Chapters   7
            12
            13
            14
            17
            19, 20   (19; 516-520
            11   527-530)
            9
            10

(20; omit 562-568
Long term memory
+ depression in CA1

571-573
Omit: Structural
Plasticity
pg 573)

NEUROSCIENCE:
EXPLORING THE BRAIN

NEUROSCIENCE: EXPLORING THE BRAIN

Mark F. Bear, Ph.D.

PROFESSOR OF NEUROSCIENCE
HOWARD HUGHES MEDICAL INSTITUTE
BROWN UNIVERSITY
PROVIDENCE, RHODE ISLAND

Barry W. Connors, Ph.D.

PROFESSOR OF NEUROSCIENCE
BROWN UNIVERSITY
PROVIDENCE, RHODE ISLAND

Michael A. Paradiso, Ph.D.

ASSOCIATE PROFESSOR OF NEUROSCIENCE
BROWN UNIVERSITY
PROVIDENCE, RHODE ISLAND

Williams & Wilkins

A WAVERLY COMPANY

BALTIMORE • PHILADELPHIA • LONDON • PARIS • BANGKOK
BUENOS AIRES • HONG KONG • MUNICH • SYDNEY • TOKYO • WROCLAW

Editor: Timothy S. Satterfield
Development Editor: Betsy Dilernia
Production Coordinator: Raymond E. Reter
Copy Editor: Bonnie Montgomery
Designer: Ashley Pound
Illustrations: J/B Woolsey Associates
Illustration Planner and Composition: Mario Fernández
Printer and Binder: Metropole Litho

351 West Camden Street
Baltimore, Maryland 21201-2436 USA
1-800-638-0672

Rose Tree Corporate Center
1400 North Providence Road
Building II, Suite 5025
Media, Pennsylvania 19063-2043 USA

Printed in Canada

Library of Congress Cataloging in Publication Data

Bear, Mark F.
 Neuroscience: exploring the brain / Mark F. Bear, Barry W. Connors, Michael A. Paradiso.
  p. cm.
 Includes bibliographical references and index.
 ISBN 0-683-00488-3
 1. Neurosciences. 2. Brain. I. Connors, Barry W. II. Paradiso, Michael A. III. Title.
 [DNLM: 1. Nervous System. 2. Neurosciences. WL 100 B368n 1995]
QP355.2.B425 1996
612.8—dc20
DNLM/DLC
for Library of Congress   95-13837
                                                             CIP

The Publishers have made every effort to trace the copyright holders for borrowed material. If they have inadvertently overlooked any, they will be pleased to make the necessary arrangements at the first opportunity.

Reprints of chapters may be purchased from Williams & Wilkins in quantities of 100 or more. Call Isabella Wise in the Special Sales Department, (800) 358-3583.

98 99
5 6 7 8 9 10

to our parents

Naomi and Firman Bear
Rose and John Connors
Marie and Nicholas Paradiso

Preface

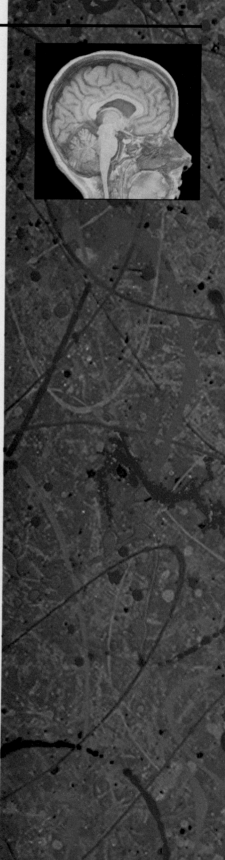

Origins of *Neuroscience: Exploring the Brain*

For over 15 years, Brown University has offered a course called "Neuroscience 1, An Introduction to the Nervous System." The architects of Neuro 1 could not have anticipated its stunning success. Enrollment now consistently tops 300, which means that approximately one in four Brown undergraduates take the course. For a few students, this is the beginning of a career in neuroscience; for others, it is the only science course he or she takes in college.

The success of Neuro 1 reflects the fascination and curiosity everyone has for how we sense, move, feel, and think. However, we believe its success also derives from the way the course is taught and what is emphasized. A cornerstone of our philosophy is that we assume only minimal prior knowledge of biology, physics, and chemistry. The details required to understand neuroscience, from the units of the metric system to the structure of water and protein molecules, are covered as the course progresses. This ensures that we can work up to advanced concepts with confidence that students are on track. We also strive to show that science is interesting, exciting, and fun. To this end we use liberal doses of common-sense metaphors, real-world examples, humor, and anecdotes. Finally, our course does not survey all of neurobiology. Instead, we focus on mammalian brains and, whenever possible, the human brain. In this sense, the course closely resembles what is taught to most second-year medical students, but without prerequisites and with less clinical emphasis. Similar courses are now offered at many colleges and universities by psychology, biology, and neuroscience departments.

We have searched over the years for a suitable textbook for Neuro 1, and for years we have been frustrated. Most books aim too high; they target advanced undergraduate science majors, graduate students, and medical students and assume sophistication in biology, chemistry, and mathematics. A few aim too low, sacrificing the neurobiological foundations of exciting new developments in molecular and cellular neuroscience. To address the needs of our students, we set out to write a textbook incorporating the subject matter and philosophy that made introductory neuroscience so successful at Brown. The result is *Neuroscience: Exploring the Brain*.

Scientific Content

Neuroscience: Exploring the Brain surveys the organization and function of the human nervous system. We present material at the cutting edge of neuroscience, but in a way that is accessible to both science and non-science students alike. The level of the material is comparable to an introductory college text in general biology.

The book is divided into three parts: (I) Foundations, (II) Sensory and Motor Systems, and (III) Brain and Behavior. We begin Part I by introducing the modern field of neuroscience and tracing some of its historical antecedents. Then we take a close look at the structure and function of individual neurons, how they communicate chemically, and how these building blocks are arranged to form a nervous system. In Part II, we break open the brain to examine the structure and function of the systems that serve the

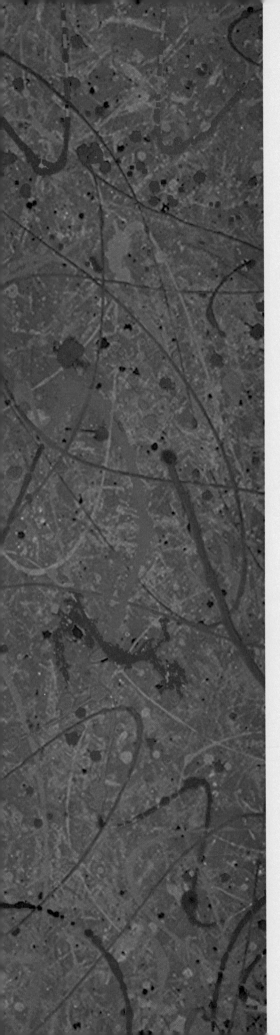

senses and command voluntary movements. Finally, in Part III we explore the neurobiology of human behavior, including mood, emotion, sleep, learning, memory, language, and attention.

The human nervous system is examined at several different scales, ranging from the molecules that determine the functional properties of neurons to the large systems in the brain that underlie cognition and behavior. Disorders of the human nervous system are introduced as the book progresses, each within the context of the specific neural system under discussion. (Indeed, many insights into the normal functions of neural systems have come from the study of diseases that cause specific malfunctions of these systems.) In addition, we discuss the actions of drugs and toxins on the brain, using this information to illustrate how different brain systems contribute to behavior and how drugs may alter brain function.

Organization of Part I: Foundations. The aim of Part I is to build a strong base of general knowledge of neurobiology. The chapters should be covered sequentially, although Chapters 1 and 6 can be skipped without loss of continuity.

In Chapter 1 we use a historical approach to review some basic principles of nervous system function, and then turn to the topic of how neuroscience research is conducted today. We directly confront the ethics of neuroscience research, particularly that which involves animals.

In Chapter 2 we focus mainly on the cell biology of the neuron. This is essential information for students inexperienced in biology, and we find that even those with a strong biology background find this review helpful. After touring the cell and its organelles, we go on to discuss the structural features that make neurons and their supporting cells unique, stressing the correlation of structure with function.

Chapters 3 and 4 are devoted to the physiology of the neuronal membrane. We cover the essential chemical, physical, and molecular properties that enable neurons to conduct electrical signals. We appeal to students' intuition throughout by employing common sense, a liberal use of metaphors, and real-life analogies. More sophisticated treatments have been placed in boxes entitled *Brain Food*, which can be covered at the instructor's discretion.

Chapters 5 and 6 cover interneuronal communication, particularly chemical synaptic transmission. Chapter 5 presents the general principles of chemical synaptic transmission; in Chapter 6 we discuss the neurotransmitters and their modes of action in greater detail. We also cover many of the modern methods used to study the chemistry of synaptic transmission. Later chapters do not assume understanding of synaptic transmission at the depth of Chapter 6, however, so this can be skipped at the instructor's discretion. Most coverage of psychopharmacology appears in Chapter 15, after the general organization of the brain and its sensory and motor systems has been presented. It is our experience that students wish to know where, in addition to how, drugs act on the nervous system and behavior.

Chapter 7 is the last of the Foundations chapters, and covers the gross anatomy of the nervous system. Here we stress the common organizational plan of the mammalian nervous system by tracing the brain's embryological development. (Cellular aspects of development are covered in Chapter 18.) We show that the specializations of the human brain are simple variations on the basic plan that applies to all mammals. We have found, contrary to what some might expect, that students love to learn about neuroanatomy. This chapter provides the canvas we use to paint the anatomy of the sensory and motor systems in Part II.

Organization of Part II: Sensory and Motor Systems. The aim of Part II is to survey the systems within the brain that control conscious sensation and voluntary movement. In general, these chapters need not be covered sequentially, except for Chapters 9 and 10 on vision and Chapters 13 and 14 on the control of movement.

We chose to begin Part II with a discussion of the chemical senses—smell and taste—in Chapter 8. The anatomical organization of these sensory systems in some respects represents the exceptions rather than the rule, but the sensory transduction mechanisms have strong parallels with other systems.

Chapters 9 and 10 cover the visual system, an essential topic for all introductory neuroscience courses. Many details of visual system organization are presented, illustrating not only the depth of current knowledge, but also principles that apply across sensory systems.

Chapters 11 and 12 cover the auditory and somatic sensory systems, respectively. These senses are such an important part of everyday life, it is hard to imagine teaching introductory neuroscience without discussing them. We point out the similarities and differences across all sensory systems.

In Chapters 13 and 14 we discuss the motor systems of the brain. Considering how much of the brain is devoted to the control of movement, this extensive treatment is clearly justified. However, we are well aware that the complexities of the motor systems are daunting to students and instructors alike. We have tried to keep our discussion sharply focused, using numerous examples to connect with personal experience.

Organization of Part III: Brain and Behavior. The aim of Part III is twofold. First, we explore how different neural systems contribute to different behaviors, emphasizing the systems where the connection between brain and behavior can be made most strongly. We cover the systems controlling visceral function and homeostasis, mood, emotion, sleep and consciousness, language, and attention. A second aim is to explore the cellular and molecular basis of brain development, and learning and memory, which represent two of the most exciting frontiers of modern neuroscience. For the most part, each chapter of Part III stands alone, but Chapters 15, 16, and 17 are closely related, as are Chapters 18, 19, and 20.

Chapters 15, 16, and 17 explore several neural systems that orchestrate widespread responses throughout the brain and body. In Chapter 15 we focus on three systems that are characterized by the long reach of their influence and their interesting neurotransmitter chemistry: the secretory hypothalamus, the autonomic nervous system, and the diffuse modulatory systems of the brain. We discuss how the behavioral manifestations of various drugs and psychiatric disorders may result from disruptions of these systems. Chapter 16 examines the neural systems believed to underlie emotional experience and expression, specifically looking at fear and anxiety, anger and aggression, reinforcement and reward. Chapter 17 explores the systems that give rise to the rhythms of the brain, ranging from the rapid electrical rhythms of the cerebrum during sleep and wakefulness to the slow circadian rhythms controlling hormones, temperature, alertness, and metabolism.

We shift gears in Chapter 18 by examining the mechanisms used during brain development to ensure that the correct connections are made between neurons. Discussion of development appears here rather than in Part I for several reasons. First, by this point in the book students fully appreciate that normal brain function depends on its precise wiring. Because we use the visual system as a concrete example, the chapter also must follow discussion of the visual pathways in Part II. Second, we explore aspects of experience-dependent development of the visual system that are regulated by the diffuse modulatory systems of the brain, so this chapter is placed after the early chapters of Part III. Finally, exploration of the role of the sensory environment in brain development in Chapter 18 is followed in the next two chapters by discussions of how experience-dependent modifications of the brain form the basis for learning and memory. We see that many of the mechanisms are similar, illustrating the unity of biology.

Chapters 19 and 20 cover learning and memory. Chapter 19 focuses on the anatomy of memory, exploring how different parts of the brain contribute to

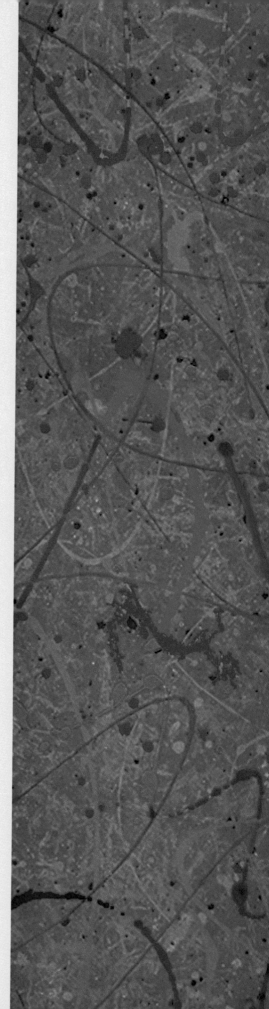

storage of different types of information. Chapter 20 takes a deeper look into the molecular and cellular mechanisms of memory, focusing on changes in synaptic connections.

Part III ends with a discussion of the neuroscience of higher brain functions. Chapter 21 illustrates how progress is being made in understanding brain mechanisms of language and attention. The exciting conclusion is that it is not too soon to begin thinking about the neurobiological bases of consciousness itself.

Capturing the Excitement

We are active neuroscience researchers as well as classroom teachers. We love the scientific work we do; it shows in our teaching and, we hope, in our writing as well.

Each fact presented in this book began as a discovery by a neuroscientist. In some cases the discovery was quite accidental, occasionally inspiration for the crucial experiment came in a dream, and the thrill of discovery often led to sleepless nights for the discoverer. We want our readers to understand the allure of research. To this end we asked a number of prominent neuroscientists to contribute a personal story about their work. (Six have been president of the Society for Neuroscience, including both its founding and current presidents.) Their stories are peppered throughout the book as boxes entitled "Paths of Discovery." We are very grateful to the Discoverers for their time, effort, and unflagging enthusiasm. The boxes serve several purposes: to give a flavor of the thrill of discovery, to show the importance of toil and patience as well as serendipity and intuition, to reveal the human side of science, and to entertain and amuse. These contributors are listed below in their order of appearance:

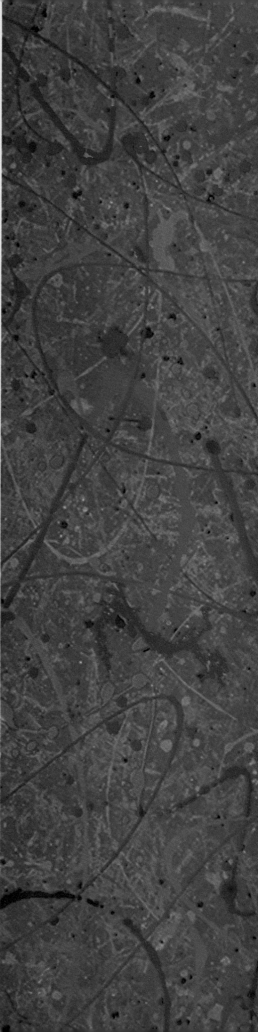

A.J. Hudspeth, M.D., Ph.D. . . . . *The Ear's Gears*
University of Texas, Dallas

James Simmons, Ph.D. . . . . *Bat Echolocation*
Brown University

Thomas Woolsey, M.D. . . . . *Cortical Barrels*
Washington University School of Medicine

Vernon Mountcastle, Ph.D. . . . *My Adventures in the Posterior Parietal Cortex*
Johns Hopkins University

Lorne Mendell, Ph.D. . . . . *Synapse Search*
SUNY, Stony Brook

Apostolos Georgopoulos, M.D. . . . . *Reaching a Goal*
VA Medical Center, Minneapolis

Paul MacLean, M.D. . . . . *Origins of the Limbic System*
National Institute of Mental Health

J. Allan Hobson, M.D. . . . . *Dreams and the Brain*
Harvard Medical School

Carla Shatz, Ph.D. . . . . *Waves in the Eye*
University of California, Berkeley

Larry Squire, Ph.D. . . . . *Memory Recollections*
University of California, San Diego

Tim Bliss, Ph.D. and Terje Lømo, M.D., Ph.D. . . . . *Discovering LTP*
National Institute for Medical Research, London
and University of Oslo

Michael I. Posner, Ph.D. . . . . *Localizing Cognitive Operations*
University of Oregon

Helping Students Learn

Our goal in writing *Neuroscience: Exploring the Brain* was not to cover the field in encyclopedic detail, but to produce a readable textbook that communicates the important principles of neuroscience clearly and effectively. *This book was written for our students, not for our colleagues.* To help students learn neuroscience, we included a number of features designed to enhance comprehension.

- **Chapter outlines, introductory and concluding remarks**. These preview the organization of each chapter, set the stage, and place the material into broader perspective.
- **Key terms list and glossary**. Neuroscience has a language of its own, and to comprehend it one must learn the vocabulary. In the text of each chapter important terms are highlighted in boldface type. To facilitate review, these terms appear again in a key term list at the end of each chapter, in the order in which they appeared in the text. The same terms are assembled at the end of the book, with definitions, in an extensive glossary.
- **Review questions**. At the end of each chapter we include 6–8 review questions. These are specifically designed to provoke thought and help students integrate the material.

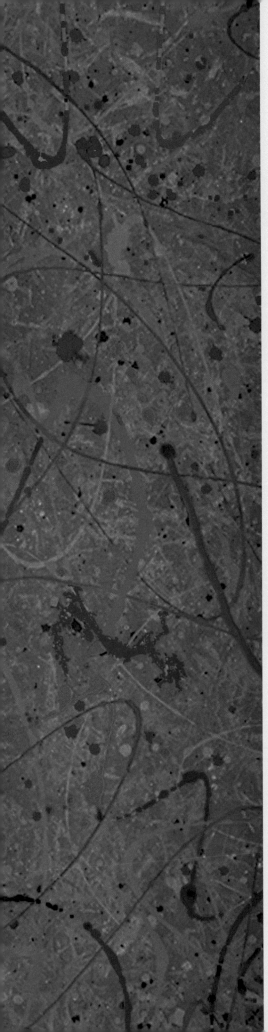

- **Review of neuroanatomical terms**. In Chapter 7, where brain anatomy is discussed, the narrative is interrupted periodically with reviews to enhance understanding.
- **Additional readings**. To guide study beyond the scope of the textbook, we provide selected readings that will lead the student into the research literature associated with each chapter. Rather than putting numerous citations into the body of the chapters, where they would greatly decrease the readability of the text, the additional readings are organized by chapter and listed at the end of the book.
- **Numerous simple illustrations in full color**. We believe in the power of illustrations; not those that "speak a thousand words," but rather those that each make a single point. We were especially fortunate to work with John Woolsey and his associates on the art. John is the son of one famous neuroscientist, Clinton Woolsey, and the brother of another, Thomas Woolsey. His understanding of neuroscience was invaluable in developing the hundreds of beautiful illustrations that appear in this book.
- **Of Special Interest boxes**. These boxes appear in all chapters to stimulate interest and help learning by enhancing the connection with real life. Many of the boxes deal with disorders of the nervous system relevant to the chapter contents: for example, multiple sclerosis, myasthenia gravis, herpes, and narcolepsy. Other boxes cover new methods, such as magnetic resonance imaging and positron emission tomography.

Acknowledgments

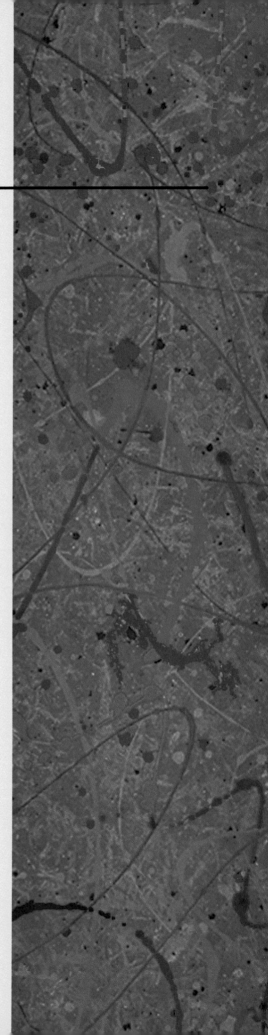

There are many people and organizations who inspired and helped us in this project. First, we thank the architects of the undergraduate neuroscience curriculum at Brown University, the finest of its kind in the world. We thank Mitchell Glickstein, Ford Ebner, James McIlwain, Leon Cooper, James Anderson, Leslie Smith, and John Donoghue for all they did to get undergraduate neuroscience off the ground at Brown. We thank the staff of Williams and Wilkins for believing in this project and guiding it towards its successful completion. Thanks in particular go to Kevin Thibodeau, Tim Satterfield, Pat Coryell, Nancy Evans, Crystal Taylor, Bonnie Montgomery, Ray Reter, and Mary Finch. We are forever indebted to our Developmental Editor, Betsy Dilernia, who guided, counseled, and sometimes humbled us with her purple pencil. The clarity and consistency in the writing is attributable to her remarkable efforts. We thank the staff of J/B Woolsey Associates for the superb art program, especially John Woolsey and Caitlin Duckwall. We thank Suzanne Meagher and Bill Ingino for keeping us organized and out of trouble with our publisher. We are especially grateful for the research support provided to us over the years by the National Institutes of Health, the Whitehall Foundation, the Alfred P. Sloan Foundation, the Klingenstein Fund, the Charles A. Dana Foundation, the National Science Foundation, the Office of Naval Research, and the Howard Hughes Medical Institute. Without the support of these organizations, we would not be neuroscientists. We thank our colleagues in the Brown University Department of Neuroscience for their support of this project and helpful advice; especially John Donoghue and Jerome Sanes. We thank the anonymous, but very helpful colleagues at other institutions who reviewed the first draft of our manuscript. We thank our loved ones for standing by us despite the countless weekends and evenings lost to preparing this book. Last, but not least, we thank the thousands of students we have had the privilege to teach neuroscience over the past 16 years.

NEUROSCIENCE: EXPLORING THE BRAIN

Mark F. Bear
Barry W. Connors
Michael A. Paradiso

PART I:
FOUNDATIONS

CHAPTER 1

Introduction to Neuroscience

INTRODUCTION

ORIGINS OF NEUROSCIENCE

NEUROSCIENCE TODAY

CONCLUDING REMARKS

REVIEW QUESTIONS

Men ought to know that from nothing else but the brain come joys, delights, laughter and sports, and sorrows, griefs, despondency, and lamentations. And by this, in an especial manner, we acquire wisdom and knowledge, and see and hear and know what are foul and what are fair, what are bad and what are good, what are sweet and what are unsavory. . . . And by the same organ we become mad and delirious, and fears and terrors assail us. . . . All these things we endure from the brain when it is not healthy. . . . In these ways I am of the opinion that the brain exercises the greatest power in the man.

—Hippocrates, *On the Sacred Disease* (Fourth century B.C.)

It is human nature to be curious about how we see and hear; why some things feel good and others hurt; how we move; how we reason, learn, remember, and forget; the nature of anger and madness. These mysteries are starting to be unraveled by basic neuroscience research, and the conclusions of this research are the subject of this textbook.

The word "neuroscience" is young. The Society for Neuroscience, an association of professional neuroscientists, was founded as recently as 1970. The study of the brain, however, is as old as science itself. Historically, the scientists that devoted themselves to an understanding of the nervous system came from different scientific disciplines: medicine, biology, psychology, physics, chemistry, mathematics. The neuroscience revolution came when these scientists realized that the best hope for understanding the workings of the brain came from an *interdisciplinary* approach, combining the traditional approaches to yield a new synthesis, a new perspective. Today, most people involved in the scientific investigation of the nervous system regard themselves as neuroscientists. Indeed, while the course you are now taking might be sponsored by the psychology or biology department at your university or college, and might be called biopsychology or neurobiology, you can bet that your instructor is a neuroscientist. Today, the Society for Neuroscience is the largest and fastest-growing association of professional scientists in all of experimental biology. Far from being overly specialized, this field is as broad as nearly all of natural science, with the nervous system serving as the common point of focus. Understanding how the brain works requires knowledge about many things, from the structure of the water molecule, to the electrical and chemical properties of the brain, to why Pavlov's dog salivated when a bell was rung. In this book, we will explore the brain with this broad perspective.

We begin the adventure with a brief tour of neuroscience. What have scientists thought about the brain over the ages? Who are the neuroscientists of today, and how do they approach studying the brain?

ORIGINS OF NEUROSCIENCE

You probably already know that the nervous system—the brain, spinal cord, and nerves of the body—is crucial for life and allows you to sense, move, and think. How did this view arise?

There is evidence that even our prehistoric ancestors appreciated that the brain was vital to life. The archeological record is rife with examples of hominid skulls, dating back a million years and more, bearing signs of fatal cranial damage, presumably inflicted by other hominids. As early as 10,000 years ago, people were boring holes in each other's skulls (a process called trepanation), evidently not with the aim to kill, but to cure (Figure 1.1). The skulls show signs of healing after the operation, indicating that this procedure was carried out on live subjects and was not merely some ritual conducted after death. Some individuals apparently survived multiple skull surgeries. What early surgeons hoped to accomplish is not clear, although it has been speculated that this procedure may have been used to treat headaches or mental disorders, perhaps by giving the evil spirits an escape route.

Recovered writings from the early physicians of ancient Egypt, dating back almost 5000 years, indicate that they were well aware of many symptoms of brain damage. However, it is also very clear that the heart, not the brain, was considered to be the seat of the soul and the repository of memories. Indeed, while the rest of the body was carefully preserved for the afterlife, the brain of the deceased was simply scooped out through the nostrils and discarded! The view that the heart was the seat of consciousness and thought was not seriously challenged until the time of Hippocrates.

Views of the Brain in Ancient Greece

Consider the notion that the different parts of your body look different because they perform different functions. The structures of the feet and hands are very different, and they perform very different functions: We walk on our feet and manipulate objects with our hands. Thus, we can say that there appears to be a very clear *correlation between structure and function*. Differences in appearance predict differences in function.

What can we glean about function from the structure of the head? Quick inspection and a few simple experiments (like closing your eyes) reveals that the head is specialized for sensing the environment. In the head are your eyes and ears, your nose and tongue. Even crude dissection shows that the nerves from these organs can be traced through the skull into the brain. What would you conclude about the brain from these observations?

If your answer is that the brain is the organ of sensation, then you have reached the same conclusion as several Greek scholars of the fourth century B.C. The most influential scholar was Hippocrates (460–379 B.C.), the father of Western medicine, who stated his belief that the brain not only was involved in sensation, but was also the seat of intelligence.

However, this view was not universally accepted. The famous Greek philosopher Aristotle (384–322 B.C.) clung to the belief that the heart was the center of intellect. What function did Aristotle reserve for the brain? He proposed it to be a radiator for the cooling of blood that was overheated by the seething heart. The rational temperament of humans was thus explained by the large cooling capacity of our brain.

Views of the Brain During the Roman Empire

The most important figure in Roman medicine was the Greek physician and writer Galen (A.D. 130–200), who embraced the Hippocratic view of brain function. As physician to the gladiators, he must have witnessed the

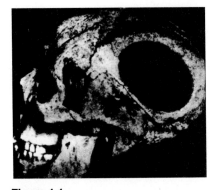

Figure 1.1.
Evidence of prehistoric brain surgery.
This skull of a man, over 5000 years old, was surgically opened when he was still alive. (Source: Finger, 1994, Fig.1.1.)

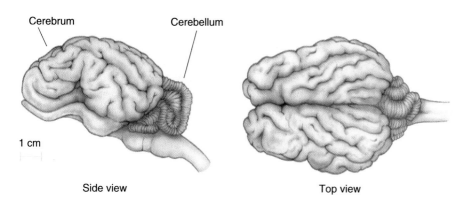

Cerebrum　　　Cerebellum

1 cm

Side view　　　　　　　　　　　Top view

Figure 1.2.
The brain of a sheep. Notice the location and appearance of the cerebrum and the cerebellum.

unfortunate consequences of spinal and brain injury. However, Galen's opinions about the brain probably were influenced more by his many careful animal dissections. Figure 1.2 is a drawing of the brain of a sheep, one of Galen's favorite subjects. Two major parts are evident: the *cerebrum* in the front and the *cerebellum* in the back. (The structure of the brain is the subject of Chapter 7.) Just as we were able to deduce function from the structure of the hands and feet, Galen tried to deduce function from the structure of the cerebrum and the cerebellum. Poking the freshly dissected brain with a finger reveals the cerebellum to be rather hard and the cerebrum to be rather soft. From this observation, Galen suggested that the cerebrum must be the recipient of sensations and the cerebellum must command the muscles. Why did he propose this distinction? He recognized that to form memories, sensations must be imprinted onto the brain. Naturally, this must occur in the doughy cerebrum.

As improbable as his reasoning may seem, Galen's deductions were not that far from the truth. The cerebrum, in fact, is largely concerned with sensation and perception and the cerebellum is primarily a movement control center. Moreover, the cerebrum is a repository of memory. We will see that this is not the only example in the history of neuroscience where the right general conclusions were reached for the wrong reasons.

How does the brain receive sensations and move the limbs? Galen cut open the brain and found that it is hollow (Figure 1.3). In these hollow spaces, called *ventricles* (like the similar chambers in the heart), there is fluid. To Galen, this discovery fit perfectly with the prevailing theory that the body functioned according to a balance of four vital fluids, or humors. Sensations were registered and movements initiated by the movement of humors to or from the brain ventricles via the nerves, which were believed to be hollow tubes, like the blood vessels.

Views of the Brain from the Renaissance to the Nineteenth Century

Galen's view of the brain prevailed for almost 1500 years. More detail was added to the structure of the brain by the great anatomist Andreas Vesalius (1514–1564) during the Renaissance (Figure 1.4). However, ventricular localization of brain function remained essentially unchallenged. Indeed, the whole concept was strengthened in the early seventeenth century when French inventors began developing hydraulically controlled

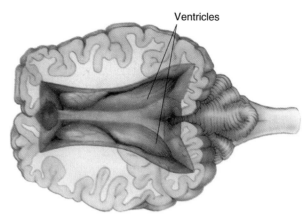

Figure 1.3
Dissected sheep brain showing the ventricles.

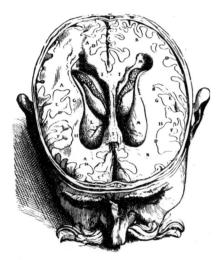

Figure 1.4
Human brain ventricles, depicted during the Renaissance. This drawing is from *De humani corporis fabrica* by Vesalius (1543). The subject was probably a decapitated criminal. Great care was taken to be anatomically correct in depicting the ventricles. (Source: Finger, 1994, Fig. 2.8.)

mechanical devices. These devices supported the notion that the brain could be machinelike in its function: Fluid forced out the ventricles through the nerves might literally "pump you up" and cause the movement of the limbs. After all, don't the muscles bulge when they contract?

A chief advocate of this fluid-mechanical theory of brain function was the French mathematician and philosopher René Descartes (1596–1650). However, while he thought this theory could explain the brain and behavior of other animals, it was inconceivable to Descartes that it could account for the full range of *human* behavior. He reasoned that unlike other animals, people possess intellect and a God-given soul. Thus, Descartes proposed that brain mechanisms control human behavior only to the extent that this behavior resembles that of the beasts. Uniquely human mental capabilities exist outside the brain in the "mind." Descartes believed that the mind is a spiritual entity that receives sensations and commands movements by communicating with the machinery of the brain via the pineal gland (Figure 1.5). There are still those today who believe that there is a "mind-brain problem," that somehow the human mind is distinct from the brain. However, as we shall see in Chapter 21, modern neuroscience research supports another conclusion: The mind has a physical basis, which is the brain.

Fortunately, other scientists during the seventeenth and eighteenth centuries broke away from Galen's tradition of focusing on the ventricles and began to give the substance of the brain a closer look. One of their observations was that brain tissue is divided into two parts: the *gray matter* and the *white matter* (Figure 1.6). What structure-function relationship did they propose? White matter, because it was continuous with the nerves of the body, was correctly believed to contain the fibers that bring information to and from the gray matter.

Another breakthrough was the observation that the same pattern of bumps (called gyri) and grooves (called sulci and fissures) on the surface of the cerebrum could be identified in every individual (Figure 1.7). This pattern, which enables the parceling of the cerebrum into "lobes," was the basis for speculation that different functions might be localized to the different bumps on the brain. The stage was now set for the era of cerebral localization.

Nineteenth-Century Views of the Brain

Let's review the state of understanding of the nervous system at the end of the eighteenth century:

- Injury to the brain can disrupt sensations, movement, and thought, and can cause death.

- The brain communicates with the body via the nerves.

- The brain has different identifiable parts, which probably perform different functions.

- The brain (if not the mind) operates like a machine and follows the laws of nature.

During the next 100 years, more would be learned about the function of the brain than had been learned in all of previously recorded history. This work provided the solid foundation on which twentieth-century neuroscience rests. Below we'll review four key insights gained during this period.

Nerves As Wires. In 1751, Benjamin Franklin published a pamphlet entitled *Experiments and Observations on Electricity*, which heralded a new understanding of electrical phenomena. By the turn of the century, Italian scientist Luigi Galvani and German biologist Emil du Bois-Reymond had shown that muscles can be caused to twitch when nerves are stimulated electrically, and that the brain itself can generate electricity. This discovery finally displaced the notion that nerves communicate with the brain by the movement of fluid. The new concept was that the nerves are wires that conduct electrical signals to and from the brain.

Unresolved was the question of whether the signals to the muscles causing movement use the same wires as those that register sensations from the skin. Bidirectional communication along the same wires was suggested by the observation that when a nerve in the body is cut, there is usually a loss of both sensation and movement in the affected region. However, it was also known that within each nerve of the body there are many thin filaments or *nerve fibers,* each one of which could serve as an individual wire carrying information in a different direction.

This question was answered around 1810 by Scottish physician Charles Bell and French physiologist François Magendie. A curious anatomical fact is that just before the nerves attach to the spinal cord, the fibers divide into two branches or roots. The dorsal root enters toward the back of the spinal cord and the ventral root enters toward the front (Figure 1.8). Bell tested the

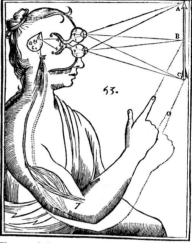

Figure 1.5
The brain according to Descartes. This drawing appeared in a 1662 publication by Descartes. Hollow nerves from the eyes project to the brain ventricles. The mind influences the motor response by controlling the pineal gland (H), which works like a valve to control the movement of animal spirits through the nerves that inflate the muscles. (Source: Finger, 1994, Fig. 2.16.)

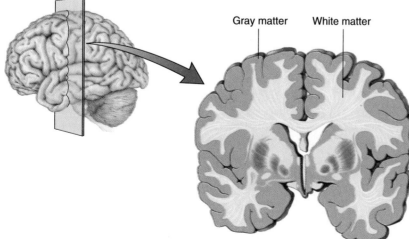

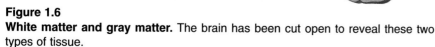

Figure 1.6
White matter and gray matter. The brain has been cut open to reveal these two types of tissue.

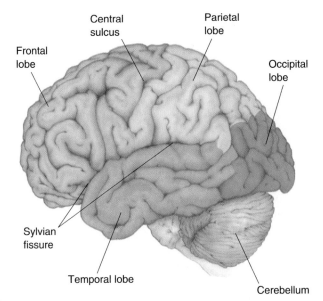

Figure 1.7
The lobes of the cerebrum. Notice the deep Sylvian fissure, dividing the frontal lobe from the temporal lobe, and the central sulcus, dividing the frontal lobe from the parietal lobe. The occipital lobe lies at the back of the brain. These landmarks can be found on all human brains.

possibility that these two spinal roots carry information in different directions by cutting each root separately and observing the consequences in experimental animals. He found that only cutting the ventral roots caused muscle paralysis. Later, Magendie was able to show that the dorsal roots carry sensory information into the spinal cord. Bell and Magendie concluded that, within each nerve, there is a mixture of many wires, some of which bring information into the brain and spinal cord and others that send information out to the muscles. In each sensory and motor nerve fiber, transmission is strictly one-way. The two kinds of fibers are bundled together for most of their length, but they are anatomically segregated when they enter or exit the spinal cord.

Localization of Specific Functions to Different Parts of the Brain. If different functions are localized in different spinal roots, then perhaps different functions also are localized in different parts of the brain. In 1811, Bell proposed that the origin of the motor fibers is the cerebellum and the destination of the sensory fibers is the cerebrum.

How would you test this proposal? One way is to use the same approach that Bell and Magendie had employed to identify the functions of the spinal roots: to destroy these parts of the brain and test for sensory and motor deficits. This approach, in which parts of the brain are systematically destroyed to determine their function, is called the *experimental ablation method*. In 1823, the esteemed French physiologist Marie-Jean-Pierre Flourens used this method in a variety of animals (particularly birds) to show that the cerebellum does indeed play a role in the coordination of movement. He also concluded that the cerebrum is involved in sensation and perception, as Bell and Galen before him had suggested. Unlike his predecessors, however, Flourens provided solid experimental support for the theory that different parts of the brain perform different functions.

What about all those bumps on the brain's surface? Do they perform different functions as well? The idea that they do was irresistible to a young Austrian medical student named Franz Joseph Gall. Believing that bumps

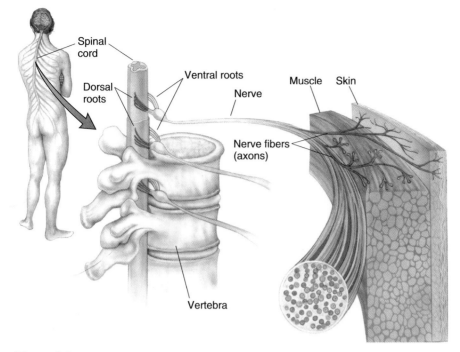

Figure 1.8
Spinal nerves and spinal nerve roots. Thirty-one pairs of nerves leave the spinal cord to supply the skin and the muscles. Cutting a spinal nerve leads to a loss of sensation and a loss of movement in the affected region of the body. Incoming sensory fibers and outgoing motor fibers divide into spinal roots where the nerves attach to the spinal cord. Bell and Magendie found that the ventral roots contain only motor fibers and the dorsal roots contain only sensory fibers.

on the surface of the skull reflect the bumps on the surface of the brain, Gall proposed in 1809 that the propensity for certain personality traits—such as generosity, secretiveness, and destructiveness—could be related to the dimensions of the head (Figure 1.9). To support his claim, Gall and his followers collected and carefully measured the skulls of hundreds of people representing the extensive range of personality types, from the very gifted to the criminally insane. This new "science" of correlating the structure of the head with personality traits was called *phrenology*. Although the claims of the phrenologists were never taken seriously by the mainstream scientific community, they did capture the popular imagination of the time. In fact, a textbook on phrenology published in 1827 sold over 100,000 copies.

One of the most vociferous critics of phrenology was Flourens, the same man who had shown experimentally that different parts of the brain perform different functions. His grounds for criticism were sound. For one thing, the dimensions of the skull are not correlated with the dimensions of the brain. In addition, Flourens performed experimental ablations showing that particular traits are not isolated to the portions of the cerebrum specified by phrenology. Flourens also concluded, however, that all regions of the cerebrum participate equally in all cerebral functions, a conclusion that later was shown to be erroneous.

The person usually credited with tilting the scales of scientific opinion back toward localization of function in the cerebrum was French neurologist Paul Broca (Figure 1.10). Broca was presented with a patient who could understand language but could not speak. Following the man's death in 1861, Broca carefully examined his brain and found a lesion in the left frontal lobe (Figure 1.11). Based on this case and several others like it, Broca

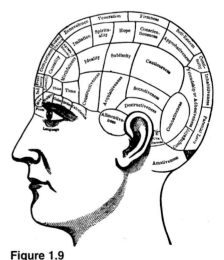

Figure 1.9
A phrenological map. According to Gall and his followers, different behavioral traits could be related to the size of different parts of the skull. (Source: Clarke and O'Malley, 1968, Fig. 118.)

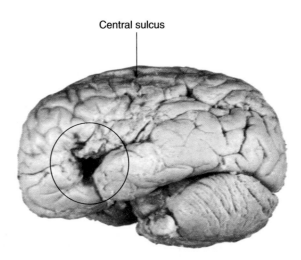

Central sulcus

Figure 1.11
The brain that convinced Broca of localization of function in the cerebrum.
This is the preserved brain of a patient who had lost the ability to speak before he
died in 1861. The lesion that produced this deficit is circled. (Source: Corsi, 1991,
Fig. III,4.)

Figure 1.10
Paul Broca (1824–1880). By the
careful study of the brain of a man
who had lost the faculty of speech
after a brain lesion (see Figure 1.11),
Broca became convinced that different
functions could be localized to different parts of the cerebrum. (Source:
Clarke and O'Malley, 1968, Fig. 121.)

concluded that this region of the human cerebrum was specifically responsible for the production of speech.

Solid experimental support for cerebral localization in animals quickly
followed. German physiologists Gustav Fritsch and Eduard Hitzig showed
in 1870 that applying small electrical currents to a circumscribed region of
the exposed surface of the brain of a dog could elicit discrete movements.
Scottish neurologist David Ferrier repeated these experiments with monkeys. And, in 1881, he showed that removal of this same region of the cerebrum causes paralysis of the muscles. Similarly, German physiologist
Hermann Munk presented evidence using experimental ablation that the
occipital lobe of the cerebrum was specifically required for vision.

As you will see in Part II of this book, we now know that there is a very
clear division of labor in the cerebrum, with different parts performing very
different functions. Today's maps of the functional divisions of the cerebrum rival even the most elaborate of those produced by the phrenologists.
The big difference is that, unlike the phrenologists, scientists today require
solid experimental evidence before attributing a specific function to a portion of the brain. All the same, Gall seems to have had the right idea. It is
natural to wonder why Flourens, the pioneer of brain localization of function, was misled into believing that the cerebrum acted as a whole and
could not be subdivided. There are many reasons that this gifted experi-
mentalist may have missed cerebral localization, but it seems clear that one
reason was his visceral reaction against Gall and phrenology. He couldn't
bring himself to agree even remotely with Gall, whom he viewed as a
lunatic. This reminds us that science, for better or worse, was and still is a
distinctly human endeavor.

Evolution of Nervous Systems. In 1859, English biologist Charles
Darwin (Figure 1.12) published *On the Origin of Species*. In this landmark
work, he articulated a theory of evolution: that species of organisms
evolved from a common ancestor. According to his theory, differences
among species arise by a process Darwin called *natural selection*. As a result

of the mechanisms of reproduction, the physical traits of the offspring are sometimes different from those of the parents. If these traits represent an advantage for survival, the offspring themselves will be more likely to reproduce, thus increasing the likelihood that the advantageous traits are passed on to the next generation. Over the course of many generations, this process has led to the development of traits that distinguish species today: flippers on harbor seals, paws on dogs, hands on raccoons, and so on. This single insight revolutionized biology. Today, scientific evidence ranging from archaeology to molecular genetics overwhelmingly supports the theory of evolution by natural selection.

Darwin included behavior among the heritable traits that could evolve. For example, he noticed that many mammalian species show the same reaction when frightened: The pupils of the eyes get bigger, the heart races, hairs stand on end. This is as true for a human as it is for a dog. To Darwin, the similarities of this response pattern indicated that these different species evolved from a common ancestor, which possessed the same behavioral trait (advantageous presumably because it facilitated escape from predators). Because behavior reflects the activity of the nervous system, we can infer that the brain mechanisms that underlie this fear reaction may be similar, if not identical, across these species.

The idea that the nervous systems of different species evolved from common ancestors and may have common mechanisms is the rationale for relating the results of animal experiments to humans. Thus, for example, many of the details of electrical impulse conduction along nerve fibers were worked out first in the squid, but are now known to apply equally well to humans. Most neuroscientists today use *animal models* of the process they wish to understand in humans. For example, rats show clear signs of addiction if they are given the chance to self-administer cocaine. Consequently, rats are a valuable animal model for research focused on understanding how psychoactive drugs exert their effects on the nervous system.

On the other hand, many behavioral traits are highly specialized for the environment (or niche) a species normally occupies. For example, monkeys swinging from branch to branch have a keen sense of sight, while rats slinking through underground tunnels have poor vision but a highly evolved sense of touch using the whiskers on the snout. Adaptations are reflected in the structure and function of the brain of every species. By comparing the specializations of the brains of different species, neuroscientists have been able to identify which parts of the brain are specialized for different behavioral functions. Examples for monkeys and rats are shown in Figure 1.13.

The Neuron: The Basic Functional Unit of the Brain. Refinement of the microscope in the early 1800s gave scientists their first opportunity to examine animal tissues at high magnifications. In 1839, German zoologist Theodor Schwann proposed what came to be known as the *cell theory*: All tissues are composed of microscopic units called cells.

Although cells in the brain had been identified and described, there was still controversy about whether the individual "nerve cell" was actually the basic unit of brain function. Nerve cells usually have a number of thin projections, or processes, that extend from a central cell body (Figure 1.14). Initially, scientists could not decide whether the processes from different cells fuse together like the blood vessels of the circulatory system. If this were true, then the "nerve net" of connected nerve cells would represent the elementary unit of brain function.

Chapter 2 presents a brief history of how this issue was resolved. Suffice it to say that by 1900, the individual nerve cell, now called the neuron, was recognized to be the basic functional unit of the nervous system.

Figure 1.12
Charles Darwin (1809–1882). Darwin proposed his theory of evolution, explaining how species evolve through the process of natural selection. (Source: The Bettman Archive.)

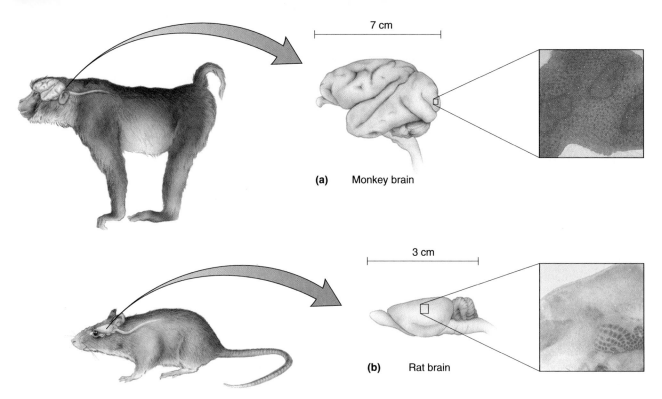

(a) Monkey brain

(b) Rat brain

Figure 1.13
Evolution of different brain specializations in monkeys and rats. (a) The brain of the macaque monkey has a highly evolved sense of sight. The boxed region receives information from the eyes. When this region is sliced open and stained to show metabolically active tissue, a mosaic of "blobs" appears. The neurons within the blobs are highly specialized to analyze colors in the visual world. **(b)** The brain of a rat has a highly evolved sense of touch to the face. The boxed region receives information from the whiskers. When this region is sliced open and stained to show the location of the neurons, a mosaic of "barrels" appears. Each barrel is specialized to receive input from a single whisker on the rat's face. (Photomicrographs courtesy of Dr. S. H. C. Hendry.)

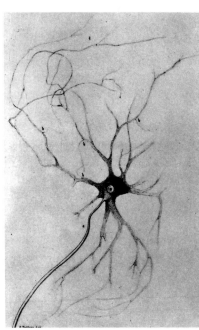

Figure 1.14
Early depiction of a nerve cell. Published in 1865, this drawing by German anatomist Otto Deiters shows a nerve cell, or neuron, and its many projections, called neurites. For a time it was thought that the neurites from different neurons might fuse together like the blood vessels of the circulatory system. We now know that neurons are distinct entities that communicate using chemical signals. (Source: Clarke and O'Malley, 1968, Fig. 16.)

NEUROSCIENCE TODAY

The history of twentieth-century neuroscience is still being written, and the accomplishments to date form the basis for this textbook. We will discuss the most recent developments throughout the book. Now let's take a look at how brain research is conducted today and why it is important to society that it continues.

Levels of Analysis

If history has taught us nothing else, it has clearly shown that understanding how the brain works is a big challenge. To reduce the complexity of the problem, neuroscientists break it into smaller pieces for systematic experimental analysis. This is called the *reductionist approach*. The size of the unit of study defines what is often called the *level of analysis*. In ascending order of complexity, these levels are molecular, cellular, systems, behavioral, and cognitive.

Molecular Neuroscience. The brain has been called the most complex piece of matter in the universe. Brain matter consists of a fantastic variety of molecules, many of which are unique to the nervous system. These different molecules play many different roles that are crucial for brain function: messengers that allow neurons to communicate with one another, sentries that control what materials can enter or leave neurons, conductors that orchestrate neuron growth, archivists of past experiences. The study of the brain at this most elementary level is called *molecular neuroscience*.

Cellular Neuroscience. The next level of analysis is cellular neuroscience, which focuses on studying how all those molecules work together to give the neuron its special properties. Among the questions asked at this level are: How many different types of neurons are there, and how do they differ in function? How do neurons influence other neurons? How do neurons become "wired together" during fetal development? How do neurons perform computations?

Systems Neuroscience. Constellations of neurons form complex circuits that perform a common function: vision, for example, or voluntary movement. Thus, we can speak of the "visual system" and the "motor system," each of which possesses its own distinct circuitry within the brain. At this level of analysis, called *systems neuroscience*, neuroscientists study how different neural circuits analyze sensory information, form perceptions of the external world, make decisions, and execute movements.

Behavioral Neuroscience. How do neural systems work together to produce integrated behaviors? For example, are different forms of memory accounted for by different systems? Where in the brain do "mind-altering" drugs act, and what is the normal contribution of these systems to the regulation of mood and behavior? What neural systems account for gender-specific behaviors? Where in the brain do dreams come from? These questions are studied in *behavioral neuroscience*.

Cognitive Neuroscience. Perhaps the greatest challenge of neuroscience is understanding the neural mechanisms responsible for the higher levels of human mental activity, such as self-awareness, mental imagery, and language. Research at this level, called *cognitive neuroscience*, studies how the activity of the brain creates the mind.

Neuroscientists

"Neuroscientist" sounds impressive, kind of like "rocket scientist." But we were all students once, just like you. For whatever reason—maybe our eyesight was poor and we wanted to know why, or perhaps a family member suffered a loss of speech after a stroke and we wanted to know why—we came to share a thirst for knowledge of how the brain works. Perhaps you will, too.

Being a neuroscientist is rewarding, but it doesn't come easily. Many years of training are required. One begins by earning an advanced degree, either a Ph.D. or an M.D. (or both). This usually is followed by several years of postdoctoral research to learn new techniques or ways of thinking under the direction of an established neuroscientist. Finally, the "young" neuroscientist is ready to set up shop at a university, institute, or hospital.

Broadly speaking, neuroscience research (and neuroscientists) may be divided into two types: *clinical* and *experimental*. Clinical research is mainly conducted by physicians (M.D.s). The medical specialties associated with the human nervous system are neurology, psychiatry, neurosurgery, and neuropathology (Table 1.1). Many who conduct clinical research continue in the tradition of Broca, attempting to deduce from the behavioral effects of brain damage the functions of different parts of the brain. Others conduct studies to assess the benefits and risks of new types of treatment.

Despite the obvious value of clinical research, the foundation for all medical treatments of the nervous system was and continues to be laid by experimental neuroscientists, who may hold either an M.D. or a Ph.D. The experimental approaches to studying the brain are so broad that they include almost every conceivable methodology. Thus, despite the interdisciplinary nature of neuroscience, expertise in a particular methodology is still often used to distinguish one neuroscientist from another. Thus, there are *neuroanatomists*, who use sophisticated microscopes to trace connections in the brain; *neurophysiologists*, who use electrodes, amplifiers, and oscilloscopes to measure the brain's electrical activity; *neuropharmacologists*, who use "designer drugs" to study the chemistry of brain function; *molecular neurobiologists*, who probe the genetic material of neurons to find clues about the structure of brain molecules; and so on. Table 1.2 lists some of the types of experimental neuroscientists. Ask your instructor what type(s) he or she is.

The Scientific Method

Neuroscientists of all stripes endeavor to establish truths about the nervous system. Regardless of the level of analysis they choose, they work according to the scientific method, which consists of four essential steps: observation, replication, interpretation, and verification.

Observation. Observations typically are made during experiments that are designed to test a particular hypothesis. For example, Bell hypothesized that the ventral roots contain the nerve fibers that control the muscles. To test this idea, he performed the experiment in which he cut these fibers and then observed whether or not muscular paralysis resulted. Other types of observation derive from human clinical cases. For example, Broca's careful observations led him to correlate left frontal lobe damage with the loss of the ability to speak.

Replication. Regardless of whether the observation is experimental or clinical, it is essential that it be replicated before it can be accepted by the scientist as fact. Replication simply means repeating the experiment on different subjects, or making similar observations in different patients, as many times as necessary to rule out the possibility that the observation occurred by chance.

Interpretation. Once the scientist believes the observation is correct, he or she makes an interpretation. Interpretations depend on the state of knowledge (or ignorance) at the time the observation was made, and on the preconceived notions (the "mind set") of the scientist who made it. As such, interpretations do not always withstand the test of time. For example, at the time he made his observations, Flourens was unaware that the cerebrum of a bird is fundamentally different from that of a mammal. Thus, he wrongly concluded from experimental ablations in birds that there was no localization of certain functions in the cerebrum of mammals. Moreover, as mentioned before, his profound distaste for Gall surely also colored his interpretation. The point is that the correct interpretation often remains unrecognized until long after the original observations were made. Indeed, major breakthroughs are sometimes made when old observations are reinterpreted in a new light.

Table 1.1
Medical Specialists Associated with the Nervous System

Specialist	Description
Neurologist	An M.D. trained to diagnose and treat diseases of the nervous system.
Psychiatrist	An M.D. trained to diagnose and treat disorders of mood and personality.
Neurosurgeon	An M.D. trained to perform surgery on the brain and spinal cord.
Neuropathologist	An M.D. or Ph.D. trained to recognize the changes in nervous tissue that result from disease.

Table 1.2
Types of Experimental Neuroscientists

Type	Description
Computational neuroscientist	Uses computers to construct models of brain functions.
Developmental neurobiologist	Analyzes the development and maturation of the brain.
Molecular neurobiologist	Uses the genetic material of neurons to understand the structure and function of brain molecules.
Neuroanatomist	Studies the structure of the nervous system.
Neurochemist	Studies the chemistry of the nervous system.
Neuroethologist	Studies the neural basis of species-specific animal behaviors in natural settings.
Neuropharmacologist	Examines the effects of drugs on the nervous system.
Neurophysiologist	Measures the electrical activity of the nervous system.
Physiological psychologist (Biological psychologist, psychobiologist)	Studies the biological basis of behavior.
Psychophysicist	Quantitatively measures perceptual abilities.

Verification. The final step of the scientific method is verification. This process is distinct from the replication performed by the original observer. Verification means that the observation is sufficiently robust that it will be seen by any competent scientist who precisely follows the protocols of the original observer. Successful verification generally means that the observation is accepted as fact. However, not all observations can be verified. Sometimes this is due to inaccuracies in the original report or insufficient replication. But failure to verify usually stems from the fact that additional variables contributed to the original result, such as temperature or time of day. Thus, the process of verification, if affirmative, establishes new scientific fact, and, if negative, suggests new interpretations for the original observation.

Occasionally, one reads in the popular press about a case of "scientific fraud." Researchers face keen competition for limited research funds and feel considerable pressure to "publish or perish." In the interest of expediency, a few have actually published "observations" that were never made. Fortunately, such instances of fraud are rare, thanks to the scientific method. It doesn't take long before other scientists find that they are unable to verify the fraudulent observations and begin to question how they could have been obtained in the first place. The material you will learn in this book stands as a strong testament to the success of the scientific method.

Use of Animals in Neuroscience Research

Most of what we know about the nervous system has come from experiments on animals. In most cases, the animals are killed so their brains can be examined neuroanatomically, neurophysiologically, and/or neurochemically. The fact that animals are sacrificed for the pursuit of human knowledge raises questions about the ethics of animal research.

The Animals. Let's begin by putting the issue in perspective. Throughout history, humans have considered animals and animal products as renewable natural resources to be used for food, clothing, transportation, recreation, sport, and companionship. The animals used for research, education, and testing have always represented a small fraction of the total used for other purposes. For example, in the United States today, the number of animals used for all types of biomedical research is less than 1% of the number that are killed for food alone.[a] The number used specifically for neuroscience research is much smaller still.

Neuroscience experiments are conducted using many different species, ranging from snails to monkeys. The choice of animal species is generally dictated by the question under investigation, the level of analysis, and the extent to which the knowledge gained at this level can be related to humans. As a rule, the more basic the process under investigation, the more distant can be the evolutionary relationship with humans. Thus, experiments aimed at understanding the molecular basis of nerve impulse conduction could be carried out on a distantly related species such as the squid. On the other hand, understanding the neural basis of movement and perceptual disorders in humans has required experiments on more closely related species, such as the macaque monkey. Today, more than half the animals used for neuroscience research are rodents—mice and rats—that are bred specially for this purpose.

Animal Welfare. In the developed world today, most educated adults have a concern for animal welfare. Neuroscientists share this concern and work to ensure that animals are well treated. It must also be appreciated, however, that society has not always placed such value on animal welfare, as reflected in some of the scientific practices of the past. For example, in his experiments early in the nineteenth century, Magendie used unanesthetized puppies (for which he was later criticized by his scientific rival Bell). Before passing judgment, consider that the philosophy of Descartes was very influential in French society at that time. Animals of all types were believed to be simple automata, biological machines that lacked any semblance of emotion. As disturbing as this now seems, it is also worth bearing in mind that humans scarcely had any more respect for one another than they did for animals during this period (slavery was still practiced in the United States, for example). Fortunately, some things have changed quite dramatically since then. Animal welfare now is important to society in general and to scientists in particular. Unfortunately, other things have changed little. Humans around the world continue to abuse one another in myriad ways (child abuse, violent crime, ethnic cleansing, and so on). Clearly, a better understanding of the brain mechanisms of aggression is urgently needed.

Today, neuroscientists accept certain moral responsibilities toward their animal subjects:

1. Animals are used only for worthwhile experiments that promise to advance our knowledge of the nervous system.

2. All necessary steps are taken to minimize pain and distress experienced by the experimental animals (use of anesthetics, analgesics, etc.).

3. All possible alternatives to the use of animals are considered.

Adherence to this ethical code is monitored in a number of ways. First, research proposals must pass a review by the Institutional Animal Care and Use Committee (IACUC). Members of this committee include a veterinari-

---

[a] According to the National Academy of Sciences Institute of Medicine, 1991.

an, scientists in other disciplines, and nonscientist community representatives. After passing the IACUC review, proposals are evaluated for scientific merit by a panel of expert neuroscientists. This step ensures that only the most worthwhile projects are carried out. Then, when neuroscientists attempt to publish their observations in the professional journals, the papers are carefully reviewed by other neuroscientists for scientific merit and for animal welfare concerns. Reservations about either issue can lead to rejection of the papers, which in turn can lead to a loss of funding for the research. In addition to these monitoring procedures, federal law sets strict standards for the housing and care of laboratory animals.

Animal Rights. Most people accept the necessity for animal experimentation to advance knowledge, as long as it is performed humanely and with the proper respect for the animal's welfare. However, a vocal and increasingly violent minority seeks the total abolition of animal use for human purposes, including experimentation. These people subscribe to a philosophical position called "animal rights." According to this way of thinking, animals have the same legal and moral rights as humans do.

If you are an animal lover, you may be sympathetic to this position. But think again. Are you willing to deprive yourself and your family of medical procedures that were developed using animals? Is the death of a mouse equivalent to the death of a human being? Is keeping a pet the moral equivalent of slavery? Is eating meat the moral equivalent of murder? Is it unethical to take the life of a pig to save the life of a child? Is controlling the rodent population in the sewers or the roach population in your home morally equivalent to the Holocaust? If your answer is no to any of these questions, then you do not subscribe to the philosophy of animal rights. A*nimal welfare*—a concern that all responsible people share—must not be confused with animal rights.

The animal rightists have vigorously pursued their agenda against animal research, sometimes with alarming success. They have manipulated public opinion with repeated allegations of cruelty in animal experiments that are grossly distorted or blatantly false. They have vandalized laboratories, destroying years of hard-won scientific data and hundreds of thousands of dollars of equipment (that you, the taxpayer, had purchased). Using threats of violence, they have driven some researchers out of science altogether.

Fortunately, the tide is turning. Thanks to the efforts of a number of people, scientists and nonscientists alike, the false claims of the extremists have been exposed, and the benefits to humankind of animal research have been extolled (Figure 1.15). Considering the staggering toll in terms of human suffering that results from disorders of the nervous system, neuroscientists take the position that it is immoral not to wisely use all the resources nature has provided, including animals, to gain an understanding of how the brain functions in health and in disease.

The Cost of Ignorance: Nervous System Disorders

Modern neuroscience research is expensive, but the cost of ignorance about the brain is far greater. Table 1.3 lists some of the disorders that affect the nervous system. It is likely that your family has felt the impact of one or more of these. Let's look at a few brain disorders and examine their impact on society.[b]

Alzheimer's disease and Parkinson's disease both are characterized by a progressive degeneration of specific neurons in the brain. Parkinson's disease results in a crippling impairment of voluntary movement and currently affects approximately 500,000 Americans. Alzheimer's disease leads to

---

[b] Source of statistics for disorders in this section: U.S. Office of Science and Technology Policy, 1991.

18 | **CHAPTER 1**
Introduction to
Neuroscience

Thanks to animal research, they'll be able to protest 20.8 years longer.

According to the U.S. Department of Health and Human Services, animal research has helped extend our life expectancy by 20.8 years. Of course, how you choose to spend those extra years is up to you.

Foundation for Biomedical Research

Figure 1.15
Our debt to animal research. This ad was used to counter the claims of animal rights activists by raising public awareness of the benefits of animal research. (Source: National Foundation for Biomedical Research.)

dementia, a state of confusion characterized by the loss of ability to learn new information and to recall previously acquired knowledge. The U.S. National Institutes of Health estimate that dementia affects 10% of people over 65 and 50% of people over 85. The number of Americans with dementia totals well over 3 million. Indeed, it is now recognized that senility is not an inevitable outcome of aging, as was once believed, but is a sign of brain disease. Alzheimer's disease progresses mercilessly, robbing its victims first of their minds, then of control over basic bodily functions, and finally of their lives; the disease is always fatal. In the United States, the annual cost of care for people with dementia is approximately $90 billion.

Depression and schizophrenia are disorders of mood and thought. Depression is characterized by overwhelming feelings of dejection, worthlessness, and guilt. Fifteen million Americans will experience a major

Table 1.3
Some Major Disorders of the Nervous System

Disorder	Description
Alzheimer's disease	A progressive degenerative disease of the brain leading to senility and dementia.
Cerebral palsy	A motor disorder in children caused by damage to the cerebrum at the time of birth.
Depression	A serious disorder of mood characterized by insomnia, loss of appetite, and feelings of dejection.
Epilepsy	A condition characterized by periodic disturbances of brain electrical activity that can lead to seizures, loss of consciousness, and sensory disturbances.
Multiple sclerosis	A progressive disease that affects nerve conduction, characterized by episodes of weakness, lack of coordination, and speech disturbance.
Parkinson's disease	A progressive disease of the brain that leads to difficulty in initiating voluntary movement.
Schizophrenia	A severe psychotic illness characterized by delusions, hallucinations, and bizarre behavior.
Spinal paralysis	A loss of feeling and movement caused by traumatic damage to the spinal cord.
Stroke	A loss of brain function caused by disruption of the blood supply, usually leading to permanent sensory, motor, or cognitive deficit.

depressive illness at some time in their lives. Depression is the leading cause of suicide, which claims 30,000 lives each year in the United States. Schizophrenia is a severe personality disorder, characterized by delusions, hallucinations, and bizarre behavior. This disease often strikes at the prime of life—adolescence or early adulthood—and can persist for life. Over 2 million Americans suffer from schizophrenia. The National Institute of Mental Health estimates that mental disorders, such as depression and schizophrenia, cost the United States in excess of $130 billion annually.

Stroke is the third leading cause of death in the United States. Those stroke victims who do not die, some 100,000 per year, are likely to be permanently disabled. The annual cost of stroke is $25 billion. Alcohol and drug addiction affects virtually every family in the United States. The cost in terms of treatment, lost wages, and other consequences approaches $150 billion per year. These few examples only scratch the surface. *More Americans are hospitalized with neurological and mental disorders than with any other major disease group, including heart disease and cancer.*

The economic costs of brain dysfunction are enormous, but they pale in comparison with the staggering emotional toll on victims and their families. The prevention and treatment of brain disorders require an understanding of normal brain function, and this basic understanding is the goal of neuroscience. Neuroscience research has already contributed to the development of increasingly effective treatments for Parkinson's disease, depression, and schizophrenia. New strategies are being tested to rescue dying neurons in people with Alzhemier's disease and those who have suffered a stroke. Major progress has been made in understanding how drugs and alcohol affect the brain and how they lead to addictive behavior. The material in this book demonstrates that a lot is known about the function of the brain. But what we know is insignificant compared with what's still left to be learned.

CONCLUDING REMARKS

In this chapter, we have emphasized that neuroscience is a distinctly human endeavor. The historical foundations of neuroscience were laid by many people over many generations. Men and women today are working at all levels of analysis, using all types of technology, to shed light on the workings of the brain. The fruits of this labor form the basis for this textbook.

The goal of neuroscience is to understand how nervous systems function. Many important insights can be gained from a vantage point outside the head. Because the brain's activity is reflected in behavior, careful behavioral measurements inform us of the capabilities and limitations of brain function. Computer models that reproduce the brain's computational properties can help us understand how these properties might arise. From the scalp we can measure brain waves, which tell us something about the electrical activity of different parts of the brain during different behavioral states. New computer-assisted imaging techniques enable researchers to examine the structure of the living brain as it sits in the head. And, using even more sophisticated imaging methods, we are beginning to see what different parts of the human brain become active under different conditions. But none of these noninvasive methods, old or new, can substitute for experimentation with living brain tissue. We cannot make sense of remotely detected signals without being able to see how they are generated and what their significance is. To understand *how* the brain works, we must open the head and examine what's inside—neuroanatomically, neurophysiologically, and neurochemically.

The pace of neuroscience research today is truly breathtaking, and raises hopes that soon we will have new treatments for the wide range of nervous system disorders that debilitate and cripple millions of people annually. In recognition of the progress and promise of brain research, the U.S. Congress designated the 1990s the "Decade of the Brain." Despite this progress (both scientific and political), we still have a long way to go before we fully understand how the brain performs all of its amazing feats. But this is the fun of being a neuroscientist: Because our ignorance of brain function is so vast, a startling new discovery lurks around virtually every corner.

✓ REVIEW QUESTIONS

1. What are brain ventricles, and what functions have been ascribed to them over the ages?

2. What experiment did Bell perform to show that the nerves of the body contain a mixture of sensory and motor fibers?

3. What did Flourens' experiments suggest were the functions of the cerebrum and the cerebellum?

4. What is the meaning of the term animal model?

5. A region of the cerebrum is now called Broca's area. What function do you think this region performs, and why?

6. What are the different levels of analysis in neuroscience research? What types of question do researchers ask at each level?

7. What are the steps in the scientific method? Describe each one.

Neurons and Glia

All organs in the body consist of cells. The specialized functions of cells and how they interact specify the functions of organs. The brain is an organ—to be sure, the most sophisticated and complex organ that nature has devised. But the basic strategy for unraveling its function is no different from that used to investigate the pancreas or the lung. We must begin by learning how brain cells work individually, and then see how they are assembled to work together. In neuroscience, there is no need to separate *mind* from *brain*; once we fully understand the individual and concerted actions of brain cells, we will understand the origins of creative thought. The organization of this book reflects this "neurophilosophy." We begin with the cells of the nervous system—their structure, function, and means of communication. In later chapters, we will explore how these cells are assembled into circuits that mediate sensation, perception, movement, speech, and emotion.

In this chapter, we focus on the structure of the different types of cells in the nervous system: *neurons* and *glia*. "Neurons" and "glia" are broad categories. Within each group are many types of cells that differ based on their structure, chemistry, and function. Nonetheless, the distinction between neurons and glia is important. Although there are many neurons in the human brain (about 100 billion, give or take a few hundred million), glia outnumber neurons by tenfold. Based on these numbers, it might appear that we should focus our attention on glia for insights into the cellular functions of the nervous system. However, neurons are the most important cells for the unique functions of the brain. It is the neurons that sense changes in the environment, communicate these changes to other neurons, and command the body's responses to these sensations. Glia are thought to contribute to brain function mainly by insulating, supporting, and nourishing neighboring neurons. If the brain were a chocolate-chip cookie and the neurons were chocolate chips, the glia would be the cookie dough that fills all the other space and ensures that the chips are suspended in their appropriate locations. Indeed, the word "glia" is derived from the Greek word for "glue," giving the impression that the main function of these cells is to keep the brain from running out of our ears! As we shall see later in this chapter, the simplicity of this view is probably a good indication of the depth of our ignorance about glial function. However, we still are confident that neurons perform the bulk of information-processing in the brain. Therefore, we will focus 90% of our attention on 10% of brain cells: the neurons.

Neuroscience, like other fields, has a language all its own. In order to use this language, you must learn the vocabulary. After you have read this chapter, take a few minutes to review the key terms list and make sure you understand the meaning of each term. Each term appears in boldface type in the text. Your neuroscience vocabulary will grow as you work your way through the book.

THE NEURON DOCTRINE

In order to study the structure of brain cells, scientists have had to overcome several obstacles. The first was their small size. Most cells range from 0.01 to 0.05 mm in diameter. The tip of an unsharpened pencil lead is about 2 mm across; neurons are 40–200 times smaller. (For a review of the metric system, see Table 2.1.) This size is at or beyond the limit of what can be seen by the naked eye. Therefore, progress in cellular neuroscience was not possible before the development of the compound microscope in the late seventeenth century. Even then, obstacles remained. To observe brain tissue using a microscope, it was necessary to make very thin slices, ideally not much thicker than the diameter of the cells. However, brain tissue has a consistency like a bowl of Jello: not firm enough to make thin slices. Thus, the study of the anatomy of brain cells had to await the development of a method to harden the tissue without disturbing its structure, and an instrument that could produce very thin slices. Early in the nineteenth century, scientists discovered how to harden or "fix" tissues by immersing them in formaldehyde, and they developed a special device called a microtome to make very thin slices.

The compound microscope and the development of methods to slice organs and fix tissues spawned a new field called **histology**, the microscopic study of the structure of tissues. But scientists studying brain structure faced yet another obstacle. Freshly prepared brain has a uniform, cream-colored appearance under the microscope; the tissue has no differences in pigmentation to enable histologists to resolve individual cells. Thus, the final breakthrough in neurohistology was the introduction of stains that could selectively color some, but not all, parts of the cells in brain tissue.

One stain, still used today, was introduced by German neurologist Franz Nissl in the late nineteenth century. Nissl showed that a class of basic dyes would stain the nuclei of all cells and also stain clumps of material surrounding the nuclei of neurons (Figure 2.1). These clumps are called *Nissl bodies*, and the stain is known as the **Nissl stain**. The Nissl stain is extremely useful for two reasons. First, it distinguishes neurons and glia from one another. Second, it enables histologists to study the arrangement, or **cytoarchitecture**, of neurons in different parts of the brain. The study of cytoarchitecture led to the realization that the brain consists of many specialized regions. We now know that each region performs a different function.

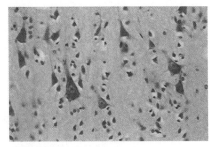

Figure 2.1
Nissl stained neurons. A thin slice of brain tissue has been stained with Cresyl violet, a Nissl stain. The clumps of deeply stained material around the cell nuclei are Nissl bodies. (Source: Hammersen, 1980, Fig. 493.)

Table 2.1
Units of Size in the Metric System

Unit	Abbreviation	Meter Equivalent	Real World Equivalent
Kilometer	km	10^3 m	About two-thirds of a mile.
Meter	m	1 m	About 3 feet.
Centimeter	cm	10^{-2} m	Thickness of your little finger
Millimeter	mm	10^{-3} m	Thickness of your toenail.
Micrometer	μm	10^{-6} m	Near the limit of resolution for the light microscope.
Nanometer	nm	10^{-9} m	Near the limit of resolution for the electron microscope.

The Golgi Stain

The Nissl stain, however, does not tell the whole story. A Nissl-stained neuron looks like little more than a lump of protoplasm containing a nucleus. Neurons are much more than that, but how much more was not recognized until the work of Italian histologist Camillo Golgi (Figure 2.2). In 1873, Golgi discovered that by soaking brain tissue in a silver chromate solution, now called the **Golgi stain**, a small percentage of neurons became darkly colored in their entirety (Figure 2.3). This revealed that the neuronal cell body, the region of the neuron around the nucleus that is shown with the Nissl stain, is actually only a small fraction of the total structure of the neuron. Notice in Figures 2.1 and 2.3 how different histological stains can provide strikingly different views of the same tissue. Today, neurohistology remains an active field in neuroscience, along with its credo: "The gain in brain is mainly in the stain."

The Golgi stain shows that neurons have at least two distinguishable parts: a central region that contains the cell nucleus, and numerous thin tubes that radiate away from the central region. The swollen region containing the cell nucleus has several names that are used interchangeably: **cell body**, **soma** (plural: somata), and **perikaryon** (plural: perikarya). The thin tubes that radiate away from the soma are called **neurites** and are divided into two types: **axons** and **dendrites** (Figure 2.4). The cell body usually gives rise to a single axon. The axon is of uniform diameter throughout its length, and, if it branches, the branches generally extend at right angles. Because axons can travel over great distances in the body (a meter or more), it was immediately recognized by the histologists of the day that axons must act like "wires" that carry the output of the neurons. Dendrites, on the other hand, rarely extend more than 2 mm in length. Many dendrites extend from the cell body, and they generally taper to a fine point. Because dendrites come in contact with many axons, early histologists recognized that they must act as the antennae of the neuron to receive incoming signals, or input.

Figure 2.2
Camillo Golgi (1843–1926). (Source: Finger, 1994, Fig. 3.22.)

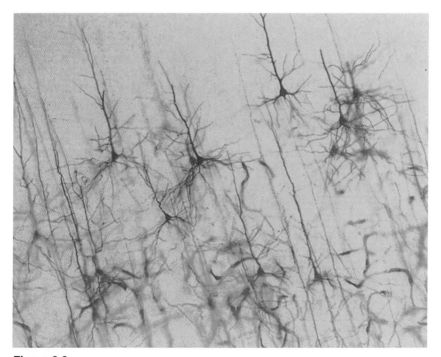

Figure 2.3
Golgi-stained neurons. (Source: Hubel, 1988, p. 126.)

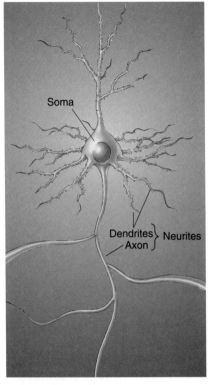

Figure 2.4
The basic parts of a neuron.

Cajal's Contribution

Golgi invented the stain, but it was a Spanish contemporary of Golgi's who used it to greatest effect. Santiago Ramón y Cajal was a skilled histologist and artist who learned about Golgi's method in 1888 (Figure 2.5). In a remarkable series of publications over the next 25 years, Cajal used the Golgi stain to work out the circuitry of many regions of the brain (Figure 2.6). Ironically, Golgi and Cajal drew completely opposite conclusions about neurons. Golgi championed the view that the neurites of different cells are fused together to form a continuous reticulum, or network, similar to the arteries and veins of the circulatory system. According to this reticular theory, the brain is an exception to the cell theory, which states that the individual cell is the elementary functional unit of all animal tissues. Cajal, on the other hand, argued forcefully that the neurites of different neurons are not continuous with one another, and must *communicate by contact, not continuity*. This idea that the neuron adhered to the cell theory came to be known as the **neuron doctrine**. Although Golgi and Cajal shared the Nobel Prize in 1906, they remained rivals to the end.

The scientific evidence over the next 50 years weighed heavily in favor of the neuron doctrine, but final proof had to wait until the development of the electron microscope in the 1950s (Box 2.1). With the increased resolving power of the electron microscope, it was finally possible to show that the neurites of different neurons are not continuous with one another. Thus, our starting point in the exploration of the brain must be the individual neuron.

THE PROTOTYPICAL NEURON

As we have seen, the neuron consists of several parts: the soma, the dendrites, and the axon. The inside of the neuron is separated from the outside by the limiting skin, the neuronal membrane, which lies like a circus tent on an intricate internal scaffolding, giving each part of the cell its spe-

---

Box 2.1	OF SPECIAL INTEREST

Advances in Microscopy

The human eye can distinguish two points only if they are separated by more than about one-tenth of a millimeter (100 μm). Thus, we can say that 100 μm is near the *limit of resolution* for the unaided eye. Neurons have a diameter of about 20 μm, and neurites can measure as small as a fraction of a micrometer. The light microscope, therefore, was a necessary development before neuronal structure could be studied. But this type of microscopy has a theoretical limit imposed by the properties of microscope lenses and visible light. With the standard light microscope, the limit of resolution is about 0.1 μm. However, the space between neurons measures only 0.02 μm (20 nm). No wonder two esteemed scientists, Golgi and Cajal, disagreed about whether or not neurites were continuous. This question could not be answered until the electron microscope was developed and applied to biological specimens, which only occurred within the last 50 years or so. The electron microscope uses an electron beam instead of light to form images, dramatically increasing the resolving power. The limit of resolution for an electron microscope is about 0.1 nm—a million times better than the unaided eye. Our insights into the fine structure of the inside of neurons—the *ultrastructure*—have all come from electron microscopic examination of the brain. Today, microscopes on the cutting edge use laser beams to illuminate the tissue and computers to create digital images. Unlike the traditional methods of light and electron microscopy, these new techniques give neuroscientists their first chance to peer into brain tissue that is still alive.

cial three-dimensional appearance. Let's explore the inside of the neuron and learn about the functions of the different parts.

The Soma

We begin our tour at the soma, the roughly spherical central part of the neuron (Figure 2.7). The cell body of the typical neuron is about 20 µm in diameter. The watery fluid inside the cell is called the **cytosol**, a salty, potassium-rich solution. The membrane, about 5 nm thick, separates the inside from the outside of the cell. Within the soma are a number of membrane-enclosed structures collectively called **organelles**.

The cell body of the neuron contains the same organelles that are found in all animal cells. The important ones are the nucleus, the rough endoplasmic reticulum, the smooth endoplasmic reticulum, the Golgi apparatus, and the mitochondria. Everything contained within the confines of the cell membrane, excluding the nucleus, is referred to collectively as the **cytoplasm** of the cell.

The Nucleus. Derived from the Latin word for "nut," the **nucleus** of the cell is spherical, centrally located, and measures about 5–10 µm across. It is contained within a double membrane called the *nuclear envelope*. The nuclear envelope is interrupted periodically by pores that measure about 0.1 µm across.

Within the nucleus are **chromosomes**, which contain the heritable material, **DNA (deoxyribonucleic acid)**. Your DNA was passed on to you from your parents, and it contains the blueprint for your entire body. The DNA in each of your neurons is the same, and it is the same as the DNA in the cells of your liver and kidney. What distinguishes a neuron from a liver cell is not the DNA *per se*, but rather which parts of the DNA are used to assemble the cell.

Each chromosome contains an uninterrupted double-stranded braid of DNA, 2 nm wide. If the DNA from the 46 human chromosomes were laid out straight, end to end, it would measure more than 2 m in length. If we were to consider this total length of DNA as analogous to the string of letters that makes up this book, the genes would be analogous to the individual words. Each gene is a segment of DNA that can measure anywhere from 0.1 µm to several micrometers in length.

The "reading" of the DNA is called **gene expression**, and the final product of gene expression is the synthesis of remarkable molecules called **proteins**. Proteins exist in a wide variety of shapes and sizes, perform many different functions, and bestow neurons with virtually all of their unique characteristics. **Protein synthesis**, the assembly of protein molecules, occurs in the cytoplasm of the soma. Because the DNA never leaves the nucleus, there must be an intermediary that carries the genetic message to the sites of protein synthesis in the cytoplasm. This function is performed by another long molecule called **messenger ribonucleic acid**, or **mRNA**. Messenger RNA consists of four different nucleic acids strung together in various sequences to form a chain. The detailed sequence of the nucleic acids in the chain represents the information in the gene, just as the sequence of letters gives meaning to a written word. The process of assembling a piece of mRNA that contains the information of a gene is called **transcription**, and the resulting mRNA is called the *transcript* (Figure 2.8).

Messenger RNA transcripts emerge from the nucleus via pores in the nuclear envelope and travel to the sites of protein synthesis elsewhere in the neuron. At these sites, a protein molecule is assembled much as the mRNA molecule was: by linking together many small molecules into a chain. In the case of protein, the building blocks are the **amino acids**, of which there are

Figure 2.5
Santiago Ramón y Cajal (1852–1934).
(Source: Finger, 1994, Fig. 3.26.)

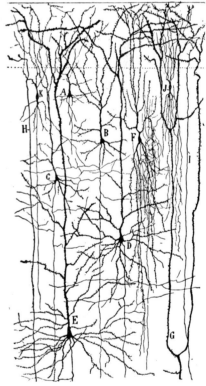

Figure 2.6
One of Cajal's many drawings of brain circuitry. The letters label the different elements Cajal identified in an area of the human cerebral cortex that controls voluntary movement. We will learn more about this part of the brain in Chapter 14. (Source: DeFelipe and Jones, 1988, Fig. 90.)

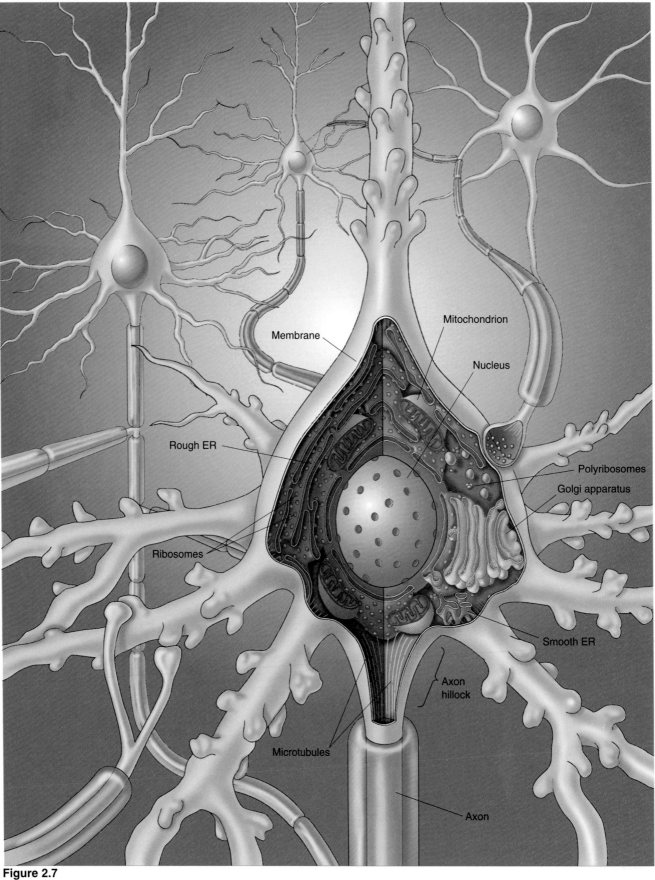

Figure 2.7
A tour of the neuron.

20 different kinds. This assembling of proteins from amino acids under the direction of the mRNA is called **translation**.

The scientific study of this process, which begins with the DNA of the nucleus and ends with the synthesis of protein molecules in the cell, is called *molecular biology*. The "central dogma" of molecular biology is summarized as:

$$\text{DNA} \xrightarrow{\text{Transcription}} \text{mRNA} \xrightarrow{\text{Translation}} \text{Protein}$$

An emerging new field within neuroscience is called *molecular neurobiology*. Molecular neurobiologists use the information contained in the genes to determine the structure and functions of neuronal proteins.

Rough Endoplasmic Reticulum. Not far from the nucleus, we encounter enclosed stacks of membrane dotted with dense globular structures called **ribosomes**, which measure about 25 nm in diameter. The stacks are called **rough endoplasmic reticulum**, or **rough ER** (Figure 2.9). Rough ER abounds in neurons, far more than in glia or most other non-neuronal cells. In fact, we have already been introduced to rough ER by another name: Nissl bodies. This organelle is stained with the dyes that Nissl introduced 100 years ago.

Rough ER is a major site of protein synthesis in neurons. RNA transcripts bind to the ribosomes, and the ribosomes translate the instructions contained in the mRNA to assemble a protein molecule. Thus, ribosomes take raw material in the form of amino acids and manufacture proteins using the blueprint provided by the mRNA (Figure 2.10a).

As we look around the cytosol, we see that not all ribosomes are attached to the rough ER. Many are freely floating and are called free ribo-

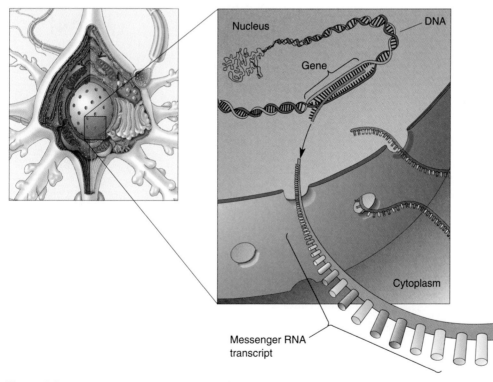

Figure 2.8
Gene transcription. Messenger RNA molecules carry the genetic instructions for protein assembly from the nucleus to the cytoplasm.

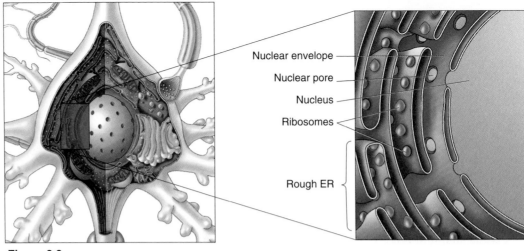

Figure 2.9
Rough endoplasmic reticulum, or rough ER.

Nuclear envelope
Nuclear pore
Nucleus
Ribosomes
Rough ER

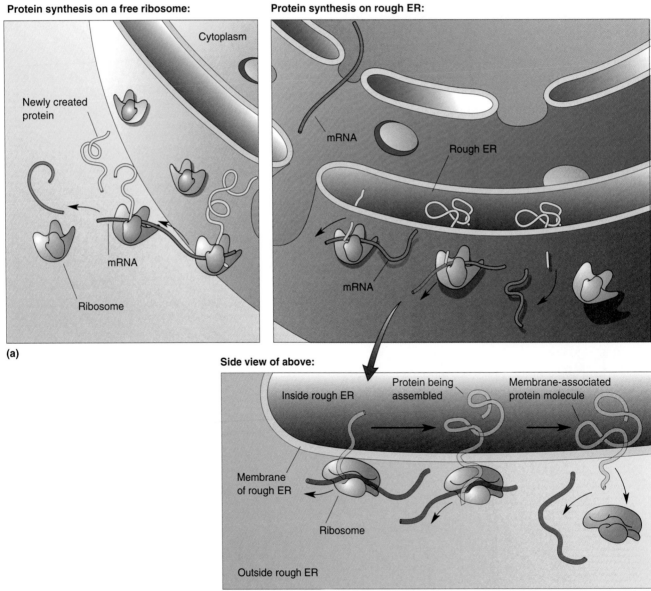

Protein synthesis on a free ribosome:

Cytoplasm

Newly created protein

mRNA

Ribosome

(a)

Protein synthesis on rough ER:

mRNA

Rough ER

mRNA

Side view of above:

Inside rough ER

Protein being assembled

Membrane-associated protein molecule

Membrane of rough ER

Ribosome

Outside rough ER

(b)

Figure 2.10
Protein synthesis on a free ribosome and on rough ER. mRNA binds to a ribosome, initiating protein synthesis. **(a)** Proteins synthesized on free ribosomes are destined for the cytosol. **(b)** Proteins synthesized on the rough ER are destined to be enclosed by or inserted in membrane. Membrane-associated proteins are inserted into the membrane as they are assembled.

somes. Several free ribosomes may appear to be attached by a thread; these are called **polyribosomes**. The thread is a single strand of mRNA, and the associated ribosomes are working on it to make multiple copies of the same protein.

What is the difference between proteins synthesized on the rough ER and those synthesized on the free ribosomes? The answer appears to lie in the intended fate of the protein molecule. If it is destined to reside within the cytosol of the neuron, then the protein's mRNA transcript shuns the ribosomes of the rough ER and gravitates toward the free ribosomes. However, if the protein is destined to become inserted into the membrane of the cell or an organelle, then it is synthesized on the rough ER. As the protein is being assembled, it is threaded back and forth through the membrane of the rough ER, where it is trapped (Figure 2.10b). It is not surprising that neurons are so well endowed with rough ER because, as we shall see in later chapters, special membrane proteins are what give these cells their remarkable information-processing abilities.

Smooth Endoplasmic Reticulum and the Golgi Apparatus. The remainder of the cytosol of the soma is crowded with stacks of membranous organelles that look a lot like rough ER without the ribosomes, so much so that one type is called **smooth endoplasmic reticulum**, or **smooth ER**. Smooth ER is actually quite heterogeneous and performs different functions in different locations. Some smooth ER is continuous with rough ER and is believed to be a site where the proteins that jut out from the membrane are carefully folded, giving them their three-dimensional structure. Other types of smooth ER play no direct role in the processing of protein molecules but function instead to regulate the internal concentrations of substances such as calcium. (This organelle is particularly prominent in muscle cells, where it is called sarcoplasmic reticulum, as we will see in Chapter 13.)

The stack of membrane-enclosed disks in the soma that lies furthest away from the nucleus is the **Golgi apparatus**, first described in 1898 by Camillo Golgi (Figure 2.11). This is a site of extensive "post-translational" chemical processing of proteins. One important function of the Golgi apparatus is believed to be the sorting of certain proteins that are destined for delivery to different parts of the neuron, such as the axon and the dendrites.

The Mitochondrion. Another very abundant organelle in the soma is the **mitochondrion** (plural: mitochondria). In neurons, these sausage-shaped

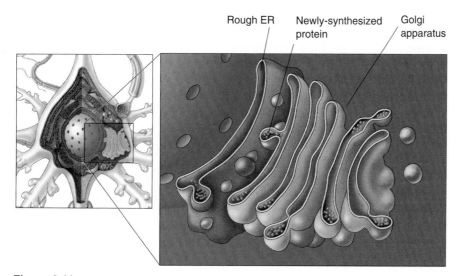

Rough ER Newly-synthesized protein Golgi apparatus

Figure 2.11
The Golgi apparatus.

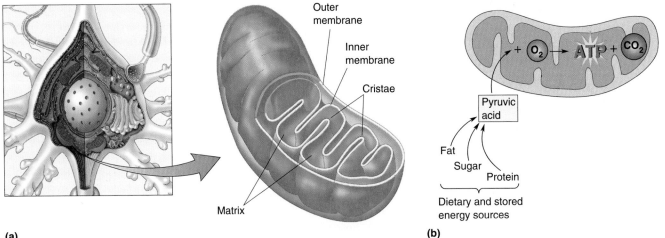

(a)

(b)

Figure 2.12
(a) The mitochondrion. (b) Cellular respiration. ATP is the energy currency that fuels biochemical reactions in neurons.

structures measure about 1 μm in length. Within the enclosure of their outer membrane are multiple folds of inner membrane called cristae (singular: crista). Between the cristae is an inner space called matrix.

The mitochondria are the site of *cellular respiration* (Figure 2.12). When a mitochondrion "takes a breath," it pulls inside pyruvic acid (derived from sugars and digested proteins and fats) and oxygen, both of which are floating in the cytosol. Within the inner compartment of the mitochondrion, pyruvic acid enters into a complex series of biochemical reactions called the *Krebs cycle*, named after the German-British scientist Hans Krebs, who first proposed it in 1937. The biochemical products of the Krebs cycle provide energy that, within the cristae, results in the addition of phosphate to adenosine diphosphate (ADP), yielding **adenosine triphosphate**, or **ATP**, the cell's energy source. When the mitochondrion "exhales," 17 ATP molecules are released for every molecule of pyruvic acid that had been taken in.

ATP is the energy currency of the cell. The chemical energy stored in ATP is used to fuel most of the biochemical reactions of the neuron. For example, as we shall see in Chapter 3, special proteins in the neuronal membrane use the energy released by the breakdown of ATP into ADP to pump certain substances across the membrane to establish concentration differences between the inside of the neuron and the outside.

The Neuronal Membrane

The **neuronal membrane** serves as a barrier to enclose the cytoplasm inside the neuron, and to exclude certain substances that float in the fluid that bathes the neuron. The membrane is about 5 nm thick, and is studded with proteins. As mentioned above, some of the membrane-associated proteins pump substances from the inside to the outside. Others form pores that regulate which substances can gain access to the inside of the neuron. An important characteristic of neurons is that the protein composition of the membrane varies depending on whether it is in the soma, the dendrites, or the axon.

The function of neurons cannot be understood without an understanding of the structure and function of the membrane and its associated proteins. In fact, this topic is so important that we'll spend a good deal of the next four chapters looking at how the membrane endows neurons with the remarkable ability to transfer electrical signals throughout the brain and body.

The Cytoskeleton

Earlier we compared the neuronal membrane to a circus tent that was draped onto an internal scaffolding. This scaffolding is called the **cytoskeleton**, and it gives the neuron its characteristic shape. The "bones" of the cytoskeleton are the microtubules, microfilaments, and neurofilaments (Figure 2.13). By drawing an analogy with a scaffolding, we do not mean that the cytoskeleton is static. On the contrary, elements of the cytoskeleton are dynamically regulated and are very likely in continual motion. Your neurons are probably squirming around in your head even as you read this sentence.

Microtubules. **Microtubules** are big, measuring 20 nm in diameter, and run longitudinally down neurites. A microtubule appears like a straight, thick-walled hollow pipe. The wall of the pipe is composed of smaller strands that are braided like rope around the hollow core. Each of the smaller strands consists of the protein *tubulin*. A single tubulin molecule is small and globular; the strand consists of tubulins stuck together like meatballs on a barbeque skewer. The process of joining small proteins to form a long strand is called *polymerization*; the resulting strand is called a *polymer*. Polymerization and depolymerization of microtubules, and therefore neuronal shape, can be regulated by various signals within the neuron.

One class of proteins that participate in the regulation of microtubule assembly and function are *microtubule-associated proteins*, or *MAPs*. Among other functions (many of which are unknown), MAPs anchor the microtubules to one another and to other parts of the neuron.

Microfilaments. **Microfilaments** are small, measuring only 5 nm in diameter (approximately the same thickness as the cell membrane). They

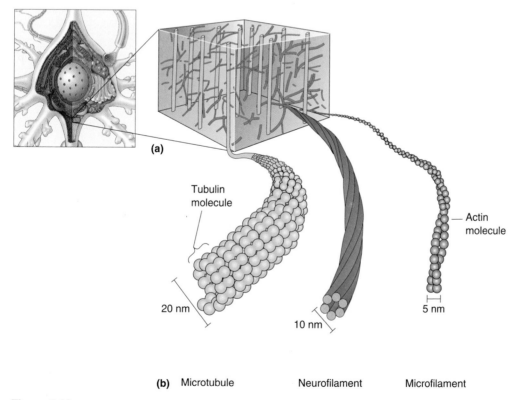

Tubulin molecule

Actin molecule

20 nm

10 nm

5 nm

(a)

(b) Microtubule   Neurofilament   Microfilament

Figure 2.13
(a) Cytoskeleton. (b) Elements of the cytoskeleton. The arrangement of microtubules, neurofilaments, and microfilaments gives the neuron its characteristic shape.

are found throughout the neuron but are particularly numerous in the neurites. Microfilaments are braids of two thin strands, and the strands are polymers of the protein *actin*. Actin is one of the most abundant proteins in cells of all types, including neurons, and is believed to play a role in changing cell shape. Indeed, as we shall see in Chapter 13, actin filaments are critically involved in the mechanism of muscle contraction.

Like microtubules, actin microfilaments are constantly undergoing assembly and disassembly, and this process is regulated by signals in the neuron. Besides running longitudinally down the core of the neurites like microtubules, microfilaments are also closely associated with the membrane. They are anchored to the membrane by attachments with a meshwork of fibrous proteins that line the inside of the membrane like a spider web.

Neurofilaments. With a diameter of 10 nm, **neurofilaments** are intermediate in size between microtubules and microfilaments. In fact, they exist in all cells of the body with the name *intermediate filament*; only in neurons are they called neurofilaments. The difference in name actually does reflect subtle differences in structure from one tissue to the next. An example of an intermediate filament from another tissue is keratin, which, when bundled together, makes up hair.

Of the types of fibrous structure we have discussed, neurofilaments most closely resemble the bones and ligaments of the skeleton. A neurofilament consists of multiple subunits (building blocks) that are organized like a chain of sausages. The internal structure of each subunit consists of three protein strands that are woven together. Unlike the microfilaments and microtubules, these strands consist of individual long protein molecules, each of which is coiled in a tight, springlike configuration. This structure makes the neurofilaments mechanically very strong.

Another distinction of neurofilaments is that their length and organization in the cell appear not to be as dynamic as microfilaments and microtubules. However, this stable situation is dramatically disrupted in Alzheimer's disease, in which neurons become stuffed with neurofilaments to the point that they finally die. The impenetrable thicket of neurofilaments actually stays in the brain long after the remains of the dead neuron have been cleared away, leaving what is called a *neurofibrillary tangle*, a characteristic pathology of this disease.

The Axon

So far, we've explored the soma, organelles, membrane, and cytoskeleton. However, nothing we saw is unique to neurons. These are structural features of all the cells in our body. But now we encounter the axon, a structure which is found only in neurons, and is highly specialized for the transfer of information over a distance in the nervous system.

The axon begins with a region called the **axon hillock**, which tapers to form the initial segment of the axon proper (Figure 2.14). Two noteworthy features distinguish the axon from the soma. First, no rough ER extends into the axon, and there are few, if any, free ribosomes. Second, the protein composition of the axon membrane is fundamentally different from that of the soma membrane. These structural differences translate into functional distinctions. Because there are no ribosomes, there is no protein synthesis in the axon. This means that all proteins in the axon must originate in the soma. And it is the different proteins in the axonal membrane that enable it to serve as the "telegraph wire" that sends information over great distances.

Axons may extend from less than a millimeter to over a meter long. An axon originating in, and coursing *away* from, a particular neuron is called an

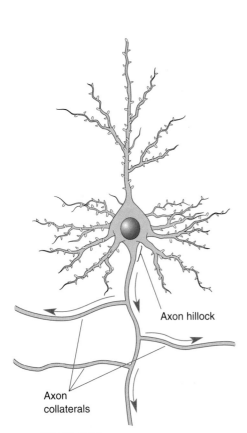

Figure 2.14
The axon and axon collaterals. The axon functions like a telegraph wire to send electrical impulses to distant sites in the nervous system. The direction of information flow is indicated by the arrows.

Axon hillock

Axon collaterals

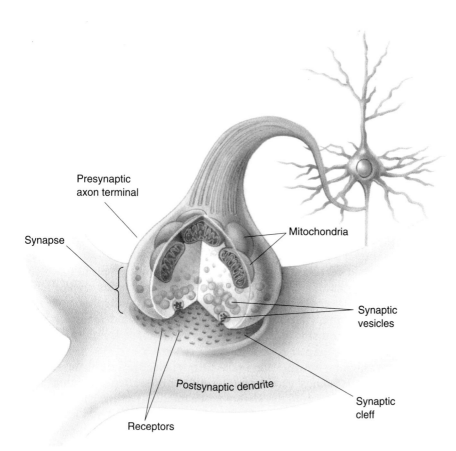

Presynaptic
axon terminal

Synapse

Mitochondria

Synaptic
vesicles

Postsynaptic dendrite

Synaptic
cleff

Receptors

Figure 2.15
The axon terminal and the synapse. Axon terminals form synapses with the dendrites or somata of other neurons. When a nerve impulse arrives in the presynaptic
axon terminal, neurotransmitter molecules are released from synaptic vesicles into
the synaptic cleft. Neurotransmitter then binds to specific receptor proteins, causing
the generation of electrical or chemical signals in the postsynaptic cell.

efferent. An axon coursing toward, and providing input to, a particular
neuron is called an **afferent**. Axons often branch, and these branches are
called **axon collaterals**. Occasionally, an efferent axon will branch, and the
collateral will return and be an afferent to the same cell that gave rise to the
axon, or to the dendrites of neighboring cells. These axon branches are
called *recurrent collaterals*.

The diameter of an axon is variable, ranging from less than 1 μm to about
25 μm in humans, and to as large as 1 mm in the squid. This variation in axon
size is important. As will be explained in Chapter 4, the speed of the electrical signal that sweeps down the axon—the *nerve impulse*—varies depending
on axonal diameter. The fatter the axon, the faster the impulse travels.

The Axon Terminal. All axons have a beginning (the axon hillock), a
middle (the axon proper), and an end. The end is called the **axon terminal**
or the **terminal bouton** (French for "button"), which reflects the fact that it
usually appears as a swollen disk (Figure 2.15). The terminal is a site where
the axon comes in contact with other neurons (or other cells) and passes
information on to them. This point of contact is called the **synapse**, a word
derived from the Greek, meaning "to fasten together." Sometimes axons
have many branches at their ends, and each branch forms a synapse on dendrites or cell bodies in the same region. These branches are collectively
called the **terminal arbor**. Sometimes axons form synapses at swollen

regions along their length and then continue on to terminate elsewhere. Such swellings are called **boutons en passant** ("buttons in passage"). In either case, when a neuron makes synaptic contact with another cell, it is said to **innervate** that cell, or to provide innervation.

The cytoplasm of the axon terminal differs from that of the axon in several ways. First, microtubules do not extend into the terminal. Second, the terminal contains numerous small bubbles of membrane, called **synaptic vesicles**, that measure about 50 nm in diameter. Third, the inside surface of the membrane that faces the synapse has a particularly dense covering of proteins. Another characteristic of the axon terminal is that it has numerous mitochondria, indicating a high energy demand.

The Synapse. Although Chapters 5 and 6 are devoted entirely to how information is transferred from one neuron to another at the synapse, we provide a preview here.

The synapse has two sides: *presynaptic* and *postsynaptic*. These names indicate the usual direction of information flow, which is from "pre" to "post." The presynaptic side generally consists of an axon terminal, while the postsynaptic side could be the dendrite or soma of another neuron. The space between the presynaptic and postsynaptic membranes is called the **synaptic cleft**. The transfer of information at the synapse from one neuron to another is called **synaptic transmission**.

At most synapses, information in the form of electrical impulses traveling down the axon is converted in the terminal into a chemical signal that crosses the synaptic cleft. On the postsynaptic membrane, this chemical signal is converted again into an electrical one. The chemical signal is called a **neurotransmitter**, and it is stored in and released from the synaptic vesicles within the terminal. As we will see, different neurotransmitters are used by different types of neurons.

This electrical-to-chemical-to-electrical transformation of information makes possible many of the brain's computational abilities. Modification of this process is involved in memory and learning, and synaptic transmission dysfunction accounts for certain mental disorders. The synapse is also the site of action for nerve gas and for most psychoactive drugs.

Axoplasmic Transport. As mentioned, one feature of the cytoplasm of axons, including the terminal, is the absence of ribosomes. Since ribosomes are the protein factories of the cell, their absence means that the proteins of the axon must be synthesized in the soma and then shipped down the axon. This movement of material down the axon is called **axoplasmic transport**.

Axoplasmic transport was first demonstrated by the experiments of American neurobiologist Paul Weiss and his colleagues in the 1940s. They found that if they tied a thread around an axon, material accumulated on the side of the axon closest to the soma. When the knot was untied, the accumulated material continued down the axon at a rate of 1–10 mm per day.

This was a remarkable discovery, but it is not the whole story. If all material moved down the axon by this transport mechanism alone, it wouldn't reach the ends of the longest axons for at least half a year—too long a wait to feed hungry synapses. In the late 1960s, methods were developed to track the movements of protein molecules down the axon into the terminal. These methods entailed injecting the somata of neurons with radioactive amino acids. Recall that amino acids are the building blocks of proteins. The "hot" amino acids were assembled into proteins, and the arrival of radioactive proteins in the axon terminal was measured to calculate the rate of transport. Bernice Grafstein of Rockefeller University discovered that this *fast axoplasmic transport* (so named to distinguish it from *slow axoplasmic transport* described by Weiss) occurred at a rate as high as 1000 mm per day (Box 2.2).

Much is now known about how axoplasmic transport works. Material is enclosed within vesicles, which then "walk down" the microtubules of the axon. The "legs" are provided by a protein called *kinesin*, and the process is fueled by ATP (Figure 2.16). Kinesin moves material only from the soma to the terminal. All movement of material in this direction is called **anterograde transport**.

In addition to anterograde transport, from the soma to the terminal, there is also a mechanism for the movement of material up the axon from

---

Box 2.2 | **PATH OF DISCOVERY**

Fast Axoplasmic Transport
by Bernice Grafstein

The beginning of 1967 was an exciting time for me. I had recently returned to full-time work in my laboratory at Rockefeller University following the birth of my son, and had resumed my studies on the regeneration of the goldfish optic nerve. I had begun these studies a few months earlier, in the expectation that while I was pregnant I would not be able to continue with the physically demanding electrophysiological experiments that until then had been the staple of my research. I had therefore deliberately embarked on a kind of work that would not be limited in its rate of progress by what I personally could do, but could be carried forward with the assistance of other hands. Having previously been a member of the nearby laboratory of Paul Weiss, I was familiar with the use by him and his associates of the technique of radioactive isotope labeling, based on methods published by Bernard Droz and Charles Leblond, to demonstrate the movement of proteins along axons of the CNS. I thought that this technique could be used to trace the trajectory of regenerating axons in the optic nerve of the goldfish, following the demonstration by Roger Sperry that these axons had an astonishing capacity for regeneration, unlike mammalian optic axons. Sperry had claimed that the regenerating axons could find their way to their correct targets in the brain [see Box 18.2], and it seemed to me that labeling the axons might make it possible to test this provocative hypothesis.

I began by injecting radioactive leucine (an amino acid) into the eye, where it was taken up by retinal neurons and incorporated into protein. The progress of the labeled protein along the axons of the retinal ganglion cell could then be followed by measuring the amount of radioactivity in successive slices of optic nerve. I found that a wave of radioactivity was moving at a rate of a fraction of a millimeter per day,

Bernice Grafstein

consistent with the "axoplasmic flow" originally postulated by Weiss. However, I was startled to find that the brain was already labeled long before this slow-moving wave of radioactivity could have reached it. The fast-moving material ended up in optic nerve axon terminals, whereas the slower-moving wave of radioactivity did not emerge from the optic axon trunks. This led me to the idea that two different kinds of cell constituents were moving down the axon at different rates.

I prepared this work for publication, and sent out an announcement that I was going to present it at the informal research-in-progress seminar for neuroscientists at Rockefeller. Almost immediately I received a telephone call from Bruce McEwen, whom I had never met, but whose name I knew as having recently joined the Rockefeller faculty. "I think we ought to talk," he said. That was a very good idea, as it turned out, since there he was, at the other end of the Rockefeller campus, likewise engaged in experiments with radioactive labeling of goldfish (can you believe it?) optic nerve proteins. What might have become an uncomfortable competition, however, turned instead into a profoundly satisfying collaboration, as we explored together the composition of different components of axonal transport.

Although I did go on to study the characteristics of axonal transport in regenerating goldfish optic axons, I never did get around to doing the experiments on specificity of regeneration that I had originally envisaged.

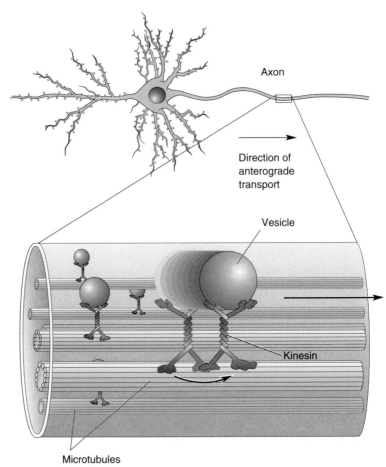

Axon

Direction of
anterograde
transport

Vesicle

Kinesin

Microtubules

Figure 2.16
Mechanism for movement of material on the microtubules of the axon. Trapped
in membrane-enclosed vesicles, material is transported from the soma to the axon
terminals by the action of the protein kinesin, which "walks" along microtubules at
the expense of ATP.

the terminal to the soma. This process is believed to provide signals to the
soma about changes in the metabolic needs of the axon terminal. Movement
in this direction, from terminal to soma, is called **retrograde transport**. The
molecular mechanism is similar to anterograde transport, except the "legs"
for retrograde transport are provided by a different protein called *dynein*.
Both anterograde and retrograde transport mechanisms have been exploited by neuroscientists to trace connections in the brain (Box 2.3).

Dendrites

The word "dendrite" is derived from the Greek for "tree," reflecting the fact
that these neurites resemble the branches of a tree as they extend from the
soma. The dendrites of a single neuron are collectively called the **dendritic tree**;
each branch of the tree is called a *dendritic branch*. Dendritic trees come in a wide
variety of shapes and sizes that are used to classify different groups of neurons.

Because dendrites function as the antennae of the neuron, they are covered with thousands of synapses (Figure 2.17). The dendritic membrane
under the synapse (the *postsynaptic* membrane) has many specialized protein molecules called **receptors** that detect the neurotransmitters in the
synaptic cleft.

The dendrites of some neurons are covered with specialized structures called **dendritic spines** that receive some types of synaptic input. Spines look like little punching bags that hang off the dendrite (Figure 2.18). They are believed to isolate various chemical reactions that are triggered by some types of synaptic activation. The true functions of spines remain a mystery because their small size (about 0.2 μm^3) has made direct examination in the living brain very difficult. This situation will soon change with the advent of new imaging techniques.

For the most part, the cytoplasm of dendrites resembles that of axons. It is filled with cytoskeletal elements and mitochondria. One interesting difference is that polyribosomes can be observed in dendrites, often right under a spine, suggesting that synaptic transmission can actually direct local protein synthesis in some neurons.

CLASSIFYING NEURONS

It is unlikely that we can ever hope to understand how each of the hundred billion neurons in the nervous system uniquely contributes to the function of the brain. But what if we could show that all the neurons in the brain can be divided into a small number of categories, and that within each category, all neurons function identically? The complexity of the problem might then be reduced to understanding the unique contribution of each category rather than each cell. It is with this hope that neuroscientists have devised schemes for classifying neurons.

Classification Based on the Number of Neurites

Neurons can be classified according to the total number of neurites (axons and dendrites) that extend from the soma (Figure 2.19). A neuron

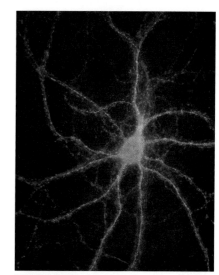

Figure 2.17
Dendrites receiving synaptic inputs from axon terminals. A neuron has been made to fluoresce green using a method that reveals the distribution of a microtubule-associated protein. Axon terminals have been made to fluoresce orange-red using a method to reveal the distribution of synaptic vesicles. The axons and cell bodies that contribute these axon terminals are not visible in this photomicrograph. (Source: Neuron 10 [Suppl.], 1993, Cover Fig.)

| Box 2.3 | **OF SPECIAL INTEREST** |

Hitching a Ride on "Retrorail"

Fast anterograde transport of proteins in axons was shown by injecting the soma with radioactive amino acids. The success of this method immediately suggested a way to trace connections in the brain. For example, to determine where neurons in the eye send their axons, the eye was injected with radioactive proline, an amino acid. The proline was incorporated into proteins in the somata that were then transported to the axon terminals. By using a technique called autoradiography, the location of radioactive axon terminals could be detected, thereby revealing the extent of the connection between the eye and the brain.

It was subsequently discovered that retrograde transport could also be exploited to work out connections in the brain. Strangely enough, the enzyme horseradish peroxidase (HRP) is selectively taken up by axon terminals and then transported retrogradely to the soma. A chemical reaction can then be initiated to visualize the location of the HRP in slices of brain tissue. This method is commonly used to trace connections in the brain.

Some viruses also exploit retrograde transport to infect neurons. For example, the oral type of herpesvirus enters axon terminals in the lips and mouth and is then transported back to the parent cell bodies. Here the virus usually remains dormant until physical or emotional stress occurs (as on a first date), at which time it replicates and returns to the nerve ending, causing a painful cold sore. Similarly, the rabies virus enters the nervous system by retrograde transport through axons in the skin. However, once inside the soma, the virus wastes no time in replicating madly, killing its neuronal host. The virus is then taken up by other neurons within the nervous system, and the process repeats itself again and again, usually until the victim dies.

Figure 2.18
Dendritic spines. This is a computer reconstruction of a segment of dendrite showing the variable shapes and sizes of spines. Each spine is postsynaptic to one or two axon terminals. (Source: Harris and Stevens, 1989, Cover Fig.)

that has a single neurite is said to be **unipolar**. If there are two neurites, the cell is called **bipolar**. If there are three or more, the cell is **multipolar**. The vast majority of neurons in the brain are multipolar.

Classification Based on Dendrites

Dendritic trees can vary wildly from one type of neuron to another. Sometimes they have inspired, elegant names like "double bouquet cells." At other times, the names have been less interesting; for example, "alpha cells." Classification is often unique to a particular part of the brain. For example, in the cerebral cortex (the structure that lies just under the surface of the cerebrum), there are two broad classes: **pyramidal cells** (pyramid shaped) and **stellate cells** (star shaped) (Figure 2.20).

Another simple way to classify neurons is according to whether their dendrites have spines. Those that do are called **spiny** and those that do not are called **aspinous**. These dendritic classification schemes can overlap. For example, in the cerebral cortex, all pyramidal cells are spiny. Stellate cells, on the other hand, can be either spiny or aspinous.

Classification Based on Connections

Information is delivered to the nervous system by neurons that have neurites in the sensory surfaces of the body, such as the skin and the retina of the eye. Cells with these connections are called **primary sensory neurons**.

Soma

Unipolar

Bipolar

Multipolar

Figure 2.19
Classification of neurons based on the number of neurites.

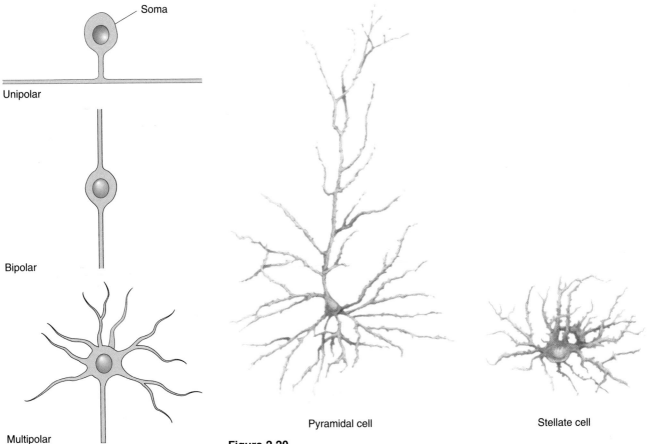

Pyramidal cell

Stellate cell

Figure 2.20
Classification of neurons based on dendritic tree structure. Pyramidal cells and stellate cells, distinguished by the arrangement of their dendrites, are two neuronal classes found in the cerebral cortex.

Other neurons have axons that form synapses with the muscles and command movements; these are called **motor neurons**. But most neurons in the nervous system form connections only with other neurons. According to this classification scheme, these cells are all called **interneurons**.

Classification Based on Axon Length

Some neurons have long axons that extend from one part of the brain to the other; these are called **Golgi Type I** or projection neurons. Other neurons have short axons that do not extend beyond the vicinity of the cell body; these are called **Golgi Type II** or local circuit neurons. In the cerebral cortex, for example, pyramidal cells usually have long axons that extend to other parts of the brain and are therefore Golgi Type I. In contrast, stellate cells have axons that never extend beyond the cerebral cortex and are therefore Golgi Type II.

Classification Based on Neurotransmitter

The classification schemes presented so far are based on the morphology of neurons as revealed by a Golgi stain. Newer methods that enable neuroscientists to identify which neurons contain specific neurotransmitters have resulted in a scheme for classifying neurons based on their chemistry. For example, the motor neurons that command voluntary movements all release the neurotransmitter *acetylcholine* at their synapses. These cells are therefore also classified as cholinergic, meaning that they use this particular neurotransmitter. Collections of cells that use a common neurotransmitter make up the brain's neurotransmitter systems (see Chapters 6 and 15).

GLIA

We have devoted most of our attention in this chapter to the neurons. While this decision is justified by the state of current knowledge, some neuroscientists consider glia to be the "sleeping giants" of neuroscience. One day, they suppose, it will be shown that glia contribute much more importantly to information-processing in the brain than is currently appreciated. At present, however, the evidence indicates that glia mainly contribute to brain function by supporting neuronal functions. Although their role may be subordinate, without glia, the brain could not function properly.

Figure 2.21
An astrocyte. Astrocytes fill most of the space in the brain that is not occupied by neurons and blood vessels.

Astrocytes

The most numerous glia in the brain are called **astrocytes** (Figure 2.21). These cells fill the space between neurons. The space that remains between the neurons and the astrocytes in the brain measures only about 20 nm wide. Consequently, astrocytes probably influence whether a neurite can grow or retract. And, when we speak of fluid "bathing" neurons in the brain, it is more like a sponge bath than a session in a hot tub.

An essential role of astrocytes is regulating the chemical content of this *extracellular space*. For example, astrocytes envelop synaptic junctions in the brain, thereby restricting the spread of neurotransmitter molecules that have been released. Astrocytes also have special proteins in their membranes that actively remove many neurotransmitters from the synaptic cleft. A recent and unexpected discovery is that astrocytic membranes also possess neurotransmitter receptors that, like the receptors on neurons, can trigger electrical and biochemical events inside the glial cell. Besides regulating neurotransmitters, astrocytes also tightly control the extracellular concentration of several substances that could potentially interfere with proper neuronal function. For example, astrocytes regulate the concentration of potassium ions in the extracellular fluid.

Myelinating Glia

Unlike astrocytes, the function of **oligodendroglia** and **Schwann cells** is clear. These glia provide layers of membrane that insulate axons. Boston University anatomist Alan Peters, a pioneer in the electron microscopic study of the nervous system, showed that this wrapping, called **myelin**, spirals around axons in the brain (Box 2.4). Because the axon fits inside the spiral wrapping like a sword in its scabbard, the name *myelin sheath* describes the entire covering. The sheath is interrupted periodically, leaving a short length where the axonal membrane is exposed. This region is called a **node of Ranvier** (Figure 2.22).

We will see in Chapter 4 that myelin serves to speed the propagation of nerve impulses down the axon. Oligodendroglia and Schwann cells differ based on their location and some other characteristics. For example, oligodendroglia are found only in the central nervous system (brain and spinal cord), while Schwann cells are found only in the peripheral nervous system (parts outside the skull and vertebral column). Another difference is that

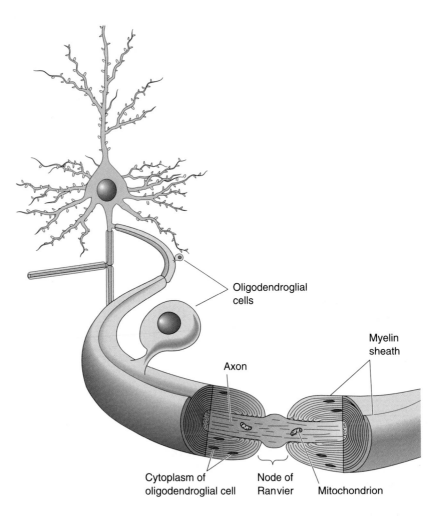

Figure 2.22

An oligodendroglial cell. Like the Schwann cells found in the nerves of the body, oligodendroglia provide myelin sheaths around axons in the brain and spinal cord. The myelin sheath of an axon is interrupted periodically at the nodes of Ranvier.

Box 2.4 PATH OF DISCOVERY

Myelin Wraps
by Alan Peters

Perhaps the discovery that excited me most was the first one I made using electron microscopy. In 1956 I went to Edinburgh as a postdoctoral fellow to continue my studies on the chemistry of silver staining of nervous tissue. At that time, silver staining was one of the most important techniques for examining nervous tissue by light microscopy. But the more my preparations improved, the more I realized how limited this method was. This led me to consider what other methods might be used to study nervous tissue, and since my arrival at the Department of Anatomy in Edinburgh coincided with the arrival of a new electron microscope, the emerging technique of electron microscopy seemed the one to try next. Fortunately, Dr. Alan Muir in the department had just returned from a sabbatical leave in the United States where he had learned the fundamentals of electron microscopy, and he was willing to teach me the necessary techniques. At that time the tissue was embedded in a soft plastic, methyl methacrylate, and the sections were all cut using glass knives. In our case, this glass had been obtained from the windows of trams that were being scrapped in Glasgow.

I decided that the first ultrastructural study I wanted to undertake was an examination of the development and structure of the myelin sheaths of axons in the central nervous system (CNS). My interest was stimulated by Betty Geren, who in 1954 had shown that the myelin sheaths in peripheral nerves are formed from lamellae that spiral around the enclosed axons to produce a multilayered wrapping. In contrast, the structure of central sheaths had been reported to be composed of concentric lamellae, like the rings of a tree. It was difficult for me to envisage how this concentric structure could arise, and, indeed, the evidence in support of this model for CNS myelin was not strong. The pictures appearing in the literature were difficult to interpret because fixation had not been optimal. Another problem, as I soon learned myself, is that the sheaths in CNS tracts are closely packed together and therefore can be difficult to discern from one another.

Fortunately, Dr. Mike Gaze introduced me to Xenopus tadpoles, which he was using for nerve regeneration studies. These tadpoles are so transparent that their optic nerves (a part of their CNS) are visible right through the head. Using potassium

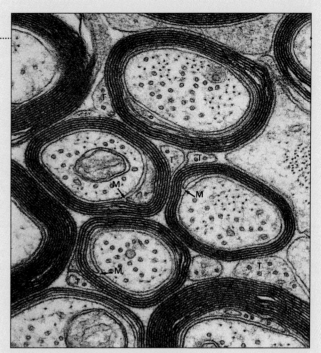

Figure A. Myelinated optic nerve fibers cut in cross-section.

permanganate, which Robertson (1957) had shown to produce good fixation of peripheral nerves, together with the use of the newly introduced hard epoxy resins for embedding, I was able to produce some reasonable electron micrographs of Xenopus optic nerves. Since the tadpoles were still developing, the myelinated nerve fibers in the optic nerves were not crowded together, and various stages in the process of myelination could be observed. Consequently, I was able to show that central myelin sheaths also formed by spiraled lamellae (Figure A). Interestingly, when my results were published in 1960, they were published alongside an article by Maturana, who had also solved the problem of the structure of central myelin using a different species of frog.

As methods of fixation and preparation improved, more has been learned about the structure and interrelations of cells in the nervous system. The last portion of the neuron to be identified and described on the basis of electron microscopic preparations was the axon initial segment, in 1968. Since all parts of the neuron then had been described, Sanford Palay, Henry deF. Webster, and I decided to write our book, The Fine Structure of the Nervous System, which was published first in 1970, and most recently as the third edition in 1991. To this day, I continue to marvel at the complexity and the beauty of nervous tissue as it is revealed by the electron microscope.

one oligodendroglial cell will contribute myelin to several axons, while each Schwann cell myelinates only a single axon.

Other Non-Neuronal Cells

Even if we eliminated every neuron, every astrocyte, and every oligo-dendroglial cell, other cells would still remain in the brain. For the sake of completeness, we must mention these other cells. First, special cells called **ependymal cells** would remain. These cells provide the lining of fluid-filled ventricles within the brain, and they also play a role in directing cell migration during brain development. Second, a class of cells called **microglia** would be present. These cells function as phagocytes to remove debris left by dead or degenerating neurons and glia. Finally, the vasculature of the brain—arteries, veins, and capillaries—would still be there.

CONCLUDING REMARKS

Learning about the structural characteristics of the neuron provides insight into how neurons and their different parts work, because structure correlates with function. For example, the absence of ribosomes in the axon correctly predicts that proteins in the axon terminal must be provided by the soma via axoplasmic transport. A large number of mitochondria in the axon terminal correctly predicts a high energy demand. The elaborate structure of the dendritic tree appears ideally suited for the receipt of incoming information, and, indeed, this is where most of the synapses are formed with the axons of other neurons.

From the time of Nissl, it has been recognized that an important feature of neurons is the rough ER. What does this tell us about neurons? We have said that rough ER is a site of the synthesis of proteins destined to be inserted into the membrane. We will now see how the various proteins in the neuronal membrane give rise to the unique capabilities of neurons to transmit, receive, and store information.

KEY TERMS

The Neuron Doctrine

histology

Nissl stain

cytoarchitecture

Golgi stain

cell body

soma

perikaryon

neurite

axon

dendrite

neuron doctrine

The Prototypical Neuron

cytosol

organelle

cytoplasm

nucleus

chromosome

DNA (deoxyribonucleic acid)

gene

gene expression

protein

protein synthesis

mRNA (messenger ribonucleic acid)

transcription

amino acid

translation

ribosome

rough endoplasmic reticulum (rough ER)

polyribosome

smooth endoplasmic reticulum (smooth ER)

Golgi apparatus

mitochondrion

ATP (adenosine triphosphate)

neuronal membrane

cytoskeleton

microtubule	**Classifying Neurons**
microfilament	unipolar neuron
neurofilament	bipolar neuron
axon hillock	multipolar neuron
efferent	pyramidal cell
afferent	stellate cell
axon collateral	spiny neuron
axon terminal	aspinous neuron
terminal bouton	primary sensory neuron
synapse	motor neuron
terminal arbor	interneuron
bouton en passant	Golgi Type I neuron
innervate	Golgi Type II neuron
synaptic vesicle	
synaptic cleft	**Glia**
synaptic transmission	astrocyte
neurotransmitter	oligodendroglial cell
axoplasmic transport	Schwann cell
anterograde transport	myelin
retrograde transport	node of Ranvier
dendritic tree	ependymal cell
receptor	microglia
dendritic spine	

REVIEW QUESTIONS

1. State the neuron doctrine in a single sentence. To whom is this insight credited?
2. Which parts of a neuron are shown by a Golgi stain that are not shown by a Nissl stain?
3. What are three physical characteristics that distinguish axons from dendrites.
4. Which of the following structures are unique to neurons and which are not: nucleus, mitochondria, rough ER, synaptic vesicle, Golgi apparatus?
5. What are the steps by which the information in the DNA of the nucleus directs the synthesis of a membrane-associated protein molecule?
6. Colchicine is a drug that causes microtubules to break apart (depolymerize). What effect would this drug have on anterograde transport? What would happen in the axon terminal?
7. Classify the cortical pyramidal cell based on (a) the number of neurites, (b) the presence or absence of dendritic spines, (c) connections, and (d) axon length.
8. What is myelin? What does it do? Which cells provide it in the central nervous system?

The Neuronal Membrane at Rest

Consider the problem your nervous system confronts when you step on a thumbtack (Figure 3.1). Your reactions are automatic: You shriek with pain as you pick up your foot. In order for this simple response to occur, breaking of the skin must be translated into neural signals that travel rapidly and reliably up the long sensory nerves of your leg. In the spinal cord, these signals are transferred to interneurons. Some of these neurons connect with the parts of your brain that interpret the signals as being painful. Others connect to the motor neurons that control the leg muscles that withdraw your foot. Thus, even this simple reflex requires the nervous system to *collect*, *distribute*, and *integrate* information. A goal of cellular neurophysiology is to understand the biological mechanisms that underlie these functions.

The neuron solves the problem of conducting information over a distance by using electrical signals that sweep along the axon. In this sense, axons act like telephone wires. However, the analogy stops here, because the type of signal used by the neuron is constrained by the special environment of the nervous system. In a copper telephone wire, information can be transferred over long distances at a high rate (about half the speed of light) because telephone wire is a superb conductor of electrons, is well insulated, and is suspended in air (air being a poor conductor of electricity). Electrons will, therefore, move within the wire instead of radiating away. In contrast, electrical charge in the cytosol of the axon is carried by electrically charged atoms (ions) instead of free electrons. This makes cytosol far less conductive than copper wire. Also, the axon is not especially well insulated and is bathed in salty extracellular fluid, which conducts electricity. Thus, like water flowing down a leaky garden hose, electrical current passively conducting down the axon would not go very far before it would leak out.

Fortunately, the axonal membrane has properties that enable it to conduct a special type of signal—the nerve impulse, or **action potential**—that overcomes these biological constraints. In contrast to passively conducted electrical signals, action potentials do not diminish over distance; they are signals of fixed size and duration. Information is encoded in the frequency of action potentials of individual neurons, as well as in the distribution and number of neurons firing action potentials in a given nerve. This type of code is partly analogous to Morse code sent down a telegraph wire; information is encoded in the pattern of electrical impulses. Cells capable of generating and conducting action potentials, which include both nerve and muscle cells, are said to have **excitable membrane**. The "action" in action potentials occurs at the membrane.

When a cell with excitable membrane is not generating impulses, it is said to be at rest. In the resting neuron, the cytosol along the inside surface of the membrane has a negative electrical charge compared to the outside. This difference in electrical charge across the membrane is called the **resting membrane potential**. The action potential is simply a brief reversal of this situation so that for an instant—about a thousandth of a second—the inside of the membrane becomes positively charged with respect to the outside. Therefore, to understand how neurons signal one another, we must learn how the neuronal membrane at rest separates electrical charge, how electrical charge can be rapidly redistributed across the membrane during the action potential, and how the impulse can propagate reliably along the axon.

In this chapter, we begin our exploration of neuronal signaling by tackling the first question: How does the resting membrane potential arise? Understanding the resting potential is very important because it forms the foundation for understanding the rest of neuronal physiology. Knowledge of neuronal physiology is central to understanding the capabilities and limitations of brain function.

THE CAST OF CHEMICALS

We begin our discussion of the resting potential by introducing the three main players: the salty fluids on either side of the membrane, the membrane itself, and the proteins that span the membrane. Each of these has certain chemical properties that contribute to establishing the resting potential.

Cytosol and Extracellular Fluid

Water is the main ingredient of the fluid inside the neuron, the intracellular fluid or cytosol, and the fluid that bathes the neuron, the extracellular

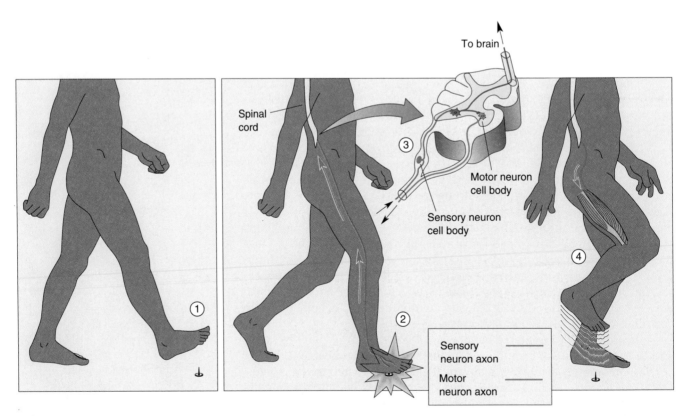

Figure 3.1
A simple reflex. (1) A person steps on a thumbtack. (2) The breaking of the skin is translated into signals that travel up sensory nerve fibers (direction of information flow indicated by the arrows). (3) In the spinal cord, the information is distributed to interneurons. Some of these neurons send axons to the brain where the painful sensation is registered. Others synapse on motor neurons, which send descending signals to the muscles. (4) The motor commands lead to muscle contraction and withdrawal of the foot.

fluid. Electrically charged atoms called ions are dissolved in this water, and these ions are responsible for the resting and action potentials.

Water. For our purpose here, the most important property of the water molecule (H$_2$O) is its uneven distribution of electrical charge (Figure 3.2a). The two hydrogen atoms and the oxygen atom are bonded together covalently, which is to say they share electrons. The oxygen atom, however, has a greater affinity for electrons than does the hydrogen atom. As a result, the shared electrons will spend more time associated with the oxygen atom than with the two hydrogen atoms. Therefore, the oxygen atom acquires a net negative charge (because it has extra electrons), and the hydrogen atoms acquire a net positive charge. Thus, H$_2$O is said to be a polar molecule, held together by *polar covalent bonds*. This electrical polarity makes water an effective solvent of other charged or polar molecules; that is, other polar molecules tend to dissolve in water.

Ions. Atoms or molecules that have a net electrical charge are called **ions**. Table salt is a crystal of sodium (Na$^+$) and chloride (Cl$^-$) ions held together by the electrical attraction of oppositely charged atoms. This attraction is called an ionic bond. Salt dissolves readily in water because the charged portions of the water molecule have a stronger attraction for the ions than they have for each other. As each ion breaks away from the crystal, it is surrounded by a sphere of water molecules. Each positively charged ion (Na$^+$, in this case) will be covered by water molecules oriented so that the oxygen atom (the negative pole) will be facing the ion. Likewise, each negatively charged ion (Cl$^-$) will be surrounded by the hydrogen atoms of the water molecules. These clouds of water that surround each ion are called spheres of hydration, and they effectively insulate the ions from one another (Figure 3.2b).

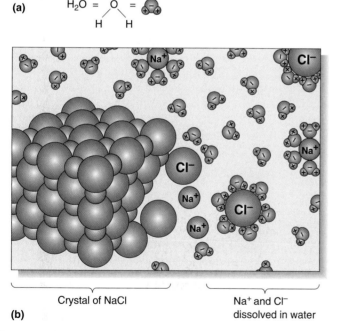

Figure 3.2
Water is a polar solvent. (a) Different representations of the atomic structure of the water molecule. The oxygen atom has a net negative electrical charge and the hydrogen atoms have a net positive electrical charge, making water a polar molecule. **(b)** A crystal of sodium chloride (NaCl) dissolves in water because the polar water molecules have a stronger attraction for the electrically charged sodium and chloride ions than the ions do for one another.

The electrical charge of an atom depends on the difference between the number of protons and electrons. When this difference is 1, the ion is said to be *monovalent*; when the difference is 2, the ion is *divalent*; and so on. Ions with a net positive charge are called **cations**; ions with a negative charge are called **anions**. Remember that ions are the major charge carriers involved in the conduction of electricity in biological systems, including the neuron. The ions of particular importance for cellular neurophysiology are the monovalent cations Na^+ (sodium) and K^+ (potassium), the divalent cation Ca^{2+} (calcium), and the monovalent anion Cl^- (chloride).

Phospholipid Membrane

As we have seen, substances with uneven electrical charges will dissolve in water because of the polarity of the water molecule. These substances, including ions and polar molecules, are said to be "water-loving," or *hydrophilic*. However, compounds whose atoms are bonded by *nonpolar covalent bonds* have no basis for chemical interactions with water. A nonpolar covalent bond occurs when the shared electrons are distributed evenly in the molecule so that no portion acquires a net electrical charge. Such compounds will not dissolve in water and are said to be "water-fearing," or *hydrophobic*. One familiar example of a hydrophobic substance is olive oil, and, as you know, oil and water don't mix. Another example is *lipid*, a class of water-insoluble biological molecules important to the structure of membranes. The lipids of the neuronal membrane contribute to the resting and action potentials by forming a barrier to water-soluble ions and, indeed, to water itself.

Phospholipid Bilayer. The main chemical building blocks of cell membranes are *phospholipid*s. Like other lipids, phospholipids contain long nonpolar chains of carbon atoms bonded to hydrogen atoms. In addition, however, a phospholipid has a polar phosphate group (a phosphorus atom bonded to three oxygen atoms) attached to one end of the molecule. Thus, phospholipids are said to have a polar "head" (containing phosphate) that is hydrophilic, and a nonpolar "tail" (containing hydrocarbon) that is hydrophobic.

The neuronal membrane consists of a sheet of phospholipids, two molecules thick. A cross section through the membrane, shown in Figure 3.3, reveals that the hydrophilic heads face the outer and inner watery environments and the hydrophobic tails face each other. This stable arrangement is called a **phospholipid bilayer**, and it effectively isolates the cytosol of the neuron from the extracellular fluid.

Protein

The type and distribution of protein molecules distinguish neurons from other types of cells. The *enzymes* that catalyze chemical reactions in the neuron, the *cytoskeleton* that gives a neuron its special shape, the *receptors* that are sensitive to neurotransmitters—all are made up of protein molecules. The resting and action potentials depend on special proteins that span the phospholipid bilayer. These proteins provide routes for ions to cross the neuronal membrane.

Protein Structure. In order to serve their many functions in the neuron, different proteins have widely different shapes, sizes, and chemical characteristics. To understand this diversity, it is necessary to briefly review protein structure.

As mentioned in Chapter 2, proteins are assembled from 20 different amino acids. The basic structure of an amino acid is shown in Figure 3.4a. All amino acids have a central carbon atom (the alpha carbon), which is covalently bonded to four molecular groups: a hydrogen atom, an amino

Figure 3.3
The phospholipid bilayer. The phospholipid bilayer is the core of the neuronal cell membrane and forms a barrier to water-soluble ions.

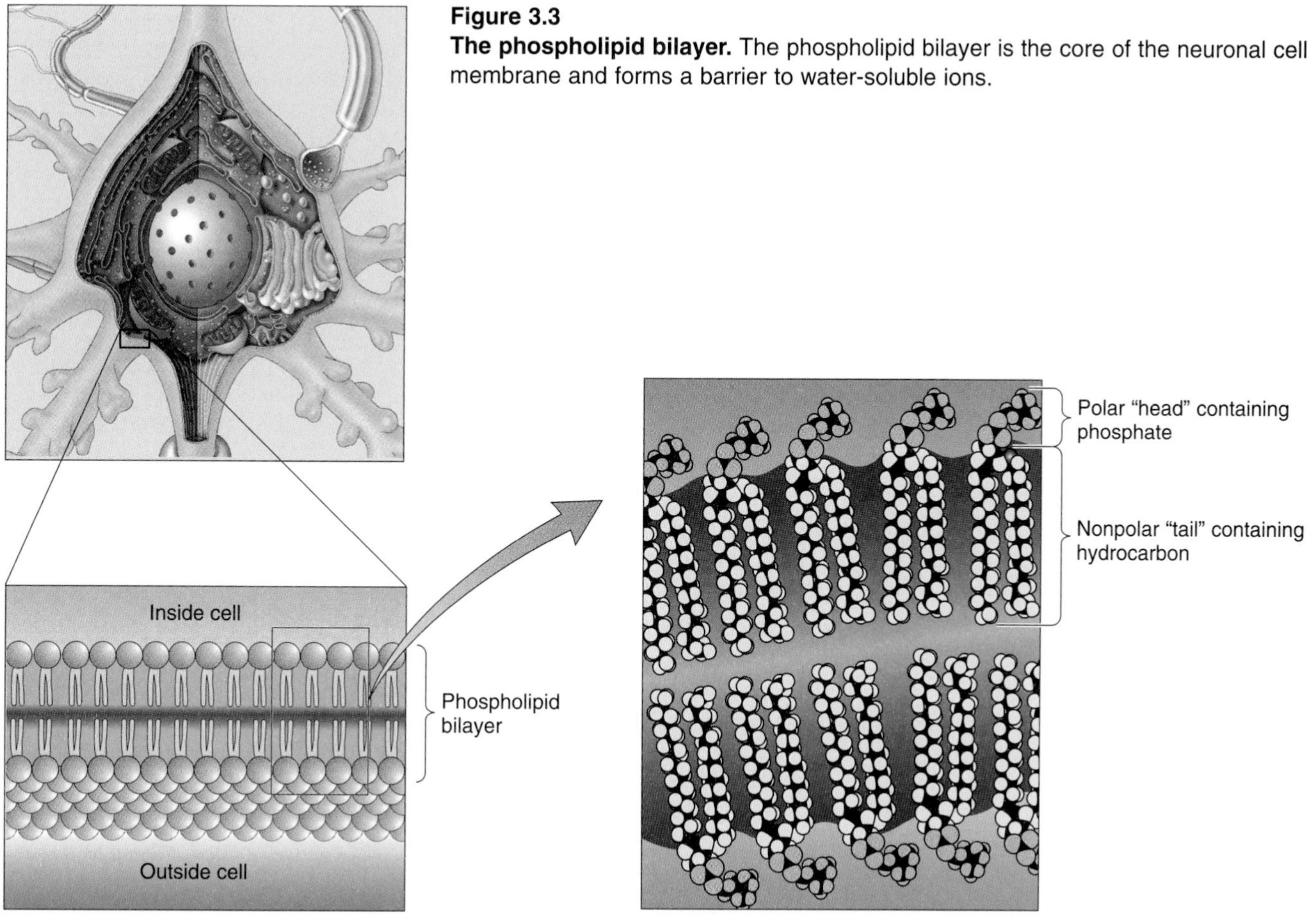

group (NH_3^+), a carboxyl group (COO^-), and a variable group called the *R group* (R for residue). The differences between amino acids result from differences in the size and nature of these R groups (Figure 3.4b). The properties of the R group determine the chemical relationships in which each amino acid can participate.

Proteins are synthesized in the ribosomes of the neuronal cell body. In this process, amino acids assemble into a chain connected by **peptide bonds**, which join the amino group of one amino acid to the carboxyl group of the next (Figure 3.5a). Proteins made of a single chain of amino acids are also called **polypeptides** (Figure 3.5b).

The four levels of protein structure are shown in Figure 3.6. The *primary structure* is like a chain, in which the amino acid R groups are linked together by peptide bonds. As a protein molecule is being synthesized, however, the polypeptide chain can coil into a spiral-like configuration called an *alpha helix*. The alpha helix is an example of what is called the *secondary structure* of a protein molecule. Interactions among the R groups can cause the molecule to change its three-dimensional conformation even further. In this way, proteins can bend, fold, and assume a globular shape. This shape is called *tertiary structure*. Finally, different polypeptide chains can bond together to form a larger molecule; such a protein is said to have *quaternary structure*. Each of the different polypeptides contributing to a protein with quaternary structure is called a *subunit*.

Channel Proteins. The exposed surface of a protein may be chemically heterogeneous. Regions where nonpolar R groups are exposed will be hydrophobic and will tend to associate readily with lipid. Regions with exposed polar R groups will be hydrophilic and will tend to avoid a lipid environment. Therefore, it is not difficult to imagine classes of rod-shaped proteins with polar groups exposed at either end, but with only hydrophobic groups showing on their middle surfaces. This type of protein could be suspended in a phospholipid bilayer, with its hydrophobic portion inside the membrane and its hydrophilic ends exposed to the watery environments on either side.

(a)

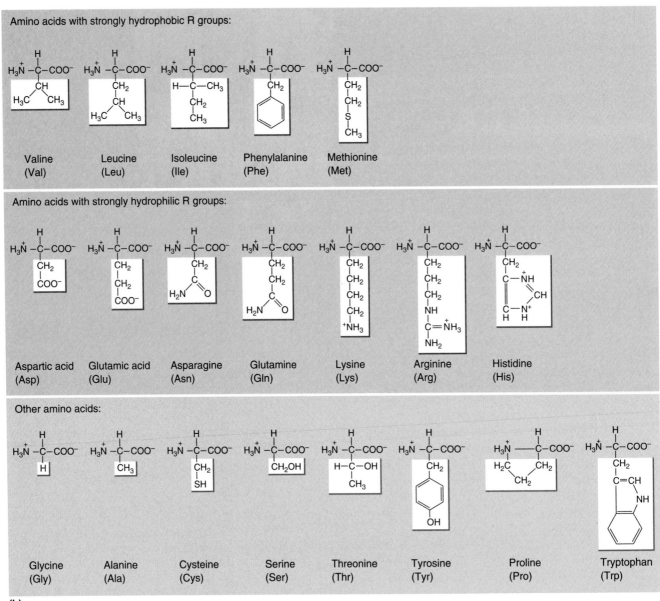

(b)

Figure 3.4
Amino acids, the building blocks of protein. (a) Every amino acid has in common a central alpha carbon, an amino group (NH₃⁺), and a carboxyl group (COO⁻). Amino acids differ from one another based on a variable R group. **(b)** The 20 different amino acids that are used by neurons to make proteins.

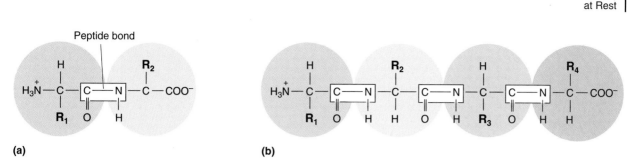

(a)   **(b)**

Figure 3.5
The peptide bond and a polypeptide. (a) Peptide bonds attach amino acids together. The bond forms between the carboxyl group of one amino acid and the amino group of another. **(b)** A polypeptide is a single chain of amino acids.

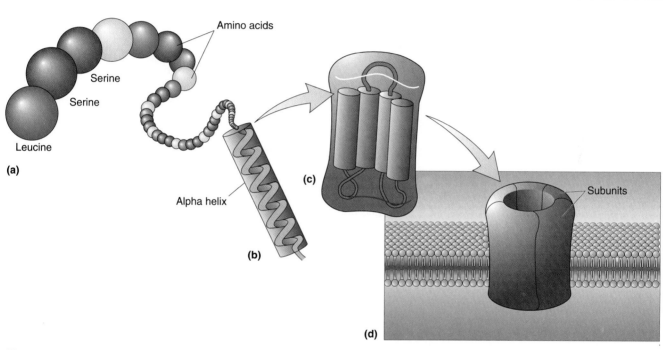

Figure 3.6
Protein structure. (a) Primary structure: the sequence of amino acids in the polypeptide. **(b)** Secondary structure: coiling of a polypeptide into an alpha helix. **(c)** Tertiary structure: three-dimensional folding of a polypeptide. **(d)** Quaternary structure: different polypeptides bonded together to form a larger protein.

Ion channels are made from just these sorts of membrane-spanning protein molecules. Typically, a functional channel across the membrane requires that 4–6 similar protein molecules assemble to form a pore between them (Figure 3.7). The subunit composition varies from one type of channel to the next, and this is what specifies their different properties. One important property of most ion channels, specified by the diameter of the pore and the nature of the R groups lining it, is **ion selectivity**. There are potassium channels that are selectively permeable to K^+. Likewise, sodium channels are permeable only to Na^+, calcium channels to Ca^{2+}, and so on. Another important property of many channels is **gating**. Channels with this property can be opened and closed—gated—by changes in the local microenvironment of the membrane. You will learn much more about channels as you work your way through this book. *Understanding ion channels in the neuronal membrane is key to understanding cellular neurophysiology.*

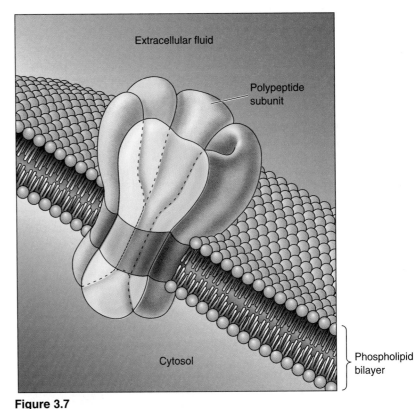

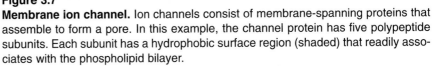

Figure 3.7
Membrane ion channel. Ion channels consist of membrane-spanning proteins that assemble to form a pore. In this example, the channel protein has five polypeptide subunits. Each subunit has a hydrophobic surface region (shaded) that readily associates with the phospholipid bilayer.

Ion Pumps. In addition to those that form channels, other membrane-spanning proteins come together to form **ion pumps**. These molecules actually are enzymes that use the energy released by the breakdown of ATP to transport certain ions across the membrane. We will see that these pumps play a critical role in neuronal signaling by transporting Na^+ and Ca^{2+} from the inside of the neuron to the outside.

MOVEMENT OF IONS

A channel across a membrane is like a bridge across a river (or, in the case of a gated channel, like a drawbridge): It provides a path to cross from one side to the other. The existence of a bridge does not necessarily compel us to cross it, however. The bridge we cross during the weekday commute may lie unused on the weekend. The same can be said of membrane ion channels. The existence of an open channel in the membrane does not necessarily mean that there will be a net movement of ions across the membrane. Such movement also requires that external forces be applied to drive them across. Because the functioning nervous system requires the movement of ions across the neuronal membrane, it is important that we understand these forces. Ionic movements through channels are influenced by two factors: diffusion and electricity.

Diffusion

Ions and molecules dissolved in water are in constant motion. This temperature-dependent, random movement will tend to distribute the ions evenly throughout the solution. In this way, there will be a net movement of ions from regions of high concentration to regions of low concentration; this movement is called **diffusion**. As an example, consider adding a teaspoon of milk to a cup of hot tea. The milk tends to spread evenly through the tea solution. If the thermal energy of the solution is reduced, as with iced tea, the diffusion of milk molecules will take noticeably longer.

Although ions typically will not pass through a phospholipid bilayer directly, diffusion will cause ions to be pushed through channels in the membrane. For example, if NaCl is dissolved in the fluid on one side of a permeable membrane (i.e., with channels that permit Na^+ and Cl^- passage), the Na^+ and Cl^- ions will cross until they are evenly distributed in the solutions on both sides (Figure 3.8). As with the previous example, the net movement is from the region of high concentration to the region of low concentration. (For a review of how concentrations are expressed, see Box 3.1). Such a difference in concentration is called a **concentration gradient**. Thus, we say that ions will flow down a concentration gradient. Driving ions across the membrane by diffusion, therefore, requires (1) that the membrane possesses channels permeable to the ions, and (2) that there be a concentration gradient across the membrane.

Electricity

Besides diffusion down a concentration gradient, another way to induce a net movement of ions in a solution is to use an electrical field, because ions are electrically charged particles. Consider the situation in Figure 3.9, where wires from the two terminals of a battery are placed in a solution containing dissolved NaCl. Remember, *opposite charges attract and like charges repel*. Consequently, there will be a net movement of Na^+ toward the negative terminal (the cathode) and of Cl^- to the positive terminal (the anode). The rate of movement of electrical charge is called **electrical current**, represented by the symbol I and measured in units called amperes (amps). According to the convention established by Benjamin Franklin, current is defined as being positive in the direction of positive-charge movement. In this example, therefore, positive current flows in the direction of Na^+ movement, from the anode to the cathode.

Two important factors determine how much current will flow: electrical potential and electrical conductance. **Electrical potential**, also called **voltage**, is the force exerted on a charged particle, and it reflects the difference in charge between the anode and the cathode. More current will flow as this difference is increased. Voltage is represented by the symbol V and is measured in units called volts. As an example, the difference in electrical potential between the terminals of a car battery is 12 volts; that is, the electrical potential at one terminal is 12 volts more positive than that at the other.

Electrical conductance is a measure of the ease with which an electrical charge can migrate from one point to another. It is represented by the symbol g and measured in units called siemens (S). Conductance depends on the number of particles available to carry electrical charge, and the ease with which these particles can travel through space. A term that expresses the same property in a different way is **electrical resistance**, a measure of how difficult it is for electrical charge to migrate. It is represented by the symbol R and measured in units called ohms (Ω). Resistance is simply the inverse of conductance (i.e., $R = 1/g$).

There is a simple relationship between potential (V), conductance (g), and the amount of current (I) that will flow. This relationship, known as **Ohm's law**, may be written $I = gV$: current is the product of the conductance and the potential difference. Notice that if the conductance is zero, no cur-

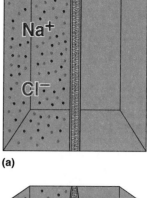

(a)

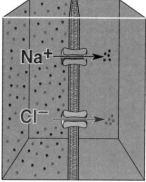

(b)

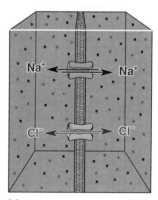

(c)

Figure 3.8
Diffusion. (a) NaCl has been dissolved on the left side of an impermeable membrane. The size of the letters Na^+ and Cl^- indicate the relative concentration of these ions. **(b)** Channels are inserted in the membrane that allow passage of Na^+ and Cl^-. Because there is a large concentration gradient across the membrane, there will be a net movement of Na^+ and Cl^- from the region of high concentration to the region of low concentration, from left to right. **(c)** In the absence of any other factors, the net movement of Na^+ and Cl^- across the membrane ceases when they are equally distributed on both sides of the permeable membrane.

rent will flow even when the potential difference is very large. Likewise, when the potential difference is zero, no current will flow even when the conductance is very large.

Consider the situation illustrated in Figure 3.10a, in which NaCl has been dissolved in equal concentrations on either side of a phospholipid bilayer. If we drop wires from the two terminals of a battery into the solution on either side, we will generate a large potential difference across this membrane. No current will flow, however, because there are no channels to allow migration of Na$^+$ and Cl$^-$ across the membrane; the conductance of the membrane is zero. Driving an ion across the membrane electrically, therefore, requires (1) that the membrane possesses channels permeable to that ion, and (2) that there be an electrical potential difference across the membrane (Figure 3.10b).

The stage is now set. We have electrically charged ions in solution on either side of the neuronal membrane. Ions can cross the membrane only by way of protein channels. The protein channels can be highly selective for specific ions. The movement of any ion through its channel depends on the concentration gradient and the difference in electrical potential across the membrane. Let's use this knowledge to explore the resting membrane potential.

THE IONIC BASIS OF THE RESTING MEMBRANE POTENTIAL

The *membrane potential*—or membrane voltage—of a neuron is represented by the symbol V_m. This V_m can be measured by inserting a microelectrode into the cytosol. A typical **microelectrode** is a thin glass tube with an extremely fine tip (diameter 0.5 μm) that will penetrate the membrane of a neuron with minimal damage. It is filled with an electrically conductive solution and connected to a device called a voltmeter. The voltmeter measures the electrical potential difference between the tip of this microelectrode and a wire placed outside the cell (Figure 3.11). This method reveals that electrical charge is unevenly distributed across the neuronal membrane. The inside of the neuron is electrically negative with respect to the outside. This steady difference, the resting potential, is maintained whenever a neuron is not generating impulses.

The resting potential of a typical neuron is about –65 millivolts (1 mV = 0.001 volts). Stated another way, for a neuron at rest, V_m = –65 mV. This negative resting membrane potential inside the neuron is an absolute requirement for a functioning nervous system. In order to understand the negative membrane potential, we look to the ions that are available and how they are distributed inside and outside the neuron.

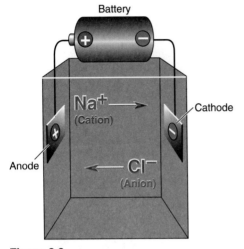

Figure 3.9
Movement of ions under the influence of an electrical field.

Review of Moles and Molarity

Concentrations of substances are expressed as the number of molecules per liter of solution. The number of molecules is usually expressed in *moles*. One mole is 6.02 x 10^{23} molecules. A solution is said to be 1 Molar (M) if it has a concentration of 1 mole per liter. A 1 millimolar (mM) solution has 0.001 moles per liter. The abbreviation for concentration is a pair of brackets. Thus, we read [NaCl] = 1 mM as: "The concentration of the NaCl solution is 1 millimolar."

Equilibrium Potentials

Consider a hypothetical "cell" where the inside is separated from the outside by a pure phospholipid membrane with no proteins. Inside this cell, a concentrated potassium salt solution is dissolved, yielding K^+ and A^- (A^- for anion, any molecule with a negative charge). Outside the cell is a solution with the same salt, but diluted twentyfold with water. Although a large concentration gradient exists between the inside of the cell and the outside, there will be no net movement of ions because the phospholipid bilayer, having no channel proteins, is impermeable to charged, hydrophilic atoms. Under these conditions, a microelectrode would record no potential difference between the inside and the outside of the cell. In other words, V_m would be equal to 0 mV because the ratio of K^+ to A^- on each side of the membrane equals 1; both solutions are electrically neutral (Figure 3.12a).

Consider how this situation would change if potassium channels were inserted into the phospholipid bilayer. Because of the selective permeability of these channels, K^+ would be free to pass across the membrane, but A^- would not. Initially, diffusion rules: K^+ ions pass through the channels out of the cell, down the steep concentration gradient. Because A^- is left behind, however, the inside of the cell immediately begins to acquire a net negative charge, and an electrical potential difference is established across the membrane (Figure 3.12b). As the inside fluid acquires more and more net negative charge, the electrical force starts to pull positively charged K^+ ions back through the channels into the cell. When a certain potential difference is reached, the electrical force pulling K^+ ions inside exactly counterbalances the force of diffusion pushing them out. Thus, an *equilibrium* state is reached in which the diffusional and electrical forces are equal and opposite, and the

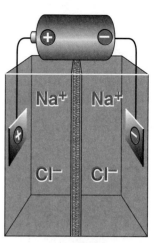

(a)   No current

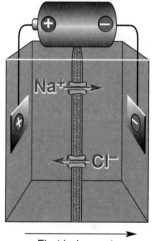

Electrical current

(b)

Figure 3.10
Electrical current flow across a membrane. (a) A voltage applied across a phospholipid bilayer causes no electrical current because there are no channels to allow the passage of electrically charged ions from one side to the other; the conductance of the membrane is zero. **(b)** Insertion of channels in the membrane allows ions to cross. Electrical current flows in the direction of cation movement (from left to right, in this example).

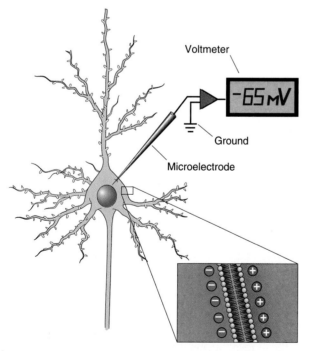

Figure 3.11
Measurement of the resting membrane potential. A voltmeter measures the difference in electrical potential between the tip of a microelectrode inside the cell and a wire placed in the extracellular fluid. Typically, the inside of the neuron is about −65 mV with respect to the outside. This potential is caused by the uneven distribution of electrical charge across the membrane (enlargement).

Inside   Outside
"cell"   "cell"

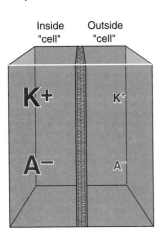

(a)

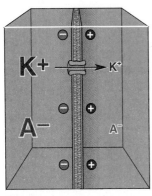

(b)

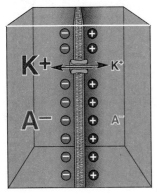

(c)

Figure 3.12
**Establishing equilibrium in a selec-
tively permeable membrane. (a)** An
impermeable membrane separates two
regions: one of high salt concentration
("inside") and the other of low salt concentration ("outside"). **(b)** Insertion of a
channel that is selectively permeable
to K⁺ into the membrane initially results in
a net movement of K⁺ ions down their
concentration gradient, from left to
right. **(c)** A net accumulation of positive
charge on the outside and negative
charge on the inside retards the movement of positively charged K⁺ ions from
the inside to the outside. An equilibrium
is established such that there is no net
movement of ions across the membrane, leaving a charge difference
between the two sides.

net movement of K^+ across the membrane ceases (Figure 3.12c). The electrical potential difference that exactly balances an ionic concentration gradient
is called an **ionic equilibrium potential**, and it is represented by the symbol
E_{ion}. In this example, the equilibrium potential will be about –80 mV.

The example in Figure 3.12 shows that it is a relatively simple matter to
generate a steady electrical potential difference across a membrane. All that
is required is an ionic concentration gradient and selective ionic permeability. Before moving on to the situation in real neurons, however, we can use
this example to make four important points.

1. *Large changes in membrane potential are caused by minuscule changes in ionic
   concentrations.* In Figure 3.12, channels were inserted, and K^+ ions flowed
   out of the cell until the membrane potential went from 0 mV to the equilibrium potential of –80 mV. How much does this ionic redistribution
   affect the K^+ concentration on either side of the membrane? Not very
   much. For a cell with a 50 μm diameter, containing 100 mM K^+, it can be
   calculated that the concentration change required to take the membrane
   from 0 to –80 mV is about 0.00001 mM. That is, when the channels were
   inserted and the K^+ flowed out until equilibrium was reached, the internal K^+ concentration went from 100 mM to 99.99999 mM—a negligible
   difference in concentration.

2. *The net difference in electrical charge occurs at the inside and outside surfaces of
   the membrane.* Because the phospholipid bilayer is so thin (less than 5 nm
   thick), it is possible for ions on one side to interact electrostatically with
   ions on the other side. Thus, the negative charges inside the neuron and
   the positive charges outside the neuron tend to be mutually attracted to
   the cell membrane. Consider how, on a warm summer evening, mosqui-
   toes are attracted to the outside face of a window pane when the inside
   lights are on. Similarly, the net negative charge inside the cell is not distributed evenly in the cytoplasm, but rather is localized at the inner face
   of the membrane (Figure 3.13). In this way, the membrane is said to store
   electrical charge, a property called capacitance.

3. *Ions are driven across the membrane at a rate proportional to the difference
   between the membrane potential and the equilibrium potential.* Notice from our
   example in Figure 3.12 that, when the channels were inserted, there was
   a net movement of K^+ only as long as the electrical membrane potential
   differed from the equilibrium potential. The difference between the real
   membrane potential and the equilibrium potential ($V_m - E_{ion}$) for a particular ion is called the **ionic driving force**. We'll talk more about this in
   Chapters 4 and 5 when we discuss the movement of ions across the membrane during the action potential and synaptic transmission.

4. *If the concentration difference across the membrane is known, an equilibrium
   potential can be calculated for any ion.* In our example in Figure 3.12, we
   assumed that K^+ was more concentrated inside the cell. From this knowledge, we were able to deduce that the equilibrium potential would be negative if the membrane were selectively permeable to K^+. Let's consider
   another example in which Na^+ is more concentrated *outside* the cell (Figure
   3.14). If the membrane contained sodium channels, Na^+ would flow down
   the concentration gradient *into* the cell. The entry of positively charged ions
   would cause the cytosol on the inner surface of the membrane to acquire a
   net positive charge. The positively charged interior of the cell would now
   repel Na^+ ions, tending to push them back out through their channels. At a
   certain potential difference, the electrical force pushing Na^+ ions out would
   exactly counterbalance the force of diffusion pushing them in. In this example, the membrane potential at equilibrium would be positive on the inside.

The examples in Figures 3.12 and 3.14 illustrate that if we know the ionic concentration difference across the membrane, we can figure out the equilibrium potential for any ion. Prove it to yourself. Assume that Ca^{2+} is more concentrated on the outside of the cell and that the membrane is selectively permeable to Ca^{2+}. See if you can figure out whether the inside of the cell would be positive or negative at equilibrium. Try it again, assuming that the membrane is selectively permeable to Cl^- and that Cl^- is more concentrated outside the cell. (Pay attention here; note the charge of the ion.)

The Nernst Equation. The preceding examples show that each ion has its own equilibrium potential—the steady electrical potential that would be achieved if the membrane were permeable only to that ion. Thus, we can speak of the potassium equilibrium potential, E_K; the sodium equilibrium potential, E_{Na}; the calcium equilibrium potential, E_{Ca}; and so on. And knowing the electrical charge of the ion and the concentration difference across the membrane, we can easily deduce whether the inside of the cell would be positive or negative at equilibrium. In fact, the *exact* value of an equilibrium potential in mV can be calculated using an equation derived from the principles of physical chemistry, the **Nernst equation**, which takes into consideration the charge of the ion, the temperature, and the ratio of the external and internal ion concentrations. Using the Nernst equation, we can calculate the value of the equilibrium potential for any ion. For example, if K^+ is concentrated twentyfold on the inside of a cell, the Nernst equation tells us that $E_K = -80$ mV (Box 3.2).

Distribution of Ions Across the Membrane

It should now be clear that the neuronal membrane potential depends on the ionic concentrations on either side of the membrane. Estimates of these concentrations appear in Figure 3.15. The important point is that *K^+ is more concentrated on the inside, and Na^+ and Ca^{2+} are more concentrated on the outside.*

How do these concentration gradients arise? Ionic concentration gradients are established by the actions of ion pumps in the neuronal membrane. Two ion pumps are especially important in cellular neurophysiology: the **sodium-potassium pump** and the **calcium pump**. The sodium-potassium pump is an enzyme that hydrolyzes ATP in the presence of internal Na^+. The chemical energy released by this reaction drives the pump, which exchanges internal Na^+ for external K^+. The actions of this pump ensure that K^+ is concentrated inside the neuron and that Na^+ is concentrated outside. Notice that the pump pushes these ions across the membrane against their concentration gradients (Figure 3.16). This work requires the expenditure of metabolic energy. Indeed, it has been estimated that the sodium-potassium pump expends as much as 70% of the total amount of ATP utilized by the brain.

The **calcium pump** is also an enzyme that actively transports Ca^{2+} out of the cytoplasm across the cell membrane. Additional mechanisms decrease intracellular $[Ca^{2+}]$ to a very low level (0.0002 mM); these include intracellular calcium-binding proteins and organelles, such as mitochondria and types of endoplasmic reticulum, that sequester cytoplasmic calcium ions.

Ion pumps are the unsung heroes of cellular neurophysiology. They work in the background to ensure that the ionic concentration gradients are established and maintained. These proteins may lack the glamour of a gated ion channel, but without ion pumps, the brain would not function.

Relative Ion Permeabilities of the Membrane at Rest

The pumps establish ionic concentration gradients across the neuronal membrane. With knowledge of these ionic concentrations, we can use the

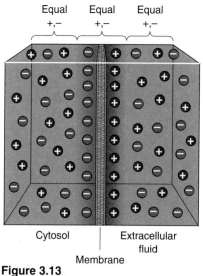

Equal +,−   Equal +,−   Equal +,−

Cytosol   Extracellular fluid

Membrane

Figure 3.13
Distribution of electrical charge across the membrane. The uneven charges inside and outside the neuron line up along the membrane because of electrostatic attraction across this very thin barrier. Notice that the bulk of the cytosol and extracellular fluid is electrically neutral.

Inside
"cell" Outside
"cell"

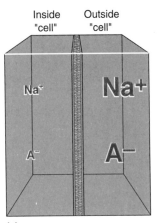

(a)

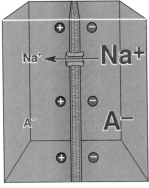

(b)

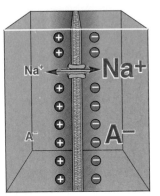

(c)

Figure 3.14
Another example of establishing equilibrium in a selectively permeable membrane. (a) An impermeable membrane separates two regions: one of high salt concentration ("outside") and the other of low salt concentration ("inside"). **(b)** Insertion of a channel that is selectively permeable to Na^+ into the membrane initially results in a net movement of Na^+ ions down their concentration gradient, from right to left. **(c)** A net accumulation of positive charge on the inside and negative charge on the outside retards the movement of positively charged Na^+ ions from the outside to the inside. An equilibrium is established such that there is no net movement of ions across the membrane, leaving a charge difference between the two sides; in this case, the inside of the cell is positively charged with respect to the outside.

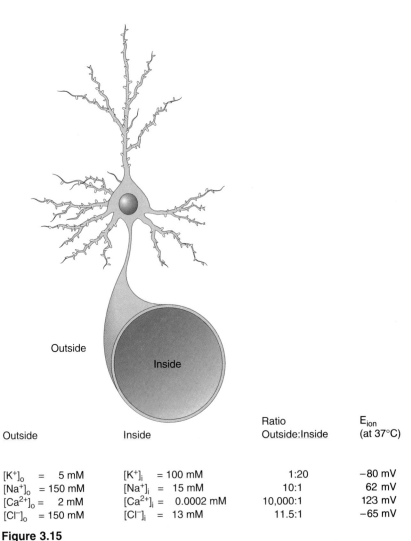

		Ratio Outside:Inside	E_{ion} (at 37°C)
$[K^+]_o$ = 5 mM	$[K^+]_i$ = 100 mM	1:20	–80 mV
$[Na^+]_o$ = 150 mM	$[Na^+]_i$ = 15 mM	10:1	62 mV
$[Ca^{2+}]_o$ = 2 mM	$[Ca^{2+}]_i$ = 0.0002 mM	10,000:1	123 mV
$[Cl^-]_o$ = 150 mM	$[Cl^-]_i$ = 13 mM	11.5:1	–65 mV

Figure 3.15
Approximate ion concentrations on either side of a neuronal membrane. E_{ion} is the membrane potential that would be achieved (at body temperature) if the membrane were selectively permeable to that ion.

Nernst equation to calculate equilibrium potentials for the different ions (see Figure 3.15). Remember, though, that an equilibrium potential for an ion is the membrane potential that results if a membrane is *selectively permeable* to that ion. In reality, however, neurons are not permeable to only a single type of ion. How do we incorporate this detail into our thinking?

Let's consider a few scenarios involving K^+ and Na^+. If the membrane of a neuron were permeable only to K^+, the membrane potential would equal E_K, which, according to Figure 3.15, is –80 mV. On the other hand, if the membrane of a neuron were permeable only to Na^+, the membrane potential would equal E_{Na}, 62 mV. If the membrane were equally permeable to K^+ and $Na+$, however, the resulting membrane potential would be some average of E_{Na} and E_K. What if the membrane were 40 times more permeable to K^+ than it is to Na^+? The membrane potential again would be between E_{Na} and E_K, but much closer to E_K than to E_{Na}. This approximates the situation in real neurons. The resting membrane potential of –65 mV approaches, but does not achieve, the potassium equilibrium potential of –80 mV. This difference arises because, although the membrane at rest is highly permeable to K^+, there is also a steady leak of Na^+ into the cell.

The resting membrane potential can be calculated using the **Goldman equation**, a mathematical formula that takes into consideration the relative permeability of the membrane to different ions. If we concern ourselves only with K^+ and Na^+, use the ionic concentrations in Figure 3.15, and assume that the resting membrane permeability to K^+ is fortyfold greater than it is to Na^+, then the Goldman equation predicts a resting membrane potential of –65 mV, the observed value (Box 3.3).

The Importance of Regulating the External Potassium Concentration. Because the neuronal membrane at rest is mostly permeable to K^+, the membrane potential is close to E_K. Another consequence of high K^+ permeability is that the membrane potential is particularly sensitive to changes in the concentration of extracellular potassium. This relationship is shown in Figure 3.17. A tenfold change in the external K^+ concentration, $[K^+]_o$, from 5 to 50 mM, would take the membrane potential from –65 to –17 mV. When the membrane potential becomes less negative, the membrane is said to be **depolarized**. Therefore, increasing extracellular potassium depolarizes neurons.

The sensitivity of the membrane potential to $[K^+]_o$ has led to the evolution of mechanisms that tightly regulate extracellular potassium concentra-

---

Box 3.2 | **BRAIN FOOD**

The Nernst Equation

The equilibrium potential for an ion can be calculated using the Nernst equation:

$$E_{ion} = 2.303 \frac{RT}{zF} \log \frac{[ion]_o}{[ion]_i}$$

where

E_{ion} = ionic equilibrium potential
R = gas constant
T = absolute temperature
z = charge of the ion
F = Faraday's constant
log = base 10 logarithm
$[ion]_o$ = ionic concentration outside the cell
$[ion]_i$ = ionic concentration inside the cell

The Nernst equation can be derived from the basic principles of physical chemistry. Let's see if we can make some sense of it.

Remember that equilibrium is the balance of two influences: diffusion, which pushes an ion down its concentration gradient, and electricity, which causes an ion to be attracted to opposite charges and repelled by like charges. Increasing the thermal energy of each particle increases diffusion and will therefore increase the potential difference achieved at equilibrium. Thus, E_{ion} is proportional to T. On the other hand, increasing the electrical charge of each

particle will decrease the potential difference needed to balance diffusion. Therefore, E_{ion} is inversely proportional to the charge of the ion (z). We need not worry about R and F in the Nernst equation because they are constants.

At body temperature (37°C), the Nernst equation for the important ions—K^+, Na^+, Cl^-, and Ca^{2+}—simplifies to:

$$E_K = 61.54 \text{ mV} \log \frac{[K^+]_o}{[K^+]_i}$$

$$E_{Na} = 61.54 \text{ mV} \log \frac{[Na^+]_o}{[Na^+]_i}$$

$$E_{Cl} = -61.54 \text{ mV} \log \frac{[Cl^-]_o}{[Cl^-]_i}$$

$$E_{Ca} = 30.77 \text{ mV} \log \frac{[Ca^{2+}]_o}{[Ca^{2+}]_i}$$

Therefore, in order to calculate the equilibrium potential for a certain type of ion at body temperature, all we need to know is the ionic concentrations on either side of the membrane. For instance, in the example we used in Figure 3.12, we stipulated that K^+ was twentyfold more concentrated inside the cell:

$$\text{If } \frac{[K^+]_o}{[K^+]_i} = \frac{1}{20}$$

$$\text{and } \log \frac{1}{20} = -1.3$$

$$\text{then } E_K = 61.54 \text{ mV} \times -1.3$$
$$= -80 \text{ mV}.$$

tions in the brain. One of these is the blood-brain barrier, a specialization of the walls of brain capillaries that limits the movement of potassium (and other bloodborne substances) into the extracellular fluid of the brain.

Glia, particularly astrocytes, also possess efficient mechanisms to take up extracellular K^+ whenever concentrations rise, as they normally do during periods of neural activity. Remember, astrocytes fill most of the space between neurons in the brain. Astrocytes have membrane potassium pumps that concentrate K^+ in their cytosol, and they also have potassium channels.

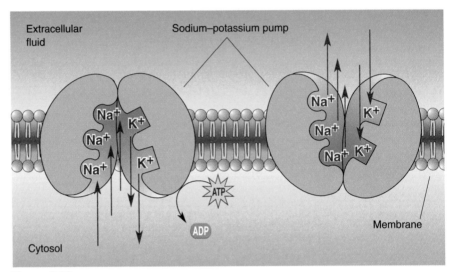

Figure 3.16
The sodium-potassium pump. This is a membrane-associated protein that transports ions across the membrane against their concentration gradients at the expense of metabolic energy.

Box 3.3 | **BRAIN FOOD**

The Goldman Equation

If the membrane of a real neuron were permeable only to K^+, the resting membrane potential would equal E_K, about –80 mV. But it does not; the measured resting membrane potential of a typical neuron is about –65 mV. This discrepancy is explained because real neurons at rest are not exclusively permeable to K^+; there is also some Na^+ permeability. Stated another way, the *relative permeability* of the resting neuronal membrane is quite high to K^+ and low to Na^+. If the relative permeabilities are known, it is possible to calculate the membrane potential at equilibrium by using the Goldman equation. Thus, for a membrane permeable only to Na^+ and K^+ at 37°C:

$$V_m = 61.54 \text{ mV} \log \frac{P_K [K^+]_o + P_{Na} [Na^+]_o}{P_K [K^+]_i + P_{Na} [Na^+]_i}$$

where V_m is the membrane potential, P_K and P_{Na} are the relative permeabilities to K^+ and Na^+, respectively, and the other terms are the same as for the Nernst equation.

If the resting membrane ion permeability to K^+ is 40 times greater than it is to Na^+, then solving the Goldman equation using the concentrations in Figure 3.15 yields:

$$V_m = 61.54 \text{ mV} \log \frac{40 (5) + 1 (150)}{40 (100) + 1 (15)}$$

$$= 61.54 \text{ mV} \log \frac{350}{4015}$$

$$= -65 \text{ mV}$$

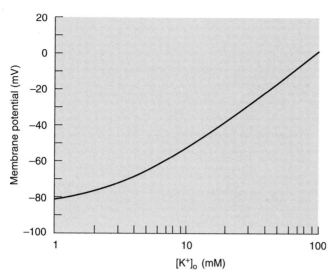

Figure 3.17
Dependence of membrane potential on external potassium concentration.
Because the neuronal membrane at rest is mostly permeable to potassium, a ten-fold change in $[K^+]_o$, from 5 to 50 mM, causes a 48 mV depolarization of the membrane. This function was calculated using the Goldman equation (see Box 3.3).

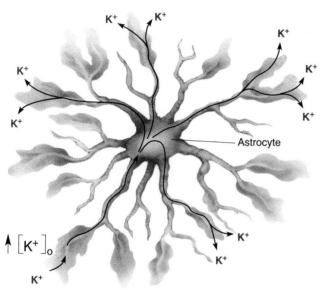

Figure 3.18
Potassium spatial buffering by astrocytes. When brain $[K^+]_o$ increases as a result of local neural activity, K^+ enters astrocytes via membrane channels. The extensive network of astrocytic processes helps dissipate the K^+ over a large area.

When $[K^+]_o$ increases, K^+ ions enter the astrocyte through the potassium channels, causing the astrocyte membrane to depolarize. In the late 1960s, Bruce Ransom and Sidney Goldring, then at Washington University in St. Louis, discovered that the membrane potential of brain astrocytes is actually a very sensitive measure of the extracellular potassium concentration in the brain (see Box 3.4). Entry of K^+ ions increases $[K^+]_i$, which is believed to be dissipated over a large area by the extensive network of astrocytic processes. This mechanism for the regulation of $[K^+]_o$ by astrocytes is called *potassium spatial buffering* (Figure 3.18).

Box 3.4   **PATH OF DISCOVERY**

The Membrane Potential of Glial Cells

by Bruce Ransom

Bruce Ransom

My experiments on the determinants of glial cell membrane potential were the first of my career. In recollecting these studies, I rediscovered why I am a scientist. I began my graduate studies in the summer of 1969 at Washington University in St. Louis. My thesis advisor was Dr. Sidney Goldring, a neurosurgeon, who was interested in the effects of direct electrical stimulation on the cerebral cortex. It became my job to study the physiology of glial cells in cat cerebral cortex in an effort to better understand how they might respond to electrical stimulation. The first question was to determine the origin of their membrane potential.

The idea was to make intracellular recordings from individual glial cells in the cortex of living cats and determine which ions controlled their membrane potential by changing the composition of the solution bathing the cortical surface. How hard could that be? As he explained more about my thesis project, my advisor didn't mention that I would find glial cells by blindly probing for them with fragile glass micropipettes that typically broke as they were pressed against the pulsating cortical surface. These, however, were my first two discoveries about glial cells in the cortex: you couldn't see them, and they moved with every beat of the heart.

Most of what was known at that time about the physiology of glial cells derived from the elegant studies of Stephen Kuffler and his colleagues on submammalian glia; reading about this work was exciting, and I began my experiments with enthusiasm. My initial efforts to impale cortical glial cells, however, proved very frustrating. I found the construction of usable glass microelectrodes for intracellular recording especially problematic. The method I used initially consisted of pulling a heated glass pipette until it had a very fine tip, filling it with methanol boiled under high negative pressure (using a cantankerous mechanical pump), and then replacing the methanol with 3 M KCl. This fabrication process was one-third method, one-third superstitious ceremony,

and one-third luck. The labs in which I worked were some of the first to use such micropipettes to record from cortical cells. I remember my astonishment on encountering a "museum specimen" micropipette on a dusty back shelf that had been pulled by hand several years earlier by Mrs. Tasaki, the wife of the famous neurophysiologist, I. Tasaki. Apparently she had a special talent for hand-pulling intracellular recording pipettes, and assisted her husband with his experiments. Fortunately for me, the science of microelectrode construction advanced rapidly, and I soon adopted a faster and more reliable method for making them.

As I learned more about glial cells, it was clear that determining which ions controlled the membrane potential of mammalian glial cells was important, even if it was not easy. This would shed light on why they depolarized when the cortical surface was stimulated. It was hypothesized that cortical stimulation produced an increase in $[K^+]_o$ that depolarized these cells; if glial membrane potential was *not* sensitive to small changes in $[K^+]_o$, this theory would need to be amended. The relationship between membrane potential and $[K^+]_o$ was also crucial for evaluating a proposed function of glial cells termed potassium spatial buffering. So I persisted.

Finally, on a September afternoon in 1970, everything worked and I got my first usable data. The experiment was straightforward. While recording from an impaled glial cell in the cortex, I changed the $[K^+]$ of the bathing solution, which diffused into the tissue. As predicted, application of K^+ caused my glial cell to depolarize in a completely reversible fashion!

I was hooked on science! Having been on such an extreme form of intermittent reinforcement, my "experimenting" behavior became very durable. In fact, I consider myself lucky to have experienced all the initial frustration for this very reason. In science, perhaps more than in other pursuits, persistence is terribly important for success.

Box 3.5 | **OF SPECIAL INTEREST**

Death by Lethal Injection

On June 4, 1990, Dr. Jack Kevorkian shocked the medical profession by assisting in the suicide of Janet Adkins. Adkins, a 54-year-old, happily married mother of three had been diagnosed with Alzheimer's disease, a progressive brain disorder that always results in senile dementia and death. Mrs. Adkins had been a member of the Hemlock Society, which advocates euthanasia as an alternative to death by terminal illness. Dr. Kevorkian agreed to help Mrs. Adkins take her own life. In the back of a 1968 Volkswagen van at a campsite in Oakland County, Michigan, she was hooked to an intravenous line that infused a harmless saline solution. To choose death, Mrs. Adkins switched the solution to one that contained an anesthetic solution, followed automatically by potassium chloride. The anesthetic caused Mrs. Adkins to become unconscious by suppressing the activity of neurons in part of the brain called the reticular formation. However, cardiac arrest and death were caused by the KCl injection. The ionic basis of the resting potential explains why the heart stopped beating.

Recall that the proper functioning of excitable cells (including those of cardiac muscle) requires that their membranes be maintained at the resting potential whenever they are not generating impulses. The negative resting potential is a result of selective ionic permeability to K^+ and to the metabolic pumps that concentrate potassium inside the cell. However, as Figure 3.17 shows, membrane potential is very sensitive to changes in the extracellular concentration of potassium. A tenfold rise in extracellular K^+ would wipe out the resting potential. Although neurons in the brain are somewhat protected from large changes in $[K^+]_o$, other excitable cells in the body, such as muscle cells, are not. Without negative resting potentials, cardiac muscle cells can no longer generate the impulses that lead to contraction, and the heart immediately stops beating. Intravenous potassium chloride is, therefore, a lethal injection.

It is important to recognize that not all excitable cells are protected from increases in potassium. Muscle cells, for example, do not have a blood-brain barrier or glial buffering mechanisms. Consequently, although the brain is relatively protected, elevations of $[K^+]$ in the blood can still have serious consequences on body physiology (Box 3.5).

We have now explored the resting membrane potential. The activity of the sodium-potassium pump produces and maintains a large K^+ concentration gradient across the membrane. The neuronal membrane at rest is highly permeable to K^+, owing to the presence of membrane potassium channels. The movement of K^+ ions across the membrane, down their concentration gradient, leaves the inside of the neuron negatively charged.

The electrical potential difference across the membrane can be thought of as a battery whose charge is maintained by the work of the ion pumps. In the following chapter we'll see how this battery runs our brain.

CONCLUDING REMARKS

KEY TERMS

Introduction

action potential

excitable membrane

resting membrane potential

The Cast of Chemicals

ion

cation

anion

phospholipid bilayer

peptide bond

polypeptide

ion channel

ion selectivity

gating

ion pump

Movement of Ions

diffusion

concentration gradient

electrical current

electrical potential

voltage

electrical conductance

electrical resistance

Ohm's law

The Ionic Basis of the Resting Membrane Potential

microelectrode

ionic equilibrium potential

ionic driving force

Nernst equation

sodium-potassium pump

calcium-pump

Goldman equation

depolarize

blood-brain barrier

✔ REVIEW QUESTIONS

1. What two functions do proteins in the neuronal membrane serve to establish and maintain the resting membrane potential?
2. On which side of the neuronal membrane are Na^+ ions more abundant?
3. When the membrane is at the potassium equilibrium potential, in which direction (in or out) is there a net movement of potassium ions?
4. There is a much greater K^+ concentration inside the cell than outside. Why, then, is the resting membrane potential negative?
5. When the brain is deprived of oxygen, the mitochondria within neurons cease producing ATP. What effect would this have on the membrane potential? Why?

The Action Potential

Now we come to the signal that conveys information over distances in the nervous system—the action potential. As we saw in Chapter 3, the cytosol in the neuron at rest is negatively charged with respect to the extracellular fluid. The action potential is a rapid reversal of this situation such that, for an instant, the inside of the membrane becomes positively charged with respect to the outside. The action potential is also often called a spike, nerve impulse, or a discharge.

The action potentials generated by a cell are all similar in size and duration, and they do not diminish as they are conducted down the axon. Keep in mind the big picture: The frequency and pattern of action potentials constitute the code used by neurons to transfer information from one location to another. In this chapter, we discuss the mechanisms that are responsible for the action potential and how it propagates down the axonal membrane.

PROPERTIES OF THE ACTION POTENTIAL

Action potentials have certain universal properties, features that are shared by axons in the nervous systems of every beast, from a lowly squid to a college student. Let's begin by exploring some of these properties. What does the action potential look like? How is it initiated? How rapidly can a neuron generate action potentials?

The Ups and Downs of an Action Potential

In Chapter 3, we saw that the membrane potential, V_m, can be determined by inserting a microelectrode in the cell. A voltmeter is used to measure the electrical potential difference between the tip of this intracellular microelectrode and another placed outside the cell. At rest, the voltmeter reads a steady potential difference of about –65 mV. During the action potential, however, the membrane potential briefly becomes positive. Because this occurs so rapidly—100 times faster than the blink of an eye—a special type of voltmeter, called an *oscilloscope*, is used to study action potentials. The oscilloscope gives a record of voltage as it changes over time (Box 4.1).

| Box 4.1 | **BRAIN FOOD** |

Methods of Recording Action Potentials

Methods used to study nerve impulses may be broadly divided into two types: intracellular and extracellular. Intracellular recording requires impaling the neuron or axon with a microelectrode. The small size of most neurons makes this method challenging (recall Box 3.4), explaining why so many of the early studies on action potentials were performed on the neurons of invertebrates, which can be 50 to 100 times larger than mammalian neurons. Fortunately, recent technical advances have made even the smallest vertebrate neurons accessible to intracellular recording methods, and these studies have confirmed that much of what was learned in invertebrates is directly applicable to humans. The goal of intracellular recording is simple: to measure the potential difference between the tip of the intracellular electrode and another electrode placed in the solution bathing the neuron (continuous with the earth, and thus called ground). The intracellular electrode is filled with a concentrated salt solution (often KCl) having a high electrical conductivity. The electrode is connected to an amplifier that compares the potential difference between this electrode and ground. This potential difference can be displayed using an oscilloscope. The oscilloscope sweeps a beam of electrons from left to right across a phosphor screen. Vertical deflections of this beam can be read as changes in voltage. The oscilloscope is really just a sophisticated voltmeter that can record rapid changes in voltage (such as an action potential).

As we shall see in this chapter, the action potential is characterized by a sequence of ionic movements across the neuronal membrane. These electrical currents can be detected without impaling the neuron by placing an electrode near the membrane. This is the principle behind extracellular recording. Again, we measure the potential difference between the tip of the recording electrode and ground. The electrode can be a fine glass capillary filled with electrolytes, but it is often simply a thin insulated metal wire. Normally, in the absence of neural activity, the potential difference between the extracellular recording electrode and ground is zero. However, when the action potential arrives at the recording position, positive charges flow away from the recording electrode, into the neuron. Then, as the action potential passes by, positive charges flow out across the membrane toward the recording electrode. Thus, the extracellular action potential is characterized by a brief, alternating voltage difference between the recording electrode and ground. These changes in voltage can be seen using an oscilloscope, but they can also be *heard* by connecting the output of the amplifier to a loudspeaker. Each impulse makes a distinctive "pop" sound. Indeed, recording the activity of an active sensory nerve sounds just like making popcorn.

An action potential, as it would appear on the display of an oscilloscope, is shown in Figure 4.1. This graph represents a plot of membrane potential versus time. Notice that the action potential has certain identifiable parts. The first part, called the **rising phase**, is characterized by a rapid depolarization of the membrane. This change in membrane potential continues until V_m reaches a peak value of about 40 mV. The part of the action potential where the inside of the neuron is positively charged with respect to the outside is called the **overshoot**. The **falling phase** of the action potential is a rapid repolarization until the membrane is actually more negative than the resting potential. The last part of the falling phase is called the **undershoot**, or **after-hyperpolarization**. Finally, there is a gradual restoration of the resting potential. From beginning to end, the action potential lasts about 2 milliseconds (msec).

Generation of an Action Potential

In Chapter 3, we said that breaking of the skin by a thumbtack was sufficient to generate action potentials in a sensory nerve. Let's use this example to see how an action potential begins.

The perception of sharp pain when a thumbtack enters your foot is caused by the generation of action potentials in certain nerve fibers in the skin. (We'll learn more about pain in Chapter 12.) The membrane of these fibers is believed to possess a type of gated sodium channel that opens when the nerve ending is stretched. The initial chain of events is therefore: (a) thumbtack enters the skin, (b) membrane of the nerve fibers in the skin is stretched, (c) Na^+-permeable channels open. Because of the large concentration gradient and the negative charge of the cytosol, Na^+ ions enter the fiber through these channels. The entry of Na^+ depolarizes the membrane; that is, the cytosol of the fiber becomes less negative. If this depolarization, called a *generator potential*, achieves a critical level, the membrane will generate an action potential. The critical level of depolarization that must be crossed in order to trigger an action potential is called **threshold**. *Action potentials are caused by depolarization of the membrane beyond threshold.*

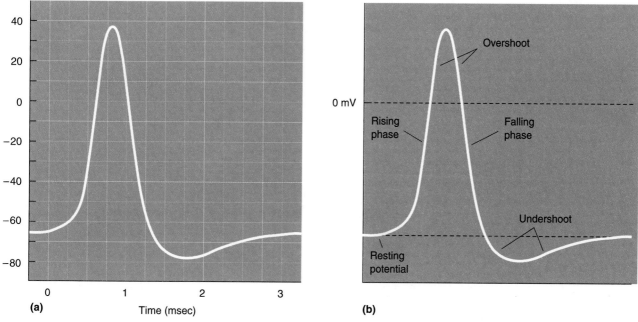

Figure 4.1
An action potential. (a) The action potential as it would be displayed by an oscilloscope. **(b)** The parts of an action potential.

The depolarization that causes action potentials arises in different ways in different neurons. In our example above, depolarization was caused by the entry of Na^+ through specialized ion channels that were sensitive to membrane stretching. In interneurons, depolarization is usually caused by Na^+ entry through channels that are sensitive to neurotransmitters released by other neurons. In addition to these natural routes, neurons can be depolarized by injecting electrical current through a microelectrode, a method commonly used by neuroscientists to study action potentials in different cells.

Generating an action potential by depolarizing a neuron is something like taking a photograph by pressing the shutter button on a camera. Applying increasing pressure on the button has no effect until it crosses a threshold value, and then "click"—the shutter opens and one frame of film is exposed. Applying increasing depolarization to a neuron has no effect until it crosses threshold, and then "pop"—one action potential. For this reason, action potentials are said to be "all-or-none."

Generation of Multiple Action Potentials

In the above example, we likened the generation of an action potential by depolarization to taking a photograph by pressing the shutter button on a camera. But what if the camera is one of those fancy motor-driven models that fashion and sports photographers use? In that case, continued pressure on the shutter button beyond threshold would cause the camera to shoot frame after frame. The same thing is true for a neuron. If, for example, we pass continuous depolarizing current into a neuron through a microelectrode, we will generate not one, but many action potentials in succession (Figure 4.2).

The rate of action potential generation depends on the magnitude of the continuous depolarizing current. If we pass enough current through a microelectrode to depolarize just to threshold, but not far beyond, we might

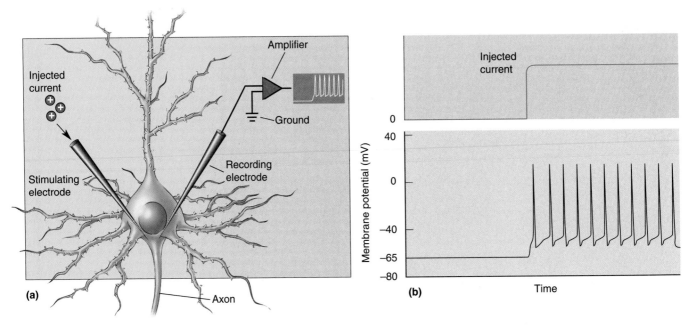

Figure 4.2
The effect of injecting positive charge into a neuron. (a) The axon hillock is impaled by two electrodes, one for recording the membrane potential relative to ground and the other for stimulating the neuron with electrical current. **(b)** When electrical current is injected into the neuron (top trace), the membrane is depolarized sufficiently to fire action potentials (bottom trace).

find that the cell generates action potentials at a rate of something like one per second, or one hertz (1 Hz). If we crank up the current a little bit more, however, we will find that the rate of action potential generation increases, say, to 50 impulses per second (50 Hz). Thus, the "firing frequency" of action potentials reflects the magnitude of the depolarizing current. This is one way that stimulation intensity is encoded in the nervous system (Figure 4.3).

Although firing frequency increases with the amount of depolarizing current, there is a limit to the rate at which a neuron can generate action potentials. The maximum firing frequency is about 1000 Hz; once an action potential is initiated, it is impossible to initiate another for about 1 msec. This period of time is called the **absolute refractory period**. In addition, it can be relatively difficult to initiate another action potential for several milliseconds after the end of the absolute refractory period. During this **relative refractory period**, the amount of current required to depolarize the neuron to action potential threshold is elevated over normal.

We will now take a look at how the movement of ions across the membrane through specialized protein channels causes a neural signal with these properties.

THE ACTION POTENTIAL, IN THEORY

The action potential is a dramatic redistribution of electrical charge across the membrane. *Depolarization of the cell during the action potential is caused by the influx of sodium ions across the membrane, and repolarization is caused by the efflux of potassium ions.* Let's apply some of the concepts introduced in Chapter 3 to help us understand how ions are driven across the membrane, and how these ionic movements affect the membrane potential.

Membrane Currents and Conductances

Consider the ideal neuron illustrated in Figure 4.4. The membrane of this cell has three types of protein molecules: sodium-potassium pumps,

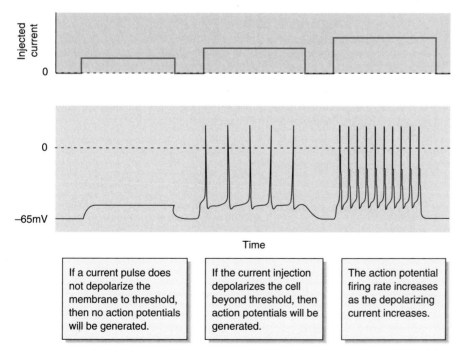

| If a current pulse does not depolarize the membrane to threshold, then no action potentials will be generated. | If the current injection depolarizes the cell beyond threshold, then action potentials will be generated. | The action potential firing rate increases as the depolarizing current increases. |

Figure 4.3
Dependence of action potential firing frequency on the level of depolarization.

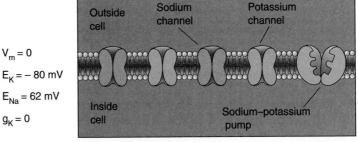

$V_m = 0$

$E_K = -80$ mV

$E_{Na} = 62$ mV

$g_K = 0$

(a) $I_K = g_K (V_m - E_K) = 0$

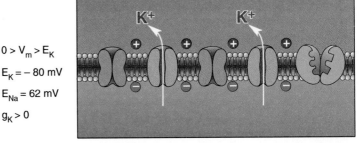

$0 > V_m > E_K$

$E_K = -80$ mV

$E_{Na} = 62$ mV

$g_K > 0$

(b) $I_K = g_K (V_m - E_K) > 0$

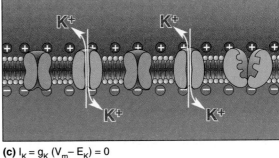

$V_m = -80$ mV

$E_K = -80$ mV

$E_{Na} = 62$ mV

$g_K > 0$

(c) $I_K = g_K (V_m - E_K) = 0$

Figure 4.4
Membrane currents and conductances. Here is an ideal neuron with sodium-potassium pumps, potassium channels, and sodium channels. The pumps establish ionic concentration gradients so that K^+ is concentrated inside the cell and Na^+ is concentrated outside the cell. **(a)** Initially, we assume that all channels are closed and the membrane potential equals 0 mV. **(b)** Now we open the potassium channels, and K^+ flows out of the cell. This movement of K^+ is an electrical current, I_K, and it flows as long as the membrane conductance to K^+ ions, g_K, is greater than zero, and the membrane potential is not equal to the potassium equilibrium potential. **(c)** At equilibrium, there is no net potassium current because, although $g_K > 0$, the membrane potential at equilibrium equals E_K.

potassium channels, and sodium channels. The pumps work continuously to establish and maintain concentration gradients. As in all our previous examples, we'll assume that K^+ is concentrated twentyfold inside the cell and that Na^+ is concentrated tenfold outside the cell. According to the Nernst equation, at 37°C $E_K = -80$ mV and $E_{Na} = 62$ mV. Let's use this cell to explore the factors that govern the movement of ions across the membrane.

We begin by assuming that both the potassium channels and the sodium channels are closed, and that the membrane potential, V_m, is equal to 0 mV (Figure 4.4a). Now let's open the potassium channels only (Figure 4.4b). As we learned in Chapter 3, K^+ ions will flow out of the cell, down their concentration gradient, until the inside becomes negatively charged, and $V_m = E_K$

(Figure 4.4c). Here we want to focus our attention on the movement of K^+ that took the membrane potential from 0 mV to –80 mV. Consider these three points:

1. The net movement of K^+ ions across the membrane is an electrical current. We can represent this current using the symbol I_K.

2. The number of open potassium channels is proportional to an electrical conductance. We can represent this conductance by the symbol g_K.

3. Membrane potassium current, I_K, will flow only as long as $V_m \neq E_K$. The driving force on K^+ is defined as the difference between the real membrane potential and the equilibrium potential, and it can be written as $V_m - E_K$.

There is a simple relationship between the ionic driving force, ionic conductance, and the amount of ionic current that will flow. For K^+ ions, this may be written $I_K = g_K (V_m - E_K)$. More generally, we write

$$I_{ion} = g_{ion} (V_m - E_{ion}).$$

If this sounds familiar, that is because it is simply an expression of Ohm's law, $I = gV$, which we learned about in Chapter 3.

Now let's take another look at this example. Initially we began with $V_m = 0$ mV, and no ionic membrane permeability (Figure 4.4a). There is a large driving force on K^+ ions because $V_m \neq E_K$; in fact, $(V_m - E_K) = 80$ mV. However, because the membrane is impermeable to K^+, the potassium conductance, g_K, equals zero. Consequently, $I_K = 0$. Potassium current only flows when we stipulate that the membrane has open potassium channels, and therefore $g_K > 0$. Now K^+ ions flow out of the cell—as long as the membrane potential differs from the potassium equilibrium potential (Figure 4.4b). Note that the current flow is in the direction that takes V_m toward E_K. When $V_m = E_K$, the membrane is at equilibrium, and no net current will flow. In this condition, although there is a large potassium conductance, g_K, there is no longer any driving force on the K^+ ions (Figure 4.4c).

The Ins and Outs of an Action Potential

Let's pick up the action where we left off in the last section. The membrane of our ideal neuron is permeable only to K^+, and $V_m = E_K = -80$ mV. What's happening with the Na^+ ions concentrated outside the cell? Because the membrane potential is so negative with respect to the sodium equilibrium potential, there is a very large driving force on Na^+ ($[V_m - E_{Na}] = [-80$ mV $- 62$ mV$] = -142$ mV). Nonetheless, there can be no net Na^+ current as long as the membrane is impermeable to Na^+. But now let's open the sodium channels and see what happens to the membrane potential.

At the instant we change the ionic permeability of the membrane, g_{Na} is high, and, as we said above, there is a large driving force pushing on Na^+. Thus, we have what it takes to generate a large sodium current, I_{Na}, across the membrane. Na^+ ions pass through the membrane sodium channels in the direction that takes V_m toward E_{Na}; in this case, the sodium current, I_{Na}, is inward across the membrane. Assuming the membrane permeability is now far greater to sodium than it is to potassium, this influx of Na^+ depolarizes the neuron until V_m approaches E_{Na}, 62 mV.

Notice that something remarkable happened here. Simply by switching the dominant membrane permeability from K^+ to Na^+, we were able to rapidly reverse the membrane potential. In theory, then, the rising phase of the action potential could be explained if, in response to depolarization of the membrane beyond threshold, membrane sodium channels opened. This would allow Na^+ to enter the neuron, causing a massive depolarization until the membrane potential approached E_{Na}.

How could we account for the falling phase of the action potential? Simply assume that sodium channels quickly close and potassium channels open, so the dominant membrane ion permeability switches back from Na^+ to K^+. Then K^+ would flow out of the cell until the membrane potential again equals E_K.

Our model for the ins and outs, ups and downs of the action potential in an ideal neuron is shown in Figure 4.5. The rising phase of the action potential is explained by an inward sodium current, and the falling phase is explained by an outward potassium current. The action potential therefore could be accounted for simply by the movement of ions through channels that are gated by changes in the membrane potential. If you understand this concept, you understand a lot about the ionic basis of the action potential. What's left now is to see how this actually happens—in a real neuron.

THE ACTION POTENTIAL, IN REALITY

Let's quickly review our theory of the action potential. When the membrane is depolarized to threshold, there is a transient increase in g_{Na}. The increase in g_{Na} allows entry of Na^+ ions, which depolarizes the neuron. And the increase in g_{Na} must be brief in duration to account for the short duration of the action potential. Restoring the negative membrane potential is further aided by a subsequent increase in g_K, which allows K^+ ions to leave the depolarized neuron faster.

Testing this theory is simple enough in principle. All one has to do is measure the sodium and potassium conductances of the membrane during the action potential. In practice, however, such a measurement proved to be quite difficult in real neurons. The key technical breakthrough was the introduction of a device called a **voltage clamp**, invented by American physiologist Kenneth C. Cole, and the key experiments using it were performed by Cambridge University physiologists Alan Hodgkin and Andrew Huxley around 1950. The voltage clamp enabled Hodgkin and Huxley to "clamp" the membrane potential of an axon at any value they chose. By measuring the currents that flowed across the membrane, they could deduce the changes in membrane conductance at different membrane potentials. In an elegant series of experiments, Hodgkin and Huxley showed that the rising phase of the action potential was indeed caused by a transient increase in g_{Na} and an influx of Na^+ ions, and that the falling phase was associated with an increase in g_K and an efflux of K^+ ions. Their accomplishments were recognized with the Nobel Prize in 1963.

To account for the transient changes in g_{Na}, Hodgkin and Huxley proposed the existence of sodium "gates" in the axonal membrane. They hypothesized that these gates are "activated"—opened—by depolarization above threshold, and "inactivated"—closed and locked—when the membrane acquires a positive membrane potential. These gates are "deinactivated"—enabled to be opened again—only after the membrane potential returns to a negative value.

It is a tribute to Hodgkin and Huxley that their hypotheses about membrane gates predated by over 20 years the direct demonstration of voltage-gated channel proteins in the neuronal membrane. We have a new understanding of gated membrane channels, thanks to two recent scientific breakthroughs. First, new molecular biological techniques have enabled neuroscientists to determine the detailed structure of these proteins. Second, new neurophysiological techniques have enabled neuroscientists to measure the ionic currents that pass through single channels. We will now explore the action potential from the perspective of these membrane ion channels.

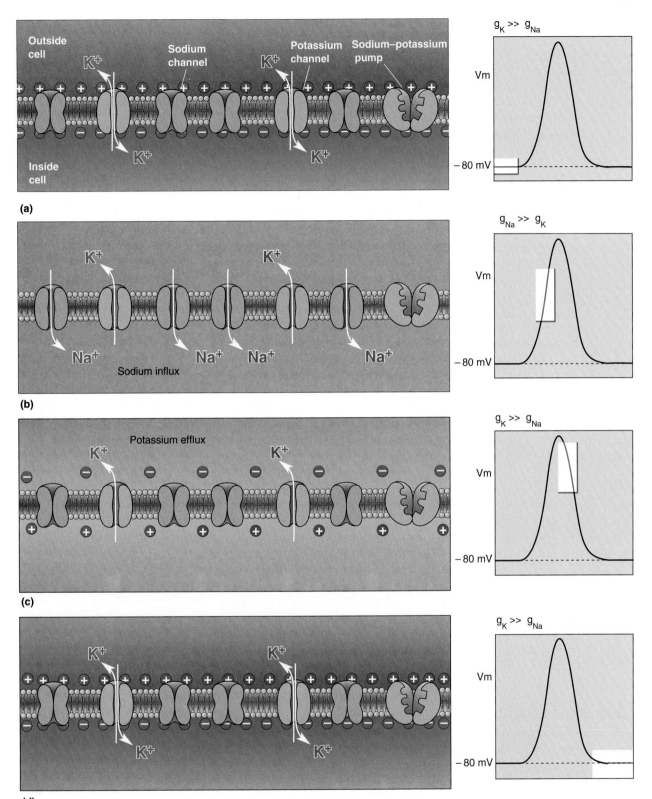

Figure 4.5
Flipping the membrane potential by changing the relative ionic permeability of the membrane. (a) The ideal neuron, introduced in Figure 4.4. We begin by assuming that the membrane is permeable only to K^+ and that $V_m = E_K$. **(b)** We now stipulate that the membrane sodium channels open so that $g_{Na} \gg g_K$. There is a large driving force on Na^+, so Na^+ ions rush into the cell, taking V_m toward E_{Na}. **(c)** Now we close the sodium channels and open more potassium channels so that $g_K \gg g_{Na}$. Because the membrane potential is positive, there is a large driving force on K^+ ions. The efflux of K^+ takes V_m back toward E_K. **(d)** The resting state is restored where $V_m = E_K$.

The Voltage-Gated Sodium Channel

The **voltage-gated sodium channel** is aptly named. The protein forms a pore in the membrane that is highly selective to Na$^+$ ions, and the pore is opened and closed by changes in the electrical potential of the membrane.

Sodium Channel Structure. The voltage-gated sodium channel is created from a single long polypeptide. The molecule has four distinct domains, each of which consists of six transmembrane alpha helices (Figure 4.6a). The four domains are believed to clump together to form a pore between them (Figure 4.6b). The pore is closed at the negative resting membrane potential. When the membrane is depolarized to threshold, however, the molecule is believed to twist into a configuration that allows the passage of Na$^+$ through the pore (Figure 4.7).

Functional Properties of the Sodium Channel. It is now possible, using a new method called the **patch clamp**, to study the ionic currents passing through individual ion channels (Box 4.2). The patch-clamp method entails sealing the tip of an electrode to a very small *patch* of neuronal membrane. This patch then can be torn away from the neuron, and the ionic currents across it can be measured as the membrane potential is *clamped* at any value the experimenter selects. With luck, the patch will contain only a single channel, and the behavior of this channel can be studied. Patch clamping has enabled neuroscientists to study the functional properties of the voltage-gated sodium channel.

Changing the membrane potential of a patch of axonal membrane from −80 to −65 mV has little effect on the voltage-gated sodium channels. They remain closed because depolarization of the membrane has not yet reached threshold. Changing the membrane potential from −65 to −40 mV, however, causes these channels to pop open. As shown in Figure 4.8, voltage-gated sodium channels have a characteristic pattern of behavior:

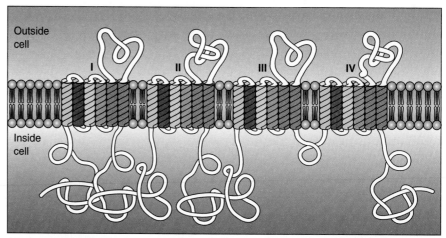

Outside cell

I   II   III   IV

Inside cell

(a)

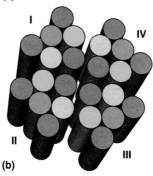

I

IV

II

III

(b)

Figure 4.6
Structure of the voltage-gated sodium channel.
(a) A depiction of how the sodium channel polypeptide chain is believed to be woven into the membrane. The molecule consists of four domains, I–IV. Each domain consists of six alpha helices (represented by the colored cylinders), which pass back and forth across the membrane. **(b)** A view of the molecule from above shows how the domains may arrange themselves to form a pore between them. (Source: Adapted from Noda et al., 1986.)

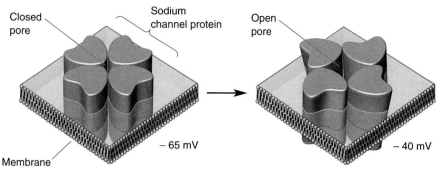

Figure 4.7
Hypothetical model for changing the configuration of the sodium channel by depolarizing the membrane.

1. They open with little delay.

2. They stay open for about 1 msec, and then close (inactivate).

3. They cannot be opened again by depolarization until the membrane potential returns to a negative value near threshold.

A hypothetical model for how conformational changes in the voltage-gated sodium channel could account for these properties is illustrated in Figure 4.8c.

A single channel does not an action potential make. The membrane of an axon may contain thousands of sodium channels per μm^2, and the concerted action of all these channels is required to generate what we measure as an action potential. Nonetheless, it is interesting to see how many of the properties of the action potential can be explained by the properties of the voltage-gated sodium channel. For example, the fact that single channels do not open until a critical level of membrane depolarization is reached explains the action potential threshold. The rapid opening of the channels in response to depolarization explains why the rising phase of the action potential occurs so quickly. And the short time the channels stay open before inactivating (about 1 msec) partly explains why the action potential is so brief. Furthermore, inactivation of the channels can account for the absolute refractory period: another action potential cannot be generated until the channels are deinactivated.

Currents through the sodium channel, and action potentials, can be completely blocked with **tetrodotoxin (TTX)**. This virulent toxin clogs the Na^+-permeable pore by binding tightly to a specific site on the outside of the channel. As we shall see in later chapters, this compound is commonly used for experiments that require the blocking of impulses in a nerve or muscle. TTX, originally isolated from the ovaries of the Japanese puffer fish, is usually fatal if ingested. Nonetheless, puffer fish are considered a delicacy in Japan. Sushi chefs train for years, and are licensed by the government, to prepare puffer fish in such a way that eating them causes numbness around the mouth. Now that's eating dangerously!

Voltage-Gated Potassium Channels

Hodgkin and Huxley's experiments indicated that the falling phase of the action potential was explained only partly by the inactivation of g_{Na}. They found there was also a transient increase in g_K that functioned to speed the restoration of a negative membrane potential after the spike. They

Box 4.2   **BRAIN FOOD**

The Patch-Clamp Method

The very existence of voltage-gated channels in the membrane was merely conjecture until the development of methods to study individual channel proteins. A revolutionary new method, the patch clamp, was developed by German neuroscientists Bert Sakmann and Erwin Neher in the mid-1970s. In recognition of their contribution, Sakmann and Neher were awarded the 1991 Nobel Prize.

Patch clamping enables one to record ionic currents through single channels. The first step is gently lowering the fire-polished tip of a glass recording electrode, 1–5 μm in diameter, onto the membrane of the neuron (Figure Aa) and then applying suction through the electrode tip (Figure Ab). A tight seal forms between the walls of the electrode and the underlying patch of membrane. This "gigaohm" seal (so named because of its high electrical resistance, $>10^9 \Omega$) leaves the ions in the electrode only one path

to take, through the channels in the underlying patch of membrane. If the electrode is then withdrawn from the cell, the membrane patch can be torn away (Figure Ac), and ionic currents can be measured as steady voltages are applied across the membrane (Figure Ad). With a little luck, one can resolve currents flowing through single channels. If the patch contains a voltage-gated sodium channel, for example, then changing the membrane potential from –65 to –40 mV will cause the channel to open, and current will flow through it (Figure Ae). The amplitude of the measured current at a constant membrane voltage reflects the channel conductance, and the duration of the current reflects the time the channel is open.

Patch-clamp recordings reveal that most channels flip between two conductance states that can be interpreted as open or closed. The time they remain open can vary, but the single-channel conductance value stays the same and is therefore said to be unitary. Ions can pass through single channels at an astonishing rate: well over a million per second.

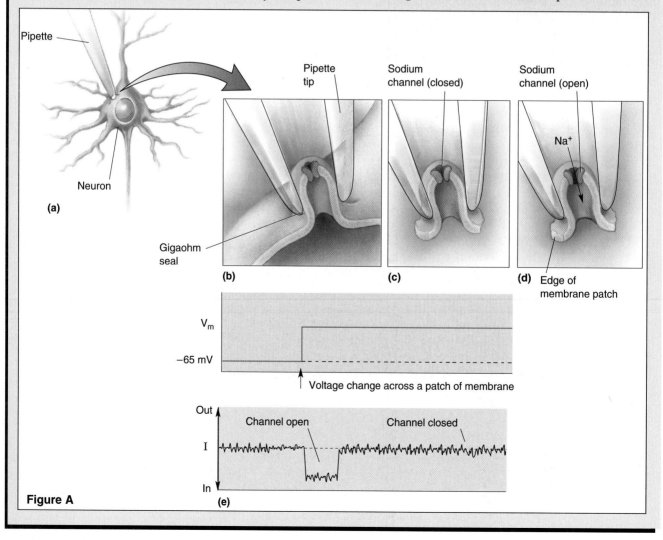

Figure A

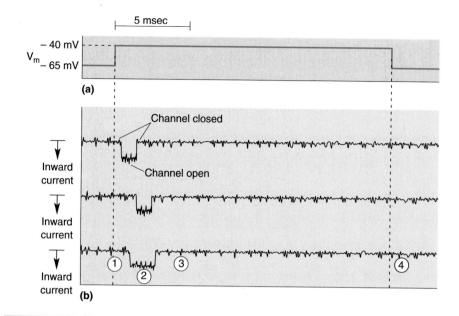

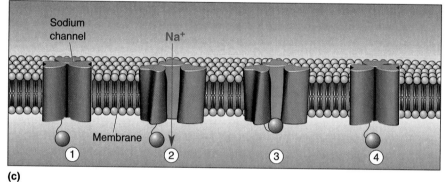

Figure 4.8
Sodium channels open and then close when depolarized. (a) This trace shows the electrical potential across a patch of membrane. When the membrane potential is changed from −65 to −40 mV, the sodium channels pop open. **(b)** These traces show how three different channels respond to the voltage step. Each line is a record of the electrical current that flows through a single channel. (1) At −65 mV, the channels are closed, so there is no current. (2) When the membrane is depolarized to −40 mV, the channels briefly open and current flows inward, represented by the downward deflection in the current traces. Although there is some variability from channel to channel, all of them open with little delay and stay open for less than 1 msec. Notice that after they have opened once, they close and stay closed as long as the membrane is maintained at a depolarized V_m. (3) The closure of the sodium channel by steady depolarization is called inactivation. (4) To deinactivate the channels, the membrane must be returned to −65 mV again. **(c)** A model for how changes in the conformation of the sodium channel protein might yield its functional properties. (1) The closed channel (2) opens upon membrane depolarization. (3) Inactivation occurs when a globular portion of the protein swings up and occludes the pore. (4) Deinactivation occurs when the globular portion swings away and the pore closes by movement of the transmembrane domains.

proposed the existence of membrane potassium gates that, like sodium gates, open in response to depolarization of the membrane. Unlike sodium gates, however, potassium gates do not open immediately upon depolarization; it takes about 1 msec for them to open. Because of this delay, and because this potassium conductance serves to rectify, or reset, the membrane potential, they called this conductance the *delayed rectifier*.

We now know that there are many different types of **voltage-gated potassium channels**. Most of them open when the membrane is depolarized and function to dampen any further depolarization by giving K^+ ions a path to leave the cell across the membrane. The known voltage-gated potassium channels have a similar structure. The channel proteins consist of four separate polypeptide subunits that come together to form a pore between them. Like the sodium channel, these proteins are sensitive to changes in the electrical field across the membrane. When the membrane is depolarized, the subunits are believed to twist into a shape that allows K^+ ions to pass through the pore.

Putting the Pieces Together

We can now use what we've learned about ions and channels to explain the key properties of the action potential (Figure 4.9):

- *Threshold*. Threshold is the membrane potential at which enough voltage-gated sodium channels pop open so that the relative ionic permeability of the membrane favors sodium over potassium.

- *Rising phase*. When the inside of the membrane has a negative electrical potential, there is a large driving force on Na^+ ions. Therefore, Na^+ ions rush into the cell through the open sodium channels, causing the membrane to rapidly depolarize.

- *Overshoot*. Because the relative permeability of the membrane greatly favors sodium, the membrane potential goes to a value close to E_{Na}, which is greater than 0 mV.

- *Falling phase*. The behavior of two types of channel contributes to the falling phase. First, the voltage-gated sodium channels inactivate. Second, the voltage-gated potassium channels finally pop open (triggered to do so 1 msec earlier by the depolarization of the membrane). There is a great driving force on K^+ ions when the membrane is strongly depolarized. Therefore, K^+ ions rush out of the cell through the open channels, causing the membrane potential to become negative again.

- *Undershoot*. The open voltage-gated potassium channels add to the resting potassium membrane permeability. Because there is very little sodium permeability, the membrane potential goes towards E_K, causing a hyperpolarization relative to the resting membrane potential until the voltage-gated potassium channels close again.

- *Absolute refractory period*. Sodium channels inactivate when the membrane becomes strongly depolarized. They cannot be activated again, and another action potential cannot be generated, until the membrane potential goes sufficiently negative to deinactivate the channels.

- *Relative refractory period*. The membrane potential stays hyperpolarized until the voltage-gated potassium channels close. Therefore, more depolarizing current is required to bring the membrane potential to threshold.

We've seen that channels and the movement of ions through them can explain the properties of the action potential. But it is important to remember that the sodium-potassium pump also is working quietly in the background. Imagine that the entry of Na^+ during each action potential is like a wave coming over the bow of a boat making way in heavy seas. Like the continuous action of the boat's bilge pump, the sodium-potassium pump works all the time to transport Na^+ back across the membrane. The pump

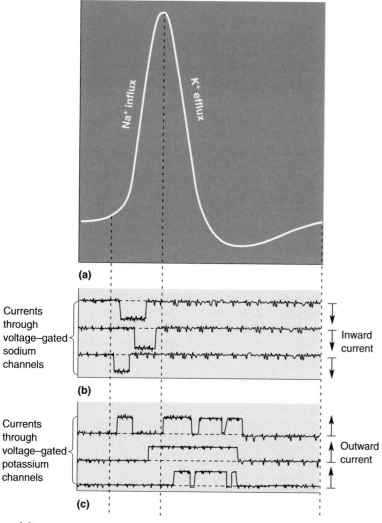

Figure 4.9

The molecular basis of the action potential. (a) The membrane potential as it changes in time during an action potential. The rising phase of the action potential is caused by the influx of Na$^+$ ions through hundreds of voltage-gated sodium channels. The falling phase is caused by sodium channel inactivation and the efflux of K$^+$ ions through voltage-gated potassium channels. **(b)** The inward currents through three representative voltage-gated sodium channels. Each channel opens with little delay when the membrane is depolarized to threshold. The channels stay open for no more than a msec and then inactivate. **(c)** The outward currents through three representative voltage-gated potassium channels. Voltage-gated potassium channels open about 1 msec after the membrane is depolarized to threshold and stay open as long as the membrane is depolarized. The high potassium permeability causes the membrane to hyperpolarize briefly. When the voltage-gated potassium channels close, the membrane potential relaxes back to the resting value, around −65 mV.

maintains the ionic concentration gradients that drive Na$^+$ and K$^+$ through their channels during the action potential.

ACTION POTENTIAL CONDUCTION

In order to transfer information from one point to another in the nervous system, it is necessary that the action potential, once generated, be conducted down the axon. This process is like the burning of a fuse. Imagine your favorite cartoon character holding a stick of dynamite with a burning match held under the fuse. The fuse ignites when it gets hot enough (beyond some threshold). The flame heats up the segment of fuse immediately ahead of it until it ignites. In this way, the flame steadily works its way down the fuse. Note that the fuse lit at one end only burns in one direction; the flame cannot turn back on itself because the combustible material just behind it is spent.

Propagation of the action potential along the axon is similar to the propagation of the flame along the fuse. When the axon is depolarized sufficiently to reach threshold, voltage-gated sodium channels open, and the action potential is initiated. The influx of positive charge depolarizes the segment of membrane immediately before it until it reaches threshold and generates its own action potential (Figure 4.10). In this way, the action potential works its way down the axon until it reaches the axon terminal, thereby initiating synaptic transmission (the subject of Chapter 5). An action potential initiated at one end of an axon only propagates in one direction; it does not turn back on itself. This is because the membrane just behind it is refractory, due to inactivation of the sodium channels. But, just like the fuse, an action potential can be generated by depolarization at either end of the axon and therefore propagate in either direction (although normally, action potentials conduct only in one direction). Note that because the axonal membrane is excitable (capable of generating action potentials) along its entire length, the impulse will propagate without decrement. The fuse works the same way, because it is combustible along its entire length. Note, however, that unlike the fuse, the axon can regenerate its firing ability.

Action potential conduction velocities vary, but 10 m/sec is a typical rate. Remember, from start to finish the action potential lasts about 2 msec. From this, we can calculate the length of membrane that is engaged in the action potential at any instant in time:

$$10 \text{ m/sec} \times 2 \times 10^{-3} \text{sec} = 2 \times 10^{-2} \text{m}.$$

Therefore, an action potential traveling at 10 m/sec occurs over a 2-cm length of axon.

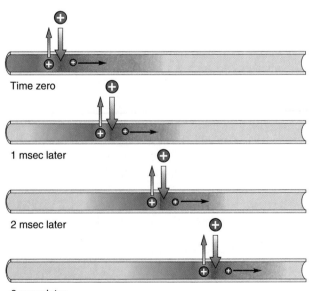

Time zero

1 msec later

2 msec later

3 msec later

Figure 4.10
Action potential conduction. The entry of positive charge during the action potential causes the membrane just ahead to depolarize to threshold.

Factors Influencing Conduction Velocity

Remember that the inward Na^+ current during the action potential depolarizes the patch of membrane just ahead. If this patch reaches threshold, it will fire an action potential, and the action potential will "burn" on down the membrane. The speed with which the action potential propagates down the axon depends on how far the depolarization ahead of the action potential spreads down the axon, which in turn depends on certain physical characteristics of the axon.

Imagine that the influx of positive charge into an axon during the action potential is like turning on the water to a leaky garden hose. There are two paths the water can take: one, down the inside of the hose; the other, across the hose through the leaks. How much water goes along each path depends on their relative resistance; most of the water will take the path of least resistance. If the hose is narrow and the leaks are numerous and large, most of the water will flow out through the leaks. If the hose is wide and the leaks are few and tiny, most of the water will flow down the inside of the hose. The same principles apply to positive current spreading down the axon ahead of the action potential. There are two paths that positive charge can take: one, down the inside of the axon; the other, across the axonal membrane. If the axon is narrow and there are many open membrane pores, most of the current will flow out across the membrane. If the axon is wide and there are few open membrane pores, most of the current will flow down inside the axon. The farther the current goes down the axon, the farther ahead of the action potential the membrane will be depolarized, and the faster the action potential will propagate. As a rule, therefore, action potential conduction velocity increases with increasing axonal diameter.

As a consequence of this relationship between axonal diameter and conduction velocity, neural pathways that are especially important for survival have evolved unusually large axons. An example is the giant axon of the squid. Originating from neurons in the squid stellate ganglion, the giant axon innervates the musculature of the mantle. It is part of a pathway that mediates an escape reflex in response to strong sensory stimulation. The squid giant axon can be 1 mm in diameter, so large that originally it was thought to be part of the squid circulatory system. Neuroscience owes a debt to British zoologist J. Z. Young who, in 1939, called attention to the squid giant axon as a preparation for studying the biophysics of the neuronal membrane. Hodgkin and Huxley used this preparation to elucidate the ionic basis of the action potential, and the giant axon continues to be used today for a wide range of neurobiological studies.

Myelin and Saltatory Conduction. The good thing about fat axons is that they conduct action potentials faster; the bad thing about them is they take up a lot of space. If all the axons in your brain were the diameter of a squid giant axon, your head would be too big to fit through most doorways. Fortunately, vertebrates evolved another solution for increasing action potential conduction velocity: ensheathing the axon with insulation called myelin (see Chapter 2). Myelin consists of many wraps of membrane provided by glial support cells—Schwann cells in the peripheral nervous system (outside the brain and spinal cord), and oligodendroglia in the central nervous system. Just as wrapping the leaky garden hose with duct tape facilitates water flow down the inside of the hose, myelin facilitates current flow down the inside of the axon, thereby increasing action potential conduction velocity (Box 4.3).

The myelin sheath does not extend continuously along the entire length of the axon. There are breaks in the insulation where ions cross the membrane, to generate action potentials. As you'll recall from Chapter 2, these breaks in the myelin sheath are the nodes of Ranvier (Figure 4.11). Voltage-gated sodium channels are concentrated in the membrane of the nodes. The distance between nodes is usually 0.2–2.0 mm, depending on the size of the axon (fatter axons have larger internodal distances).

Imagine that the action potential traveling along the axon membrane is like you traveling down a sidewalk. Action potential conduction without myelin is like walking down the sidewalk in small steps, heel-to-toe, using every inch of the sidewalk to creep along. Conduction with myelin, in contrast, is like skipping down the sidewalk. In myelinated axons, action potentials skip from node to node (Figure 4.12). This type of action potential propagation is called **saltatory conduction** (from the Latin meaning "to leap").

ACTION POTENTIALS, AXONS, AND DENDRITES

Action potentials of the type discussed in this chapter are a feature mainly of axons. As a rule, membrane of dendrites and neuronal cell bodies does not generate sodium-dependent action potentials. This is because this membrane has very few voltage-gated sodium channels. Only membrane that contains these specialized protein molecules is capable of generating action potentials, and this type of excitable membrane is usually found only in axons. Therefore, the part of the neuron where an axon originates from

---

Box 4.3 ▸ **OF SPECIAL INTEREST**

Multiple Sclerosis, a Demyelinating Disease

The critical importance of myelin for the normal transfer of information in the human nervous system is revealed by the neurological disorder known as multiple sclerosis (MS). Victims of MS often complain of weakness, lack of coordination, and impairments of vision and speech. The disease is capricious, usually marked by remissions and relapses that occur over a period of many years. Although the precise cause of MS is still poorly understood, the cause of the sensory and motor disturbances is now quite clear. MS attacks the myelin sheaths of bundles of axons in the brain, spinal cord, and optic nerves. The name is derived from the Greek word for "hardening," which describes the lesions that develop around bundles of axons; and the sclerosis is multiple because the disease attacks many sites in the nervous system at the same time.

Lesions in the brain can now be viewed noninvasively using new methods such as magnetic resonance imaging (MRI). However, neurologists have been able to diagnose MS for many years by taking advantage of the fact that myelin serves the nervous system by speeding axonal conduction velocity. One simple test involves stimulating the eye with a checkerboard pattern, then measuring the time that elapses until an electrical response is measured from the scalp over the part of the brain that is a target of the optic nerve. People who have MS are characterized by a marked slowing of the conduction velocity of their optic nerve.

Another, more common demyelinating disease is called Guillain-Barré syndrome, which attacks the myelin of the peripheral nerves that innervate muscle and skin. This disease may follow minor infectious illnesses and inoculations, and it appears to result from an anomalous immunological response against one's own myelin. The symptoms stem directly from the slowing and/or failure of action potential conduction in the axons that innervate the muscles. This conduction deficit can be demonstrated clinically by stimulating the peripheral nerves electrically through the skin, then measuring the time it takes to evoke a response (a twitch of a muscle, for instance). Both MS and Guillain-Barré syndrome are characterized by a profound slowing of the response time, because saltatory conduction is disrupted.

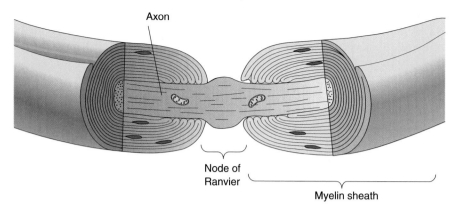

Figure 4.11
The myelin sheath and node of Ranvier. The electrical insulation provided by myelin helps speed action potential conduction from node to node. Voltage-gated sodium channels are concentrated in the axonal membrane at the nodes of Ranvier.

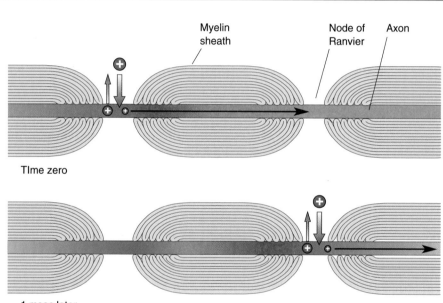

Time zero

1 msec later

Figure 4.12
Saltatory conduction. Myelin allows current to spread farther and faster between nodes, thus speeding action potential conduction. Compare this figure with Figure 4.10.

the soma, the axon hillock, is often also called the **spike-initiation zone**. In a typical neuron in the brain or spinal cord, the depolarization of the dendrites and soma caused by synaptic input from other neurons leads to the generation of action potentials if the membrane of the *axon hillock* is depolarized beyond threshold (Figure 4.13a). In most sensory neurons, however, the spike-initiation zone occurs near the *sensory nerve endings* where the depolarization caused by sensory stimulation leads to the generation of action potentials that propagate up the sensory nerves (Figure 4.13b).

In Chapter 2, we learned that axons and dendrites differ in their morphology. We now see that they are functionally different, and that this difference in function is specified at the molecular level by the type of protein in the neuronal membrane. Differences in the types and density of mem-

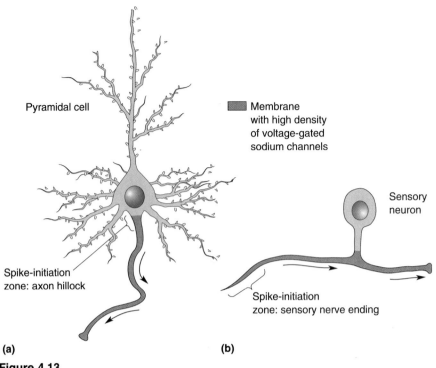

Pyramidal cell

Membrane
with high density
of voltage-gated
sodium channels

Sensory
neuron

Spike-initiation
zone: axon hillock

Spike-initiation
zone: sensory nerve ending

(a)

(b)

Figure 4.13
Membrane proteins specify the function of different parts of the neuron.
Depicted here are **(a)** a cortical pyramidal neuron and **(b)** a primary sensory neuron.
Despite the diversity of neuronal structure, the axonal membrane can be identified
at the molecular level by its high density of voltage-gated sodium channels. This
molecular distinction enables axons to generate and conduct action potentials. The
region of membrane where action potentials are normally generated is called the
spike-initiation zone. The arrows indicate the normal direction of action potential
propagation in these two types of neuron.

brane ion channels also can account for the characteristic electrical properties of different types of neuron (Box 4.4).

CONCLUDING REMARKS

Let's return briefly to the example in Chapter 3 of stepping on a thumbtack. The breaking of the skin caused by the tack stretches the sensory nerve
endings of the foot. Special ion channels that are sensitive to the stretching
of the membrane open and allow positively charged sodium ions to enter
the nerve endings. This influx of positive charge depolarizes the membrane
of the spike-initiation zone to threshold, and the action potential is generated. The positive charge that enters during the rising phase of the action
potential spreads down the axon and depolarizes the membrane ahead to
threshold. In this way, the action potential is continuously regenerated as it
sweeps like a wave up the sensory axon. We now come to the step where
this information is distributed and integrated by other neurons in the central nervous system. This transfer of information from one neuron to another is called *synaptic transmission*, the subject of the next two chapters.

The Eclectic Electrical Behavior of Neurons

Neurons are not all alike; they vary in shape, size, and connections. Neurons also differ from one another in their electrical properties. A few examples of the diverse behavior of neurons are shown in Figure A. The cerebral cortex has two major types of neurons, as defined by morphology: aspinous stellate cells and spiny pyramidal cells. A stellate cell typically responds to a steady depolarizing current injection into its soma by firing action potentials at a relatively steady frequency throughout the stimulus (Figure Aa). However, most pyramidal cells cannot sustain a steady firing rate. Instead, they fire rapidly at the beginning of the stimulus and then slow down, even if the stimulus remains strong (Figure Ab). This slowing over time is called *adaptation*, and it is a very common property among excitable cells. Another firing pattern is the burst, a rapid cluster of action potentials followed by a brief pause. Some cells, including a particular subtype of large pyramidal neuron in the cortex, can even respond to a steady input with rhythmic, repetitive bursts (Figure Ac). Variability of firing patterns is not confined to the cerebral cortex. Surveys of many areas of the brain imply that neurons have just as large an assortment of electrical behaviors as they do morphologies.

What accounts for the diverse behavior of different types of neurons? Ultimately, each neuron's physiology is determined by the properties and numbers of the ion channels in its membrane. There are many more types of ion channels than the few described in this chapter, and each has distinctive properties. For example, some potassium channels activate only very slowly. A neuron with a high density of these will show adaptation because during a prolonged stimulus, more and more of the slow potassium channels will open, and the outward currents they progressively generate will tend to hyperpolarize the membrane. When you realize that a single neuron may express more than a dozen types of ion channels, the source of diverse firing behavior becomes clear. It is the complex interactions between multiple ion channels that create the eclectic electric signature of each class of neuron.

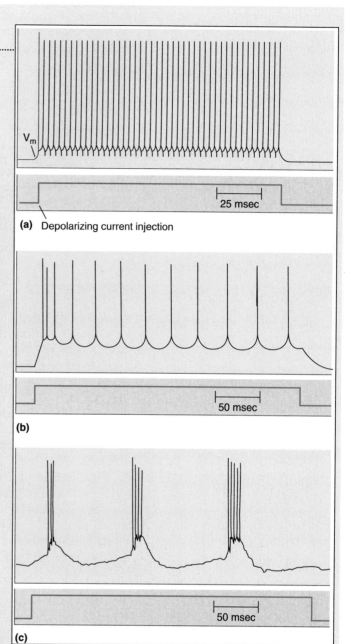

(a) Depolarizing current injection

25 msec

(b)

50 msec

(c)

50 msec

Figure A (Source: Adapted from Agmon and Connors, 1992.)

KEY TERMS

Properties of the Action Potential

rising phase

overshoot

falling phase

undershoot

after-hyperpolarization

threshold

absolute refractory period

relative refractory period

voltage-gated sodium channel

patch clamp

tetrodotoxin (TTX)

voltage-gated potassium channel

Action Potential Conduction

saltatory conduction

Action Potentials, Axons, and Dendrites

spike-initiation zone

The Action Potential, in Reality

voltage clamp

✔ REVIEW QUESTIONS

1. Define membrane potential (V_m) and sodium equilibrium potential (E_{Na}). Which of these, if any, changes during the course of an action potential?
2. What ions carry the early inward and late outward currents during the action potential?
3. Why is the action potential referred to as "all-or-none"?
4. Some voltage-gated K^+ channels are known as delayed rectifiers because of the timing of their opening during an action potential. What would happen if these channels took much longer than normal to open?
5. Imagine we have labeled tetrodotoxin (TTX) so that it can be seen using a microscope. If we wash this TTX onto a neuron, what parts of the cell would you expect to be labeled? What would be the consequence of applying TTX to this neuron?
6. How does action potential conduction velocity vary with axonal diameter? Why?

Synaptic Transmission

In Chapters 3 and 4, we discussed how mechanical energy, such as a thumbtack entering your foot, can be converted into a neural signal. First, specialized ion channels of the sensory nerve endings allow positive charge to enter the axon. If this depolarization reaches threshold, then action potentials are generated. Because the axonal membrane is excitable and has voltage-gated sodium channels, action potentials can propagate without decrement up the long sensory nerves. For this information to be integrated by the rest of the nervous system, it is necessary that these neural signals be passed on to other neurons—for example, the motor neurons that control muscle contraction, as well as neurons in the brain and spinal cord that lead to a coordinated reflex response. By the end of the nineteenth century, it was recognized that this transfer of information from one neuron to another occurs at specialized sites of contact. In 1897, English physiologist Charles Sherrington gave these sites their name: synapses. The process of information transfer at a synapse is called *synaptic transmission.*

The physical nature of synaptic transmission was argued for almost a century. One attractive hypothesis, especially considering the speed of synaptic transmission, was that it was simply electrical current flowing from one neuron to the next. The existence of such **electrical synapses** was finally proven, in 1959, by Harvard University physiologists Edwin Furshpan and David Potter. We now know, however, that electrical synapses account for only a very small fraction of the total number of synapses in the brain.

An alternative hypothesis, also dating back to the 1800s, was that chemical neurotransmitters transfer information from one neuron to another at the synapse. Solid support for the concept of **chemical synapses** was provided in 1921 by Otto Loewi, who then was the head of the Pharmacology Department at the University of Graz in Austria. Loewi showed that electrical stimulation of axons innervating the frog's heart caused the release of a chemical, and that this chemical could mimic the effects of neuron stimulation on the heartbeat (Box 5.1). Later, Bernard Katz and his colleagues at University College, London, conclusively demonstrated that fast transmission at the synapse between a motor neuron axon and skeletal muscle was chemically mediated. By 1951, John Eccles of the Australian National University was able to study the physiology of synaptic transmission within the mammalian central nervous system (CNS) using a new tool, the glass microelectrode. These experiments indicated that many CNS synapses also use a chemical transmitter. We now know that chemical synapses make up the vast majority of synapses in the brain. In the last decade there has been a revolution in our understanding of chemical synaptic transmission, thanks to new methods of studying the structure and function of the molecules involved.

Synaptic transmission is a large and fascinating topic. The actions of psychoactive drugs, the causes of mental disorders, the neural bases of learning and memory—indeed, all the operations of the nervous system—cannot be understood without knowledge of synaptic transmission. Therefore, we've devoted several chapters to this topic, mainly focusing on chemical synapses. In this chapter, we begin by exploring the basic mechanisms of synaptic transmission. What do different types of synapse look like? How are neurotransmitters synthesized and stored, and how are they released in response to an action potential in the axon terminal? How do neurotransmitters act on the postsynaptic membrane? How do single neurons integrate the inputs provided by the thousands of synapses that impinge upon them?

TYPES OF SYNAPSES

We introduced the synapse in Chapter 2. A synapse is the specialized junction where an axon terminal contacts another neuron or cell type. The normal direction of information flow is from the axon terminal to the target neuron; thus, the axon terminal is said to be *presynaptic* and the target neuron is said to be *postsynaptic*. Let's take a closer look at the different types of synapse.

Electrical Synapses

As we have said, most mammalian synapses are chemical, but there is a very simple and evolutionarily ancient form of electrical synapse that allows the direct transfer of ionic current from one cell to the next. Electrical synapses occur at specialized sites called **gap junctions**. At a gap junction, the pre- and postsynaptic membranes are separated by only about 3 nm, and this narrow gap is spanned by special proteins called *connexons*. The connexons form channels that allow ions to pass directly from the cytoplasm of one cell to the cytoplasm of the other (Figure 5.1). Because electrical current can pass through these channels, cells connected by gap junctions are said to be *electrotonically coupled*. The pore of the channel formed by connexons is among the largest known. Its diameter is about 2 nm, big enough for all the major cellular ions, and many small organic molecules, to pass through.

Box 5.1	OF SPECIAL INTEREST

Otto Loewi and *Vagusstoff*

One of the more colorful stories in the history of neuroscience was contributed by Otto Loewi, who, working in Austria in the 1920s, showed definitively that synaptic transmission between nerve and heart was chemically mediated. The heart is supplied with two types of innervation; one type speeds the beating of the heart, and the other slows it. The latter type of innervation is supplied by the vagus nerve. Loewi isolated a frog heart with the vagal innervation left intact, stimulated the nerve electrically, and observed the expected effect, the slowing of the heartbeat. The critical demonstration that this effect was chemically mediated came when he took the solution that bathed this heart, applied it to a second isolated frog heart, and found that the beating of this one also slowed.

The idea for this experiment had actually come to Loewi in a dream. Below is his own account:

In the night of Easter Sunday, 1921, I awoke, turned on the light, and jotted down a few notes on a tiny slip of paper. Then, I fell asleep again. It occurred to me at six o'clock in the morning that during the night I had written down something most important, but I was unable to decipher the scrawl. That Sunday was the most desperate day in my whole scientific life. During the next night, however, I awoke again, at three o'clock, and I remembered what it was. This time I did not take any risk; I got up immediately, went to the laboratory, made the experiment on the frog's heart, described above, and at five o'clock the chemical transmission of the nervous impulse was conclusively proved. . . . Careful consideration in daytime would undoubtedly have rejected the kind of experiment I performed, because it would have seemed most unlikely that if a nervous impulse released a transmitting agent, it would do so not just in sufficient quantity to influence the effector organ, in my case the heart, but indeed in such an excess that it could partly escape into the fluid which filled the heart, and could therefore be detected. Yet the whole nocturnal concept of the experiment was based on this eventuality, and the result proved to be positive, contrary to expectation. (Loewi, 1953, pp. 33, 34)

The active compound, which Loewi called *vagusstoff*, turned out to be acetylcholine. As we shall see in this chapter, acetylcholine is also a transmitter at the synapse between nerve and skeletal muscle. Here, unlike at the heart, acetylcholine causes excitation and contraction of the muscle.

Transmission at electrical synapses is very fast and, in many cases, fail-safe. Thus, an action potential in the presynaptic neuron can produce, almost instantaneously, an action potential in the postsynaptic neuron. In invertebrate species, electrical synapses are commonly found between sensory and motor neurons in neural pathways mediating escape reflexes. In the adult mammalian CNS, electrical synapses are mainly found in specialized locations where normal function requires that the activity of neighboring neurons be highly synchronized.

Although gap junctions are relatively rare between adult mammalian neurons, they are very common in a large variety of non-neural cells, including glia, epithelial cells, smooth and cardiac muscle cells, liver cells, and some glandular cells. They also occur frequently between neurons during early embryonic stages. There is evidence that during brain development, gap junctions allow neighboring cells to share both electrical and chemical signals that may help coordinate their growth and maturation.

Chemical Synapses

As a rule, synaptic transmission in the mature human nervous system is chemical, so we will now focus exclusively on chemical synapses. Before we discuss the different types of chemical synapse, let's take a look at some of their universal characteristics (Figure 5.2).

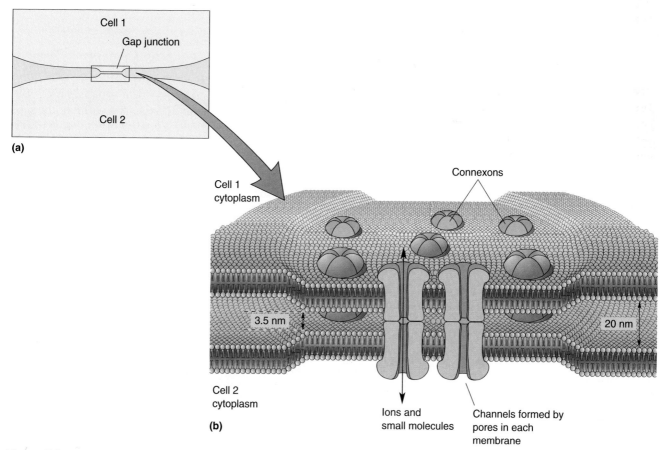

Figure 5.1
A gap junction. (a) Neurites of two cells connected by a gap junction. **(b)** An enlargement showing connexons, the channel proteins that bridge the cytoplasm of the two cells. Ions and small molecules can pass in both directions through these channels.

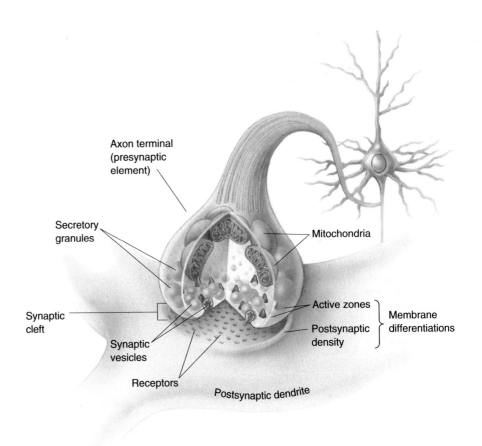

Axon terminal
(presynaptic
element)

Secretory
granules

Mitochondria

Synaptic
cleft

Active zones

Membrane
differentiations

Postsynaptic
density

Synaptic
vesicles

Receptors

Postsynaptic dendrite

Figure 5.2
Parts of a chemical synapse.

The pre- and postsynaptic membranes at chemical synapses are separated by a *synaptic cleft* that is 20–50 nm wide, 10 times the width of the separation at gap junctions. The cleft is filled with a matrix of fibrous extracellular protein that adheres the pre- and postsynaptic membranes to each other. The presynaptic side of the synapse, also called the *presynaptic element*, is usually an axon terminal. The terminal typically contains dozens of small membrane-enclosed spheres, about 50 nm in diameter, called *synaptic vesicles*. These vesicles store neurotransmitter, the chemical used to communicate with the postsynaptic neuron. Many axon terminals also contain larger vesicles, about 100 nm in diameter, called **secretory granules**. Secretory granules contain soluble protein that appears dark in the electron microscope, so they are sometimes called **large, dense-core vesicles**.

There are dense accumulations of protein in the membranes on either side of the synaptic cleft that are collectively called **membrane differentiations**. On the *presynaptic* side, proteins jutting into the cytoplasm of the terminal along the intracellular face of the membrane look like a field of pyramids. The pyramids, and the membrane associated with them, are the actual sites of neurotransmitter release, called **active zones**. Synaptic vesicles are clustered in the cytoplasm adjacent to the active zones.

The protein spanning the thickness of the *postsynaptic* membrane is called the **postsynaptic density**. The postsynaptic density contains the neurotransmitter receptors, which convert the *intercellular* chemical signal into an *intracellular* signal in the postsynaptic cell. As we shall see, the nature of this postsynaptic response can be quite varied, depending on the type of protein receptor that is activated by the neurotransmitter.

CNS Synapses. In the CNS, different types of synapse may be distinguished by which part of the neuron is postsynaptic to the axon terminal. If the postsynaptic membrane is on a dendrite, the synapse is said to be *axodendritic*. If the postsynaptic membrane is on the cell body, the synapse is said to be *axosomatic*. In some cases the postsynaptic membrane is on another axon, and these synapses are called *axoaxonic* (Figure 5.3). In certain specialized neurons, *dendrites* actually form synapses with one another; these are called *dendrodendritic* synapses.

CNS synapses may be further divided into two general categories based on the appearance of their presynaptic and postsynaptic membrane differentiations. Synapses in which the membrane differentiations are of similar thickness are called symmetrical, or **Gray's type II synapses**; those in which the membrane differentiation on the postsynaptic side is thicker than that on the presynaptic side are called asymmetrical, or **Gray's type I synapses** (Figure 5.4). As we shall see later in the chapter, these structural differences can be correlated with functional differences. Gray's type I synapses are usually excitatory, while Gray's type II synapses are usually inhibitory.

Neuromuscular Junction. Synaptic junctions also exist outside the brain and spinal cord. For example, axons of the autonomic nervous system innervate glands, smooth muscle, and the heart. Chemical synapses also occur between the axons of motor neurons of the spinal cord and skeletal muscle. Such a synapse is also called a **neuromuscular junction**, and it has many of the structural features of chemical synapses in the CNS.

Neuromuscular synaptic transmission is fast and reliable. An action potential in the motor axon normally always causes an action potential in the muscle cell it innervates. This reliability is accounted for, in part, by structural specializations of the neuromuscular junction. For example, the presynaptic terminal contains a large number of active zones. In addition, the postsynaptic membrane, also called the **motor end-plate**, contains a series of shallow folds (Figure 5.5). The presynaptic active zones are precisely aligned with these junctional folds, and the postsynaptic membrane of the folds is packed with neurotransmitter receptors. These specializations ensure that a lot of neurotransmitter molecules are focally released onto a large chemoreceptive membrane surface.

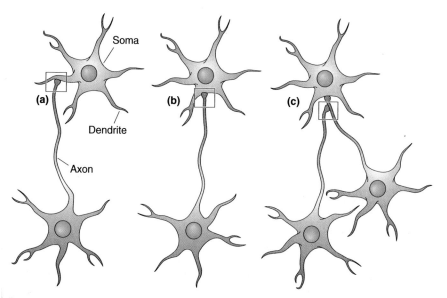

Figure 5.3
Synaptic arrangements in the CNS. (a) An axodendritic synapse, **(b)** an axosomatic synapse, and **(c)** an axoaxonic synapse.

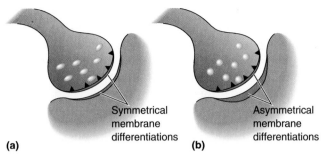

Figure 5.4
Two categories of CNS synaptic membrane differentiations. (a) A Gray's type II synapse is symmetrical and usually inhibitory. **(b)** A Gray's type I synapse is asymmetrical and usually excitatory.

Because neuromuscular junctions are more accessible than CNS synapses, much of what we know about the mechanisms of synaptic transmission was first established here. Neuromuscular junctions are also of considerable clinical significance; diseases, drugs, and poisons that interfere with this chemical synapse have direct effects on vital bodily functions.

PRINCIPLES OF CHEMICAL SYNAPTIC TRANSMISSION

Consider the basic requirements of chemical synaptic transmission. There must be a mechanism for synthesizing and replenishing neurotransmitter in the synaptic vesicles, a mechanism causing vesicles to spill their contents into the synaptic cleft in response to a presynaptic action potential, a mechanism for producing an electrical or biochemical response to neurotransmitter in the postsynaptic neuron, and a mechanism for removing neurotransmitter from the synaptic cleft. And, to be useful for sensation, perception, and the control of movement, all these things must occur very rapidly. No wonder physiologists initially were skeptical about the existence of chemical synapses in the brain!

Fortunately, thanks to several decades of research on the topic, we have learned a lot about how many of these aspects of synaptic transmission are so efficiently carried out. Here we'll present a general survey of the basic principles as they are currently understood. In Chapter 6, we will take a more detailed look at the individual neurotransmitters and their modes of postsynaptic action.

Neurotransmitters

Since the discovery of chemical synaptic transmission, the search has been on to identify neurotransmitters in the brain. Our current understanding is that most neurotransmitters fall into one of three chemical categories: (1) *amino acids*, (2) *amines*, and (3) *peptides* (Table 5.1). Some representatives of these categories are shown in Figure 5.6. The amino acid and amine neurotransmitters are all small organic molecules containing a nitrogen atom, and they are stored in and released from synaptic vesicles. Peptide neurotransmitters are large molecules stored in and released from secretory granules. As mentioned above, secretory granules and synaptic vesicles are frequently observed in the same axon terminals. Consistent with this observation, peptides often exist in the same axon terminals that contain amine or amino acid neurotransmitters. And, as we'll discuss in a moment, these different neurotransmitters are released under different conditions.

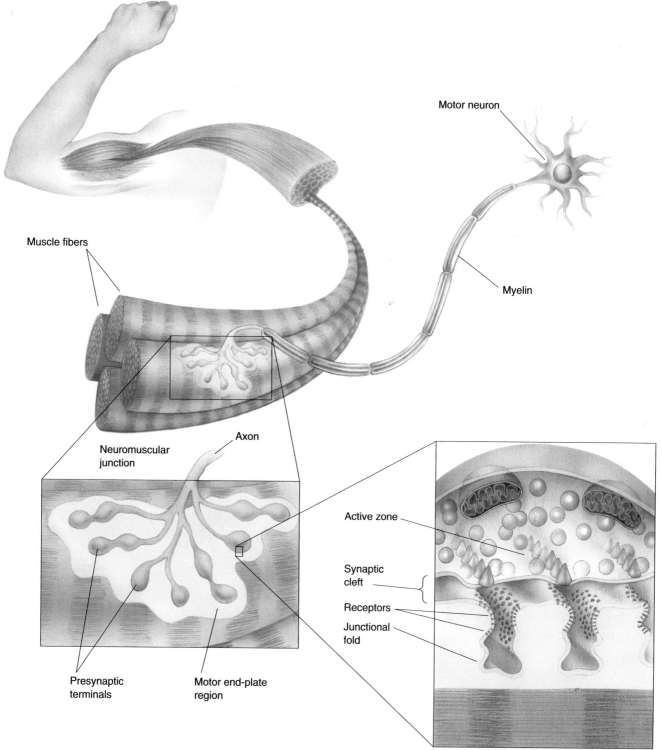

Motor neuron

Myelin

Muscle fibers

Axon

Neuromuscular
junction

Active zone

Synaptic
cleft

Receptors

Junctional
fold

Presynaptic
terminals

Motor end-plate
region

Figure 5.5
The neuromuscular junction. The postsynaptic membrane, called the motor end-plate, contains junctional folds with numerous neurotransmitter receptors.

Table 5.1
The Major Neurotransmitters

Amino Acids	Amines	Peptides
Gamma-amino butyric acid (GABA)	Acetylcholine (ACh)	Cholecystokinin (CCK)
Glutamate (Glu)	Dopamine (DA)	Dynorphin
Glycine (Gly)	Epinephrine	Enkephalins (Enk)
	Histamine	*N*-acetylaspartylglutamate (NAAG)
	Norepinephrine (NE)	Neuropeptide Y
	Serotonin (5-HT)	Somatostatin
		Substance P
		Thyrotropin-releasing hormone
		Vasoactive intestinal polypeptide (VIP)

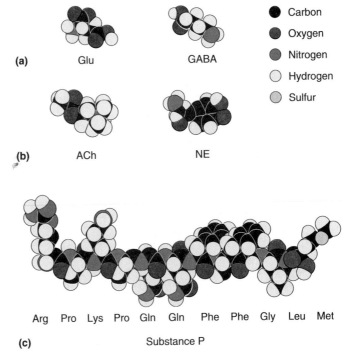

Figure 5.6
Representative neurotransmitters. (a) The amino acid neurotransmitters glutamate and GABA, **(b)** the amine neurotransmitters acetylcholine and norepinephrine, and **(c)** the peptide neurotransmitter substance P. (For the abbreviations and chemical structures of amino acids in substance P, see Figure 3.4b.)

Different neurons in the brain release different neurotransmitters. Fast synaptic transmission at most CNS synapses is mediated by the amino acids **glutamate (Glu)** and **gamma-aminobutyric acid (GABA)**. The amine **acetylcholine (ACh)** mediates fast synaptic transmission at all neuromuscular junctions. Slower forms of synaptic transmission in the CNS and in the periphery are mediated by transmitters of all three types.

Neurotransmitter Synthesis and Storage

Chemical synaptic transmission requires that neurotransmitters be synthesized and ready for release. Different neurotransmitters are synthesized in different ways. For example, glutamate is among the 20 amino acids that are the building blocks of protein (see Figure 3.4b); consequently, it is abundant in all cells of the body, including neurons. In contrast, GABA and the amines are made only by the neurons that release them. These neurons contain specific enzymes that synthesize the neurotransmitters from various metabolic precursors. The synthetic enzymes for both amino acid and amine neurotransmitters are transported to the axon terminal, where they locally and rapidly direct transmitter synthesis.

Once synthesized in the cytosol of the axon terminal, the amino acid and amine neurotransmitters must be taken up by the synaptic vesicles. Concentrating these neurotransmitters inside the vesicle is the job of **transporters**, special proteins embedded in the vesicle membrane.

The synthesis and storage of peptides in secretory granules are quite different. As we learned in Chapters 2 and 3, peptides are formed when amino acids are strung together by the ribosomes of the cell body. In the case of peptide neurotransmitters, this occurs in the rough ER. Generally, the peptide synthesized in the rough ER is cleaved in the Golgi apparatus to yield the active neurotransmitter. Secretory granules containing the peptide bud off from the Golgi apparatus and are carried to the axon terminal by axoplasmic transport. Figure 5.7 compares the synthesis and storage of amine and amino acid neurotransmitters with peptide neurotransmitters.

Neurotransmitter Release

Neurotransmitter release is triggered by the arrival of an action potential in the axon terminal. The depolarization of the terminal membrane causes **voltage-gated calcium channels** in the active zones to open. These membrane channels are very similar to the sodium channels we discussed in Chapter 4, except they are permeable to Ca^{2+} ions instead of Na^+ ions. There is a large inward driving force on Ca^{2+}. Remember that the internal calcium ion concentration, $[Ca^{2+}]_i$, at rest is very low, only 0.0002 mM; so Ca^{2+} will flood the axon terminal as long as the calcium channels are open. The resulting elevation in the internal calcium ion concentration, $[Ca^{2+}]_i$, is the signal that causes neurotransmitter to be released from synaptic vesicles.

The vesicles release their contents by a process called **exocytosis**. The membrane of the synaptic vesicle fuses to the presynaptic membrane at the active zone, allowing the contents of the vesicle to spill out into the synaptic cleft (Figure 5.8). New York University Medical Center neuroscientist Rodolfo Llinás, using a "giant synapse" in the squid nervous system, showed that exocytosis can occur very rapidly—within 0.2 msec of the Ca^{2+} influx into the terminal (Box 5.2). Subsequent work by Llinás and others has shown that exocytosis is fast because Ca^{2+} enters at the active zone, precisely where synaptic vesicles are ready and waiting to release their contents. In this local "microdomain" around the active zone, calcium can achieve very high concentrations (about 0.3 mM).

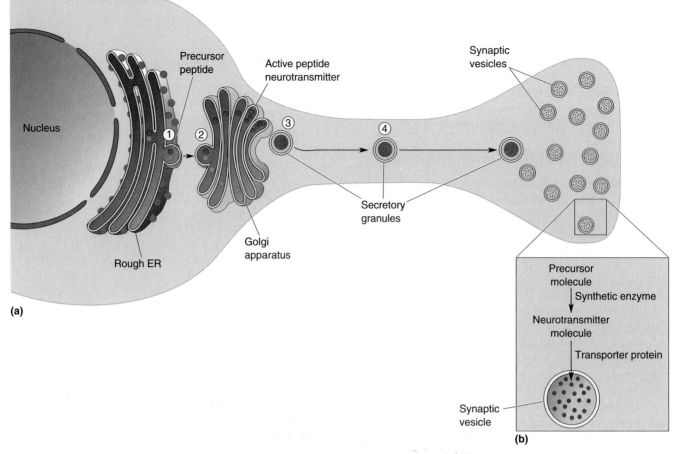

(a)

Figure 5.7
Synthesis and storage of different types of neurotransmitter. (a) Peptides: (1) A precursor peptide is synthesized in the rough ER; (2) the precursor peptide is cleaved in the Golgi apparatus to yield the active neurotransmitter; (3) secretory vesicles containing the peptide bud off from the Golgi apparatus; (4) the secretory granules are transported down the axon to the terminal where the peptide is stored. **(b)** Amine and amino acid neurotransmitters: (1) Enzymes convert precursor molecules into neurotransmitter molecules in the cytosol; (2) transporter proteins load the neurotransmitter into synaptic vesicles in the terminal, where they are stored.

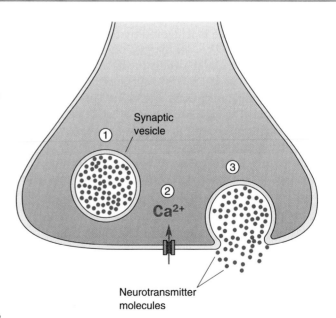

Figure 5.8
Release of neurotransmitter by exocytosis. (1) A synaptic vesicle loaded with neurotransmitter, in response to (2) an influx of Ca^{2+} through voltage-gated calcium channels, (3) releases its contents into the synaptic cleft by the fusion of the vesicle membrane with the presynaptic membrane.

The precise mechanism for how increased $[Ca^{2+}]_i$ stimulates exocytosis is poorly understood, but it is currently under intensive investigation. The speed of neurotransmitter release suggests that the vesicles involved are those already "docked" at the active zones. Docking is believed to involve interactions between proteins in the synaptic vesicle membrane and the active zone. In the presence of high $[Ca^{2+}]_i$, these proteins alter their conformation so that the lipid bilayers of the vesicle and presynaptic membranes fuse, forming a pore that allows the neurotransmitter to escape into the cleft. The mouth of this exocytotic fusion pore continues to expand until the membrane of the vesicle is fully incorporated into the presynaptic membrane (Figure 5.9). The vesicle membrane is later recovered by the process of endocytosis, and the recycled vesicle is refilled with neurotransmitter. During periods of prolonged stimulation, vesicles are mobilized from a "reserve pool" that is bound to the cytoskeleton of the axon terminal. The release of these vesicles from the cytoskeleton, and their docking to the active zone, is also triggered by elevations of $[Ca^{2+}]_i$.

Storage granules also release peptide neurotransmitters by exocytosis, in a calcium-dependent fashion, but typically not at the active zones. Because the sites of exocytosis occur at a distance from the sites of Ca^{2+} entry, peptide neurotransmitters are usually not released in response to every action potential invading the terminal. Instead, the release of peptides

---

Box 5.2	**PATH OF DISCOVERY**

Sleepless in Woods Hole
by Rodolfo Llinás

Doing research into the workings of the squid giant synapse I find absolutely thrilling, even after 26 summers at the Marine Biological Laboratory in Woods Hole, Massachusetts. The thrill that such research generates can only be compared to such experiences as one's first solo flight in a small plane, a view of the cave paintings at Altamira, or being given an unexpected present that, in addition, one clearly does not deserve.

While many moments come to mind when thinking about the work in squid synaptic transmission, three remain pure excitement. The first occurred in 1966 when working with Bloedel, Gage, and Quastel. After having penetrated with microelectrodes the pre- and postsynaptic fibers in the squid synapse, and having blocked the presynaptic action potential with TTX, we found that direct depolarization of the terminal evoked transmitter release. I vividly remember being incapable of sleeping that night. Synaptic transmission and nerve conduction were truly different events! The second came about when performing the first voltage-clamp studies of the presynaptic terminal with Walton. We found that we

could actually measure the Ca^{2+} current responsible for transmitter release. More important, however, we discovered that a delay of only

Rodolfo Llinás

180 μsec separated presynaptic Ca^{2+} entry and neurotransmitter release. Again I could not sleep that night. The transmission machinery had to be packed very densely and with an incredible level of structural order to display such rapid responsiveness. The third time came recently when Sugimori, Silver, and I performed experiments using n-aequorin-J, a jellyfish protein that emits light when it binds Ca^{2+}. After injecting aequorin into the presynaptic fiber, we were able to see from the rapidly flickering spots of light that Ca^{2+} entry is highly localized to the active zones, as expected from theory. Synaptic transmission had a general coinage, that of dynamic microcompartments. Once again, the excitement would not allow me to sleep. And so, one looks with expectation to more sleepless nights!

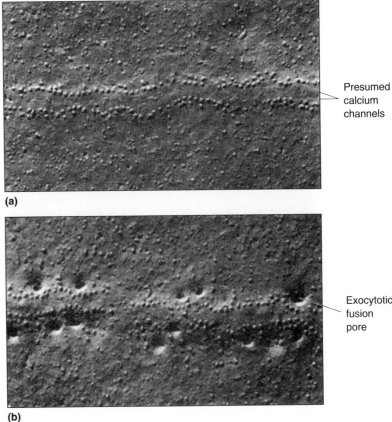

Presumed
calcium
channels

(a)

Exocytotic
fusion
pore

(b)

Figure 5.9
A "receptor's eye" view of neurotransmitter release. (a) This is a view of the extracellular surface of the active zone at the frog neuromuscular junction. The particles are believed to be calcium channels. **(b)** In this view, the presynaptic terminal had been stimulated to release neurotransmitter. The exocytotic fusion pores are where synaptic vesicles have fused with the presynaptic membrane and released their contents. (Source: Heuser and Reese, 1973.)

requires high-frequency trains of action potentials, so that the $[Ca^{2+}]_i$ in the terminal can build to the level required to trigger release away from the active zones. Unlike the fast release of amino acid and amine neurotransmitters, the release of peptides is a leisurely process, taking 50 msec or more.

Neurotransmitter Receptors and Effectors

Neurotransmitters released in the synaptic cleft affect the postsynaptic neuron by binding to thousands of specific receptor proteins that are embedded in the postsynaptic density. The binding of neurotransmitter to the receptor is like inserting a key in a lock; this causes conformational changes in the protein. Although there are well over 100 different neurotransmitter receptors, they can be divided into two types: transmitter-gated ion channels and G-protein-coupled receptors.

Transmitter-Gated Ion Channels. Transmitter-gated ion channels are membrane-spanning proteins consisting of five subunits that come together to form a pore between them (Figure 5.10). In the absence of neurotransmitter, the pore is closed. When neurotransmitter binds to specific sites on the extracellular region of the channel, it induces a conformational change—just a slight twist of the subunits—which within microseconds causes the pore to open. The functional consequence of this depends on which ions can pass through the pore.

Transmitter-gated channels generally do not show the same degree of ion selectivity as do voltage-gated channels. For example, the ACh-gated ion channels at the neuromuscular junction are permeable to both Na$^+$ and K$^+$ ions. Nonetheless, as a rule, if the open channels are permeable to Na$^+$, the net effect will be to depolarize the postsynaptic cell from the resting membrane potential (Box 5.3). Because it tends to bring the membrane potential toward threshold for generating action potentials, this effect is said to be *excitatory*. A transient postsynaptic membrane depolarization caused by presynaptic release of neurotransmitter is called an **excitatory postsynaptic potential** (**EPSP**) (Figure 5.11). Synaptic activation of ACh-gated and glutamate-gated ion channels causes EPSPs.

If the transmitter-gated channels are permeable to Cl$^-$, the net effect will be to hyperpolarize the postsynaptic cell from the resting membrane potential (because the chloride equilibrium potential is negative; see Chapter 3). Because it tends to bring the membrane potential away from threshold for generating action potentials, this effect is said to be *inhibitory*. A transient hyperpolarization of the postsynaptic membrane potential caused by the presynaptic release of neurotransmitter

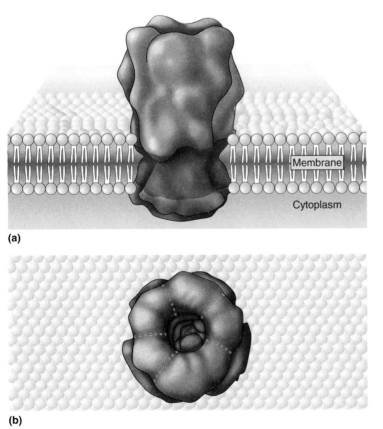

(a)

(b)

Figure 5.10
Structure of a transmitter-gated ion channel. (a) Side view of an ACh-gated ion channel, as it is believed to appear. **(b)** Top view of the channel, showing the pore. Dashed lines indicate the approximate borders between the five subunits. (Source: Adapted from Unwin, 1993.)

is called an **inhibitory postsynaptic potential (IPSP)** (Figure 5.12). Synaptic activation of GABA-gated ion channels causes an IPSP. We'll discuss EPSPs and IPSPs more shortly, when we explore the principles of synaptic integration.

G-Protein-Coupled Receptors. Fast chemical synaptic transmission is mediated by amino acid and amine neurotransmitters acting on transmitter-gated ion channels. However, all three types of neurotransmitter, acting on **G-protein-coupled receptors**, can also have slower, longer-lasting,

---

Reversal Potentials

In Chapter 4, we saw that when the membrane voltage-gated sodium channels open during the action potential, Na$^+$ ions enter the cell, causing the membrane potential to rapidly depolarize until it approaches the sodium equilibrium potential, E_{Na}, about 40 mV. Unlike the voltage-gated channels, however, many transmitter-gated ion channels are not permeable to a single type of ion. For example, the ACh-gated ion channel at the neuromuscular junction is permeable to both Na$^+$ and to K$^+$. Let's explore the functional consequence of activating these channels.

In Chapter 3, we learned that the membrane potential, V_m, can be calculated using the Goldman equation, which takes into account the relative permeability of the membrane to different ions (recall Box 3.3). If the membrane were equally permeable to Na$^+$ and K$^+$, as it would be if the ACh-gated channels were open, then V_m would have a value between E_{Na} and E_K, around 0 mV. Therefore, ionic current would flow through the channels in a direction that brings the membrane potential toward 0 mV. If the membrane potential was < 0 mV before ACh was applied, as is usually the case, the direction of net current flow through the ACh-gated ion channels would be *inward*, causing a depolarization. However, if the membrane potential was > 0 mV before ACh was applied, the direction of net current flow through the ACh-gated ion channels would be *outward*, causing the membrane potential to become less positive. Ionic current flow at different membrane voltages can be graphed, as shown in Figure A. Such a graph is called an I-V plot (I: current; V: voltage). The critical value of membrane potential at which the direction of current flow reverses is called the *reversal potential*. In this case, the reversal potential would be 0 mV. The experimental determination of a reversal

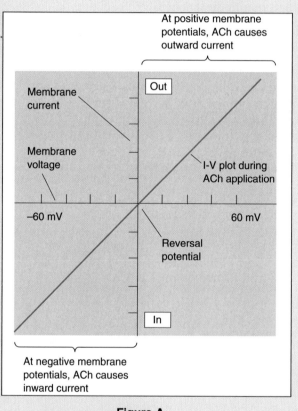

Figure A

potential, therefore, helps tell us which types of ions the membrane is permeable to.

If a neurotransmitter, by changing the relative permeability of the membrane to different ions, causes V_m to move toward a value that is more positive than action potential threshold, the neurotransmitter action would be termed excitatory. As a rule, neurotransmitters that open a channel permeable to Na$^+$ are excitatory. If a neurotransmitter causes V_m to take on a value that is more negative than action potential threshold, the neurotransmitter action would be termed inhibitory. Neurotransmitters that open a channel permeable to Cl$^-$ are inhibitory, as are neurotransmitters that open a channel permeable only to K$^+$.

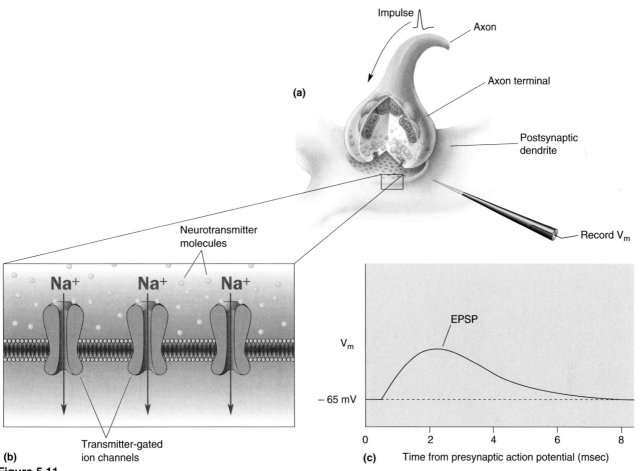

Figure 5.11
Generation of an EPSP. (a) An impulse arriving in the presynaptic terminal causes the release of neurotransmitter. **(b)** The molecules bind to transmitter-gated ion channels in the postsynaptic membrane. If Na$^+$ enters the postsynaptic cell through the open channels, the membrane will become depolarized. **(c)** The resulting change in membrane potential (V_m), as recorded by a microelectrode in the cell, is the EPSP.

and much more diverse postsynaptic actions. This type of transmitter action involves three steps:

1. Neurotransmitter molecules bind to receptor proteins embedded in the postsynaptic membrane.

2. The receptor proteins activate small protein molecules, called **G-proteins**, that are free to move along the intracellular face of the postsynaptic membrane.

3. The activated G-proteins activate "effector" proteins.

The effector proteins can be G-protein-gated ion channels in the membrane (Figure 5.13a), or they can be enzymes that synthesize molecules called **second messengers** that diffuse away in the cytosol (Figure 5.13b). Second messengers can activate additional enzymes in the cytosol that can regulate ion channel function and alter cellular metabolism. Because G-protein-coupled receptors can trigger widespread metabolic effects, they are often referred to as **metabotropic receptors**.

We'll discuss the different neurotransmitters, their receptors, and their effectors in more detail in Chapter 6. However, you should be aware that the same neurotransmitter can have different postsynaptic actions, depending on what receptors it binds to. An example of this diversity is the effect of

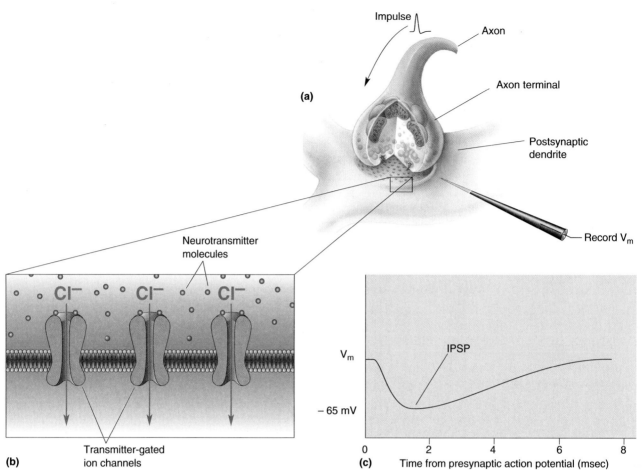

Figure 5.12
Generation of an IPSP. (a) An impulse arriving in the presynaptic terminal causes the release of neurotransmitter. **(b)** The molecules bind to transmitter-gated ion channels in the postsynaptic membrane. If Cl⁻ enters the postsynaptic cell through the open channels, the membrane will become hyperpolarized. **(c)** The resulting change in membrane potential (V_m), as recorded by a microelectrode in the cell, is the IPSP.

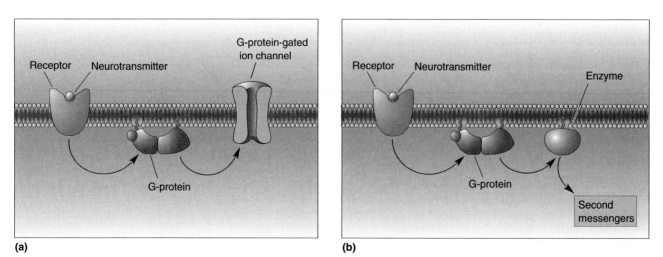

Figure 5.13
Transmitter actions at G-protein-coupled receptors. The binding of neurotransmitter to the receptor leads to activation of G-proteins. Activated G-proteins activate effector proteins, which may be **(a)** ion channels, or **(b)** enzymes that generate intracellular second messengers.

ACh on the heart and on skeletal muscles. ACh slows the rhythmic contractions of the heart by causing a slow hyperpolarization of the cardiac muscle cells. In contrast, in skeletal muscle, ACh induces contraction by causing a rapid depolarization of the muscle fibers. These different actions are explained by different receptors. In the heart, the ACh receptor is coupled by a G-protein to a potassium channel. The opening of the potassium channel hyperpolarizes the cardiac muscle fibers. In skeletal muscle, the receptor is an ACh-gated ion channel permeable to Na^+. The opening of this channel depolarizes the muscle fibers.

Autoreceptors. Neurotransmitter receptors, besides being a part of the postsynaptic density, are also commonly found in the membrane of the presynaptic axon terminal. Presynaptic receptors that are sensitive to the neurotransmitter released by the presynaptic terminal are called **autoreceptors**. Typically, autoreceptors are G-protein-coupled receptors that stimulate second messenger formation. The consequences of activating these receptors varies, but a common effect is inhibition of neurotransmitter release and, in some cases, neurotransmitter synthesis. Such autoreceptors appear to function as a sort of safety valve to reduce release when the concentration of neurotransmitter in the synaptic cleft gets too high.

Neurotransmitter Recovery and Degradation

Once the released neurotransmitter has interacted with postsynaptic receptors, it must be cleared from the synaptic cleft to permit another round of synaptic transmission. One way this happens is by simple diffusion of the transmitter molecules away from the synapse. For most of the amino acid and amine neurotransmitters, however, diffusion is aided by reuptake of the neurotransmitter into the presynaptic axon terminal. Reuptake occurs by the action of specific neurotransmitter transporter proteins located in the presynaptic membrane. Once inside the cytosol of the terminal, the transmitters may be enzymatically destroyed, or they may be reloaded into synaptic vesicles. Neurotransmitter transporters also exist in the membranes of glia surrounding the synapse, which assist in the removal of neurotransmitter from the cleft.

Another way neurotransmitter action can be terminated is by enzymatic destruction in the synaptic cleft itself. This is how ACh is removed at the neuromuscular junction, for example. The enzyme acetylcholinesterase (AChE) is deposited in the cleft by the muscle cells. AChE cleaves the ACh molecule, rendering it inactive at the ACh receptors.

The importance of transmitter removal from the cleft should not be underestimated. At the neuromuscular junction, for example, uninterrupted exposure to high concentrations of ACh after several seconds leads to a process called *desensitization*, in which, despite the continued presence of ACh, the transmitter-gated channels close. This desensitized state can persist for many seconds even after the neurotransmitter is removed. The rapid destruction of ACh by AChE normally prevents desensitization from occurring. However, if the AChE is inhibited, as it is by various nerve gases, the ACh receptors will become desensitized and neuromuscular transmission will fail.

Neuropharmacology

Each of the steps of synaptic transmission we've discussed so far—neurotransmitter synthesis, loading into synaptic vesicles, exocytosis, binding and activation of receptors, reuptake, and degradation—is chemical, and therefore can be affected by specific drugs and toxins (Box 5.4). The study of drug effects on nervous tissue is called **neuropharmacology**.

Above, we mentioned that nerve gases can interfere with synaptic transmission at the neuromuscular junction by inhibiting the enzyme AChE. This interference represents one class of drug action, which is to *inhibit* the normal function of specific proteins involved in synaptic transmission; such drugs are called **inhibitors**. Inhibitors of neurotransmitter receptors, called **receptor antagonists**, bind to the receptors and block (antagonize) the normal action of the transmitter. An example of a receptor antagonist is curare, an arrow-tip poison traditionally used by South American Indians to paralyze their prey. Curare binds tightly to the ACh receptors on skeletal muscle cells and blocks the actions of ACh, thereby preventing muscle contraction.

Other drugs bind to receptors, but instead of inhibiting them, they mimic the actions of the naturally occurring neurotransmitter. These drugs are called **receptor agonists**. An example of a receptor agonist is nicotine, derived from the tobacco plant. Nicotine binds to, and activates, the ACh receptors in skeletal muscle. In fact, the ACh-gated ion channels in muscle are also called **nicotinic ACh receptors**, to distinguish them from other types of ACh receptors, such as those in the heart, that are not activated by nicotine.

The immense chemical complexity of synaptic transmission makes it especially susceptible to the medical corollary of Murphy's Law, which states that if a physiological process can go wrong, it will go wrong. When chemical synaptic transmission goes wrong, the nervous system malfunc-

---

| Box 5.4 | **OF SPECIAL INTEREST** |

Bacteria, Spiders, Snakes, and You

What do the bacteria *Clostridium botulinum*, black widow spiders, cobras, and humans have in common? They all produce toxins that attack the chemical synaptic transmission that occurs at the neuromuscular junction. Botulism is caused by the neurotoxin botulin, which is produced by the growth of *C. botulinum* in improperly canned foods. ("Botulism" comes from the Latin word for "sausage" because of the early association of the disease with poorly preserved meat.) Botulin is a very potent blocker of neuromuscular transmission; it has been estimated that as few as 10 molecules of the toxin are enough to inhibit a cholinergic synapse. It is believed that botulin interferes with the mobilization of transmitter by Ca^{2+} and thus inhibits the release of ACh at the neuromuscular junction.

Although the mechanism of action is different, black widow spider venom also exerts its deadly effects by affecting transmitter release. The venom first increases, and then eliminates, ACh release at the neuromuscular junction. Electron microscopic examination of synapses poisoned with black widow spider venom reveals that the axon terminals are swollen and the synaptic vesicles are missing. The explanation appears to be that the venom, a protein molecule, associates with the presynaptic membrane to form a pore. This pore depolarizes the membrane by allowing both Na^+ and Ca^{2+} to enter the terminal, leading rapidly to a total depletion of transmitter.

The bite of the cobra also results in the blockade of neuromuscular transmission in its victim, but by yet another mechanism. The active compound in the snake's venom, called cobratoxin, is a peptide molecule that binds tightly to the postsynaptic nicotinic receptors and prevents their activation by ACh.

We humans have synthesized a large number of chemicals that poison synaptic transmission at the neuromuscular junction. Originally motivated by the search for chemical warfare agents, this effort led to the development of a new class of compounds called organophosphates. These are irreversible inhibitors of AChE, and by preventing the degradation of ACh, they probably kill their victims by causing a desensitization of ACh receptors. The organophosphates used today as insecticides, like parathion, are toxic to humans only in high doses.

tions. Defective neurotransmission is believed to be the root cause of a large number of neurological and mental disorders. The good news is that, thanks to our growing knowledge of the neuropharmacology of synaptic transmission, clinicians have new and increasingly effective therapeutic drugs to treat these disorders. We'll discuss the synaptic basis of some mental disorders, and their neuropharmacological treatment, in Chapter 15.

PRINCIPLES OF SYNAPTIC INTEGRATION

Most CNS neurons receive thousands of synaptic inputs that activate different combinations of transmitter-gated ion channels and G-protein-coupled receptors. The postsynaptic neuron integrates all these complex ionic and chemical signals and gives rise to a simple form of output: action potentials. The transformation of many synaptic inputs to a single neuronal output constitutes a neural computation. The brain performs billions of neural computations every second we are alive. As a first step toward understanding how neural computations are performed, let's explore some basic principles of synaptic integration.

Integration of EPSPs

The most elementary postsynaptic response is the opening of a single transmitter-gated channel (Figure 5.14). Inward current through these channels depolarizes the postsynaptic membrane, causing the EPSP. The postsynaptic membrane contains thousands of transmitter-gated channels; how many are activated during synaptic transmission depends on how much neurotransmitter is released.

Quantal Analysis of EPSPs. The elementary unit of neurotransmitter release is the contents of a single synaptic vesicle. Vesicles each contain about the same number of transmitter molecules (several thousand); the total amount of transmitter released is some multiple of this number. Consequently, the amplitude of the postsynaptic EPSP is some multiple of the response to the contents of a single vesicle. Stated another way, postsynaptic EPSPs at a given synapse are *quantized*; they are multiples of an indivisible unit, the *quantum*, that reflects the number of transmitter molecules in a single synaptic vesicle and the number of postsynaptic receptors available at the synapse.

At many synapses, exocytosis of vesicles occurs at some low rate in the absence of presynaptic stimulation. The size of the postsynaptic response to this spontaneously released neurotransmitter can be measured electrophys-

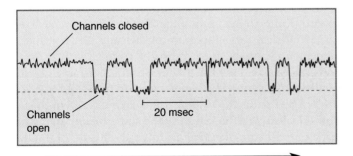

Application of neurotransmitter to membrane patch

Figure 5.14
A patch-clamp recording of a transmitter-gated ion channel. Ionic current passes through the channels when the channels are open. In the presence of neurotransmitter, they rapidly alternate between open and closed states. (Source: Adapted from Neher and Sakmann, 1992.)

iologically. This tiny response is a **miniature postsynaptic potential**, often called simply a "mini." The amplitude of the postsynaptic EPSP evoked by a presynaptic action potential, then, is simply an integer multiple (i.e., 1×, 2×, 3×, etc.) of the mini amplitude.

Quantal analysis, a method of comparing the amplitudes of miniature and evoked postsynaptic potentials, can be used to determine how many vesicles release neurotransmitter during normal synaptic transmission. Quantal analysis of transmission at the neuromuscular junction reveals that a single action potential in the presynaptic terminal triggers the exocytosis of about 200 synaptic vesicles, causing an EPSP of 40 mV or more. At many CNS synapses, in striking contrast, the contents of only a *single vesicle* is released in response to a presynaptic action potential, causing an EPSP of only a few tenths of a millivolt.

EPSP Summation. The difference between excitatory transmission at neuromuscular junctions and CNS synapses is not surprising. The neuromuscular junction has evolved to be fail-safe; it needs to work every time, and the best way to ensure this is to generate an EPSP with a huge size. On the other hand, if every CNS synapse were, by itself, capable of triggering an action potential in its postsynaptic cell (as the neuromuscular junction can), then a neuron would be little more than a simple relay station. Instead, neurons seem to perform more sophisticated computations, requiring that EPSPs add together to produce a significant postsynaptic depolarization. This is what is meant by integration of EPSPs.

EPSP summation represents the simplest form of synaptic integration in the CNS. There are two types of summation: spatial and temporal. **Spatial summation** is the adding together of EPSPs generated simultaneously at many different synapses on a dendrite. **Temporal summation** is the adding together of EPSPs generated at the same synapse if they occur in rapid succession, within 5–15 msec of one another (Figure 5.15).

Contribution of Dendritic Properties to Synaptic Integration

Even with the summation of EPSPs out on a dendrite, the depolarization still may not be enough to cause a neuron to fire an action potential. The current entering at the sites of synaptic contact must spread down the dendrite, through the soma, and cause the membrane of the spike-initiation zone to be depolarized beyond threshold, before an action potential can be generated. The effectiveness of an excitatory synapse in triggering an action potential, therefore, depends on how far the synapse is from the spike-initiation zone and on the properties of the dendritic membrane.

Dendritic Cable Properties. To simplify the analysis of how dendritic properties contribute to synaptic integration, let's assume that dendrites function as cylindrical cables that are electrically passive; that is, lacking voltage-gated ion channels (in contrast, of course, with axons). Using an analogy introduced in Chapter 4, imagine that the influx of positive charge into the dendrite at a synapse is like turning on the water to a leaky garden hose. There are two paths the water can take: One is down the inside of the hose; the other is through the leaks. By the same token, there are two paths that synaptic current can take: One is down the inside of the dendrite; the other is across the dendritic membrane. At some distance from the site of current influx, the EPSP amplitude goes to zero because of the dissipation of the current across the membrane.

The decrease in depolarization as a function of distance along a dendritic cable is plotted in Figure 5.16. Notice that the amount of depolarization falls off exponentially with increasing distance. Depolarization of the membrane at a given distance (V_x) can be described by the equation

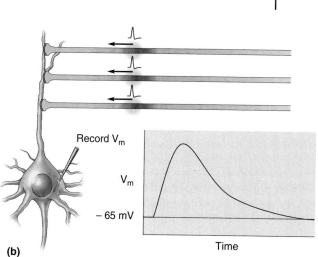

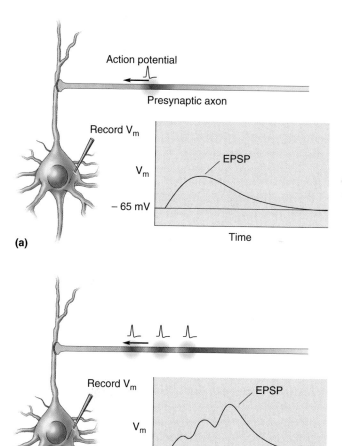

Figure 5.15
EPSP summation. (a) A presynaptic action potential triggers a small EPSP in a postsynaptic neuron. **(b)** Spatial summation of EPSPs: When two or more presynaptic inputs are active at the same time, their individual EPSPs sum. **(c)** Temporal summation of EPSPs: When the same presynaptic fiber fires action potentials in quick succession, the individual EPSPs sum.

$V_x = V_o / e^{x/\lambda}$, where V_o is depolarization at the origin (just under the synapse), e (= 2.718 . . .) is the base of natural logarithms, x is distance from the synapse, and λ is a constant that depends on the properties of the dendrite. Notice that when $x = \lambda$, then $V_x = V_o / e$. Put another way, $V_\lambda = 0.37 (V_o)$. This distance λ, where the depolarization is 37% of that at the origin, is called the dendritic length constant.

The **length constant** is an index of how far depolarization can spread down a dendrite. The longer the length constant, the more likely it is that EPSPs generated at distant synapses will depolarize the membrane at the axon hillock. The value of λ depends on two factors: (1) the resistance to current flowing longitudinally down the dendrite, called the **internal resistance** (r_i), and (2) the resistance to current flowing across the membrane, called the **membrane resistance** (r_m). Most current will take the path of least resistance; therefore, the value of λ will increase as membrane resistance increases because more depolarizing current will flow down the inside of the dendrite. The value of λ will decrease as internal resistance increases because more current will flow across the membrane. Just as water will flow farther down a wide hose with few leaks, synaptic current will flow farther down a wide dendrite (low r_i) with few open membrane channels (high r_m).

The internal resistance depends only on the diameter of the dendrite and the electrical properties of the cytoplasm; consequently, it is relatively constant in a mature neuron. The membrane resistance, in contrast,

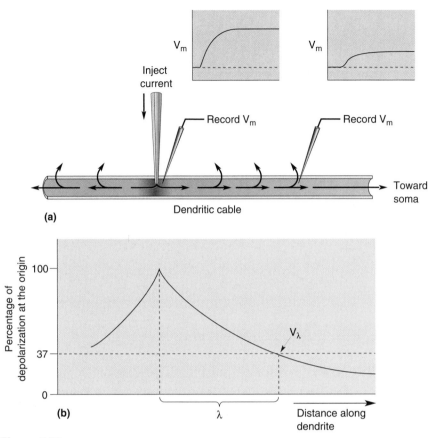

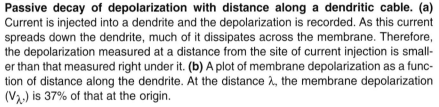

Figure 5.16
Passive decay of depolarization with distance along a dendritic cable. (a)
Current is injected into a dendrite and the depolarization is recorded. As this current spreads down the dendrite, much of it dissipates across the membrane. Therefore, the depolarization measured at a distance from the site of current injection is smaller than that measured right under it. **(b)** A plot of membrane depolarization as a function of distance along the dendrite. At the distance λ, the membrane depolarization (V_λ) is 37% of that at the origin.

depends on the number of open ion channels, which changes from moment to moment depending on what other synapses are active. The dendritic length constant, therefore, isn't constant at all! As we will see in a moment, fluctuations in the value of λ are an important factor in synaptic integration.

Excitable Dendrites. Our analysis of dendritic cable properties made an important assumption: The dendrite's membrane is electrically passive and lacks voltage-gated channels. Some dendrites in the brain actually have passive and inexcitable membranes, and thus follow the simple cable equations. The dendrites of spinal motor neurons, for example, are very close to passive. However, many other neuronal dendrites are decidedly not passive (Figure 5.17). A variety of neurons have dendrites with significant numbers of voltage-gated sodium and/or calcium channels. There are not enough of these dendritic channels to generate fully propagating action potentials, as in axons. But the voltage-gated channels in dendrites can act as important amplifiers of small postsynaptic potentials generated far out on dendrites. EPSPs that would diminish to near nothingness in a long, passive dendrite may nevertheless be large enough to trigger the opening of voltage-gated sodium channels, which in turn would add current to boost the synaptic signal along toward the soma.

Inhibition

So far, we've seen that whether an EPSP contributes to the action potential output of a neuron or not depends on several factors, including the number of coactive excitatory synapses, the distance the synapse is from the spike-initiation zone, and the properties of the dendritic membrane. Of course, not all synapses in the brain are excitatory. The action of some synapses is to take the membrane potential away from action potential threshold; these are said to be inhibitory synapses. Inhibitory synapses exert powerful control over a neuron's output.

IPSPs and Shunting Inhibition. The postsynaptic receptors under most inhibitory synapses are very similar to those under excitatory synapses; they're transmitter-gated ion channels. The only important differences are that they bind different neurotransmitters and that they allow different ions to pass through their channels. The transmitter-gated channels of most inhibitory synapses are permeable to only one natural ion, Cl⁻. Opening of the chloride channel allows Cl⁻ ions to cross the membrane in a direction that brings the membrane potential toward the chloride equilibrium potential, E_{Cl}, about −65 mV. If the membrane potential was greater than −65 mV when the transmitter was released, activation of these channels would cause a hyperpolarizing IPSP.

Notice that if the membrane potential was already −65 mV, no IPSP would be visible after chloride channel activation because the value of the

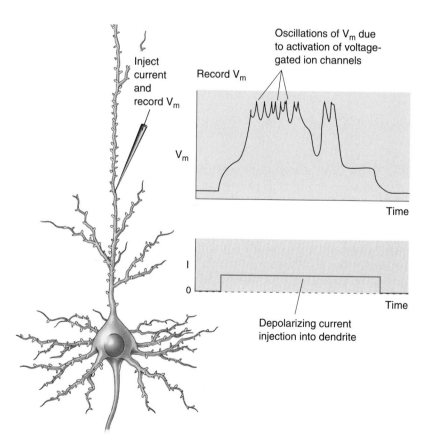

Figure 5.17
An excitable dendrite. Pyramidal cells in the cerebral cortex often have long, vertically oriented apical dendrites. When these dendrites are injected with depolarizing current, they generate complex oscillations of membrane potential because of the voltage-gated sodium and calcium channels in their membrane.

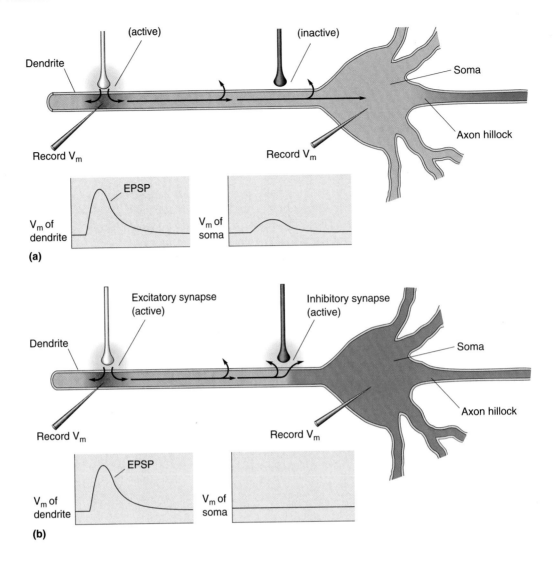

Figure 5.18
Shunting inhibition. Illustrated is a neuron receiving one excitatory and one inhibitory input. **(a)** Stimulation of the excitatory input causes inward postsynaptic current that spreads to the soma, where it can be recorded as an EPSP. **(b)** When the inhibitory and excitatory inputs are stimulated together, the depolarizing current leaks out before it reaches the soma.

membrane potential would already equal E_{Cl} (i.e., the reversal potential for that synapse; see Box 5.3). If there is no visible IPSP, is the neuron really inhibited? The answer is yes. Consider the situation illustrated in Figure 5.18, with an excitatory synapse on a distal segment of dendrite and an inhibitory synapse on a proximal segment of dendrite, near the soma. Activation of the excitatory synapse leads to the influx of positive charge into the dendrite. This current depolarizes the membrane as it flows toward the soma. At the site of the active inhibitory synapse, however, the membrane potential is approximately equal to E_{Cl}, –65 mV. Positive current therefore flows outward across the membrane at this site to bring V_m to –65 mV. This synapse acts as an electrical shunt, preventing the current from flowing through the soma to the axon hillock. This type of inhibition is called **shunting inhibition**. The actual physical basis of shunting inhibition is the *inward movement of negatively charged chloride ions*, which is formally equivalent to *outward positive current flow*. Shunting inhibition is like cutting a big hole in the leaky garden hose—all the water flows down this path of least resistance, out of the hose, before it gets to the nozzle where it can "activate" the flowers in your garden.

Thus, you can see that the action of inhibitory synapses also contributes to synaptic integration. The IPSPs can be subtracted from EPSPs, making the postsynaptic neuron less likely to fire action potentials. In addition, shunting inhibition acts to drastically reduce r_m and consequently λ, thus allowing positive current to flow out across the membrane instead of internally down the dendrite toward the spike-initiation zone.

Geometry of Excitatory and Inhibitory Synapses. Inhibitory synapses in the brain that use GABA as a neurotransmitter always have a morphology characteristic of Gray's type II (see Figure 5.4). This structure contrasts with excitatory synapses use glutamate, which always have a Gray's type I morphology. This correlation between structure and function has been useful for working out the geometrical relationships among excitatory and inhibitory synapses on individual neurons. In addition to being spread over the dendrites, inhibitory synapses on many neurons are found clustered on the soma and near the axon hillock, where they are in an especially powerful position to influence the activity of the postsynaptic neuron.

Modulation

Most of the postsynaptic mechanisms we've discussed so far involve transmitter receptors that are, themselves, ion channels. To be sure, synapses with transmitter-gated channels carry the bulk of the specific information that is processed by the nervous system. However, there are many synapses with G-protein-coupled neurotransmitter receptors that are not directly associated with an ion channel. Synaptic activation of these receptors does not directly evoke EPSPs and IPSPs, but instead *modulates* the effectiveness of EPSPs generated by other synapses with transmitter-gated channels. This type of synaptic transmission is called **modulation**. We'll give you a taste for how modulation influences synaptic integration by exploring the effects of activating one type of G-protein-coupled receptor in the brain, the norepinephrine beta (β) receptor.

The binding of the amine neurotransmitter **norepinephrine** (**NE**) to the β receptor triggers a cascade of biochemical events within the cell. In short, the β receptor activates a G-protein that, in turn, activates an effector protein, the intracellular enzyme adenylyl cyclase. **Adenylyl cyclase** catalyzes the chemical reaction that converts adenosine triphosphate (ATP), the product of oxidative metabolism in the mitochondria, into a compound called **cyclic adenosine monophosphate**, or **cAMP**, that is free to diffuse within the cytosol. Thus, the *first* chemical message of synaptic transmission (the release of NE into the synaptic cleft) is converted by the β receptor into a *second* message (cAMP); cAMP is an example of a second messenger.

The effect of cAMP is to stimulate another enzyme called a protein kinase. **Protein kinases** catalyze a chemical reaction called **phosphorylation**, the transfer of phosphate groups (PO_3) from ATP to specific sites on cell proteins (Figure 5.19). The significance of phosphorylation is that it can change the conformation of a protein, thereby changing that protein's activity.

In some neurons, one of the proteins that is phosphorylated when cAMP rises is a particular type of potassium channel in the dendritic membrane. Phosphorylation causes this channel to close, thereby reducing the membrane K^+ conductance. By itself, this does not cause any dramatic effects on the neuron. But consider the wider consequence: *Decreasing the K^+ conductance increases the dendritic membrane resistance and therefore increases the length constant.* It is like wrapping the leaky garden hose in duct tape; more water can flow down the inside of the hose and less leaks out the sides. As a consequence of increasing λ, distant or weak excitatory synapses will become more effective in depolarizing the spike-initiation zone beyond

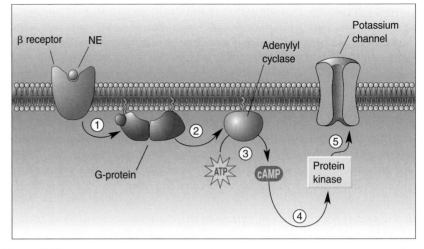

Figure 5.19
Modulation by the NE ß receptor. (1) The binding of NE to the receptor activates a G-protein in the membrane. (2) The G-protein activates the enzyme adenylyl cyclase. (3) Adenylyl cyclase converts ATP into the second messenger cAMP. (4) cAMP activates a protein kinase. (5) The protein kinase causes a potassium channel to close by attaching a phosphate group to it.

threshold; the cell will become *more excitable*. Thus, the binding of NE to β receptors produces little change in membrane potential but greatly increases the response produced by another neurotransmitter at an excitatory synapse. Because this effect involves several biochemical intermediaries, it can last far longer than the presence of the modulatory transmitter itself.

We have described one particular G-protein-coupled receptor and the consequences of activating it in one type of neuron. But it is important to recognize that other types of receptor can lead to the formation of other types of second messenger molecule. Activation of each of these receptor types will initiate a distinct cascade of biochemical reactions in the postsynaptic neuron that do not always include phosphorylation and decreases in membrane conductance. In fact, cAMP in a different cell type with different enzymes may produce functionally opposite changes in the excitability of cells. In Chapter 6, we will describe more examples of synaptic modulation and their mechanisms. However, you can already see that modulatory forms of synaptic transmission offer an almost limitless number of ways that information encoded by presynaptic impulse activity can be transformed and used by the postsynaptic neuron.

CONCLUDING REMARKS

In this chapter, we have covered the basic principles of chemical synaptic transmission. The action potential that arose in the sensory nerve when you stepped on that thumbtack in Chapter 3, and swept up the axon in Chapter 4, has now reached the axon terminal in the spinal cord. The depolarization of the terminal triggered the presynaptic entry of Ca^{2+} through voltage-gated calcium channels, which then stimulated exocytosis of the contents of synaptic vesicles. The liberated transmitter diffused across the synaptic cleft and bound to specific receptors in the postsynaptic membrane. The transmitter (probably glutamate) caused transmitter-gated channels to open, which allowed positive charge to enter the postsynaptic dendrite. Because the sensory nerve was firing action potentials at a high rate, and because many synapses were activated at the same time, the EPSPs summed to bring the spike-initiation zone of the postsynaptic neuron to threshold, and this cell then generated action potentials. If the cell were a motor neuron, this activity would cause the release of ACh at the neuromuscular junction and muscle contraction. If the postsynaptic cell were an interneuron that used GABA as a neurotransmitter, the activity of the cell would result in inhibition of its synaptic targets. If this cell used a modulatory transmitter such as NE, the activity could cause lasting changes in the excitability or metabolism of its synaptic targets. It is this rich diversity of chemical synaptic interactions that allows complex behaviors (such as shrieking with pain as you pick up your foot) to emerge from simple stimuli (such as a thumbtack).

Although we surveyed chemical synaptic transmission in this chapter, we did not cover the *chemistry* of synaptic transmission in any detail. In Chapter 6, we'll take a closer look at the chemical "nuts and bolts" of different neurotransmitter systems. In Chapter 15, after we've examined the sensory and motor systems in Part II, we'll explore the contributions of several different neurotransmitters to nervous system function and behavior. You'll see that the chemistry of synaptic transmission warrants all this attention because defective neurotransmission is the basis for many neurological and psychiatric disorders. And virtually all psychoactive drugs, both therapeutic and illicit, exert their effects at chemical synapses.

In addition to providing explanations for aspects of neural information-processing and the effects of drugs, knowledge of chemical synaptic transmission is also the key to understanding the neural basis of learning and memory. Memories of past experiences are established by modification of the effectiveness of chemical synapses in the brain. Material discussed in this chapter suggests possible sites of modification, ranging from changes in presynaptic Ca^{2+} entry and neurotransmitter release to alterations in postsynaptic receptors or excitability. As we shall see in Chapter 20, all of these changes are likely to contribute to the storage of information by the nervous system.

KEY TERMS

Introduction
electrical synapse
chemical synapse

Types of Synapses
gap junction
secretory granule
large, dense-core vesicle
membrane differentiation
active zone
postsynaptic density
Gray's type II synapse
Gray's type I synapse
neuromuscular junction
motor end-plate

Principles of Chemical Synaptic Transmission
glutamate (Glu)
gamma-aminobutyric acid (GABA)
acetylcholine (ACh)
transporter
voltage-gated calcium channel
exocytosis
transmitter-gated ion channel
excitatory postsynaptic potential (EPSP)
inhibitory postsynaptic potential (IPSP)

G-protein-coupled receptor
G-protein
second messenger
metabotropic receptor
autoreceptor
neuropharmacology
inhibitor
receptor antagonist
receptor agonist
nicotinic ACh receptor

Principles of Synaptic Integration
miniature postsynaptic potential
quantal analysis
EPSP summation
spatial summation
temporal summation
length constant
internal resistance
membrane resistance
shunting inhibition
modulation
norepinephrine (NE)
adenylyl cyclase
cyclic adenosine monophosphate (cAMP)
protein kinase
phosphorylation

✔ REVIEW QUESTIONS

1. What is meant by quantal release of neurotransmitter?
2. You apply ACh and activate nicotinic receptors on a muscle cell. Which way will current flow through the receptor channels when V_m = –60 mV? When V_m = 0 mV? When V_m = 60 mV? Why?
3. In this chapter we discussed a GABA-gated ion channel that is permeable to Cl^-. GABA also activates a G-protein-coupled receptor, called the $GABA_B$ receptor, which causes potassium-selective channels to open. What effect would $GABA_B$ receptor activation have on the membrane potential?
4. You are studying the neuronal response to glycine, an amino acid transmitter. The reversal potential for postsynaptic responses evoked by glycine is –60 mV. Is this neurotransmitter excitatory or inhibitory? Why?
5. A drug called strychnine, isolated from the seeds of a tree native to India and commonly used as rat poison, blocks the effects of glycine. Is strychnine an agonist or an antagonist of the glycine receptor?
6. How does nerve gas cause respiratory paralysis?
7. Why is an excitatory synapse on the soma more effective in evoking action potentials in the postsynaptic neuron than an excitatory synapse on the tip of a dendrite?
8. What are the steps that lead to increased excitability in a neuron when NE is released presynaptically?

Neurotransmitter Systems

The normal human brain is an orderly set of chemical reactions. As we have seen, some of the brain's most important chemical reactions are those associated with synaptic transmission. Chapter 5 introduced the general principles of chemical synaptic transmission, using a few specific neurotransmitters as examples. In this chapter, we will explore in more depth the variety and elegance of the major neurotransmitter systems.

Neurotransmitter systems begin with neurotransmitters. In Chapter 5 we discussed the three major classes: *amino acids*, *amines*, and *peptides*. Even a partial list of the known transmitters, such as that appearing in Table 5.1, has over 20 molecules. Each of these molecules can define a particular transmitter system. In addition to the molecule itself, a neurotransmitter system includes all the molecular machinery responsible for transmitter synthesis, vesicular packaging, reuptake and degradation, and transmitter action (Figure 6.1).

The first molecule positively identified as a neurotransmitter by Otto Loewi in the 1920s was acetylcholine, or ACh (see Box 5.1). To describe the cells that produce and release ACh, British pharmacologist Henry Dale introduced the term *cholinergic*. (Dale shared the 1936 Nobel Prize with Loewi, in recognition of his neuropharmacological studies of synaptic transmission.) Dale termed the neurons that use the amine neurotransmitter norepinephrine (NE) *noradrenergic* (NE is known as noradrenaline in Great Britain). The convention of using the suffix *-ergic* continued when additional transmitters were identified. Therefore, today we speak of *glutamatergic* synapses that use glutamate, *GABAergic* synapses that use GABA, *peptidergic* synapses that use peptides, and so on. These adjectives are also used to identify the various neurotransmitter systems. Thus, for example, ACh and all the molecular machinery associated with it are collectively called the *cholinergic system*.

With this terminology in hand, we can begin our exploration of the neurotransmitter systems. We start with a discussion of the experimental strategies that have been used to study neurotransmitter systems. Then we look at the synthesis and metabolism of specific neurotransmitters and explore how these molecules exert their postsynaptic effects. In Chapter 15, after we have learned more about the structural and functional organization of the nervous system, we'll take another look at specific neurotransmitter systems in the context of their individual contributions to the regulation of brain function and behavior.

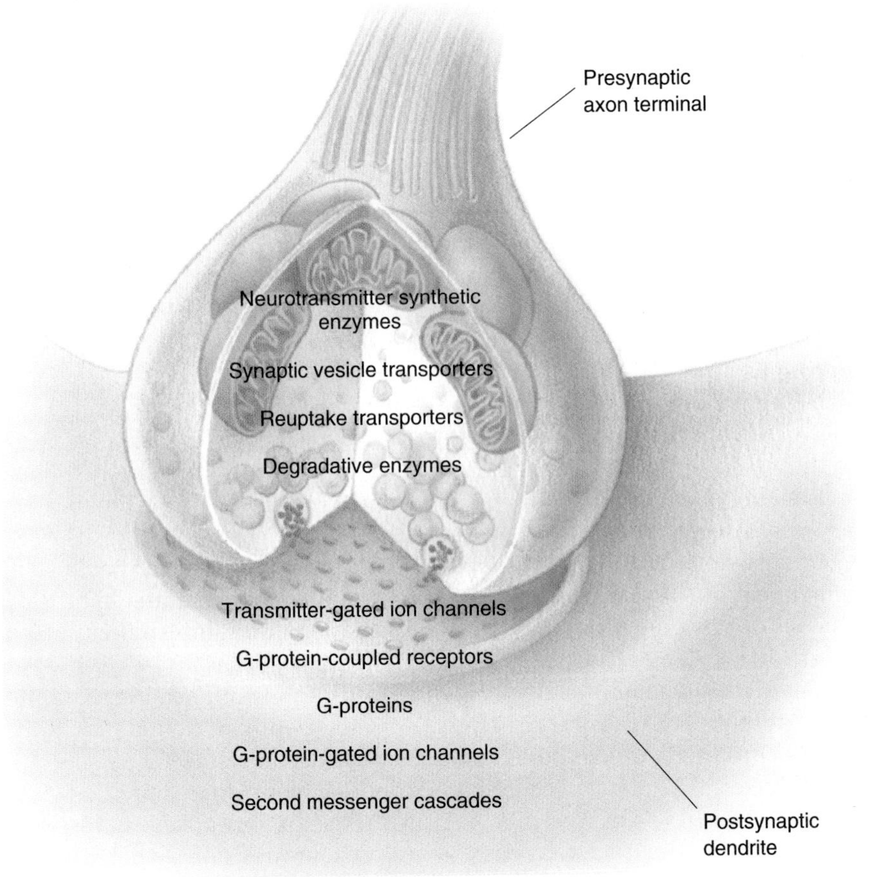

Figure 6.1
Elements of neurotransmitter systems.

STUDYING NEUROTRANSMITTER SYSTEMS

The first step in studying a neurotransmitter system is usually identifying the neurotransmitter. This is no simple task; the brain contains uncountable chemicals. How can we decide which few chemicals are used as transmitters?

Over the years, neuroscientists have established certain criteria that must be met for a molecule to be considered a neurotransmitter:

1. The molecule must be synthesized and stored in the presynaptic neuron.

2. The molecule must be released by the presynaptic axon terminal upon stimulation.

3. The molecule must produce a response in the postsynaptic cell that mimics the response produced by the release of neurotransmitter from the presynaptic neuron.

Let's begin by exploring some of the strategies and methods that are used to satisfy these criteria.

Localization of Transmitters and Transmitter Synthetic Enzymes

The investigator often begins with little more than a hunch that a particular molecule may be a neurotransmitter. This idea may be based on observing that the molecule is concentrated in brain tissue or that the application of the molecule to certain neurons alters their action potential firing rate. Whatever the inspiration, the first step in confirming the hypothesis is to show that the molecule is, in fact, localized in, and synthesized by, particular neurons. Many methods have been used to satisfy this criterion for different neurotransmitters. Two of the most important techniques used today are immunocytochemistry and *in situ* hybridization.

Immunocytochemistry. The method of **immunocytochemistry** is used to anatomically localize particular molecules to particular cells. The principle behind the method is quite simple (Figure 6.2). Once the neurotransmitter candidate has been chemically purified, it is injected into the bloodstream of an animal, where it stimulates an immune response. (Often, to evoke a response, the molecule is chemically coupled to a larger molecule.) One aspect of the immune response is the generation of antibodies that can bind tightly to specific sites on the foreign molecule—in this case, the transmitter candidate. These specific antibody molecules can be recovered from a blood sample of the immunized animal and chemically tagged with a colorful marker that can be seen in a microscope. When these labeled antibodies are applied to a section of brain tissue, they will color just those cells that contain the transmitter candidate (Figure 6.3). Immunocytochemistry can be used to localize any molecule for which a specific antibody can be generated, including the synthetic enzymes for transmitter candidates. Demonstration that the transmitter candidate and its synthetic enzyme are contained in the same neuron—or better yet, in the same axon terminal—can satisfy the criterion that the molecule be localized in, and synthesized by, a particular neuron.

In Situ *Hybridization*. The method called ***in situ* hybridization** is useful for confirming that a cell synthesizes a particular protein or peptide. Recall from Chapter 2 that proteins are assembled by the ribosomes according to instructions from specific mRNA molecules. There is a unique mRNA molecule for every polypeptide synthesized by a neuron. The mRNA transcript consists of four different nucleic acids linked together in various sequences to form a strand. Each nucleic acid has the unusual property that it will bind tightly to one other complementary nucleic acid. Thus, if the

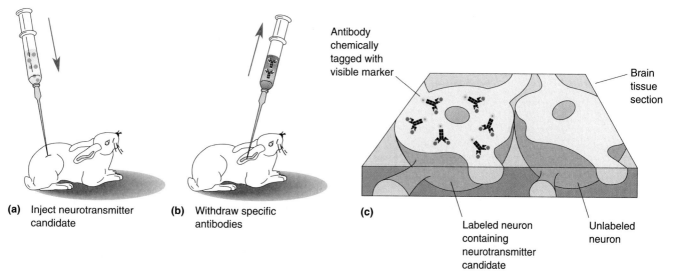

(a) Inject neurotransmitter candidate

(b) Withdraw specific antibodies

(c)

Antibody chemically tagged with visible marker

Brain tissue section

Labeled neuron containing neurotransmitter candidate

Unlabeled neuron

Figure 6.2
Immunocytochemistry. This method uses labeled antibodies to identify the location of molecules within cells. **(a)** The molecule of interest (a neurotransmitter candidate) is injected into an animal, causing an immune response and the generation of antibodies. **(b)** Blood is withdrawn from the animal, and the antibodies are isolated from the serum. **(c)** The antibodies are tagged with a visible marker and applied to sections of brain tissue. The antibodies label those cells that contain the neurotransmitter candidate.

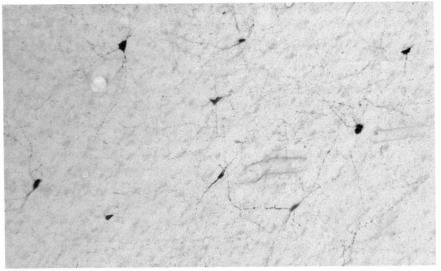

Figure 6.3
Immunocytochemical localization of a peptide neurotransmitter in neurons.
(Source: Courtesy of Dr. S. H. C. Hendry)

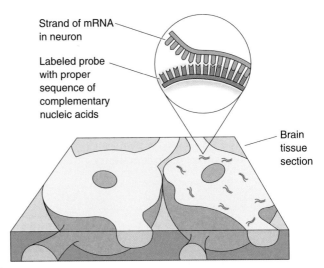

Strand of mRNA in neuron

Labeled probe with proper sequence of complementary nucleic acids

Brain tissue section

Figure 6.4
In situ hybridization. Strands of mRNA consist of nucleotides arranged in a specific sequence. Each of these nucleotides will stick to one other complementary nucleotide. In the method of *in situ* hybridization, a synthetic probe is constructed that contains a sequence of complementary nucleotides that will allow it to stick to the mRNA. If the probe is labeled, the location of cells containing the mRNA will be revealed.

sequence of nucleic acids in a strand of mRNA is known, it is possible to construct in the lab a complementary strand that will stick, like a strip of Velcro, to the mRNA molecule. The complementary strand is called a *probe*, and the process by which the probe bonds to the mRNA molecule is called *hybridization* (Figure 6.4). In order to see if the mRNA for a particular peptide is localized in a neuron, we chemically label the appropriate probe so it can be detected, apply it to a section of brain tissue, and then search for neurons that contain the label.

For *in situ* hybridization, the probes are usually labeled by making them radioactive. Since we cannot see radioactivity, hybridized probes are detected by laying the brain tissue on a sheet of special film that is sensitive to radioactive emissions. After exposure to the tissue, the film is developed like a photograph, and negative images of the radioactive cells are visible (Figure 6.5). This technique for viewing the distribution of radioactivity is called **autoradiography**.

In summary, immunocytochemistry is a method for viewing the location of specific molecules, including proteins, in sections of brain tissue. *In situ* hybridization is a method for localizing specific mRNA transcripts for proteins. Together, these methods enable us to see if a neuron contains and synthesizes a transmitter candidate.

Studying Transmitter Release

Once we are satisfied that a transmitter candidate is synthesized by a neuron and localized to the presynaptic terminal, we must show that it is actually released upon stimulation. In some cases, a specific set of cells or axons can be stimulated while taking samples of the fluids bathing their synaptic targets. The biological activity of the sample can then be tested to see if it mimics the effect of the intact synapses, and then the sample can be chemically analyzed to reveal the structure of the active molecule. This general approach helped Loewi and Dale identify ACh as a transmitter at many peripheral synapses.

Unlike the peripheral nervous system (PNS, the nervous system outside the brain and spinal cord), most regions of the CNS contain a diverse mixture of intermingled synapses using different neurotransmitters. Thus, it is often impossible to stimulate a single population of synapses, containing only a single neurotransmitter. Researchers must be content with stimulating many synapses in a region of the brain and collecting and measuring all the chemicals that are released. One way this is done is by using brain slices that are kept alive *in vitro*. To stimulate release, the slices are bathed in a solution containing a high K^+ concentration. This treatment causes a large membrane depolarization (see Figure 3.17), thereby stimulating transmitter release from the axon terminals in the tissue. Because transmitter release requires entry of Ca^{2+} into the axon terminal, it must also be shown that the

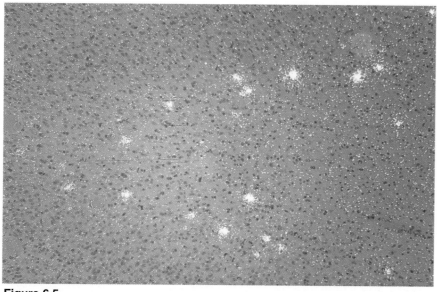

Figure 6.5
In situ hybridization of the mRNA for a peptide neurotransmitter in neurons.
(Source: Courtesy of Dr. S. H. C. Hendry.)

release of the neurotransmitter candidate from the tissue slice after depolarization occurs only when Ca^{2+} ions are present in the bathing solution.

Even when it has been shown that a transmitter candidate is released upon depolarization in a calcium-dependent fashion, however, we still cannot be sure that the molecules collected in the fluids were released from the axon terminals; they may have been released as a secondary consequence of synaptic activation. These technical difficulties make the second criterion—that a transmitter candidate must be released by the presynaptic axon terminal upon stimulation—the most difficult to satisfy unequivocally in the CNS.

Studying Synaptic Mimicry

Establishing that a molecule is localized in, synthesized by, and released from a neuron is still not sufficient to qualify it as a neurotransmitter. A third criterion must be met: The molecule must evoke the same response as that produced by the release of naturally occurring neurotransmitter from the presynaptic neuron.

In order to assess the postsynaptic actions of a transmitter candidate, a method called **microionophoresis** is often used. Most neurotransmitter candidates can be dissolved in solutions that will cause them to acquire a net electrical charge. A glass pipette with a very fine tip, just a few μm across, is filled with the ionized solution. The tip of the pipette is carefully positioned next to the postsynaptic membrane of the neuron, and the transmitter-candidate is ejected in very small amounts by passing electrical current through the pipette. A microelectrode in the postsynaptic neuron can be used to measure the effects of the transmitter-candidate on the membrane potential (Figure 6.6).

If ionophoretic application of the molecule causes electrophysiological changes that mimic the effects of transmitter released at the synapse, *and* if the other criteria of localization, synthesis, and release have been met, then the molecule and the transmitter usually are considered to be the same chemical.

Studying Receptors

Each neurotransmitter exerts its postsynaptic effects by binding to specific receptors. As a rule, no two neurotransmitters bind to the same receptor; however, the same neurotransmitter can bind to many different receptors. Each of the different receptors that a neurotransmitter binds to is called a **receptor subtype**. For example, in Chapter 5 we learned that ACh acts on two different cholinergic receptor *subtypes*: one in skeletal muscle and another in heart muscle.

Researchers have tried almost every method of biological and chemical analysis to study the different receptor subtypes of the various neurotransmitter systems. Three approaches have proved to be particularly useful: neuropharmacological analysis of synaptic transmission, ligand-binding methods, and, most recently, molecular analysis of receptor proteins.

Neuropharmacological Analysis. Much of what we know about receptor subtypes was first learned using neuropharmacological analysis. For instance, the different subtypes of cholinergic receptor at skeletal muscle and heart muscle respond differently to various drugs. *Nicotine*, derived from the tobacco plant, is a receptor agonist in skeletal muscle but has no effect in the heart. On the other hand, *muscarine*, derived from a poisonous species of mushroom, has no effect on skeletal muscle but is an agonist at the cholinergic receptor subtype in the heart. (You may recall that ACh slows the heart; ingestion of muscarine is poisonous because it causes a precipitous drop in heart rate and blood pressure.) Thus, two ACh receptor subtypes could be distinguished by these different drug actions. In fact, these agonists gave the subtypes their names: **nicotinic receptors** in skele-

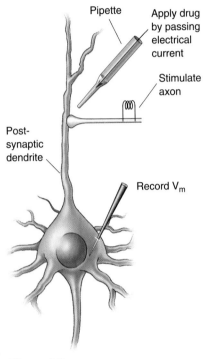

Pipette

Apply drug by passing electrical current

Stimulate axon

Postsynaptic dendrite

Record V_m

Figure 6.6
Microionophoresis. This method enables a researcher to apply drugs or neurotransmitter candidates in very small amounts to the surface of neurons. The responses generated by the drug can be compared to those generated by synaptic stimulation.

tal muscle and **muscarinic receptors** in the heart. There are also selective antagonists that act at these two ACh receptor subtypes. The arrow-tip poison *curare* inhibits the action of ACh at nicotinic receptors (therefore causing paralysis), and *atropine*, derived from belladonna plants, antagonizes ACh at muscarinic receptors (Figure 6.7). (The eyedrops an ophthalmologist uses to dilate your pupils are related to atropine.)

Different drugs were also used to distinguish several subtypes of glutamate receptors. Three subtypes are **AMPA receptors**, **NMDA receptors**, and **kainate receptors**, each named for a different chemical agonist. (AMPA stands for α-amino-3-hydroxy-5-methyl-4-isoxazole propionate, and NMDA stands for N-methyl-D-aspartate.) Glutamate activates all three subtypes, but AMPA acts only at the AMPA receptor, NMDA acts only at the NMDA receptor, and so on (Figure 6.8).

Similar pharmacological analyses were used to split the NE receptors into two subtypes, α and β, and to divide GABA receptors into GABA$_A$ and GABA$_B$ subtypes. The same can be said for virtually all the neurotransmitter systems. Thus, selective drugs have been extremely useful for categorizing receptor subclasses (Table 6.1). In addition, neuropharmacological

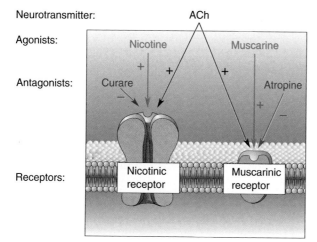

Figure 6.7
Neuropharmacology of cholinergic synaptic transmission.

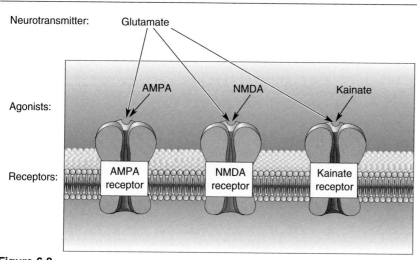

Figure 6.8
Neuropharmacology of glutaminergic synaptic transmission.

analysis has been invaluable for assessing the contributions of neurotransmitter systems to brain function.

Ligand-Binding Methods. As we said, the first step in studying a neurotransmitter system is usually identifying the neurotransmitter. However, with the discovery in the 1970s that many drugs interact selectively with neurotransmitter receptors, researchers realized that they could use these compounds to begin analyzing receptors, even before the neurotransmitter itself had been identified. The pioneer of this approach was Solomon Snyder at Johns Hopkins University (Box 6.1).

The question Snyder and his student Candace Pert originally set out to answer was how heroin, morphine, and other opiate compounds exert their effects on the brain. They hypothesized that opiates might be agonists at specific receptors in neuronal membranes. To test this idea, they radioactively labeled opiate compounds and applied them in small quantities to neuronal membranes that had been isolated from different parts of the brain. If receptors existed in the membrane, the labeled opiates should bind tightly to them. This is just what they found. The radioactive drugs labeled specific sites on the membranes of some, but not all, neurons in the brain (Figure 6.9). Following the discovery of opiate receptors, the search was on to identify endogenous opiates or *endorphins*, the naturally occurring neurotransmitters that act on these receptors. Soon it was discovered that a peptide, *enkephalin*, is a neurotransmitter that acts on these receptors. Subsequent neuropharmacological analysis showed that enkephalin acts on several subtypes of opiate receptor.

Any chemical compound that binds to a specific site on a receptor is called a *ligand* for that receptor (from the Latin word meaning "to bind"). The technique of studying receptors using radioactively labeled ligands is called the **ligand-binding method**. Notice that a ligand for a receptor can be an agonist, an antagonist, or even the chemical neurotransmitter itself. Ligand-binding methods have been enormously valuable for mapping the anatomical distribution of different neurotransmitter receptors in the brain.

Molecular Analysis. There has been an explosion of information about neurotransmitter receptors in the last 10 years, thanks to the new methods for studying protein molecules. Information obtained with these methods has enabled us to divide the neurotransmitter receptor proteins into two groups: transmitter-gated ion channels and G-protein-coupled (metabotropic) receptors (see Chapter 5).

Molecular neurobiologists have determined the structure of the polypeptides that make up many proteins, and these studies have led to

Table 6.1
Neuropharmacology of Some Receptor Subtypes

Neurotransmitter	Receptor Subtype	Agonist	Antagonist
Acetylcholine (ACh)	Nicotinic receptor	Nicotine	Curare
	Muscarinic receptor	Muscarine	Atropine
Norepinephrine	α receptor	Phenylephrine	Phenoxybenzamine
	β receptor	Isoproterenol	Propranolol
Glutamate	AMPA	AMPA	CNQX
	NMDA	NMDA	AP5
GABA	$GABA_A$	Muscimol	Bicuculline
	$GABA_B$	Baclofen	Phaclofen

Box 6.1 | **PATH OF DISCOVERY**

Finding Opiate Receptors
by Solomon H. Snyder

Solomon H. Snyder

Like so many events in science, identifying the opiate receptors was not simply an intellectual feat accomplished in an ethereal pursuit of pure knowledge. Instead, it all began with Richard Nixon and his "war on drugs" in 1971, at the height of very well-publicized use of heroin by hundreds of thousands of American soldiers in Vietnam. To combat all this, Nixon appointed as czar of drug abuse research Dr. Jerome Jaffe, a psychiatrist who had pioneered in methadone treatment for heroine addicts. Jaffe was to coordinate the several billions of federal dollars in agencies ranging from the Department of Defense to the National Institutes of Health.

Jerry, a good friend, pestered me to direct our research toward the "poor soldiers" in Vietnam. So I began wondering how opiates act. The notion that drugs act at receptors, specific recognition sites, had been appreciated since the turn of the century. In principle, one could identify such receptors simply by measuring the binding of radioactive drugs to tissue membranes. However, countless researchers had applied this strategy to opiates with no success.

About this time, a new Johns Hopkins faculty member, Pedro Cuatrecasas, located his laboratory adjacent to mine, and we became fast friends. Pedro had recently attained fame for his discovery of receptors for insulin. His success depended upon seemingly simple, but important technical advances. Past efforts to identify receptors for hormones had failed because hormones can bind to many nonspecific sites, comprising proteins, carbohydrates, and lipids. The numbers of these nonspecific sites would likely be millions of times greater than the number of specific receptors. To identify the "signal" of insulin receptors binding above the "noise" of nonspecific interactions, Pedro developed a simple filtration assay. Since insulin should adhere more tightly to its receptors than to nonspecific sites, he incubated liver membranes with radioactive insulin and poured the mixture over filters attached to a vacuum that rapidly sucked away the incubation fluid, leaving the membrane with attached insulin stuck to the filters. He then "washed" the filters with large volumes of saline, but did this very rapidly so as to preserve insulin bound to receptors while washing away nonspecific binding.

Despite Pedro's proximity, it did not immediately occur to me that the insulin success could be transferred to the opiate receptor problem. Instead, I had read a paper on nerve growth factor, showing that its amino acid sequence closely resembled that of insulin. Pedro and I soon collaborated in a successful search for the nerve growth factor receptor. Only then did I marshal the courage to extend this approach from proteins such as insulin and nerve growth factor to much smaller molecules such as opiates. Candace Pert, a graduate student in my laboratory, was eager to take on a new research project. We obtained a radioactive drug and monitored its binding to brain membranes using Pedro's magic filter machine. The very first experiment, which took only about two hours, was successful.

Within a few months, we were able to characterize many features of opiate receptors. Knowing the exact sites where receptors are concentrated in the brain explained all the major actions of opiates, such as euphoria, pain relief, depression of breathing, and pupillary constriction. The properties of opiate receptors resembled very much what one would expect for neurotransmitters. Accordingly, we used similar approaches to search for receptors for neurotransmitters in the brain, and within a few years had identified receptors for most of them.

These findings raised an obvious question: Why do opiate receptors exist? Humans were not born with morphine in them. Might the opiate receptor be a receptor for a new transmitter that regulates pain perception and emotional states? We and other groups attempted to isolate the hypothesized normally occurring morphinelike neurotransmitters. John Hughes and Hans Kosterlitz in Aberdeen, Scotland, were the first to succeed. They isolated and obtained the chemical structures of the first "endorphins," which are called the enkephalins. In our own laboratory, Rabi Simantov and I obtained the structure of the enkephalins soon after the published success of the Scottish group.

From the first experiments identifying opiate receptors until the isolation of the enkephalins, only three years elapsed—an interval of frantic, exhilarating work that changed profoundly how we think about drugs and the brain.

some startling conclusions. Receptor subtype diversity was expected from the actions of different drugs, but the extent of the diversity was not appreciated until researchers determined how many different polypeptides could serve as subunits of functional receptors.

Consider as an example the $GABA_A$ receptor, a transmitter-gated chloride channel. The channel is constructed of five major subunit proteins, designated α, β, γ, δ, and ϵ. However, at least six different polypeptides (designated $\alpha1$–6) can substitute for one another as an α subunit. Three different polypeptides (designated $\beta1$–3) can substitute as a β subunit, and two different polypeptides ($\gamma1$ and $\gamma2$) can be used as a γ subunit. Although this is probably not the final tally, let's use these numbers to make an interesting calculation. If it takes five subunits to form a $GABA_A$ receptor-gated channel and there are 13 possible subunits to choose from, then there are 74,269 possible combinations of subunits. This means there are 74,269 potential subtypes of $GABA_A$ receptors!

It is important to recognize that most of the possible subunit combinations are never manufactured by neurons and, even if they were, they would not work properly. Nonetheless, it is clear that receptor classifications like those appearing in Table 6.1, while still useful, underestimate the diversity of possible receptor subtypes in the brain.

NEUROTRANSMITTER CHEMISTRY

Research using methods such as those discussed above has led to the conclusion that the major neurotransmitters are amino acids, amines, and peptides. Evolution is conservative and opportunistic, and it often puts common and familiar things to new uses. This seems to be true about the evolution of neurotransmitters. For the most part, they are similar or identical to the basic chemicals of life, the same substances that cells in all species, from bacteria to giraffes, use for metabolism. Amino acids, the building blocks of protein, are essential to life. Most of the known neurotransmitter molecules are either (1) amino acids, (2) amines derived from amino acids, or (3) peptides constructed from amino acids. ACh is an exception; but it is derived from acetyl CoA, a ubiquitous product of cellular respiration in mitochondria, and choline, which is important for fat metabolism throughout the body.

Amino acid and amine transmitters are generally each stored in and released by separate sets of neurons. The convention established by Henry Dale classifies neurons into mutually exclusive groups by neurotransmitter (cholinergic, glutaminergic, GABAergic, and so on). The idea that a neuron has a unique identity with respect to neurotransmitter is often called **Dale's principle**. Strictly speaking, many peptide-containing neurons violate Dale's principle, because these cells usually release more than one neurotransmitter: an amino acid or amine *and* a peptide. Nonetheless, the amino acid and amine neurotransmitters can be used to divide most neurons into distinct, nonoverlapping classes. Let's take a look at the biochemical mechanisms that differentiate these neurons.

Cholinergic Neurons

ACh is the neurotransmitter at the neuromuscular junction and therefore is synthesized by all the motor neurons in the spinal cord. Other cholinergic cells contribute to the functions of specific circuits in the PNS and CNS, as we will see in Chapter 15.

ACh synthesis requires a specific enzyme, *choline acetyltransferase* (*ChAT*). Like all presynaptic proteins, ChAT is manufactured in the soma

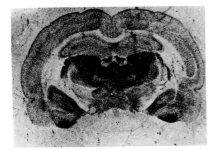

Figure 6.9
Opiate receptor binding to a slice of rat brain. Special film was exposed to a brain section that had radioactive opiate receptor ligands bound to it. The dark regions contain more receptors. (Source: Snyder, 1986, p. 44.)

and transported to the axon terminal. Only cholinergic neurons contain ChAT, so this enzyme is a good marker for cells that use ACh as a neurotransmitter. ChAT synthesizes ACh in the cytosol of the axon terminal, and the neurotransmitter is concentrated in synaptic vesicles by the actions of an ACh transporter.

ChAT transfers an acetyl group from acetyl CoA to choline (Figure 6.10a). The source of choline is the extracellular fluid, where it exists in low micromolar concentrations. Choline is taken up by the cholinergic axon terminals via a specific transporter. Because the availability of choline limits how much ACh can be synthesized in the axon terminal, the transport of choline into the neuron is said to be the **rate-limiting step** in ACh synthesis. For certain diseases in which a deficit in cholinergic synaptic transmission has been noted, dietary supplements of choline are sometimes prescribed to boost ACh levels in the brain.

Cholinergic neurons also manufacture the ACh degradative enzyme *acetylcholinesterase (AChE)*. AChE is secreted into the synaptic cleft and is associated with cholinergic axon terminal membranes. However, AChE is also manufactured by some noncholinergic neurons, so this enzyme is not as useful a marker for cholinergic synapses as ChAT.

AChE degrades ACh into choline and acetic acid (Figure 6.10b). Much of the resulting choline is taken up by the cholinergic axon terminal and reused for ACh synthesis. In Chapter 5, we mentioned that AChE is the target of many nerve gases and insecticides. Inhibition of AChE prevents the breakdown of ACh, disrupting transmission at cholinergic synapses on skeletal muscle and heart muscle. Acute effects include marked decreases in heart rate and blood pressure; however, death from the irreversible inhibition of AChE is typically a result of respiratory paralysis.

Catecholaminergic Neurons

The amino acid tyrosine is the precursor for three different amine neurotransmitters that contain a chemical structure called a *catechol* (Figure 6.11a). These neurotransmitters are collectively called **catecholamines**. The

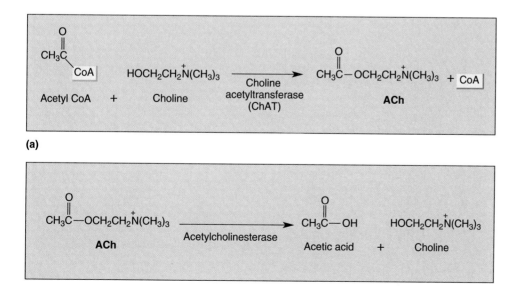

(a)

(b)

Figure 6.10
(a) The synthesis and **(b)** degradation of ACh.

catecholamine neurotransmitters are **dopamine** (**DA**), **norepinephrine** (**NE**), and **epinephrine** (Figure 6.11b). Catecholaminergic neurons are found in regions of the nervous system involved in the regulation of movement, mood, attention, and visceral function (discussed further in Chapter 15).

All catecholaminergic neurons contain the enzyme *tyrosine hydroxylase* (*TH*), which catalyzes the first step in catecholamine synthesis, the conversion of tyrosine to a compound called **dopa** (L-dihydroxyphenylalanine) (Figure 6.12a). The activity of TH is rate-limiting for catecholamine synthesis. The enzyme's activity is regulated by various signals in the cytoplasm of the axon terminal. For example, decreased catecholamine release by the axon terminal causes the catecholamine concentration in the cytosol to rise, thereby inhibiting TH. This type of regulation is called *end-product inhibition*. On the other hand, during periods when catecholamines are released at a high rate, the elevation in $[Ca^{2+}]_i$ that accompanies neurotransmitter release triggers an increase in the activity of TH, so transmitter supply keeps up with demand. In addition, prolonged periods of stimulation actually cause the synthesis of more mRNA for the enzyme.

(a) Catechol

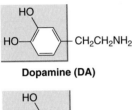

Dopamine (DA)

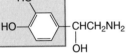

Norepinephrine (NE)

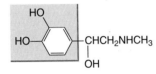

Epinephrine

(b) Catecholamines

Figure 6.11
(a) A catechol group and **(b)** the catecholamine neurotransmitters.

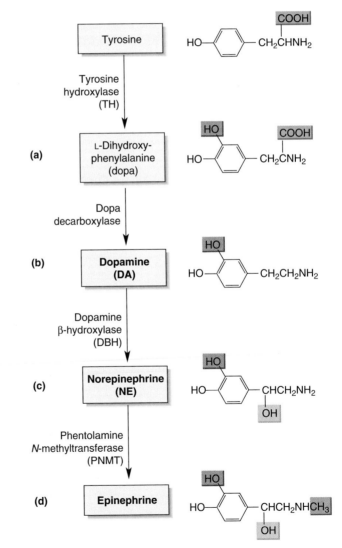

Figure 6.12
The synthesis of catecholamines from tyrosine. The catecholamine neurotransmitters are in boldface type.

Dopa is converted into the neurotransmitter dopamine by the enzyme *dopa decarboxylase* (Figure 6.12b). Dopa decarboxylase is abundant in catecholaminergic neurons, so the amount of dopamine synthesized primarily depends on the amount of dopa available. In a movement disorder called *Parkinson's disease*, dopaminergic neurons in the brain slowly degenerate and eventually die. A strategy for treating Parkinson's disease is the administration of dopa, which causes an increase in DA synthesis in the surviving neurons, increasing the amount of DA available for release. (We will learn more about dopamine and movement in Chapter 14.)

Neurons that use NE as a neurotransmitter contain, in addition to TH and dopa decarboxylase, the enzyme *dopamine ß-hydroxylase (DBH)*, which converts dopamine to norepinephrine (Figure 6.12c). It is interesting to note that DBH is not found in the cytosol, but instead is located within the synaptic vesicles. Thus, in noradrenergic axon terminals, DA is transported from the cytosol to the synaptic vesicles, and there it is made into NE.

The last in the line of catecholamine neurotransmitters is epinephrine, also called **adrenaline**. Adrenergic neurons contain the enzyme *phentolamine N-methyltransferase (PNMT)* that converts NE to epinephrine (Figure 6.12d). Curiously, PNMT is in the cytosol of adrenergic axon terminals. Thus, NE must first be synthesized in the vesicles, released into the cytosol for conversion into epinephrine, and then the epinephrine must again be transported into vesicles for release. In addition to serving as a neurotransmitter in the brain, epinephrine is released by the adrenal gland into the bloodstream. As we shall see in Chapter 15, circulating epinephrine acts at receptors throughout the body to produce a coordinated visceral response.

The actions of catecholamines in the synaptic cleft are terminated by selective uptake of the neurotransmitters back into the axon terminal. This step is sensitive to a number of different drugs. For example, amphetamine and cocaine block catecholamine uptake, therefore prolonging the actions of the neurotransmitter in the cleft. Once inside the axon terminal, the catecholamines may be reloaded into synaptic vesicles for reuse, or they may be enzymatically destroyed by the action of *monoamine oxidase (MAO)*, an enzyme found on the outer membrane of mitochondria.

Serotonergic Neurons

The amine neurotransmitter **serotonin**, also called *5-hydroxytryptamine* and abbreviated **5-HT**, is derived from the amino acid tryptophan. Serotonergic neurons are relatively few in number, but, as we shall see in Part III, they appear to play an important role in the brain systems that regulate mood, emotional behavior, and sleep.

Synthesis of serotonin occurs in two steps, just like the synthesis of dopamine (Figure 6.13). Tryptophan is converted first into an intermediary called 5-HTP (5-hydroxytryptophan) by the enzyme *tryptophan hydroxylase*. The 5-HTP is then converted to 5-HT by the enzyme *5-HTP decarboxylase*. Serotonin synthesis appears to be limited by the availability of tryptophan in the extracellular fluid bathing neurons. The source of brain tryptophan is the blood, and the source of blood tryptophan is the diet. Thus, a deficiency of tryptophan in the diet can quickly lead to the depletion of serotonin in the brain.

Following release from the axon terminal, 5-HT is removed from the synaptic cleft by the action of a specific transporter. The process of serotonin reuptake, like catecholamine reuptake, is sensitive to a number of different drugs. For example, several clinically useful antidepressant drugs, including fluoxetine (trade name Prozac), are selective inhibitors of serotonin reuptake. Once it is back in the cytosol of the serotonergic axon terminal, the transmitter is either reloaded into synaptic vesicles or degraded by MAO.

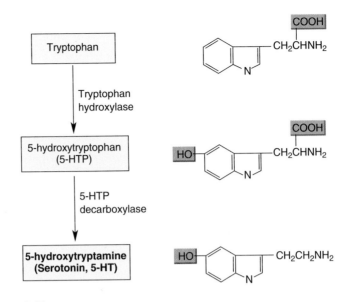

Figure 6.13
The synthesis of serotonin from tryptophan.

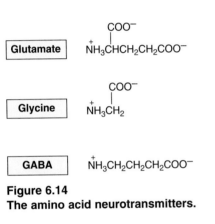

Figure 6.14
The amino acid neurotransmitters.

Amino Acidergic Neurons

The amino acids **glutamate (Glu)**, **glycine (Gly)**, and **gamma-aminobutyric acid (GABA)** serve as neurotransmitters at most CNS synapses (Figure 6.14). Of these, only GABA is unique to those neurons that use it as a neurotransmitter; the others are among the 20 amino acids that make up proteins.

Glutamate and glycine are synthesized from glucose and other precursors using enzymes that exist in all cells. Differences among neurons in the synthesis of these amino acids are therefore quantitative rather than qualitative. For example, the average glutamate concentration in the cytosol of glutaminergic axon terminals has been estimated to be about 20 mM, two or three times higher than that in nonglutaminergic cells. The more important distinction between glutaminergic and nonglutaminergic neurons, however, is the transporter that loads the synaptic vesicles. In glutaminergic axon terminals, but not in other types, the glutamate transporter concentrates glutamate until it reaches a value of about 50 mM in the synaptic vesicles.

Because GABA is not one of the 20 amino acids used to construct proteins, it is synthesized only by those neurons that use it as a neurotransmitter. The precursor for GABA is glutamate, and the key synthetic enzyme is *glutamic acid decarboxylase* (*GAD*) (Figure 6.15). GAD, therefore, is a good marker for GABAergic neurons. Immunocytochemical studies have shown that GABAergic neurons are distributed widely in the brain. GABAergic neurons are the major source of synaptic inhibition in the nervous system.

The synaptic actions of the amino acid neurotransmitters are terminated by selective uptake into the presynaptic terminals and glia. Inside the terminal or glial cell, GABA is metabolized by the enzyme *GABA transaminase*.

Other Neurotransmitter Candidates and Intercellular Messengers

In addition to the amines and amino acids, a few other small molecules may serve as chemical messengers between neurons. For instance, researchers have recently focused on ATP, a key molecule in cellular metabolism, as a possible neurotransmitter. ATP is concentrated in vesicles at many synapses in the CNS and PNS, and it is released into the cleft in a Ca^{2+}-dependent manner. ATP directly excites some neurons by gating a cation channel. In this sense, some of the neurotransmitter functions of ATP may be similar to that of glutamate.

The newest and most exotic chemical messenger to be proposed for intercellular communication is actually a gaseous molecule, **nitric oxide (NO)**. Carbon monoxide (CO) has also been suggested as a messenger, although evidence thus far is meager. These are the same nitric oxide and carbon monoxide that are major air pollutants from internal combustion engines. NO is synthesized from the amino acid arginine by many cells of the body, and it has powerful biological effects. In the nervous system, NO may have unique functions. It seems to be released from postsynaptic neurons, by nonvesicular means, and acts on presynaptic terminals. Communication in this direction, from "post" to "pre," is called *retrograde communication*; thus, NO is said to be a **retrograde messenger**. Because NO is small and membrane-permeable, it can diffuse much more freely than other transmitter molecules, even penetrating through one cell to affect another beyond it. Its influence may spread throughout a small region of local tissue, rather than being confined to the site of the cells that released them. On the other hand, NO is evanescent and breaks down very rapidly. The functions of gaseous transmitters are being extensively studied and hotly debated.

Before leaving the topic of neurotransmitter chemistry, we wish to point out, once again, that many of the chemicals we call neurotransmitters may also be present in high concentrations in non-neural parts of the body. A chemical may serve dual purposes, mediating communication in the nervous system but something entirely different elsewhere. Amino acids, of course, are used to make proteins throughout the body. ATP is the energy source for all cells. Nitric oxide is released from endothelial cells and causes the smooth muscle of blood vessels to relax. (One consequence in males is penile erection.) The cells with the highest levels of ACh are not in the brain but in the cornea of the eye, where there are no ACh receptors. Likewise, the highest serotonin levels are not in neurons, but in blood platelets. These observations underscore the importance of rigorous analysis before a chemical is assigned a neurotransmitter role.

The operation of a neurotransmitter system is like a play with two acts. Act I is presynaptic and culminates in the transient elevation of neurotransmitter concentration in the synaptic cleft. We are now ready to move on to Act II, the generation of electrical and biochemical signals in the postsynaptic neuron. The main players are transmitter-gated channels and G-protein-coupled receptors.

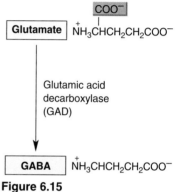

Figure 6.15
The synthesis of GABA from glutamate.

TRANSMITTER-GATED CHANNELS

In Chapter 5, we learned that ACh and the amino acid neurotransmitters mediate fast synaptic transmission by acting on transmitter-gated ion channels. These channels are magnificent minuscule machines. A single channel can be a sensitive detector of chemicals and voltage, it can regulate the flow of surprisingly large currents with great precision, it can sift and select between very similar ions, and it can be regulated by other receptor systems. Yet each channel is only about 11 nm long, just barely visible with the best computer-enhanced electron microscopic methods. Although the secrets of these channels are now being revealed, we still have much to learn.

Basic Structure of Transmitter-Gated Channels

The most thoroughly studied transmitter-gated ion channel is the nicotinic ACh receptor in skeletal muscle. It is a pentamer, an amalgam of five protein subunits arranged like the staves of a barrel to form a single pore through the membrane. Four different polypeptides are used as subunits for the nicotinic receptor, and they are designated α, β, γ, and δ. A complete mature channel is made from two α units, and one each of β, γ, and δ (abbreviated $\alpha_2\beta\gamma\delta$). There is one ACh binding site on each of the α subunits; the simultaneous binding of ACh to both sites is required for the channel to open (Figure 6.16).

Although each subunit has a different primary structure, there are stretches where the different polypeptide chains have a similar sequence of amino acids. For example, each subunit polypeptide has four separate segments that will coil into alpha helices (see Figure 3.6). Because the amino acid residues of these segments are hydrophobic, the four alpha helices are believed to be where the polypeptide is threaded back and forth across the membrane.

The primary structures of the subunits of many other transmitter-gated channels in the brain have now been deduced, and there are obvious similarities (Figure 6.17). All contain the four membrane-spanning segments (although subunits of glutamate-gated channels may have an extra one), while keeping them in about the same places. Most of the channels are thought to be pentameric complexes. Perhaps more interesting than the similarities, however, are the variations among channel structures. Different transmitter binding sites let one channel respond to Glu while another responds to GABA; certain amino acids around the narrow ion pore allow only Na^+ and K^+ to flow through some channels, Ca^{2+} through others, and only Cl^- through yet others.

Amino Acid-Gated Channels

Amino acid-gated channels mediate most of the fast synaptic transmission in the CNS. Let's take a closer look at their functions, because they are central to topics as disparate as sensory systems, memory, and disease. Several properties of these channels distinguish them from one another and define their functions within the brain.

- The *pharmacology* of their binding sites describes which transmitters affect them and how drugs interact with them.

- The *kinetics* of the transmitter binding process and channel gating determine the duration of their effect.

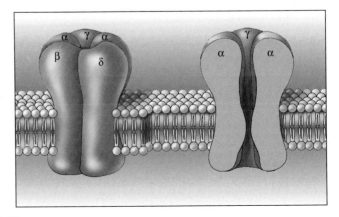

Figure 6.16
Subunit arrangement of the nicotinic ACh receptor.

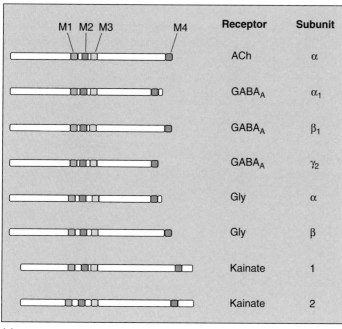

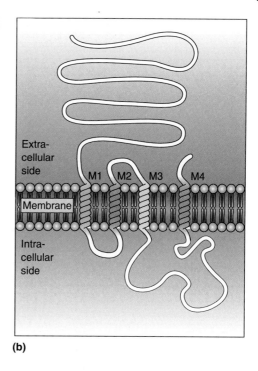

(a) **(b)**

Figure 6.17
Similarities in the structure of subunits for different transmitter-gated ion channels. (a) If the polypeptides for various channel subunits were stretched out in a line, this is how they would compare. They have in common the four regions called M1–M4, which are segments where the polypeptides will coil into alpha helices to span the membrane. **(b)** M1–M4 regions of the ACh α subunit as they are believed to be threaded through the membrane.

- The *selectivity* of the ion channels determines whether they produce excitation or inhibition, and whether Ca^{2+} enters the cell in significant amounts.

- The *conductance* of the open channels helps determine the magnitude of their effects.

All these properties are a direct result of the molecular structure of the channels.

Glutamate-Gated Channels. As we discussed previously, three glutamate receptor subtypes bear the names of their selective agonists: AMPA, NMDA, and kainate. Each of these is a glutamate-gated ion channel. The AMPA-gated and NMDA-gated channels mediate the bulk of fast excitatory synaptic transmission in the brain. Kainate receptors also exist throughout the brain, but their functions are poorly understood.

AMPA-gated channels are permeable to both Na^+ and K^+. The net effect of activating them at normal, negative membrane potentials is to admit Na^+ ions into the cell, causing a rapid and large depolarization. Thus, AMPA receptors at CNS synapses mediate excitatory transmission in much the same way as nicotinic receptors mediate synaptic excitation at neuromuscular junctions.

AMPA receptors coexist with NMDA receptors at many synapses in the brain, so most glutamate-mediated EPSPs have components contributed by both (Figure 6.18). NMDA-gated channels also cause excitation of a cell by admitting Na^+, but they differ from AMPA receptors in two very important ways: (1) NMDA-gated channels are permeable to Ca^{2+}, and (2) inward ionic current through NMDA-gated channels is voltage-dependent. We'll discuss each of these properties, in turn.

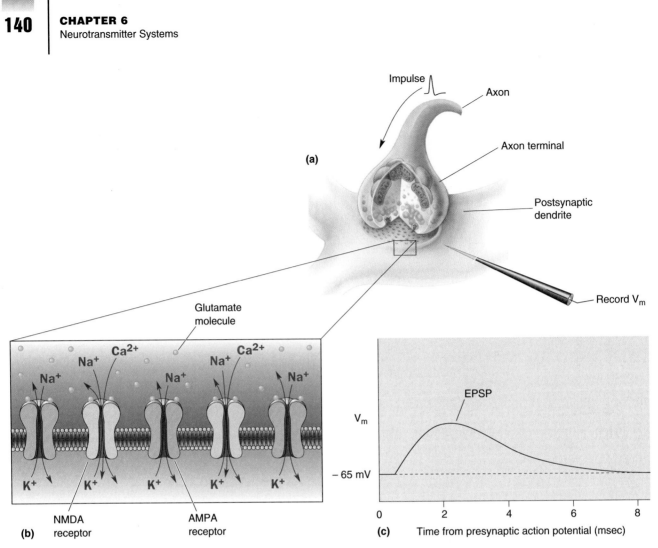

Figure 6.18
Coexistence of NMDA and AMPA receptors in the postsynaptic membrane of a CNS synapse. (a) An impulse arriving in the presynaptic terminal causes the release of glutamate. **(b)** Glutamate binds to AMPA-gated and NMDA-gated channels in the postsynaptic membrane. **(c)** Entry of Na+ through the AMPA channels, and Na+ and Ca2+ through the NMDA channels, causes an EPSP.

It is hard to overstate the importance of intracellular Ca^{2+} to cell functions. We have already seen that Ca^{2+} can trigger presynaptic neurotransmitter release. Postsynaptically, Ca^{2+} can activate many enzymes, regulate the opening of a variety of channels, and affect gene expression; in excessive amounts, Ca^{2+} can even trigger the death of a cell (Box 6.2). Thus, activation of NMDA receptors can, in principle, cause widespread and lasting changes in the postsynaptic neuron. Indeed, as we will see in Chapter 20, Ca^{2+} entry through NMDA-gated channels may cause the changes that lead to long-term memory.

When the NMDA-gated channel opens, Ca^{2+} and Na^+ enter the cell (and K^+ leaves), but the magnitude of this inward ionic current depends on the postsynaptic membrane potential in an unusual way, for an unusual reason. When glutamate binds to the NMDA receptor, the pore opens as usual. However, at normal negative resting membrane potentials, the channel becomes clogged by Mg^{2+} ions, and the "magnesium block" prevents other ions from passing freely into the cell. Mg^{2+} pops out of the pore only when the membrane is depolarized, which usually follows activation of AMPA channels at the same and neighboring synapses. Thus, inward ionic current through the NMDA channel is *voltage-dependent*, in addition to being transmitter-gated. Both glutamate and depolarization must coincide before the channel will pass current (Figure 6.19). This property has a significant impact on synaptic integration at many locations in the CNS.

GABA-Gated and Glycine-Gated Channels. GABA mediates most synaptic inhibition in the CNS, and glycine mediates most of the rest. Both the GABA$_A$ receptor and the glycine receptor gate a Cl$^-$ channel. Surprisingly, inhibitory GABA$_A$ and glycine receptors have a structure very similar to that of excitatory nicotinic ACh receptors, despite the fact that the first two are selective for anions while the last is selective for cations. Both the glycine-gated channel and the ACh-gated channel receptor have α subunits that bind the transmitter and β subunits that do not. GABA can bind to either the α or the β subunit of its receptor.

Synaptic inhibition must be tightly regulated in the brain. Too much causes loss of consciousness and coma; too little leads to a seizure. The need

---

Box 6.2 | **OF SPECIAL INTEREST**

The Brain's Exciting Poisons

Neurons of the brain do not regenerate, so each dead neuron is one less we have to think with. One of the fascinating ironies of neuronal life and death is that glutamate (Glu), the most essential neurotransmitter in the brain, is also one of the biggest killers of neurons. A large percentage of the brain's synapses release glutamate, and it is stored in large quantities. Even the cytosol of nonglutaminergic neurons has a very high glutamate concentration, greater than 3 mM. An ominous observation is that when you apply this same amount of glutamate to isolated neurons, they die within minutes.

The voracious metabolic rate of the brain demands a continuous supply of oxygen and glucose. If blood flow ceases, as in cardiac arrest, neural activity will stop within seconds, and permanent damage will result within a few minutes. Disease states such as cardiac arrest, stroke, brain trauma, seizures, and oxygen deficiency can initiate a vicious cycle of excess glutamate release. Whenever neurons cannot generate enough ATP to keep their ion pumps working hard, membranes depolarize, and Ca^{2+} leaks into cells. Entry of Ca^{2+} triggers the synaptic release of glutamate. Glutamate further depolarizes neurons, which further raises intracellular Ca^{2+} and causes still more glutamate to be released. When glutamate reaches high concentrations, it kills neurons by overexciting them, a process called *excitotoxicity*. Glutamate simply activates its several types of receptors, which allow excessive amounts of Na$^+$, K$^+$, and

Ca^{2+} to flow across the membrane. The NMDA subtype of glutamate-gated channel is a critical player in excitotoxicity, because it is the main route for Ca^{2+} entry. Neuron damage or death occurs because of swelling due to water uptake and the stimulation by Ca^{2+} of intracellular enzymes that degrade proteins, lipids, and nucleic acids. Neurons literally digest themselves.

Excitotoxicity has recently been implicated in several progressive neurodegenerative human diseases such as amyotrophic lateral sclerosis (ALS, also known as Lou Gehrig's disease), in which spinal motor neurons slowly die, and Alzheimer's disease, in which brain neurons slowly die. The effects of various environmental toxins mimic aspects of these diseases. Eating large quantities of a certain type of chick pea can cause lathyrism, a degeneration of motor neurons. The pea contains an excitotoxin called β-oxalylaminoalanine, which activates glutamate receptors. A toxin called domoic acid, found in contaminated mussels, is also a glutamate receptor agonist. Ingesting small amounts of domoic acid causes seizures and brain damage. And another plant excitotoxin, β-methylaminoalanine, may cause a hideous condition that combines signs of ALS, Alzheimer's disease, and Parkinson's disease in individual patients on the island of Guam.

As research sorts out the tangled web of excitotoxins, receptors, enzymes, and neurological disease, new strategies for treatment emerge. Already, glutamate receptor antagonists that can obstruct these excitotoxic cascades, and minimize neuronal suicide, show clinical promise. Genetic manipulations may eventually thwart neurodegenerative conditions in susceptible people.

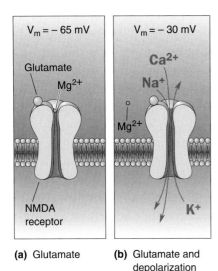

(a) Glutamate

(b) Glutamate and depolarization

Figure 6.19
Inward ionic current through the NMDA-gated channel. (a) Glutamate alone causes the channel to open, but at the resting membrane potential, the pore becomes blocked by Mg^{2+} ions. **(b)** Depolarization of the membrane relieves the magnesium block and allows Na^+ and Ca^{2+} to enter.

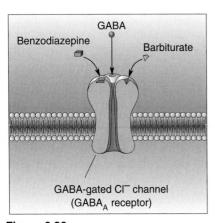

Figure 6.20
Binding of drugs to the $GABA_A$ receptor.

to control it may explain why the $GABA_A$ receptor has, in addition to its GABA binding site, several other sites where chemicals can dramatically modulate its function. For example, two classes of drugs, **benzodiazepines** (one of which is the tranquilizer diazepam, or Valium) and **barbiturates** (such as phenobarbital), bind to their own specific site on the outside face of the $GABA_A$ channel (Figure 6.20). By themselves these drugs do very little to the channel. But when GABA is present, benzodiazepines increase the frequency of channel openings, while barbiturates increase the duration of channel openings. The result in each case is more inhibitory Cl^- current, stronger IPSPs, and the behavioral consequences of enhanced inhibition.

But surely the $GABA_A$ receptor did not evolve specialized binding sites just for the benefit of our modern drugs. This sort of logic has motivated researchers to find endogenous ligands, natural chemicals that bind to benzodiazepine and barbiturate sites and serve as regulators of inhibition. There is now substantial evidence that such ligands exist.

G-PROTEIN-COUPLED RECEPTORS AND EFFECTORS

There are multiple subtypes of G-protein-coupled receptor in every known neurotransmitter system. In Chapter 5, we learned that transmission at these receptors involves three steps: (1) binding of the neurotransmitter to the receptor protein, (2) activation of G-proteins, and (3) activation of effector systems. Let's focus on each of these steps.

Basic Structure of G-Protein-Coupled Receptors

Most G-protein-coupled receptors are simple variations on a common plan, consisting of a single polypeptide containing seven membrane-spanning alpha helices (Figure 6.21). Two of the extracellular loops of the polypeptide form the transmitter binding sites. Structural variations in this region determine which neurotransmitters, agonists, and antagonists bind to the receptor. Two of the intracellular loops can bind to and activate G-proteins. Structural variations here determine which G-proteins and, consequently, which effector systems are activated in response to transmitter binding.

A partial list of G-protein-coupled receptors appears in Table 6.2. About 100 such receptors have been described. Most of these were unknown only a few years ago, before the powerful methods of molecular neurobiology were applied to the problem.

The Ubiquitous G-Proteins

G-proteins are the common link in most signaling pathways that start with a neurotransmitter receptor and end with effector proteins. G-protein is short for guanosine triphosphate (GTP)-binding protein, which is actually a diverse family of about 20 types. There are many more transmitter receptors than G-proteins, so some types of G-proteins can be activated by many receptors.

G-proteins all have the same basic mode of operation (Figure 6.22):

1. Each G-protein has three subunits, termed α, β, and γ. In the resting state, a guanosine diphosphate (GDP) molecule is bound to the G_α subunit, and the whole complex floats around on the inner surface of the membrane.

2. If this GDP-bound G-protein bumps into the proper type of receptor *and* that receptor has a transmitter molecule bound to it, the G-protein releases its GDP and exchanges it for a GTP that it picks up from the cytosol.

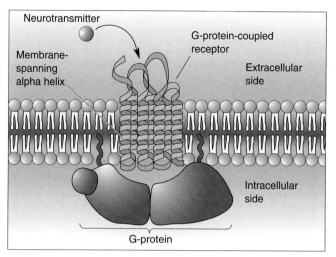

Figure 6.21
The basic structure of a G-protein-coupled receptor. Most metabotropic receptors have seven membrane-spanning alpha helices, a transmitter binding site on the extracellular side, and a G-protein binding site on the intracellular side.

Table 6.2
Some G-Protein-Coupled Neurotransmitter Receptors

Neurotransmitter	Receptor(s)
Acetylcholine (ACh)	Muscarinic receptors (M_1, M_2, M_3, M_4, M_5)
Glutamate (Glu)	Metabotropic glutamate receptors (mGluR1–7)
GABA	$GABA_B$ receptor
Serotonin (5-HT)	$5\text{-}HT_{1(A, B, C, D\alpha, D\beta, E, F)}$ $5\text{-}HT_{2, 2F}$, $5\text{-}HT_4$, $5\text{-}HT_{5\alpha, 5\beta}$
Dopamine (DA)	D_1, $D_{2S, 2I}$, D_3, D_4, D_5
Norepinephrine (NE)	α_1, α_2, β_1, β_2, β_3
Enkephalin	μ, δ, κ

3. The activated GTP-bound G-protein splits into two parts: the G_α subunit plus GTP, which can go on to stimulate various effector proteins, and the $G_{\beta\gamma}$ complex.

4. The G_α subunit is itself an enzyme that breaks down GTP into GDP. Therefore, G_α eventually terminates its own activity by converting the bound GTP to GDP.

5. The G_α and $G_{\beta\gamma}$ subunits come back together, allowing the cycle to begin again.

The first G-proteins that were discovered had the effect of stimulating effector proteins. Subsequently, it was found that other G-proteins could inhibit these same effectors. Thus, the simplest scheme for subdividing the G-proteins is G_S, designating that the G-protein is stimulatory, and G_i, designating that the G-protein is inhibitory.

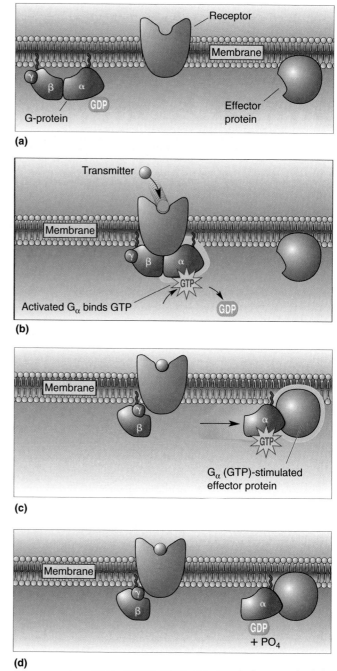

Figure 6.22
Basic mode of operation of G-proteins. (a) In its inactive state, the α subunit of the G-protein binds GDP. **(b)** When activated by a G-protein-coupled receptor, the GDP is exchanged for GTP. **(c)** The activated G-protein splits, and the G_α (GTP) subunit activates effector proteins. **(d)** The G_α subunit slowly removes phosphate (PO_4) from GTP, converting GTP to GDP and terminating its own activity.

G-Protein-Coupled Effector Systems

In Chapter 5, we learned that activated G-proteins exert their effects via two types of effector proteins: G-protein-gated ion channels and G-protein-activated enzymes. Because the effects do not involve any other chemical intermediaries, the first route is sometimes called the shortcut pathway.

The Shortcut Pathway. A variety of neurotransmitters use the shortcut pathway, from receptor to G-protein to ion channel. One example is the muscarinic receptors in the heart. These ACh receptors are coupled via G-proteins to potassium channels, explaining why ACh slows the heart rate (Figure 6.23). Another example is neuronal GABA$_B$ receptors, also coupled by the shortcut pathway to potassium channels.

Shortcut pathways are relatively fast, with responses beginning within 30–100 msec of neurotransmitter binding. Although not quite as fast as a transmitter-gated channel, which uses no intermediary between receptor and channel, this is faster than the second messenger cascades we will describe next. The shortcut pathway is also very localized, compared with other effector systems. As the G-protein diffuses within the membrane, it apparently cannot move very far, so only channels nearby can be affected.

Second Messenger Cascades. G-proteins can also exert their effects by directly activating certain enzymes. Activation of these enzymes can trigger an elaborate series of biochemical reactions, a cascade that often ends in the activation of "downstream" enzymes that alter neuronal function. Between the first enzyme and the last are several chemical intermediaries often collectively called *second messengers*. The whole process that couples the neurotransmitter, via multiple steps, to activation of a downstream enzyme is called a **second messenger cascade** (Figure 6.24).

In Chapter 5, we introduced the cAMP second messenger cascade initiated by activation of the NE β receptor (Figure 6.25a). It begins with the β receptor activating the stimulatory G-protein, G$_S$, which proceeds to stimulate the membrane-bound enzyme adenylyl cyclase. Adenylyl cyclase converts ATP to cAMP. The subsequent rise of cAMP in the cytosol activates a specific downstream enzyme called **protein kinase A (PKA)**.

Many biochemical processes are regulated with a push-pull method, one to stimulate them and one to inhibit them, and cAMP production is no

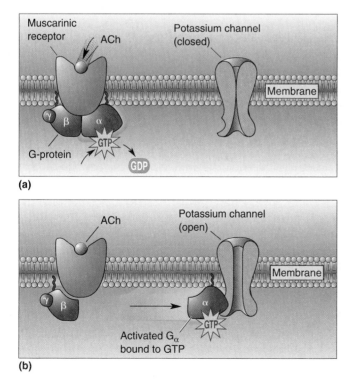

(a)

(b)

Figure 6.23
The shortcut pathway. (a) G-proteins in heart muscle are activated by ACh binding to muscarinic receptors. **(b)** The activated G-protein directly gates a potassium channel.

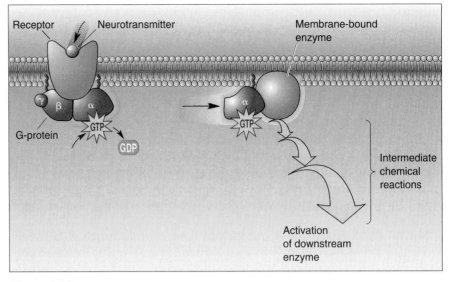

Figure 6.24
Components of a second messenger cascade.

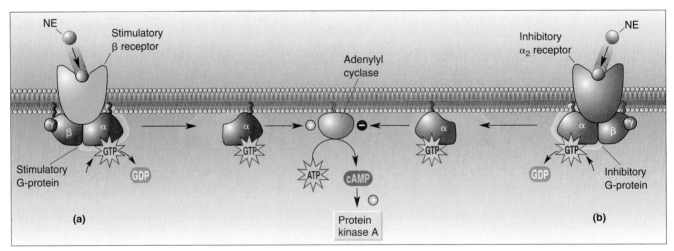

Figure 6.25
Stimulation and inhibition of adenylyl cyclase by different G-proteins. (a) Binding of NE to the ß receptor activates G_s, which in turn activates adenylyl cyclase. Adenylyl cyclase generates cAMP, which activates the downstream enzyme protein kinase A. **(b)** Binding of NE to the α_2 receptor activates G_i, which inhibits adenylyl cyclase.

exception. Activation of a second type of NE receptor, called the α_2 receptor, leads to activation of G_i (the inhibitory G-protein). G_i suppresses the activity of adenylyl cyclase, and this effect can take precedence over the stimulatory system (Figure 6.25b).

Some messenger cascades can branch. Figure 6.26 shows how activation of various G-proteins can stimulate **phospholipase C (PLC)**, an enzyme that floats in the membrane like adenylyl cyclase. PLC acts on a membrane phospholipid (PIP$_2$, or phosphatidylinositol-4,5-bisphosphate), splitting it to form the two molecules that serve as second messengers: **diacylglycerol (DAG)** and **inositol-1,4,5-triphosphate (IP$_3$)**. DAG, which is lipid-soluble, stays within the plane of the membrane where it activates a downstream enzyme, **protein kinase C (PKC)**. At the same time, the water-soluble IP$_3$ diffuses away in the cytosol and binds to specific receptors on the smooth ER and other membrane-enclosed organelles in the cell. These receptors are

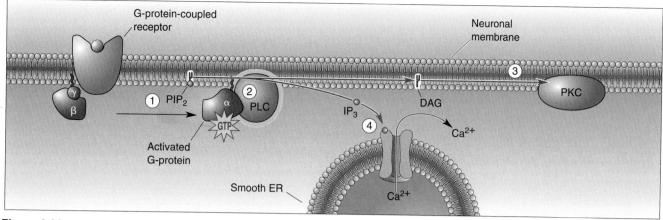

Figure 6.26
Second messengers generated by the breakdown of PIP$_2$, a membrane phospholipid. (1) Activated G-proteins stimulate the enzyme phospholipase C (PLC). (2) PLC splits PIP$_2$ into DAG and IP$_3$. (3) DAG stimulates the downstream enzyme protein kinase C (PKC). (4) IP$_3$ stimulates the release of Ca^{2+} from intracellular stores. The Ca^{2+} can go on to stimulate various downstream enzymes.

IP$_3$-gated calcium channels, so IP$_3$ causes the organelles to discharge their stores of Ca^{2+}. As we have said, elevations in cytoplasmic Ca^{2+} can trigger widespread and long-lasting effects. One effect is activation of the enzyme **calcium-calmodulin-dependent protein kinase**, or **CaMK**.

Phosphorylation and Dephosphorylation. The preceding examples show that key downstream enzymes in many of the second messenger cascades are *protein kinases* (PKA, PKC, CaMK). Protein kinases transfer phosphate from ATP floating in the cytosol to proteins, a reaction called *phosphorylation.* The addition of phosphate groups to a protein changes its conformation slightly, thereby changing its biological activity. Phosphorylation of ion channels, for example, can strongly influence the probability that they will open or close.

Consider the consequence of activating β receptors on cardiac muscle cells. The subsequent rise in cAMP activates PKA, which phosphorylates the cell's voltage-gated calcium channels, and this *enhances* their activity. More Ca^{2+} flows, and the heart beats more strongly. By contrast, stimulation of β adrenergic receptors in many neurons seems to have no effect on calcium channels, but instead causes *inhibition* of certain potassium channels. Reduced K$^+$ conductance causes a slight depolarization, increases the length constant, and makes the neuron more excitable (see Chapter 5).

If transmitter-stimulated kinases were allowed to phosphorylate without some method of reversing the process, all proteins would quickly become saturated with phosphates, and further regulation would become impossible. Enzymes called **protein phosphatases** save the day, because they act rapidly to remove phosphate groups. The degree of channel phosphorylation at any moment therefore depends on the dynamic balance of phosphorylation by kinases and dephosphorylation by phosphatases (Figure 6.27).

The Function of Signal Cascades. Synaptic transmission using transmitter-gated channels is simple and fast. Transmission involving G-protein-coupled receptors is complex and slow. What are the advantages of having such long chains of command? One important advantage is *signal amplification*: Activation of one G protein-coupled receptor can lead to the activation of not one, but many, ion channels (Figure 6.28).

Signal amplification can occur at several places in the cascade. A single neurotransmitter molecule, bound to one receptor, can activate perhaps 10–20 G-proteins; each G-protein can activate an adenylyl cyclase, which can make

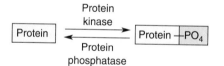

Figure 6.27
Protein phosphorylation and dephosphorylation.

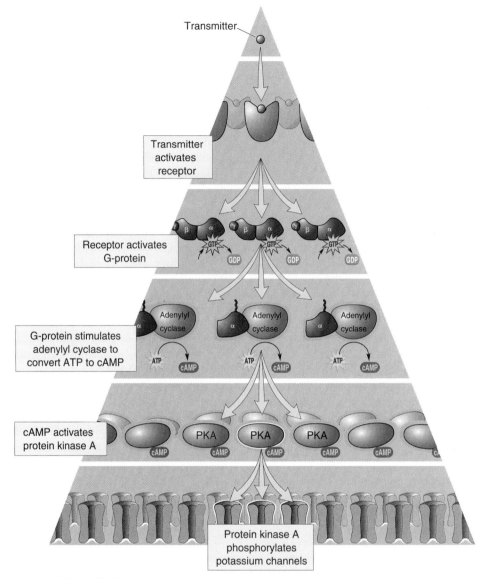

Figure 6.28
Signal amplification by G-protein-coupled second messenger cascades. When a transmitter activates a G-protein-coupled receptor, there can be amplification of the messengers at several stages of the cascade, so that ultimately many channels are affected.

many cAMP molecules that can spread to activate many kinases; each kinase can then phosphorylate many channels. If all cascade components were tied together in a clump, signaling would be severely limited. The use of small messengers that can diffuse quickly (such as cAMP) also allows signaling at a distance, over a wide stretch of cell membrane. Signal cascades also provide many sites for further regulation, as well as interaction between cascades. Finally, signal cascades can generate very long-lasting chemical changes in cells, which may form the basis for, among other things, a lifetime of memories.

DIVERGENCE AND CONVERGENCE IN NEUROTRANSMITTER SYSTEMS

Glutamate is the most common excitatory neurotransmitter in the brain, while GABA is the pervasive inhibitory neurotransmitter. But this is only part of the story, because any single neurotransmitter can have many dif-

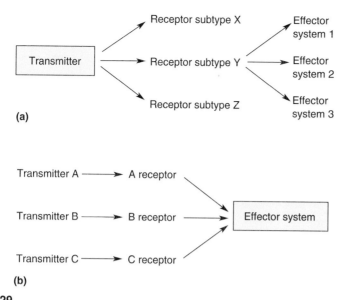

Figure 6.29
Principles of (a) divergence and (b) convergence in neurotransmitter systems.

ferent effects. A molecule of glutamate can bind to any of several kinds of glutamate receptors, and each of these can mediate a different effect. The ability of one transmitter to activate more than one subtype of receptor, and cause more than one type of postsynaptic response, is called *divergence*.

Divergence is the rule among neurotransmitter systems. Every known neurotransmitter can activate multiple receptor subtypes, and recent experience indicates that the number of receptors will continue to escalate as the powerful methods of molecular neurobiology are applied to each system. Because of the multiple receptor subtypes, one transmitter can affect different neurons (or even different parts of the same neuron) in very different ways. Divergence also occurs at points beyond the receptor level, depending on which G-proteins and which effector systems are activated. Divergence may occur at any stage in the cascade of transmitter effects (Figure 6.29a).

Neurotransmitters can also exhibit *convergence* of effects. Multiple transmitters, each activating their own receptor type, can converge to affect the same effector systems (Figure 6.29b). Convergence in a single cell can occur at the level of the G-protein, the second messenger cascade, or the type of ion channel.

CONCLUDING REMARKS

Neurotransmitters are the essential links between neurons, and between neurons and other effector cells such as muscle cells and glandular cells. But it is important to think of transmitters as one link in a chain of events, inciting chemical changes both fast and slow, divergent and convergent. You can envision the many signaling pathways onto and within a single neuron as a kind of information network. This network is in delicate balance, shifting its effects dynamically as the demands on a neuron vary with changes in the organism's behavior.

The signaling network within a single neuron resembles in some ways the neural networks of the brain itself. It receives a variety of inputs, in the form of transmitters bombarding it at different times and places. These inputs cause an increased drive through some signal pathways and a decreased drive through others, and the information is recombined to yield a particular output that is more than a simple summation of the inputs. Signals regulate signals, chemical changes can leave lasting traces of their history, drugs can shift the balance of signaling power—and, in a literal sense, the brain and its chemicals are one.

KEY TERMS

Studying Neurotransmitter Systems

immunocytochemistry

in situ hybridization

autoradiography

microionophoresis

receptor subtype

nicotinic receptor

muscarinic receptor

AMPA receptor

NMDA receptor

kainate receptor

ligand-binding method

Neurotransmitter Chemistry

Dale's principle

rate-limiting step

catecholamines

dopamine (DA)

norepinephrine (NE)

epinephrine

dopa

adrenaline

serotonin (5-HT)

glutamate (Glu)

glycine (Gly)

gamma-aminobutyric acid (GABA)

nitric oxide (NO)

retrograde messenger

Transmitter-Gated Channels

benzodiazepine

barbiturate

G-Protein-Coupled Receptors and Effectors

second messenger cascade

protein kinase A (PKA)

phospholipase C (PLC)

diacylglycerol (DAG)

inositol-1,4,5-triphosphate (IP_3)

protein kinase C (PKC)

calcium-calmodulin-dependent protein kinase (CaMK)

protein phosphatase

✔ REVIEW QUESTIONS

1. If you could place microelectrodes into both a presynaptic and a postsynaptic neuron, how would you determine whether the synapse between them was chemically or electrically mediated?

2. List the criteria that are used to determine if a chemical serves as a neurotransmitter. What are the various experimental strategies you could use to show that ACh fulfills the criteria of a neurotransmitter at the neuromuscular junction?

3. What are three methods that could be used to show that a neurotransmitter receptor is synthesized or localized in a particular neuron?

4. Compare and contrast the properties of (a) AMPA and NMDA receptors, and (b) $GABA_A$ and $GABA_B$ receptors.

5. Synaptic inhibition is an important feature of the circuitry in the cerebral cortex. How would you determine whether GABA or Gly, or both, or neither, is the inhibitory neurotransmitter of the cortex?

6. Glutamate activates a number of different metabotropic receptors. The consequence of activating one subtype is the *inhibition* of cAMP formation. A consequence of activating a second subtype is *activation* of protein kinase C. Propose mechanisms for these different effects.

7. Do convergence and divergence of neurotransmitter effects occur in single neurons?

8. Ca^{2+} ions are considered to be second messengers. Why?

Structure of the Nervous System

Now that we have seen how individual neurons function and communicate, we are ready to assemble them into a nervous system that sees, hears, feels, moves, remembers, and dreams. Just as an understanding of neuronal structure was necessary to understand neuronal function, we must understand nervous system structure in order to understand brain function.

Neuroanatomy has challenged generations of students, and for good reason: The human brain is extremely complicated. However, our brain is merely a variation on a plan that is common to the brains of all mammals (Figure 7.1). The human brain appears complicated because it is distorted (twisted, literally) as a result of the selective growth of some parts within the confines of the skull, and because we stand on two legs instead of four. But once the basic mammalian plan is understood, these specializations of the human brain become transparent.

We begin by introducing the general organization of the mammalian brain and the terms used to describe it. Then we take a look at how the three-dimensional structure of the brain arises during embryological and fetal development. Following the course of development makes it easier to understand how the parts of the adult brain fit together. Finally, we explore the cerebral neocortex, a structure that is unique to mammals and most highly evolved in humans.

The neuroanatomy presented in this chapter will provide the canvas on which we will paint the sensory and motor systems in Chapters 8–14. Because you will be introduced to a lot of new terms, there are occasional breaks in the chapter to provide an opportunity for review.

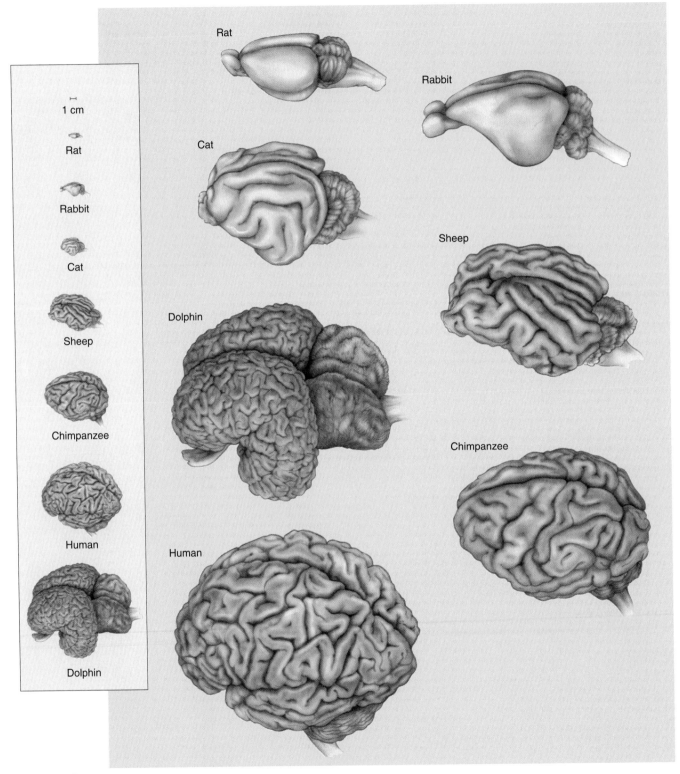

Figure 7.1
Mammalian brains. Despite differences in complexity, the brains of all these species have many features in common. The brains have been drawn to appear approximately the same size; their relative sizes are shown in the inset to the left.

GROSS ORGANIZATION OF THE MAMMALIAN NERVOUS SYSTEM

The nervous system of all mammals has two divisions: the central nervous system (CNS) and the peripheral nervous system (PNS). In this section, we identify some of the important components of the CNS and the PNS. We will also discuss the fluid-filled ventricles within the brain and the membranes that surround the brain. But first, we need to review some anatomical terminology.

Anatomical References

Getting to know your way around the brain is like getting to know your way around a city. To describe your location in the city, you would use points of reference such as north, south, east, and west, up and down. The same is true for the brain, only the terms—called *anatomical references*—are different.

Consider the nervous system of a rat (Figure 7.2a). In the head lies the brain, and the spinal cord runs down inside the backbone toward the tail. The direction pointing toward the rat's nose is called **anterior** or **rostral** (from the Latin for "beak"). The direction pointing toward the rat's tail is called **posterior** or **caudal** (from the Latin for "tail"). The direction pointing up is called **dorsal** (from the Latin for "back") and the direction pointing down is called **ventral** (from the Latin word for "belly"). Thus, the rat spinal cord runs anterior to posterior. The top side of the spinal cord is the dorsal side, and the bottom side is the ventral side.

If we look down on the nervous system, we see that it may be divided into two equal halves (Figure 7.2b). The right side of the brain and spinal cord is the mirror image of the left side. This characteristic is known as *bilateral symmetry*. With just a few exceptions, most structures within the nervous system come in pairs, one on the right side and the other on the left. The line running down the middle of the nervous system is called the **mid-**

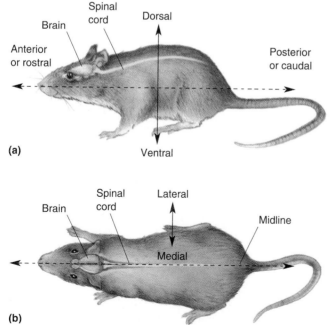

(a)

(b)

Figure 7.2
Basic anatomical references in the nervous system of a rat. (a) Side view. **(b)** Top view.

line, and this gives us another way to describe direction. Structures closer to the midline are **medial**; structures farther away from the midline are **lateral**. In other words, the nose is medial to the eyes, the eyes are medial to the ears, and so on. In addition, two structures that are on the same side are said to be **ipsilateral** to each other; for example, the right ear is ipsilateral to the right eye. If the structures are on opposite sides of the midline, they are said to be **contralateral** to each other; the right ear is contralateral to the left ear.

To view the internal structure of the brain, it is usually necessary to slice it up. In the language of anatomists, a slice is called a *section*; to slice is *to section*. Although one could imagine an infinite number of ways we might cut into the brain, the standard approach is to make cuts parallel to one of the three *anatomical planes of section*. The plane of the section resulting from splitting the brain into equal right and left halves is called the **midsagittal plane** (Figure 7.3a). Sections parallel to the midsagittal plane are in the **sagittal plane**.

The two other anatomical planes are perpendicular to the sagittal plane and to one another. The **horizontal plane** is parallel to the ground (Figure 7.3b). A single section in this plane could pass through both the eyes and the ears. Thus, horizontal sections split the brain into dorsal and ventral parts. The **coronal plane** is perpendicular to the ground and to the sagittal plane (Figure 7.3c). A single section in this plane could pass through both eyes or both ears, but not through all four at the same time. Thus, the coronal plane splits the brain into anterior and posterior parts.

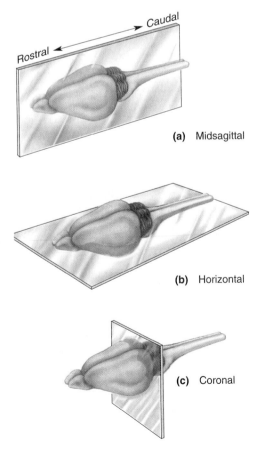

(a) Midsagittal

(b) Horizontal

(c) Coronal

Figure 7.3
Anatomical planes of section.

Take a few moments right now and be sure you understand the meaning of these terms:

anterior   ipsilateral
rostral   contralateral
posterior   midsagittal plane
caudal   sagittal plane
dorsal   horizontal plane
ventral   coronal plane
midline
medial
lateral

The Central Nervous System

The **central nervous system**, or **CNS**, consists of the parts of the nervous system that are encased in bone: the **brain** and the **spinal cord**. The brain lies entirely within the skull. A side view of the rat brain reveals three parts that are common to all mammals: the cerebrum, the cerebellum, and the brain stem (Figure 7.4a).

Cerebrum. The rostral-most and largest part of the brain is the **cerebrum**. Figure 7.4b shows the rat cerebrum as it appears when viewed from above. Notice that it is clearly split down the middle into two **cerebral hemispheres**, separated by the deep *sagittal fissure*. In general, the *right* cerebral hemisphere receives sensations from, and controls movements of, the *left*

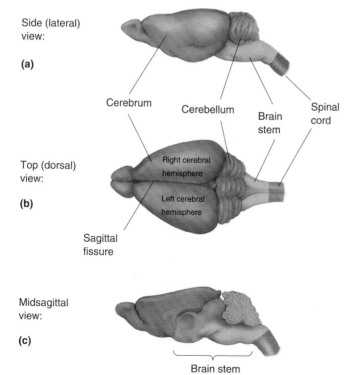

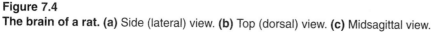

Figure 7.4
The brain of a rat. (a) Side (lateral) view. **(b)** Top (dorsal) view. **(c)** Midsagittal view.

side of the body. Similarly, the *left* cerebral hemisphere is concerned with sensations and movements on the *right* side of the body.

Cerebellum. Lying behind the cerebrum is the **cerebellum**. The word "cerebellum" is derived from the Latin for "little brain." While the cerebellum is in fact dwarfed by the large cerebrum, it actually contains as many neurons as both cerebral hemispheres combined. The cerebellum is primarily a movement control center having extensive connections with the cerebrum and the spinal cord. In contrast to the cerebral hemispheres, the left cerebellar hemisphere is concerned with movements of the left side of the body, and the right cerebellar hemisphere is concerned with movements of the right side.

Brain Stem. The remaining part of the brain is the brain stem, best observed in a midsagittal view of the brain (Figure 7.4c). The **brain stem** forms the stalk from which the cerebral hemispheres and the cerebellum sprout. The brain stem is a complex nexus of fibers and cells that in part serves to relay information from the cerebrum to the spinal cord and cerebellum, and vice versa. However, the brain stem is also the site where vital functions are regulated, such as breathing, consciousness, and the control of body temperature. Indeed, while the brain stem is considered the most primitive part of the mammalian brain, it is also the most important to life. One can survive damage to the cerebrum and cerebellum, but damage to the brain stem usually means rapid death.

Spinal Cord. The spinal cord is encased in the bony vertebral column and is attached to the brain stem. The spinal cord is the major conduit of information from the skin, joints, and muscles of the body to the brain, and vice versa. A transection of the spinal cord results in anesthesia (lack of feeling) in the skin and paralysis of the muscles in parts of the body caudal to the cut. Paralysis in this case does not mean that the muscles cannot function, rather that they cannot be controlled by the brain.

The spinal cord communicates with the body via the **spinal nerves**, which are part of the peripheral nervous system (discussed below). Spinal nerves exit the spinal cord through notches between each vertebra of the vertebral column. Each spinal nerve attaches to the spinal cord by means of two branches, the **dorsal root** and the **ventral root** (Figure 7.5). Recall from Chapter 1 that François Magendie showed that the dorsal root contains axons that bring information *into* the spinal cord, for example, those that signal the accidental entry of a thumbtack into your foot (see Figure 3.1). Charles Bell showed that the ventral root contains axons that carry information *away from* the spinal cord, for example, to the muscles that withdraw your foot in response to the pain of the thumbtack.

The Peripheral Nervous System

All the parts of the nervous system other than the brain and spinal cord comprise the **peripheral nervous system**, or **PNS**. The PNS is divided into two parts: the somatic PNS and the visceral PNS.

Somatic PNS. All the spinal nerves that innervate the skin, the joints, and the muscles that are under voluntary control are part of the **somatic PNS**. The somatic motor axons, which command muscle contraction, derive from motor neurons in the ventral spinal cord. The cell bodies of the motor neurons lie within the CNS, but their axons are mostly in the PNS.

The somatic sensory axons, which innervate and collect information from the skin, muscles, and joints, enter the spinal cord via the dorsal roots. The somata of these neurons lie outside the spinal cord in clusters called **dorsal root ganglia**. There is a dorsal root ganglion for each spinal nerve (Figure 7.5).

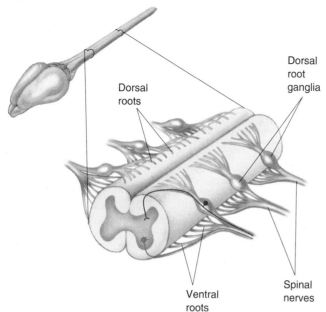

Figure 7.5
The spinal cord. The spinal cord runs inside the vertebral column. Axons enter and exit the spinal cord via the dorsal and ventral roots, respectively. These roots come together to form the spinal nerves that course through the body.

Visceral PNS. The **visceral PNS**, also called the involuntary, vegetative, or **autonomic nervous system** (**ANS**), consists of the neurons that innervate the internal organs, blood vessels, and glands. Visceral sensory axons bring information about visceral function to the CNS, such as the pressure and oxygen content of the blood in the arteries. Visceral motor fibers command the contraction and relaxation of muscles that form the walls of the intestines and the blood vessels (called smooth muscles), the rate of cardiac muscle contraction, and the secretory function of various glands. For example, the visceral PNS controls blood pressure by regulating the diameter of the blood vessels and the heart rate.

We will return to the structure and function of the ANS in Chapter 15. For now, remember that when one speaks of an emotional reaction that is beyond voluntary control—like "butterflies in the stomach" or blushing—it usually is mediated by the visceral PNS (the ANS).

The Cranial Nerves

In addition to the nerves that arise from the spinal cord and innervate the body, there are 12 pairs of **cranial nerves** that arise from the brain stem and innervate (mostly) the head. As shown in Table 7.1, each cranial nerve has a name and a number associated with it (originally numbered by Galen, about 1800 years ago, from anterior to posterior).

Some of the cranial nerves are part of the CNS, others are part of the somatic PNS, and still others are part of the visceral PNS. For example, cranial nerve II—the optic nerve—is part of the CNS. It carries visual information from the retina to the brain stem. Cranial nerve V—the trigeminal nerve—is part of the somatic PNS. It carries somatic sensory (touch) information from the skin of the face to the brain stem. Cranial nerve X—the vagus nerve—is part of the visceral PNS. The vagus nerve descends within the neck to innervate many internal organs of the chest and abdomen. ("Vagus" is from the Latin for "wandering.") Recall from Box 5.1 that Otto

Table 7.1
The Cranial Nerves

Number	Name	Main Function
I	Olfactory nerve	Special sensory (smell)
II	Optic nerve	Special sensory (vision)
III	Oculomotor nerve	Somatic motor
IV	Trochlear nerve	Somatic motor
V	Trigeminal nerve	Somatic sensory and motor
VI	Abducens nerve	Somatic motor
VII	Facial nerve	Somatic motor and special sensory (taste)
VIII	Auditory vestibular nerve	Special sensory (hearing and balance)
IX	Glossopharyngeal nerve	Somatic sensory and motor
X	Vagus nerve	Visceral sensory and motor
XI	Spinal accessory nerve	Somatic motor
XII	Hypoglossal nerve	Somatic motor

Loewi used the vagal innervation of the heart to show that ACh is a neurotransmitter.

The Meninges

The CNS, that part of the nervous system encased in the skull and vertebral column, does not come in direct contact with the overlying bone. It is protected by three membranes collectively called the **meninges** (singular: meninx), from the Greek for "covering." The three membranes are the dura mater, the arachnoid membrane, and the pia mater (Figure 7.6).

The outermost covering is the **dura mater**, from the Latin words meaning "hard mother," an accurate description of the dura's leatherlike consistency. The dura forms a tough, inelastic bag that surrounds the brain and spinal cord. Just under the dura lies the **arachnoid membrane**. This meningeal layer has an appearance and a consistency resembling a spider web ("arachnoid" comes from the Greek word for "spider"). While there normally is no space between the dura and the arachnoid, if the blood vessels passing through the dura are ruptured, blood can collect here and form what is called a *subdural hematoma*. The buildup of fluid in this subdural space can disrupt brain function by compressing parts of the CNS. The disorder is treated by drilling a hole in the skull and draining the blood.

The **pia mater**, the "gentle mother," is a thin membrane that adheres closely to the surface of the brain. Along the pia run many blood vessels that ultimately dive into the substance of the underlying brain. The pia is separated from the arachnoid by a fluid-filled space. This *subarachnoid space* is filled with salty clear liquid called **cerebrospinal fluid**, or **CSF**. Thus, in a sense, the brain floats inside the head in this thin layer of CSF.

The Ventricular System

In Chapter 1, we noted that the brain is hollow. The fluid-filled caverns and canals inside the brain constitute the **ventricular system**. The fluid that runs in this system is CSF, the same as the fluid in the subarachnoid space. CSF is produced by a special tissue, called the choroid plexus, in the ventricles of the cerebral hemispheres. CSF flows from the paired ventricles of the cerebrum to a series of connected, unpaired cavities at the core of the brain stem (Figure 7.7). CSF exits the ventricular system and enters the subarachnoid space by way of small openings, or apertures, located near where the cerebellum attaches to the brain stem. In the subarachnoid space, CSF is

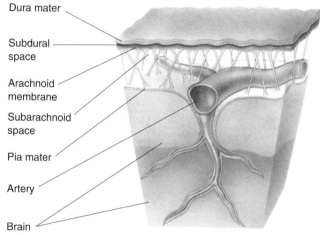

Figure 7.6
The meninges. The three meningeal layers protecting the brain and spinal cord are the dura mater, the arachnoid membrane, and the pia mater.

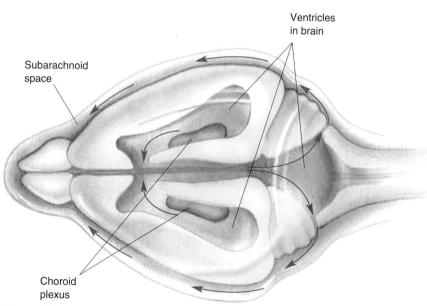

Figure 7.7
The ventricular system. CSF is produced in the ventricles of the paired cerebral hemispheres and flows through a series of unpaired ventricles at the core of the brain stem. CSF escapes into the subarachnoid space via small apertures near the base of the cerebellum. In the subarachnoid space, CSF is absorbed into the blood.

absorbed by the blood vessels at special structures called arachnoid villi. If the normal flow of CSF is disrupted, brain damage can result (Box 7.1).

We will return to fill in some details about the ventricular system in a moment. As we will see, understanding the organization of the ventricular system holds the key to understanding how the mammalian brain is organized.

☑ REVIEW

Take a few moments right now and be sure you understand the meaning of these terms:

central nervous system (CNS)   somatic PNS
brain   dorsal root ganglia
spinal cord   visceral PNS
cerebrum   autonomic nervous system (ANS)
cerebral hemispheres   cranial nerve
cerebellum   meninges
brain stem   dura mater
spinal nerve   arachnoid membrane
dorsal root   pia mater
ventral root   cerebrospinal fluid (CSF)
peripheral nervous system   ventricular system
  (PNS)

UNDERSTANDING CNS STRUCTURE THROUGH DEVELOPMENT

The entire CNS is derived from the walls of a fluid-filled tube that is formed at an early stage of embryological development. The tube itself becomes the adult ventricular system. Thus, by examining how this tube changes during the course of fetal development, we can understand how the brain is organized and how the different parts fit together. In this section, we focus on development as a way to understand the structural organization of the brain. In Chapter 18, we will revisit the topic of development to see how neurons are born, how they find their way to their final locations

---

Box 7.1 | **OF SPECIAL INTEREST**

Water on the Brain

If the flow of CSF from the choroid plexus through the ventricular system to the subarachnoid space is impaired, the fluid will back up and cause a swelling of the ventricles. This condition is called *hydrocephalus*, meaning "water head."

Occasionally babies are born with hydrocephalus. However, because the skull is soft and not completely formed, the head will expand in response to the increased intracranial fluid, sparing the brain from damage. Often this condition goes unnoticed until the size of the head reaches enormous proportions.

In adults, hydrocephalus is a much more serious situation because the skull cannot expand, and intracranial pressure increases as a result. The soft brain tissue is then compressed, impairing function and leading to death if left untreated. Typically, this "obstructive" hydrocephalus is also accompanied by severe headache, caused by the distention of nerve endings in the meninges. Treatment consists of inserting a tube into the swollen ventricle and draining off the excess fluid.

in the CNS, and how they make the appropriate synaptic connections with one another.

As you work your way through this section, and through the rest of the book, you will encounter many different names used by anatomists to refer to groups of related neurons and axons. Some common names used to describe collections of neurons and axons are given in Tables 7.2 and 7.3, respectively. Take a few moments to familiarize yourself with these new terms before continuing.

Anatomy, by itself, can be pretty dry. It really comes alive only after the functions of different structures are understood. The remainder of this book is devoted to understanding the functional organization of the nervous system. However, we have punctuated this section with a preview of some structure-function relationships to provide you with a general sense for how the different parts contribute, individually and in concert, to the function of the CNS.

Formation of the Neural Tube

The embryo begins as a flat disk with three distinct layers of cells called *endoderm*, *mesoderm*, and *ectoderm*. The endoderm ultimately gives rise to the lining of many of the internal organs (viscera). From the mesoderm arises

Table 7.2
Collections of Neurons

Name	Description/Example
gray matter	A generic term for a collection of neuronal cell bodies in the CNS. When a freshly dissected brain is cut open, neurons appear gray.
cortex	A collection of neurons that forms a thin sheet, usually at the brain's surface. "Cortex" is Latin for "bark." Example: *cerebral cortex*, the sheet of neurons found just under the surface of the cerebrum.
nucleus	A clearly distinguishable mass of neurons, usually deep in the brain (*not* to be confused with the nucleus of a cell). "Nucleus" is from the Latin word for "nut." Example: the *lateral geniculate nucleus*, a cell group in the brain stem that relays information from the eye to the cerebral cortex.
substantia	A group of related neurons deep within the brain, but usually with less distinct borders than those of nuclei. Example: the *substantia nigra* (from the Latin for "black substance"), a brain stem cell group involved in the control of voluntary movement.
locus (plural: loci)	A small, well-defined group of cells. Example: the *locus coeruleus* (Latin for "blue spot"), a brain stem cell group involved in the control of wakefulness and behavioral arousal.
ganglion (plural: ganglia)	A collection of neurons in the PNS. "Ganglion" is from the Greek for "knot." Example: the *dorsal root ganglia*, which contain the cell bodies of sensory axons entering the spinal cord via the dorsal roots. Only one cell group in the CNS goes by this name, the *basal ganglia*, which are structures lying deep within the cerebrum that control movement.

Table 7.3
Collections of Axons

Name	Description/Example
nerve	A bundle of axons in the PNS. Only one collection of CNS axons is called a nerve: the *optic nerve.*
white matter	A generic term for a collection of CNS axons. When a freshly dissected brain is cut open, axons appear white.
tract	A collection of CNS axons having a common site of origin and a common destination. Example: the *corticospinal tract*, which originates in the cerebral cortex and ends in the spinal cord.
bundle	A collection of axons that run together but that do not necessarily have the same origin and destination. Example: the *medial forebrain bundle*, which connects cells scattered within the cerebrum and brain stem.
capsule	A collection of axons that connect the cerebrum with the brain stem. Example: the *internal capsule*, which connects the brain stem with the cerebral cortex.
commissure	Any collection of axons that connect one side of the brain with the other side.
lemniscus	A tract that meanders through the brain like a ribbon. Example: the *medial lemniscus*, which brings touch information from the spinal cord through the brain stem.

the bones of the skeleton and the muscles. *The nervous system and the skin derive entirely from the ectoderm.*

Our focus is on changes in that part of the ectoderm that gives rise to the nervous system, called the *neural plate*. At this early stage (about 3 weeks in the human), the brain consists only of a flat sheet of cells (Figure 7.8a). The next event of interest is the formation of a groove in the neural plate that runs rostral to caudal, called the *neural groove* (Figure 7.8b). The walls of the groove are called *neural folds*, which subsequently move together and fuse dorsally forming the **neural tube** (Figure 7.8c). *The entire central nervous system develops from the walls of the neural tube.* As the neural folds come together, some neural ectoderm is pinched off and comes to lie just lateral to the neural tube. This tissue is called the **neural crest** (Figure 7.8d). *All neurons of the peripheral nervous system derive from the neural crest.*

The neural crest develops in close association with the underlying mesoderm. The mesoderm at this stage in development forms prominent bulges on either side of the neural tube called *somites*. From these somites, the 33 individual vertebrae of the spinal column and the related skeletal muscles will develop. The nerves that innervate the skeletal muscles are therefore called *somatic* motor nerves.

Three Primary Brain Vesicles

The process whereby structures become more elaborate and specialized during development is called **differentiation**. The first step in the differentiation of the brain is the development, at the rostral end of the neural tube, of three swellings called the primary vesicles (Figure 7.9). *The entire brain derives from the three primary vesicles of the neural tube.*

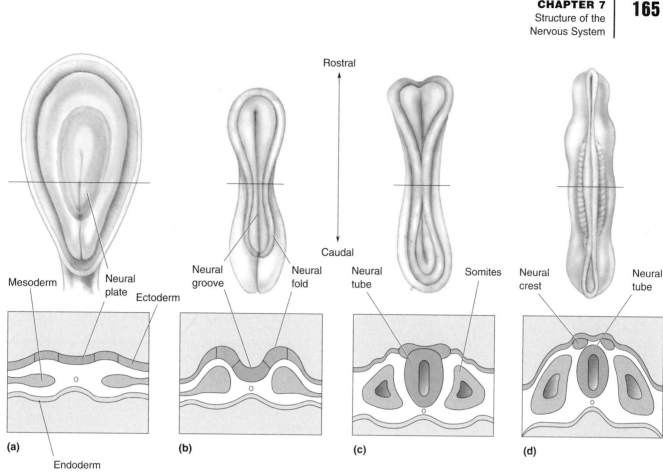

Figure 7.8
Formation of the neural tube and crest. These schematic illustrations follow the early development of the nervous system in the embryo. The drawings above are dorsal views of the embryo; those below are cross-sections. **(a)** The primitive embryonic central nervous system begins as a thin sheet of ectoderm. **(b)** The first important step in the development of the nervous system is the formation of the neural groove. **(c)** The walls of the groove, called neural folds, come together and fuse, forming the neural tube. **(d)** The bits of neural ectoderm that are pinched off when the tube rolls up is called the neural crest, from which the PNS will develop. The somites are mesoderm that will give rise to much of the skeletal system and the muscles.

The rostral-most vesicle is called the *prosencephalon. Pro* is Greek for "before"; "encephalon" is derived from the Greek for "brain." Thus, the prosencephalon is also called the **forebrain**. Behind the prosencephalon lies another vesicle called the *mesencephalon* or **midbrain**. Caudal to this is the third primary vesicle, the *rhombencephalon* or **hindbrain**. The rhombencephalon connects with the caudal neural tube, which gives rise to the spinal cord.

Differentiation of the Forebrain

The next important developments occur in the forebrain, where secondary vesicles sprout off on both sides of the prosencephalon. The secondary vesicles are the *optic vesicles* and the *telencephalic vesicles*. The unpaired structure that remains after the secondary vesicles have sprouted off is called the **diencephalon**, or "between brain" (Figure 7.10). Thus, the forebrain at this stage consists of the two optic vesicles, the two telencephalic vesicles, and the diencephalon.

The optic vesicles grow and invaginate to form the optic stalks and the optic cups, which will ultimately become the *optic nerves* and the two *retinas* in the adult (Figure 7.11). The important point is that the retina at the back of the eye, and the optic nerve connecting the eye to the diencephalon, are part of the brain, not the PNS.

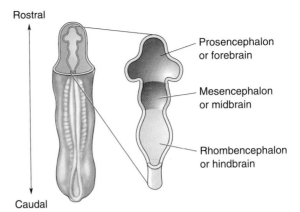

Figure 7.9
The three primary brain vesicles. The rostral end of the neural tube differentiates to form the three vesicles that will give rise to the entire brain. This view is from above, and the vesicles have been cut horizontally so that we can see the inside of the neural tube.

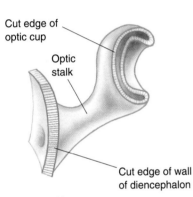

Figure 7.11
Early development of the eye. The optic vesicle differentiates into the optic stalk and the optic cup. The optic stalk will become the optic nerve, and the optic cup will become the retina.

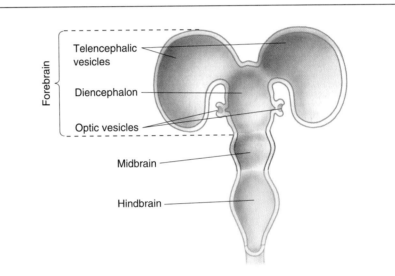

Figure 7.10
The secondary brain vesicles of the forebrain. The forebrain differentiates into the paired telencephalic and optic vesicles, and the diencephalon. The optic vesicles develop into the eyes.

Differentiation of the Telencephalon and Diencephalon. The telencephalic vesicles, together, form the **telencephalon**, or "endbrain," consisting of the two cerebral hemispheres. The telencephalon continues to develop in four ways. (1) The telencephalic vesicles grow posteriorly so that they lie over and lateral to the diencephalon (Figure 7.12a). (2) Another pair of vesicles sprout off the ventral surfaces of the cerebral hemispheres, giving rise to the **olfactory bulbs** and related structures that participate in the sense of smell (Figure 7.12b). (3) The cells of the walls of the telencephalon divide and differentiate into various structures. (4) White matter systems develop, carrying axons to and from the neurons of the telencephalon.

Figure 7.13 shows a coronal section through the primitive mammalian forebrain, to view how the different parts of the telencephalon and diencephalon differentiate and fit together. Notice that the two cerebral hemispheres lie above and on either side of the diencephalon, and that the ventral-medial surfaces of the hemispheres have fused with the lateral surfaces of the diencephalon (Figure 7.13a).

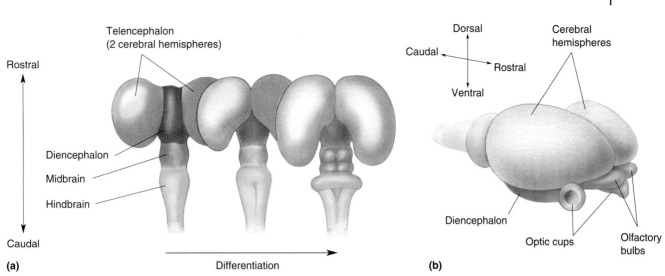

Figure 7.12
Differentiation of the telencephalon. (a) As development proceeds, the cerebral hemispheres swell and grow posteriorly and laterally to envelop the diencephalon. **(b)** The olfactory bulbs sprout off the ventral surfaces of each telencephalic vesicle.

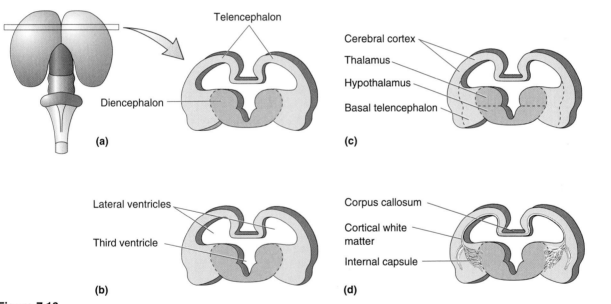

Figure 7.13
Structural features of the forebrain. (a) A coronal section through the primitive forebrain, showing the two main divisions: the telencephalon and the diencephalon. **(b)** Ventricles of the forebrain. **(c)** Gray matter of the forebrain. **(d)** White matter systems of the forebrain.

The fluid-filled spaces within the cerebral hemispheres are called the **lateral ventricles**, and the space at the center of the diencephalon is called the **third ventricle** (Figure 7.13b). The paired lateral ventricles are a key landmark in the adult brain: Whenever you see paired fluid-filled ventricles in a brain section, you know that the tissue surrounding them is in the telencephalon. The elongated, slitlike appearance of the third ventricle in cross section is also a useful feature for identifying the diencephalon.

Notice in Figure 7.13 that the walls of the telencephalic vesicles appear swollen due to the proliferation of neurons. These neurons form two different types of gray matter in the telencephalon: the **cerebral cortex** and the basal telencephalon. Likewise, the diencephalon differentiates into two structures: the **thalamus** and the **hypothalamus** (Figure 7.13c). The thalamus, nestled deep inside the forebrain, gets its name from the Greek word for "inner chamber."

The neurons of the developing forebrain extend axons to communicate with other parts of the nervous system. These axons bundle together to form three major white matter systems: the cortical white matter, the corpus callosum, and the internal capsule (Figure 7.13d). The **cortical white matter** contains all the axons that run to and from the neurons in the cerebral cortex. The **corpus callosum** is continuous with the cortical white matter and forms an axonal bridge that links cortical neurons of the two cerebral hemispheres. The cortical white matter is also continuous with the **internal capsule**, which links the cortex with the brain stem, particularly the thalamus.

Forebrain Structure-Function Relationships. The forebrain is the seat of perceptions, conscious awareness, cognition, and voluntary action. All this depends on extensive interconnections with the sensory and motor neurons of the brain stem and spinal cord.

Arguably the most important structure in the forebrain is the cerebral cortex. As we will see later in this chapter, the cortex is the brain structure that has expanded the most over the course of human evolution. Cortical neurons receive sensory information, form perceptions of the outside world, and command voluntary movements.

Neurons in the olfactory bulbs sense chemicals in the nose (odors) and relay this information caudally to a part of the cerebral cortex for further analysis. Information from the eyes, ears, and skin is also brought to the cerebral cortex for analysis. However, each of the sensory pathways serving vision, audition, and somatic sensation relays (i.e., synapses upon neurons) in the thalamus en route to the cortex. Thus, the thalamus is often referred to as the gateway to the cerebral cortex (Figure 7.14).

Thalamic neurons send axons to the cortex via the internal capsule. As a general rule, the axons of each internal capsule carry information to the cortex about the contralateral side of the body. Therefore, if a thumbtack entered the *right* foot, it would be relayed to the *left* cortex by the *left* thalamus via axons in the *left* internal capsule. But how is the right foot to know what the left foot is doing? One important way is by communication between the hemispheres via the axons in the corpus callosum.

Cortical neurons also send axons through the internal capsule, back to the brain stem. Some cortical axons course all the way to the spinal cord, forming the corticospinal tract. This is one important way cortex can command voluntary movement. Another way is by communicating with neurons in the basal ganglia, a collection of cells in the basal telencephalon. The word "basal" is used to describe structures deep in the brain, and the basal ganglia lie deep within the cerebrum. The functions of the basal ganglia are poorly understood, but it is known that damage to these structures disrupts the ability to initiate voluntary movement. Other structures, contributing to other brain functions, are also present in the basal telencephalon. For example, in Chapter 16 we'll discuss a structure called the amygdala that is involved in fear and emotion.

Although the hypothalamus lies just under the thalamus, functionally it is more closely related to certain telencephalic structures, like the amygdala. The hypothalamus performs many primitive functions and therefore has not changed much over the course of mammalian evolution. "Primitive" does not mean unimportant or uninteresting, however. The hypothalamus controls the visceral (autonomic) nervous system, which regulates bodily functions in response to the needs of the organism. For example, when you are faced with a threatening situation, the hypothalamus orchestrates the

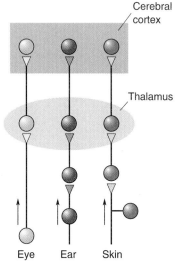

Figure 7.14
The thalamus: gateway to the cerebral cortex. The sensory pathways from the eye, ear, and skin all relay in the thalamus before terminating in the cerebral cortex. The arrows indicate the direction of information flow.

body's visceral fight-or-flight response. Hypothalamic commands to the ANS will lead to (among other things) an increase in the heart rate, increased blood flow to the muscles for escape, and even the standing of hair on end. Conversely, when you're relaxing after Sunday brunch, the hypothalamus ensures that the brain is well-nourished via commands to the ANS, which will increase peristalsis (movement of material through the gastrointestinal tract) and redirect blood to your digestive system. The hypothalamus also plays a key role in motivating animals to find food, drink, and sex in response to their needs. Aside from its connections to the ANS, the hypothalamus also directs bodily responses via connections with the pituitary gland located below the diencephalon. This gland communicates to many parts of the body by releasing hormones into the bloodstream.

REVIEW ✔

Listed below are derivatives of the forebrain that we have discussed. Be sure you know what each of these terms means.

Primary Vesicle	Secondary Vesicle	Some Adult Derivatives
Forebrain (prosencephalon)	Optic vesicle	Retina Optic nerve
	Thalamus (diencephalon)	Dorsal thalamus Hypothalamus Third ventricle
	Telencephalon	Olfactory bulb Cerebral cortex Basal telencephalon Corpus callosum Cortical white matter Internal capsule

Differentiation of the Midbrain

Unlike the forebrain, the midbrain differentiates relatively little during subsequent brain development (Figure 7.15). The dorsal surface of the mesencephalic vesicle becomes a structure called the **tectum** (Latin for "roof"). The floor of the midbrain becomes the **tegmentum**. The fluid-filled space in between constricts into a narrow channel called the **cerebral aqueduct**. The aqueduct connects rostrally with the third ventricle of the diencephalon. Because it is small and circular in cross section, the cerebral aqueduct is a good landmark for identifying the midbrain.

Midbrain Structure-Function Relationships. For such a seemingly simple structure, the functions of the midbrain are remarkably diverse. Besides serving as a conduit for information passing from the spinal cord to the forebrain and vice versa, the midbrain contains neurons that contribute to sensory systems, the control of movement, and several other functions.

The midbrain contains axons descending from the cerebral cortex to the brain stem and the spinal cord. For example, the corticospinal tract courses through the midbrain en route to the spinal cord. Damage to this tract in the midbrain on one side produces a loss of voluntary control of movement on the opposite side of the body.

The tectum differentiates into two structures: the superior colliculus and the inferior colliculus. The superior colliculus receives direct input from the

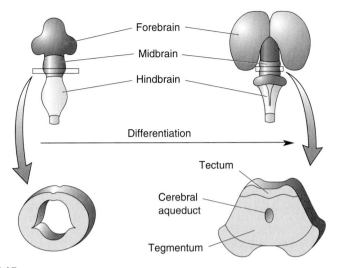

Figure 7.15
Differentiation of the midbrain. The midbrain differentiates into the tectum and the tegmentum. The CSF-filled space at the core of the midbrain is the cerebral aqueduct. (Drawings are not to scale.)

eye, so it is also called the optic tectum. One function of the optic tectum is to control eye movements, which it does via synaptic connections with the motor neurons that innervate the eye muscles. Some of the axons that supply the eye muscles originate in the midbrain, bundling together to form cranial nerves III and IV (see Table 7.1).

The inferior colliculus also receives sensory information, but from the ear instead of the eye. The inferior colliculus serves as an important relay station for auditory information en route to the thalamus.

The tegmentum is one of the most colorful regions of the brain because it contains both the substantia nigra (the black substance) and the red nucleus. These two cell groups are involved in the control of voluntary movement. Other cell groups scattered in the midbrain have axons that project widely throughout much of the CNS and function to regulate consciousness, mood, pleasure, and pain.

Differentiation of the Hindbrain

The hindbrain differentiates into three important structures: the **cerebellum**, the **pons** and the **medulla**. The cerebellum and pons derive from the rostral half of the hindbrain (called the metencephalon); the medulla derives from the caudal half (called the myelencephalon). The fluid-filled tube becomes the **fourth ventricle**, which is continuous with the cerebral aqueduct of the midbrain.

At the three-vesicle stage, the rostral hindbrain in cross section is a simple tube. In subsequent weeks, the tissue along the dorsal-lateral wall of the tube, called the rhombic lip, grows dorsally and medially until it fuses with its twin on the other side. The resulting flap of brain tissue grows into the cerebellum. The ventral wall of the tube differentiates and swells to form the pons (Figure 7.16).

Less dramatic changes occur during the differentiation of the caudal half of the hindbrain into the medulla. The ventral and lateral walls of this region swell, leaving the roof covered only with a thin layer of non-neuronal ependymal cells (Figure 7.17). Along the ventral surface of each side of the medulla runs a major white matter system. Cut in cross section, these bundles of axons appear somewhat triangular in shape, explaining why they are called the *medullary pyramids*.

Hindbrain Structure-Function Relationships. Like the midbrain, the hindbrain is an important conduit for information passing from the forebrain to the spinal cord, and vice versa. In addition, neurons of the hindbrain contribute to the processing of sensory information, control of voluntary movement, and regulation of the autonomic nervous system.

The cerebellum, the "little brain," is an important movement control center. It receives massive axonal inputs from the spinal cord and the pons. The spinal cord inputs provide information about where the body is in space. The inputs from the pons relay information from the cerebral cortex, specifying the goals of intended movements. The cerebellum compares these types of information and calculates the sequences of muscle contractions that are required to achieve the movement goals. Damage to the cerebellum results in uncoordinated and inaccurate movements.

Of the descending axons passing through the midbrain, over 90%—about 20 million axons in the human—synapse on neurons in the pons. The pontine cells relay all this information to the cerebellum on the opposite site. So the pons serves as a massive switchboard connecting the cerebral cortex to the cerebellum. (The word "pons" is from the Latin word for "bridge.") The pons bulges out from the ventral surface of the brain stem to accommodate all this circuitry.

The axons that do not terminate in the pons continue caudally and enter the medullary pyramids. Most of these axons originate in the cerebral cortex and are part of the corticospinal tract. Thus, "pyramidal tract" is often used as a synonym for corticospinal tract. Near where the medulla joins

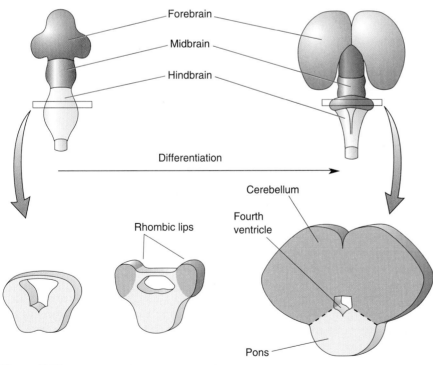

Figure 7.16
Differentiation of the rostral hindbrain. The rostral hindbrain differentiates into the cerebellum and pons. The cerebellum is formed by the growth and fusion of the rhombic lips. The CSF-filled space at the core of the hindbrain is the fourth ventricle. (Drawings are not to scale.)

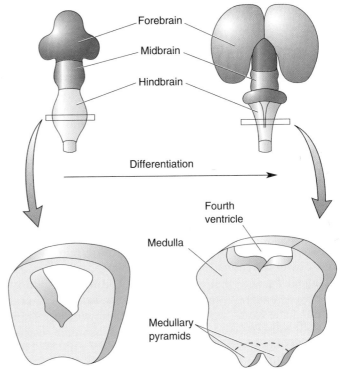

Figure 7.17
Differentiation of the caudal hindbrain. The caudal hindbrain differentiates into
the medulla. The medullary pyramids are bundles of axons coursing caudally toward
the spinal cord. The CSF-filled space at the core of the medulla is the fourth ventricle. (Drawings are not to scale.)

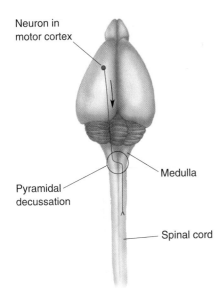

Figure 7.18
The pyramidal decussation. The corticospinal tract crosses from one side
to the other in the medulla.

with the spinal cord, each pyramidal tract crosses from one side of the
midline to the other. A crossing of axons from one side to the other is
called a decussation, and this one is called the pyramidal decussation.
The crossing of axons in the medulla explains why the cortex of one side
of the brain controls movements on the opposite side of the body
(Figure 7.18).

In addition to the white matter systems passing through, the medulla contains neurons that perform many different sensory and motor
functions. For example, the axons of the auditory nerves, bringing
auditory information from the ears, synapse on cells in the cochlear
nuclei of the medulla. The cochlear nuclei project axons to a number
of different structures, including the tectum of the midbrain (inferior
colliculus, discussed above). Damage to the cochlear nuclei leads to
deafness.

Other sensory functions of the medulla include touch and taste. The
medulla contains neurons that relay somatic sensory information from
the spinal cord to the thalamus. Destruction of these cells leads to
anesthesia (loss of feeling). Other neurons relay gustatory information
from the tongue to the thalamus. And among the motor neurons in the
medulla are cells that control the tongue muscles via cranial nerve XII.
(So think of the medulla next time you stick out your tongue!)

Listed below are derivatives of the midbrain and hindbrain that we have discussed. Be sure you know what each of these terms means.

Primary Vesicle	Some Adult Derivatives
Midbrain (mesencephalon)	Tectum
	Tegmentum
	Cerebral aqueduct
Hindbrain (rhombencephalon)	Cerebellum
	Pons
	Fourth ventricle
	Medulla

Differentiation of the Spinal Cord

As shown in Figure 7.19, the transformation of the caudal neural tube into the spinal cord is straightforward compared to the differentiation of the brain. With the expansion of the tissue in the walls, the lumen of the tube constricts to form the tiny **spinal canal**.

Cut in cross section, the gray matter of the spinal cord (where the neurons are) has the appearance of a butterfly. The upper part of the butterfly's

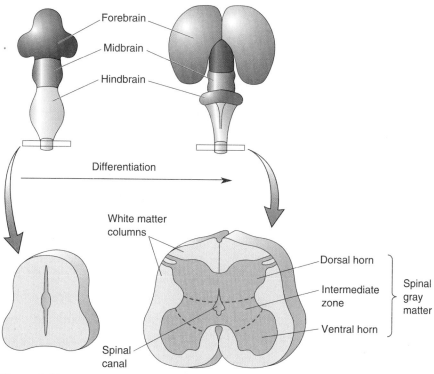

Figure 7.19
Differentiation of the spinal cord. The butterfly-shaped core of the spinal cord is gray matter, divisible into dorsal and ventral horns, and an intermediate zone. Surrounding the gray matter are white matter columns running rostrocaudally, up and down the cord. The narrow CSF-filled channel is the spinal canal. (Drawings are not to scale.)

wing is the **dorsal horn**, and the lower part is the **ventral horn**. The gray matter between the dorsal and ventral horns is called the intermediate zone. Everything else is white matter, consisting of columns of axons that run up and down the spinal cord. Thus, the bundles of axons running along the dorsal surface of the cord are called the *dorsal columns*, the bundles of axons lateral to the spinal gray matter on each side are called the *lateral columns*, and the bundles on the ventral surface are called the *ventral columns*.

Spinal Cord Structure-Function Relationships. As a general rule, dorsal horn cells receive sensory inputs from the dorsal root fibers, ventral horn cells project axons into the ventral roots that innervate muscles, and intermediate zone cells are interneurons that shape motor outputs in response to sensory inputs and descending commands from the brain.

The large dorsal column contains axons that carry somatic sensory (touch) information up the spinal cord toward the brain. It is a superhighway that speeds information from the ipsilateral side of the body up to nuclei in the medulla. The postsynaptic neurons in the medulla give rise to axons that decussate and ascend to the thalamus on the contralateral side. This crossing of axons in the medulla is why touching the left side of the body is sensed by the right side of the brain.

The lateral column contains the axons of the descending corticospinal tract, which also crossed from one side to the other in the medulla. These axons innervate the neurons of the intermediate zone and ventral horn, and communicate the signals that control voluntary movement.

There are at least a half-dozen other tracts that run in the columns of each side of the spinal cord. Most are one-way and bring information to or from the brain. Thus, the spinal cord is the major conduit of information from the skin, joints, and muscles to the brain, and vice versa. However, the spinal cord is also much more than that. The neurons of the spinal gray matter begin the analysis of sensory information, play a critical role in coordinating movements, and orchestrate simple reflexes (like foot withdrawal from a thumbtack).

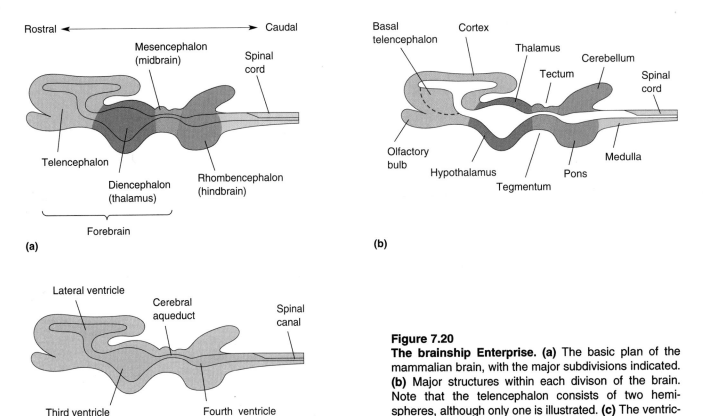

(a)

(b)

(c)

Figure 7.20
The brainship Enterprise. (a) The basic plan of the mammalian brain, with the major subdivisions indicated. **(b)** Major structures within each divison of the brain. Note that the telencephalon consists of two hemispheres, although only one is illustrated. **(c)** The ventricular system.

Putting the Pieces Together

We have discussed the development of different parts of the CNS: the telencephalon, diencephalon, midbrain, hindbrain, and spinal cord. Now let's put all the individual pieces together to make a whole central nervous system.

Figure 7.20 is a highly schematic illustration that captures the basic organizational plan of the CNS of all mammals, including humans. The paired hemispheres of the telencephalon surround the lateral ventricles. Dorsal to the lateral ventricles, at the surface of the brain, lies the cortex. Ventral and lateral to the lateral ventricles, lies the basal telencephalon. The lateral ventricles are continuous with the third ventricle of the diencephalon. Surrounding this ventricle is the thalamus and the hypothalamus. The third ventricle is continuous with the cerebral aqueduct. Dorsal to the aqueduct is the tectum. Ventral to the aqueduct is the midbrain tegmentum. The aqueduct connects with the fourth ventricle that lies at the core of the hindbrain. Dorsal to the fourth ventricle sprouts the cerebellum. Ventral to the fourth ventricle lie the pons and the medulla.

You should see by now that finding your way around the brain is easy if you can identify which parts of the ventricular system are in the neighborhood (Table 7.4). Even in the complicated human brain, the ventricular system holds the key to understanding brain structure.

Special Features of the Human CNS

So far, we've explored the basic plan of the CNS as it applies to all mammals. Figure 7.21 compares the brains of the rat and the human. You can see immediately that there are indeed many similarities, but also some obvious differences.

Let's start by reviewing the similarities. The dorsal view of both brains reveals the paired hemispheres of the telencephalon (Figure 7.21a). A midsagittal view of the two brains shows the telencephalon extending rostrally from the diencephalon (Figure 7.21b). The diencephalon surrounds the third ventricle, the midbrain surrounds the cerebral aqueduct, and the cerebellum, pons, and medulla surround the fourth ventricle. Notice how the pons swells below the cerebellum, and how structurally elaborate the cerebellum is.

Now let's consider some of the structural differences between the rat and human brains. Figure 7.21a reveals a striking one: the many convolutions on the surface of the human cerebrum. The grooves in the surface of the cerebrum are called **sulci** (singular: sulcus), and the bumps are called **gyri** (singular: gyrus). Remember, the thin sheet of neurons that lies just under the surface of the cerebrum is the cerebral cortex. Sulci and gyri result from the tremendous expansion of the surface area of the cerebral

Table 7.4
The Ventricular System of the Brain

Component	Related Brain Structures
Lateral ventricles	Cerebral cortex Basal telencephalon
Third ventricle	Thalamus Hypothalamus
Cerebral aqueduct	Tectum Midbrain tegmentum
Fourth ventricle	Cerebellum Pons Medulla

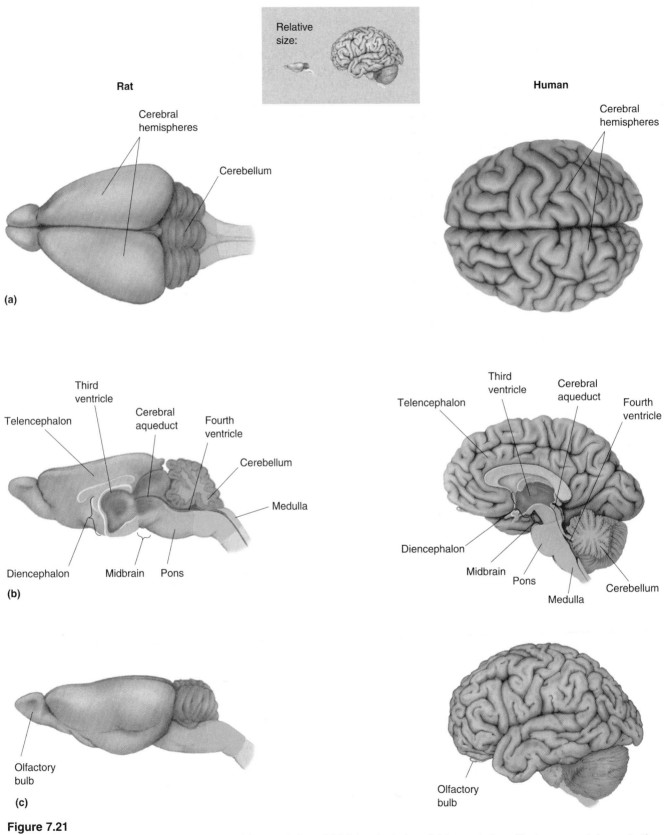

Figure 7.21
The rat brain and human brain compared. (a) Dorsal view. **(b)** Midsagittal view. **(c)** Lateral view. (Brains are not drawn to the same scale.)

cortex during human fetal development. The adult human cortex, measuring about 1.5 ft^2, must fold and wrinkle to fit within the confines of the skull. This increase in cortical surface area is one of the "distortions" of the human brain. Clinical and experimental evidence has implicated the cortex as the seat of uniquely human reasoning and cognition. Without cerebral cortex, a person would be blind, deaf, dumb, and unable to initiate voluntary movement. We will take a closer look at the structure of the cerebral cortex in a moment.

The side views of the rat and human brains in Figure 7.21c reveal further differences in the forebrain. One is the small size of the olfactory bulb in the human relative to the rat. On the other hand, notice again the growth of the cerebral hemisphere in the human. See how the cerebral hemisphere of the human brain arcs posteriorly, ventrolaterally, and then anteriorly to resemble a ram's horn. The tip of the "horn" lies right under the temporal bone (temple) of the skull, so this part of the brain is called the **temporal lobe**. Three other lobes, also named after skull bones, are also used to describe the human cerebrum. The part of the cerebrum lying just under the frontal bone of the forehead is called the **frontal lobe**. The deep **central sulcus** marks the posterior border of the frontal lobe, caudal to which lies the **parietal lobe**, under the parietal bone. Caudal to that, at the back of the cerebrum under the occipital bone, lies the **occipital lobe** (Figure 7.22).

It is important to realize that, despite the disproportionate growth of the cerebrum, the human brain still follows the basic mammalian brain plan laid out during early development. Again, the ventricles are key. Although the ventricular system is distorted, particularly by the growth of the temporal lobes, the relationships that relate the brain to the different ventricles still hold (Figure 7.23).

A GUIDE TO THE CEREBRAL CORTEX

Considering its prominence in the human brain, the cerebral cortex deserves further description. As we will see repeatedly in subsequent chapters, the systems in the brain that are responsible for sensations, perceptions, voluntary movement, learning, speech, and cognition all converge in this remarkable organ.

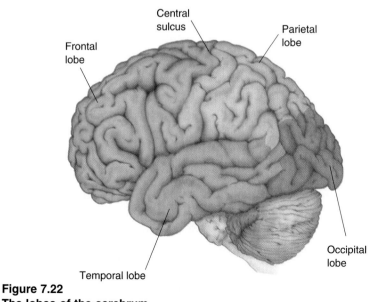

Figure 7.22
The lobes of the cerebrum.

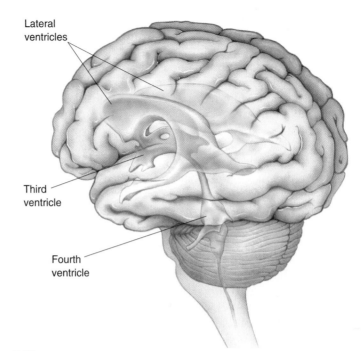

Lateral
ventricles

Third
ventricle

Fourth
ventricle

Figure 7.23
The human ventricular system. Although the ventricles are distorted by the growth
of the brain, the basic relationships of the ventricles to the surrounding brain are the
same as those illustrated in Figure 7.20.

Types of Cerebral Cortex

Cerebral cortex in the brain of all vertebrate animals has several common features, as shown in Figure 7.24. First, the cell bodies of cortical neurons are always arranged in layers, or sheets, that usually lie parallel to the surface of the brain. Second, the layer of neurons closest to the surface (the most superficial cell layer) is separated from the pia by a zone that lacks neurons; it is called the molecular layer, or simply *layer I*. Third, at least one cell layer contains pyramidal cells that emit large dendrites, called *apical dendrites*, that extend up to layer I, where they form multiple branches. Thus, we can say that the cerebral cortex has a characteristic cytoarchitecture that distinguishes it, for example, from the nuclei of the basal telencephalon or the thalamus.

Figure 7.25 shows a Nissl-stained coronal section through the caudal telencephalon of a rat brain. You don't need to be Cajal to see that different types of cortex can also be discerned based on cytoarchitecture. Medial to the lateral ventricle is a piece of cortex that is folded onto itself in a peculiar shape. This structure is called the **hippocampus**, which, despite its bends, has only a single cell layer. ("Hippocampus" is from the Greek word meaning "seahorse.") Connected to the hippocampus ventrally and laterally is another type of cortex that has only two cell layers. It is called the **olfactory cortex**, because it is continuous with the olfactory bulb, which sits further anterior. The olfactory cortex is separated by a sulcus, called the *rhinal fissure*, from another more elaborate type of cortex that has many cell layers. This remaining cortex is called **neocortex**. Unlike the hippocampus and olfactory cortex, *neocortex is found only in mammals*. Thus, when we said previously that the cerebral cortex has expanded over the course of human evolution, we really meant that the neocortex has expanded. Similarly, when we said that the thalamus is the gateway to the cortex, we meant that it is the gateway to the neocortex. Most neuroscientists are such neocortical

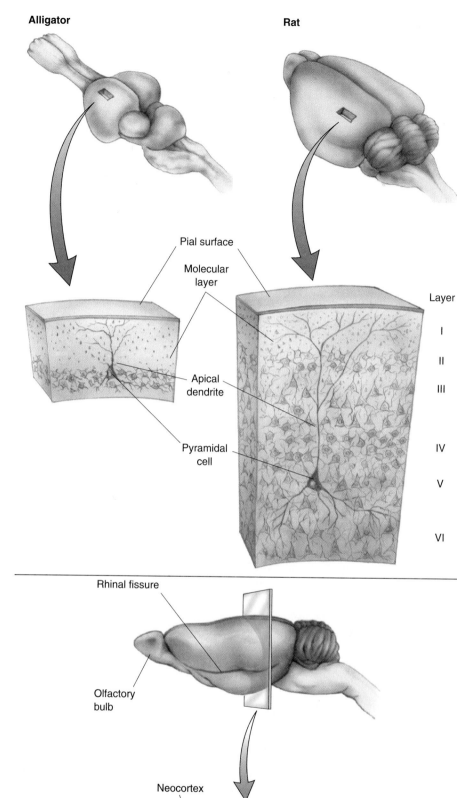

Alligator

Rat

Pial surface

Molecular
layer

Apical
dendrite

Pyramidal
cell

Layer

I

II

III

IV

V

VI

Figure 7.24
General features of cerebral cortex.
On the left is the structure of cortex in an
alligator; on the right, a rat. In both
species, the cortex lies just under the pia
of the cerebral hemisphere, contains a
molecular layer, and has pyramidal cells
arranged in layers.

Rhinal fissure

Olfactory
bulb

Figure 7.25
Three types of cortex in a mammal.
In this section of a rat brain, the lateral
ventricles lie between the neocortex
and the hippocampus on each side.
The ventricles are not obvious
because they are very long and thin in
this region. Below the telencephalon
lies the brain stem. What region of
brain stem is this, based on the
appearance of the fluid-filled space at
its core?

Neocortex

Hippocampus

Lateral
ventricle

Rhinal
fissure

Brain stem

Olfactory
cortex

snobs (ourselves included) that the term cortex, if left unqualified, is usually intended to refer to the cerebral neocortex.

In Chapter 8, we will discuss the olfactory cortex in the context of the sense of smell. Further discussion of the hippocampus is reserved until Part III when we explore its role in the limbic system (Chapter 16) and in memory and learning (Chapters 19 and 20). The neocortex will figure prominently in our discussions of vision, audition, somatic sensation, and the control of voluntary movement in Part II, so let's examine its structure in more detail.

Areas of Neocortex

Just as cytoarchitecture can be used to distinguish cerebral cortex from the basal telencephalon, and the neocortex from the olfactory cortex, it can be used to divide the neocortex up into different zones. This is precisely what the famous German neuroanatomist Korbinian Brodmann did at the beginning of the twentieth century. He constructed a **cytoarchitectural map** of the neocortex (Figure 7.26). In this map, each area of cortex having a common cytoarchitecture is given a number. Thus, we have "area 17" at the tip of the occipital lobe, "area 4" just anterior to the central sulcus in the frontal lobe, and so on.

What Brodmann guessed, but could not show, was that cortical areas that look different perform different functions. We now have evidence that this is true. For instance, we can say that area 17 is visual cortex, because it receives signals from a nucleus of the thalamus that is connected to the retina at the back of the eye. Indeed, without area 17, a human is blind. Similarly, we can say that area 4 is motor cortex, because neurons in this area project axons directly to the motor neurons of the ventral horn that command muscles to contract. Notice that the different functions of these two areas are specified by their different connections.

Neocortex Structure-Function Relationships. Neocortical areas fall into four main categories: primary sensory areas, motor areas, secondary sensory areas, and association areas. The primary sensory areas of cortex are those that are first to receive signals from the ascending sensory pathways. For example, area 17 is *primary* visual cortex, also called V1, because it receives input from the eyes via a direct path, retina to thalamus to cortex. Motor areas of cortex are intimately concerned with the control of voluntary move-

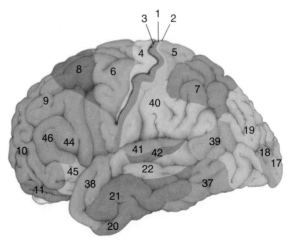

Figure 7.26
Brodmann's cytoarchitectural map of the human cerebral cortex.

ment. These cortical areas receive inputs from thalamic nuclei that relay information from the basal telencephalon and the cerebellum, and they send outputs to motor control neurons in the brain stem and spinal cord. Because cortical area 4 sends outputs directly to motor neurons in the ventral horn of the spinal cord, it is designated as primary motor cortex, or M1.

Figure 7.27 shows views of the brain of a rat, a cat, and a human, with the primary sensory and motor areas identified. It is plain to see that when we speak of the expansion of the cortex in mammalian evolution, what has expanded is the region that lies in between these areas. Recent research reveals that much of this "in-between" cortex is also devoted to the analysis of sensory information. For example, there are secondary sensory areas of cortex in the temporal and parietal lobes that receive their major input from V1 and therefore are "visual."

Even after we have assigned primary sensory, motor, and secondary sensory functions to large regions of cortex, a considerable amount of area remains in the human brain, particularly in the frontal and temporal lobes. These are the association areas of cortex, which appear to be responsible for uniquely human traits like spoken and written language, reasoning, and abstract thinking.

When examining the structure of the human brain, it is natural to wonder whether intellectual ability is correlated with the amount of neural circuitry in the association areas of the cerebral cortex. At the microscopic level, this correlation may exist. Marian Diamond, of the University of California at Berkeley, investigated the microscopic structure of an association area of the parietal lobe in autopsy specimens from several normal human brains and one exceptional one: Albert Einstein's (Figure 7.28). She discovered that Einstein's brain had significantly more glial cells per neuron. Because glia support the dendrites and axons of cortical neurons, the presence of more of them per neuron was interpreted to mean that Einstein's brain had more elaborate cortical circuitry (Box 7.2).

Does this mean we are intellectual hostages to the cortical circuits we have inherited from our parents? To some extent, yes. However, we can take heart in the fact that the complexity of cortical circuitry can be modified throughout life by the environment. For example, Diamond and her colleagues found that if rats are reared in an enriched environment, filled with

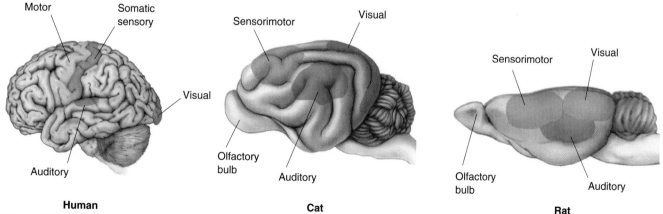

Figure 7.27
Lateral view of the cerebral cortex in three species. Notice the expansion of the human cortex that is neither strictly primary sensory nor motor.

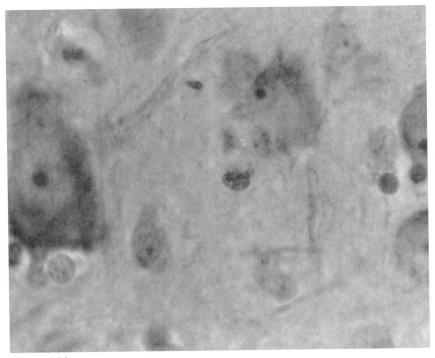

Figure 7.28
Einstein's brain. This is a Nissl stain of neurons and glia in the parietal association cortex of Albert Einstein's brain. (Source: Courtesy of Dr. M. C. Diamond.)

toys and playmates, their neocortex will develop more glia per neuron than littermates reared in the usual laboratory cages. When it comes to the complexity of cortical circuitry, it's a case of "use it or lose it." We will return to the role of the environment in brain development in Chapter 18.

CONCLUDING REMARKS

Although we have covered a lot of new ground in this chapter, we have only scratched the surface of neuroanatomy. Clearly, the brain deserves its status as the most complex piece of matter in the universe. What we have presented here is a shell, or scaffold, of the nervous system and some of its contents.

Understanding neuroanatomy is necessary for understanding how the brain works. This statement is just as true for an undergraduate first-time neuroscience student as it is for a neurologist or a neurosurgeon. In fact, neuroanatomy has taken on a new relevance with the advent of methods to image the living brain. These methods include positron emission tomography (PET) and magnetic resonance imaging (MRI), as shown in Figure 7.29. These techniques have given neuroscientists their first glimpse at how neural activity is distributed in the brain during mental acts. You can appreciate that even the most sophisticated computer images of the brain are meaningless unless you know what you are looking at. (Imaging methods will be discussed in more detail in Chapter 21.)

In Part II "Sensory and Motor Systems," the anatomy presented here will come alive as we explore how the brain goes about the tasks of smelling, seeing, hearing, sensing touch, and moving.

Box 7.2   PATH OF DISCOVERY

Marian C. Diamond

Einstein's Brain
by Marian C. Diamond

The excitement of discovery is infectious. One new enticing finding leads continuously on and on. What discovery led me to study the ratio of glial cells to neurons in Einstein's brain? The answer is not found in a single, simple statement but is based on decades of work on how the environment affects the anatomy of the brain.

Reconstructing the series of events that led up to the final cell counts was a challenge, but roughly here's the sequence: (1) a comment by a friendly professor 25 years ago, (2) the results showing that brains from rats living in enriched conditions possessed more glial cells per neuron than brains from rats living in impoverished conditions, (3) a photograph from the journal *Science* showing a cardboard box sitting beside a desk with a caption stating that Einstein's brain was in that box, (4) a quiet afternoon during which everyone appeared to be busy but me so I had time to think, and (5) superb technical and statistical assistance. In general, one could now say simply that $1 + 2 + 3 + 4 + 5 =$ the difference in the glial/neuron ratio in Einstein's brain compared with the average glial/neuron ratio in a sample of other human male brains. Let me expand upon each part of the equation.

1. One day in my teaching laboratory, Professor Gerhardt von Bonin mentioned that he thought the inferior parietal cortex was more highly evolved than the prefrontal cortex. Since the number of glial cells per neuron increases as one ascends the phylogenetic tree, I reasoned that the more highly evolved area should have more glial cells per neuron. From 11 human, male, preserved brains, I removed a sugar cube-size piece from right and left prefrontal and inferior parietal cortex, giving 44 samples of tissue. The glial/neuron ratios showed the prefrontal cortex to have more glial cells per neuron compared to the inferior parietal lobe. Thus, we had a database of human cortical glial/neuron ratios.

2. In the early 1960s, the laboratories both of Joseph Altman at Purdue and ours had found that the cerebral cortex from rats living in enriched environments had more glial cells per neuron than the cortex from rats living in impoverished environments. Active cortical neurons needed more support cells, namely glial cells, since cortical nerve cells do not divide after birth and glial cells do.

3. The graduate students had clipped the photo of the box with Einstein's brain from *Science* and attached it to the laboratory wall. The caption mentioned that the brain was in Kansas. A daily exposure to its message was available for all who entered the laboratory.

4. I was sitting by myself in my husband's office at UCLA one day with nothing to do but think. I wondered if I could only obtain the four pieces of Einstein's brain comparable to the ones already studied. I picked up the phone and called the University of Kansas Department of Anatomy and learned that Thomas Harvey in Weston, Missouri, had Einstein's brain. After three years of negotiating by calling every six months, I received the four precious pieces of tissue.

5. With the help of an excellent technician and statistician (a scientist rarely works alone), we learned that in all four areas, Einstein had more glial cells per neuron than the average man, but in only the left inferior parietal area did he have statistically significantly more. The differences were unusually large, but we only had one Einstein to compare with 11 males. The findings would be more valid if we had 11 Einsteins, but at least the study was a first step that no one had taken previously!

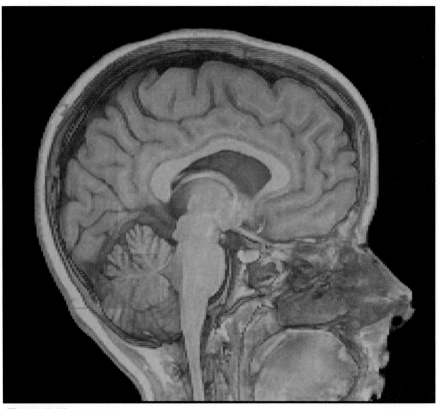

Figure 7.29
An MRI scan of the living human brain. How many structures can you label?
(Source: Courtesy of Dr. J. N. Sanes.)

KEY TERMS

**Gross Organization of the
Mammalian Nervous System**

anterior

rostral

posterior

caudal

dorsal

ventral

midline

medial

lateral

ipsilateral

contralateral

midsagittal plane

sagittal plane

horizontal plane

coronal plane

central nervous system (CNS)

brain

spinal cord

cerebrum

cerebral hemispheres

cerebellum

brain stem

spinal nerve

dorsal root

ventral root

peripheral nervous system (PNS)

somatic PNS

dorsal root ganglion

visceral PNS

autonomic nervous system (ANS)

cranial nerves

meninges

dura mater

arachnoid membrane

pia mater

cerebrospinal fluid (CSF)

ventricular system

Understanding CNS Structure Through Development

neural tube

neural crest

differentiation

forebrain

midbrain

hindbrain

diencephalon

telencephalon

olfactory bulb

lateral ventricle

third ventricle

cerebral cortex

basal telencephalon

thalamus

hypothalamus

cortical white matter

corpus callosum

internal capsule

tectum

tegmentum

cerebral aqueduct

pons

medulla

fourth ventricle

spinal canal

dorsal horn

ventral horn

sulcus

gyrus

temporal lobe

frontal lobe

central sulcus

parietal lobe

occipital lobe

A Guide to the Cerebral Cortex

hippocampus

olfactory cortex

neocortex

cytoarchitectural map

REVIEW QUESTIONS ✔

1. Are the dorsal root ganglia in the central or peripheral nervous system?
2. Is the myelin sheath of optic nerve axons provided by Schwann cells or oligodendroglia? Why?
3. Imagine that you are a neurosurgeon, about to remove a tumor lodged deep inside the brain. The top of the skull has been removed. What now lies between you and the brain? Which layer(s) must be cut before you reach the CSF?
4. What is the fate of tissue derived from the embryonic neural tube? Neural crest?
5. Name the four main parts of the brain stem. Which of these are also part of the forebrain?
6. Name the three main parts of the hindbrain. Which of these are also part of the brain stem?
7. Where is CSF produced? What path does it take before it is absorbed into the bloodstream? Name the parts of the CNS it will pass through in its voyage from brain to blood.
8. What are three features that characterize the structure of cerebral cortex?

PART II:
SENSORY & MOTOR SYSTEMS

CHAPTER

8

Chemical Senses

INTRODUCTION

TASTE

SMELL

CONCLUDING REMARKS

KEY TERMS

REVIEW QUESTIONS ✔

Life evolved in a sea of chemicals. From the beginning, organisms have floated or swum in water containing chemical substances that signal food, poison, or sex. In this sense, things have not changed much in three billion years. Animals, including humans, depend on the chemical senses to help identify nourishment (the sweetness of honey, the aroma of pizza), noxious substances (the bitterness of plant alkaloids), or the suitability of a potential mate. Chemical sensation is the oldest and most common of the sensory systems. Even brainless bacteria can detect, and tumble toward, a favorable food source.

Multicellular organisms must detect chemicals in both their internal and external environments. The variety of chemical detection systems has expanded considerably over the course of evolution. Humans live in a sea of air, full of volatile chemicals; we put chemicals into our mouths for a variety of reasons, and we carry a complex sea within us in the form of blood and the other fluids that bathe our cells. We have specialized detection systems for the chemicals in each milieu. The mechanisms of chemical sensation that originally evolved to detect environmental substances now form the basis for chemical communication between cells and organs, using hormones and neurotransmitters. Virtually every cell in every organism is responsive to many chemicals.

In this chapter we consider the most familiar of our chemical senses: taste, or **gustation**, and smell, or **olfaction**. Although taste and smell reach our awareness most often, they are not the only important chemical senses we have. Many types of chemically sensitive cells, or **chemoreceptors,** are distributed through our body. For example, some nerve endings in skin and mucous membranes warn us of irritating chemicals. A wide range of chemoreceptors report subconsciously and consciously about our internal state: nerve endings in the digestive organs detect many types of ingested substances, receptors in arteries of the neck measure carbon dioxide and oxygen levels in our blood, and sensory endings in muscles respond to acidity, giving us the burning feeling that comes with exertion and oxygen debt.

Gustation and olfaction have a similar task: the detection of environmental chemicals. In fact, only by using both senses can the nervous system perceive flavor. Gustation and olfaction have unusually strong and direct connections with our most basic internal needs, including thirst, hunger, emotion, sex, and certain forms of memory. However, the systems of gustation and olfaction are separate and different, from the structures and mechanisms of their chemoreceptors, to the gross organization of their central connections, to their effects on behavior. The neural information from each system is processed in parallel and is merged only at rather high levels in the cerebral cortex.

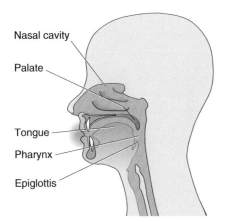

Figure 8.1
Anatomy of the mouth, throat, and nasal passages. Taste is primarily a function of the tongue, but regions of the pharynx, palate, and epiglottis have some sensitivity. Notice how the nasal passages are located so that odors from ingested food can enter through the nose or the pharynx, thereby easily contributing to perceptions of flavor through olfaction.

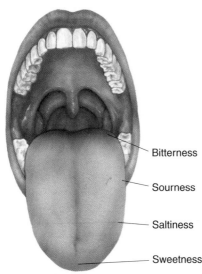

Figure 8.2
Tongue map regions of the lowest thresholds for the basic tastes. The center of the tongue lacks taste buds and is relatively taste-insensitive

TASTE

Humans evolved as omnivores (from the Latin *omnis*, "all," and *vorare*, "to eat"), opportunistically eating the plants and animals they could gather, scavenge, or kill. A sensitive and versatile system of taste was necessary to distinguish between new sources of food and potential toxins. Some of our taste preferences are inborn. We have an innate preference for sweetness, satisfied by mother's milk. Bitter substances are instinctively rejected and, indeed, many kinds of poisons are bitter. However, experience can strongly modify our instincts, and we can learn to tolerate and even enjoy the bitterness of such substances as coffee and quinine. The body also has the capacity to recognize a deficiency of certain key nutrients and induce an appetite for them. For example, when deprived of essential salt, we may crave salty foods.

The Basic Tastes

Although the number of different chemicals is endless and the variety of flavors seems immeasurable, it is likely that we can recognize only a few basic tastes. Most neuroscientists put the number at four or five. The four basic taste qualities are saltiness, sourness, sweetness, and bitterness. The more unfamiliar fifth taste quality is *umami*, meaning "delicious" in Japanese; it is defined by the taste of the amino acid glutamate (monosodium glutamate, or MSG, being the familiar culinary form). There are several kinds of evidence for the basic tastes, the most compelling of which is behavioral. When people are asked to taste many pairs of chemicals and compare them, most substances fall into one of four categories: salt, sour, sweet, and bitter.

The correspondence between chemistry and taste is obvious in some cases. Most acids taste sour, and most salts taste salty. But the chemistry of substances can vary considerably, while their basic taste remains the same. Many substances are sweet, from familiar sugars (like fructose, present in fruits and honey, and sucrose, which is white table sugar) to certain proteins (monellin) to artificial sweeteners (saccharin and aspartame, the second of which is made from two amino acids). Surprisingly, sugars are the least sweet of all of these; gram for gram, the artificial sweeteners and proteins are 10,000–100,000 times sweeter than sucrose. Bitter substances range from simple ions like K^+ (KCl actually evokes both bitter and salty tastes) and Mg^{2+} to complex organic molecules such as quinine and caffeine.

How, then, do we perceive the countless flavors of food such as chocolate, strawberries, and barbecue sauce? First, each food activates a different combination of the basic tastes, helping make it unique. Second, most foods have a distinctive flavor as a result of their taste *and* smell, occurring simultaneously. For example, without the sense of smell, a bite of onion is easily mistaken for an apple. Third, other sensory modalities may contribute to a unique food-tasting experience. Texture and temperature are important, and pain sensations are essential to the hot, spicy flavor of foods laced with *capsaicin*, the key ingredient in hot peppers.

The Organs of Taste

Experience tells us that we taste with our tongue; but other areas of the mouth, such as the palate, pharynx, and epiglottis, are also involved (Figure 8.1). Odors from the food we are eating can also pass, via the pharynx, into the nasal cavity, where they can be detected by olfactory receptors. The tip of the tongue is most sensitive to sweetness, the back of it to bitterness, and the sides to saltiness and sourness, as shown in Figure 8.2. This does not mean, however, that we taste sweetness only with the tip of our tongue. Most of the tongue is sensitive to all basic tastes. The tongue map implies only that certain regions of the tongue are more sensitive to the basic tastes than other regions.

Scattered about the surface of the tongue are small projections called **papillae** (Latin for "bumps"; singular: papilla), which are shaped like ridges (*foliate papillae*), pimples (*vallate papillae*), or mushrooms (*fungiform papillae*) (Figure 8.3a). Facing a mirror, stick your tongue out and shine a flashlight on it, and you will see your papillae easily—small, rounded ones at the front and sides, large ones in the back. Each papilla has from one to several hundred **taste buds**, visible only with a microscope (Figure 8.3b). Each taste bud has 50–150 **taste receptor cells,** or taste cells, arranged within the bud like the sections of an orange. Taste cells are only about 1% of the tongue epithelium. Taste buds also have basal cells that surround the taste cells, plus a set of gustatory afferent axons (Figure 8.3c). A person usually has 2000–5000 taste buds, although exceptional cases range from 500 to 20,000.

Using tiny droplets, it is possible to expose a single papilla on a person's tongue to low concentrations of various basic taste stimuli (something almost purely sour, like vinegar, or almost purely sweet, like sucrose). Concentrations too low will not be tasted, but at some critical concentration,

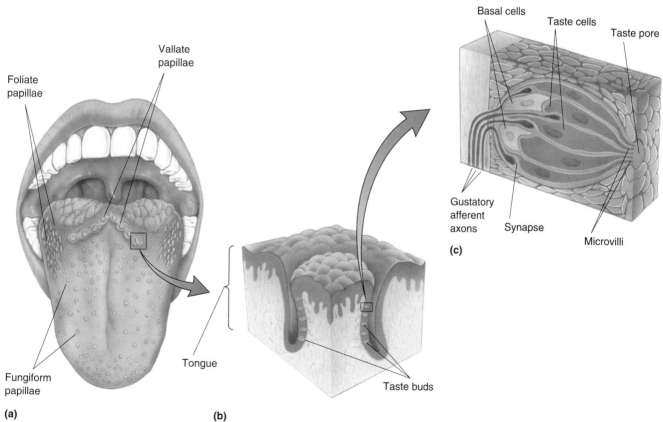

Figure 8.3
The tongue, its papillae, and its taste buds. (a) Papillae are the taste-sensitive structures. The largest and most posterior are the vallate papillae. Foliate papillae are elongated. Fungiform papillae are relatively large toward the back of the tongue and much smaller along the sides and tip. **(b)** A cross-sectional view of a vallate papilla, showing the locations of taste buds. **(c)** A taste bud is a cluster of taste cells (the receptor cells), gustatory afferent axons and their synapses with taste cells, and basal cells. Microvilli at the apical end of the taste cells extend into the taste pore, the site where chemicals dissolved in saliva can interact directly with taste cells.

the stimulus will evoke a perception of taste—this is the *threshold* concentration. At concentrations just above threshold, most papillae tend to be sensitive to only one basic taste; there are sour-sensitive papillae and sweet-sensitive papillae, for example. When the concentrations of the taste stimuli are increased, however, most papillae become less selective. Whereas a papilla might have responded only to sweet when all stimuli were weak, it may also respond to sour and salt if they are made stronger. This lack of specificity is a common phenomenon in sensory systems. Most sensory receptors are surprisingly indiscriminate about the things that excite them. This presents a paradox: If single taste receptors show only small differences in response to ice cream and bananas, how can we distinguish reliably between differences as subtle as two brands of chocolate? The answer lies in the brain, as we will discuss below.

Taste Receptor Cells

The chemically sensitive part of a taste receptor cell is its small membrane region, called the *apical end,* near the surface of the tongue. The apical ends have thin extensions called *microvilli* that project into the *taste pore,* a small opening on the surface of the tongue where the taste cell is exposed to the contents of the mouth (Figure 8.3c). Taste receptor cells are not neurons, by standard histological criteria. However, they do form synapses with the endings of the gustatory afferent axons near the bottom of the taste bud. Taste receptor cells also make both electrical and chemical synapses onto some of the basal cells; some basal cells synapse onto the sensory axons, and these may form a simple information-processing circuit within each taste bud. Cells of the taste bud undergo a constant cycle of growth, death, and regeneration; the lifespan of one taste cell is about 2 weeks. This process depends on an influence of the sensory nerve, because if the nerve is cut, taste buds will degenerate.

When a taste receptor cell is activated by an appropriate chemical, its membrane potential changes, either depolarizing or hyperpolarizing. This voltage shift is called the **receptor potential** (Figure 8.4a). If the receptor potential is depolarizing and large enough, some taste receptor cells, like neurons, may fire action potentials. In any case, depolarization of the receptor membrane causes voltage-gated calcium channels to open; Ca^{2+} enters the cytoplasm, triggering the release of transmitter molecules. This is basic synaptic transmission, from taste cell to sensory axon. The identity of the taste receptor's transmitter is unknown, but we do know that it excites the postsynaptic sensory axon and causes it to fire action potentials (Figure 8.4b), which communicate the taste signal into the brain stem.

More than 90% of receptor cells respond to two or more of the basic tastes, emphasizing that even the first cell in the taste process is relatively unselective to chemical types. An example is Cell 2 in Figure 8.4a, which gives strong depolarizing responses to both salt (NaCl) and sour (HCl) stimuli. Taste cells and their gustatory axons differ widely in their response preferences, however. Each of the gustatory axons in Figure 8.4b responds to three or four basic tastes, but each has a clearly different bias. Figure 8.5 shows the results of similar recordings from four gustatory axons in a rat. One responds strongly only to salt, one only to sweet, and two to all but sweet. Why will one cell respond only to a single chemical type, while another responds to three or four categories of chemicals? The answer is that the responses depend on the particular transduction mechanisms present in each cell.

Mechanisms of Taste Transduction

The process by which an environmental stimulus causes an electrical response in a sensory receptor cell is called **transduction** (from the Latin

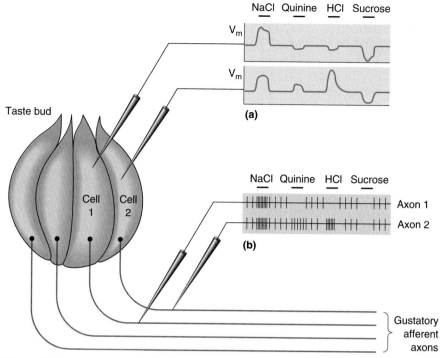

Figure 8.4
Taste responsiveness of taste cells and gustatory axons. (a) Two different cells
were exposed to a salt (NaCl), bitter (quinine), sour (HCl), and sweet (sucrose) stimulus, and their membrane potential was recorded with an electrode. Notice the different sensitivities of the two cells. **(b)** In this case, the action potential discharge of
the sensory axons was recorded. This is an example of extracellular recording of
action potentials. Each vertical deflection in the record is a single action potential.

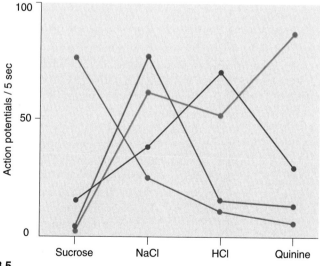

Figure 8.5
**The action potential firing rates of four different primary gustatory nerve
axons in a rat.** The taste stimuli were sweet (sucrose), salt (NaCl), sour (HCl), and
bitter (quinine). Notice the differences in selectivity. (Source: Adapted from Sato,
1975, p. 23.)

transducere, "to lead across"). The nervous system has myriad transduction mechanisms, which make it sensitive to chemicals, pressures, sounds, and light. The nature of the transduction mechanism determines the specific sensitivity of a sensory system. We see because our eyes have photoreceptors. If our tongues had photoreceptors, we might see with our mouth.

Some sensory systems have a single basic type of receptor cell that uses one transduction mechanism (e.g., the auditory system). However, taste transduction involves several different processes, and each basic taste may use one or more of these mechanisms. Taste stimuli, or *tastants*, may (1) directly pass through ion channels (salt and sour), (2) bind to and block ion channels (sour and bitter), (3) bind to and open ion channels (some sweet amino acids), or (4) bind to membrane receptors that activate second messenger systems that in turn open or close ion channels (sweet and bitter). These are familiar processes, the functional building blocks of signaling in all neurons and synapses.

As explained in Chapter 1, much of what we know about how our brain works comes from studying animals other than humans. Our understanding of taste transduction is no exception. Important clues have come from the cells of catfish, mudpuppies (a large, voracious salamander), mice, and rats. Each animal has certain experimental advantages (very large taste cells, for example), and the information they provide tells us where to start looking when we are able to study human cells directly. In the case of taste (and much of the rest of neuroscience), the human story is still sparse and must be supplemented by animal stories.

Saltiness. The prototypical salty chemical is table salt (NaCl), which, apart from water, is the major component of the ocean, blood, and chicken soup. The taste of salt is mostly the taste of the cation Na^+. Salt-sensitive taste cells have a Na^+-selective channel that is common in other epithelial cells, and which is blocked by the drug amiloride (Figure 8.6a). The amiloride-sensitive sodium channel is quite different from the sodium channel that generates action potentials; the taste channel is insensitive to voltage, and it stays open at rest. When you sip chicken soup, the Na^+ concentration rises outside the receptor cell, and the gradient for Na^+ across the membrane is made steeper. Na^+ then diffuses down its concentration gradient, which means it flows into the cell, and the resulting inward current causes the membrane to depolarize. This is similar to the process that causes the rising phase of the action potential, except in that case the concentration gradient for Na^+ stays constant, while the conductance for Na^+ (the number of open sodium channels) increases temporarily.

The anions of salts affect the taste of the cations. For example, NaCl tastes saltier than Na acetate, apparently because the larger an anion is, the more it *inhibits* the salt taste of the cation. The mechanisms of anion inhibition are poorly understood. Another complication is that as the anions become larger, they tend to take on tastes of their own. Sodium saccharin tastes sweet, because the Na^+ concentrations are far too low to taste the saltiness.

Sourness. Foods taste sour because of their high acidity (otherwise known as low pH). Acids, such as HCl, dissolve in water to generate hydrogen ions (protons, or H^+). Thus, protons are the causative agents of acidity and sourness, and they are known to affect sensitive taste receptors in two ways (Figure 8.6b). First, H^+ can permeate the amiloride-sensitive sodium channel, the same channel that mediates the taste of salt. This causes an inward H^+ current and depolarizes the cell. (Note that the cell would not be able to distinguish a H^+ ion from a Na^+ ion if this was the only transduction mechanism available to it.) Second, H^+ ions can bind to and *block* a K^+-selective channel. When the K^+ permeability of a membrane is decreased, it depolarizes. These may not be the only mechanisms of sour taste transduction, because changes of pH can affect virtually all cellular processes.

Sweetness. Molecules become sweet when they bind to specific receptor sites and activate a cascade of second messengers in certain taste cells (Figure 8.7). The cascade is similar to that caused by the activation of the norepinephrine receptor in some neurons (see Chapter 5). It involves a G-protein-coupled membrane receptor triggering the formation of cAMP within the cytoplasm, which activates protein kinase A (PKA), which phosphorylates a K^+-selective channel (apparently a different one from that involved in sourness), causing a blockade. Once again, the blocking of K^+ channels causes

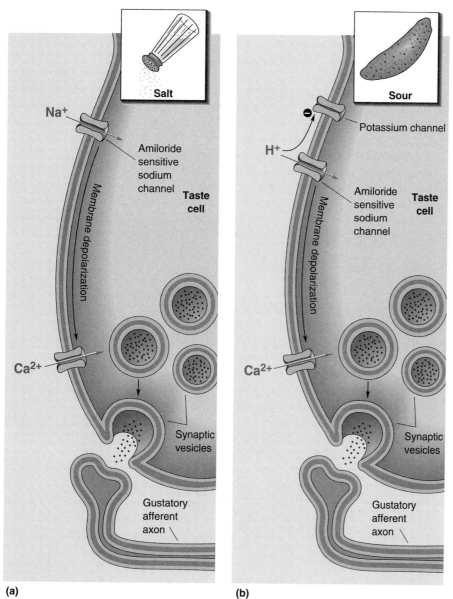

(a)   **(b)**

Figure 8.6
Transduction mechanisms of (a) salt and (b) sour tastants. Tastants can interact directly with ion channels, either by passing through them (Na^+ and H^+) or blocking them (H^+ blocking the potassium channel). Then, membrane voltage influences calcium channels on the basal membrane, which in turn influence the intracellular $[Ca^{2+}]$ and transmitter release.

depolarization of the receptor cell. There may also be a second transduction mechanism for sweetness that does not involve second messengers. In this case, a set of cation channels may be gated directly by sugars.

Bitterness. Bitter taste receptors are poison detectors. Perhaps because poisons are so chemically diverse, there are several different mechanisms for bitter taste transduction (Figure 8.8). Some bitter compounds, such as calcium and quinine (Bitter$_1$ in Figure 8.8), can bind directly to K$^+$-selective channels and block them (similar to the sourness mechanism). There are also specific membrane receptor proteins for bitter substances (as in Bitter$_2$ in Figure 8.8), which activate G-protein-coupled second messenger cascades that are different from those of the sweetness mechanism. One type of bitter receptor triggers an increased production of the intracellular messenger inositol triphosphate (IP$_3$). IP$_3$ pathways are ubiquitous signaling systems in cells throughout the body, as described in Chapter 6. An unusual feature of the IP$_3$ bitterness system is that it modulates transmitter release without changing the membrane potential of the receptor cell, because it directly triggers the release of Ca^{2+} from intracellular storage sites.

Umami. "Amino acids" may not be the answer at the tip of your tongue when asked to list your favorite tastes, but recall that proteins are made from amino acids, and that they are also excellent energy sources. In short, amino acids are the foods your mother would want you to eat. Many amino acids also taste good, although some taste bitter. The umami taste may be generated when glutamate (and perhaps certain other amino acids, such as arginine) bind to and activate a cation-permeable channel, causing depolarization (Figure 8.9). This is another common process in the brain, because glutamate-gated channels are part of the most prevalent excitatory neurotransmitter system in the brain. Other amino acids taste bitter (leucine, for example), and they may trigger either cAMP- or IP$_3$-mediated messenger systems.

Central Taste Pathways

The main flow of taste information is from taste buds, to the primary gustatory axons, into the brain stem, up to the thalamus, to the cerebral cortex (Figure 8.10). Three cranial nerves carry primary gustatory axons and bring taste information to the brain. The anterior two-thirds of the tongue and the palate send axons into cranial nerve VII, the *facial nerve*; the posterior third of the tongue is innervated by cranial nerve IX, the *glossopharyngeal nerve*; the regions around the throat, including the glottis, epiglottis, and pharynx, send taste axons to cranial nerve X, the *vagus nerve*. These nerves are involved in a variety of other sensory and motor functions, but their taste axons all enter the brain stem, bundle together, and synapse within the slender **gustatory nucleus** in the medulla.

From the gustatory nucleus, taste pathways diverge. The conscious experience of taste is presumably mediated by the cerebral cortex. The path to the neocortex via the thalamus is a common one for sensory information. Neurons of the gustatory nucleus synapse on a subset of small neurons in the **ventral posterior medial (VPM) nucleus**, which is a portion of the thalamus that deals with sensory information from the head. The VPM taste neurons then send axons to the **primary gustatory cortex** (mainly Brodmann's area 43), in the ventral parietal lobe. Lesions within the VPM thalamus or the gustatory cortex—as a result of a stroke, for example—can cause *ageusia*, the loss of taste perception.

Gustation is important to basic behaviors such as the control of feeding and digestion, both of which involve additional taste pathways. Gustatory nucleus cells project to a variety of brain stem regions, largely in the medulla, that are involved in swallowing, salivation, gagging, vomiting, and basic

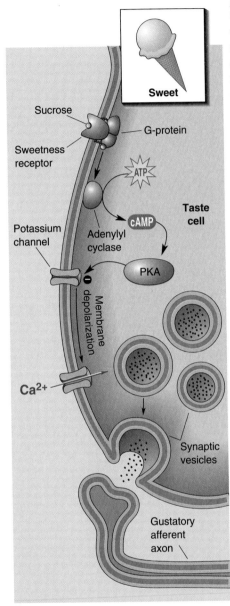

Figure 8.7
Transduction mechanisms for sweet tastants. Tastants bind directly to G-protein-coupled membrane receptors and trigger the synthesis of cAMP, which leads to the blockade of a potassium channel, depolarization, Ca^{2+} entry, and transmitter release.

physiological functions such as digestion and respiration. In addition, gustatory information is distributed to the hypothalamus and related parts of the basal telencephalon (structures of the *limbic system*, described in Chapter 16). These structures seem to be involved in the palatability of foods and the forces that motivate us to eat. Localized lesions of the hypothalamus or amygdala, a nucleus of the basal telencephalon, can cause an animal to either chronically overeat or ignore food, or alter preferences for food types.

Neural Coding of Taste

If you were going to design a system for coding tastes, you might begin with many specific taste receptors for many basic tastes (e.g., sweet, sour, salty, bitter, chocolate, banana, mango, beef, Swiss cheese, etc.). Then you

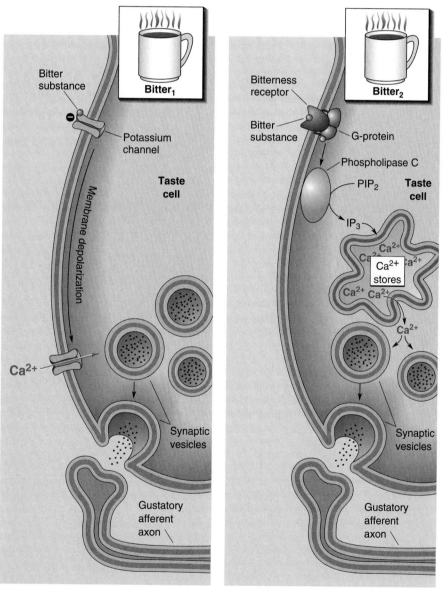

Figure 8.8
Transduction mechanisms for bitter tastants. Tastants can either block a potassium channel (Bitter$_1$) or bind directly to a G-protein-coupled membrane receptor (Bitter$_2$) that triggers IP$_3$ synthesis and the release of Ca^{2+} from internal storage sites. PIP$_2$ is the membrane lipid phosphatidylinositol-4,5-bisphosphate.

would connect each receptor type, by separate sets of axons, to neurons in the brain that also responded to only one specific taste. All the way up to the cortex you would expect to find specific neurons responding to "sweet" and "chocolate," and the flavor of chocolate ice cream would involve the rapid firing of these cells and very few of the "salty," "sour," and "banana" cells. This concept is the *labeled line hypothesis*, and it seems simple and rational. Unfortunately, it is incompatible with several facts. As we have seen, individual taste receptor cells tend to be *broadly tuned* to stimuli; that is, they are not very specific in their responses. Primary taste axons are even less specific than receptor cells, and most central taste neurons continue to be broadly responsive all the way into the cortex. In other words, the response of a single taste cell is often ambiguous about the food being tasted; the labels on the taste lines are uncertain rather than distinct.

There are several reasons that cells in the taste system are broadly tuned. If one taste receptor cell has several different transduction mechanisms, it will respond to several types of tastants (although it may still respond more strongly to one or two). In addition, there is convergence of receptor cell input onto afferent axons. Each receptor synapses onto a primary taste axon that also receives input from several other receptors, in that papilla as well as its neighbors. This means that one axon may combine the taste information from several papillae. If one of those receptors is mostly sensitive to sour stimuli and another to salt stimuli, then the axon will respond to salt *and* sour. This pattern continues into the brain: neurons of the solitary nucleus receive synapses from many axons of different taste specificities, and they may become less selective for tastes than the primary taste axons.

All of this mixing of taste information might seem like an inefficient way to design a coding system. When you taste chocolate ice cream, how does the brain sort through its apparently ambiguous information about the flavor to make clear distinctions between chocolate and thousands of other possibilities? The likely answer is a scheme that includes features of roughly labeled lines and *population coding*, in which the responses of a large number of broadly tuned neurons, rather than a small number of precisely tuned neurons, are used to specify the properties of a particular stimulus, such as a taste. Population coding schemes seem to be used throughout the sensory and motor systems of the brain, as we shall see in later chapters. In the case of taste, receptors are not sensitive to all tastes; most respond broadly: to salt and sour but not to bitter and sweet, for example. Only with a large population of taste cells, with different response patterns, can the brain distinguish between alternative tastes. One food activates a certain subset of neurons, some of them firing very strongly, some moderately, some not at all, others perhaps even inhibited below their spontaneous firing rates (i.e., their nonstimulated rates); a second food excites some of the cells activated by the first food, but also others; and the overall patterns of discharge rates will be distinctly different. The relevant population may even include neurons activated by the olfactory, temperature, and textural features of a food; certainly the creamy coldness of chocolate ice cream contributes to your ability to distinguish it from chocolate cake.

SMELL

Olfaction brings both good news and bad news. It combines with taste to help us identify foods, and it increases our enjoyment of many of them. But it can also warn of potentially harmful substances (spoiled meat) or places (smoke-filled rooms). In olfaction, the bad news may outweigh the good; by some estimates, we can smell several hundred thousand substances, but only about 20% of them smell pleasant. Practice helps in olfaction, and pro-

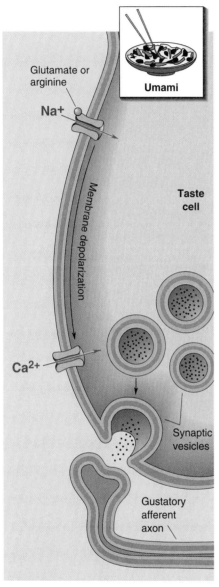

Figure 8.9
Transduction mechanisms for umami tastants (glutamate). Some amino acid tastants can bind to a cation-permeable channel, leading to a change in membrane current and potential.

Labels in figure: Glutamate or arginine; Umami; Na⁺; Membrane depolarization; Taste cell; Ca²⁺; Synaptic vesicles; Gustatory afferent axon

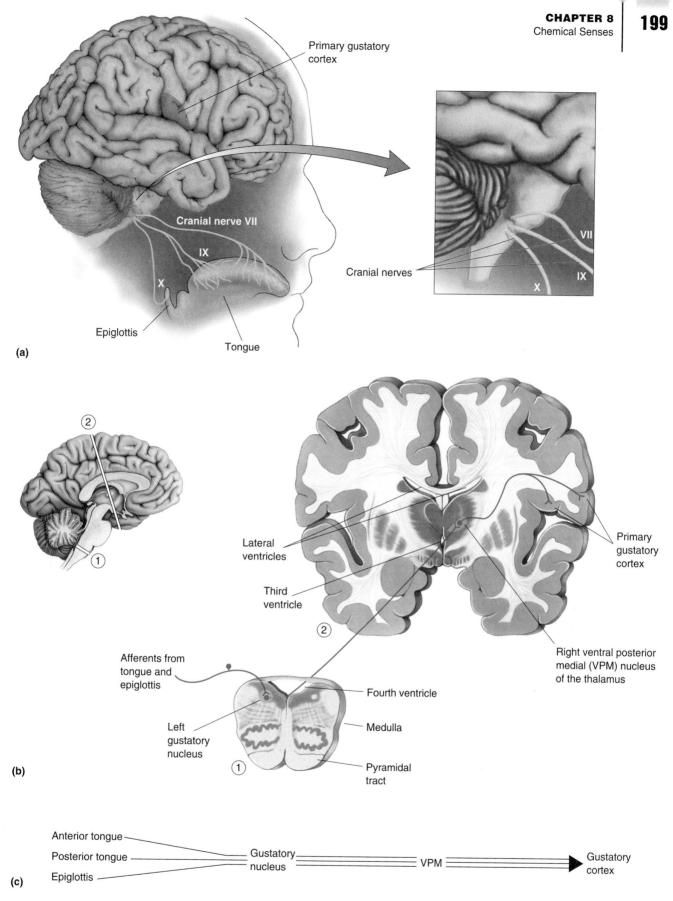

Figure 8.10
Central taste pathways. (a) Taste information from different regions of the tongue and mouth cavity is carried by three cranial nerves, VII, IX, and X, which enter the medulla. **(b)** Gustatory axons enter the small gustatory nucleus located within the medulla. Axons from the gustatory nucleus synapse upon neurons of the thalamus, and these in turn project to regions of the cerebral cortex that include the postcentral gyrus. The insets show the locations of the planes of section through the medulla (1) and the forebrain (2). **(c)** A summary of the central taste pathways.

fessional perfumers and whiskey blenders can actually distinguish between as many as 100,000 different odors.

Smell is also a mode of communication. Chemicals released by the body, called **pheromones** (from the Greek *pherein*, "to carry," and *horman*, "to excite"), are important signals for reproductive behaviors, and they may also mark territories, identify individuals, and indicate aggression or submission. Although systems of pheromones are well developed in many animals, their importance to humans is not clear (Box 8.1).

Organs of Smell

We do not smell with our nose. Rather, we smell with a small, thin sheet of cells high up in the nasal cavity called the **olfactory epithelium** (Figure 8.11a). The olfactory epithelium has three main cell types (Figure 8.11b). *Olfactory receptor cells* are the site of transduction. Unlike taste receptor cells, olfactory receptors are genuine neurons, with axons of their own that penetrate into the central nervous system. *Supporting cells* are similar to glia; among other things, they help produce mucus. *Basal cells* are the source of new receptor cells. Olfactory receptors (similar to taste receptors) continually grow, die, and regenerate, in a cycle that lasts about 4–8 weeks. In fact,

---

| Box 8.1 | OF SPECIAL INTEREST |

Human Pheromones?

Smells are surer than sounds and sights to make your heart-strings crack.

—*Rudyard Kipling*

Odors can certainly sway emotions and arouse memories, but just how important are they to human behavior? Each of us has a distinctive set of odors, which mark our identity just as surely as do our fingerprints or genes. In fact, variations in body odor are probably genetically determined. Bloodhounds have great difficulty distinguishing between the smells of identical twins, but not between those of fraternal siblings. For some animals, odor identity is an essential fact: when her lamb is born, the ewe establishes a long-term memory of its specific smell and develops an enduring bond based largely on olfactory cues. In a newly inseminated female mouse, the smell of a strange male (but not the smell of her recent mate, which she remembers) will trigger an abortion of the pregnancy.

Humans have the ability to recognize the scents of other humans. Infants as young as 6 days old show a clear preference for the smell of their own mother's breast over that of other nursing mothers. The mothers, in turn, can usually identify the odor of their own infant from among several choices. Women who spend a lot of time together (college roommates, for example) often find that their menstrual cycles synchronize. The effect may be mediated by a phero-

mone; in experiments, the daily odors of one woman's underarm sweat alone have been shown to synchronize the cycles of other women.

Many subhuman animals use the *accessory olfactory system* to detect pheromones and mediate a variety of social behaviors involving mother, mating, territory, and food. The accessory system runs parallel to the primary olfactory system. It consists of a separate chemically sensitive region in the nasal cavity, the *vomeronasal organ*, which projects to the *accessory olfactory bulb*, and from there provides input to the hypothalamus. It had been thought for a long time that the vomeronasal organ in mature humans was absent or vestigial, but recent research indicates that it is present in adults. Its precise function in humans is not yet clear.

Napoleon Bonaparte once wrote to his love Josephine, asking her not to bathe for the two weeks until they would next meet, so he could enjoy her natural aromas. The scent of a woman may indeed be a source of arousal for sexually experienced males, presumably because of learned associations. But there is not yet any hard evidence for human pheromones that might mediate sexual attraction (for members of either sex) via innate mechanisms. However, considering the commercial implications of such a substance, you can be sure the search will continue.

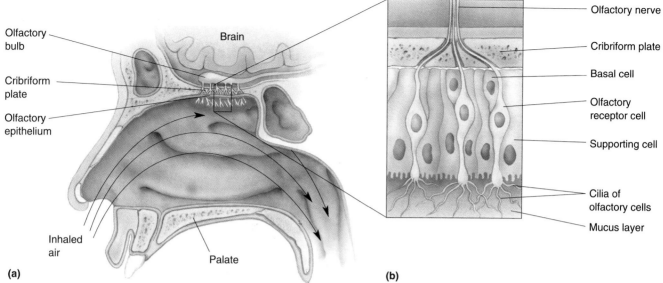

Figure 8.11

(a) Location and (b) structure of the olfactory epithelium. The epithelium consists of a layer of olfactory receptor cells, supporting cells, and basal cells. Odorants dissolve in the mucus layer and contact the cilia of the olfactory cells. Axons of the olfactory cells penetrate the bony cribriform plate, on their way to the CNS.

olfactory receptor cells are the only *neurons* in the nervous system that are regularly replaced throughout life.

Sniffing brings air through the convoluted nasal passages, but only a small percentage of that air passes over the olfactory epithelium. The epithelium exudes a thin coating of mucus, which flows constantly and is replaced about every 10 minutes. Chemical stimuli in the air, or *odorants*, dissolve in the mucus layer before they reach receptor cells. Mucus consists of a water base with dissolved mucopolysaccharides (long chains of sugars); a variety of proteins including antibodies, enzymes, and odorant binding proteins; and salts. The antibodies are critical because olfactory cells can be a direct route by which some viruses (such as the rabies virus) and bacteria enter the brain.

The size of the olfactory epithelium is one indicator of an animal's olfactory acuity. Humans are relatively weak smellers (although even we can detect some odorants at concentrations as low as a few parts per trillion). The surface area of the human olfactory epithelium is only about 10 cm². The olfactory epithelium of certain dogs may be over 170 cm², and dogs have over 100 times more receptors in each square centimeter than humans. By sniffing the aromatic air above the ground, dogs can detect the few molecules left by someone walking there hours before. Humans may only be able to smell the dog when he licks their face.

Olfactory Receptor Neurons

Olfactory receptor neurons have a single, thin dendrite that ends with a small knob at the surface of the epithelium (see Figure 8.11b). Waving from the knob, within the mucus layer, are several long, thin cilia. Odorants dissolved in the mucus bind to the surface of the cilia and activate the transduction process. On the opposite side of the olfactory receptor cell is a very thin, unmyelinated axon. Collectively the olfactory axons constitute the *olfactory nerve* (or cranial nerve I). After leaving the epithelium, the axons penetrate a thin sheet of bone called the *cribriform plate,* then course into the

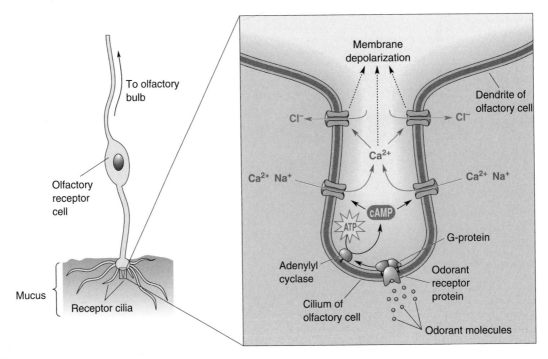

Figure 8.12
Transduction mechanisms of vertebrate olfactory receptor cells. This drawing shows a single cilium of an olfactory receptor cell, and the signaling molecules of olfactory transduction.

olfactory bulb (see Figure 8.11a). The olfactory axons are fragile, and during traumatic injury, such as a blow to the head, the forces between the cribriform plate and surrounding tissue can permanently sever the olfactory axons. The result is *anosmia*, the inability to smell.

Olfactory Transduction. Although taste receptors use many different molecular signaling systems for transduction, olfactory receptors apparently use only one or two (Figure 8.12). All of the transduction molecules are located in the thin cilia. The best-known olfactory pathway can be summarized as follows:

Odorants →
  Binding to specific membrane receptor proteins →
    G-protein stimulation →
      Activation of adenylyl cyclase →
        Formation of cyclic AMP →
          Binding of cAMP to specific cation channel →
            Opening of cation channels and influx of Ca^{2+} →
              Opening of calcium-activated chloride channels →
                Current flow and membrane depolarization (receptor potential).

Once the cation-selective cAMP-gated channels open, current flows inward, and the membrane of the olfactory neuron depolarizes (Figures 8.12 and 8.13). Besides Na^+, the cAMP-gated channel allows substantial amounts of Ca^{2+} to enter the cilia. Recent work by biophysicists Graeme Lowe and Geoffrey Gold, at the Monell Chemical Senses Center in Philadelphia, suggests that a Ca^{2+}-activated chloride current may amplify the olfactory receptor potential. (This is a switch from the usual effect of Cl^- currents, which inhibit neurons; in olfactory cells, the internal Cl^- concentration must be unusually high so that a Cl⁻ current tends to depolarize rather than hyperpolarize the membrane.) If the resulting receptor potential is large enough, it will exceed the threshold for action potentials in the cell body, and spikes will propagate out along the axon into the CNS (Figure 8.13). The olfactory response may terminate for several reasons. Odorants diffuse away, scavenger enzymes in the mucus layer often break them down, and cAMP in the receptor cell may activate other signaling pathways that end the transduction process.

There are two unusual features of this signaling pathway: the receptor binding proteins at the beginning and the cAMP-gated channels near the end. The receptor proteins have odorant binding sites on their extracellular surface. Researchers Linda Buck and Richard Axel, working at Columbia University, recently found that there are 500–1000 different odorant binding protein genes, making it the largest family of genes yet discovered! Presumably, each receptor protein can bind only certain types of odorants, but the details are not known. It now seems likely that each olfactory cell expresses only one of each receptor type. With hundreds of possible binding proteins to choose from, it is easy to imagine how neurons with widely different odorant preferences might be assembled.

Olfactory receptor proteins belong to a protein "superfamily" whose members all have seven transmembrane alpha helices. This superfamily also includes a variety of neurotransmitter receptors (see Chapter 6). All proteins in this superfamily are coupled to G-proteins, which in turn relay a signal to other second messenger systems within the cell. In most olfactory cells, this second messenger is cAMP. There may also be an alternative pathway that uses the messenger IP_3.

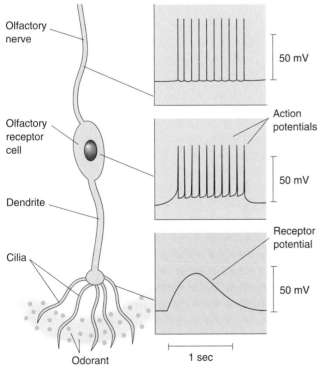

Figure 8.13
Voltage recordings from an olfactory receptor cell during stimulation. Odorants generate a slow receptor potential in the cilia; the receptor potential propagates down the dendrite and triggers a series of action potentials within the soma of the olfactory receptor cell; finally, the action potentials (but not the receptor potential) propagate continuously down the olfactory nerve axon.

In neurons, cAMP is a common second messenger, but the way it acts in olfactory transduction is quite unusual. Tadashi Nakamura and Geoffrey Gold, working at Yale University in 1987, showed that a population of channels in the cilia of olfactory cells responds directly to cAMP; that is, the channels are cAMP-gated. In Chapter 9, we will see that cyclic nucleotide-gated channels are also used for visual transduction. This is another demonstration that biology is conservative, that evolution recycles its good ideas: Smelling and seeing use some very similar molecular mechanisms (Box 8.2).

Box 8.2 | **PATH OF DISCOVERY**

Channels of Vision and Smell

by Geoffrey Gold

The discovery of cyclic nucleotide-gated ion channels in olfactory receptor cells provides illuminating examples of how scientific orthodoxy can inhibit progress. Ironically, the story begins with work on vision. Studies of visual transduction began in earnest following the discovery, in 1971, that light caused the breakdown of cyclic guanosine monophosphate (cGMP) in photoreceptors. Yet it was not until 1985 that the patch-clamp method was used to demonstrate a direct effect of cGMP on ion channels from photoreceptors. This long delay was not due to lack of interest, because there were at least a dozen labs working on the visual transduction mechanism. Rather, I think the widespread acceptance of protein phosphorylation as the mechanism of action of cyclic nucleotides in most cells effectively suppressed curiosity about other (direct) effects of cyclic nucleotides on ion channels. The cGMP-gated channel of photoreceptors was discovered by a group in the former Soviet Union, perhaps because these scientists were less influenced by the reigning dogma in Western countries.

The discovery of the olfactory cyclic nucleotide-gated channel by Tadashi Nakamura and myself also emphasizes the importance of listening to the beat of your own drum. After the odorant-stimulated adenylyl cyclase was discovered in 1985, only a few months after the discovery of the photoreceptor cGMP-gated channel, we (and probably others) thought that olfactory cilia might contain a cyclic nucleotide-gated channel. This was because the biochemical similarities between visual and olfactory

Geoffrey Gold

transduction suggested an evolutionary relationship between photoreceptors and olfactory receptor cells. Thus, we hypothesized that if the biochemical reactions of sensory transduction were conserved throughout evolution, ion channels might also be conserved. However, we knew that the transduction process was located in the cilia, and structures as small as cilia, which are about 0.2 µm in diameter, had never before been studied with the patch-clamp technique. Indeed, most people I talked to thought it would be impossible to excise patches from the cilia. Nevertheless, we reasoned that it should be possible if only we could make patch pipettes with tip openings smaller than the ciliary diameter. This proved easy to accomplish; it only required fire polishing (melting) the tips of the patch pipettes slightly longer than was customary. Once we obtained high-resistance seals on cilia, patch excision and current recording were done in the conventional way.

Perhaps the most ironic thing about this story is that the photoreceptor channel was discovered by a group led by E. E. Fesenko, whose previous (and present) work was on olfactory receptor proteins, whereas our work prior to the discovery of the olfactory channel was on phototransduction. That just goes to show how useful it can be for people to move into new areas. I like to point out that ours was a project that never would have been funded by the conventional grant review process because it was so unlikely to work.

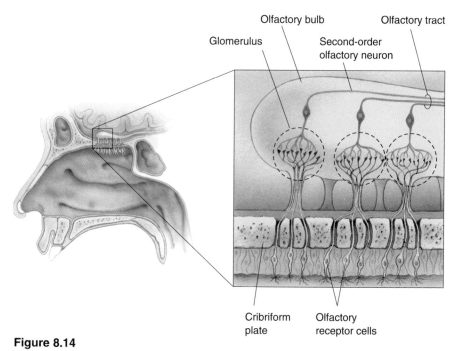

Olfactory bulb

Olfactory tract

Glomerulus

Second-order
olfactory neuron

Cribriform
plate

Olfactory
receptor cells

Figure 8.14
The structure of an olfactory bulb. Axons of olfactory receptor cells penetrate the cribriform plate and penetrate the olfactory bulb. After multiple branching, olfactory axons synapse upon second-order neurons within spherical glomeruli. The second-order neurons send axons through the olfactory tract further into the brain.

Central Olfactory Pathways

Olfactory receptor neurons send axons into the two olfactory bulbs (Figure 8.14). The bulbs are a neuroscientist's wonderland, full of fascinating dendritic arrangements, unusual reciprocal synapses, and high levels of exotic neurotransmitters. The input layer of each bulb contains numerous spherical structures called **glomeruli** (singular: glomerulus). Within a glomerulus, about 25,000 primary olfactory axons (axons from the receptor cells) converge and terminate on the dendrites of about 100 second-order olfactory neurons. Olfactory information is modified by inhibitory and excitatory interactions within and between the bulbs, and by modulation from systems of axons descending from higher areas of the brain. The bulbs send axons out through the olfactory tracts. While it is obvious that the elegant circuitry of the olfactory bulbs has important functions, it is not yet clear what those functions are.

The anatomy of olfaction is unique. The olfactory tract projects directly into the primitive regions of the cerebral cortex, and only then into the thalamus and on to the neocortex. All other sensory systems *first* pass through the thalamus before projecting to the cerebral cortex. Among other things, the olfactory arrangement gives an unusually direct and widespread influence on the parts of the forebrain that have roles in emotion, motivation, and certain kinds of memory (see Chapters 16 and 19). A partial summary of olfactory pathways begins with axons from the olfactory bulbs, which project to several areas of the olfactory cortex, which in turn relay the information on to other structures. There are so many structures that receive olfactory connections that listing those parts of the brain to which olfactory projections do *not* have access might be easier. As a general principle, several parallel pathways mediate different olfactory functions, including odor discrimination and conscious perception, motivational and emotional features, behaviors as disparate as reproduction and feeding, and imprinting

Olfactory
bulbs

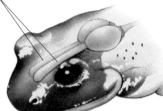

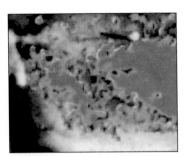

(a)

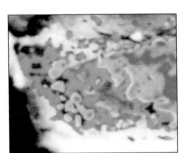

(b)

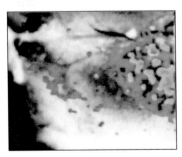

(c)

Figure 8.16
Maps of neural activation of the olfactory bulb. The activity of many olfactory neurons in a salamander olfactory bulb was recorded with a specialized optical method. The cells were stained with dyes that are sensitive to membrane voltage, and neural activity was then signaled by changes in the amount of light emitted by the dye. The colors on the maps represent differing levels of neural activity; hotter colors imply more activity. Different olfactants evoked different spatial patterns of neural activation in the bulb: **(a)** amyl acetate (pear), **(b)** limonene (citrus), **(c)** ethyl-n-butyrate (pineapple). (Source: Adapted from Kauer, 1991, p. 82.)

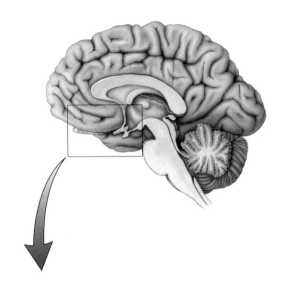

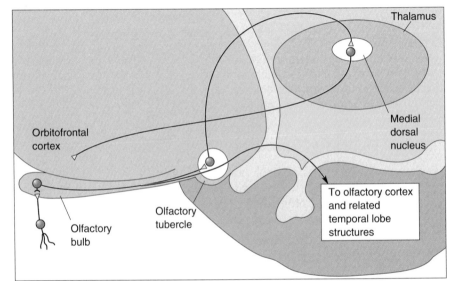

Figure 8.15
Central olfactory pathways. Axons of the olfactory tract branch and enter many regions of the forebrain, including the olfactory cortex. The neocortex is reached only by a pathway that synapses in the medial dorsal nucleus of the thalamus.

and memory. Conscious perceptions of smell may be mediated by a path from the *olfactory tubercle*, to the *medial dorsal nucleus* of the thalamus, to the *orbitofrontal cortex* (situated right behind the eyes) (Figure 8.15).

Spatial and Temporal Coding of Olfactory Information

In olfaction, we are faced with a paradox similar to the one in gustation. Individual receptors are broadly tuned to their stimuli; each cell is sensitive to a wide variety of chemicals. However, when we smell many of those same chemicals, we can easily tell them apart. How is the whole brain doing what single olfactory cells cannot?

As in gustation, a key point is that the olfactory system uses the simultaneous responses of a large population of receptors to encode a specific stimulus. Microelectrode recordings show that many receptor neurons will respond to the presentation of a single odorant and that these cells are distributed across a wide area of the olfactory epithelium. The axons of each receptor cell synapses upon many neurons of the olfactory bulb; therefore, one chemical also activates a very large population of bulb neurons. You can make a map of the distribution of bulb neurons activated by one chemical stimulus (some strongly, others moderately or weakly, others even inhibited). Maps confirm that many, but not all, bulb neurons are involved in each type of smell, and together the active cells form complex *spatial* patterns, as shown in Figure 8.16. The smell of a particular chemical is converted into a specific map within "neural space."

Could neural odor maps be used to distinguish between chemicals? Although each receptor cell is broadly tuned, there can be many types of tuning because there are many types of membrane receptor proteins. If we imagine that receptor cells with distinctive tunings are distributed across the olfactory epithelium differently, then we would predict that each odor will generate a different map on the epithelial surface. Recent work by Buck and her colleagues shows, in fact, that the epithelium has several zones, each with a population of receptor cells that expresses a different subset of receptor binding proteins (Figure 8.17). Within each zone, individual receptors form a random mosaic of protein (and presumably odorant) specificity. Maps within the epithelium seem to generate maps within the olfactory bulb. Several million olfactory receptor neurons converge onto only a few thousand glomeruli in the olfactory bulb, and the neural activation maps of the olfactory bulb show that different odors produce different patterns (see Figure 8.16). With practice, we might be able to read the "alphabet" of odors on the surface of the olfactory bulb, which may roughly approximate the functions of cortical regions of the olfactory system. The idea that odors are encoded by spatial maps, and that this information is used by the brain, is only one possibility, not yet confirmed. The *temporal* patterns of activity in olfactory neurons may also be essential features of odorant coding. The importance of timing is evident even from the odor maps, as they sometimes change shape during the presentation of a single odor.

You will see in subsequent chapters that every sensory system uses neural maps, for many different purposes. In most cases, the maps correspond obviously to features of the sensory world. For example, in the visual system, there are maps of visual space, and in the somatic sensory system, there are maps of the body surface. The maps of the chemical senses are unusual in that the stimuli themselves have no meaningful spatial properties. Although seeing a skunk walking in front of you may tell you *what* and *where* it is, smell by itself can reveal only the *what*. (By moving the head about, it is only possible to localize smells crudely.) Because the olfactory system does not have to map the spatial pattern of an odor in the same way that the visual system has to map the spatial patterns of light, neural odor maps may be available for other purposes, such as discrimination between a huge number of different chemicals.

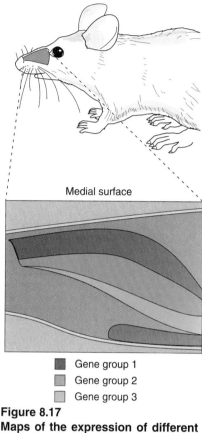

Medial surface

■ Gene group 1
■ Gene group 2
□ Gene group 3

Figure 8.17
Maps of the expression of different olfactory receptor proteins on the olfactory epithelium of a mouse. Three different groups of genes were mapped in this case, and each had a different, nonoverlapping zone of distribution. (Source: Adapted from Ressler et al., 1993, p. 602.)

CONCLUDING REMARKS

The chemical senses are a good place to begin learning about sensory systems. The functions of taste and smell are (usually) obvious, and most of their basic mechanisms are utilized (with modification) by other sensory systems. More significantly, the molecular pathways of gustatory and olfactory transduction are very similar to the signaling systems used in every cell of the body for functions as diverse as neurotransmission and fertilization. Similarities among sensory systems also extend to the level of neural systems. In the chapters that follow, we will see trends in the anatomy and physiology of systems dealing with light, sound, and pressure.

KEY TERMS

Introduction

gustation

olfaction

chemoreceptor

Taste

papilla

taste bud

taste receptor cell

receptor potential

transduction

gustatory nucleus

ventral posterior medial (VPM) nucleus

primary gustatory cortex

Smell

pheromone

olfactory epithelium

olfactory bulb

glomerulus

✔ REVIEW QUESTIONS

1. Most tastes in the real world are some combination of the five basic tastes. What other sensory factors can help define the specific perceptions associated with a particular food?

2. The transduction of saltiness is accomplished, in part, by a Na^+-specific channel. Why would a sugar-specific membrane channel be a poor mechanism for the transduction of sweetness?

3. Some tastants apparently cause a hyperpolarization of the receptor cell membrane. Do the transduction mechanisms outlined in this chapter suggest ways that the cells might hyperpolarize? If not, name two possible mechanisms that might account for a negative shift in the membrane potential.

4. Why would the size of an animal's olfactory epithelium (and consequently the number of receptor cells) be related to its olfactory acuity?

5. Receptor cells of the gustatory and olfactory systems undergo a constant cycle of growth, death, and maturation. Therefore, the connections they make with the brain must be continually renewed as well. Can you propose a set of mechanisms that would allow the connections to be remade in a specific way, again and again, over the course of an entire lifetime?

6. If the olfactory system does use some kind of spatial mapping to encode specific odors, how might the rest of the brain read the map?

CHAPTER
9

The Eye

INTRODUCTION

PROPERTIES OF LIGHT

STRUCTURE OF THE EYE

**IMAGE FORMATION BY
THE EYE**

**MICROSCOPIC ANATOMY
OF THE RETINA**

PHOTOTRANSDUCTION

RETINAL PROCESSING

RETINAL OUTPUT

CONCLUDING REMARKS

KEY TERMS

REVIEW QUESTIONS ✔

The adaptive value of **vision** in animal evolution is obvious. The detection of light enables an individual to detect prey, predators, and mates, even at a great distance. The success of the visual process requires that light reflected off distant objects be localized in reference to the individual and the environment; that some form of object identification take place based on size, shape, color, and past experience; that the movement of objects be detected; and that the detection of objects is possible in the full range of lighting conditions normally experienced by the individual in its habitat.

Light is electromagnetic energy that is emitted in the form of waves. We can safely say that we live in a turbulent sea of electromagnetic radiation. Like any ocean, this sea has large waves and small waves, short wavelets and long rollers. The waves crash into objects and are absorbed, scattered, reflected, and bent. Because of the nature of electromagnetic waves and their interactions with the environment, the visual system can extract information about the world. This is a big job, and it requires a lot of neural machinery. However, the mastery of vision over the course of vertebrate evolution has had surprising rewards. It has provided new ways to communicate, given rise to brain mechanisms for predicting the trajectory of objects and events in time and space, allowed for new forms of mental imagery and abstraction, and led to the creation of a world of art. The significance of vision is perhaps best attested to by the fact that about half of the human cerebral cortex is devoted to the analysis of the visual world.

The mammalian visual system begins with the eye. At the back of the eye is the **retina**, which contains photoreceptors specialized to convert light energy into neural activity. The rest of the eye acts like a camera and forms crisp, clear images of the world on the retina. Like a quality 35-mm camera, the eye automatically adjusts to differences in illumination to maximize the "depth of field" and automatically focuses itself on objects of interest. The eye has some additional features not yet available on cameras, such as the ability to track moving objects (by eye movement) and the ability to keep its transparent surfaces clean (by tears and blinking).

While much of the eye might be considered to be analogous to a camera, the retina is much more than film. In fact, as mentioned in Chapter 7, the retina is actually part of the brain. (Think about that the next time you look deeply into someone's eyes.) In a sense, each eye has two overlapping retinas: one specialized for low light levels that we encounter from dusk to dawn; and another specialized for higher light levels, and for the detection of color, from sunrise to sunset. Regardless of the time of day, however, the output of the retina is not a faithful reproduction of the intensity of the light falling on it. Rather, the retina is specialized to detect *differences* in the intensity of light falling on different parts of it, (almost) regardless of the absolute intensity. This capability probably accounts for the fact that print looks black on a white page both outside on a sunny day and inside a dimly lit room. Image processing is well under way in the retina, before any visual information reaches the rest of the brain.

Axons of retinal neurons are bundled into optic nerves, which distribute visual information (in the form of action potentials) to several brain structures that perform different functions. Some targets of the optic nerves are involved in regulating biological rhythms, which are synchronized with the light-dark diurnal cycle; others are involved in the control of eye position and optics. However, the first synaptic relay in the pathway that serves visual perception occurs in a cell group of the dorsal thalamus called the *lateral geniculate nucleus*, or *LGN*. From the LGN, visual information ascends to the cerebral cortex, where it is interpreted and remembered.

In this chapter, we explore the eye and the retina. We shall see how light carries information to our visual system, how the eye forms images on the retina, and how the retina converts light energy into neural signals that can be used to extract information about luminance and color differences. In Chapter 10, we will pick up the visual pathway at the back of the eye and take it through the thalamus to the cerebral cortex.

PROPERTIES OF LIGHT

The visual system uses light to form images of the world around us. Let's briefly review the physical properties of light and its interactions with the environment that are important for an understanding of vision.

Light

Light is the electromagnetic radiation that is visible to our eyes. Electromagnetic radiation can be described as a wave of energy. Like any wave, electromagnetic radiation can be characterized in terms of wavelength, the distance between successive waves; frequency, the number of waves per second; and amplitude, the difference between wave trough and peak (Figure 9.1).

The energy content of electromagnetic radiation is proportional to its frequency. Radiation emitted at a high frequency (short wavelengths) has the highest energy content; examples are gamma radiation and X-rays, with wavelengths less than 10^{-9} m (< 1 nm). Conversely, radiation emitted at lower frequencies (longer wavelengths) has less energy; examples are radar and radio waves, having wavelengths greater than 1 mm. Only a small part of the electromagnetic spectrum is detectable by our visual system; visible light consists of wavelengths of 400–700 nm (Figure 9.2). As first shown by Isaac Newton early in the eighteenth century, the mix of wavelengths in this range emitted by the sun appears to humans as white, whereas light of a single wavelength appears as one of the colors of the rainbow. It is interesting to note that a "hot" color like red or orange consists of light with a longer wavelength, and hence has *less* energy, than a "cool" color like blue

Figure 9.1
Characteristics of electromagnetic radiation.

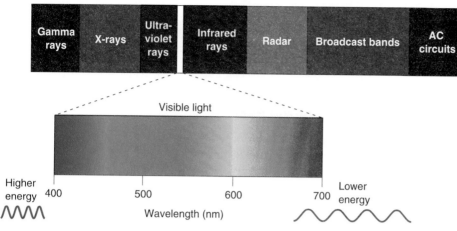

Figure 9.2
The electromagnetic spectrum. Only electromagnetic radiation with wavelengths of 400–700 nm is visible to the naked human eye. Within this visible spectrum, different wavelengths appear as different colors.

or violet. Clearly, colors are themselves "colored" by the brain, based on our subjective experiences.

Optics

In a vacuum, a wave of electromagnetic radiation will travel in a straight line and thus can be described as a *ray*. Light rays in our environment also travel in straight lines until they interact with the atoms and molecules of the atmosphere and objects on the ground. These interactions include reflection, absorption, and refraction (Figure 9.3). The study of light rays and their interactions is called *optics*.

Reflection is the bouncing of light rays off a surface. The manner in which a ray of light is reflected depends on the angle at which it strikes the surface. A ray striking a mirror perpendicularly is reflected 180° back upon itself, a ray striking the mirror at a 45° angle is reflected 90°, and so on. Most of what we see is light that has been reflected off objects in our environment.

Absorption is the transfer of light energy to a particle or surface. You can feel this energy transfer on your skin on a sunny day, as visible light is absorbed and warms you up. Surfaces that appear black absorb the energy of all visible wavelengths. Some compounds absorb light energy only in a limited range of wavelengths, then reflect the remaining wavelengths. This property is the basis for the colored pigments of paints. For example, a blue pigment absorbs long wavelengths but reflects a range of short wavelengths centered on 480 nm that are perceived as blue. As we will see in a moment, light-sensitive photoreceptor cells in the retina contain pigments and use the energy absorbed from light to generate changes in membrane potential.

A physical property of light that is important for image formation by the eye is **refraction**, the bending of light rays that can occur when they travel from one transparent medium to another. Consider a ray of light passing from the air into a pool of water. If the ray strikes the water surface perpendicularly, it will pass through in a straight line. However, if light strikes the surface at an angle, it will bend toward a line that is perpendicular to the surface. This bending of light occurs because the speed of light differs in the two media; light passes through air more rapidly than through water. The greater the difference between the speed of light in the two media, the greater the angle of refraction. The transparent media in the eye bend light rays to form images on the retina.

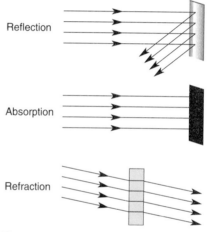

Figure 9.3
Some interactions between light and the environment.

STRUCTURE OF THE EYE

The eye is an organ specialized for the detection, localization, and analysis of light. Here we introduce the structure of this remarkable organ in terms of its gross anatomy, ophthalmoscopic appearance, and cross-sectional anatomy.

Gross Anatomy of the Eye

When you look into someone's eyes, what are you really looking at? The main structures are shown in Figure 9.4. The **pupil** is the opening that allows light to enter the eye and reach the retina; it appears dark because of the light-absorbing pigments in the retina. The size of the pupil is controlled by a circular muscle, the **iris**, whose pigmentation represents what we call the eye's color. The pupil and iris are covered by the glassy transparent external surface of the eye, the **cornea**. The cornea lacks blood vessels and is nourished by the fluid behind it, the **aqueous humor**, and, on its outside surface, by the tear film that is continuously replenished by the blinking of the eyelids. The cornea is continuous with the **sclera**, the "white of the eye," which forms the tough wall of the eyeball. Inserted into the sclera are three pairs of muscles, the **extraocular muscles**, that move the eyeball in the bony orbits of the skull. These muscles normally are not visible because they lie behind the **conjunctiva**, a membrane that folds back from the inside of the eyelids and attaches to the sclera. The **optic nerve**, carrying axons from the retina, exits the eye at the back, passes through the orbit, and reaches the brain at its base, near the pituitary gland.

Ophthalmoscopic Appearance of the Eye

Another view of the eye is afforded by the ophthalmoscope, a device that enables one to peer into the eye through the pupil to the retina (Figure 9.5). The most obvious feature of the retina viewed through an ophthalmoscope is the blood vessels on its surface. These retinal vessels originate from a pale circular region called the **optic disk**, which is also where the optic nerve fibers exit the retina. The optic disk is also sometimes called the optic nerve head.

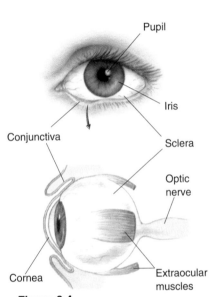

Figure 9.4
Gross anatomy of the human eye.

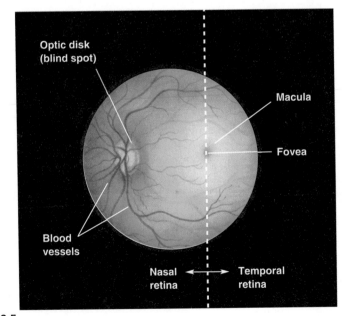

Figure 9.5
Ophthalmoscopic appearance of the retina.

It is interesting to note that the sensation of light cannot occur at the optic disk because there are no photoreceptors here, nor can it occur where the large blood vessels exist because the vessels cast shadows on the retina. And yet, our perception of the visual world appears seamless. We are not aware of any holes in our field of vision because the brain fills in our perception of these areas. (They are not *actually* filled in—we fool ourselves into thinking they are.) There are tricks by which we can demonstrate the "blind" retinal regions (Box 9.1).

At the middle of each retina is a darker colored region with a yellowish hue. This is the **macula** (from the Latin word for "spot"), the part of the retina for central (as opposed to peripheral) vision. Besides its color, the macula is distinguished by the relative absence of large blood vessels. Notice in

---

Box 9.1 | **OF SPECIAL INTEREST**

Demonstrating the Blind Regions of Your Eye

A look through an ophthalmoscope leads to the startling revelation that there is a sizable hole in the retina. The region where the optic nerve axons exit the eye and the retinal blood vessels enter the eye, the optic disk, is devoid of any photoreceptors at all. Moreover, the blood vessels coursing across the retina are opaque and block the light from falling on photoreceptors beneath them. Although we normally do not notice them, these blind regions can be demonstrated. Look at Figure A. Hold the book about 1.5 ft away, close your right eye, and fixate on the cross with your left eye. Move the book (or your head) around slightly, and eventually you will find a position where the black circle disappears. At this position, the spot is imaged on the optic disk of the left eye. This region of visual space is called the blind spot for the left eye.

The blood vessels are a little tricky to demonstrate, but give this a try. Get a standard household flashlight. In a dark or dimly lit room, close your left eye (it helps to hold the eye closed with your finger so you can open your right eye further). Look straight ahead with the open right eye, and shine the flashlight at an angle into the corner of the eye from the side. Jiggle the light back and forth, up and down. If you're lucky, you'll see an image of your own retinal blood vessels. This is possible because the illumination of the eye at this oblique angle causes the retinal blood vessels to cast long shadows on the adjacent regions of retina. For the shadows to be visible, they must be swept back and forth on the retina, hence the jiggling of the light.

If we have all these light-insensitive regions in the retina, why does the visual world appear uninterrupted and seamless? The detailed answer to this question is still unknown, but mechanisms in the visual cortex apparently "fill in" the missing regions. Filling in can be demonstrated with the stimulus shown in Figure B. Again, fixate on the cross with your left eye. You will find that at the place where the break in the line falls is imaged on the blind spot, your perception will be of a continuous, uninterrupted line.

Figure A.

Figure B.

Figure 9.5 that the vessels arc from the optic disk to the macula; this is also the trajectory of the optic nerve fibers from the macula en route to the optic nerve head. The relative absence of large blood vessels in this region of the retina is one of the specializations that improves the quality of central vision. Another specialization of the central retina can sometimes be discerned with the ophthalmoscope: a dark spot, about 2 mm in diameter, called the **fovea**. "Fovea" is from the Latin for "pit," and the retina is thinner in the fovea than elsewhere. Because it marks the center of the retina, the fovea is a convenient anatomical reference point. Thus, the part of the retina that lies closer to the nose than the fovea is called nasal, the part that lies further away is called temporal, the part of the retina above the fovea is called superior, and that below it is called inferior.

Cross-Sectional Anatomy of the Eye

A cross-sectional view of the eye shows the path taken by light as it passes through the cornea toward the retina (Figure 9.6). This view reveals the **lens**, another transparent surface located behind the iris. The lens is suspended by ligaments (called zonule fibers) attached to the **ciliary muscles**, which are attached to the sclera and form a ring inside the eye. When they contract, the middle of the ring gets smaller, and the tension on the suspensory ligaments decreases. When the tension of these ligaments decreases, the lens tends to become fatter because of its natural elasticity. Conversely, relaxation of the ciliary muscles causes the ring to get bigger, and, consequently, the tension on the suspensory ligaments increases. This has the effect of stretching the lens into a flat shape. As we shall see, these changes in the shape of the lens enable our eyes to adjust their focus to different viewing distances. The lens also divides the interior of the eye into two compartments containing slightly different fluids. The aqueous humor, introduced previously, is the watery fluid that lies between the cornea and the lens. The more viscous, jellylike **vitreous humor** lies between the lens and the retina; its pressure serves to keep the globe of the eye spherical.

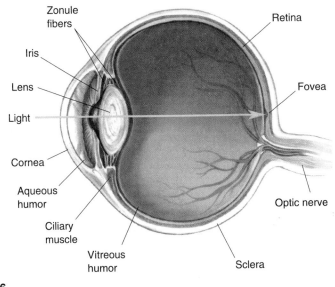

Figure 9.6
The eye in cross section.

IMAGE FORMATION BY THE EYE

The eye collects the light rays emitted by or reflected off objects in the environment, and focuses them onto the retina to form images. Let's see how the different structures in the eye contribute to this function.

Refraction by the Cornea

Consider the light emitted from a distant source, perhaps a bright star at night. We see the star as a point of light because the eye focuses the light into a **point image** on the retina. The light rays striking the surface of the eye from a distant star are virtually parallel, so they must be bent by the process of refraction.

Recall that as light passes into a medium where its speed is slowed, it will bend toward a line that is perpendicular to the border between the media. This is precisely the situation as light strikes the cornea and passes from the air into the aqueous humor. As shown in Figure 9.7, the light rays that strike the curved surface of the cornea bend so that they converge on the back of the eye; those that enter the center of the eye pass straight to the retina. The distance from the refractive surface to the point where parallel light rays converge is called the *focal distance*. Focal distance depends on the curvature of the cornea—the tighter the curve, the shorter the focal distance. The reciprocal of the focal distance in meters is a unit of measurement called the **diopter**. The cornea has a refractive power of about 42 diopters, which means that parallel light rays striking the corneal surface will be focused 0.024 m (2.4 cm) behind it, about the distance from cornea to retina.

Remember that refractive power depends on the slowing of light at the air-cornea interface. If we replace air with a medium that passes light at about the same speed as the eye, the refractive power of the cornea will be eliminated. This is why things look blurry when you open your eyes underwater; the water-cornea interface has very little focusing power. A scuba mask restores the air-cornea interface and, consequently, the refractive power of the eye.

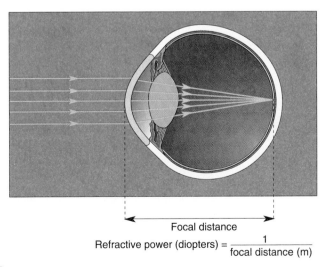

Focal distance

$$\text{Refractive power (diopters)} = \frac{1}{\text{focal distance (m)}}$$

Figure 9.7
Refraction by the cornea.

Accommodation by the Lens

Although the cornea performs most of the eye's refraction, the lens also contributes another dozen or so diopters to the formation of a sharp image at a distant point. However, the lens is involved more importantly in the formation of point images closer than about 9 m. As objects approach, the light rays originating at a point can no longer be considered to be parallel. Rather, these rays diverge, and greater refractive power is required to bring them into focus on the retina. This additional focusing power is provided by changing the shape of the lens, a process called **accommodation** (Figure 9.8).

During accommodation, the ciliary muscles contract (relieving the tension on the suspensory ligaments), and the lens, because of its natural elasticity, becomes rounder. This rounding increases the curvature of the lens surfaces, thereby increasing their refractive power. The ability to accommodate changes with age. An infant can focus objects just beyond his or her nose, whereas many middle-aged adults cannot focus objects closer than about arm's length. Fortunately, this and other defects of the eye's optics can be corrected with artificial lenses (Box 9.2).

Pupillary Light Reflex

In addition to the cornea and the lens, the pupil also contributes to the optical qualities of the eye by continuously adjusting for different ambient light levels. Check this for yourself by standing before a bathroom mirror with the lights out for a few seconds, and then watch your pupils change size when the light goes on. This **pupillary light reflex** involves connections between the retina and neurons in the brain stem that synapse on the motor neurons that control the iris muscle. An interesting property of this reflex is that it is *consensual*; that is, shining a light into only one eye causes the constriction of the pupils of both eyes. It is unusual, indeed, when the pupils are not the same size; the lack of a consensual pupillary light reflex is often taken as a sign of a serious neurological disorder involving the brain stem.

Constriction of the pupil has the effect of increasing the depth of focus, just like decreasing the aperture size (increasing the f-stop) on a camera lens. To understand why this is true, consider two points in space, one close and the other far away. When the eye accommodates to the closer point, the image of the farther point no longer forms a point on the retina but rather a

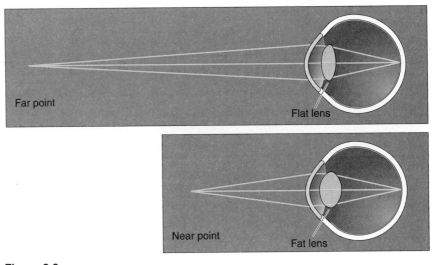

Figure 9.8
Accommodation: refraction by the lens.

Box 9.2	OF SPECIAL INTEREST

Corrective Lenses

When the ciliary muscles are relaxed and the lens is flat, the eye is said to be *emmetropic* if parallel light rays from a distant point source are focused sharply on the back of the retina. ("Emmetropic" is from the Greek words *emmetros*, "in proper measure," and *ope*, "sight.") Stated another way, the emmetropic eye focuses parallel light rays on the retina without the need for accommodation (Figure A). Now consider what happens when the eyeball is too short from front to back (Figure B). The light rays are focused at some point *behind* the retina. Without accommodation, all the retina can see of the point image is a blurry circle. This condition is known as *hyperopia*, or farsightedness, because much of the accommodative power of the lens is needed to image far points and the lens cannot accommodate enough to image near

points. Farsightedness can be corrected, however, by placing a convex glass or plastic lens in front of the eye (Figure C). The curved front edge of the lens, like the cornea, bends light toward the center of the retina. Because the light again passes from glass into air, so does the back of the lens (light going from glass to air speeds up and is bent *away* from the perpendicular). If the eyeball is too long rather than too short, parallel rays will converge before the retina, cross, and again be imaged on the retina as a blurry circle (Figure D). This condition is known as *myopia*, or nearsightedness. Thus, for the nearsighted eye to see distant points clearly, artificial concave lenses must be used to move the point image back onto the retina (Figure E). Some eyes have irregularities in the curvature of the cornea or lens that lead to different amounts of refraction depending on whether light rays enter the eye in a horizontal or a vertical axis.

continued on page 220

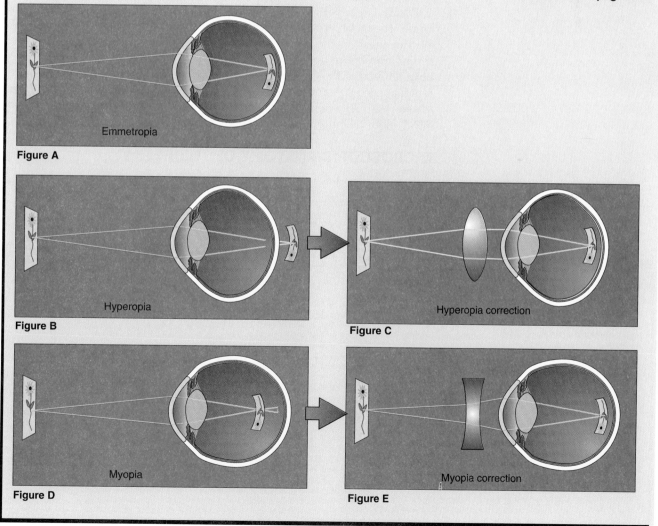

Emmetropia

Figure A

Hyperopia

Figure B

Hyperopia correction

Figure C

Myopia

Figure D

Myopia correction

Figure E

blurred circle. Decreasing the aperture—constricting the pupil—reduces the size of this blurred circle so that its image more closely approximates a point. In this way, distant objects appear to be less out of focus.

The Visual Field

The structure of the eyes, and where they sit in our head, limits how much of the world we can see at any one time. Let's investigate the extent of the space seen by one eye. Holding a pencil horizontally in your right hand, close your left eye and look at a point straight ahead. Keeping your eye fixated on this point, slowly move the pencil to the right (toward your right ear) across your field of view until the whole pencil disappears. Repeat this exercise, moving the pencil to the left, and then up and down. The points where you can no longer see the pencil mark the limits of the **visual field** for your right eye. Now look at the middle of the pencil as you hold it horizontally in front of you. Figure 9.9 shows how the light reflected off this pencil falls on your retina. Notice that the image is inverted; the left visual field is imaged on the right side of the retina, and the right visual field is imaged on the left side of the retina.

Visual Acuity

The ability of the eye to distinguish between two nearby points is called **visual acuity**. Acuity depends on several factors, but especially the spacing of photoreceptors in the retina and the precision of the eye's refraction.

Distance across the retina can be described in terms of degrees of **visual angle**. The moon, for example, subtends an angle of about 0.5° (Figure 9.10). We can speak of the eye's ability to resolve points that are separated by a certain number of degrees of visual angle. The Snellen eye chart, which we have all read at the doctor's office, tests our ability to discriminate characters that each subtend various angles at a viewing distance of 20 feet. Your vision is 20/20 when you can recognize a letter that subtends an angle of 0.083° (equivalent to 5 minutes of arc, where 1 minute is 1/60 of a degree).

MICROSCOPIC ANATOMY OF THE RETINA

Now that we have an image formed on the retina, we can get to the neuroscience of vision: the conversion of light energy into neural activity. To

---

Box 9.2 **OF SPECIAL INTEREST**

continued from page 219

This condition is called *astigmatism*, and it can also be corrected by using artificial lenses.

Even if you are fortunate enough to have perfectly shaped eyeballs and a symmetrical refractive system, you probably will not escape *presbyopia*. This condition is a hardening of the lens that accompanies the aging process and is thought to be explained by the fact that new lens cells are generated throughout

life, while none are lost. The hardening denies the lens of its elasticity, leaving it unable to thicken sufficiently during accommodation and unable to flatten sufficiently during relaxation. The correction for presbyopia, first introduced by none other than Benjamin Franklin, is a bifocal lens. These lenses are concave on top to assist far vision and convex on the bottom to assist near vision.

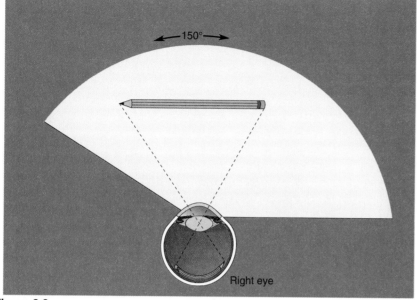

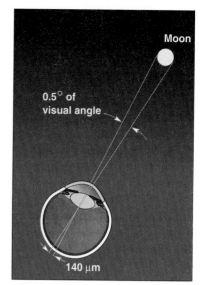

Figure 9.9
The visual field for one eye. The visual field is the total amount of space that can be viewed by the retina when the eye is fixated straight ahead. Notice how the image of an object in the visual field (pencil) is inverted on the retina.

Figure 9.10
Visual angle. Distances across the retina can be expressed as degrees of visual angle.

begin our discussion of image processing in the retina, we must introduce the cellular architecture of this bit of brain.

The basic system of retinal information processing is shown in Figure 9.11. The most direct pathway for visual information flow is from **photoreceptors** to **bipolar cells** to **ganglion cells**. The ganglion cells fire action potentials in response to light, and these impulses propagate down the optic nerve to the rest of the brain. Besides the cells in this direct path from photoreceptor to brain, retinal processing is influenced by two additional cell types. **Horizontal cells** receive input from the photoreceptors and project neurites laterally to influence surrounding bipolar cells. **Amacrine cells** receive input from bipolar cells and project laterally to influence surrounding ganglion cells.

There are two important points to remember here. First, *the only light-sensitive cells in the retina are the photoreceptors;* all other cells are influenced by light only via direct and indirect synaptic interactions with the photoreceptors. Second, *the ganglion cells are the only source of output from the retina;* no other retinal cell type projects an axon through the optic nerve. Now let's take a look at how the different cell types are arranged in the retina.

Laminar Organization of the Retina

Figure 9.12 shows that the cells of the retina are organized in layers. (In other words, the retina has a *laminar organization*.) Notice that the layers are seemingly inside-out: Light must pass from the vitreous humor through the ganglion cells and bipolar cells *before* it reaches the photoreceptors. Because these cells are relatively transparent, image distortion is minimal.

The cell layers of the retina are named in reference to the middle of the eyeball. Thus, the innermost layer is the **ganglion cell layer,** which contains the cell bodies of the ganglion cells. Next is the **inner nuclear layer,** which contains the cell bodies of the bipolar cells, the horizontal and amacrine cells. The next layer is the **outer nuclear layer,** which contains the cell bod-

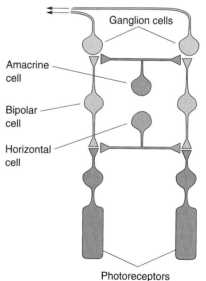

Figure 9.11
The basic system of retinal information processing. Information about light flows from the photoreceptors to bipolar cells to ganglion cells, which project axons out of the eye in the optic nerve. Horizontal cells and amacrine cells modify the responses of bipolar cells and ganglion cells via lateral connections.

Light

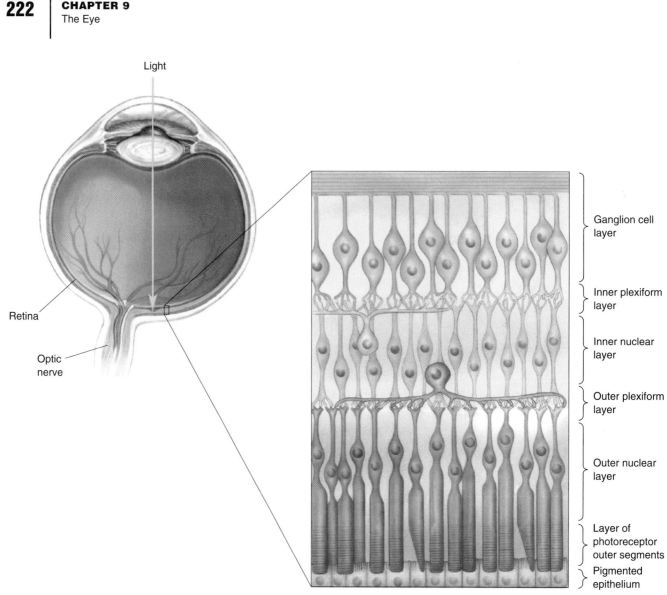

Retina

Optic
nerve

Ganglion cell
layer

Inner plexiform
layer

Inner nuclear
layer

Outer plexiform
layer

Outer nuclear
layer

Layer of
photoreceptor
outer segments

Pigmented
epithelium

Figure 9.12
Laminar organization of the retina. Notice how light must pass through several cell layers before it reaches the photoreceptors at the back of the retina.

ies of the photoreceptors. Finally, the **layer of photoreceptor outer segments** contains the light-sensitive elements of the retina. The outer segments are embedded in a pigmented epithelium that is specialized to absorb any light that passes entirely through the retina. Interspersed between the ganglion cell layer and the inner nuclear layer is the **inner plexiform layer**, which contains the axons and dendrites of ganglion cells, bipolar cells, and amacrine cells. Between the outer and inner nuclear layers is the **outer plexiform layer**, which contains the axons and dendrites of bipolar cells and horizontal cells, and the synaptic terminals of the photoreceptors.

Photoreceptor Structure

The conversion of electromagnetic radiation into neural signals occurs in the 125 million photoreceptors at the back of the retina. Every photoreceptor has four regions: an outer segment, an inner segment, a cell body, and a synaptic terminal. The outer segment contains a stack of membranous disks. Light-sensitive photopigments in the disk membranes absorb light, thereby triggering changes in the photoreceptor membrane potential (discussed below). Figure 9.13 shows the two types of photoreceptor in the retina, easily distinguished by the appearance of their outer segments. **Rod photoreceptors** have a long, cylindrical outer segment, containing many disks. **Cone photoreceptors** have a shorter, tapering outer segment, containing relatively few membranous disks.

The structural differences between rods and cones correlate with important functional differences. For example, the greater number of disks and higher photopigment concentration in rods makes them 1000 times more sensitive to light than cones. Indeed, under nighttime lighting, or *scotopic* conditions, only rods contribute to vision. Conversely, under daytime lighting, or *photopic* conditions, cones do the bulk of the work. For this reason, the retina is said to be *duplex*: a scotopic retina using only rods, and a photopic retina using mainly cones.

Rods and cones differ in other respects as well. All rods contain the same photopigment, but there are three types of cone, each containing a different pigment. The variations among pigments make the different cones sensitive to different wavelengths of light. As we shall see in a moment, only the cones, not the rods, are responsible for our ability to see color.

Regional Differences in Retinal Structure

Retinal structure varies from the fovea to the retinal periphery. In general, the peripheral retina has more rods and fewer cones (Figure 9.14). It also has a larger ratio of photoreceptors to ganglion cells. The combined effect of this arrangement is that the peripheral retina is more sensitive to light because (1) rods are specialized for low light and (2) there are more photoreceptors feeding information to each ganglion cell. You can prove this to yourself on a starry night. (It's fun; try it with a friend.) First, spend about 20 minutes in the dark getting oriented, and then gaze at a bright star. Fixating on this star, search your peripheral vision for a dim star. Then move your eyes to look at this dim star. You will find that the faint star disappears when it is imaged on the central retina (when you look straight at it) but reappears when it is imaged on the peripheral retina (when you look slightly to the side of it).

The same characteristics that enable the peripheral retina to detect faint stars at night make it relatively poor at resolving fine details in daylight. This is because daytime vision requires cones, and because good visual acuity requires a low ratio of photoreceptors to ganglion cells. The region of retina most highly specialized for high-resolution vision is the fovea. Recall that the fovea is a thinning of the retina at the center of the macula. In cross section, the fovea appears as a pit in the retina. Its pitlike appearance is due to the lateral displacement of ganglion cells from the fovea, allowing light to strike the photoreceptors without passing through the other retinal cell layers (Figure 9.15). The fovea also is unique because it contains no rods; all the photoreceptors are cones.

PHOTOTRANSDUCTION

The photoreceptors convert, or *transduce*, light energy into changes in membrane potential. We begin our discussion of phototransduction with rods, which outnumber cones in the human retina by 20 to 1. Most of what has been learned about phototransduction by rods has proven to be applicable to cones as well.

Phototransduction in Rods

As we discussed in Part I, one way information is represented in the nervous system is as changes in the membrane potential of neurons. Thus, we look for a mechanism by which the absorption of light energy can be trans-

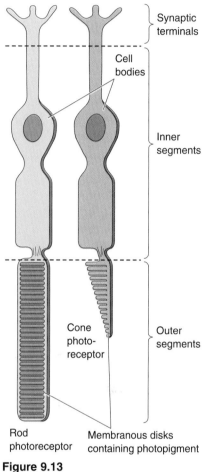

Synaptic terminals

Cell bodies

Inner segments

Cone photoreceptor

Outer segments

Rod photoreceptor

Membranous disks containing photopigment

Figure 9.13
A rod and a cone.

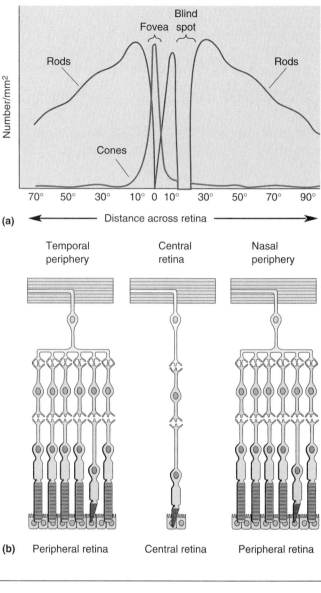

(a)

Figure 9.14
Regional differences in retinal structure. (a) Cones are found primarily in the central retina, within 10° of the fovea. Rods are absent from the fovea and are found mainly in the peripheral retina. **(b)** In the central retina, relatively few photoreceptors feed information directly to a ganglion cell; in the peripheral retina, many photoreceptors provide input. This arrangement makes the peripheral retina better at detecting dim light but the central retina better for high-resolution vision.

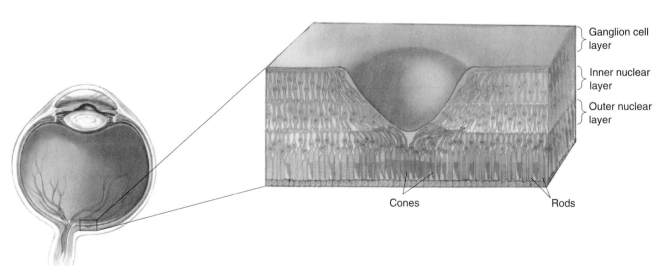

Figure 9.15
The fovea in cross section. The ganglion cell layer and the inner nuclear layer are displaced laterally to allow light to strike the foveal photoreceptors directly.

duced into a change in the photoreceptor membrane potential. In many respects, this process is analogous to the transduction of chemical signals into electrical signals that occurs during synaptic transmission. At a G-protein-coupled neurotransmitter receptor, for example, the binding of transmitter to the receptor activates G-proteins in the membrane, which in turn stimulate various effector enzymes (Figure 9.16a). These enzymes alter the intracellular concentration of cytoplasmic second messenger molecules, which (directly or indirectly) change the conductance of membrane ion channels, thereby altering membrane potential. Similarly, in the photoreceptor, light stimulation of the photopigment activates G-proteins, which in turn activate an effector enzyme that changes the cytoplasmic concentration of a second messenger molecule. This change causes a membrane ion channel to close, and the membrane potential is thereby altered (Figure 9.16b). One of the interesting functional consequences of using such a biochemical cascade for transduction is signal *amplification*. Many G-proteins are activated by each photopigment molecule, and many enzyme molecules are activated by each G-protein. This amplification gives our visual system the ability to detect as little as a single photon, the elementary unit of light energy.

Recall from Chapter 3 that a typical neuron at rest has a membrane potential of about –65 mV, close to the equilibrium potential for K^+. In contrast, in complete darkness, the membrane potential of the rod outer segment is about –30 mV. This depolarization is caused by the steady influx of Na^+ through special channels in the outer segment membrane (Figure

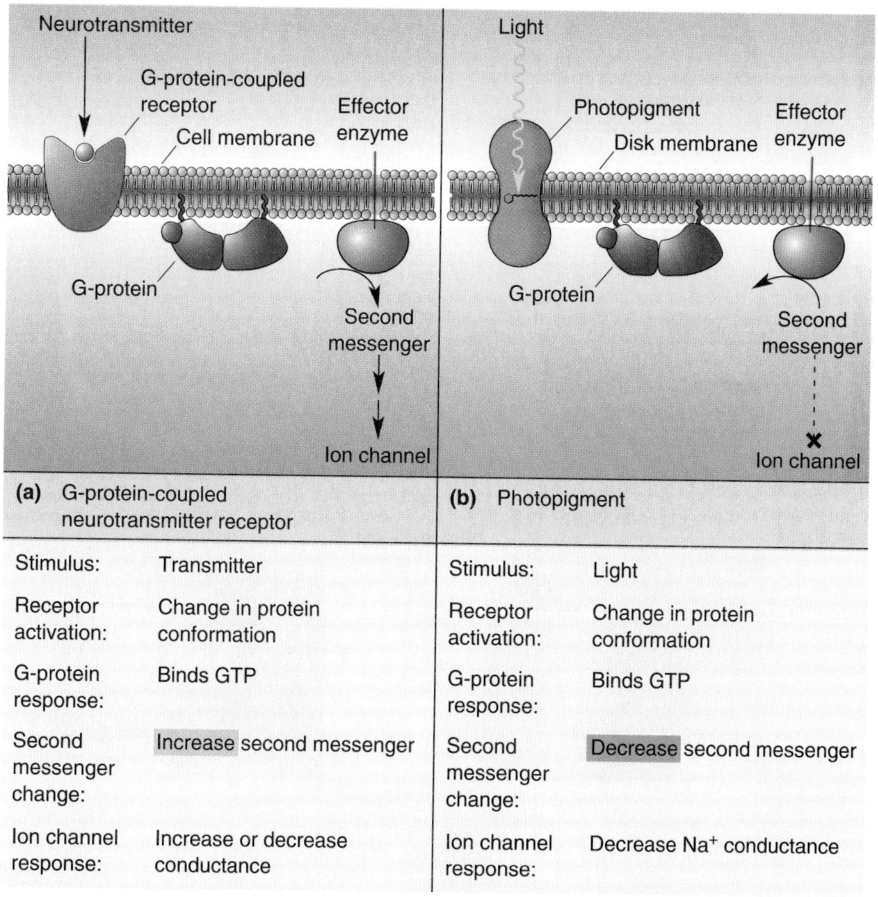

Figure 9.16
Comparison of events triggered by activation of (a) a G-protein-coupled neurotransmitter receptor and (b) a photopigment.

9.17a). This movement of positive charge across the membrane is called the **dark current**. In 1985, a team of Russian scientists led by Evgeniy Fesenko found that these sodium channels are stimulated to open—are gated—by an intracellular second messenger called **cyclic guanosine monophosphate**, or **cGMP** (see Box 8.2). Evidently, cGMP is continually produced in the photoreceptor by the enzyme guanylate cyclase, keeping the Na⁺ channels open. Light reduces cGMP; this causes the Na⁺ channels to close, and the membrane potential becomes *more negative* (Figure 9.17b). Thus, *photoreceptors hyperpolarize in response to light*.

The hyperpolarizing response to light is initiated by the absorption of electromagnetic radiation by the photopigment in the membrane of the stacked disks in the rod outer segments. In the rods, this pigment is called **rhodopsin**. Rhodopsin can be thought of as a receptor protein with a prebound chemical agonist. The receptor protein is called *opsin*, and it has the

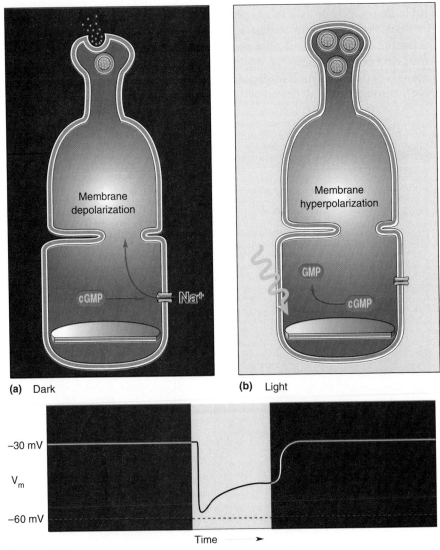

Figure 9.17
Hyperpolarization of photoreceptors in response to light. Photoreceptors are continuously depolarized in the dark because of an inward sodium current, the dark current. **(a)** Sodium enters the photoreceptor through a cGMP-gated channel. **(b)** Light leads to the activation of an enzyme that destroys cGMP, thereby shutting off the Na⁺ current and hyperpolarizing the cell.

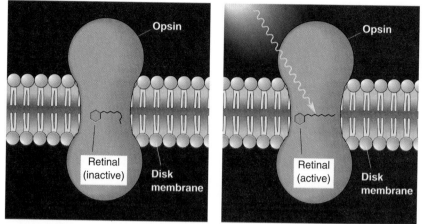

Figure 9.18
Activation of rhodopsin by light. Rhodopsin consists of a protein with seven trans-membrane alpha helices, called an opsin, and a small molecule derived from vitamin A, called retinal. Retinal undergoes a change in atomic conformation when it absorbs light, activating the opsin.

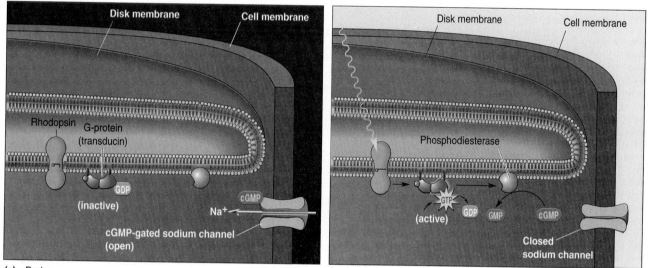

(a) Dark
(b) Light

Figure 9.19
The light-activated biochemical cascade in a photoreceptor. (a) In the dark, cGMP gates a sodium channel, causing an inward Na^+ current and depolarization of the cell. **(b)** Activation of rhodopsin by light energy causes the G-protein (transducin) to exchange GDP for GTP (see Chapter 6), which in turn activates the enzyme phosphodiesterase (PDE). PDE breaks down cGMP and shuts off the dark current.

seven transmembrane alpha helices typical of G-protein-coupled receptors throughout the body. The prebound agonist is called *retinal*, a derivative of vitamin A. Absorption of light causes a change in the atomic conformation of retinal so that it activates the opsin (Figure 9.18). This process is called bleaching because it changes the wavelengths absorbed by the rhodopsin (the photopigment literally changes color from purple to yellow). The bleaching of rhodopsin stimulates a G-protein called **transducin** in the disk membrane, which in turn activates the effector enzyme **phosphodiesterase (PDE)**, which breaks down the cGMP that is normally present in the cytoplasm of the rod (in the dark). The reduction in cGMP causes the Na^+ channels to close and the membrane to hyperpolarize.

The complete sequence of events of phototransduction in rods is illustrated in Figure 9.19.

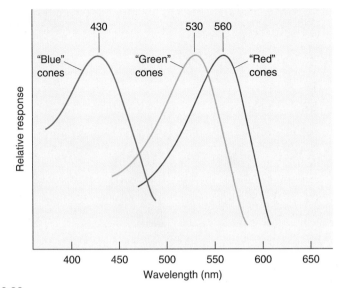

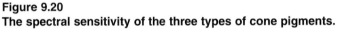

Figure 9.20
The spectral sensitivity of the three types of cone pigments.

Phototransduction in Cones

Prolonged illumination of the rods causes cGMP levels to fall to the point where the response to light becomes *saturated*; additional light causes no more hyperpolarization. This is the situation in bright sunlight. Thus, vision during the day depends entirely on the cones, whose photopigments require more energy to become bleached. The process of phototransduction in cones is virtually the same as in rods; the only major difference is in the type of opsins in the membranous disks of the cone outer segments. The cones in our retinas contain one of three opsins that give the photopigments different spectral sensitivities. Thus, we can speak of "blue" cones that are maximally activated by light with a wavelength of about 430 nm, "green" cones that are maximally activated by light with a wavelength of about 530 nm, and "red" cones that are maximally activated by light with a wavelength of about 560 nm (Figure 9.20).

Color Detection. The color that we perceive is largely determined by the relative activation of blue, green, and red cones. The fact that our visual system detects colors in this way was actually predicted almost 200 years ago by British physicist Thomas Young. Young showed in 1802 that all the colors of the rainbow, including white, could be created by mixing the proper ratio of red, green, and blue light (Figure 9.21). He proposed, quite correctly, that at each point in the retina there exists a cluster of three receptor types, each type being sensitive to either blue, green, or red. Young's ideas were later championed by Hermann von Helmholtz, an influential nineteenth-century German physiologist. (Among his accomplishments is the invention of the ophthalmoscope in 1851.) This theory of color vision came to be known as the **Young-Helmholtz trichromacy theory**. According to the theory, the brain assigns colors based on a comparison of the readout of the three cone types. When all types of cones are equally active, as in broad-spectrum light, we perceive "white." Color blindness results when one or more of the cone photopigment types is missing. The genes for two of the three cone opsins ("red" and "green") are on the X chromosome. Because males have only one X chromosome, genetic defects of color vision occur more often in males than in females. The congenital absence of one type of cone opsin is actually fairly common (1–2% of males) and results in the confusion of shades of red and green.

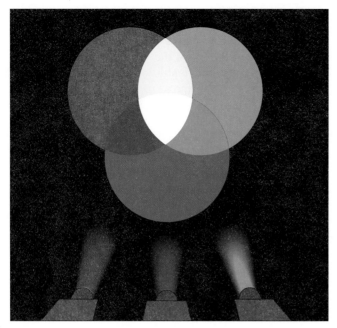

Figure 9.21
Mixing colored lights. Mixing of red, green, and blue light causes equal activation of the three types of cones, and the perception of "white" results.

According to the previous discussion, without the activity of the three cone types, we would not perceive color differences. You can prove this to yourself. Go outside on a dark night and try to distinguish the colors of different objects. It is difficult to detect colors at night because only the rods, with a single type of photopigment, are activated under dim lighting conditions. The peak sensitivity of the rods is to a wavelength of about 500 nm, perceived as blue-green (under photopic conditions). This fact is the basis for two points of view about the design of automobile dashboard indicator lights. According to one view, the lights should be dim blue-green to take advantage of the spectral sensitivity of the rods. According to the other view, the lights should be bright red because this wavelength affects mainly cones, leaving the rods unsaturated, resulting in better night vision.

Dark and Light Adaptation

This transition from all-cone daytime vision to all-rod nighttime vision is not instantaneous; it takes about 20–25 minutes (hence the time needed to get oriented in the star-gazing exercise above). This phenomenon is called **dark adaptation**, or getting used to the dark. Sensitivity to light actually increases a millionfold during this period. Dark adaptation is explained by a number of factors, including dilation of the pupils, the regeneration of unbleached rhodopsin, and an adjustment of the functional circuitry of the retina so that information from more rods is available to each ganglion cell. Because of this tremendous increase in sensitivity, when the dark-adapted eye goes back into bright light, it is temporarily saturated. This explains what happens when you first go outside on a bright day. Over the next 5–10 minutes, the eyes undergo **light adaptation,** reversing the changes in the retina that accompanied dark adaptation. This light-dark adaptation in the duplex retina gives our visual system the ability to operate in light intensities ranging from moonless midnight to bright high noon.

RETINAL PROCESSING

Well before the discovery of how photoreceptors work, researchers were able to explain some of the ways the retina processes visual images. Since about 1950, neuroscientists have studied the action potential discharges of retinal ganglion cells as the retina is stimulated with light. The pioneers of this approach were neurophysiologists Stephen Kuffler and Horace Barlow, Kuffler working in the United States and Barlow working in England (Box 9.3). Their research uncovered which aspects of a visual image were encoded as ganglion cell output. Early studies of horseshoe crabs and frogs gave way to research on cats and monkeys. It soon became clear that the less sophisticated the brain (and central visual system), the more sophisticated the analysis that takes place in the retina. Nonetheless, there are similar principles involved in retinal processing across a wide range of species.

Progress in understanding how ganglion cell properties are generated by synaptic interactions in the retina has been slower. This is because *only ganglion cells fire action potentials*; all other cells in the retina (except some amacrine cells) respond to stimulation with graded changes in membrane potential. Detection of such graded changes requires technically challenging intracellular recording methods, whereas action potentials can be detected using simple extracellular recording methods (see Box 4.1). It was not until the early 1970s that John Dowling and Frank Werblin, of Harvard University, were able to show how ganglion responses are built from the interactions of horizontal and bipolar cells.

The direct path for information flow in the retina is from photoreceptor to bipolar cell to ganglion cell. At each synaptic relay, the responses are modified by the lateral connections of horizontal cells and amacrine cells. We'll first focus on how information is transformed as it passes from photoreceptors to bipolar cells, and then explore ganglion cell output in the next section.

Transformations in the Outer Plexiform Layer

Photoreceptors, like other neurons, release neurotransmitter when depolarized. As we have seen, photoreceptors are depolarized in the dark and are *hyperpolarized* by light. We thus have the counterintuitive situation in which photoreceptors actually release fewer transmitter molecules in the light than in the dark. However, we can reconcile this apparent paradox if we take the point of view that *dark* rather than *light* is the preferred stimulus for a photoreceptor. Thus, when a shadow passes across a photoreceptor, it responds by depolarizing and releasing neurotransmitter. The transmitter substance of the photoreceptor is thought to be the amino acid glutamate.

In the outer plexiform layer, each photoreceptor is in synaptic contact with two types of retinal neuron: bipolar cells and horizontal cells. Remember, bipolar cells create the direct pathway from photoreceptors to ganglion cells; horizontal cells feed information laterally in the outer plexiform layer to influence the activity of neighboring bipolar cells.

Bipolar Cell Receptive Fields. Bipolar cells can be divided into two categories based on their responses to the glutamate released by photoreceptors. On one class of bipolar cell, glutamate-gated cation channels mediate a classical depolarizing EPSP from the influx of Na^+ ions; these cells are called **OFF bipolar cells**. A second class of bipolar cell responds to glutamate by hyperpolarizing (evidently using G-protein-coupled receptors); these are called **ON bipolar cells**. Notice that the names OFF and ON refer to whether these cells depolarize in response to light *off* (more glutamate) or to light *on* (less glutamate).

The Fly and the Frog's Eye

by Horace Barlow

What I find most exciting is when a new concept links together and joins up what were previously remote and unconnected sets of facts. This is crucially important in neuroscience, for it has many well-worked fields that have developed almost independently of each other, but all of them are relevant to understanding how the brain and the mind work. We need to know what the brain is made of, its structure, how it evolved, how it develops, what it does, and how it works—all of this on a spatial scale from nanometers to meters, and on a temporal scale from less than a millisecond to months to the lifetime of a species. The sheer mass of facts is overwhelming, and it grows at an overwhelming rate—but you must refuse to be overwhelmed! The secret is to ask questions and seek connections, temporarily ignoring what you don't want to know, and with luck you'll find a pattern that makes a lot of the facts fit together and become comprehensible.

I started research in vision almost 50 years ago, because it seemed the right place to pursue Sherrington's notion of the integrative action of the system. He had shown how layers of semiautomatic reflex mechanisms provide a coherent substrate for pain avoidance, coordinated movement, and balance, but all his examples were on the motor side. The coherent nature of our perceptions shows that sensory messages are also integrated, and it seemed likely to me that analysis of the visual system would enable us to see furthest into the brain mechanisms that do this.

I had the good luck to start research at a moment when initially unrelated developments were occurring, one inside neurophysiology and a second in animal behavior. The first was the ability to record from single neurons in sensory pathways. E. D. Adrian and his colleagues had done this with peripheral nerve fibers in the 1920s and 1930s, but the exciting developments around 1940 were the demonstrations, by R. Granit and K. Hartline, that one could record from single retinal ganglion cells. These were at a level where one would expect some sensory integration to have occurred, because there are at least two layers of synaptic convergence and interaction on the path from the photoreceptors to the ganglion cells. But what might sensory integration consist of?

Horace Barlow

Here the second development had something to say. In about 1950, the subject of animal behavior was enlivened by ethologists Konrad Lorenz and Niko Tinbergen's semipopular books about the ideas of fixed action patterns and releasing stimuli. All the first phrase, "fixed action patterns," really means is that animals do not have unlimited, nor necessarily a very large, range of possible behavioral reactions to stimuli but make selections from a relatively limited repertoire. On the sensory side, the second phrase, "releasing stimuli," suggests that they respond to a relatively restricted range of specially effective sensory stimuli, such as the red spot on the herring gull's bill that evokes gaping for a hatchling gull, the oscillatory movements that make a frog or fish lunge at its prey, or the various signs by which potential mates advertise their availability. One task for integration might be the detection and discrimination of these special releasers. I have a keen recollection of the night when I tested this idea.

For many weeks I had been attempting to test a quantitative theory about summation in isolated retinal ganglion cells of the frog's retina. One evening, after a frustrating start to an experiment, I abandoned it and decided to try something new. I thought it would be interesting to see how intact frogs reacted to ganglion cells being vigorously activated, so I searched for an easily delivered stimulus that produced a continuously high rate of firing in one particular type of cell. I ended up with a disk of paper subtending two or three degrees, twiddled on the end of a piece of wire. Then I descended to the froggery in the basement. To my disappointment, in the first batch of frogs I tested, the appropriately sized stimuli did nothing whatever. They were stolid English frogs, presumably satiated with a rich diet from the Fenland ditches. Fortunately there was a second batch, recently imported from Hungary for Huxley and Staempfli's experiments on saltatory conduction, and these were transformed by my little twiddling disks from their normal torpid state into voraciously active beasts. It appeared that they mis-

continued on page 232

Each bipolar cell receives direct synaptic input from a cluster of photoreceptors. The number of photoreceptors in this cluster ranges from very few near the fovea (less than a half-dozen) to thousands in the peripheral retina. In addition to these direct connections with photoreceptors, bipolar cells also are connected via horizontal cells to a circumscribed ring of photoreceptors that surrounds this central cluster. The **receptive field** of a bipolar cell (or any other cell in the visual system) is *the area of retina that, when stimulated with light, changes the cell's membrane potential*. The receptive field of a bipolar cell consists of two parts: a circular area of retina providing direct photoreceptor input, called the *receptive field center*, and a surrounding area of retina providing input via horizontal cells, called the *receptive field surround* (Figure 9.22). The receptive field dimensions can be measured in millimeters across the retina or, more commonly, in degrees of visual angle. One millimeter on the retina corresponds to a visual angle of about 3.5°. Bipolar cell receptive field diameters range from a fraction of a degree in the central retina to several degrees in the peripheral retina.

The response of a bipolar cell's membrane potential to light in the receptive field center is opposite to that of light in the surround. For example, if illumination of the center causes depolarization of the bipolar cell (an ON response), then illumination of the surround will cause an antagonistic hyperpolarization of the bipolar cell. Likewise, if the cell is depolarized by a dark spot falling on the center of its receptive field (an OFF response), it will be hyperpolarized by the same stimulus applied to the surround. Thus, these cells are said to have antagonistic **center-surround receptive fields**.

The center-surround receptive field organization is passed on from bipolar cells to ganglion cells via synapses in the inner plexiform layer. The lateral connections of the amacrine cells in the inner plexiform layer also contribute to the elaboration of ganglion cell receptive fields, but so far the precise contributions of these connections remain poorly understood.

RETINAL OUTPUT

The sole source of output from the retina to the rest of the brain is the action potentials arising from the million or so ganglion cells. The activity of these cells can be recorded electrophysiologically not only in the retina, but also in the optic nerve where their axons are.

| Box 9.3 | **PATH OF DISCOVERY** |

continued from page 231

took my disks for their favorite bugs, and this gave an early hint that sensory neurons are likely to be opened to give a burst of impulses by a particular key pattern that we called the *trigger feature*. Since then, the visual pathways of invertebrates, cats, monkeys, and rabbits have given other examples.

I find it very difficult to undertake a set of careful observations without having a quantitative theory to test (like my theory of summation in frog retinal ganglion cells). But confirming the theory is a relatively dull outcome. It is almost always the unexpected finding that is important (like the discovery of trigger features). Notice, however, that you must have the theory to know what is unexpected!

So ask questions, form theories to test, do not confine your reading to this textbook or the latest neuroscience journals—and when doing experiments, keep your eyes open for the unexpected!

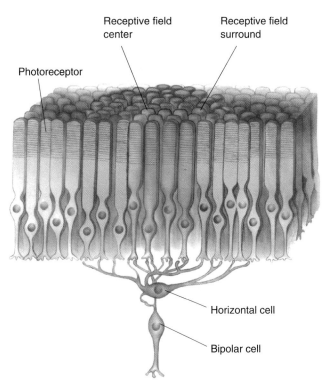

Photoreceptor

Receptive field
center

Receptive field
surround

Horizontal cell

Bipolar cell

Figure 9.22
Direct and indirect pathways from photoreceptor to bipolar cell. Bipolar cells receive direct synaptic input from a cluster of photoreceptors, constituting the receptive field center. In addition, they receive indirect input from surrounding photoreceptors via horizontal cells.

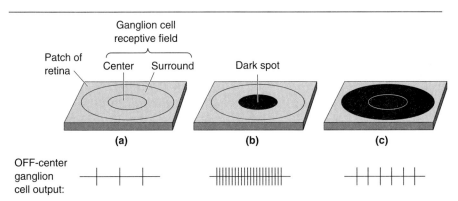

Ganglion cell
receptive field

Patch of
retina

Center   Surround

Dark spot

(a)   (b)   (c)

OFF-center
ganglion
cell output:

Figure 9.23
A center-surround ganglion cell receptive field. (a, b) An OFF-center ganglion cell responds with a barrage of action potentials when a dark spot is imaged on its receptive field center. **(c)** If the spot is enlarged to include the receptive field surround, the response is greatly reduced.

Ganglion Cell Receptive Fields

Most retinal ganglion cells have the concentric center-surround receptive field organization discussed above for bipolar cells. Thus, an ON-center ganglion cell will be depolarized and respond with a barrage of action potentials when a small spot of light is projected onto the middle of its receptive field. Likewise, an OFF-center cell will respond to a small dark spot presented to the middle of its receptive field. However, in both types of cell, the response to stimulation of the center is canceled by the response to stimulation of the surround (Figure 9.23). The surprising implication is

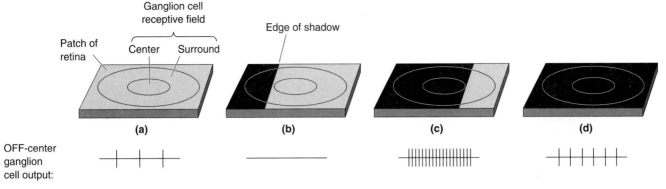

Figure 9.24
Ganglion cell output in response to a light-dark border falling within its receptive field.

that most retinal ganglion cells are not particularly responsive to changes in illumination that include both the receptive field center and the receptive field surround. Rather, it appears that the ganglion cells are mainly responsive to *differences* in illumination that occur within their receptive fields.

To illustrate this point, consider the response generated by an OFF-center cell as a dark shadow crosses its receptive field (Figure 9.24). Remember that in such a cell, dark in the center of the receptive field causes the cell to depolarize, whereas dark in the surround causes the cell to hyperpolarize. In uniform illumination, the center and surround cancel to yield some low level of response (Figure 9.24a). When the shadow enters the surround region of the receptive field without encroaching on the center, it has the effect of hyperpolarizing the neuron, leading to a decrease in the cell's firing rate (Figure 9.24b). As the shadow begins to include the receptive field center, however, the partial inhibition by the surround is overcome, and the cell response increases (Figure 9.24c). But when the shadow finally fills the entire surround, the center response is again canceled (Figure 9.24d). Notice that the cell response in this example is only slightly different in uniform light and in uniform dark; the response is modulated mainly by the presence of a light-dark border in its receptive field.

Now let's consider the output of *all* the OFF-center ganglion cells that are stimulated by a stationary light-dark border imaged on the retina. The responses will fall into the same four categories illustrated in Figure 9.24. Thus, the cells that will register the presence of this shadow are those with receptive field centers and surrounds that are differentially affected by the border. The population of cells with receptive field centers "viewing" the light side of the border will be inhibited (Figure 9.24b). The population of cells with receptive field centers "viewing" the dark side of the border will be excited (Figure 9.24c). In this way, the difference in illumination at a light-dark border is not faithfully represented by the difference in the output of ganglion cells on either side of the border. Instead, *the center-surround organization of the receptive fields leads to a neural response that exaggerates the contrast at borders*.

From the organization of ganglion cell receptive fields we can infer that our visual system is specialized to detect local spatial variations rather than the absolute magnitudes of light falling on the retina. Thus, the perception of lightness or darkness is not absolute, but relative. This is demonstrated by the two framed boxes in Figure 9.25. Even though the central boxes are shaded identically, the left box *appears* darker than the right box because the left one is framed by a lighter background shading. Similarly, the print on this page appears black on a white background both in a dimly lit room and

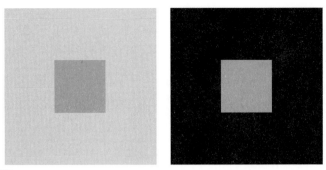

Figure 9.25
Distortion of the perception of light and dark by ganglion cell contrast enhancement. The central boxes are identical shades of gray, but because the border is lighter on the left, the left box appears darker.

outside at high noon. This perceptual constancy cannot be explained by the absolute light intensities reflected by the print and the page. In fact, the "white" page reflects less light in the dorm than the "black" print does outside at high noon. Again, what matters is not the absolute light intensity, but the *difference* in light intensities that arise from the print and the page, and this difference remains constant despite the change in illumination.

Types of Ganglion Cell

Most ganglion cells in the mammalian retina have a center-surround receptive field with either an ON or an OFF center. They can be further subdivided based on their appearance, connectivity, and electrophysiological properties. For example, in the cat retina, the ganglion cells can be divided into three categories (α, β, and γ) based on the size of their somata and dendritic arbors. Another classification scheme (W, X, Y) was developed for the cat based on the visual response properties of ganglion cells. As it turned out, α cells and Y cells were one and the same, as were β cells and X cells. Studies of the cat retina helped guide subsequent research on the primate retina. In the macaque monkey, whose retina is almost identical to our own, two major types of ganglion cells are distinguished: large **M-type ganglion cells** and smaller **P-type ganglion cells**. (M stands for "magno," from the Latin for "large"; P stands for "parvo," from the Latin for "small.") M cells constitute about 10% of the ganglion cell population; P cells constitute most of the rest.

The visual response properties of M cells differ from those of P cells in several ways. They have larger receptive fields, they conduct action potentials more rapidly in the optic nerve, and they are more sensitive to low-contrast stimuli. In addition, M cells respond to stimulation of their receptive field centers with a transient burst of action potentials, while P cells respond with a sustained discharge as long as the stimulus is on (Figure 9.26). The prevailing view is that M cells are particularly important for the detection of stimulus movement, while P cells are more sensitive to stimulus form and fine detail.

Color-Opponent Ganglion Cells. Another important distinction between M cells and P cells is that *P cells are sensitive to differences in the wavelength of light.* The majority of these P cells are called **color-opponent cells**, reflecting the fact that the response to one wavelength in the receptive field center is canceled by showing another wavelength in the receptive field surround. The color pairs that cancel each another are red-green and blue-yellow. Consider, for example, a cell with a red ON center and a green surround

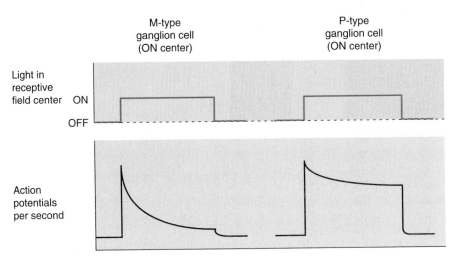

Figure 9.26
Different responses to light of M-type and P-type ganglion cells.

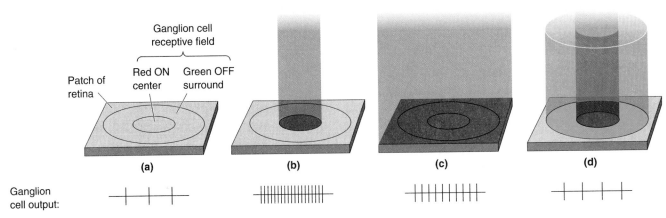

Figure 9.27
A color-opponent center-surround receptive field of a P-type ganglion cell.

(Figure 9.27). The center of the receptive field is fed mainly by red cones; therefore, the cell responds to red light by firing action potentials. Note that even a red light that bathes the entire receptive field is an effective stimulus, because the surround is less sensitive to this wavelength. The response to red is only canceled by green light on the surround. Shorthand notation for such a cell is R⁺G⁻, meaning simply that it responds to red in the receptive field center, and this response is inhibited by green in the surround. What would be the response to white light on the entire receptive field? Because white light contains all visible wavelengths, both center and surround would be equally activated, thereby canceling the response of the cell.

Blue-yellow color opponency works the same way. Consider a cell with a yellow ON center and a blue surround (Y⁺B⁻). Yellow light activates both red and green cones that feed the receptive field center, while blue cones feed the surround. Again, diffuse yellow light would be an effective stimulus for this cell, but blue on the surround would cancel the response, as would diffuse white light.

Perceived color is based on the relative activity of ganglion cells whose receptive field centers receive input from red, green, and blue cones. Demonstrate this to yourself by staring at the cross in the middle of the red box in Figure 9.28 for a minute or so. This will have the effect of fatiguing

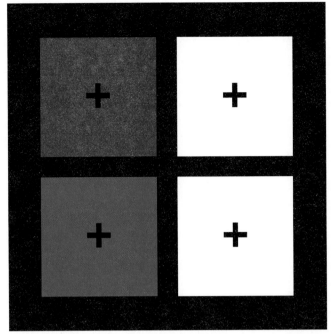

Figure 9.28
Color opponency revealed. Stare at the cross in the red box on the left for 60 seconds, then shift your gaze to the cross in the white box. What color do you see? Try it again with the blue box.

some of the red cones. Then look at the white box. The activation of the green cones by the white light is unopposed, and you see a green square. Similarly, if you stare at the blue box, you will see yellow when you shift your gaze to the white box. Thus, it appears that the ganglion cells provide a stream of information to the brain that is involved in the spatial comparison of three different opposing processes: light versus dark, red versus green, and blue versus yellow.

Parallel Processing

One of the important concepts that emerges from our discussion of the retina is the idea of **parallel processing** in the visual system. Here's why. First, we view the world with not one, but two eyes that provide two parallel streams of information. In the central visual system these streams are compared to give information about *depth*, the distance of an object from the observer. Second, there appear to be independent streams of information about light and dark that arise from the ON-center and OFF-center ganglion cells in each retina. Third, ganglion cells of both ON and OFF varieties have different types of receptive fields and response properties. M cells can detect subtle contrasts over their large receptive fields and are likely to contribute to low-resolution vision. P cells have small receptive fields that are well suited for the discrimination of fine detail, and may be further divided according to their color sensitivity.

CONCLUDING REMARKS

In this chapter, we have seen how light emitted by or reflected off objects in space can be imaged by the eye onto the retina. Light energy is first converted into membrane potential changes in the mosaic of photoreceptors. It is interesting to note that the transduction mechanism in photoreceptors is very similar to that in olfactory receptor cells, both of which involve cyclic nucleotide-gated ion channels. Photoreceptor membrane potential is converted into a chemical signal (the neurotransmitter glutamate), which is again converted into membrane potential changes in the postsynaptic bipolar and horizontal cells. This process of electrical-to-chemical-to-electrical signaling repeats again and again, until the presence of light or dark or color is finally converted to a change in the action potential firing frequency of the ganglion cells.

The information from the 125 million photoreceptors is funneled into 1 million ganglion cells. In the central retina, particularly the fovea, relatively few photoreceptors feed each ganglion cell, whereas in the peripheral retina, thousands of receptors do. Thus, the mapping of visual space onto the array of optic nerve fibers is not uniform. Rather, in "neural space," there is an overrepresentation of the central few degrees of visual space. This specialization ensures high acuity in central vision but also requires that the eye move to bring the images of objects of interest onto the fovea.

As we shall see in the next chapter, there is good reason to believe that the different types of information that arise from different types of ganglion cells are, at least in the early stages, processed independently. Parallel streams of information—for example, from the right and left eyes—remain segregated at the first synaptic relay in the lateral geniculate nucleus of the thalamus. The same can be said for the M-cell and P-cell synaptic relays in the LGN. In the visual cortex, it appears that parallel paths may process different visual attributes. It is possible that each of the more than two-dozen visual cortical areas is specialized for the analysis of a different type of retinal output.

KEY TERMS

Introduction
vision
retina

Properties of Light
refraction

Structure of the Eye
pupil
iris
cornea
aqueous humor
sclera
extraocular muscle
conjunctiva
optic nerve
optic disk
macula
fovea
lens
ciliary muscle
vitreous humor

Image Formation by the Eye
point image
diopter
accommodation
pupillary light reflex
visual field
visual acuity
visual angle

Microscopic Anatomy of the Retina
photoreceptor
bipolar cell

ganglion cell
horizontal cell
amacrine cell
ganglion cell layer
inner nuclear layer
outer nuclear layer
layer of photoreceptor outer segments
inner plexiform layer
outer plexiform layer
rod photoreceptor
cone photoreceptor

Phototransduction
dark current
cyclic guanosine monophosphate (cGMP)
rhodopsin
transducin
phosphodiesterase (PDE)
Young-Helmholtz trichromacy theory
dark adaptation
light adaptation

Retinal Processing
OFF bipolar cell
ON bipolar cell
receptive field
center-surround receptive field

Retinal Output
M-type ganglion cell
P-type ganglion cell
color-opponent cell
parallel processing

1. What physical property of light is most closely related to the perception of color?
2. Name eight structures in the eye that light passes through before it strikes the photoreceptors.
3. Why is a scuba mask necessary for clear vision under water?
4. What is myopia, and how is it corrected?
5. Give three reasons explaining why visual acuity is best when images fall on the fovea.
6. How does the membrane potential change in response to a spot of light in the receptive field center of a photoreceptor? Of an ON bipolar cell? Of an OFF-center ganglion cell? Why?
7. What happens in the retina when you "get used to the dark"? Why can't you see color at night?
8. In what way is retinal output *not* a faithful reproduction of the visual image falling on the retina?

The Central Visual System

lthough our visual system provides us with a unified picture of the world around us, this picture has multiple facets. Objects we see have shape and color. They have position in the world, and sometimes they move. For us to see each of these facets, neurons somewhere in the visual system must be sensitive to them. Moreover, because we have two eyes, we actually have two visual images in our head, and somehow they must be merged.

In Chapter 9, we saw that in many ways the eye acts like a camera. But starting with the retina, the rest of the visual system is far more elaborate, far more interesting, and capable of doing far more than any camera. For example, we saw that the retina does not simply pass along information about the patterns of light and dark that fall on it. Rather, the retina *extracts* information about different facets of the visual image. What we perceive about the world around us, therefore, depends on what information is extracted by the retina and how this information is analyzed and interpreted by the rest of the CNS. An example is color. There is no such thing as color in the physical world; there is simply a spectrum of visible wavelengths of light that are reflected by objects around us. Based on the information extracted by the three types of cone photoreceptors, however, our brain somehow synthesizes a rainbow of color and fills our world with it.

In this chapter, we explore how the information extracted by the retina is analyzed by the central visual system. The pathway serving conscious visual perception includes the *lateral geniculate nucleus* (*LGN*) of the thalamus and the primary visual cortex, also called *area 17*, *V1*, or *striate cortex*. We will see that the information funneled through this geniculocortical pathway is segregated into separate parallel processing channels that are specialized for the analysis of different stimulus attributes. The striate cortex then feeds this information to over two dozen different extrastriate cortical areas in the temporal and parietal lobes, and many of these appear to be specialized for different types of analysis.

Much of what we know about the central visual system was first worked out in the domestic cat and then extended to the rhesus monkey, *Macaca mulatta*. The macaque monkey, as it is also called, relies heavily on vision for survival in its habitat, as do we humans. In fact, tests of the performance of this primate's visual system show that in virtually all respects, it rivals that of humans. Thus, although most of this chapter concerns the organization of the macaque visual system, most neuroscientists agree that it approximates very closely the situation in our own brain.

Although visual neuroscience cannot yet explain many aspects of visual perception (some interesting examples are shown in Figure 10.1), significant progress has been made in answering a more basic question: How do neurons represent the different facets of the visual world? By examining those stimuli that make different neurons in the visual cortex respond, and how these response properties arise, we begin to see how the brain portrays the visual world around us.

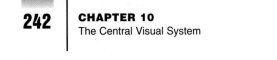

(a)   (b)

Figure 10.1
Perceptual illusions. (a) The two table tops are of identical dimensions and are imaged on similar sized patches of retina, but the perceived sizes are quite different. **(b)** This is an illusory spiral. Try tracing it with your finger.

THE RETINOFUGAL PROJECTION

The neural pathway that leaves the eye, beginning with the optic nerve, is often referred to as the **retinofugal projection**. The suffix *-fugal* is from the Latin word meaning "to put to flight" and is commonly used in neuroanatomy to describe a pathway that is directed away from a structure. Thus, a centrifugal projection goes away from the center, a corticofugal projection goes away from the cortex, and the retinofugal projection goes away from the retina.

We begin our tour of the central visual system by looking at how the retinofugal projection courses from each eye to the brain stem on each side, and how the task of analyzing the visual world initially is divided among, and organized within, certain structures of the brain stem. Then we'll focus on the major arm of the retinofugal projection that mediates conscious visual perception.

The Optic Nerve, Optic Chiasm, and Optic Tract

The ganglion cell axons "fleeing" the retina pass through three structures before they form synapses in the brain stem. The components of this retinofugal projection are, in order, the optic nerve, the optic chiasm, and the optic tract (Figure 10.2). The **optic nerves** exit the left and right eyes at the optic disks, travel through the fatty tissue behind the eyes in their bony orbits, then pass through holes in the floor of the skull. The optic nerves from both eyes combine to form the **optic chiasm**, which lies at the base of the brain, just anterior to where the pituitary gland dangles down. At the optic chiasm, the axons originating in the nasal retinas cross from one side to the other. The crossing of a fiber bundle from one side of the brain to the other is called a *decussation*. Because only the axons originating in the nasal retinas cross, we say that a **partial decussation** of the retinofugal projection occurs at the optic chiasm. Following the partial decussation at the optic chiasm, the axons of the retinofugal projections form the **optic tracts,** which run just under the pia along the lateral surfaces of the diencephalon.

Right and Left Visual Hemifields

To understand the significance of the partial decussation of the retinofugal projection at the optic chiasm, let's review the concept of the visual field introduced in Chapter 9. The full visual field is the entire region of space (measured in degrees of visual angle) that can be seen with both eyes looking straight ahead. Fix your gaze on a point straight ahead. Now imagine a vertical line passing through the fixation point, dividing the visual field into left and right halves. By definition, objects appearing to the left of the midline are in the left **visual hemifield,** and objects appearing to the right of the midline are in the right visual hemifield (Figure 10.3). By looking straight ahead with both eyes open and then alternately closing one eye and then the other, you will see that the central portion of both visual hemifields is viewed by *both* retinas. This region of space is therefore called the **binocular visual field**. Notice that objects in the binocular region of the left visual hemifield will be imaged on the nasal retina of the left eye and on the temporal retina of the right eye. Because the fibers from the nasal portion of the left retina cross to the right side at the optic chiasm, all the information about the left visual hemifield is directed to the right side of the brain. Remember this rule of thumb: Optic nerve fibers cross in the optic chiasm so that *the left visual hemifield is "viewed" by the right hemisphere and the right visual hemifield is "viewed" by the left hemisphere.*

Targets of the Optic Tract

A small number of optic tract axons peel off to form synaptic connections with cells in the hypothalamus, and another 10% or so continue past the thalamus to innervate the midbrain. But most of them innervate the **lateral geniculate nucleus (LGN)** of the dorsal thalamus. The neurons in the LGN give rise to axons that project to the primary visual cortex. This projection from LGN to cortex is called the **optic radiation**. Lesions anywhere in the retinofugal projection from eye to LGN to visual cortex cause blindness in humans. Therefore, we know that it is this pathway that mediates conscious visual perception (Figure 10.4).

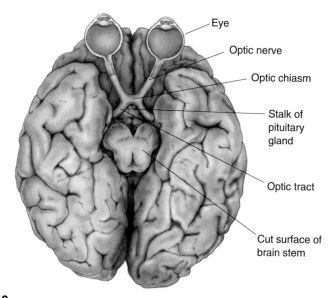

Eye

Optic nerve

Optic chiasm

Stalk of
pituitary
gland

Optic tract

Cut surface of
brain stem

Figure 10.2
The retinofugal projection. View of the base of the brain showing the optic nerves, optic chiasm, and optic tracts.

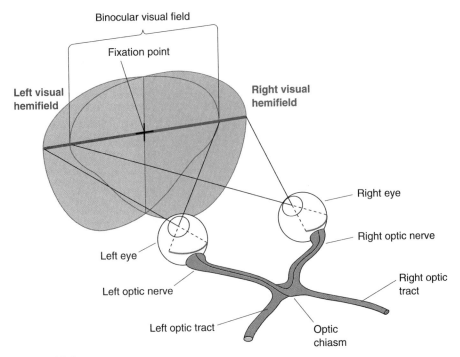

Figure 10.3
Right and left visual hemifields. Ganglion cells in both retinas that are responsive to visual stimuli in the right visual hemifield project axons into the left optic tract. Similarly, ganglion cells "viewing" the left visual hemifield project into the right optic tract.

From our knowledge of how the visual world is represented in the retinofugal projection, we can predict the types of perceptual deficits that would result from its destruction at different levels, as might occur from a traumatic injury to the head, a tumor, or an interruption of the blood supply. As shown in Figure 10.5, while a transection of the left optic *nerve* would render a person blind in the left eye only, a transection of the left optic *tract* would lead to blindness in the right visual field as viewed through either eye. A midline transection of the optic chiasm would affect only the fibers that cross the midline. Because these fibers originate in the nasal portions of both retinas, blindness would result in the regions of the visual field viewed by the nasal retinas, that is, the peripheral visual fields on both sides (Box 10.1). Because unique deficits result from lesions at different sites, neurologists and neuro-ophthalmologists can locate sites of damage by assessing visual field deficits.

Nonthalamic Targets of the Optic Tract. As we have said, some retinal ganglion cells send axons to innervate structures other than the LGN. Direct projections to part of the hypothalamus play an important role in synchronizing a variety of biological rhythms, including sleep and wakefulness, with the daily dark-light cycle (see Chapter 17). Direct projections to part of the midbrain, called the *pretectum*, control the size of the pupil and certain types of eye movement. And about 10% of the ganglion cells in the retina project to a part of the midbrain tectum called the **superior colliculus** (Latin for "mound") (Figure 10.6). While 10% may not sound like much of a projection, bear in mind that in the primate this is about 150,000 neurons, equivalent to the *total* number of retinal ganglion cells in the cat! In fact, the tectum of the midbrain is the major target of the retinofugal projection in all nonmammalian vertebrates (fish, amphibians, birds, and reptiles). In these vertebrate groups, the superior colliculus is called the **optic tectum**. This is why the projection from the retina to the superior colliculus is often called the **retinotectal projection** even in mammals.

Retinotopy. The retinotectal projection illustrates a general organizational feature of the central visual system called retinotopy. **Retinotopy** is an organization whereby *neighboring cells in the retina feed information to neighboring places in their target structures*, in this case the superior colliculus. In this way, the two-dimensional surface of the retina is *mapped* onto the two-dimensional surface of the colliculus.

There are two important points to remember about retinotopy. First, the mapping of the visual field onto a retinotopically organized structure is

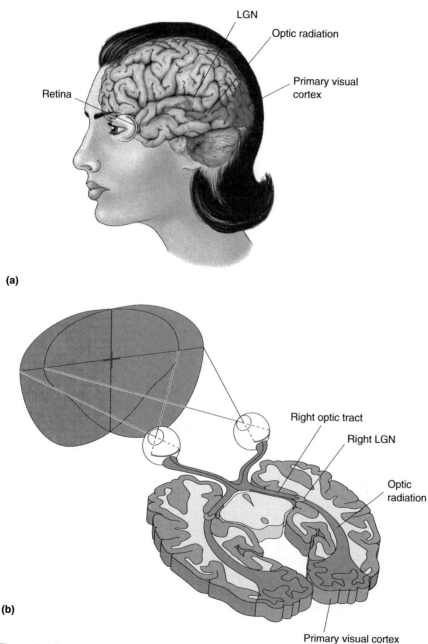

(a)

(b)

Figure 10.4
The visual pathway that mediates conscious visual perception. (a) Side view of the human brain with the retinogeniculocortical pathway shown inside. **(b)** A horizontal slab through the brain exposing the same pathway.

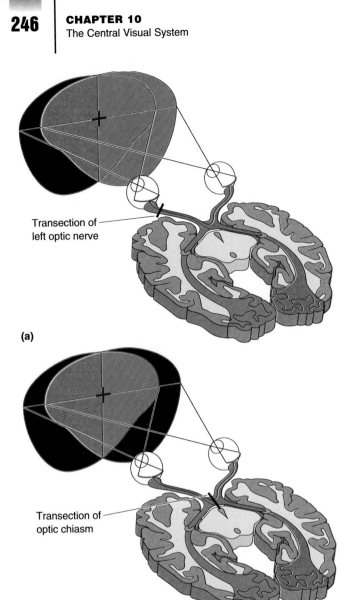

(a)

Transection of
left optic nerve

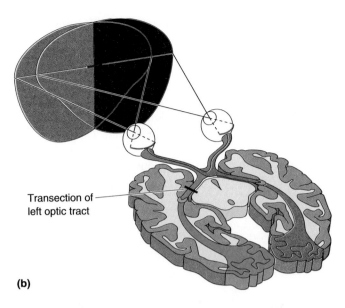

(b)

Transection of
left optic tract

Transection of
optic chiasm

(c)

Figure 10.5
Visual field deficits from lesions in the retinofugal projection. (a) If the optic nerve on the left side is cut, vision will be lost completely in the left eye. **(b)** If the optic tract on the left side is cut, vision will be lost in the right visual field of either eye. **(c)** If the optic chiasm is split down the middle, only the crossing fibers will be damaged, and peripheral vision will be lost in both eyes.

often distorted, because visual space is not sampled uniformly by the cells in the retina. Recall from Chapter 9 that there are many more ganglion cells with receptive fields in or near the fovea than in the periphery. Thus, the representation of the visual field is distorted in the superior colliculus: The central few degrees of the visual field are overrepresented, or *magnified*, in the retinotopic map (Figure 10.7).

The second point to remember is that a discrete point of light can activate many cells in the retina, and often many more cells in the target structure, due to the overlap of receptive fields. Thus, the image of a point of light on the retina actually activates a large population of superior colliculus neurons; every neuron that contains that point in its receptive field would be activated. Thus, a position on the superior colliculus in retinotopic coordinates is the site of a peak in the *distribution* of postsynaptic activity that arises when the retina is stimulated with a point of light at the corresponding retinal location.

In the superior colliculus, a patch of neurons activated by a point of light, via indirect connections with motor neurons in the brain stem, commands eye and head movements to bring the image of this point in space onto the fovea. Thus, this branch of the retinofugal projection is involved in orienting the eyes in response to new stimuli in the visual periphery.

We will return to the superior colliculus when we discuss motor systems in Chapter 14. However, as we shall now see, the basic principles of retinotopy also apply to the LGN and primary visual cortex.

THE LATERAL GENICULATE NUCLEUS

The right and left lateral geniculate nuclei, located in the dorsal thalamus, are the major targets of the two optic tracts. Viewed in cross section, each LGN appears to be arranged in six distinct layers of cells (Figure 10.8). By convention, the layers are numbered 1 through 6, starting with the most ventral layer, layer 1. In three dimensions, the layers of the LGN are arranged like a stack of six pancakes, one on top of the other. The pancakes do not lie flat, however; they are bent around the optic tract like a knee joint. This shape explains the name "geniculate," from the Latin *geniculatus*, meaning "bent like a knee."

The LGN is the gateway to the visual cortex and, therefore, to conscious visual perception. Here we will explore the structure and function of this thalamic nucleus.

Segregation of Input by Eye and by Ganglion Cell Type

LGN neurons receive synaptic input from the retinal ganglion cells, and most geniculate neurons project an axon to primary visual cortex via the optic radiation. The segregation of LGN neurons into layers suggests that

---

Box 10.1	OF SPECIAL INTEREST

David and Goliath

Many of you are familiar with the famous story of David and Goliath, which appears in the Hebrew scriptures (Old Testament). The armies of the Philistines and the Israelites were gathered for battle when Goliath, a Philistine, came forth and challenged the Israelites to settle the dispute by sending out their best man to face him in a fight to the death. Goliath, it seems, was a man of great proportions, measuring over "six cubits" in height. If you consider that a cubit is the distance from the elbow to the tip of the middle finger, about 20 inches, this guy was over 10 feet tall! Goliath was armed to the teeth with body armor, a javelin, and a sword. To face this giant, the Israelites sent David, a young and diminutive shepherd, armed only with a sling and five smooth stones. Here's how the action is described in the Revised Standard Version of the Bible (1 Samuel 17: 48):

> When the Philistine arose and came and drew near to meet David, David ran quickly toward the battle line to meet the Philistine. And David put his hand in his bag and took out a stone, and slung it, and struck the Philistine on his forehead; the stone sank into his forehead, and he fell on his face to the ground.

Now why, you might ask, are we giving a theology lesson in a neuroscience textbook? The answer is that our understanding of the visual pathway offers an explanation, in addition to divine intervention, for why Goliath was at a disadvantage in this battle. Body size is regulated by the secretion of growth hormone from the anterior lobe of the pituitary gland. In some cases, the anterior lobe becomes hypertrophied (swollen) and produces excessive amounts of the hormone, resulting in body growth to unusually large proportions. Such individuals are called pituitary giants and can measure well over 8 feet tall. Pituitary hypertrophy also disrupts normal vision. Recall that the optic nerve fibers from the nasal retinas cross in the optic chiasm, which butts up against the stalk of the pituitary. Any enlargement of the pituitary compresses these crossing fibers and results in a loss of peripheral vision called bitemporal hemianopia, or tunnel vision. (See if you can figure out why this is so from what you know about the visual pathway.) We can speculate that David was able to draw close and smite Goliath because, when David raced to the battle line, the pituitary giant had completely lost sight of him!

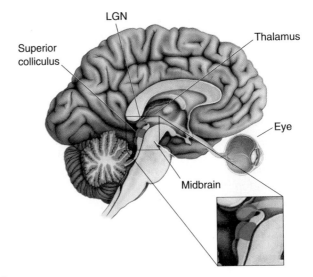

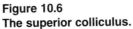

Figure 10.6
The superior colliculus.

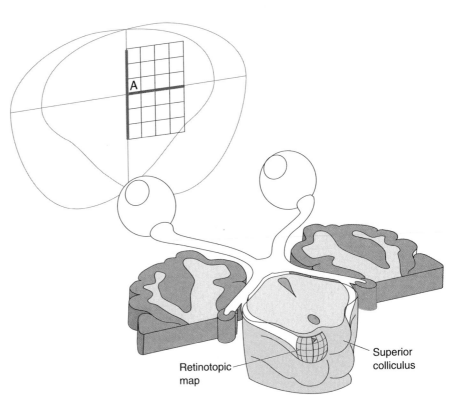

Figure 10.7
Retinotopic map of the superior colliculus. Notice that there is a map of visual space in the superior colliculus, but also that the map is distorted, with more tissue devoted to analysis of the central visual field. Similar maps are found in the LGN and primary visual cortex.

different types of retinal information are being kept separate at this synaptic relay, and indeed this is the case: Axons arising from the two major classes of ganglion cells in the two retinas synapse on cells in different LGN layers.

Recall from our rule of thumb that the *right* LGN receives information about the *left* visual field. The left visual field is viewed by both the nasal left retina and the temporal right retina. At the LGN, input from the two eyes is kept separate. In the right LGN, the right eye (ipsilateral) axons synapse on LGN cells in layers 2, 3, and 5. The left eye (contralateral) axons synapse on cells in layers 1, 4, and 6 (Figure 10.9).

A complete retinotopic map exists in each LGN layer, just as in the superior colliculus. And, despite the segregation of left-eye and right-eye fibers, these different maps occur in perfect register. In other words, if we stuck a toothpick perpendicularly through the layers of the LGN, it would pass through cells responding to the same points in visual space viewed alternately by the right and left eyes.

A closer look at the LGN in Figure 10.8 reveals that the two ventral layers, 1 and 2, contain larger neurons, and the four more dorsal layers, 3 through 6, contain smaller cells. The ventral layers are therefore called the **magnocellular LGN layers**, and the dorsal layers are called **parvocellular LGN layers**. Recall from Chapter 9 that ganglion cells in the retina may also be split into magnocellular and parvocellular groups. As it turns out, P-type ganglion cells in the retina project exclusively to the parvocellular LGN, and M-type ganglion cells in the retina project entirely to the magnocellular LGN. Thus, in the LGN there is also a segregation of information arising from these two distinct classes of ganglion cells.

In addition to the neurons in the six principal layers of the LGN, there are also numerous tiny neurons that lie just ventral to each layer. Cells in these **koniocellular layers** (*konio* is from the Greek for "dust"), as they have

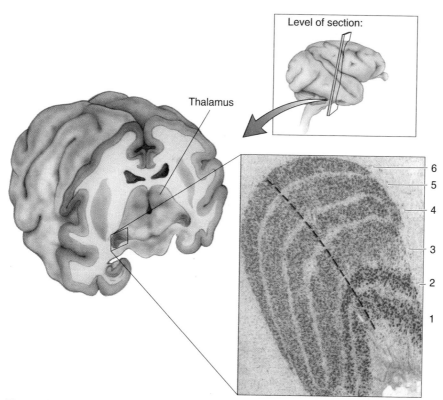

Figure 10.8
The LGN of the macaque monkey. The tissue has been stained to show cell bodies, which appear as dots. Notice particularly the six layers and the larger size of the cells in the two ventral layers (layers 1 and 2). (Source: Adapted from Hubel, 1988; p. 65.)

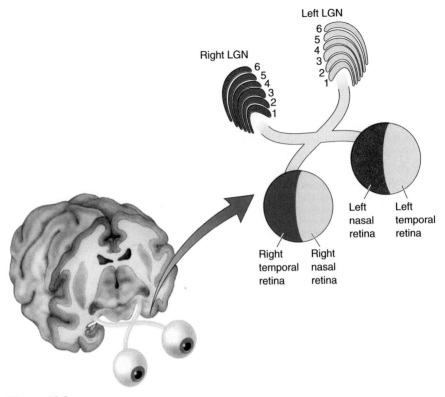

Figure 10.9
Retinal inputs to the LGN layers.

been called, also receive retinal input and project to visual cortex. Their input is believed to arise from odd retinal ganglion cells that cannot be classified as P cells or M cells.

The anatomical organization of the LGN supports the idea that the retina gives rise to streams of information that are processed in parallel. This organization is summarized in Figure 10.10.

Receptive Fields

By inserting a microelectrode into the LGN, it is possible to study the action potential discharges of geniculate neurons in response to visual stimuli, just as was done in the retina. The surprising conclusion of such studies is that the visual receptive fields of LGN neurons are almost identical to those of the ganglion cells that feed them. So, for example, magnocellular LGN neurons have relatively large center-surround receptive fields, respond to stimulation of their receptive field centers with a transient burst of action potentials, and are insensitive to differences in wavelength. All in all, they are just like M-type ganglion cells. Likewise, parvocellular LGN cells, like P-type retinal ganglion cells, have relatively small center-surround receptive fields, respond to stimulation of their receptive field centers with a sustained increase in the frequency of action potentials, and are generally very sensitive to differences in wavelength. Although the small size of the cells in the koniocellular layers has impeded progress in identifying their response properties, recent work indicates that many of these cells are also sensitive to different wavelengths of light.

We can further divide most LGN cells into ON-center and OFF-center categories and blue-yellow and red-green categories, as in the retina. These different LGN cell types receive synaptic input only from the appropriate retinal cell types (i.e., ON-center ganglion cells connect to ON-center LGN cells, etc.). However, there is not a segregation of these cells within the LGN; they are found scattered among one another in the various layers.

Nonretinal Inputs to the LGN

What makes the similarity of LGN and ganglion cell receptive fields so surprising is that the retina is not the main source of synaptic input to the LGN. The major input, constituting about 80% of the excitatory synapses, comes from primary visual cortex. Thus, one might reasonably expect that this corticofugal feedback pathway would significantly alter the qualities of the visual responses recorded in the LGN. So far, however, a role for this massive input has not been clearly identified.

The LGN also receives synaptic inputs from neurons in the brain stem whose activity is related to alertness and attentiveness (see Chapters 15 and 17). Have you ever "seen" a flash of light when you are startled in a dark room? This perceived flash might be a result of direct activation of LGN neurons by this pathway. Usually, however, this input does not directly evoke action potentials in LGN neurons. But it can powerfully modulate the magnitude of LGN responses to visual stimuli. (Recall modulation from Chapters 5 and 6.) Thus, the LGN is more than a simple relay from retina to cortex; it is the first site in the ascending visual pathway where what we see is influenced by how we feel.

ANATOMY OF THE STRIATE CORTEX

The LGN has a single major synaptic target: primary visual cortex. Recall from Chapter 7 that the cortex may be divided into a number of distinct areas based on their connections and cytoarchitecture. Primary visual cortex is Brodmann's **area 17** and is located in the occipital lobe of the pri-

Retinal output		LGN	
Eye	Ganglion cell type	LGN cell type	Layer
Contralateral	P-type →	Parvocellular	6
	? →	Koniocellular	
Ipsilateral	P-type →	Parvocellular	5
	? →	Koniocellular	
Contralateral	P-type →	Parvocellular	4
	? →	Koniocellular	
Ipsilateral	P-type →	Parvocellular	3
	? →	Koniocellular	
Ipsilateral	M-type →	Magnocellular	2
	? →	Koniocellular	
Contralateral	M-type →	Magnocellular	1
	? →	Koniocellular	

Figure 10.10
Summary of ganglion cell inputs to the different LGN layers.

mate brain. Much of area 17 lies on the medial surface of the hemisphere, surrounding the calcarine fissure (Figure 10.11). Other terms used interchangeably to describe the primary visual cortex are **V1** and **striate cortex**.

We have seen that the axons of different types of retinal ganglion cells synapse on anatomically segregated neurons in the LGN. In this section we look at the anatomy of the striate cortex, and then trace the connections different LGN cells make with cortical neurons. In a later section we will use microelectrodes to explore how this information is analyzed by cortical neurons. As we did in the LGN, in striate cortex we'll see a close correlation of structure with function.

Lamination of Striate Cortex

The neocortex in general, and striate cortex in particular, has neuronal cell bodies arranged into about a half-dozen layers. These layers can be seen clearly in a Nissl stain of the cortex, which, as described Chapter 2, leaves a deposit of dye (usually blue or violet) in the soma of each neuron. Starting at the white matter (containing the cortical input and output fibers), the cell layers are named by Roman numerals VI, V, IV, III, and II. Layer I, just under the pia mater, is largely devoid of neurons and consists almost entirely of axons and dendrites of cells in other layers. The full thickness of the striate cortex from white matter to pia is about 2 mm, the height of the letter m in mm (Figure 10.12).

As Figure 10.12 shows, the lamination of striate cortex has been somewhat forced into a six-layer scheme. There are actually at least nine distinct layers of neurons. But, to maintain Brodmann's convention that neocortex has six layers, we combine three sublayers into layer IV, labeled IVA, IVB, and IVC. Layer IVC is further divided into two tiers called IVCα and IVCβ. The anatomical segregation of neurons into layers suggests that there is a division of labor in the cortex, similar to what we saw in the LGN. We can learn a lot about how the cortex handles visual information by examining the structure and connections of its different layers.

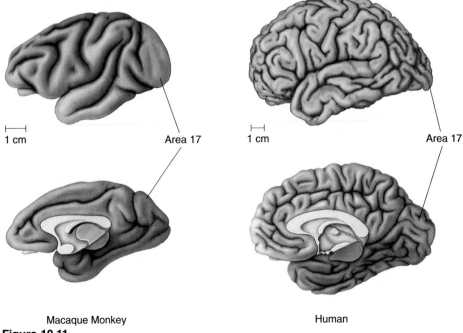

1 cm   Area 17   1 cm   Area 17

Macaque Monkey   Human

Figure 10.11
Cortical area 17. Top views are lateral; bottom views are medial.

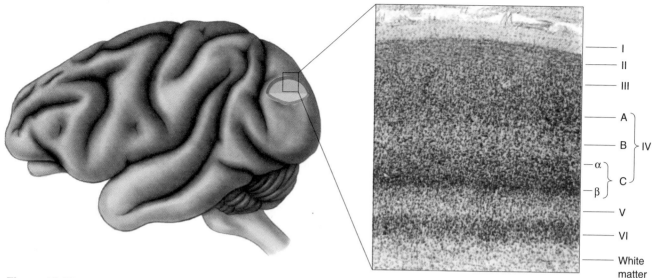

Figure 10.12
The cytoarchitecture of the striate cortex. The tissue has been Nissl stained to show cell bodies, which appear as dots. (Source: Adapted from Hubel, 1988; p. 97.)

The Cells of Different Layers. Many different neuronal shapes have been identified in striate cortex, but here we'll focus on two principal types, defined by the appearance of their dendritic trees (Figure 10.13). *Spiny stellate cells* are small neurons with spine-covered dendrites that radiate out from the cell body (recall dendritic spines from Chapter 2). They are seen primarily in the two tiers of layer IVC. Outside of layer IVC, there are many *pyramidal cells.* These neurons are characterized by a single, thick apical dendrite that branches as it ascends to the pia, and by multiple basal dendrites that extend horizontally. (Pyramidal cell dendrites are also covered with spines.) Notice that a pyramidal cell in one layer may have dendrites extending into other layers. It is important to remember that *only pyramidal cells send axons out of striate cortex* to form connections with other parts of the brain. The axons of stellate cells only make local connections within the cortex.

Input-Output Organization of Different Layers. Axons from the LGN terminate in several different cortical layers, but mainly in layer IVC. This axonal projection maintains the retinotopy of the LGN, so neighboring cells in layer IV receive their inputs (via the LGN relay) from neighboring regions of the retina (Figure 10.14). As in the LGN and superior colliculus, many more cortical neurons receive input from the central part of the retina than from more peripheral parts.

Layer IVC stellate cells send axons up mainly to layers IVB and III. Here, an axon may form synapses with the dendrites of pyramidal cells of all layers. As previously mentioned, the pyramidal cells send axons out of striate cortex into the white matter. The pyramidal cells in different layers innervate different structures. Layer III and layer IVB pyramidal cells send their axons to other cortical areas. Layer V pyramidal cells send axons all the way down to the superior colliculus and pons. Layer VI pyramidal cells give rise to the massive axonal projection back to the LGN (Figure 10.15). Pyramidal cell axons in all layers also branch and form local connections in the cortex.

Most intracortical connections occur along a radial line that runs across the layers, from white matter to layer I, perpendicular to the cortical surface. This pattern of *radial connections* maintains the retinotopic organization established in layer IV. Therefore, a cell in layer VI, for example, receives

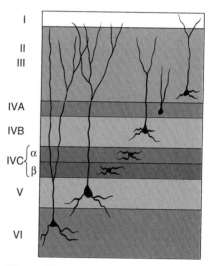

Figure 10.13
Dendritic morphology of some cells in striate cortex. Notice particularly that pyramidal cells are found in layers III, IVB, V, and VI, and that spiny stellate cells are found in layer IVC.

information from the same part of the retina as does a cell above it in layer IV (Figure 10.16a). However, the axons of some layer III pyramidal cells extend collateral branches that make *horizontal connections* within layer III (Figure 10.16b). Radial and horizontal connections play different roles in the analysis of the visual world, as we shall see later in the chapter.

LGN Input to Layer IVC

Now we want to focus on the details of how the different types of LGN neurons innervate the cells in cortical layer IVC. We've seen that the output of the LGN is divided into streams of information, for example, from the magnocellular and parvocellular layers serving the right and left eyes. These streams remain anatomically segregated in layer IVC.

Segregation of magnocellular and parvocellular LGN inputs is straightforward. Magnocellular LGN neurons project to layer IVCα, and parvocellular LGN neurons project to layer IVCβ. Imagine the two tiers of layer IVC are pancakes, stacked one (α) on top of the other (β). Because the input from

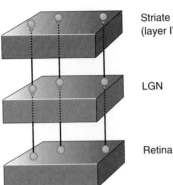

Figure 10.14
Retinotopy is preserved in the LGN projection to layer IVC.

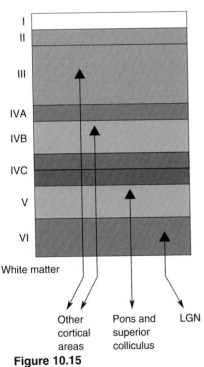

Figure 10.15
The organization of outputs from striate cortex.

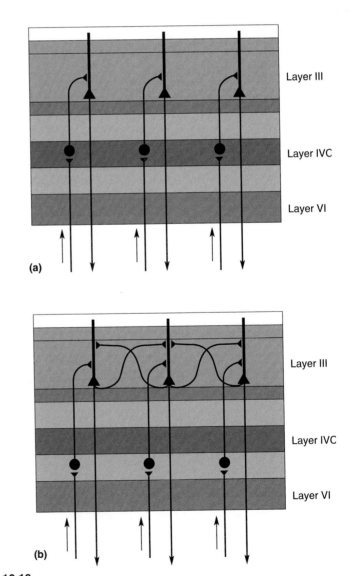

(a)

(b)

Figure 10.16
Patterns of intracortical connections. (a) Radial connections. **(b)** Horizontal connections.

the LGN to the cortex is arranged topographically, we see that layer IVC contains two overlapping retinotopic maps, one from the magnocellular LGN (IVCα) and the other from the parvocellular LGN (IVCβ).

How are the left eye and right eye LGN inputs segregated when they reach layer IVC of striate cortex? The answer was provided by a ground-breaking experiment performed in the early 1970s at Harvard Medical School by neuroscientists David Hubel and Torsten Wiesel. They injected a radioactive amino acid into one eye of a monkey (Figure 10.17). This amino acid was incorporated into proteins by the ganglion cells, and the proteins were transported down the ganglion cell axons into the LGN (recall anterograde transport in Chapter 2). Here, the radioactive proteins spilled out of the ganglion cell axon terminals and were taken up by nearby LGN neurons. But not all LGN cells took up the radioactive material; only those cells that were postsynaptic to the inputs from the injected eye incorporated the labeled protein. These cells then transported the radioactive proteins to their axon terminals in layer IVC of striate cortex. The location of the radioactive axon terminals was visualized by first placing a film of emulsion over thin sections of striate cortex and later developing the emulsion like a photograph, a process called *autoradiography* (introduced in Chapter 6). The resulting collection of silver grains on the film marked the location of the radioactive LGN inputs.

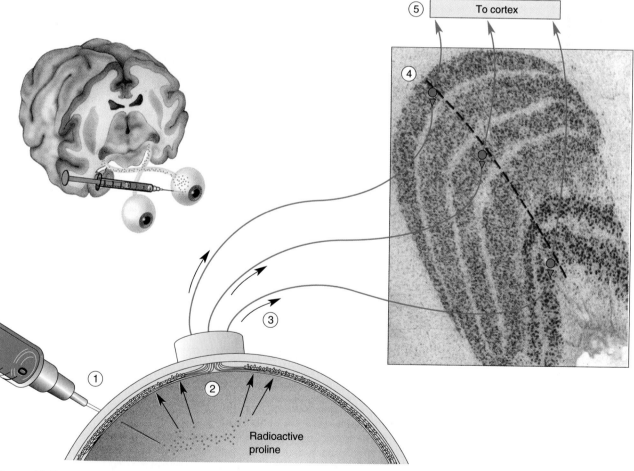

Figure 10.17
Transneuronal autoradiography. Radioactive proline is (1) injected into one eye, where it is taken up by retinal ganglion cells and (2) incorporated into proteins that are (3) transported down the axons to the LGN. Here, some radioactivity spills out of the retinal terminals (4) and is taken up by LGN neurons that then transport it to striate cortex (5). The location of radioactivity can be determined using autoradiography.

In sections cut perpendicular to the cortical surface, Hubel and Wiesel observed that the distribution of axon terminals relaying information from the injected eye was not continuous in layer IVC, but rather was split up into a series of equally spaced patches, each about 0.5 mm wide (Figure 10.18a). These patches were termed **ocular dominance columns**. In later experiments, the cortex was sectioned tangentially, parallel to layer IV. This revealed that the left eye and right eye inputs to layer IV are laid out as a series of alternating bands, like the stripes of a zebra (Figure 10.18b).

Putting the anatomy together, you can see that any chunk of layer IVC measuring 0.5 mm (the thickness of layer IVC) by 1 mm × 1 mm would contain a full complement of segregated inputs from both magnocellular and parvocellular LGN layers from both left and right eyes.

Layer IVC Innervation of Other Cortical Layers

Layer IVC neurons project axons radially up to layers IVB and III where, for the first time, information from the left eye and right eye begins to mix. Even so, there continues to be considerable anatomical segregation of the magnocellular and parvocellular processing streams. Layer IVCα, which

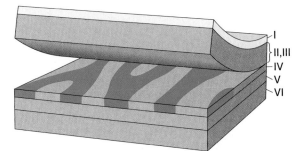

(a)

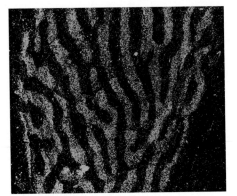

(b)

Figure 10.18
Ocular dominance columns in striate cortex. (a) The organization of ocular dominance columns in layer IV of the striate cortex of a macaque monkey. The distribution of LGN axons serving one eye is darkly shaded. In cross section, these eye-specific zones appear as patches, each about 0.5 mm wide, in layer IV. When the superficial layers are peeled back, allowing a view of the ocular dominance columns in layer IV from above, these zones take on the appearance of zebra stripes. **(b)** An autoradiograph of a histological section of layer IV viewed from above. Two weeks prior to the experiment, one eye of this monkey was injected with radioactive proline. In the autoradiograph, the radioactive LGN terminals appear bright on a dark background. (Source: LeVay et al., 1980.)

receives magnocellular LGN input, projects mainly to cells in layer IVB. Layer IVCβ, which receives parvocellular LGN input, projects mainly to layer III.

Blobs. In addition to the parvocellular input relayed from layer IVCβ, very recent research indicates that a subset of cells in layer III receives *direct* input from the LGN. These layer III cells are found in what are known as "blobs." The blob story began in 1978 when Margaret Wong-Riley, at the University of California, San Francisco, discovered a nonhomogeneity in the anatomy of cortical layer III. She used a staining procedure designed to reveal the presence of **cytochrome oxidase**, a mitochondrial enzyme used for cell metabolism, and found that it is not uniformly distributed in layers II and III. Rather, the cytochrome oxidase staining in cross sections of striate cortex appears as a colonnade, a series of pillars at regular intervals, running the full thickness of layers II and III (Figure 10.19a). When the cortex is sliced tangentially through layer III, these pillars appear like the spots of a leopard (Figure 10.19b). These pillars of cytochrome oxidase-rich neurons in layers II and III, after some debate, eventually were given the name **blobs**. The blobs are in rows, each blob centered on an ocular dominance stripe in layer IV. Between the blobs are "interblob" regions in layers II and III.

It was soon shown that both blobs and interblobs receive parvocellular LGN input via a relay in layer IVCβ. In 1994, Johns Hopkins University neuroscientists Stewart Hendry and Takashi Yoshioka discovered that the blob cells in macaque striate cortex also receive direct LGN innervation—from the mysterious koniocellular layers.

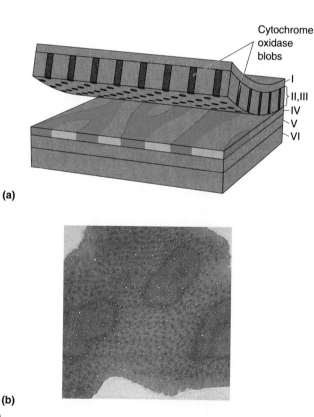

(a)

(b)

Figure 10.19
Cytochrome oxidase blobs. Distribution of the enzyme cytochrome oxidase in the cells of striate cortex. **(a)** The organization of cytochrome oxidase blobs in layer III of striate cortex in a macaque monkey. **(b)** A photograph of a histological section of layer III, stained for cytochrome oxidase and viewed from above. (Source: Courtesy of Dr. S. H. C. Hendry.)

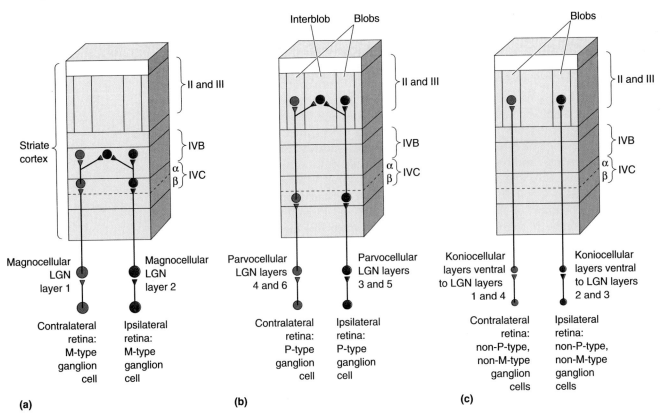

Figure 10.20
Three parallel pathways from the retina to striate cortex. (a) magnocellular pathway, **(b)** parvocellular pathway, **(c)** koniocellular pathway.

Parallel Pathways

Figure 10.20 summarizes the connections of striate cortex and its LGN input in terms of anatomically distinct, parallel pathways that arise in the retina. The magnocellular pathway begins with the M-type ganglion cells of the retina. These cells project axons to the magnocellular layers of the LGN. The magnocellular layers of the LGN project to layer IVCα of striate cortex. Layer IVCα neurons project to layer IVB (Figure 10.20a).

The parvocellular pathway originates with the P-type ganglion cells of the retina, which project to the parvocellular layers of the LGN. The parvocellular LGN sends axons to layer IVCβ of striate cortex. Layer IVCβ neurons project to layer III blob and interblob regions (Figure 10.20b).

The koniocellular pathway arises from the subset of ganglion cells that are neither M-type nor P-type. These poorly characterized cells project to the koniocellular layers of the LGN. The koniocellular LGN projects directly to the layer III blobs (Figure 10.20c). Notice that the blobs are a site of convergence of parvocellular and koniocellular inputs.

Notice also in Figure 10.20 that the information from the two eyes remains segregated all the way through layer IVC. Inputs from the two eyes begin to mix in layer IVB and in the interblob region of layer III. However, note that the blob cells receive input mainly from one eye or the other. Keep these anatomically defined parallel pathways in mind as we look at how neurons in striate cortex respond to visual stimuli in the next section.

PHYSIOLOGY OF THE STRIATE CORTEX

Beginning in the early 1960s, Hubel and Wiesel were the first to systematically explore the physiology of striate cortex with microelectrodes.

They were students of Stephen Kuffler, who was then at Johns Hopkins University. They extended Kuffler's innovative methods of receptive field mapping to the central visual pathways. After showing that LGN neurons behave much like retinal ganglion cells, they turned their attention to striate cortex, initially in cats and later in monkeys. (Here we will be concerned only with the monkey cortex.) The work that continues today on the physiology of striate cortex is built on the solid foundation provided by Hubel and Wiesel's pioneering studies (Box 10.2). Their contributions to our understanding of the cerebral cortex were recognized with the Nobel Prize in 1981.

Instead of presenting the physiological results in the order in which they were discovered, we will discuss the physiology of cortical neurons in light of the anatomy described above. The latest research suggests that, functionally, there are three relatively independent channels of visual information processing. The channel that begins with the magnocellular retinal ganglion cells and leads to layer IVB of striate cortex we will call the **magnocellular channel**, or simply the **M channel**. The channel that begins with the parvocellular retinal ganglion cells and leads to the interblob regions of layer III we will call the **parvocellular-interblob channel**, or the **P-IB channel**. The channel that converges on the blobs of layer III, and passes through the parvocellular and koniocellular layers of the LGN, we will call simply the **blob channel**. We'll see that each of these channels appears to process a different facet of vision.

The M Channel

Recall that cells in the magnocellular LGN are activated by only one eye (i.e., they are monocular), have center-surround receptive fields, respond transiently to visual stimulation, and are insensitive to the wavelength of light. At the level of the first synaptic relay in layer IVCα, we see an interesting elaboration of receptive field structure. Instead of being circular, the receptive fields of layer IVCα neurons are elongated along a particular axis, with an ON-center or OFF-center region flanked on one or both sides by an antagonistic surround (Figure 10.21). One gets the impression that these cells receive a converging input from three or more LGN cells with receptive fields that are aligned along one axis (Figure 10.21c). Hubel and Wiesel called neurons of this type **simple cells**. Simple cells respond best to a thin slit of light or dark that is aligned with the long axis of the receptive field, and they respond poorly, if at all, to a stimulus that is perpendicular to this axis. Thus, we say that layer IVCα cells possess **orientation selectivity**.

Layer IVCα cells project up to layer IVB, where the cells also have orientation-selective simple receptive fields. One important difference between the cells in the two layers is that while cells in layer IVCα respond only to stimulation of one eye or the other (just as the anatomy of ocular dominance columns predicts), many cells in layer IVB respond to stimulation of both eyes. Such cells are said to have **binocular receptive fields**. The construction of binocular receptive fields is essential in binocular animals, such as humans. Without binocular neurons, we would be unable to use the inputs from both eyes to form a single image of the world around us.

Besides binocularity, another important physiological property of layer IVB cells is **direction selectivity**. Figure 10.22 shows how a direction-selective cell might respond to a moving stimulus. Notice that the cell responds to an elongated stimulus swept across the receptive field, but only in a particular direction of movement. Sensitivity to the direction of stimulus motion is a hallmark of neurons in the M channel. For this reason, the M channel is thought to be specialized for the *analysis of object motion*.

Exploring the Visual Cortex

by David Hubel

David Hubel

Torsten Wiesel and I began what was to be a 25-year collaboration at the Johns Hopkins Hospital in the fall of 1958. For almost all of our first year we worked in the dark and dingy basement of the Wilmer Ophthalmological Institute, in the laboratory of Stephen Kuffler. It was a wonderful place despite its tawdriness; soon after we got there Ed Furshpan and David Potter arrived from Katz' lab in London, and in those days the field was smaller, more intimate, and we had daily discussions ranging from synapses to higher CNS. It was at the Wilmer that Steve Kuffler had done his work on the receptive field organization of retinal ganglion cells in the cat retina, surely one of the most beautiful pieces of work in sensory physiology of this century. Torsten and I had already been working in vision for several years, and we set out to extend Kuffler's receptive field studies into the brain. For several reasons we thought it was likely that the most interesting things would be found in the cortex, so for the time being we skipped the intermediate stage, the LGN, and went right to the primary visual cortex.

At that time recording from single CNS cells, particularly from the cortex, was in its infancy. The most interesting work by far had been done by Vernon Mountcastle (also at Hopkins) in the somatosensory cortex of cats and monkeys, where he had discovered, in 1956, the columnar organization of cortex. The curious thing was that there seemed to be no obvious differences between the cells of the somatosensory cortex and those at lower levels, either thalamus or primary sensory nerves. Nevertheless, just looking at the high complexity and high degree of order in the visual cortex was enough to give us hope that there we might find something more elaborate than the center-surround receptive fields of retinal ganglion cells.

We certainly had no preconceived ideas of what these differences might be. It may be worth remarking here that many sciences are not structured the way much of physics is, with hypothesis, leading to measurements, and then laws, often with much quantification. This dissimilarity holds especially for much of biology. Much of science is akin to geographical exploration in the days of Columbus,

where one sails off not knowing what to expect. This has no bearing on whether a particular area of science is good or bad, or revolutionary or pedestrian. One need only ask what hypotheses Darwin had in mind when he set forth on the Beagle; one need only search the *Origin of Species* for equations, to be struck by how great the differences are among different sciences. Our work began with no hypotheses, unless the vague expectation of finding something interesting constitutes a hypothesis; as for quantification, we did not even plot graphs of anything until the 1970s.

Within a few weeks of starting, we began to get results. At first the going was rough. Cells failed to respond to our large patches of light or to the small spots Kuffler had used so effectively. The break came during our third or fourth experiment. We had been recording large, stable impulses from one cell for 3 or 4 hours with no hint of any response to our stimuli. The cell was #3004 in our series, which we had decided to begin at 3000 to bolster our confidence and also to impress Mountcastle, whose series of cells generally ran into the thousands. At that time we were generating our light spots with a most unwieldy concoction designed for retinal work, for which it was ideal—in fact, it was the same stimulator that Kuffler had used so effectively in the early 1950s. To stimulate with a spot of light, one used a metal plate about the shape of a microscope slide into which a small hole had been drilled. The metal plate was inserted into a slot, and the spot was projected onto the retina with an elaborate system of lamps and mirrors. Dark spots on a light background were produced by using a glass microscope slide instead of a brass plate, onto which a tiny opaque disc had been glued. For cell #3004 neither type of stimulus seemed to work, despite hours of searching around in the retina. Suddenly, in mid-afternoon, we started to get hints of a response from the cell just as we inserted the glass slide into the slot. It seemed that what the cell wanted in order to respond was the faint shadow of the edge of the glass as it crossed a particular part of the animal's retina. After working with that one cell for 9 hours, we finally concluded that the shadow only worked over a restricted range of orientations: when the orientation was right, the cell would respond with a roar of impulses.

continued on next page

| Box 10.2 | PATH OF DISCOVERY |

continued from page 260

By now, in 1995, we have seen tens of thousands of such cells, in the visual cortex of cats and monkeys, but it took several years to convince ourselves that the great majority of V1 cells respond to line segments in specific orientations, that all orientations are represented about equally in the population, and to work out such things as the degrees of complexity, binocular properties, responses to movement, and the geometrical distributions of cells by layers and columns. Later we went on to study the effects of visual deprivation, and to compare the responses in adult and newborn animals. On one visit, Vernon remarked, "What a wonderful system! It will be five years before you get this all worked out." We thought he was exaggerating, but now, 36 years later,

the mine is still producing, in dozens of laboratories worldwide.

This might seem to be an example of the importance of luck in science. Of course luck always helps. We were surely lucky to get into such a field early—it fit our personalities not to have to begin by months of reading in the library. We were lucky, if you like, that we had the sense not to spend months or years designing elaborate calibrated stimulators, before having any idea what stimuli the cells might be interested in. We were lucky in being a bit sloppy long before any rigor or fine measurements or statistics were called for. And we were certainly lucky in being stubborn and persistent in our willingness to fight that cell for 9 hours. In retrospect, it was time well spent.

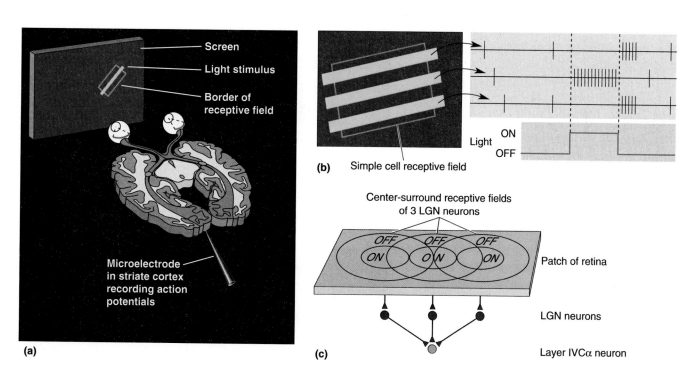

Figure 10.21
Simple cell receptive field. (a) The responses of a neuron in layer IVCα are monitored as visual stimuli are presented to its receptive field. The visual stimulus is a light bar. **(b)** The response of a simple cell to different stimuli. Notice that it responds best to an oriented bar of light, and the response can be ON or OFF depending on where the bar lies in the receptive field. **(c)** Possible construction of a simple cell receptive field by the convergence of three LGN cell axons with center-surround receptive fields.

The P-IB Channel

Neurons in the parvocellular layers of the LGN have monocular, small, circular receptive fields. About 80% of these cells show red-green or blue-yellow center-surround opponency. In all these respects, the cells of layer IVCβ act just like the LGN neurons that feed them. Recall that IVCβ cells relay information up to layers II and III, and that this projection targets both blob and interblob regions.

We will discuss blobs in a moment; for now, let's concentrate on the interblob cells in layer III. Hubel and Wiesel called most of these **complex cells**, because their receptive fields appeared to be more complex than those of simple cells. Unlike simple cells, complex cell receptive fields do not have distinct ON and OFF regions (Figure 10.23). Complex cells are mostly binocular, relatively insensitive to the wavelength of light, and highly selective to stimulus orientation (even more so than simple cells).

The analysis of stimulus orientation appears to be one of the most important functions of striate cortex. Presumably this analysis is required to discriminate and identify objects based on their shape. Because virtually all interblob cells are orientation-selective and have small receptive fields, the P-IB channel is thought to be devoted to the *analysis of object shape.*

Orientation Columns

We have mentioned that many cortical neurons exhibit orientation selectivity. However, because this property may be generated independently by both M channels and P-IB channels, you might wonder whether the orientation selectivity of neurons in different layers is related. From the earliest work of Hubel and Wiesel, the answer to this question is an emphatic yes. As a microelectrode is advanced *radially* from one layer to the next, the preferred orientation remains the same for all the selective neurons encountered. This includes

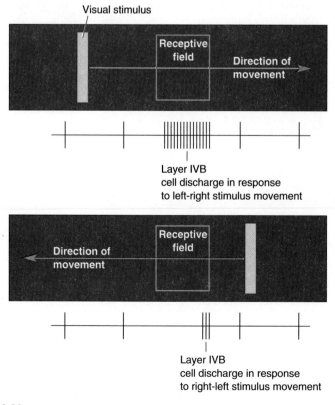

Figure 10.22
Responses of a direction-selective neuron.

all the cells of layers V and VI having both simple and complex receptive fields. Hubel and Wiesel called such a radial column of cells an **orientation column**. Perhaps predictably, as an electrode passes *tangentially* through the cortex in a single layer, the preferred orientation shifts systematically, like the sweep of the minute hand of a clock, from the top of the hour to ten past to twenty past, and so on (Figure 10.24). Hubel and Wiesel found that a complete 180° shift in preferred orientation required a traverse of about 1 mm within layer III.

Physiology of the Blobs

After Wong-Riley's discovery of the cytochrome oxidase blobs, Hubel, working with Harvard Medical School neurobiologist Margaret

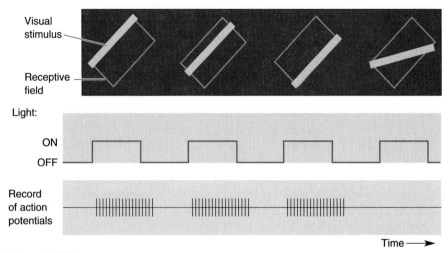

Figure 10.23
Complex cell receptive field. Like the simple cell, the complex cell responds best to an oriented bar of light. However, responses occur at both light ON and light OFF, regardless of position in the receptive field.

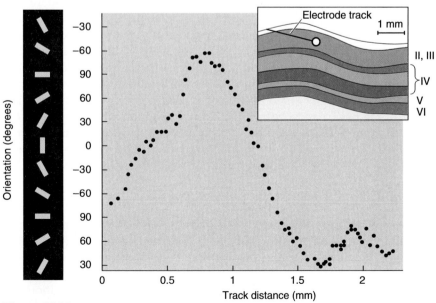

Figure 10.24
Systematic variation among orientation preferences across striate cortex. As an electrode is advanced tangentially across layer III of striate cortex, the orientation preference of the neurons encountered is recorded and plotted. Notice that there is a periodic, regular shift in preferred orientation. (Source: Adapted from Hubel and Wiesel, 1968.)

Livingstone, recorded from neurons in the blobs and found that they are very different from the interblob cells in layer III. The blob cells are wavelength-sensitive and monocular, and they lack orientation selectivity. The visual responses of blob cells resemble those of their major sources of input: the neurons in layer $IVC\beta$ and the koniocellular layers of the LGN.

The receptive fields of most blob neurons are circular. Some have the color-opponent center-surround organization observed in P-type ganglion cells and in the parvocellular layers of the LGN. Other blob cell receptive fields have red-green or blue-yellow color opponency in the center of their receptive fields, with no surround regions at all. Still other cells have both a color-opponent center *and* a color opponent-surround (called double-opponent cells). For present purposes, the most important thing to remember about blobs is that they contain the only color-sensitive neurons outside of layer IVC. Thus, the blob channels appear to be specialized for the *analysis of object color*. Without them, we would be color-blind.

Putting the Pieces Together

We have seen that the anatomy and physiology of the central visual pathways, from retina to striate cortex, are consistent with the idea that there are several, relatively independent, parallel processing channels. Each may be responsible for the analysis of a different facet of the visual scene: the M channel for the analysis of object motion, the P-IB channel for the analysis of object shape, and the blob channel for the analysis of object color.

Each point in the visual world is analyzed by a circumscribed patch of cells in the cortex. Hubel and Wiesel showed that the image of a point in space falls within the receptive fields of neurons across a traverse of about 2 mm of layer III. For a complete analysis, this 2×2 mm patch of active neurons must include representatives from each of the processing channels from right and left eyes.

Fortunately, a 2×2 mm chunk of cortex would contain two complete sets of ocular dominance columns in layer IV, 16 blobs in layer III, and, in the cells between blobs, a complete sampling (twice over) of all 180° of possible orientations. Thus, Hubel and Wiesel have argued that a 2×2 mm chunk of striate cortex is both necessary and sufficient to analyze the image of a point in space; *necessary* because its removal would leave a blind spot for this point in the visual field, and *sufficient* because it contains all the neural machinery required to analyze the participation of this point in oriented and/or colored contours viewed through either eye. Such a unit of brain tissue has come to be called a **cortical module**. Striate cortex is constructed from perhaps a thousand of these modules, and one is shown in Figure 10.25. We can think of a visual scene being simultaneously processed by these modules, each "looking" at a portion of the scene.

FROM SINGLE NEURONS TO PERCEPTION

Visual *perception*—the task of identifying and assigning meaning to objects in space—obviously requires the concerted action of many cortical modules. *How* is the simultaneous activity of widely separated cortical neurons compared, and *where* does this comparison take place?

Let's first explore some general ideas about how the information distributed across striate cortex might be combined. Then we'll examine a couple of cortical areas in the parietal and temporal lobes that depend on striate cortex for their major inputs. We will see that the processing channels that begin in the retina, and are elaborated in striate cortex, are maintained by corticocortical connections. Entirely different regions of association cor-

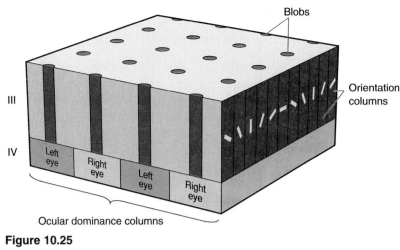

Blobs

Orientation
columns

III

IV

Left
eye

Right
eye

Left
eye

Right
eye

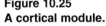

Ocular dominance columns

Figure 10.25
A cortical module.

tex appear to be dedicated to the analysis of visual motion and color, for example. We'll see that although neuroscience cannot yet answer the questions, posed above, of how and where the streams of information combine to form seamless perceptions of the world, our best hope lies beyond V1.

Communication Among Modules

Consider the pattern of activity that might arise in striate cortex when you view a simple geometric shape like a stop sign (Figure 10.26). Because the sign takes up a substantial portion of the visual field, neural activity in the cortex is distributed among many cortical modules. The readout of each module provides information about orientation and color at a fraction of the contrast border along the edge of the sign. Somehow, the activity of these modules must be compared, and a decision made, that this activity represents a single object—a stop sign—and not a jumble of line segments. Such a decision might be made by cells in higher cortical areas that are postsynaptic to the active striate cortical cells. In this case, we might expect to find "stop sign cells," neurons that only respond to an octagonal red shape. Elsewhere we might find "grandmother cells," neurons that respond only to one's grandmother, and so on.

Another scheme for perceptually linking the neurons that respond to a stop sign or one's grandmother takes advantage of the fact that the cortical modules activated by a single object will be active simultaneously. Thus, the contours of an object could be perceptually bound together if the activity of a single module were in some way sensitive to the activity of other modules that were active at precisely the same time. In fact, there is an anatomical basis for cross-talk between modules: the horizontal connections that connect the cells of layer III. Similar connections even cross from one side of the brain to the other via the corpus callosum to bind together the cells that respond to images falling on the seam between the right and left visual fields. These connections selectively wire up the cells that have common receptive field characteristics; for example, blob cells connect to other blob cells, cells responding only to horizontal lines connect to cells that share that same orientation preference, and so on. In 1989, neuroscientists Charles Gray and Wolf Singer, working together at the Max Planck Institute in Frankfurt, Germany, discovered that visual cortical cells respond with repetitive *bursts* of action potentials to their preferred stimuli, and that the bursts of activity occurring in widely separated modules were often syn-

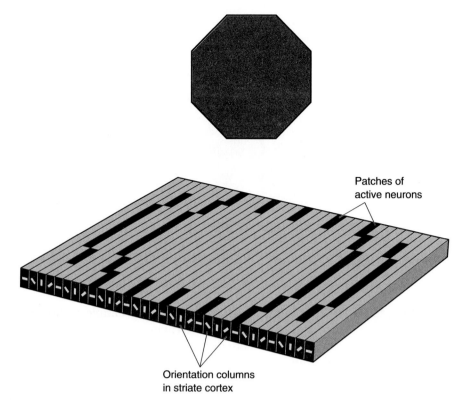

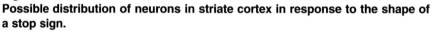

Patches of
active neurons

Orientation columns
in striate cortex

Figure 10.26
Possible distribution of neurons in striate cortex in response to the shape of a stop sign.

chronous when activated by the contours of a single common object. This work suggests that the internal neural representation of a stop sign or one's grandmother might be the synchronous activity of a large population of cells distributed across the cortex.

So is "grandmother" a cell, or a population of cells? The "grandmother cell" theory predicts that we should find neurons in higher cortical areas that are selective to, say, faces. As improbable as this might seem, there are now reports of cells in the temporal lobes of monkeys that respond only to faces! Then again, there do not seem to be neurons tuned to everything we perceive. Most neurons in higher cortical areas appear to represent the things we perceive in a much more abstract manner than the grandmother cell theory suggests. Today, most visual neuroscientists believe that the representation of objects is distributed among a population of many cortical neurons, but there is still little agreement as to the nature of this representation.

Beyond V1

The search for understanding of visual perception takes us beyond the borders of striate cortex. As already mentioned, striate cortex is also called V1, for "visual area one," because it is the first cortical area to receive information from the LGN. Beyond V1, however, lies another two dozen distinct areas of cortex, each of which contains a map of the visual world. Most of the extrastriate areas remain a wild neuroscientific frontier, and their contributions to vision are still being vigorously debated. However, the emerging picture is that there are two large-scale cortical *streams* of visual processing, one stretching from striate cortex toward the parietal lobe and the other projecting toward the temporal lobe (Figure 10.27a).

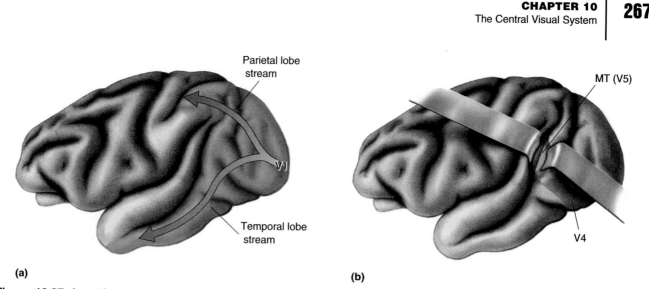

Parietal lobe
stream

MT (V5)

V1

Temporal lobe
stream

V4

(a)

(b)

Figure 10.27. Location of (a) parietal and temporal lobe visual processing streams and (b) areas MT and V4 in the macaque brain.

Area MT and the Parietal Lobe Stream. As an example of the areas found in the parietal lobe stream, we consider cortical area V5, or, as it is usually called, area MT. **Area MT** receives retinotopically organized input from a number of other cortical areas such as V2 and V3, but it also is directly innervated by cells in layer IVB of striate cortex. Recall that layer IVB is part of the magnocellular channel, which is characterized by cells with large receptive fields, transient responses to light, and direction selectivity. Area MT is most notable for the fact that almost all the cells are direction-selective, unlike areas earlier in the parietal stream, or anywhere in the temporal stream. The neurons in MT also respond to types of motion, such as drifting spots of light, that are not good stimuli for other areas. For these reasons and others, MT appears to be specialized for the analysis of motion. Consistent with this idea, lesions in MT seriously disrupt the ability of monkeys to discriminate the direction in which a stimulus moves.

Neurons in area MT have large receptive fields that respond to stimulus movement in a narrow range of directions. As with orientation-selective cells in striate cortex, these direction-selective cells are arranged into a system of columns. A complete analysis requires that each point in visual space be "viewed" by cells in columns that respond to a full 360° range of directions. Presumably, the perception of movement at any point in space then depends on a comparison of the activity across these direction columns. This idea was recently tested directly in a fascinating experiment performed by William Newsome and his colleagues at Stanford University. They used a method pioneered at the National Institutes of Health by Edward Evarts, in which electrodes are painlessly lowered into the cortex of awake monkeys (see Box 14.1). In Newsome's experiment, a monkey watched a video display of moving dots. The monkey would report the direction in which he believed the dots were moving by moving his eyes in that direction. (Monkeys, like humans, appear to enjoy video games.) Amazingly, Newsome and his colleagues discovered that by electrically stimulating the cells in a single-direction column of area MT, they could shift the monkey's perception of stimulus movement (Box 10.3). This is perhaps the closest visual neurophysiology has come to visual perception, and it bodes well for our eventual understanding of the neurophysiological basis of vision.

Beyond area MT in the parietal lobe stream are several cortical areas involved in directing eye movements. It is likely that one of the main reasons the brain exerts so much effort on analyzing motion is that it is critical

Box 10.3 | **PATH OF DISCOVERY**

Influencing a Perceptual Judgment

by William T. Newsome

William T. Newsome

By the summer of 1989, my colleagues and I were fairly sure that MT was an important source of the directional signals that enabled monkeys to discriminate the direction of motion in our visual displays. We had used microelectrode recordings to "listen" to the traffic of electrical signals in MT while monkeys worked on the task, and we knew that MT neurons were exquisitely sensitive to the motion signals in the displays. We also knew that small lesions in MT would selectively disrupt a monkey's ability to perform the task. But we lacked a "smoking gun"—an experiment that could directly show that changes in the activity of MT neurons *cause* corresponding changes in a monkey's choices in the direction discrimination task.

From the beginning of our work with the direction discrimination task, I had considered trying just such an experiment. The idea was to use electrical microstimulation to activate a column of directionally selective neurons in MT while a monkey inspected the visual display and judged the direction of motion. In other words, rather than "listen" to the natural electrical signals of MT neurons, we would use our microelectrodes to artificially "boost" the signals related to a particular direction of motion. If MT neurons actually provided the signals used by the monkey in choosing the direction of motion, stimulating a particular column should increase the probability of a choice favoring the direction of motion encoded by that column.

Though intriguing, this experiment stayed on the back burner because the chances of success seemed very slim. Columns encoding different directions of motion are interspersed among each other in MT, and it would thus be necessary to restrict activation to one or a very few columns in order to avoid "smearing" the microstimulation signal across an arbitrary collection of directions. Even if the stimulating current could be placed with sufficient precision, it seemed unlikely that activation of a *single* cortical column could exert a measurable influence on perceptual behavior.

In August of 1989, however, Daniel Salzman (a Stanford University medical student) and I gave the experiment a try. I warned Daniel not to expect much from this experiment because it was fairly far-fetched. Remarkably, the very first experiment we attempted yielded a significant effect—the monkey's choices were indeed biased toward the direction encoded by the neurons at the electrode tip! Not believing the result, I insisted that we immediately repeat the experiment with the stimulator turned off to determine whether any undetected bugs in our computer software could cause the effect. To my surprise, the result held up in that experiment and in several more experiments over the next few days.

In the tenth experiment we got our first "whopper," a term we came to apply to experiments with spectacularly large effects. In this experiment, the monkey chose the direction encoded by the stimulated neurons on 51 of the first 54 trials! It was almost as though the monkey could not help but choose that direction when we "tickled" the cortex at exactly the right place. I distinctly remember that Daniel and I had the same reaction as we watched this amazing result unfold—uncontrollable laughter. The most analogous experience I can imagine is hitting a jackpot at a Las Vegas slot machine; laughter is probably the most appropriate response to such a preposterous turn of events.

One of the most important lessons I learned from this investigation was to be suspicious of my own preconceptions. I would never have trained a monkey simply to do the microstimulation experiment; the investment of time and money seemed unwarranted, given what appeared to be a very small probability of success. Since the monkeys were already trained for the single-unit recordings, however, only a modest amount of additional effort was necessary to try a few microstimulation experiments. How many other important results are we missing even now because our most basic assumptions about the brain are somehow flawed?

for our eyes to be able to follow objects and move toward the objects when necessary.

Area V4 and the Temporal Lobe Stream. One of the most-studied areas in the temporal lobe is area V4 (Figure 10.27b). V4 receives input from the blob and interblob regions of striate cortex via a relay in V2. Neurons in V4 have larger receptive fields than cells in striate cortex, and many of the cells are both orientation-selective and color-selective. Neurons in this area also have more elaborate spatial and color interactions than those in striate cortex. Although there is a good deal of ongoing speculation concerning the function of V4, this area appears to be important for both shape and color perception. If this area is damaged, there are perceptual deficits involving both shape and color.

Beyond V4 in the temporal lobe are cortical areas that contain neurons with complicated spatial receptive fields. Although some of these neurons have been found to be particularly responsive to identifiable objects such as faces, the receptive fields of many temporal neurons are more abstract. As we will see in Chapter 19, these areas in temporal cortex appear to be important for both visual perception and visual memory.

CONCLUDING REMARKS

In this chapter, we have outlined the organization of the sensory pathway from eye to thalamus to cortex. We saw that the sensory modality we call vision actually may be divided into a number of submodalities—color, contrast, form, movement—and these are processed in parallel by different cells of the visual system. This processing of information in the visual system evidently requires a strict segregation of inputs at the thalamus, some limited convergence of information in striate cortex, and finally a massive divergence of information as it is passed on to higher cortical areas. The distributed nature of the cortical processing of visual information is underscored when you consider that the output of a million ganglion cells can recruit the activity of well over a *billion* cortical neurons throughout the occipital, parietal, and temporal lobes! Somehow, this widespread cortical activity is combined to form a single, seamless perception of the visual world.

Heed the lessons learned from the visual system. As we shall see in later chapters, the basic principles of organization in this system—parallel processing, topographic mappings of sensory surfaces, synaptic relays in the dorsal thalamus, cortical modules, and multiple cortical representations—are also features of the sensory systems devoted to hearing and touch.

KEY TERMS

The Retinofugal Projection

retinofugal projection

optic nerve

optic chiasm

partial decussation

optic tract

visual hemifield

binocular visual field

lateral geniculate nucleus (LGN)

optic radiation

superior colliculus

optic tectum

retinotectal projection

retinotopy

The Lateral Geniculate Nucleus

magnocellular LGN layer

parvocellular LGN layer

koniocellular LGN layer

Anatomy of the Striate Cortex

area 17

V1

striate cortex

ocular dominance column

cytochrome oxidase

blob

Physiology of the Striate Cortex

magnocellular channel (M channel)

parvocellular-interblob channel
(P-IB channel)

blob channel

simple cell

orientation selectivity

binocular receptive field

direction selectivity

complex cell

orientation column

cortical module

From Single Neurons to Perception

area MT

✔ R E V I E W   Q U E S T I O N S

1. Following a bicycle accident, you are disturbed to find that you cannot see anything in the left visual field. Where has the retinofugal pathway been damaged?
2. What is the source of most of the input to the *left* LGN?
3. A worm has eaten part of one lateral geniculate nucleus. You can no longer perceive color in the right visual field of the right eye. What layer(s) of which LGN have been damaged?
4. List the chain of connections that link a cone in the retina to a blob cell in striate cortex.
5. How are receptive fields transformed at each of the synaptic relays that connect a M-type retinal ganglion cell to a neuron in striate cortical layer IVB?
6. Which pathway, magnocellular or parvocellular, constitutes four-fifths of the input to striate cortex? What are two analyses of the visual world that are thought to involve mainly this pathway? What about the other pathway?
7. What is meant by parallel processing in the visual system? Give two examples.
8. If a child is born cross-eyed and the condition is not corrected before the age of 10, binocular depth perception will be lost forever. This is explained by a modification in the circuitry of the visual system. From your knowledge of the central visual system, where do you think the circuitry has been modified?

CHAPTER 11

The Auditory System

The sense of hearing, also known as **audition**, plays many important roles in our daily lives. Even if we cannot see an object, we can detect its presence by hearing a sound that it makes. Anyone who has ever hiked through the forest in an area where there are bears or snakes knows that the sound of rustling leaves can be a powerful attention-grabber. From the sound, we can often identify the source and determine its location—an important piece of information if you have to make a run for it. Aside from the ability to detect and locate sound, we are able to perceive and interpret its nuances. We can immediately identify the bark of a dog, the voice of a friend, the crash of an ocean wave. Because humans are able to produce a wide variety of sounds as well as hear them, spoken language and its reception via the auditory system have become an extremely important means of communication. Audition in humans has even evolved beyond the strictly utilitarian functions of communication and survival: In a manner analogous to artists who use visual media, musicians explore the sensations and emotions evoked by sound.

In this chapter, we explore the mechanisms within the ear and brain that translate the sounds in our environment into meaningful neural signals. We will find that this transformation is carried out in stages rather than all at once. Within the ear, auditory receptors convert the mechanical energy in sound into a neural response. At subsequent stages in the brain stem and thalamus, signals from the receptors are integrated before they ultimately reach auditory cortex. By looking at the response properties of neurons at various points in the system, we begin to get some idea of the relationship between activity in the auditory system and our perception of sound.

THE NATURE OF SOUND

Sound consists of audible variations in air pressure resulting from a great variety of processes that move air molecules; these include the whisper of the human voice, the vibration of a string on a guitar, and the explosion of a firecracker. In general, when an object moves toward a patch of air, it compresses the air, increasing the density of the molecules. Conversely, the air is rarefied (made less dense) when an object moves away. This is particularly easy to visualize in the case of a stereo speaker, in which a paper cone attached to a magnet vibrates in and out, alternately rarefying and compressing the air (Figure 11.1). These changes in air pressure are transferred away from the speaker at the speed of sound, which is about 343 m/sec (767 mph) for air at room temperature.

Many sources of sound, such as vibrating strings or a stereo speaker reproducing the sound of a stringed instrument, produce variations in air pressure that are periodic. The **frequency** of the sound is the number of compressed or rarefied patches of molecules that pass by our ears each second. One cycle of the sound is the distance between successive compressed patches; the sound frequency, expressed in units called **hertz (Hz)**, is the number of cycles per second. Because sound waves all propagate at the same speed, high-frequency sound waves have more compressed and rarefied regions packed into the same space than low-frequency waves (Figure 11.2a). Our auditory system can respond to pressure waves over the remarkable range of 20–20,000 Hz (though this audible range decreases significantly with age and exposure to noise, especially at the high-frequency end). Whether a sound is perceived to have a high or low tone, or *pitch*, is determined by the frequency. In order to relate frequency to recognizable examples, remember that a room-shaking low note on an organ is about 20

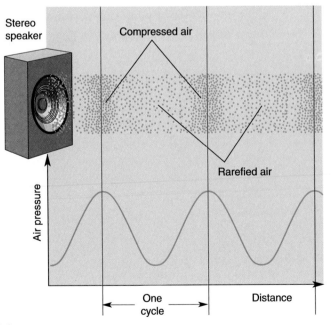

Figure 11.1
The production of sound by variations in air pressure. When the paper cone of a stereo speaker pushes out, it compresses the air; when the cone pulls in, it rarefies the air. If the push and pull are periodic, there will also be a periodic variation in the air pressure, as shown in the graph. The distance between successive compressed (high-pressure) patches of air is one cycle of the sound (indicated by the vertical lines). The sound wave propagates away from the speaker at the speed of sound.

Hz and an ear-piercing high note on a piccolo is about 10,000 Hz. Although humans can hear a great range of frequencies, there are high and low sound wave frequencies our ears cannot hear, just as there are electromagnetic waves of light our eyes cannot see (Box 11.1).

Another important property of a sound wave is its **intensity**, which is the difference in pressure between the compressed and rarefied patches of air

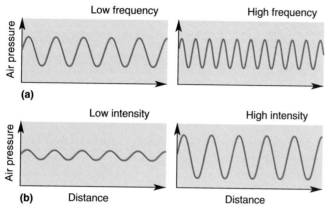

Figure 11.2
The frequency and intensity of sound waves. (a) We perceive high-frequency waves as having a higher pitch. **(b)** We perceive high-intensity waves as louder.

| Box 11.1 | OF SPECIAL INTEREST |

Infrasound

Most people are familiar with ultrasound (sound above the 20 kHz limit of our hearing) because it has everyday applications, from ultrasonic cleaners to medical imaging. Some animals can hear these high frequencies. For instance, dog whistles work because dogs can hear up to about 40 kHz. Less well known is infrasound, or sound at low frequencies, below about 20 Hz. Some animals can hear these frequencies; one is the elephant, which can detect 15 Hz tones at sound levels inaudible to humans. Whales produce low-frequency sounds, which are thought to be a means of communication over distances of kilometers. Low-frequency vibrations are also produced by the earth, and it is thought that some animals may sense an impending earthquake by hearing such sound.

Even though we usually cannot hear very low frequencies with our ears, they are present in our environment and can have unpleasant subconscious effects. Infrasound is produced by such devices as air conditioners, boilers, aircraft, and automobiles. Though even intense infrasound from these machines does not cause hearing loss, it can cause dizziness, nausea, and headache. Many cars produce low-frequency sound when they're moving at high-way speeds, making sensitive people carsick. At very high levels, low-frequency sound may also produce resonances in body cavities such as the chest and stomach, which can damage internal organs. You might want to think twice before standing directly in front of a large speaker at a concert!

In addition to mechanical equipment, our own bodies generate inaudible low-frequency sound. When muscle changes length, individual fibers vibrate, producing low-intensity sound at about 25 Hz. While we cannot normally hear these sounds, you can demonstrate them to yourself by carefully putting your thumbs in your ears and making a fist with each hand. As you tighten your fist, you can hear a low rumbling sound produced by the contraction of your forearm muscles. Other muscles, including your heart, produce inaudible sound at frequencies near 20 Hz.

It's probably just as well that we aren't more aware of infrasound. It would be hard to get any work done listening to our bodies rumble and groan in a world filled with the drone of manmade machinery.

(Figure 11.2b). Sound intensity determines the *loudness* we perceive, loud sounds having higher intensity. The range of intensities to which the human ear is sensitive is astonishing: The intensity of the loudest sound that doesn't damage our ears is about a trillion times greater than the intensity of the faintest sound that can be heard. If we were any more sensitive, we would hear a constant roar from the random movement of air molecules.

Real world sounds rarely consist of simple periodic sound waves at one frequency and intensity. It is the simultaneous combination of different frequency waves at different intensities that gives different musical instruments and human voices their unique tonal qualities.

STRUCTURE OF THE AUDITORY SYSTEM

Before exploring the details of translating variations in air pressure into neural activity, let's quickly survey the structural organization of the auditory system. The structure of the ear is shown in Figure 11.3. The visible portion of the ear consists primarily of cartilage covered by skin, forming a sort of funnel called the **pinna** (from the Latin for "wing"), which brings sound into the head. It has also been suggested that the convolutions in the pinna play a role in localizing sounds, something we will discuss later in the chapter. In humans, the pinna is more or less fixed in position, but animals such as cats and horses have considerable muscular control over the position of their pinna and can orient it toward a source of sound. The entrance to the internal ear is called the **auditory canal**, and it extends about 2.5 cm (1 inch) inside the skull before it ends at the **tympanic membrane**, also known as the *eardrum*. Connected to the medial surface of the tympanic membrane is a series of bones called **ossicles** (from the Latin for "little bones"). Located in a small air-filled chamber, the ossicles transfer movements of the tympanic membrane into movements of a second membrane covering a hole in the bone of the skull called the **oval window**. Behind the oval window is the

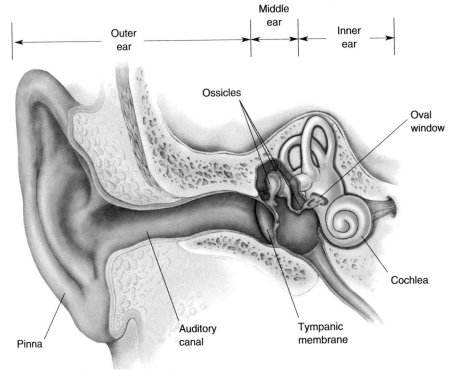

Figure 11.3
The outer, middle, and inner ear.

fluid-filled **cochlea**, which contains the apparatus for transforming the physical motion of the oval window membrane into a neuronal response. Thus, the first stages of the basic auditory pathway look like this:

Sound wave moves the tympanic membrane→
  Tympanic membrane moves the ossicles→
    Ossicles move the membrane at the oval window→
      Motion at the oval window moves fluid in the cochlea→
        Movement of fluid in the cochlea causes a response in sensory neurons.

All of the structures from the pinna inward are considered components of the ear, and it is conventional to refer to the ear as having three main divisions. The structures from the pinna to the tympanic membrane make up the **outer ear**, the tympanic membrane and the ossicles compose the **middle ear**, and the apparatus medial to the oval window is the **inner ear**.

Once a neural response to sound is generated in the inner ear, the signal is transferred to and processed by a series of nuclei in the brain stem. Output from these nuclei is sent to a relay in the thalamus, the **medial geniculate nucleus**, or **MGN**. Finally, the MGN projects to **primary auditory cortex**, or **A1**, located in the temporal lobe. In one sense, the auditory pathway is more complex than the visual pathway because there are more intermediate stages between the sensory receptors and cortex. However, the systems have analogous components, including sensory receptors, early integration stages (which in the visual system are within the retina), a thalamic relay, and sensory cortex (Figure 11.4).

THE MIDDLE EAR

The outer ear funnels sound to the middle ear, an air-filled cavity containing the first elements that move in response to sound. In the middle ear, variations in air pressure are converted into movements of the ossicles. In this section, we'll explore how this first transformation of sound energy occurs.

Components of the Middle Ear

The structures within the middle ear are the tympanic membrane and the ossicles. The tympanic membrane is somewhat conical in shape, with the point of the cone extending into the cavity of the middle ear. There are three ossicles, each named (from the Latin) after an object it slightly resembles (Figure 11.5). The ossicle attached to the tympanic membrane is the **malleus** ("hammer"), which forms a rigid connection with the **incus** ("anvil"). The incus forms a flexible connection with the **stapes** ("stirrup"). The flat bottom portion of the stapes, the *footplate*, moves in and out like a piston at the oval window, thus transmitting sound vibrations to the fluids of the cochlea in the inner ear.

The air in the middle ear is continuous with that in the mouth via the **Eustachian tube**, although this tube is usually closed by a valve. When you're in an ascending airplane or a car heading up a mountain, the pressure of the surrounding air decreases. However, as long as the valve on the Eustachian tube is closed, the air in the middle ear is at the pressure of the air before you started to climb. As a result of the pressure inside the middle ear being higher than the air pressure outside, the tympanic membrane bulges out, and you experience unpleasant pressure or pain in the ear. The pain can be relieved by yawning or swallowing, either of which opens the Eustachian tube, thereby equalizing the air pressure in the middle ear with the ambient air pressure.

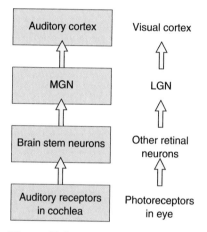

Figure 11.4
Auditory and visual pathways compared. Following the sensory receptors, both systems have early integration stages, a thalamic relay, and a projection to neocortex.

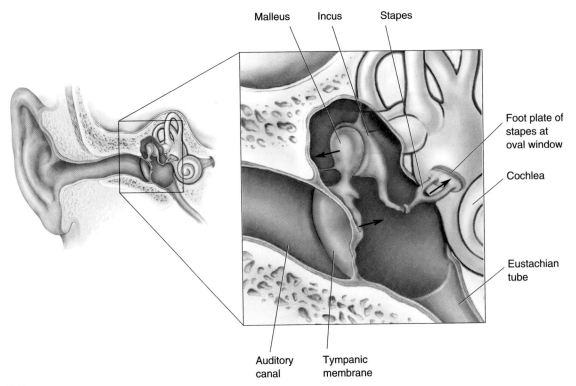

Malleus   Incus   Stapes

Foot plate of
stapes at
oval window

Cochlea

Eustachian
tube

Auditory
canal

Tympanic
membrane

Figure 11.5
The middle ear. As the arrows indicate, when the bottom of the malleus is pushed inward by the tympanic membrane, the lever action of the ossicles makes the footplate of the stapes push inward at the oval window. The pressure pushing at the oval window is greater than that at the tympanic membrane, in part because the surface area of the footplate of the stapes is smaller than the surface area of the tympanic membrane.

Sound Force Amplification by the Ossicles

Sound waves move the tympanic membrane, and the ossicles move another membrane at the oval window. Why don't sound waves simply directly move the membrane at the oval window? The answer lies in the fact that the cochlea is filled with fluid, not air. If sound waves impinged directly on the oval window, the membrane would barely move, and most of the sound would be reflected because of the pressure the cochlear fluid exerts at the back of the oval window. If you've ever noticed how quiet it is under water, you know how well water reflects sound coming from above. Because the fluid in the inner ear resists being moved much more than air does (i.e., fluid has greater inertia), more pressure is needed to vibrate the fluid. The ossicles provide this necessary amplification in pressure.

To understand the process, consider the definition of pressure. The pressure on a membrane is defined as the force pushing it divided by its surface area. The pressure at the oval window can become greater than the pressure at the tympanic membrane if the force on the oval window membrane is greater than that on the tympanic membrane, or, conversely, if the surface area of the oval window is smaller than the area of the tympanic membrane. It turns out that the middle ear increases pressure at the oval window by altering both the force and the surface area. The force at the oval window is greater because the ossicles act like levers; a large movement of the tympanic membrane is transformed into a smaller but stronger movement of the oval window. And the surface area of the oval window is much smaller than that of the tympanic membrane. These factors combine to make the pressure at the oval window about 20 times greater than that at the tympanic membrane, and this increase is sufficient to move the fluid in the inner ear.

The Attenuation Reflex

Two muscles attached to the ossicles have a significant effect on sound transmission to the inner ear. The *tensor tympani muscle* is anchored to bone in the cavity of the middle ear at one end and attaches to the malleus at the other end. The *stapedius muscle* also extends from a fixed anchor of bone and attaches to the stapes. When these muscles contract, the chain of ossicles becomes much more rigid, and sound conductance to the inner ear is greatly diminished. The onset of a loud sound causes these muscles to contract, a response called the **attenuation reflex**. Sound attenuation is much greater at low frequencies than at high frequencies.

A number of functions have been proposed for this reflex. One function may be to adapt the ear to continuous sound at high intensities. Loud sounds that would otherwise saturate the response of the receptors in the inner ear could be reduced to a level below saturation by the attenuation reflex, thus increasing the dynamic range we can hear. The attenuation reflex also protects the inner ear from loud sounds that would otherwise damage it. Unfortunately, the reflex has a delay of 50–100 msec from the time that sound reaches the ear, so it doesn't offer much protection from very sudden loud sounds; damage might already be done by the time the muscles contract. This is why, despite the best efforts of your attenuation reflex, a loud explosion (or music on your Walkman) can still damage your cochlea. Because the attenuation reflex suppresses low frequencies more than high frequencies, it tends to make high-frequency sounds easier to discern in an environment with a lot of low-frequency noise. This capability enables us to understand speech more easily in a noisy environment than we could without the reflex. It is thought that the attenuation reflex is also activated when we speak, so we don't hear our own voices as loudly as we otherwise would.

THE INNER EAR

Although considered part of the ear, not all of the inner ear is concerned with hearing. The inner ear consists of the cochlea, which is part of the auditory system, and the labyrinth, which is not. The labyrinth is an important part of the *vestibular apparatus*, which helps maintain the body's equilibrium (see Chapter 14). Here we are only concerned with the cochlea and the role it plays in transforming sound into a neural signal.

Anatomy of the Cochlea

The cochlea (from the Latin for "snail") has a spiral shape resembling a snail's shell. One can approximate the structure of the cochlea by wrapping a drinking straw 2½ times around the sharpened tip of a pencil (Figure 11.6). In the cochlea, the hollow tube (represented by the straw) has walls made of bone. The central pillar of the cochlea (represented by the pencil) is a conical bony structure called the **modiolus**. The actual dimensions are smaller than the straw-and-pencil model, the cochlea's hollow tube being about 32 mm long and 2 mm in diameter. At the base of the cochlea there are two

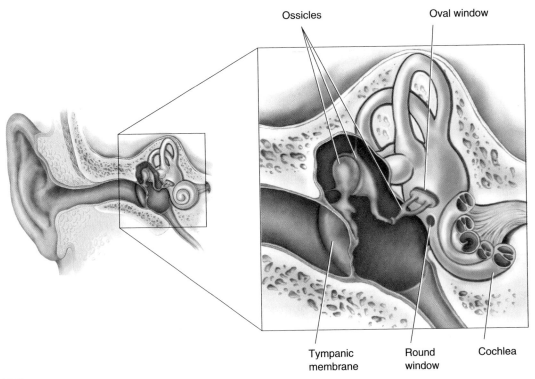

Ossicles

Oval window

Tympanic
membrane

Round
window

Cochlea

Figure 11.6
The inner ear.

membrane-covered holes; the oval window, which as we have seen is below the footplate of the stapes, and the **round window**.

If the cochlea is cut in cross section, it can be seen that the tube is divided into three chambers: the **scala vestibuli**, the **scala media**, and the **scala tympani** (Figure 11.7). The three scalae wrap around inside the cochlea like a spiral stairway ("scala" is from the Latin for "stairway"). **Reissner's membrane** separates the scala vestibuli from the scala media, and the **basilar membrane** separates the scala tympani from the scala media. Sitting upon the basilar membrane is the **organ of Corti**, which contains auditory receptor neurons; hanging over this organ is the **tectorial membrane**. At the apex of the cochlea the scala media is closed off, and the scala tympani and scala vestibuli are connected by a hole in the membranes, the **helicotrema** (Figure 11.8). Consequently, the fluid in the scala vestibuli is continuous with that in the scala tympani. At the base of the cochlea, the scala vestibuli meets the oval window and the scala tympani meets the round window.

The fluid in the scala vestibuli and scala tympani, called **perilymph**, has an ionic content similar to that of cerebrospinal fluid: low K^+ (7 mM) and high Na^+ (140 mM) concentrations. The scala media is filled with **endolymph**, which is unusual in that it has ionic concentrations similar to intracellular fluid, high K^+ (150 mM) and low Na^+ (1 mM), even though it is extracellular. The explanation for this difference is that an active process taking place at the *stria vascularis* (the endothelium lining one wall of the scala media) reabsorbs sodium and secretes potassium against their concentration gradients. Because of the ionic concentration differences and the permeability of Reissner's membrane, the endolymph has an electrical potential that is about 80 mV more positive than that of the perilymph; this is called the *endocochlear potential*. We shall see that the endocochlear potential is important because it enhances auditory transduction.

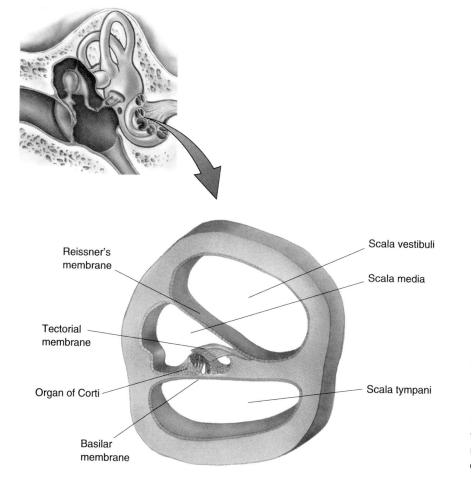

Figure 11.7
Three parallel canals within the cochlea. Viewed in cross section, the cochlea is seen to contain three smaller tubes running in parallel. These tubes, the scalae, are separated by Reissner's membrane and the basilar membrane. The organ of Corti contains the auditory receptors; it sits upon the basilar membrane and is covered by the tectorial membrane.

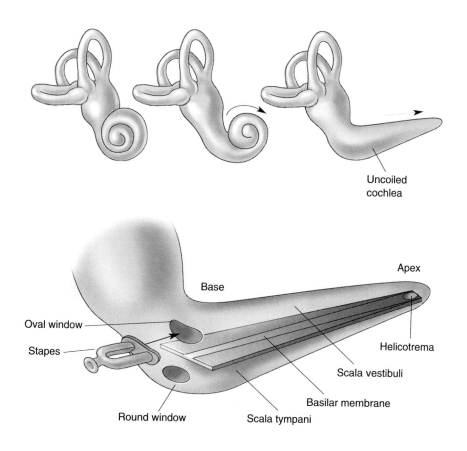

Figure 11.8
The basilar membrane in an uncoiled cochlea. Although the cochlea narrows from base to apex, the basilar membrane widens toward the apex. The helicotrema is a hole at the apex of the basilar membrane, which connects the scala vestibuli and scala tympani.

Physiology of the Cochlea

Despite its structural complexity, the basic operation of the inner ear is fairly simple. Look at Figure 11.8 and imagine what happens when the ossicles move the membrane that covers the oval window. Inward motion at the oval window pushes perilymph into the scala vestibuli. If the membranes inside the cochlea were completely rigid, then the increase in fluid pressure at the oval window would reach up the scala vestibuli, through the helicotrema, and back down the scala tympani to the round window. The membrane at the round window would bulge out in response to the inward movement of the membrane at the oval window. Although this isn't exactly what happens, the description conveys the very important point that any motion at the oval window must be accompanied by a complementary motion at the round window. Such movement must occur because the cochlea is filled with incompressible fluid held in a solid bony container. The consequence of pushing in at the oval window is a bit like pushing in one end of a tubular water balloon—the other end has to bulge out. The only reason the description of the events in the cochlea given above is not quite accurate is that the structures inside the cochlea are not rigid. Most importantly, the basilar membrane is flexible and bends in response to sound.

Response of the Basilar Membrane to Sound. The basilar membrane has two structural properties that determine the way it responds to sound. First, the membrane is wider at the apex than at the base by a factor of about 5. Second, the stiffness of the membrane decreases from base to apex, the base being about 100 times stiffer. Think of it as a flipper of the sort used for swimming, with a narrow, stiff base and a wide, floppy apex. When sound pushes the footplate of the stapes at the oval window, perilymph is displaced within the scala vestibuli, and endolymph is displaced within the scala media because Reissner's membrane is very flexible. We owe much of our understanding of the response of the basilar membrane to the research of Hungarian-American biophysicist Georg von Békésy. von Békésy determined that the movement of the endolymph makes the basilar membrane bend near its base, starting a wave that propagates toward the apex. The wave that travels up the basilar membrane is similar to the wave that runs along a rope if you hold one end in your hand and give it a snap. The distance the wave travels up the basilar membrane depends on the frequency of the sound. If the frequency is high, the stiffer base of the membrane will vibrate a good deal, dissipating most of the energy, and the wave will not propagate very far (Figure 11.9a). However, low-frequency sounds generate waves that travel all the way up to the floppy apex of the membrane before most of the energy is dissipated (Figure 11.9b). The response of the basilar membrane establishes a place code in which different locations of membrane are maximally deformed at different sound frequencies (Figure 11.9c). As we shall see, the differences in the traveling waves produced by different sound frequencies are responsible for the coding of pitch.

The Organ of Corti and Associated Structures. Everything we have discussed to this point involves the mechanical transformations of sound energy that occur in the middle and inner ear. Now we come to the point in the system where neurons are involved. The auditory receptor cells, which convert the mechanical energy into a change in membrane polarization, are located in the organ of Corti (named for the Italian anatomist who first identified it). The organ of Corti consists of hair cells, the rods of Corti, and various supporting cells.

The auditory receptors are called **hair cells** because each one has about 100 hairy-looking **stereocilia** extending from its top. The hair cells and stereocilia are shown in Figure 11.10 as they appear when viewed through

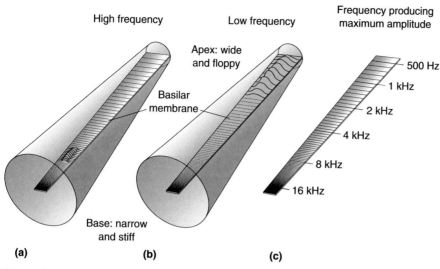

High frequency Low frequency Frequency producing maximum amplitude

Apex: wide and floppy

Basilar membrane

500 Hz
1 kHz
2 kHz
4 kHz
8 kHz
16 kHz

Base: narrow and stiff

(a)   (b)   (c)

Figure 11.9
Response of the basilar membrane to sound. The cochlea is again shown uncoiled. **(a)** High-frequency sound produces a traveling wave, which dissipates near the narrow and stiff base of the basilar membrane. **(b)** Low-frequency sound produces a wave that propagates all the way to the apex of the basilar membrane before dissipating. The bending of the basilar membrane is greatly exaggerated for the purpose of illustration. **(c)** There is a place code on the basilar membrane for the frequency that produces the maximum amplitude deflection.

a scanning electron microscope. The critical event in the transduction of sound into a neural signal is the bending of these cilia. For this reason, we want to examine the organ of Corti in more detail to see how flexing of the basilar membrane leads to bending of the stereocilia.

The hair cells are sandwiched between the basilar membrane and a thin sheet of tissue called the **reticular lamina** (Figure 11.11). The *rods of Corti* span these two membranes and provide structural support. Hair cells between the modiolus and the rods of Corti are called **inner hair cells** (there are about 3500), and cells farther out than the rods of Corti are called **outer hair cells** (there are about 15,000–20,000 arranged in 3–5 rows). The stereocilia at the tops of the hair cells extend above the reticular lamina into the endolymph, and their tips end in the gelatinous substance of the tectorial membrane. To keep the membranes within the organ of Corti straight in your mind, remember that the *basilar* is at the *base* of the organ of Corti, the *tectorial* forms a *roof* over the structure, and the *reticular* is in the *middle*, holding onto the hair cells.

Hair cells form synapses on neurons whose cell bodies are located in the **spiral ganglion** within the modiolus. Spiral ganglion cells are bipolar, with neurites extending to the bases and sides of the hair cells, where they receive synaptic input. Axons from the spiral ganglion enter the **auditory nerve** (cranial nerve VIII), which projects to the cochlear nuclei in the medulla.

Transduction by Hair Cells. When the basilar membrane moves in response to a motion at the stapes, the entire foundation supporting the hair cells moves because the basilar membrane, rods of Corti, reticular lamina, and hair cells are all rigidly connected. These structures move as a unit, pivoting up toward the tectorial membrane or away from it. When the basilar membrane moves up, the reticular lamina moves up and in toward the modiolus. Conversely, downward motion of the basilar membrane causes the reticular lamina to move down and away from the modiolus. When the

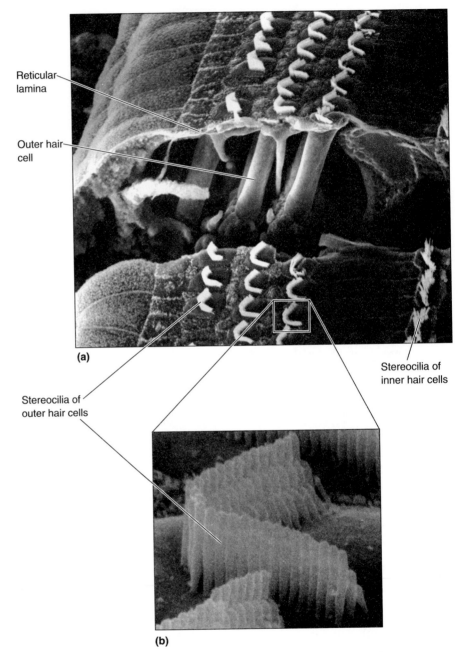

Reticular
lamina

Outer hair
cell

(a)

Stereocilia of
inner hair cells

Stereocilia of
outer hair cells

(b)

Figure 11.10
Hair cells viewed through the scanning electron microscope. (a) Hair cells and
their stereocilia. **(b)** Higher resolution view of the stereocilia on an outer hair cell. The
stereocilia are approximately 5 μm in length. (Source: Courtesy of I. Hunter-Duvar
and R. Harrison, The Hospital for Sick Children, Toronto, Ontario)

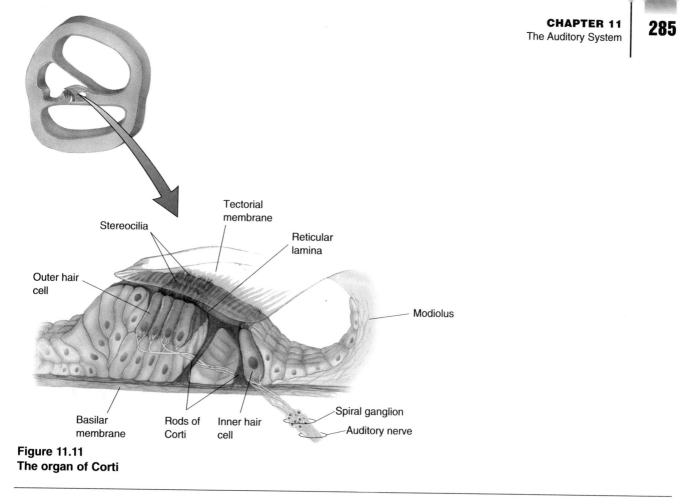

Figure 11.11
The organ of Corti

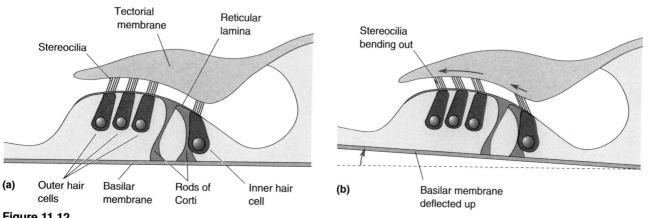

Figure 11.12
Bending of the stereocilia produced by upward motion of the basilar membrane. (a) At rest, the hair cells are held between the reticular lamina and the basilar membrane, and the tips of the outer hair cell stereocilia are attached to the tectorial membrane. **(b)** When sound causes the basilar membrane to deflect upward, the reticular membrane moves up and inward toward the modiolus, causing the stereocilia to bend outward.

reticular lamina moves inward or outward relative to the modiolus, it also moves in or out with respect to the tectorial membrane. Because the tectorial membrane holds the tips of the hair cell stereocilia, the lateral motion of the reticular lamina relative to the tectorial membrane bends the stereocilia on the hair cells one way or the other (Figure 11.12). The stereocilia contain aligned actin filaments that make them stiff, so they bend as rigid rods. Cross-link filaments connect the stereocilia on each hair cell, making all the c.lia on a hair cell move together as a unit.

Until recently, progress was slow in understanding how hair cells convert the mechanical deformation of stereocilia into neural signals. Because the cochlea is encased in bone, it is quite difficult to record from the hair cells. In the 1980s, A. J. Hudspeth and his colleagues, now at the University of Texas, pioneered a new approach, in which hair cells are isolated from the inner ear and studied *in vitro* (Box 11.2). The *in vitro* technique has revealed much about the transduction mechanism. Recordings from hair cells indicate that when the stereocilia bend in one direction, the hair cell depolarizes, and when they bend in the other direction, the cell hyperpolarizes.

Box 11.2 | **PATH OF DISCOVERY**

The Ear's Gears
by A. J. Hudspeth

A. J. Hudspeth

My research interest arose by accident. As graduate students, two colleagues and I were given the task of lecturing to the first-year medical class in neurobiology. This exercise was undoubtedly meant to build character, and incidentally freed the faculty from the necessity of lecturing on the three least palatable topics in neuroscience! We dutifully went about our preparation, then met to rehearse one another's talks. Although all the assigned subjects were challenging, the material on hearing and balance proved most opaque. We read that, in the darkness of the labyrinth, strangely named structures (basilar membranes, cupulae) performed arcane tasks, eliciting electrical signals of questionable significance (summating potentials, cochlear microphonics) in response to obscurely described stimuli. I came away from the experience with the conviction that the ear could not be as obscure as it seemed, that the unrivaled structural beauty of its cellular organization must reflect a comparable—if as yet unrevealed—precision of operation. I resolved to study how the internal ear's receptors, the hair cells, transduce mechanical stimuli into electrical responses.

The precedent of Theseus suggested that one should venture into a labyrinth only with due precautions. Indeed, the vestibular labyrinth's complexity, fragility, and position in the hardest bone of the body make experimentation on hair cells *in situ* very difficult. I therefore initiated a different approach, removing receptor organs from the internal ear and exploring the electrical responsiveness of their constituent hair cells *in vitro*. This strategy presented formidable problems. Isolating intact cells and maintaining them alive for hours was a challenge; stimulating and recording from such small cells was frustrating. It was especially difficult to achieve a mechanically stable experimental environment when working with cells sensitive to movements as small

as 0.3 nm, the diameter of a single atom. At the same time, though, experimentation with isolated hair cells offered important advantages. One could be certain from which cell a recording was made, and could employ a variety of extra- and intracellular recording techniques. Most importantly, the experimenter could confront a hair cell with stimuli of whatever amplitude, orientation, and timecourse he or she wished, without the necessity of dealing with the complex and incompletely understood mechanical and hydrodynamic properties of the intact ear.

Despite the differences between our various senses, all sensory receptors face similar challenges. Each such cell must transduce some significant form of stimulus energy into an electrical response that the nervous system can interpret. How can a receptor filter away irrelevant inputs and focus its attention on those of greatest behavioral importance? In order that the receptor organ be as sensitive as possible, the cell must detect the smallest possible amount of energy. How does the cell contrive to make such sensitive measurements without being overwhelmed by thermal, electrical, or other noise? The receptor must at the same time be able to adapt to maintained stimulation, so that its responsiveness is not impaired by large, maintained inputs. How can a receptor respond effectively to salient components of a stimulus, while ignoring background stimuli? These common demands are an important attraction of biophysical research on sensory transduction. Working in this field, one may attempt to understand how evolution has shaped a receptor to specific tasks that may be precisely defined in physical terms.

Some of the most important aspects of transduction by hair cells remain to be explored. The small **continued on page 287**

Depolarization or hyperpolarization produced by bending of the stereocilia is the receptor potential, which is above or below the resting potential of –70 mV (Figure 11.13).

These changes in cell potential result from the opening of potassium channels on the tips of the stereocilia. Figure 11.14 shows how these interesting channels are believed to function. Each channel is linked by an elastic filament to the wall of the adjacent cilium. When the cilia are straight, the tension on this filament holds the channel in a partially opened state, allowing a steady leak of K^+ from the endolymph into the hair cell. Displacement of the cilia in one direction increases tension on the linking filament, increasing the inward K^+ current. Displacement in the opposite direction relieves tension on the linking filament, thereby allowing the channel to

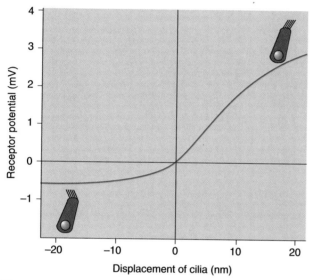

Figure 11.13
Hair cell receptor potentials. The hair cell depolarizes or hyperpolarizes depending on the direction in which the stereocilia bend. (Source: Adapted from Russel, Cody, and Richardson, 1986, Fig. 5.)

Box 11.2	**PATH OF DISCOVERY**

continued from page 286

number of cells in the internal ear has limited the application of biochemical techniques, so we know almost nothing about the molecular constituents of hair cells. It will be especially interesting to learn the structure of the hair bundle's mechanically sensitive ion channel, to see how that molecule responds to stimuli, and to ascertain whether it or a similar molecule subserves other forms of mechanoreceptivity—for example, touch. Perhaps the greatest challenge is understanding how the ear achieves its extraordinary sensitivity. Several lines of evidence suggest that hair cells are not just passive receptors of mechanical stimuli, but that they are somehow capable of producing active motions that enhance the ear's inputs. The cochlear amplifier is so powerful that most human ears actually *produce* spontaneous tones that can be detected by a sensitive microphone placed in the ear canal! The contraction of electrically stimulated hair cells and the presence of myosin in hair bundles provide hints of mechanisms by which a hair cell could augment its mechanical inputs. It will be fascinating to learn not only how amplification occurs, but also how a hair cell regulates its amplifier so as to avoid instability.

close completely, preventing inward K⁺ movement. The entry of K⁺ into the hair cell causes a depolarization, which in turn activates voltage-gated calcium channels (Figure 11.14b). The entry of Ca^{2+} triggers the release of neurotransmitter, which activates the spiral ganglion fibers lying postsynaptic to the hair cell. It is interesting that the opening of K⁺ channels produces a depolarization in the hair cell, whereas the opening of K⁺ channels *hyperpolarizes* most neurons. The reason that hair cells respond differently from neurons is the unusually high K⁺ concentration in endolymph, which yields a K⁺ equilibrium potential of 0 mV compared to the equilibrium potential of –80 mV in typical neurons.

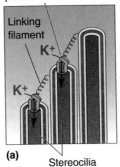
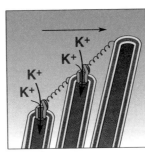

(a)

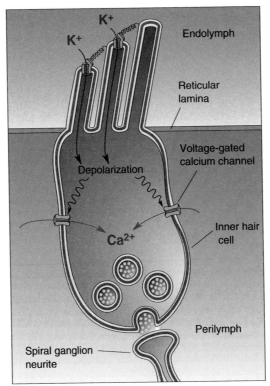

(b)

Figure 11.14
Depolarization of a hair cell. (a) Potassium channels are opened when the linking filaments joining the stereocilia are stretched. **(b)** Entry of potassium depolarizes the hair cell, which opens voltage-gated Ca^{2+} channels. Incoming calcium further depolarizes the cell, leading to the release of synaptic vessels to the postsynaptic neurite from the spiral ganglion.

The Innervation of Hair Cells. The auditory nerve consists of the axons of neurons whose cell bodies are located in the spiral ganglion. Thus, the spiral ganglion neurons, which are the first in the auditory pathway to fire action potentials, provide all the auditory information sent to the brain. For this reason, it is important to note that there is a significant difference in the spiral ganglion innervation of the inner and outer hair cells. It is estimated that the number of neurons in the spiral ganglion is in the range of 35,000–50,000. Despite the fact that inner hair cells are outnumbered by outer hair cells by a factor of 5 to 1, more than 95% of the spiral ganglion neurons communicate with the relatively small number of inner hair cells, and less than 5% receive synaptic input from the more numerous outer hair cells (Figure 11.15). Consequently, one spiral ganglion fiber receives input from only one inner hair cell; each inner hair cell feeds about 10 spiral ganglion neurites. The situation is the opposite with outer hair cells. Because they outnumber their spiral ganglion cells, an individual spiral ganglion fiber synapses with numerous outer hair cells. Simply based on these numbers, we can infer that most of the information leaving the cochlea comes from the responses of inner hair cells.

Amplification by Outer Hair Cells. Given that outer hair cells far outnumber inner hair cells, it seems paradoxical that most of the cochlear output is derived from inner hair cells. However, ongoing research suggests that outer hair cells play a very important role in sound transduction. The key to this function are *motor proteins,* found in the membranes of outer hair cells (Figure 11.16a). Motor proteins can change the length of outer hair cells, and it has been found that outer hair cells respond to sound with both a receptor potential and a change in length (Figure 11.16b). Because outer hair cells are attached to the basilar membrane and reticular lamina, when motor proteins change the length of the hair cell, the basilar membrane is pulled toward or pushed away from the reticular lamina and tectorial membrane. This is why the word *motor* is used—the outer hair cells actively change the physical relationship between the cochlear membranes.

The motor effect of outer hair cells makes a significant contribution to the traveling wave that propagates down the basilar membrane. This was demonstrated in 1991 by Mario Ruggero and Nola Rich at the University of Minnesota, who administered the chemical furosemide into experimental animals. Furosemide temporarily decreases the transduction that normally results from the bending of stereocilia on hair cells and it was found to significantly reduce the movement of the basilar membrane in response to sound (Figure 11.16 c, d). Reduction in the movement of the basilar membrane

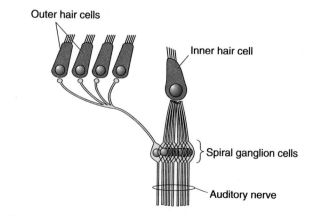

Figure 11.15
Innervation of hair cells by neurons from the spiral ganglion.

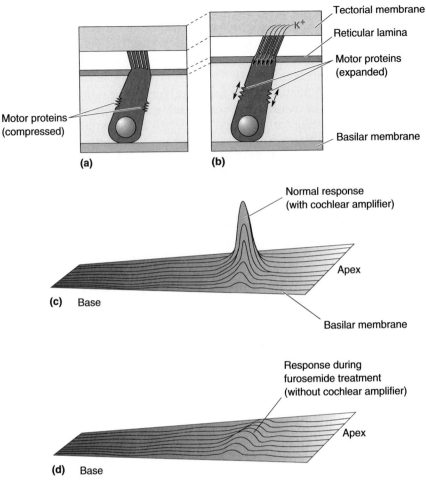

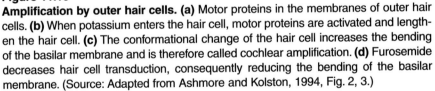

Figure 11.16
Amplification by outer hair cells. (a) Motor proteins in the membranes of outer hair cells. **(b)** When potassium enters the hair cell, motor proteins are activated and lengthen the hair cell. **(c)** The conformational change of the hair cell increases the bending of the basilar membrane and is therefore called cochlear amplification. **(d)** Furosemide decreases hair cell transduction, consequently reducing the bending of the basilar membrane. (Source: Adapted from Ashmore and Kolston, 1994, Fig. 2, 3.)

produced by furosemide is believed to result specifically from inactivation of the outer hair cell motor proteins. For this reason, it is sometimes said that the outer hair cells constitute a **cochlear amplifier**. When the outer hair cells amplify the response of the basilar membrane, the stereocilia on the inner hair cells will bend more, and the increased transduction process in the inner hair cells will produce a greater response in the auditory nerve. Through this feedback system, therefore, outer hair cells contribute significantly to the output of the cochlea. (In some situations, the cochlear amplifier can also produce sounds of its own; see Box 11.4.)

The effect of outer hair cells on the response of inner hair cells can also be influenced by neurons outside the cochlea. In addition to the spiral ganglion afferents that project from the cochlea to the brain stem, there also are about 1000 efferent fibers projecting *from* the brain stem *toward* the cochlea. These efferents diverge widely, synapsing onto outer hair cells. Stimulation of these efferents changes the shape of the outer hair cells, and, as we've already discussed, this affects the responses of inner hair cells. In this way, descending input from the brain to the cochlea can regulate auditory sensitivity.

The amplifying effect of outer hair cells explains how certain antibiotics (e.g., kanamycin) that damage hair cells can lead to deafness. After excessive exposure to antibiotics, the responses to sound of many inner hair cells are reduced. However, the antibiotic almost exclusively damages outer hair cells, not inner hair cells! For this reason, deafness produced by antibiotics is thought to be a consequence of damage to the cochlear amplifier (i.e., outer hair cells), demonstrating just how essential this amplifier is.

CENTRAL AUDITORY PROCESSES

Because there are more synapses at nuclei intermediate between the sensory organ and the cortex, the auditory pathway appears more complex than the visual pathway. Also, in contrast to the visual system, there are many more alternative pathways by which signals can get from one nucleus to the next. Nonetheless, the amount of information processing in the two systems is not that different when you consider that the cells and synapses of the auditory system in the brain stem are analogous to interactions in the layers of the retina. We will now look at auditory circuitry, focusing on the transformations of auditory information that occur along the way.

Anatomy of Auditory Pathways

Afferents from the spiral ganglion enter the brain stem in the auditory nerve. At the level of the medulla, the axons innervate the **dorsal cochlear nucleus** and **ventral cochlear nucleus** ipsilateral to the cochlea where the axons originated. Each axon branches so that it synapses on neurons in both cochlear nuclei. From this point on, the system gets more complicated, and the connections are less well understood, because there are multiple pathways. Rather than trying to describe all of these connections, we will follow one pathway from the cochlear nuclei to auditory cortex (Figure 11.17). Cells in the ventral cochlear nucleus send out axons that project to the **superior olive** (also called the superior olivary nucleus) on both sides of the brain stem. Axons of the olivary neurons ascend in the *lateral lemniscus* (a lemniscus is a collection of axons) and innervate the **inferior colliculus** of the midbrain. Many efferents of the dorsal cochlear nucleus follow a similar route as the pathway from the ventral cochlear nucleus, but the dorsal path bypasses the superior olive. Although there are other routes from the cochlear nuclei to the inferior colliculus, with additional intermediate relays, *all ascending auditory pathways converge onto the inferior colliculus*. The neurons in the colliculus send out axons to the medial geniculate nucleus (MGN) of the thalamus, which in turn projects to auditory cortex.

Before moving on to the response properties of auditory neurons, we should make several points. (1) As we have said, there are projections and brain stem nuclei, other than the ones described, that contribute to the auditory pathways. For instance, the inferior colliculus sends axons not only to the MGN but also to the superior colliculus (where integration of auditory and visual information occurs) and to the cerebellum. (2) There is extensive feedback in the auditory pathways. For instance, the brain stem neurons send axons that contact outer hair cells, and auditory cortex sends axons to the MGN and inferior colliculus. (3) It is important to note that, other than the cochlear nuclei, auditory nuclei in the brain stem receive input from both ears. This explains the clinically important fact that the only way in which brain stem damage can produce deafness in one ear is if the cochlear nuclei (or auditory nerve) on one side are destroyed.

Response Properties of Neurons in the Auditory Pathway

To understand the transformations of auditory signals that occur in the brain stem, we must first consider the nature of the input from the neurons in the spiral ganglion of the cochlea. Because most spiral ganglion cells receive input from a single inner hair cell at a particular location on the basilar membrane, they fire action potentials only in response to sound within a limited frequency range. After all, hair cells are excited by deformations of the basilar membrane, and each portion of the membrane is maximally sen-

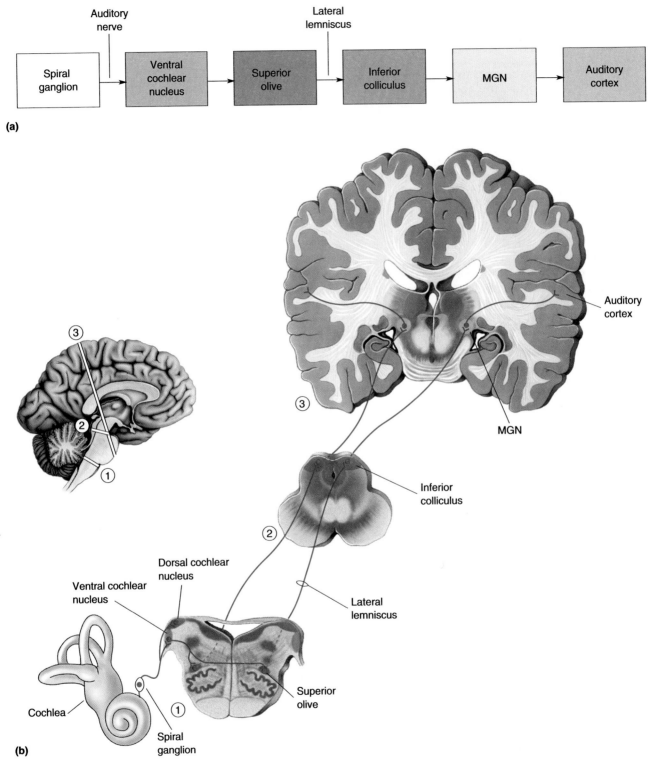

(a)

(b)

Figure 11.17
Auditory pathways. There are numerous paths by which neural signals can travel from the spiral ganglion to auditory cortex. A primary pathway is shown **(a)** schematically and **(b)** through brain stem cross sections.

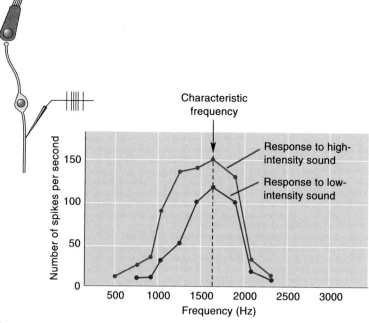

Figure 11.18
Response of an auditory nerve fiber to different sound frequencies. This neuron is frequency tuned and has greatest response at the characteristic frequency. (Source: Adapted from Rose, Hind, Anderson, and Brugge, 1971, Fig. 2.)

sitive to a particular range of frequencies. Figure 11.18 shows the results of an experiment in which action potentials were recorded from a single auditory nerve fiber (i.e., the axon of a spiral ganglion cell). The graph represents the firing rate in response to sounds at different frequencies. The neuron is most responsive to sound at one frequency, called the neuron's **characteristic frequency**, and it is less responsive at neighboring frequencies. This type of frequency tuning is typical of neurons at each of the relays from cochlea to cortex.

As one ascends the auditory pathway in the brain stem, the response properties of the cells become more diverse and complex, just as in the visual pathway. For instance, some cells in the cochlear nuclei are especially sensitive to sounds varying in frequency with time. In the MGN there are cells that respond to fairly complex sounds such as vocalizations, as well as other cells that show simple frequency selectivity as in the auditory nerve. An important development in the superior olive is that cells receive input from cochlear nuclei on both sides of the brain stem. As discussed below, such binaural neurons are probably important for sound localization.

ENCODING SOUND INTENSITY AND FREQUENCY

If you stop reading this book for a moment, you can focus on the many sounds around you. You can hear sounds you might have been ignoring, and you can selectively pay attention to different sounds occurring at the same time. There is truly an incredible diversity—from chattering people to cars to the radio to electrical noises. We cannot yet account for the perception of each of these sounds by pointing to particular neurons in the brain. However, most sounds have certain features in common, including intensity, frequency, and a location from which they emanate. Each of these features is represented in a different manner in the auditory pathway.

Stimulus Intensity

Information about sound intensity is coded in two interrelated ways: the firing rates of neurons and the number of active neurons. As a stimulus gets more intense, the basilar membrane vibrates with greater amplitude, causing the membrane potential of the activated hair cells to be more depolar-

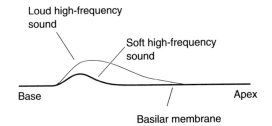

Figure 11.19
The influence of sound intensity on the response of the basilar membrane. At the same sound frequency, the maximum bending of the basilar membrane will occur closer to the apex as the intensity is increased.

ized or hyperpolarized. This causes the nerve fibers with which the hair cells synapse to fire action potentials at greater rates. In Figure 11.18, the auditory nerve fiber fires faster to the same sound frequencies when the intensity is increased. In addition, more intense stimuli produce movements of the basilar membrane over a greater distance (Figure 11.19), which leads to hair cell activation over a larger area. In a single auditory nerve fiber, this increase in the number of activated hair cells is seen as a broadening of the frequency range to which the fiber responds. The number of active neurons in the auditory nerve (and throughout the auditory pathway), and their firing rates, are thought to be the neural correlates of perceived loudness.

Stimulus Frequency and Tonotopy

As already discussed, from the hair cells in the cochlea through the various nuclei leading to auditory cortex, most neurons are sensitive to stimulus frequency and are most sensitive at the characteristic frequency. This sensitivity is largely a consequence of the mechanics of the basilar membrane because different portions of the membrane are maximally deformed by sound of different frequencies. Moving from the base to the apex of the cochlea, there is a progressive decrease in the frequency that produces a maximal deformation of the basilar membrane. There is a corresponding representation of frequency in the auditory nerve; auditory nerve fibers connected to hair cells near the apical basilar membrane have low characteristic frequencies and those connected to hair cells near the basal basilar membrane have high characteristic frequencies (Figure 11.20). When neurons in the auditory nerve synapse in the cochlear nuclei, they do so based on characteristic frequency. Nearby neurons have similar characteristic frequencies, and there is a systematic relationship between position in the cochlear nucleus and characteristic frequency. In other words, there is a map of the basilar membrane within the cochlear nuclei. Systematic organization within an auditory structure based on characteristic frequency is called **tonotopy**, and it is analogous to retinotopy in the visual system. There are tonotopic maps within each of the auditory nerve relay nuclei, the MGN, and auditory cortex.

Because of the tonotopy present throughout the auditory system, the location of active neurons in auditory nuclei is one indication of the frequency of the sound. However, there are two reasons why frequency must be coded in some way other than the site of maximal activation in tonotopic maps. One reason is that these maps do not contain neurons with very low characteristic frequencies, below about 200 Hz. As a result, the site of maximal activation might be the same for a 50 Hz tone as for a 200 Hz tone, so there must be some other way to distinguish them. The second reason that

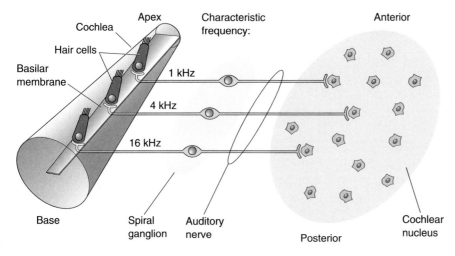

Figure 11.20
Tonotopic maps on the basilar membrane and cochlear nucleus. Going from the base to the apex of the cochlea, the basilar membrane resonates with increasingly lower frequencies. This tonotopy is preserved in the auditory nerve and cochlear nucleus. In the cochlear nucleus, there are bands of cells with similar characteristic frequencies; characteristic frequencies increase progressively from posterior to anterior.

something other than tonotopy is needed is, as we saw in Figure 11.19, that the region of the basilar membrane maximally displaced by a sound depends on its intensity in addition to its frequency. At a fixed frequency, a more intense sound will produce a maximal deformation at a point further up the basilar membrane than a less intense sound.

The main source of information about sound frequency that complements information derived from tonotopic maps is the timing of neural firing. Recordings made from neurons in the auditory nerve show **phase locking**, the consistent firing of a cell at the same phase of a sound wave (Figure 11.21). If you think of a sound wave as a sinusoidal variation in air pressure, a phase-locked neuron would fire action potentials at either the peaks, the troughs, or some other constant location on the wave. At low frequencies, some neurons fire action potentials every time the sound has a particular phase (Figure 11.21a). This makes it easy to determine the frequency of the sound; it is the same as the frequency of the neuron's action potentials.

It is important to realize that there can be phase locking even if an action potential has not fired on every cycle (Figure 11.21b). For instance, a neuron may have a response phase-locked to a 1000 Hz sound in such a way that an action potential fires on perhaps 25% of the cycles of the input. If you have a group of such neurons, each responding to different cycles of the input signal, it is possible to have a response to every cycle (by some member of the group) and thus a measure of sound frequency. The idea that intermediate frequencies are represented in the pooled activity of a number of neurons, each of which fires in a phase-locked manner, is called the **volley principle**. Phase locking occurs with sound waves up to about 4 kHz. Above this point, the action potentials fired by a neuron are at random phases of the sound wave (Figure 11.21c) because the intrinsic variability in the timing of the action potential becomes comparable to the time interval between successive cycles of the sound.

To summarize, here is how different frequencies are represented. At very low frequencies, phase locking is used; at intermediate frequencies, both phase locking and tonotopy are useful; and at high frequencies, tonotopy must be relied on to indicate sound frequency.

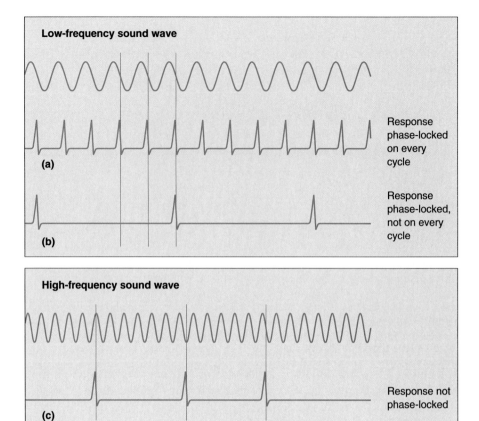

Figure 11.21
Phase locking in the response of auditory nerve fibers. Sound at a low frequency can elicit a phase-locked response, either **(a)** on every cycle of the stimulus or **(b)** on some fraction of the cycles. At high frequencies, **(c)** the response does not have a fixed phase relationship to the stimulus.

MECHANISMS OF SOUND LOCALIZATION

While the use of frequency information is essential for interpreting sounds in our environment, sound localization can be of critical importance for survival. If a predator is about to eat you, finding the source of a sudden sound and running away are much more important than analyzing the subtleties of the sound. Humans are not eaten by wild animals much anymore, but there are other situations in which sound localization can save your life. If you carelessly try to cross the street, your localization of a car's horn may be all that saves you. Our current understanding of the mechanisms underlying sound localization suggests that we use different techniques for locating sources in the horizontal plane (left–right) and vertical plane (up–down).

If you close your eyes and plug one ear, you can locate a bird singing as it flies overhead almost as well as with both ears open. But if you try to locate the horizontal position of a duck quacking as it swims across a pond, you'll find that you're much less able using only one ear. Thus, good horizontal localization requires a comparison of the sounds reaching the two ears, whereas good vertical localization does not.

Localization of Sound in the Horizontal Plane

Perhaps the most obvious cue to the location of a sound source is the time at which the sound arrives at each ear. We have two ears, and if we aren't facing a source directly, it will take the sound longer to reach one ear than the other. For instance, if a sudden noise comes at you from the right, it will reach your right ear first (Figure 11.22a); it will arrive at your left ear later, after what is known as an *interaural time delay*. If the distance between your ears is 20 cm, sound coming from the right, perpendicular to your head, will reach your left ear 0.6 msec after reaching your right ear. If the sound comes from straight ahead, there will be no interaural delay; and at angles between straight ahead and perpendicular, the delay will be between 0 and 0.6 msec (Figure 11.22b). Thus, there is a simple relationship between location and interaural delay. Detected by specialized neurons in the brain stem, the delay enables us to locate the source of the sound.

If we don't hear the onset of a sound because it is a continuous tone rather than a sudden noise, we cannot know the initial arrival times of the sound at the two ears. Thus, continuous tones pose more of a problem for sound localization because they are always present at both ears. We can still use arrival time to localize the sound but in a slightly different manner from localizing a sudden sound. The only thing that can be compared between continuous tones is the time at which the same *phase* of the sound wave reaches each ear. Imagine you are exposed to a 200 Hz sound coming from the right. At this frequency, one cycle of the sound covers 172 cm, which is much more than the 20 cm distance between your ears. After a peak in the sound wave passes the right ear, you must wait 0.6 msec, the time it takes

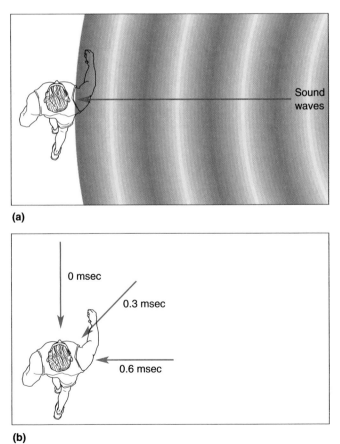

(a)

(b)

Figure 11.22
Interaural time delay as a cue to the location of sound. (a) Sound waves coming from the right side will reach the right ear first, and there will be a large interaural delay before the sound propagates to the left ear. **(b)** If the sound comes from straight ahead, there is no interaural delay. Delays for three different sound directions are shown.

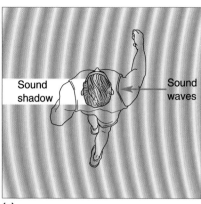

(a)

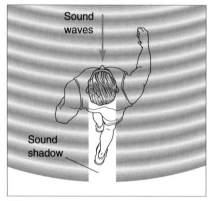

(b)

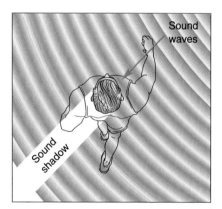

(c)

Figure 11.23
Interaural intensity difference as a cue to sound location. (a) With high-frequency sound, the head will cast a sound shadow to the left, when sound waves come from the right. Lower-intensity sound in the left ear is a cue that the sound came from the right. **(b)** If the sound comes from straight ahead, a sound shadow is cast behind the head but the sound reaches the two ears with the same intensity. **(c)** Sound coming from an oblique angle will partially shadow the left ear.

sound to travel 20 cm, before detecting a peak at the left ear. Of course, if the sound is straight ahead, peaks in the continuous tone will reach the ears simultaneously. Because the sound wave is much longer than the distance between the ears, we can reliably use the interaural delay of the peak in the wave to determine sound location.

However, things are more complicated with continuous tones at high frequencies. Suppose that the sound coming from the right now has a frequency of 20,000 Hz, which means that one cycle of the sound covers 1.7 cm. After a peak reaches the right ear, does it still take 0.6 msec before a peak arrives at the left ear? No! It takes a much shorter time because many peaks of such a high-frequency wave will fit between your ears. No longer is there a simple relationship between the direction the sound comes from and the arrival times of the peaks at the two ears. Interaural arrival time simply is not useful for locating sounds with frequencies so high that one cycle of the sound wave is smaller than the distance between your ears.

The brain has another process for sound localization at high frequencies: *interaural intensity difference*. An interaural intensity difference exists between the two ears because your head effectively casts a sound shadow (Figure 11.23). There is a direct relationship between the direction the sound comes from and the extent to which your head shadows the sound to one ear. If sound comes directly from the right, the left ear will hear a significantly lower intensity (Figure 11.23a). With sound coming from straight ahead, the same intensity reaches the two ears (Figure 11.23b), and with sound coming from intermediate directions, there are intermediate intensity differences (Figure 11.23c). Neurons sensitive to differences in intensity can use this information to locate the sound. The reason intensity information cannot be used to locate sounds at lower frequencies is that sound waves at these frequencies diffract around the head, and the intensities at the two ears are roughly equivalent. There is no sound shadow at low frequencies.

Let's summarize the two processes for localizing sound in the horizontal plane. With sounds in the range of 20–2000 Hz, the process involves *interaural time delay*. From 2000–20,000 Hz, *interaural intensity difference* is used. Together these two processes constitute the **duplex theory of sound localization**.

Sensitivity of Binaural Neurons to Sound Location. From our discussion of the auditory pathway, recall that neurons in the cochlear nuclei only receive afferents from the ipsilateral auditory nerve. Thus, all of these cells are *monaural*, meaning that they only respond to sound presented to one ear. At all later stages of processing in the auditory system, however, there are *binaural* neurons whose responses are influenced by sound at both ears. The response properties of binaural neurons imply that they play an important role in sound localization in the horizontal plane. The first structure where binaural neurons are present is the superior olive. While there is some controversy concerning the relationship between the activity of such neurons and the behavioral localization of sound, there are several compelling correlations. Neurons in the superior olive receive input from cochlear nuclei on both sides of the brain stem. Cells in the cochlear nuclei that project to the superior olive typically have responses phase-locked to lower-frequency sound input. Consequently, the olivary neuron receiving spikes from the left and right cochlear nuclei can compute interaural time delay. Recordings made in the superior olive show that the neurons typically give the greatest response to a particular interaural delay (Figure 11.24). One model for the circuitry that produces neurons sensitive to interaural delay involves *delay lines*. If the afferents reaching the superior olive are active with different delay times following sound in the left or right ear, the olivary neurons will have a preferred interaural delay (Figure 11.25). In other words, the left ear and right ear afferents are delay lines that determine the optimal interaural

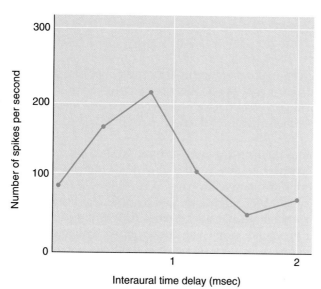

Figure 11.24
A neuron in the superior olive sensitive to interaural time delay. This neuron has an optimal delay of 1 second.

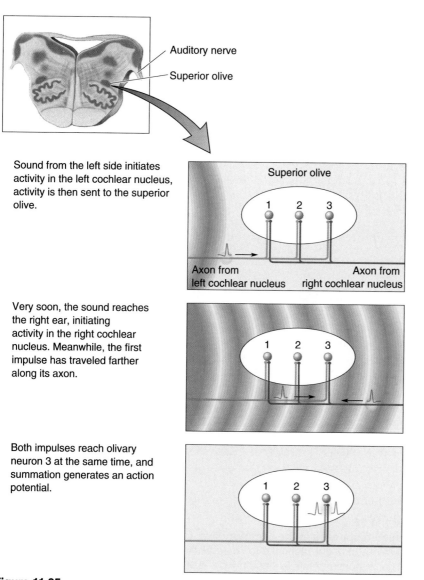

Sound from the left side initiates activity in the left cochlear nucleus, activity is then sent to the superior olive.

Very soon, the sound reaches the right ear, initiating activity in the right cochlear nucleus. Meanwhile, the first impulse has traveled farther along its axon.

Both impulses reach olivary neuron 3 at the same time, and summation generates an action potential.

Figure 11.25
Delay lines and neuronal sensitivity to interaural delay.

delay for a given neuron in the superior olive. Because phase locking only occurs in the neurons of the cochlear nucleus at relatively low frequencies, and phase locking is essential for precise comparison of the timing of inputs, it makes sense that interaural delays are useful for localization only at low frequencies.

In addition to their sensitivity to interaural delay, neurons in the superior olive are sensitive to interaural intensity. The responses of two types of binaural neurons are shown in Figure 11.26. One type of neuron (type EE) is moderately excited by sound presented to either ear but only gives a maximal response when both ears are stimulated. The other type of neuron (type EI) is excited by sound in one ear but inhibited by sound in the other ear. These EI neurons can be very sensitive to differences in intensity at the two ears because the effect of added excitation from one ear is compounded by the effect of decreased inhibition from the other ear. Presumably such a mechanism serves horizontal localization of high-frequency sound by means of differences in interaural intensity.

Localization of Sound in the Vertical Plane

As mentioned above, localizing sounds in the vertical plane is much less affected by plugging one ear than localizing sounds in the horizontal plane. In order to seriously impair vertical sound localization, one must place a tube into the auditory canal to bypass the pinna. The necessity of this step indicates the importance of the shape of the outer ear for assessing the elevation of a source of sound. The bumps and ridges of the outer ear apparently produce reflections of the entering sound. The delays between the direct path and the reflected path make vertical localization possible (Figure 11.27). Consistent with the idea that reflections off the outer ear are important is the finding that vertical localization is seriously impaired if the convolutions of the pinna are covered.

Some animals are extremely good at vertical sound localization even though they do not have a pinna. For example, a barn owl can swoop down on a squeaking mouse in the dark, locating accurately by sound, not sight. Although owls do not have a pinna, they can use the same techniques we

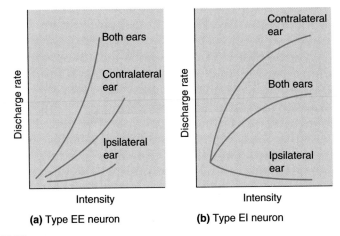

(a) Type EE neuron   **(b)** Type EI neuron

Figure 11.26
Binaural responses in the superior olive. (a) At all sound intensities, this EE neuron gives a stronger response to *bilateral* stimulation than to *unilateral* stimulation. **(b)** The response of this EI neuron increases with increasing sound intensity in the contralateral ear but decreases as intensity rises in the ipsilateral ear. The response to bilateral stimulation is reduced from the contralateral response.

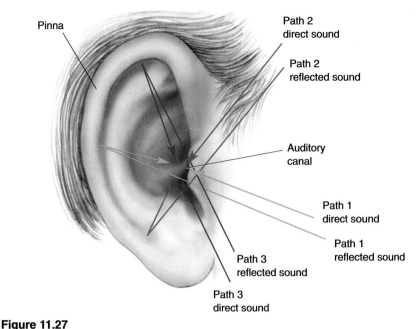

Figure 11.27
Vertical sound localization based on reflections from the pinna.

use for horizontal localization (interaural differences) because their ears are at different heights on their head. Some animals have a more "active" system for sound localization than humans and owls. Certain bats emit sounds that are reflected off objects, and these echoes are used to locate objects without sight. Emission and reception of reflected sound, which is analogous to the sonar used by ships, is used by many bats to hunt insects. In 1989, James Simmons at Brown University made the startling discovery that bats can discriminate time delays that differ by as little as 0.00001 msec (Box 11.3). This finding challenges our current understanding of how the nervous system, using action potentials lasting almost a millisecond, can perform such fine temporal discriminations.

AUDITORY CORTEX

Axons leaving the MGN project to auditory cortex via the internal capsule in an array called the *acoustic radiation*. Primary auditory cortex (A1) corresponds to Brodmann's area 41 in the temporal lobe (Figure 11.28a). The structure of A1 and the secondary auditory areas is in many ways similar to corresponding areas of the visual cortex. Layer I contains few cell bodies, and layers II and III contain mostly small pyramidal cells. Layer IV, where the medial geniculate axons terminate, is composed of densely packed granule cells. Layers V and VI contain mostly pyramidal cells that tend to be larger than those in the superficial layers. Let's look at how these cortical neurons respond to sound.

Neuronal Response Properties

In general, neurons in monkey (and presumably human) A1 are relatively sharply tuned for sound frequency and possess characteristic frequencies covering the audible spectrum of frequencies. In electrode penetrations made perpendicular to the cortical surface in monkeys, the cells encountered tend to have similar characteristic frequencies, suggesting a columnar organization on the basis of frequency. In the tonotopic representation in A1, low frequencies are represented rostrally and laterally, whereas high

Box 11.3 | **PATH OF DISCOVERY**

James Simmons

Bat Echolocation

by James Simmons

One powerful approach to learning how the brain works is to study an animal whose behavior might embody, in an exaggerated form, some general principle of brain function. The same exaggeration that makes the chosen creature seem exotic might also make the workings of the system easier to observe than in a more conventionally studied species. My colleagues and I have been very fortunate in getting a glimpse of how the brain creates the images that constitute perception from an unlikely source: the sonar system that bats use to "see" in the dark. Bats are popularly regarded as truly strange animals; they have even been demonized in myths and in horror movies. The bat's very body is unusual, with wings, tiny eyes, huge ears, and often a nose-leaf. But in reality these odd characteristics are fascinating adaptations of the animal to its nocturnal environment. Bats hunt flying insects in the dark with astonishing precision, using a form of biological sonar called *echolocation*. Bats actually *do* this—finding, tracking, and intercepting each flying insect—in only a second or so. In flight, the bat broadcasts ultrasonic chirps and perceives objects from their echoes, determining distance from the time it takes for the sound to travel out to the target and back to the ears. This mode of perception naturally places a premium on the brain's capacity to register the timing of sounds with great precision. Echolocation in bats provides an opportunity to observe how the brain uses time as a dimension of perception when it must do its utmost to be accurate.

My interest in bats began in graduate school in 1965 in the laboratory of auditory physiologist E. G. Wever, at Princeton University. This laboratory was a virtual menagerie of strange and unusual creatures with interesting auditory systems: dolphins, fish, frogs, owls, lizards—and bats. Wever provided the best of atmospheres for scientific research: support, encouragement, and a true interest in knowledge. With one of the lab's buildings to use as a sonar test range, I was told to try to find out just how accurate the bat's sonar really was. My first experiments involved training bats to sit still on a platform and "tell" me how far away objects were; the bats got mealworms as a reward. This involved catching bats in abandoned buildings, then hand-feeding them until they ate from a dish, and then getting them to sit still long enough to begin learning what I wanted them to do. Many bat-hunting afternoons were spent tottering high up on a rickety ladder, getting strange stares from passers-by. Word got around that I was an exterminator, and calls kept coming in to "please take away the bats from my house."

In the first experiments on perception of echo delay, the big brown bat could judge the distance to objects with an accuracy of about a centimeter. This is equivalent to telling a time difference of about 50 msec in echo delays from one object to another. It soon became apparent, however, that the movements of the bat while it scanned the targets were introducing a blurring effect on the results. More refined experiments with echoes delivered electronically to the bat through a loudspeaker revealed that the bat's echo-delay acuity actually was a fraction of a microsecond. Under quiet conditions the big brown bat can detect changes as small as 10 nanoseconds (0.01 microsecond) in the delay of echoes! Not surprisingly, there was a lot of skepticism about this result because it seemed biologically impossible. After all, individual neural action potentials are roughly 300,000 nsec (300 microseconds) wide, and the jitter in their timing is also hundreds of microseconds. How could bats possibly perceive changes of only 10 nsec using such inherently imprecise neural events to register echoes? Besides, aren't these animals too strange to challenge assumptions drawn from studying more conventional laboratory species, in which perception appears to emerge from representing stimulus information on maps in the brain?

But behavior doesn't lie; in fact, it is what the brain is for, so my colleagues and I searched long and hard to find the bat's neural mechanism for registering echoes with submicrosecond precision. We eventually found that the bat stretches small differences of a few microseconds in the timing of sounds into differences of hundreds of microseconds in the timing of neural events. Furthermore, with Steven Dear in the Department of Neuroscience at Brown University, we have most recently seen how the bat might set up intricate timing patterns of neural responses to create the images the bat perceives.

D. R. Griffin, one of the discoverers of echolocation, has described progress in this field as a "state of dynamic habituation to successive surprises," which seems to characterize scientific inquiry in general very accurately. The possible role of hearing in the bat's orientation was greeted with much disbelief, and each subsequent discovery has revealed capabilities that at first seemed beyond reason. What is important to science? Being critical, even skeptical, about your own experiments while keeping an open mind for what is possible to find.

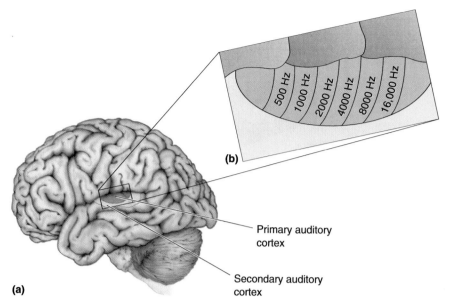

(b)

Primary auditory cortex

Secondary auditory cortex

(a)

Figure 11.28
Primary auditory cortex in humans. (a) Primary auditory cortex (purple) and secondary auditory areas (yellow) on the superior temporal lobe. **(b)** Tonotopic organization within primary auditory cortex. The numbers are characteristic frequencies.

frequencies are represented caudally and medially (Figure 11.28b). Roughly speaking, there are *isofrequency bands* running mediolaterally across A1. In other words, there are strips of neurons running across A1 in which all the neurons have fairly similar characteristic frequencies.

In the visual system, it is possible to describe large numbers of cortical neurons as having some variation on a general receptive field that is either simple or complex. So far, it has not been possible to place the diverse auditory receptive fields into a similarly small number of categories. As at earlier stages in the auditory pathway, neurons have different temporal response patterns; some have a transient response to a brief sound, and others have a sustained response. In addition to the frequency tuning that occurs in most cells, some neurons are intensity tuned, giving a peak response to a particular sound intensity. Even within a vertical column perpendicular to the cortical surface, there can be considerable diversity in the degree of tuning to sound frequency. Some neurons are sharply tuned for frequency, and others are barely tuned at all; the degree of tuning does not seem to correlate well with cortical layers. Other sounds that produce responses in cortical neurons include clicks, bursts of noise, frequency-modulated sounds, and animal vocalizations. Attempting to understand the role of these neurons that respond to seemingly complex stimuli is one of the challenges researchers currently face.

Given the wide variety of response types that neurophysiologists encounter in studying auditory cortex, you can understand why it is reassuring to see some sort of organization or unifying principle. One organizational principle already discussed is the tonotopic representation in many auditory areas. A second organizational principle is the presence in auditory cortex of columns of cells with similar binaural interaction. As at lower levels in the auditory system, one can distinguish EE and EI neurons (see Figure 11.26). Recall that EE cells respond more to stimulation of both ears than to either ear separately, while EI cells are inhibited if both ears are stimulated. If one makes an electrode penetration perpendicular to the cortical

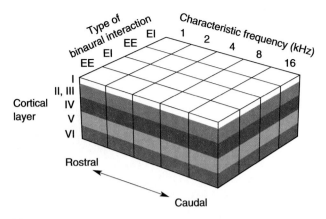

Figure 11.29
Hypothetical ice cube model of auditory cortex. In this model, characteristic frequencies increase moving rostral to caudal, and alternating bands of EE and EI cells run mediolateral.

surface, most of the cells encountered will be of one type, either EE or EI. In cat cortex, there are alternating patches of EE and EI cells within an isofrequency band, and evidence suggests that the binaural interaction columns (sometimes called *summation and suppression columns*) are roughly at right angles to the isofrequency bands (Figure 11.29). As we discussed for the superior olive, neurons sensitive to interaural time delays and interaural intensity differences probably play a role in sound localization.

In addition to A1, other cortical areas located on the superior surface of the temporal lobe respond to auditory stimuli. Some of these higher auditory areas are tonotopically organized, and others do not seem to be. As in visual cortex, there is a tendency for the stimuli that evoke a strong response to be more complex than at lower levels in the system. An example of specialization is Wernicke's area, which we will discuss in Chapter 21. Destruction of this area does not interfere with the sensation of sound, but it seriously impairs the ability to interpret spoken language.

Effects of Auditory Cortical Lesions and Ablation

While deafness results from bilateral ablation of auditory cortex, it is more often the consequence of damage to the ear (Box 11.4). A surprising degree of normal auditory function is retained after lesions of auditory cortex that are unilateral. This is in marked contrast to the visual system, in which a unilateral cortical lesion of striate cortex leads to complete blindness in one hemifield. The reason for greater preservation of function after lesions of auditory cortex is that both ears send output to cortex in both hemispheres. In humans, the primary deficit that results from unilateral loss of A1 is an inability to localize the source of a sound. It may be possible to determine which side of the head a sound comes from, but there is little ability to locate the sound more precisely. Performance on such tasks as frequency or intensity discrimination is near normal.

Studies in experimental animals indicate that smaller lesions can produce rather specific localization deficits. Because of the tonotopic organization of A1, it is possible to make a restricted cortical lesion that destroys neurons with characteristic frequencies within a limited frequency range. Interestingly, there is a deficit in localization only for sounds roughly corresponding to the characteristic frequencies of the missing cells. This finding reinforces the idea that information in different frequency bands may be processed in parallel by tonotopically organized structures.

Box 11.4   OF SPECIAL INTEREST

Auditory Disorders

Although the effects of cortical lesions provide important information about the role of auditory cortex in perception, the perceptual deficit we all associate with the auditory system, deafness, usually results from problems in or near the cochlea. Deafness is conventionally subdivided into two categories: conduction deafness and nerve deafness. Loss of hearing resulting from a disturbance in sound conductance from the external ear to the cochlea is called *conduction deafness*. This sensory deficit can result from something as simple as excessive wax in the ear to more serious diseases of the ossicles. A number of diseases cause binding of the ossicles to the bone of the middle ear, impairing the transfer of sound. Fortunately, most of the mechanical problems in the middle ear that interfere with sound conductance can be treated surgically.

Nerve deafness is deafness associated with the loss of neurons in either the auditory nerve or the hair cells of the cochlea. Nerve deafness sometimes results from tumors affecting the inner ear. It also can be the result of drugs that are toxic to hair cells, such as quinine and certain antibiotics. Exposure to loud sounds, such as explosions and loud music, is another cause of injury to the cochlea. Depending on the degree of cell loss, different treatments are possible. If the cochlea or auditory nerve on one side is completely destroyed, deafness in that ear will be absolute. However, a partial loss of hair cells is more common. In these cases, a hearing aid can be used to amplify the sound for the remaining hair cells. In more severe cases in which there is considerable damage to the cochlea but an intact auditory nerve, it is sometimes possible to restore some hearing by implanting an artificial electronic cochlea. This device has an electrode that is placed over or within the cochlea in order to electrically stimulate the auditory nerve. The artificial cochlea is wired to a microphone, which receives the sound. The frequency coding usually provided by the mechanics of the basilar membrane and hair cells can be partially achieved by the use of multiple electrodes that stimulate different portions of the auditory nerve in response to sound of various frequencies. The success of these devices is quite variable, ranging from restoration of the ability to understand speech to only a crude perception of changes in sounds.

With deafness, a person hears less sound than normal. With a hearing disorder called *tinnitus*, a person hears noises in the ears in the absence of any sound stimulus. The subjective sensation can take many forms, including buzzing, humming, and whistling. You may have experienced a mild and temporary form of tinnitus after being at a party with really loud music; the rest of your brain may have had fun, but your hair cells are in shock! Tinnitus is a relatively common disorder that can seriously interfere with concentration and work if it persists. You can imagine how distracting it would be if you constantly heard whispering or humming or the crinkling of paper. Tinnitus can be a symptom of any of a number of neurological problems. Although it frequently accompanies diseases involving the cochlea or auditory nerve, it may result from exposure to loud sounds or abnormal vasculature of the neck. Although clinical treatment of tinnitus is only partially successful, the annoyance of the noise can often be lessened by using a device to constantly produce a sound in the affected ear(s). For unknown reasons, the constant real sound is less annoying than the sound of the tinnitus that gets blocked out.

Perhaps the most interesting cause of tinnitus is sound produced by the ear itself. It is a curious fact that even normal ears not only respond to sound, they also can produce sound. Short sound clicks presented to the ear cause a faint "echo" that can be picked up with a microphone in the auditory canal. This echo is thought to be a by-product of the cochlear amplifier bending the cochlear membranes, which moves the ossicles and tympanic membrane, in the reverse order from normal sound transduction. This echo is not usually heard because it is faint and there are other sounds in the environment. However, in some cases, the ears emit sounds in the absence of any incoming sound at all. Called *spontaneous oto-acoustic emissions*, these sounds are thought to be a result of cochlear damage that makes the cochlear amplifier continuously active. As in the response of a normal ear to a click, the amplifier drives the cochlear membranes, ultimately moving the tympanic membrane and producing sound. If these spontaneous sounds are loud enough, they are heard as tinnitus.

CONCLUDING REMARKS

We have followed the auditory pathway from the ear to cerebral cortex and seen the ways in which sound information is transformed. Variations in the density of air molecules are converted to movement of the mechanical components of the middle and inner ear that is transduced into a neural response. The structure of the ear and cochlea are highly specialized for the transduction of sound. However, this fact should not blind us to the considerable similarities between the organization of the auditory system and that of other sensory systems.

Many analogies can be made between the auditory and visual systems. In the sensory receptors of both systems, a spatial code is established. In the visual system, the code in the photoreceptors is retinotopic; the activity of a given photoreceptor indicates light at a particular location. The receptors in the auditory system establish a spatial code that is tonotopic because of the unique properties of the cochlea. In each system, the retinotopy or tonotopy is preserved as signals are processed in secondary neurons, the thalamus, and finally in sensory cortex.

The convergence of inputs from lower levels produces neurons at higher levels that have more complex response properties. Combinations of LGN inputs give rise to simple and complex receptive fields in visual cortex; similarly in the auditory system, the integration of inputs tuned to different sound frequencies yields higher-level neurons that respond to complex combinations of frequencies. Another example of the increase in complexity at higher levels is that in the visual system, the convergence of inputs from the two eyes yields binocular neurons, which are important for depth perception. Analogously, in the auditory system, input from the two ears is combined to create binaural neurons, which are used for horizontal sound localization. These are just a few of the many similarities in the two systems. Principles governing one system can often help us understand other systems. Keep this in mind while reading about the somatosensory system in the next chapter, and you'll be able to predict some features of cortical organization based on the types of sensory receptors.

KEY TERMS

Introduction
audition

The Nature of Sound
frequency
hertz (Hz)
intensity

Structure of the Auditory System
pinna
auditory canal
tympanic membrane
ossicle
oval window
cochlea
outer ear
middle ear
inner ear
medial geniculate nucleus (MGN)
primary auditory cortex (A1)

The Middle Ear
malleus
incus
stapes
Eustachian tube
attenuation reflex

The Inner Ear
modiolus
round window
scala vestibuli
scala media
scala tympani
Reissner's membrane
basilar membrane
organ of Corti
tectorial membrane
helicotrema
perilymph
endolymph
hair cell

stereocilium
reticular lamina
inner hair cell
outer hair cell
spiral ganglion
auditory nerve
cochlear amplifier

Central Auditory Processes
dorsal cochlear nucleus
ventral cochlear nucleus
superior olive

inferior colliculus
characteristic frequency

Encoding Sound Intensity and Frequency
tonotopy
phase locking
volley principle

Mechanisms of Sound Localization
duplex theory of sound localization

REVIEW ✔ QUESTIONS

1. How is the conduction of sound to the cochlea facilitated by the ossicles of the middle ear?
2. Why is the round window crucial for the function of the cochlea? What would happen if it were rigid instead of a flexible membrane?
3. Why is it impossible to predict the frequency of a sound simply by looking at which portion of the basilar membrane is the most deformed?
4. Why would the transduction process in hair cells fail if the stereocilia as well as the hair cell bodies were surrounded by perilymph?
5. If inner hair cells are primarily responsible for hearing, what is the function of outer hair cells?
6. Why doesn't unilateral damage to the inferior colliculus or MGN lead to deafness in one ear?
7. What mechanisms function to localize sounds in the horizontal and vertical planes?
8. What symptoms would you expect to see in a person who had recently had a stroke affecting A1 unilaterally? How does the severity of these symptoms compare with the effects of a unilateral stroke involving V1?
9. What is the difference between nerve deafness and conduction deafness?

CHAPTER
12

The Somatic Sensory System

The somatic sensory system brings us some of life's most enjoyable experiences, as well as some of the most aggravating. **Somatic sensation** enables our bodies to feel, to ache, to chill, and to know what its parts are doing. It is sensitive to many kinds of stimuli: the pressure of objects against the skin, the position of joints and muscles, distension of the bladder, and the temperature of the limbs and of the brain itself. When stimuli become so strong that they may be damaging, somatic sensation is also responsible for the feeling that is most offensive, but vitally important: pain.

The somatic sensory system is different from other sensory systems in two interesting ways. First, its receptors are distributed throughout the body rather than being concentrated at small, specialized locations. Second, because it responds to many different kinds of stimuli, we can think of it as a group of at least four senses rather than a single one: the senses of touch, temperature, body position, and pain. In fact, those four can in turn be subdivided into many more. The somatic sensory system is really a catch-all name, a collective category for all the sensations that are *not* seeing, hearing, tasting, smelling, and the vestibular sense of equilibrium. The familiar idea that we have only five senses is obviously too simple.

If something touches your finger, you can accurately gauge the place, pressure, sharpness, texture, and duration of the touch. If it is a pinprick, there is no mistaking it for a hammer. If the touch moves from your hand to your wrist, and up your arm to your shoulder, you can track its speed and position. Assuming you are not looking, this information must be entirely described by the activity of the sensory nerves in your limb. A single sensory receptor can encode stimulus features such as intensity, duration, position, and sometimes direction. But a single stimulus usually activates many receptors. The task of the central nervous system is to interpret the activity of the vast receptor array and use it to generate coherent perceptions.

In this chapter, we describe the many somatic sensory receptors and their properties, proceed up their several pathways through the spinal cord to the brain, and try to explain why one stimulus may be pleasant while another hurts.

RECEPTORS OF SOMATIC SENSATION

Somatic sensation deals with many types of information from almost all parts of the body. It begins with sensory receptors in the skin and the body wall; in the muscles, tendons, ligaments, and connective tissue of the joints; and in the internal organs. Because the stimuli vary from the mechanical (e.g., a tickle) to the chemical (bee venom) and the physical (a cold shower), it takes many different kinds of specialized receptors to detect them. A receptor might be a simple, bare nerve ending or a complex combination of nerve with supporting cells, connective tissue, muscle cells, or even hair. A few examples are shown in Figure 12.1. This range of receptor structures is very different from all the other sensory systems, which have only one type, or a small set of subtypes, of sensory receptor.

Most of the sensory receptors in the somatic sensory system are **mechanoreceptors**, which are sensitive to physical distortion such as bending or stretching. Present throughout the body, they monitor contact with the skin, pressure in the heart and blood vessels, stretching of the digestive organs and bladder, and force against the teeth. At the heart of each mechanoreceptor are unmyelinated axon branches. These axons have *mechanosensitive ion channels;* their gating depends on stretching, or changes in tension, of the surrounding membrane. Of the major types of known ion channels, mechanosensitive channels are the least understood.

Besides mechanoreceptors, somatic sensation depends strongly on **nociceptors**, which respond to potentially damaging stimuli; **thermoreceptors**, which are sensitive to changes in temperature; **proprioceptors**, which monitor body position; and various **chemoreceptors**, which respond to certain chemicals (see Chapter 8). Let's explore how these receptors transduce different stimuli into neural signals.

Mechanoreceptors of the Skin

There are several kinds of mammalian skin. The backs and palms of your hands are examples of *hairy* and *glabrous* (hairless) skin, the two major types. Skin has an outer layer, the *epidermis,* and an inner layer, the *dermis.*

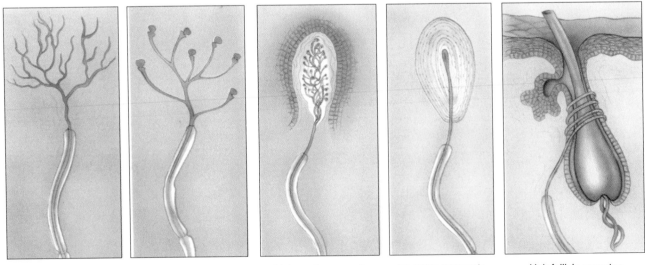

| Free nerve ending | Merkel's disk | Meissner's corpuscle | Pacinian corpuscle | Hair follicle receptor |

Figure 12.1
Specialized somatic sensory nerve endings. Each receptor has an axon, and all but the free nerve ending have non-neural tissues associated with them.

Skin performs an essential protective function, and it prevents the evaporation of body fluids into the dry environment we live in. But skin also provides our most direct contact with the world—indeed, skin is the largest sensory organ we have. Imagine the beach without the squish of sand between your toes, or consider *watching* a kiss instead of experiencing it yourself. Skin is sensitive enough that a raised dot measuring only 0.006 mm high and 0.04 mm wide can be felt when stroked by a fingertip. For comparison, a braille dot is 167 times higher!

The mechanoreceptors of the skin are shown in Figure 12.2. Most of them are named after the nineteenth century German and Italian histologists who discovered them. The largest and best studied receptor is the **Pacinian corpuscle**, which lies deep in the dermis, and can be as long as 2 mm and almost 1 mm in diameter. *Ruffini's endings,* found in both hairy and glabrous skin, resemble small Pacinian corpuscles. *Meissner's corpuscles* are about one-tenth the size of Pacinian corpuscles and are located in the ridges of glabrous skin (the raised parts of your fingerprints, for example). Located within the epidermis, *Merkel's disks* each consist of a nerve terminal and a flattened, non-neural epithelial cell. In this case it may be that the epithelial cell is the mechanically sensitive part, because it makes a synapse-like junction with the nerve terminal. In *Krause end bulbs,* which lie in the border regions of dry skin and mucous membrane (around the lips and genitals, for example), the nerve terminals look like knotted balls of string.

Skin can be vibrated, pressed, pricked, and stroked, and its hairs can be bent or pulled. These are quite different kinds of mechanical energy, yet we can feel them all and easily tell them apart. Accordingly, we have mechanoreceptors that vary in their preferred stimulus frequencies, pressures, and receptive field sizes. Swedish neuroscientist Åke Vallbo and his colleagues developed methods to record from single sensory axons in the human arm, so that they could simultaneously measure the sensitivity of mechanoreceptors in the hand *and* evaluate the perceptions produced by various mechanical stimuli (Figure 12.3a). When the stimulus probe was

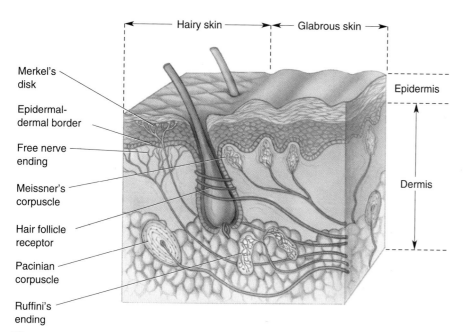

Figure 12.2
Somatic sensory receptors in the skin. Hairy and glabrous skin have a variety of sensory receptors within the dermal and epidermal layers.

touched to the surface of the skin and moved around, the receptive field of a single mechanoreceptor could be mapped. Meissner's corpuscles and Merkel's disks had small receptive fields, only a few millimeters wide, while Pacinian corpuscles and Ruffini's endings had large receptive fields that could cover an entire finger or half the palm (Figure 12.3b).

Mechanoreceptors also vary in the persistence of their responses to long-lasting stimuli. If a stimulus probe is suddenly pressed against the skin within the receptive field, some mechanoreceptors, such as Meissner's and Pacinian corpuscles, tend to respond quickly at first but then stop firing even though the stimulus continues; these receptors are said to be *rapidly adapting*. Other receptors, such as Merkel's disks and Ruffini's endings, are *slowly adapting*, and generate a more sustained response during a long stimulus. Figure 12.4 summarizes the receptive field size and adaptation rate for four mechanoreceptors of the skin.

Hairs do more than adorn our head and keep a dog warm in winter. Many hairs are part of a sensitive receptor system. To demonstrate this, brush just a single hair on the back of your arm with the tip of a pencil; it feels like an annoying mosquito. For some animals, hair is a major sensory system. Imagine a rat slinking confidently through dark passageways and alleys. The rat navigates in part by waving its facial *vibrissae* (whiskers) to sense the local environment and derive information about the texture, distance, and shape of nearby objects. Hairs grow from *follicles* embedded in the skin; each follicle is richly innervated by free nerve endings that either

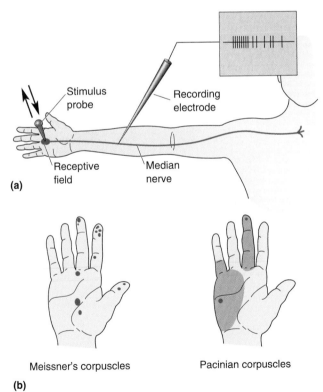

Meissner's corpuscles   Pacinian corpuscles

(b)

Figure 12.3
Testing the receptive fields of human sensory receptors. (a) By introducing a micro-electrode into the median nerve of the arm, it is possible to record the action potentials from a single sensory axon and map its receptive field on the hand with a fine stimulus probe. **(b)** Results show that receptive fields are either relatively small, as for the Meissner's corpuscles, or large, as for Pacinian corpuscles. (Source: Adapted from Vallbo and Johansson, 1984.)

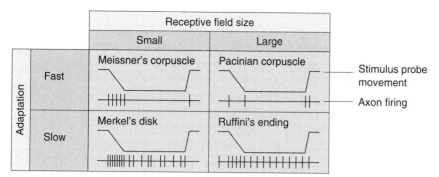

Figure 12.4
Variations among somatic sensory receptors of the skin in receptive field size and adaptation rate. (Source: Adapted from Vallbo and Johansson, 1984.)

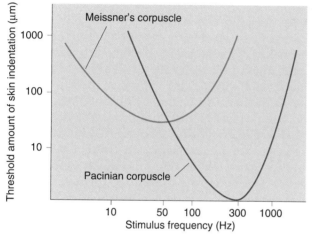

Figure 12.5
Frequency sensitivity of two rapidly adapting mechanoreceptors of the skin. Pacinian corpuscles are most sensitive to high-frequency stimuli, and Meissner's corpuscles are more sensitive to low-frequency stimuli. The skin was indented with a pressure probe, at various frequencies, while recording from the nerve. The amplitude of the stimulus was increased until it generated action potentials; threshold was measured as the amount of skin indentation in micrometers (μm). (Source: Adapted from Schmidt, 1978.)

wrap around it or run parallel to it (see Figure 12.1). There are several types of hair follicles, including some with erectile muscles (essential for mediating the strange sensations we know as goose pimples), and the details of their innervation differ. In all cases, the bending of the hair causes a deformation of the follicle and surrounding skin tissues. This, in turn, stretches, bends, or flattens the nearby nerve endings, which then increase or decrease their action potential firing frequency. The mechanoreceptors of hair follicles may be either slowly adapting or rapidly adapting.

The different mechanical sensitivities of mechanoreceptors mediate different sensations. Pacinian corpuscles are most sensitive to vibrations of about 200–300 Hz, while Meissner's corpuscles respond best around 50 Hz (Figure 12.5). Place your hand against a speaker while playing your favorite music loudly; you "feel" the music largely with your Pacinian corpuscles. If you stroke your fingertips across the coarse screen covering the speaker, each point of skin will hit the bumps at frequencies about optimal to activate Meissner's corpuscles. You feel this as a sensation of rough texture. Stimulation of even lower frequency can activate Ruffini's endings and Meissner's corpuscles, and yields a "fluttering" feeling.

Vibration and the Pacinian Corpuscle. The selectivity of a mechanoreceptive axon depends primarily on the structure of its special ending. For example, the Pacinian corpuscle has a football-shaped capsule with 20–70 concentric layers of connective tissue, arranged like the layers of an onion, with a nerve terminal in the middle (see Figure 12.1). When the capsule is compressed, energy is transferred to the nerve terminal, its membrane is deformed, and mechanosensitive channels open. Current flowing through the channels generates a receptor potential, which is depolarizing (Figure 12.6a). If the depolarization is large enough, the axon will fire an action potential. But the capsule layers are slick, with viscous fluid between them. If the stimulus pressure is maintained, the layers slip past one another and transfer the stimulus energy in such a way that the axon terminal is no longer deformed, and the receptor potential dissipates. When pressure is released, the events reverse themselves; the terminal depolarizes again and may fire another action potential.

In the 1960s, Werner Loewenstein and his colleagues, working at Columbia University, stripped away the capsule from single corpuscles and found that the naked nerve terminal became much less sensitive to vibrating stimuli and much more sensitive to steady pressure (Figure 12.6b). Clearly, it is the layered capsule (and not some property of the nerve ending itself) that makes the Pacinian corpuscle exquisitely sensitive to vibrating, high-frequency stimuli, and almost unresponsive to steady pressure.

Two-Point Discrimination. Our ability to discriminate the detailed features of a stimulus varies tremendously across the body. A simple measure of spatial resolution is the two-point discrimination test. You can do this yourself with a paper clip bent into the shape of a U. Start with the ends about an inch apart and touch them to the tip of a finger—you should have no problem telling that there are two separate points touching your finger. Then bend the wire to bring the points closer together, and touch them to your fingertip again. Repeat, and see how close the points have to be before they feel like a single point. (This test is best done with two people, one testing and the other being tested without looking.) Now try it on the back of your hand, on your lips, on your leg, and any other place that interests you. Compare your results with those shown in Figure 12.7.

Two-point discrimination varies at least twentyfold across the body. Fingertips have the highest resolution. The dots of braille are 1 mm high and 2.5 mm apart; up to six dots make a letter. An experienced braille reader can scan an index finger across a page of raised dots and read about 600 letters per minute, which is roughly as fast as someone reading aloud. There are several reasons why the fingertip is so much better than, say, the elbow for braille reading: (1) There is a much higher density of mechanoreceptors in the skin of the fingertip than on other parts of the body; (2) the fingertips are enriched in receptor types that have small receptive fields; (3) there is more brain tissue (and thus more raw computing power) devoted to the sensory information of each square millimeter of fingertip than elsewhere; and (4) there may be special neural mechanisms devoted to high-resolution discriminations.

Nociceptors

Nociceptors are free, branching, unmyelinated nerve endings that signal that body tissue is being damaged or is at risk of being damaged. (The word

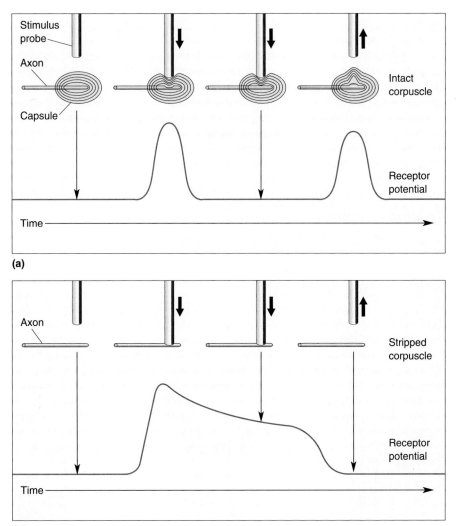

(a)

(b)

Figure 12.6
Adaptation in the Pacinian corpuscle. A single Pacinian corpuscle was isolated and stimulated by a probe that indented it briefly. The receptor potential was recorded from a nearby portion of the axon. **(a)** In the intact corpuscle, a large receptor potential was generated at the onset and offset of the stimulus; during maintained indentation, the receptor potential disappears. **(b)** The onionlike encapsulation was dissected away, leaving a bare axon ending. When indented by the probe, a receptor potential was again generated, so the capsule is not necessary for mechanoreception. But while the normal corpuscle responded only to the onset or offset of a long indentation, the stripped version gave a much more prolonged response; its adaptation rate was slowed. Apparently it is the capsule that makes the corpuscle insensitive to low-frequency stimuli.

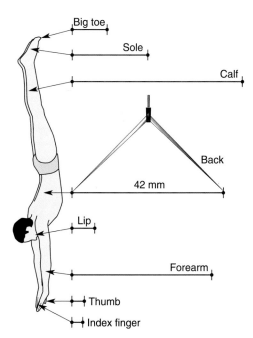

Figure 12.7
Two-point discrimination on the body surface. The pairs of dots show the minimum distance necessary to differentiate between two points touching you simultaneously. Notice the sensitivity of the fingertips compared to the rest of the body.

is from the Latin *nocere*, "to hurt.") The protective advantage of pain is clear in many cases: We quickly remove our hand from a hot stove. But as we shall see, the value of pain is not always obvious: What purpose is served by the pain of a migraine headache or the severe, chronic pain of terminal cancer?

Nociceptors are selective in their responses and can be divided into four broad categories. *Mechanical nociceptors* respond to strong pressure, in particular from sharp objects (such as that thumbtack in Chapter 3). *Thermal nociceptors* signal either burning heat (above about 45°C, when tissues begin to be destroyed) or extreme cold. *Chemically sensitive, mechanically insensitive nociceptors* respond to a variety of agents from the environment or from the tissue itself, including potassium, extremes of pH, neuroactive substances such as histamine and bradykinin, and various irritants. *Polymodal nociceptors* respond to combinations of mechanical, thermal, and chemical stimuli.

Nociceptors are present in most body tissues, including skin, bone, muscle, most internal organs, blood vessels, and the heart. They are notably absent in the brain itself, except for the meninges. This fact allows neurosurgeons to occasionally perform procedures on patients using only a local anesthetic on the scalp. In certain forms of temporal lobe epilepsy, for example, it becomes necessary to remove a piece of diseased cortex. While the patient is awake, the sensory and motor functions of different cortical areas can be tested to ensure that essential regions are not removed.

Thermoreceptors

Because the rate of a chemical reaction depends on temperature, the functioning of all cells is sensitive to temperature. Because of its complexity, the brain requires a more stable temperature than other parts of the body. It is vitally important that brain temperature stay close to 37°C. It cannot rise above 40.5°C without serious dysfunction. A drop of a just a few degrees below normal will impair movement and cognition, and will ultimately depress respiration, eventually causing death.

The fact that all neurons are sensitive to temperature does not make them all thermoreceptors. Because of specific (but unknown) membrane mechanisms, some neurons are exquisitely sensitive to temperature. For example, we can perceive changes in our average skin temperature of as little as 0.01°C! Temperature-sensitive neurons clustered in the hypothalamus and the spinal cord are important in the physiological responses that maintain stable body temperatures, but it is the thermoreceptors in the skin that apparently contribute to our perception of temperature. Although very little is known about the structure of thermoreceptive nerve endings, we do know that temperature sensitivity is not spread uniformly across the skin. It is possible to take a small cold or warm probe and map your skin's sensitivity to temperature changes. There are spots about 1 mm wide that are especially sensitive to *either* hot or cold, but not both. The locations of hot and cold sensitivity are different, showing that separate receptors encode them. There are also small areas of skin in between the hot and cold spots that are relatively insensitive to temperature.

Warm receptors begin firing above about 30°C and increase their firing rate until about 45°C (Figure 12.8a). As temperature increases further, the firing rate actually falls off steeply. Recall that 45°C is just about the point where thermal nociceptors begin firing; it is also where perception changes from hot to scalding, and tissues begin to burn. *Cold receptors* are relatively unresponsive at skin temperatures above about 35°C, but they fire faster over a broad range down to about 10°C. Below that temperature, firing ceases, and cold becomes a very effective anesthetic, as you know if you've ever had the pleasure of wading through an icy mountain stream. Strangely, some cold receptors also begin to fire when their temperature is raised *above* 45°C (Figure 12.8a). If such temperatures are applied to wide areas of skin, they are usually painful; but if they are restricted to small regions of skin innervated by a cold receptor, they produce a paradoxical feeling of cold. This emphasizes an important point: The CNS does not know *what* kind of stimulus (in this case, heat) caused the receptor to fire, but it continues to interpret all activity from its cold receptor as a response to cold.

As with other somatic sensory receptors, the responses of thermoreceptors adapt during long-duration stimuli. Figure 12.8b shows that a sudden drop in skin temperature causes a cold receptor to fire strongly, while it

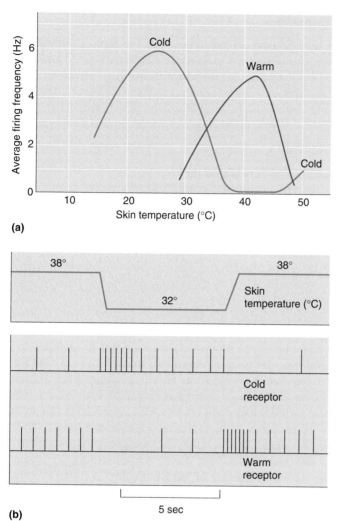

(a)

(b)

5 sec

Figure 12.8

Responses of thermoreceptors. (a) Action potential firing rate as a function of skin temperature for cold and warm receptors. Notice that the cold receptor fires not only at relatively low temperatures but also above about 45°C. **(b)** Responses of cold and warm receptors to a step reduction in skin temperature. Both receptors are most responsive to sudden changes in temperature, but they adapt over several seconds.

silences a warm receptor. After a few seconds at 32°, however, the cold receptor slows its firing (but still fires faster than it did at 38°), while the warm receptor speeds up slightly. Notice that a return to the original warm skin temperature causes opposite responses—transient silence of the cold receptor, and a burst of activity in the warm receptor—before both return to their steady, adapted rates. Thus, the differences between the response rates of warm and cold receptors are greatest during, and shortly after, temperature changes. This often mirrors our perceptions of temperature. Try a simple experiment. Fill two buckets with tap water, one cold and one hot (but not painfully hot). Then plunge your right hand for one minute into each of them in turn. Notice the striking sensations of hot and cold that occur with each change, but also notice how transient the sensations are. With thermoreception, as with most other sensory systems, it is the sudden *change* in the quality of a stimulus that generates the most intense neural and perceptual responses.

Proprioceptors

Most of the somatic sensory receptors we have examined so far give direct information about the external environment. Life would be deficient, however, without a sense of **proprioception** (from the Latin word *proprius*, "one's own"), which tells us where the parts of our bodies are in space, whether they are moving, and, if so, in which direction and how fast. Proprioception has two main functions. First, knowing the positions of our limbs as they move helps us identify things. It's much easier to recognize an object if you can handle it actively than if it is placed passively into your hand so that your skin is stimulated but you are not allowed to personally guide your fingers around it. Second, proprioceptive information is essential for guiding movements.

The muscles of voluntary movement, *skeletal muscles*, have two mechanosensitive proprioceptors: muscle spindles and Golgi tendon organs. *Muscle spindles* measure the length and rate of stretch of the muscles. *Golgi tendon organs* gauge the force generated by a muscle by measuring the tension in its tendon (we will discuss these proprioceptors further in Chapter 13). In addition, there is a variety of mechanoreceptors in the connective tissues of joints, especially within the capsules and ligaments. Many resemble Ruffini's endings, Golgi tendon organs, and Pacinian corpuscles, and others are free nerve endings. They respond to changes in the angle, direction, and velocity of movement in a joint. Most are rapidly adapting, meaning that sensory information about a *moving* joint is plentiful, but nerves encoding the *resting* position of a joint are few. We are, nevertheless, quite good at judging the position of a joint, even with our eyes closed. It seems that information from joint receptors is combined with that from muscle spindles and Golgi tendon organs, and probably from cutaneous receptors, to estimate joint angle. Removing one source of information can be compensated for by the use of the other sources. When an arthritic hip is replaced with a steel and plastic one, patients are still able to tell the angle between their thigh and their pelvis, despite the fact that all their hip joint mechanoreceptors are sitting in a jar of formaldehyde in another room.

PRIMARY AFFERENT NERVES

Impulses from somatic sensory receptors are transmitted to the CNS by primary afferent axons. The cell body of the primary afferent axon is the dorsal root ganglion cell, and the axon enters the spinal cord through the dorsal roots (Figure 12.9). In the following sections, we will describe a basic plan for the organization of the somatic sensory pathways. There are two broad types of somatic sensory information: touch and proprioception; temperature and pain. The pathways for these two types remain largely segregated as axons course into and up the spinal cord, through the brain stem and thalamus, and eventually terminate in the neocortex. *Parallel pathways* are common in sensory organization. A particularly striking example was the parvocellular and magnocellular pathways of visual information described in Chapters 9 and 10. As in vision, the two parallel pathways of somatic sensation are distinguished by the type of information they carry, the size and speed of their primary axons, the route they take through the brain, and the connections they make.

Sensory Receptors and the Size of Primary Afferent Axons

Primary afferent axons have widely varying diameters, and their size correlates with the type of sensory receptor to which they are attached. Unfortunately, the terminology approaches absurdity here because the dif-

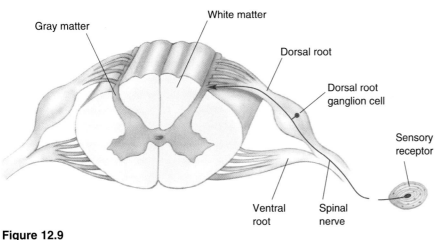

Figure 12.9
Structure of the spinal cord and its roots.

ferent sizes of axons are designated by two sets of names, using Arabic *and* Greek letters *and* Roman numerals. As shown in Figure 12.10, in order of decreasing size, axons supplying sensory receptors are usually designated Aα, Aβ, Aδ, and C; axons of similar size, but supplying the muscles and tendons, are called groups I, II, III, and IV. Note that the group of largest axons (Aα) is missing in the sensory nerves originating from the skin. Group C (or IV) axons are, by definition, unmyelinated axons, while all the rest are myelinated.

An interesting and simple point is hidden in the many axon names. Recall that the diameter of an axon, together with its myelin, determines its speed of action potential conduction. The smallest axons, the so-called C fibers, have no myelin and are less than about 1 μm in diameter. C fibers mediate some nociception and thermoreception, and they are the slowest of axons, conducting at about 0.5–1 m/sec. To appreciate just how slow this is, stretch out your arm. Imagine a bee stinging a fingertip. The painful components of the sting that are conducted by C fibers will take a second or two to reach your spinal cord, assuming you have a reach of about 1 m. Of course, it does not take a couple of seconds to appreciate the sting of a bee. Some of the mechanical nociceptors will also fire, and their Aδ axons, conducting at 5–30 m/sec, will get the message there in a few tenths of a second. The stinger will also activate cutaneous mechanoreceptors, most of which are connected to relatively large Aβ axons. These can conduct at 35–75 m/sec, so bringing that information from your fingertip will take only about 0.01–0.03 sec to reach your CNS.

Big axons are costly, in terms of the space they occupy, the protein and lipid it takes to make them, and the energy they burn. But big is swift, when it comes to axons. Aα (group I) fibers conduct at 80–120 m/sec. A good major league baseball pitcher can throw a fastball about 90 miles per hour, which is only 40 m/sec. High-speed fibers are reserved for connections between the skeletal muscles and spinal cord, both for sensory information (from muscle spindles and Golgi tendon organs) and for outgoing motor signals. There is no such urgency in transmitting some kinds of information, like the dull pain of a toothache or the temperature of a bath, and the small, slow, metabolically frugal C fibers suffice.

Segmental Organization of the Spinal Cord

The arrangement of paired dorsal and ventral roots shown in Figure 12.9 is repeated 30 times down the length of the human spinal cord. Each spinal

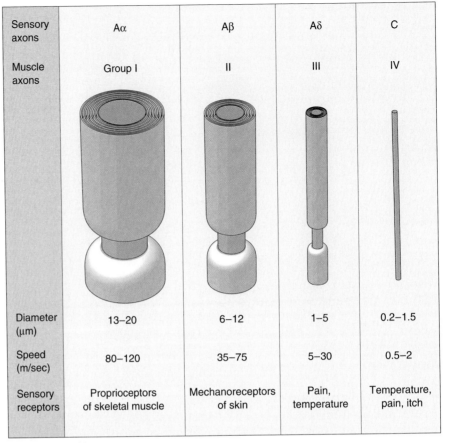

Sensory axons	Aα	Aβ	Aδ	C
Muscle axons	Group I	II	III	IV
Diameter (μm)	13–20	6–12	1–5	0.2–1.5
Speed (m/sec)	80–120	35–75	5–30	0.5–2
Sensory receptors	Proprioceptors of skeletal muscle	Mechanoreceptors of skin	Pain, temperature	Temperature, pain, itch

Figure 12.10
Various sizes of primary afferent axons. The axons are drawn to scale but are shown 2000 times life size. The diameter of an axon is correlated with its conduction velocity, and with the type of sensory receptor to which it is connected.

nerve, consisting of dorsal root and ventral root axons, passes through a notch between the vertebrae of the spinal column (the "backbone"). There are as many spinal nerves as there are notches between vertebrae. As shown in Figure 12.11, the 30 **spinal segments** are divided into four groups, and each segment is named after the vertebra from which the nerves originate: cervical (C) 1–8, thoracic (T) 1–12, lumbar (L) 1–5, and sacral (S) 1–5.

The segmental organization of spinal nerves and the sensory innervation of the skin are related. The area of skin innervated by the dorsal roots of a single spinal segment is called a **dermatome**, and thus there is a one-to-one correspondence between dermatomes and spinal segments. When mapped, the dermatomes delineate a set of stripes on the body surface, as shown in Figure 12.12. What the illustration does not show is that the dermatomes of adjacent dorsal roots overlap extensively, so that cutting only one root does not anesthetize any area of skin (but see Box 12.1).

THE TWO ASCENDING PATHWAYS OF SOMATIC SENSATION

Somatic sensory signals can take two major routes as they pass through the spinal cord and up to higher levels of the CNS: (1) the **dorsal column-medial lemniscal pathway** carries information about touch and vibration from the skin, as well as proprioceptive signals from the limbs; (2) the **spinothalamic pathway** is mainly concerned with information about pain and temperature. Figure 12.13 shows the main components of these two

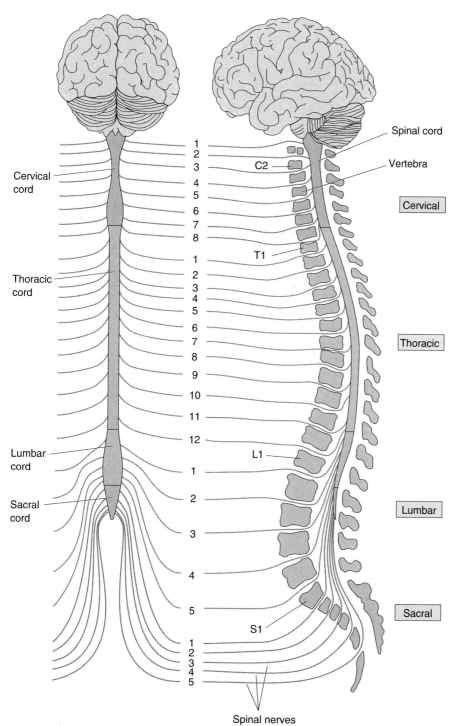

Cervical
cord

Thoracic
cord

Lumbar
cord

Sacral
cord

1
2
3
4
5
6
7
8

1
2
3
4
5
6
7
8
9
10
11
12

1
2
3
4
5

1
2
3
4
5

Spinal cord

Vertebra

C2

T1

L1

S1

Cervical

Thoracic

Lumbar

Sacral

Spinal nerves

Figure 12.11
Segmental organization of the spinal cord. The spinal cord is divided into cervical, thoracic, lumbar, and sacral divisions (left). The right side shows the spinal cord within the vertebral column. Spinal nerves are named for the level of the spinal cord from which they exit and are numbered in order from rostral to caudal.

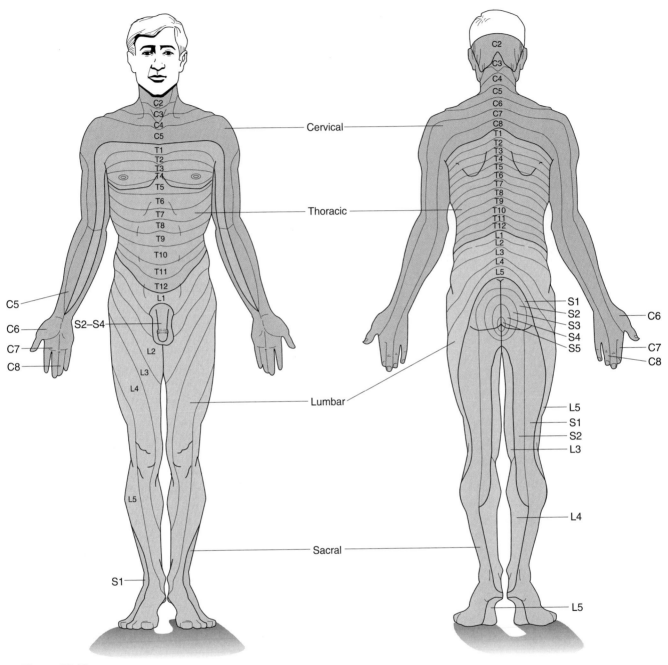

Figure 12.12
Dermatomes. This map shows the approximate boundaries of the dermatomes.

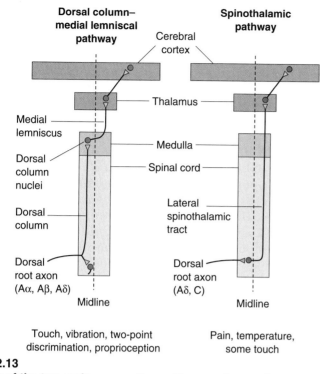

Figure 12.13
Overview of the two major ascending pathways of somatic sensation.

Herpes, Shingles, and Dermatomes

As children, most all of us were infected by the herpes zoster virus, commonly known as chicken pox. After a week or so covered with red, itchy spots on our skin, we usually recovered. Out of sight is not out of body, however. The virus remains in our primary sensory ganglion neurons, dormant but viable. Most people never hear from it again, but in some cases the virus revives decades later, wreaking havoc with the somatic sensory system. The result is *shingles*, a condition that can be agonizingly painful for periods of months or even years. The reactivated virus increases the excitability of the sensory neurons, leading to very low thresholds of firing as well as spontaneous activity. The pain is a constant burning, sometimes a stabbing sensation, and the skin is exquisitely sensitive to any stimulus. People with

shingles often shun clothes because of their hypersensitivity. The skin itself becomes inflamed and blistered, then scaly—hence, the name.

Fortunately, the herpes virus usually reactivates only in the neurons of one dorsal root ganglion. This means that the symptoms are restricted only to the axons of the affected dorsal root and the skin that it innervates. In effect, the virus performs an anatomical labeling experiment for us by clearly marking the skin territory of one dermatome. Almost any dermatome may be involved, although the thoracic and facial areas are most common. Observations of many shingles patients and their infected areas were actually useful in mapping the dermatomes (see Figure 12.12).

Neuroscientists are only now learning to use herpes, and other viruses, to their advantage. Ironically, viruses are useful in research for labeling cells, introducing new genes into them, and killing them selectively.

pathways as they wend their way from the inputs (the dorsal roots) to the terminations (the cerebral cortex). In the following sections, we will describe the anatomy and functions of these two pathways in detail.

Sensory Organization of the Spinal Cord

The basic anatomy of the spinal cord was introduced in Chapter 7. The spinal cord is comprised of an inner core of gray matter, surrounded by a thick covering of white matter tracts that are often called columns (see Figure 12.9). Each half of the spinal gray matter is divided into a *dorsal horn*, an *intermediate zone*, and a *ventral horn* (Figure 12.14).

The two pathways of somatic sensation segregate at the very margin of the spinal cord. Large-diameter fibers enter more medially, and small-diameter fibers, both myelinated and unmyelinated, enter more laterally, through the **zone of Lissauer** at the outer edge of the dorsal horn. Every axon branches numerous times, and each branch follows one of three routes: it may ascend toward the brain, it may penetrate the gray matter and synapse upon neurons at the same spinal segment, or it may descend into other spinal segments. In every case, the branch eventually synapses onto sensory neurons within the CNS; these are often called *second-order sensory neurons* because they receive input from primary sensory neurons. By this arrangement, parallel streams of sensory information are transferred quickly either directly to circuits of the spinal cord or upward to the brain. The ascending input allows perception, enabling us to form complex judgments about the information, while the direct spinal input can initiate or modify a variety of rapid and unconscious reflexes.

The two groups of primary afferent axons make their first synapses in different places. Large-diameter mechanosensory axons terminate in the deep part of the dorsal horn, within the intermediate zone, and to some extent in the ventral horn. The small-diameter axons that convey pain and temperature data ascend and descend within the zone of Lissauer, and then synapse on cells in the outer part of the dorsal horn in a region known as the **substantia gelatinosa**. From this organization, it should be clear that neurons of the dorsal horns are almost exclusively concerned with sensory functions. Cells of the intermediate zone serve, among other things, to couple the sensory functions of the dorsal horns with the motor functions of the ventral horns (discussed in Chapter 13).

The two streams of somatic sensory information take very different pathways up the spinal cord to the brain (Figure 12.14). The ascending branch of the large sensory axons (Aβ) enters the ipsilateral **dorsal column**, which is a thick white matter tract that provides a fast, direct pathway to the medulla (notice that no synapse intervenes; see Figure 12.13). By contrast, the small fibers (Aδ and C) of pain and temperature do not send a branch directly to the brain stem. Instead, axons of second-order neurons in the dorsal horn cross, or decussate, to the other side of the cord, enter the ventral part of the lateral columns, and ascend to the brain in the *lateral spinothalamic tract* (see Figure 12.13).

The central sensory pathways may cross, not cross, and even double-cross. This pattern can lead to a curious, but predictable, group of deficits when the nervous system is impaired. For example, if one half of the spinal cord is damaged, the patterns of sensory loss are determined by the differential crossing of the two ascending pathways. Because dorsal columns transfer information from the same side of the body, certain deficits of mechanosensitivity occur on the same—*ipsilateral*—side as the spinal cord damage: insensitivity to light touch, the vibrations of a tuning fork on the skin, the position of a limb. Information about pain and temperature, on the

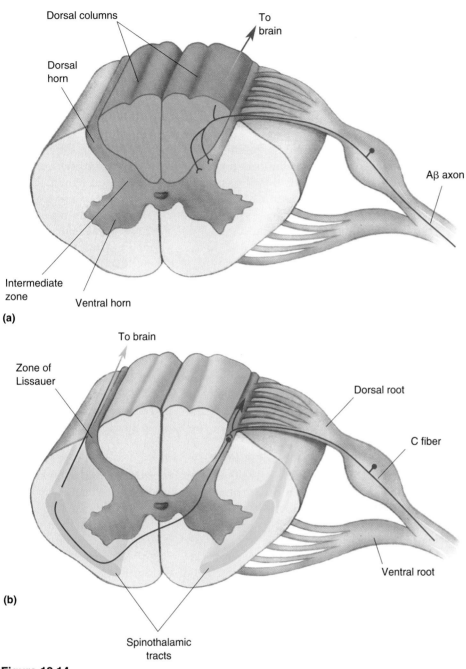

Figure 12.14
Two pathways of primary sensory information through the spinal cord. (a)
Large-diameter fibers (such as this Aβ axon) enter through a dorsal root and send
one branch to the dorsal horn and another into the dorsal column on the same side.
(b) Small-diameter fibers (such as this C axon) enter the zone of Lissauer and send
branches that synapse on neurons of the dorsal horn. The second-order neurons of
the dorsal horn then send an axon across the cord, and into the spinothalamic tract.

other hand, travels up the side of the cord that is *contralateral* to where it originates. Deficits in pain and temperature sensitivity will show up on the side of the body opposite the cord damage. Other signs, such as motor deficiency and the exact map of sensory deficits, give additional clues about the site of spinal cord damage. For example, movements will be impaired on the ipsilateral side. The constellation of sensory and motor signs following damage to one side of the spinal cord is called the *Brown-Séquard syndrome*.

The Dorsal Column-Medial Lemniscal Pathway

The dorsal column-medial lemniscal pathway is shown in Figure 12.15. The dorsal columns carry information about tactile sensation and limb proprioception. They are composed of primary sensory axons, as well as second-order axons from neurons in the spinal gray matter. Axons from the dorsal columns terminate and synapse within the **dorsal column nuclei**, at the border of the spinal cord and the medulla. At this point information is still represented ipsilaterally, that is, touch information from the *right* side of the body is represented in the activity of cells in the *right* dorsal column nuclei. Axons from cells of the dorsal column nuclei then arch toward the ventral and medial medulla, and decussate. This crossing finally brings the mechanoreceptive and proprioceptive inputs into closer register with the pain and temperature inputs, on the contralateral side of the brain.

Axons of the dorsal column nuclei ascend within a conspicuous white matter tract called the **medial lemniscus**, which meanders upward through the brain stem. The medial lemniscus rises through the medulla, pons, and midbrain, and its axons synapse upon neurons of the **ventral posterior (VP) nucleus** of the thalamus. Remember that no sensory information goes directly into the cortex without first synapsing in the thalamus (with the exception of olfactory information). Thalamic neurons of the VP nucleus then project to specific regions of **primary somatosensory cortex, or S1**.

It is tempting to assume that sensory information is simply transferred, unchanged, through nuclei in the brain stem and thalamus on its way to the cortex, with the actual processing taking place only in the cortex. In fact, this assumption is demonstrated by the term *relay nuclei*, which is often used to describe specific sensory nuclei of the thalamus such as the VP nucleus. Physiological studies prove otherwise, however. In both dorsal column and thalamic nuclei, considerable transformation of information takes place. In particular, there are inhibitory interactions between adjacent sets of inputs that serve to sharpen the contrast of a stimulus. Some synapses in these nuclei can also change their strength, depending on their recent activity. This may be one mechanism of learning. Neurons of both the thalamus and the dorsal column nuclei are also controlled by input from the cerebral cortex. Accordingly, the output of the cortex can influence the input of the cortex! As a general rule, information is altered every time it passes through a set of synapses in the brain.

The Spinothalamic Pathway

The spinothalamic pathway carries information about pain and temperature, and some touch, from the spinal cord to the brain. Dorsal root axons of this pathway make an obligatory synapse onto second-order sensory neurons within the dorsal horn of the spinal cord. Axons of the second-order neurons immediately decussate and ascend through the *spinothalamic tract* (Figure 12.16). As the name implies, these fibers project up the spinal cord and through the medulla, pons, and midbrain without synapsing, until they reach the thalamus. As the spinothalamic axons journey through the

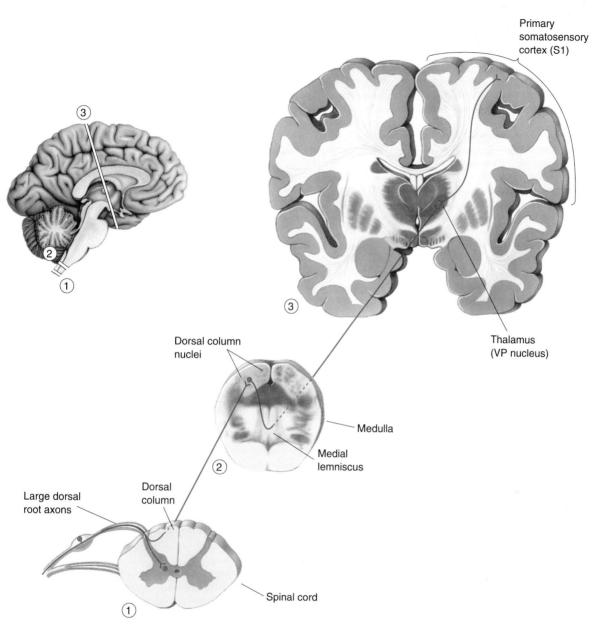

Primary
somatosensory
cortex (S1)

Thalamus
(VP nucleus)

Dorsal column
nuclei

Medulla

Medial
lemniscus

Dorsal
column

Large dorsal
root axons

Spinal cord

Figure 12.15
The dorsal column-medial lemniscal pathway. This is the major route by which touch and proprioceptive information ascend
to the cerebral cortex.

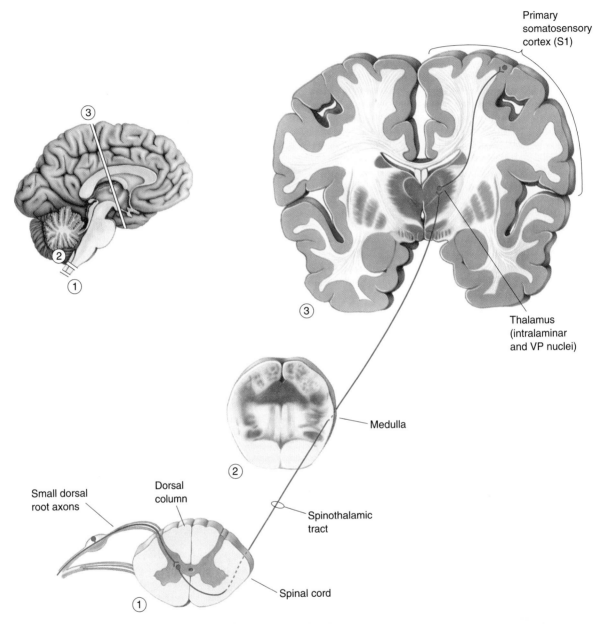

Figure 12.16
The spinothalamic pathway. This is the major route by which pain and temperature information ascend to the cerebral cortex.

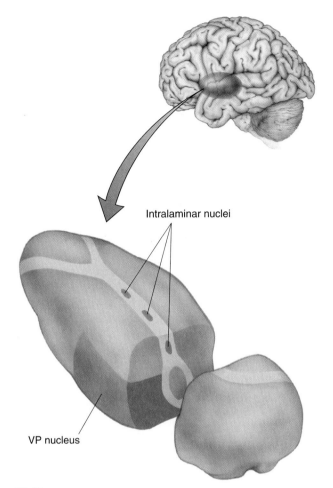

Figure 12.17
Somatic sensory nuclei of the thalamus.

brain stem, they eventually come to lie alongside the medial lemniscus, but the two groups of axons remain distinct from each other.

The spinothalamic tract synapses over a wider region of the thalamus than does the medial lemniscus. Some of its axons terminate in the VP nucleus, just as the medial lemniscal axons do, but the two systems *still* remain segregated there by occupying separate regions of the nucleus. Other spinothalamic axons end in the small *intralaminar nuclei* of the thalamus (Figure 12.17). From the thalamus, pain and temperature information is projected to various areas of the cerebral cortex. As in the thalamus, this pathway covers a much wider territory than the cortical connections of the dorsal column-medial lemniscal pathways.

In addition to the spinothalamic pathway, there are closely related pain and temperature pathways that send axons into a variety of structures at all levels of the brain stem, before they reach the thalamus and cortex. Some of these pathways are particularly important in generating sensations of slow, burning, agonizing pain, while others are involved in arousal.

The Trigeminal Pathways

Thus far we have described only the part of the somatic sensory system that enters the spinal cord. If this were the whole story, your face would be numb. Somatic sensation of the face is supplied mostly by the large **trigeminal nerves** (cranial nerve V), which enter the brain at the pons.

"Trigeminal" is from the Latin *tria*, "three," and *geminus*, "twin." There are twin trigeminal nerves, one on each side, and each breaks up into three peripheral nerves that innervate the face, mouth areas, the outer two-thirds of the tongue, and the dura mater covering the brain. Additional sensation from the skin around the ears, nasal areas, and pharynx is provided by other cranial nerves: the facial (VII), glossopharyngeal (IX), and vagus (X).

The sensory connections of the trigeminal nerve are analogous to those of the dorsal roots. The large-diameter sensory axons of the trigeminal nerve carry tactile information from skin mechanoreceptors and jaw proprioceptors (stretch receptors and Golgi tendon organs). The thinner axons carry pain and temperature signals. Just as in the spinal cord, these groups of axons take different routes into the brain stem, where they synapse onto second-order neurons in the trigeminal nuclei: the large axons in the principal sensory trigeminal nucleus, and the small axons in the spinal trigeminal nucleus (Figure 12.18).

Neurons of the trigeminal sensory nuclei send axons into the medial part of the VP nucleus of the thalamus where they synapse, and information is relayed to the somatosensory cortex. As we would predict by now, the two main streams of somatic sensory input from the face (touch and proprioception versus temperature and pain) remain separate all the way to the cortex.

SOMATOSENSORY CORTEX

As with all other sensory systems, the most complex levels of somatosensory processing occur in the cerebral cortex. Most of the cortex concerned with the somatic sensory system is located in the parietal lobe (Figure 12.19). Primary somatosensory cortex (S1) is easy to find in humans because it occupies an exposed cortical strip called the postcentral gyrus (right behind the central sulcus). Structurally, S1 is really four distinct cortical areas—Brodmann's areas 3a, 3b, 1 and 2—counting from the central sulcus back. There is also a secondary somatosensory cortex (S2), lying at the lateral end of S1, which is revealed by pulling back the temporal lobe and peeking over the auditory cortex at the lower part of the parietal lobe. Finally, the **posterior parietal cortex,** consisting of areas 5 and 7, sits just posterior to S1.

Primary Somatosensory Cortex

S1 is the *primary* somatic sensory cortex because (1) it receives dense inputs from the VP nucleus of the thalamus; (2) its neurons are very responsive to somatosensory stimuli (but not to stimuli of other sensory input); (3) lesions in S1 impair somatic sensation; and (4) electrically stimulated, it evokes somatic sensory experiences. Most inputs from the thalamus end in areas 3a and 3b, and these areas then project to areas 1 and 2, and to S2; connections within the cortex are nearly always bidirectional, so axons also go back to areas 3a and 3b. These are more examples of cortical *association pathways,* linking cortical areas, similar to the way area 17 is connected to area MT in the visual system.

The different areas of S1 have different functions. Area 3b is concerned mainly with the texture, size, and shape of objects. Its projection to area 1 sends mainly texture information, while its projection to area 2 emphasizes size and shape. Small lesions in area 1 or 2 produce predictable deficiencies in discrimination of texture, size, and shape.

Cortical Somatotopy. Electrical stimulation of the S1 surface can cause somatic sensations, which can be localized to a specific part of the body. Systematically moving the stimulator around S1 will cause the sensation to

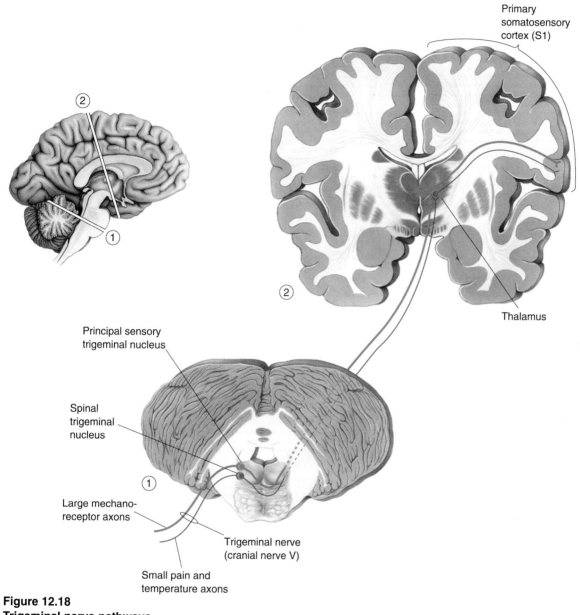

Primary
somatosensory
cortex (S1)

Thalamus

Principal sensory
trigeminal nucleus

Spinal
trigeminal
nucleus

Large mechano-
receptor axons

Trigeminal nerve
(cranial nerve V)

Small pain and
temperature axons

Figure 12.18
Trigeminal nerve pathways.

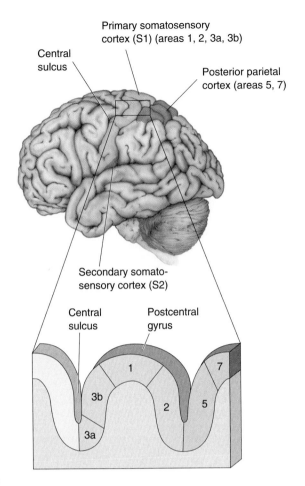

Figure 12.19
Somatic sensory areas of the cortex. All of the illustrated areas lie in the parietal lobe. The lower drawing shows that the postcentral gyrus contains S1, which consists of four different cortical areas.

move across the body. Canadian neurosurgeon Wilder Penfield, working at McGill University from the 1930s through the 1950s, actually used this method to map the cortex of locally anesthetized neurosurgical patients. Another way to map the somatosensory cortex is to record the activity of a single neuron and determine the site of its somatosensory receptive field on the body. The receptive fields of many S1 neurons produce an orderly map of the body on the cortex. The mapping of the body's surface sensations onto a structure in the brain is called **somatotopy**. We have seen previously that the brain has maps of other sensory surfaces, such as the light-sensitive retina in the eye (*retinotopy*) and the frequency-sensitive cochlea in the inner ear (*tonotopy*). Somatotopic maps generated by stimulating and recording methods are similar and roughly resemble a trapeze artist hanging upside down, his legs hooked over the top of the postcentral gyrus and dangling into the medial cortex between the hemispheres, and his head at the opposite, lower end of the gyrus (Figure 12.20). A somatotopic map is sometimes called a *homunculus* (from the Latin diminutive of "man"; the little man in the brain).

Two things are obvious about a somatotopic map. First, the map is not always continuous but can be broken up. Notice in Figure 12.20 that the representation of the hand separates that of the face and the head, while the genitals are mapped onto the most hidden part of S1, somewhere below the toes. Second, the map is not scaled like the human body. Instead, it looks like a caricature: The mouth, tongue, and fingers are absurdly large, while the trunk, arms, and legs are tiny. The relative size of cortex devoted to each body part is correlated with the *density* of sensory input received from that part. Size on the map is also related to the *importance* of the sensory input from that part of the body; information from your index finger is more useful than that from your elbow. The importance of touch information from our hands and fingers is obvious, but why throw so much cortical computing power at the mouth? Two likely reasons are that tactile sensations are important in the production of speech; and your lips and tongue (feeling, as well as tasting) are the last line of defense when deciding if a morsel is delicious, nutritious food or something that could choke you, break your tooth, or bite back.

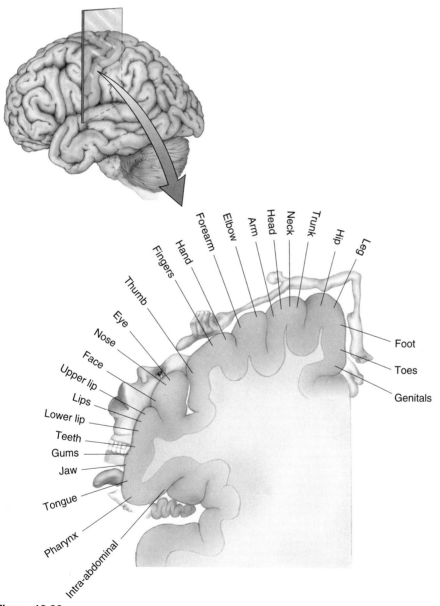

Figure 12.20
Somatotopic map of the body surface onto primary somatosensory cortex. This map is a cross section through the postcentral gyrus (shown at top). Neurons in each area are most responsive to the parts of the body illustrated above them. (Source: Adapted from Penfield and Rasmussen, 1952.)

The importance of a body part can vary greatly in different species. For example, the large facial vibrissae (whiskers) of rodents receive a huge share of the territory in S1, while the digits of the paws receive relatively little (Figure 12.21). Remarkably, the sensory signals from each vibrissa follicle go to one clearly defined cluster of S1 neurons; such clusters are called *barrels*.

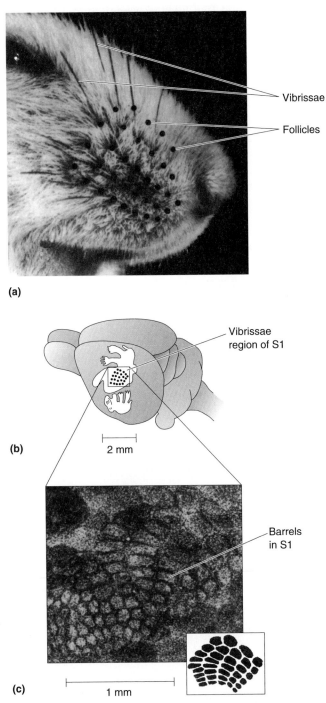

Figure 12.21
Somatotopic map of the facial vibrissae on mouse cerebral cortex. (a) Position of the major vibrissae on the face. **(b)** Somatotopic map within S1 of the mouse brain. **(c)** Barrels within S1. The cortex has been thinly sectioned parallel to the surface and Nissl stained. The inset shows the pattern of barrels, laid out in five rows; compare with the five rows of vibrissae shown in the photograph above. (Source: Adapted from Woolsey and Van der Loos, 1970.)

Thomas Woolsey and Hendrik Van der Loos, working at Johns Hopkins University, first observed that the somatotopic map of rodent vibrissae is easily seen in thin sections of S1, and that the five rows of cortical barrels precisely match the five rows of facial vibrissae (Box 12.2).

Somatotopy in the cerebral cortex is not limited to a single map. Just as the visual system builds multiple retinotopic maps, the somatic sensory system has several maps of the body. Figure 12.22 shows the detailed somatotopy of S1 in an owl monkey. Carefully compare the maps in areas 3b and 1; they map the same parts of the body, literally in parallel along adjacent strips of cortex. The two somatotopic maps are not identical, but mirror images, as an enlargement of the hand regions makes clear (Figure 12.22b). As described above, areas 3b and 1 are also concerned with different types of somatotopic information.

Printed somatotopic maps cannot represent the fact that the cortex is a dynamic structure, changing the patterns of its maps in response to changes in sensory input. Long-lasting changes of sensory experience can lead to major rearrangements of cortical maps. The mechanisms of such plasticity are not understood, but, as described in Chapter 20, they may be related to processes involved in learning and memory.

Cortical Columns. We have emphasized that various types of somatic sensory information remain segregated as they ascend through brain. This principle continues in S1, but in a different and interesting way. Neurophysiologist Vernon Mountcastle, working at Johns Hopkins University, found that each S1 neuron tends to respond to input from only one class of sensory receptors, such as rapidly adapting mechanoreceptors of the skin, the bending of a hair, or the stretch of a muscle spindle. As he recorded from one cell after another, he found that this preference was not scattered randomly, nor was it arranged according to cortical layers. Instead, Mountcastle found that cells responsive to similar inputs were stacked vertically into cortical columns that cut across all layers. Because the arrangement of cortical neurons maps the surface of the body and codes for different types of sensory input simultaneously, the different columns of S1 must form a systematically repeating series. Figure 12.23 illustrates this point schematically. Information from each fingertip in turn is mapped sequentially across the cortex. At the same time, within the small cortical map of each digit is a column of neurons responding only to slowly adapting receptors in that fingertip, and an adjacent column responding only to rapidly adapting receptors from the same fingertip. Mountcastle's discovery of cortical columns was fundamental; columnlike arrangements of neurons were subsequently observed in both visual and auditory systems. Although details vary from area to area, a general rule of cortical architecture seems to be that neurons with similar physiological properties are arranged into vertical patterns.

Posterior Parietal Cortex

Information of different sensory types cannot remain separate forever. When we feel for a key in our pocket, we do not ordinarily sense it as a list of traits: a particular size and shape, textured and smooth edges, hard and smooth flat surfaces, a certain weight. Instead, without thinking much about it, we simply confirm with our fingers "key," as opposed to "coin" or "wad of old chewing gum." Separate aspects of a stimulus come effortlessly together as a meaningful object. We have a very poor understanding of how this occurs biologically within any sensory system, much less between sensory systems. After all, many objects have a distinct look, sound, feel, *and* smell, and the melding of these sensations is necessary for the complete mental image of something like your pet cat.

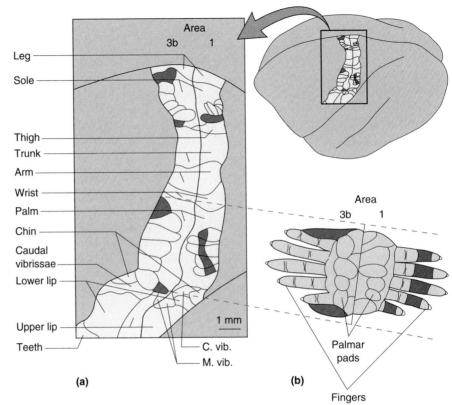

Figure 12.22
Multiple somatotopic maps in S1.
Recordings were made from areas 3b
and 1 of an owl monkey. **(a)** Results
show that each area has its own
somatotopic map. **(b)** Detailed examination of the hand area shows that the
two maps are mirror images. Shaded
regions represent the dorsal surfaces
of the hands and feet, unshaded
regions the ventral surfaces. (Source:
Adapted from Kaas et al., 1981.)

Cortical Barrels

by Thomas Woolsey

Thomas Woolsey

In the mid-1960s, I completed a physiological study of the organization of touch, hearing, and vision in the mouse brain for an undergraduate research project at the University of Wisconsin. At Wisconsin, histology was done routinely on all such brains. After my first year at medical school, I came back to look at the sections. There was something odd about cortical layer IV just where I had recorded responses to moving the whiskers: The cell bodies were distributed unevenly. This was not new. Several different authors, in nearly forgotten papers written over 50 years earlier, showed this pattern of neurons; but that was before recording was possible, so no one knew the function of this cortex.

Cortex is usually studied in sections cut perpendicular to the brain surface. It occurred to me that cutting the brain parallel to its surface, which had only been done rarely, might give a view of the entire layer IV. The late H. Van der Loos, who taught neuroanatomy, gave me a place to work at Johns Hopkins during an elective period. I prepared specimens in a way so that I could accurately position them for cutting (I knew where I usually got responses to stimulating the face) and cut thicker sections than customary. About 10:00 on a bright late spring morning, after struggling to mount the first sections on slides, I took them down a corridor to the dark student histology lab, where I had a microscope. That first look showed a stunning pattern of cells in layer IV that obviously mimicked the whiskers. There was no doubt about what I had seen; I immediately showed the sections to Van der Loos, who was the second person in the world to know that whiskers are stamped in the mouse brain. We named the cell groups barrels. Later, the hypotheses that each barrel is associated with a single whisker and that each one forms part of a functional cortical column were proven.

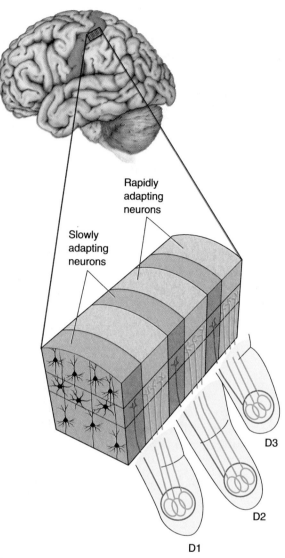

Figure 12.23
Columnar organization of S1's area 3b. Each finger (D1–D3) is represented by an adjacent area of cortex. Within the area of each finger representation are alternating columns of cells with rapidly adapting (green) and slowly adapting (red) sensory responses. (Source: Adapted from Kaas et al., 1981, Fig. 8.)

What we do know is that the character of neuronal receptive fields tends to change as information passes through the cortex, and receptive fields enlarge. For example, neurons below the cortex and in cortical areas 3a and 3b are not sensitive to the direction of stimulus movement across the skin, but cells in areas 1 and 2 are. The stimuli that neurons prefer become increasingly complex. Certain cortical areas seem to be sites where simple, segregated streams of sensory information converge to generate particularly complex neural representations. The posterior parietal cortex is such an area. Its neurons have large receptive fields, with stimulus preferences that are a challenge to characterize because they are so elaborate (Box 12.3). Moreover, the area is concerned not only with somatic sensation but also with visual stimuli, and even a person's state of attentiveness.

Damage to posterior parietal areas can yield some bizarre neurological disorders. Among these is **agnosia**, the inability to recognize objects even though simple sensory skills seem to be normal. People with *astereognosia*

Box 12.3 | **PATH OF DISCOVERY**

My Adventures in the Posterior Parietal Cortex

by **Vernon Mountcastle**

Vernon Mountcastle

The most exciting days of my experimental life were September 19–20, 1971. I had been working for two years to perfect methods of recording the electrical signs of the neuronal activity of single neurons in the postcentral somatic sensory cortex of monkeys as they executed sensory tasks—in this case, the detection of a mechanical oscillation superimposed upon a small indentation of the glabrous skin of the hand. All had gone well, and we had observed a nearly perfect correlation between the increasing probability of correct detection with increasing stimulus amplitude and the increase in the cyclic entrainment of the cortical neuronal activity at the stimulus frequency. Alas! When we examined neuronal responses to stimuli detected only 50% of the time, we found no differences between those evoked by stimuli detected and those missed! Where was the critical neural event upon which detection depended?

Thus, it was more in frustration than in a cleverly planned experimental foray that we (Robert LaMotte, now at Yale University; Carlos Acuna, now at the University of Santiago del Compestella in Spain; and I) suddenly changed the locus of recording from the postcentral somatic sensory area to the posterior parietal homotypical cortex, first on that day to area 5, and a few weeks later to area 7. What we observed on those very first days determined my experimental life for 15 years: neurons were active if and only if the animal "had a mind" to deal with the stimulus in a behaviorally meaningful way!

In area 5, neurons were active when the animal projected his arm toward a target of interest, but not during random arm movements; neurons activated when the animal worked with his fingers to extract a morsel of food from a recess, but not during other hand actions (we called them "winkle" neurons); neurons activated by joint rotation, more intensely during his active joint rotations than when the joints were passively moved by us; neurons activated by cutaneous stimulation, frequently from very large and sometimes bilateral receptive fields and often with directional selectivity. In area 7, neurons were active during visual stimuli per se; neurons were active during visually evoked but not during spontaneous saccadic movements of the eyes; neurons were active during smooth visual pursuit tracking; and neurons responded to visual stimuli, per se, but with unusual receptive fields and response properties.

The major sources of information about the putative functions of the parietal system were in the descriptions of patients with lesions in the parietal lobe. I began to attend the neurological ward rounds in the hospital and was soon shown a patient with the classical parietal lobe neglect syndrome. He complained loudly of a foreign arm in bed with him—his own! I reviewed the case histories filed under the label "parietal lobe syndrome" in the hospital. I also reviewed the relatively sparse literature on the effects of parietal lobe lesions in nonhuman primates, as well as that describing the connectivity of these regions. These reviews provided a base for a series of studies of the parietal lobe in a number of different experimental designs I then carried out with a succession of capable collaborators, in chronological order: W. H. Talbot, R. H. LaMotte, C. Acuna, J. C. Lynch, H. Sakata, A. P. Georgopoulos, T. T. C. Yin, B. C. Motter, C. J. Duffy, R. A. Andersen, M. A. Steinmetz, and A. K. Sestokas.

Several things became clear early in our studies. First, it is possible to study the higher functions of the brain in nonhuman primates, but only if the complex properties of neurons of the homotypical cortex are matched by equally complex behavioral tasks; also, quantitative studies under controlled conditions are essential. Second, although the properties of posterior parietal neurons do provide a limited positive image of the defects of the parietal lobe syndrome, the functions inferred are embedded in dynamic neuronal processes in widely distributed systems of the forebrain, in which the parietal areas are essential nodes. Third, what is important for experimental design, in any given experimental arrangement, one cannot study quantitatively all the many classes of parietal neurons—perhaps at best only two. This makes the ubiquitous sampling problem that limits all electrophysiological studies of the cerebral cortex especially acute, and emphasizes that conclusions drawn from any small sample of neurons are especially hazardous.

Model   Patient's Copy

Figure 12.24
An example of a neglect syndrome. A patient who had had a stroke in the posterior parietal cortex was asked to copy the model drawing but was unable to reproduce many of the features on the left side of the model. (Source: Springer and
Deutsch, 1989, p. 193.)

cannot recognize common objects by feeling them (a key, for example),
though their sense of touch is otherwise normal and they may have no trouble recognizing the object by sight or sound. Deficits are often limited to the
side contralateral to the damage. Parietal cortical lesions may also cause a
neglect syndrome, in which a part of the body or a part of the world (the
entire visual field left of the center of gaze, for example) is ignored or suppressed, and its very existence is denied (Figure 12.24). Neurologist Oliver
Sacks described such a patient in his essay, "The Man Who Fell Out of Bed."
After suffering a stroke that presumably damaged his cortex, the man insisted that someone was playing a macabre joke on him by hiding an amputat-
ed leg under his blanket. When he tried to remove the leg from his bed, he
and the leg ended up on the floor. Of course the leg in question was his own,

still attached, but he was unable to recognize it as part of his body. A neglect patient may ignore the food on one half of his plate, or attempt to dress only one side of his body. Neglect syndromes are most common following damage to the right hemisphere, and, happily, they usually improve or disappear with time. In general, the posterior parietal cortex seems to be essential for the perception and interpretation of spatial relationships, accurate body image, and the learning of tasks involving coordination of the body in space. This involves a complex integration of somatosensory information with that from other sensory systems, in particular vision.

PAIN AND ITS CONTROL

Nociception and Pain

Pain is a fact of life (but see Box 12.4). We have already described the peripheral receptors and some of the central pathways involved in nociception, but we have not discussed, or even defined, pain itself. Nociception and pain are not the same thing. *Pain* is the feeling, or the perception, of irritating, sore, stinging, aching, throbbing, miserable, or unbearable sensations arising from a part of the body. *Nociception* is the sensory process that provides the signals that trigger pain. While nociceptors may fire away wildly and continually, pain may come and go. The opposite may also happen. Pain may be agonizing, even without activity in nociceptors. More than any other sensory system, the cognitive qualities of nociception can be controlled from within, by the brain itself.

Box 12.4 | **OF SPECIAL INTEREST**

The Misery of Life Without Pain

Pain teaches us to avoid harmful situations. It elicits withdrawal reflexes from noxious stimuli. It exhorts us to rest an injured part of our body. Pain is vital, and the most convincing arguments are the very rare people who are born without the sensation of pain. They go through life in constant danger of destroying themselves, because they do not realize the harm they are doing. They often die young. One Canadian woman born with an indifference to painful stimuli had no other sensory deficiencies and was quite intelligent. Despite early training to avoid damaging situations, she developed progressive degeneration of her joints and spinal vertebrae, leading to skeletal deformation, degeneration, infection, and, finally, death at the age of 28. Apparently, low

levels of nociceptive activity are important during everyday tasks to tell us when a particular movement or prolonged posture is putting too much strain on our body. Even during sleep, nociception may be the prod that makes us toss and turn enough to prevent bedsores or skeletal strain.

People with a congenital absence of pain reveal that pain is a separate sensation, and not simply an excess of the other sensations. Such people usually have a normal ability to perceive other somatic sensory stimuli. The actual causes may vary. Some may have abnormally high endorphin activity, since taking a drug that blocks endorphins can lower their tolerance to noxious stimuli. Other people with the condition may actually lack certain pain-specific structures, such as the thin C fibers and their peripheral nociceptors, or the zone of Lissauer in the spinal cord. Or they may have impaired synaptic transmission in pain-mediating pathways of the CNS. In any case, life without pain is *not* a blessing.

Hyperalgesia

Nociceptors are sensory endings that respond only when stimuli are strong enough to damage tissue, or at least threaten to damage it. But we all know that skin, joints, or muscles that have *already* been damaged or inflamed are unusually sensitive. A light, sympathetic mother's touch to a burned area of her child's skin may elicit howls of excruciating pain. This phenomenon is called hyperalgesia, and it is the most familiar example of our body's ability to control its own pain. **Hyperalgesia** can be a reduced threshold for pain, an increased intensity of painful stimuli, or even spontaneous pain. *Primary hyperalgesia* occurs within the area of damaged tissue, but tissues surrounding a damaged area may become supersensitive by the process of *secondary hyperalgesia*.

There seem to be many different mechanisms involved in hyperalgesia, some in and around the peripheral receptors and others within the CNS. When skin is damaged, a variety of chemical substances are released from skin tissue itself, from blood cells, and from nerve endings. These substances trigger a set of local responses known as *inflammation*. Blood vessels become more leaky, causing tissue swelling and redness; the chemical histamine is released from local cells and directly excites nociceptors; and the spreading axon branches of the nociceptors themselves may release substances that sensitize surrounding nociceptors. Aspirin is a useful treatment for hyperalgesia because it suppresses the synthesis of prostaglandins, one of the sensitizing substances. New research also indicates that there may be many "silent" nociceptors among our small Aδ and C axons. The idea is that these insidious receptors are normally unresponsive to stimuli, even destructive ones, unless the surrounding tissue has been previously sensitized. Only then do they become responsive to mechanical or chemical stimuli, contributing strongly to hyperalgesia. It is also clear that changes within the spinal cord and brain contribute to hyperalgesia, particularly its secondary aspects.

Regulation of Pain

Pain can be modified both by *nonpainful* sensory input and by neural activity from various nuclei within the brain. Pain evoked by activity in nociceptors (Aδ and C fibers) can be *reduced* by simultaneous activity in low-threshold mechanoreceptors (Aα and Aβ fibers). Presumably this is why it feels good to rub the skin around your shin when you bruise it. This may also explain an electrical treatment for some kinds of chronic, intractable pain. Wires are taped to the skin surface, and pain is suppressed when the patient simply turns on an electrical stimulator designed to activate large-diameter sensory axons. In the 1960s, Ronald Melzack and Patrick Wall, then working at MIT, proposed a hypothesis to explain these phenomena. Their *gate theory of pain* proposes that certain neurons of the dorsal horns, which project an axon up the spinothalamic tract, are excited by both large diameter sensory axons and unmyelinated pain axons. The projection neuron is also inhibited by an interneuron, and the interneuron is both *excited* by the large sensory axon and *inhibited* by the pain axon (Figure 12.25). By this arrangement, activity in the pain axon alone maximally excites the projection neuron, allowing nociceptive signals to rise to the brain. However, if the large mechanoreceptive axon fires concurrently, it activates the interneuron and suppresses nociceptive signals.

Stories abound of soldiers, athletes, and torture victims who sustained horrible injuries but apparently felt no pain. Strong emotion, stress, or stoic determination can powerfully suppress feelings of pain. Several brain

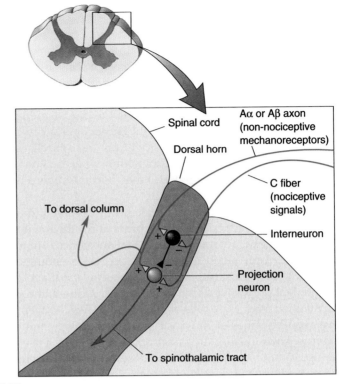

Figure 12.25
Melzack and Wall's gate theory of pain.

regions have been implicated in pain suppression (Figure 12.26). One is a zone of neurons in the midbrain called the periventricular and **periaqueductal gray matter** (**PAG**). Electrical stimulation of the PAG can cause a profound **analgesia**, or an absence of pain, which has sometimes been exploited clinically. The PAG normally receives input from several brain structures, many of them appropriate for transmitting signals related to emotional status. PAG neurons send descending axons into various midline regions of the medulla, particularly to the *raphe nuclei* (which use the neurotransmitter serotonin). These medullary neurons in turn project axons down to the dorsal horns of the spinal cord, where they can effectively depress the activity of nociceptive neurons.

Brain Chemicals and Pain

The self-regulation of pain is accomplished by a variety of unique chemical systems in the CNS. When nociceptors first synapse in the spinal cord, they release not only the excitatory neurotransmitter glutamate, as all sensory afferents do, but also small peptides such as substance P. Substance P causes a very long-lasting EPSP in nociceptive spinal cord neurons and probably helps sustain the effect of noxious stimuli. If you like spicy food, you should know that the active ingredient is *capsaicin*, which generates its piquant effect by potently activating certain nociceptors in the mouth. If you cook spicy food, you should be forewarned about the searing pain that comes from rubbing your eyes with chili-stained (i.e., capsaicin-coated) fingers. Capsaicin causes the release of substance P from nociceptors. Ironically, when applied in large quantities, it can cause analgesia because it depletes substance P from nerve terminals.

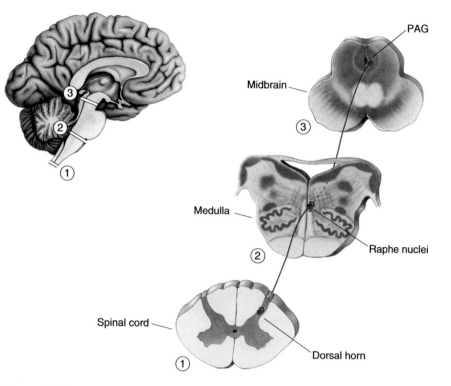

Figure 12.26
Descending pain-control pathways. A variety of brain structures, many of which are affected by behavioral state, can influence activity within the periaqueductal gray matter (PAG) of the midbrain. The PAG can influence the raphe nuclei of the medulla, which in turn can modulate the flow of nociceptive information through the dorsal horns of the spinal cord.

Endorphins. Opium was probably known to the ancient Sumerians around 4000 B.C. Their pictograph for the poppy roughly translates to "joy plant." By the seventeenth century, the therapeutic value of opium was undisputed. Opium—its active narcotic ingredients and their analogues, including morphine, codeine, and heroin—are used and abused widely today, in most cultures of the world. Opioids produce profound analgesia when taken systemically. They can also produce mood changes, drowsiness, mental clouding, nausea, vomiting, and constipation. The 1970s brought the stunning discoveries that opioids act by binding tightly and specifically to several types of **opioid receptors** in the brain, and that the brain itself manufactures endogenous morphinelike substances, collectively called **endorphins** (see Box 6.1). Endorphins are relatively small proteins, or peptides.

Endorphins and their receptors are distributed widely in the CNS, but they are particularly concentrated in areas that process or modulate nociceptive information. Small injections of morphine or endorphins into the PAG, the raphe nuclei, or the dorsal horn can produce analgesia. Because this effect is prevented by administration of the specific blocker of opioid receptors, *naloxone,* the injected drugs must have acted by binding to opioid receptors in those areas. Naloxone can also block the analgesic effects induced by electrically stimulating these areas. At the cellular level, endorphins exert multiple effects that include suppressing the release of glutamate from presynaptic terminals and inhibiting neurons by hyperpolarizing their postsynaptic membranes. In general, the PAG, the raphe nuclei, and the spinal cord can use extensive systems of endorphin-containing neurons to prevent the passage of nociceptive signals through the dorsal horn and into higher levels of the brain where the perception of pain is generated.

Itch

There are many reasons to believe that itch is similar to, but distinctly different from, pain. Both sensations are mediated by Aδ and C fibers and the spinothalamic pathways in the spinal cord, but pain can be generated in most tissues, while itch is restricted to skin and mucous membranes. Both itch and pain can be induced by local applications of histamine or capsaicin, but only pain is suppressed by opioid drugs such as morphine. A key point is that pain and itch rarely occur in the same area of skin at the same time; and of course the best way to get rid of an itch is to scratch it, which is actually a mildly noxious stimulus. It is also interesting that people born without a sense of pain also do not itch. This confusing picture has not been adequately explained. It may be that there are specific, small-diameter "pruritoceptors" (*pruritus* is Latin for "itching"), but these have never been observed. Another possibility is that itch is mediated by a subset of nociceptors having somewhat different central connections than other nociceptors. In any case itch, like pain, is strongly influenced by psychological factors. This is easily demonstrated: After reading this far, haven't you felt the urge to scratch yet?

CONCLUDING REMARKS

This concludes our descriptions of the sensory systems. Though each has evolved to be the brain's interface with a different form of environmental energy, the systems have striking similarities of organization and function. Different types of somatic sensory information are necessarily kept separate in the spinal nerves because each axon is connected only to one type of sensory receptor ending. Segregation of sensory types continues within the spinal cord and is largely maintained all the way to the cerebral cortex. In this way, the somatic sensory system repeats a theme common throughout the nervous system: Several flows of related, but distinct, information are passed in parallel through a series of neural structures. Mixing of these streams occurs along the way, but only judiciously, until higher levels of processing are reached in the cerebral cortex. We saw other examples of parallel processing of sensory information in the chemical senses, vision, and audition.

Exactly how the parallel streams of sensory data are melded into perception, images, and ideas remains the Holy Grail of neuroscience. Thus, perception of any handled object involves the seamless coordination of all facets of somatic sensory information. The bird in hand is rounded, warm, and soft, light in weight; its heartbeat flutters against your fingertips; its claws scratch; and its textured wings brush against your palm. Somehow your brain knows it's a bird, even without looking or listening, and would never mistake it for a toad. In the chapters that follow, we describe how the brain begins to use sensory information to plan and coordinate movement.

KEY TERMS

Introduction
somatic sensation

Receptors of Somatic Sensation
mechanoreceptor
nociceptor
thermoreceptor
proprioceptor
chemoreceptor
Pacinian corpuscle
proprioception

Primary Afferent Nerves
spinal segment
dermatome

The Two Ascending Pathways of Somatic Sensation
dorsal column-medial lemniscal pathway
spinothalamic pathway
zone of Lissauer
substantia gelatinosa
dorsal column

dorsal column nucleus
medial lemniscus
ventral posterior (VP) nucleus
primary somatosensory cortex
S1
trigeminal nerve

Somatosensory Cortex
posterior parietal cortex
somatotopy

agnosia
neglect syndrome

Pain and Its Control
hyperalgesia
periaqueductal gray matter (PAG)
analgesia
opioid receptor
endorphin

R E V I E W ✅
Q U E S T I O N S

1. Imagine rubbing your fingertips across a pane of smooth glass and then across a brick. What kinds of skin receptors help you distinguish the two surfaces? As far as your somatic sensory system is concerned, what is different about the two surfaces?

2. Imagine this experiment: Fill two buckets with water, one relatively cold and one hot. Fill a third bucket with water of an intermediate, lukewarm temperature. Put your left hand into the hot water, your right hand into the cold, and wait one minute. Now quickly plunge both hands into the lukewarm water. Try to predict what sensations of temperature you will feel in each hand. Will they feel the same? Why?

3. What purpose is served by the encapsulations around some sensory nerve endings in the skin?

4. If someone tossed you a hot potato and you caught it, which information would reach your CNS first: the news that the potato was hot, or that it was relatively smooth? Why?

5. At what levels of the nervous system are *all* types of somatic sensory information represented on the contralateral side: the spinal cord, the medulla, the pons, the midbrain, the thalamus, the cortex?

6. What lobe of the cortex contains the main somatic sensory areas? Where are these areas relative to the main visual and auditory areas?

7. Where within the body can pain be modulated, and what causes its modulation?

8. Where in the CNS does information about touch, shape, temperature, and pain converge?

Spinal Control of Movement

We are now ready to turn our attention to the system that, in response to the sensory information collected by the chemical, visual, auditory, and somatic sensory systems, gives rise to behavior. The motor system consists of all our muscles and the neurons that command them. The importance of the motor system is obvious; as the pioneering English neurophysiologist Charles Sherrington put it in the Linacre Lecture of 1924: "To move things is all that mankind can do . . . for such the sole executant is muscle, whether in whispering a syllable or in felling a forest." A moment's thought will convince you that the motor system is also incredibly complex. Behavior requires the coordinated action of various combinations of over 750 muscles in a changing and often unpredictable environment.

Have you ever heard the expression "running around like a chicken with its head cut off"? It is based on the observation that complex patterns of behavior (running around the barnyard) can be generated without the participation of the brain. There exists a significant amount of circuitry within the spinal cord for the coordinated control of movements, particularly stereotyped (repetitive) ones such as those associated with locomotion. This point was established early in this century by Sherrington and his English contemporary Graham Brown, who showed that rhythmic movements could be elicited in the hind legs of cats and dogs long after their spinal cords had been severed from the rest of the central nervous system. Today's view is that contained within the spinal cord are certain *motor programs* for the generation of coordinated movements, and that these programs are accessed, executed, and modified by descending commands from the brain. Thus, motor control can be divided into two parts: (1) the spinal cord's command and control of coordinated muscle contraction, and (2) the brain's command and control of the motor programs in the spinal cord.

In this chapter, we will explore the peripheral somatic motor system: the joints, skeletal muscles, and spinal motor neurons, and how they communicate with each other. In Chapter 14, we will take a look at how the brain influences the activity of the spinal cord.

THE SOMATIC MOTOR SYSTEM

The muscles in the body may be divided into two broad categories, striated and smooth, based on their appearance under the microscope. But they are also distinct in other ways. **Smooth muscle** lines the digestive tract and arteries, among other things, and is innervated by nerve fibers from the autonomic nervous system (ANS). Smooth muscle plays a role in peristalsis (movement of material through the intestines) and the control of blood pressure. **Striated muscle** is further divisible into two types, cardiac and skeletal. **Cardiac muscle** is heart muscle and contracts rhythmically in the absence of any innervation. Innervation of the heart from the ANS functions to accelerate or slow down the heart rate. (Recall Otto Loewi's experiment in Chapter 5.) **Skeletal muscle** constitutes the bulk of the muscle mass of our bodies and functions to move bones around joints, to move the eyes within the head, to control respiration, to control facial expression, and to produce speech. Each skeletal muscle is enclosed in a connective tissue sheath that, at the ends of the muscle, forms the tendons. Within each muscle are hundreds of **muscle fibers**—the cells of skeletal muscle—and each fiber is innervated by a single axon from the CNS (Figure 13.1). Because skeletal muscle is derived embryologically from 33 paired somites (see

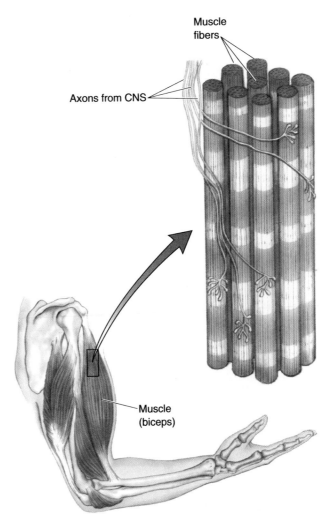

Muscle fibers

Axons from CNS

Muscle (biceps)

Figure 13.1
Structure of skeletal muscle. Each muscle fiber is innervated by a single axon.

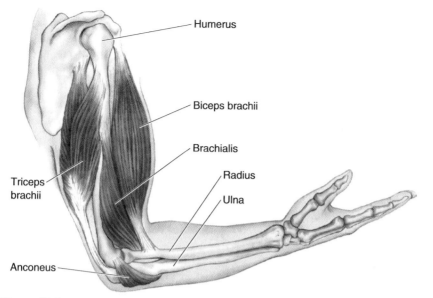

Figure 13.2
Major muscles of the elbow joint. The biceps and triceps are antagonistic muscles. Contraction of the biceps causes flexion, and contraction of the triceps causes extension of the elbow.

Chapter 7), these muscles, and the parts of the nervous system that control them, are collectively called the **somatic motor system**. Here we will focus our attention on this system because it is under voluntary control and is responsible for generating behavior. (The visceral motor system of the ANS will be discussed in Chapter 15.)

Consider the elbow joint (Figure 13.2). This is formed where the humerus, the upper arm bone, is bound by fibrous ligaments to the radius and ulna, the bones of the lower arm. The joint functions like a hinge on a pocket knife. Movement in the direction that closes the knife is called **flexion**, and movement in the direction that opens the knife is called **extension**. The major muscle that causes flexion is the brachialis, whose tendons insert into the humerus at one end and into the ulna at the other. Two other muscles cause flexion at this joint, the biceps brachii and the coracobrachialis. Together, these muscles are called **flexors** of the elbow joint, and, because the three muscles all work together, they are called **synergists** of one another. The two synergistic muscles that cause extension of the elbow joint are the triceps brachii and the anconeus; these two muscles are called **extensors**. Because the flexors and extensors pull on the joint in opposite directions, they are called **antagonists** to one another. Note that muscles only pull on a joint; they cannot push. Even the simple flexion of the elbow joint requires the coordinated contraction of the synergistic flexor muscles *and* the relaxation of the antagonistic extensor muscles.

Other terms to note about somatic musculature refer to the location of the joints they act on. The muscles that are responsible for movements of the trunk are called **axial muscles**; those that move the shoulder, elbow, pelvis, and knee are called **proximal** (or **girdle**) **muscles**; and those that move the hands, feet, and digits are called the **distal muscles**. The axial musculature is very important for maintaining posture, the proximal musculature is critical for locomotion, and the distal musculature, particularly of the hands, is specialized for the manipulation of objects.

THE LOWER MOTOR NEURON

The somatic musculature is innervated by the somatic motor neurons in the ventral horn of the spinal cord (Figure 13.3). These cells are sometimes called "lower motor neurons" to distinguish them from the higher-order "upper motor neurons" of the brain that supply input to the spinal cord. Remember, only the lower motor neurons directly command muscle contraction. Sherrington called these cells the *final common pathway* for the control of behavior.

Segmental Organization of Lower Motor Neurons

The axons of lower motor neurons bundle together to form ventral roots; each ventral root joins with a dorsal root to form a spinal nerve that exits the cord through the notches between vertebrae. Recall from Chapter 12 that there are as many spinal nerves as there are notches between vertebrae; in humans, this adds up to 30 on each side. Because they contain sensory and motor fibers, they are called the *mixed spinal nerves*. The motor neurons that provide fibers to one spinal nerve are said to belong to a spinal segment, named for the vertebra where the nerve originates. The segments are cervical (C) 1–8, thoracic (T) 1–12, lumbar (L) 1–5, and sacral (S) 1–5.

Skeletal muscles are not distributed evenly throughout the body, nor are lower motor neurons distributed evenly within the spinal cord. For example, innervation of the more than 50 muscles of the arm originates entirely from spinal segments C3–T1. Thus, in this region of the spinal cord, the ventral horns appear swollen to accommodate the large number of motor neurons that control the arm musculature (Figure 13.4). Similarly, spinal segments L1–S3 have a swollen ventral horn because this is where the motor neurons controlling the leg musculature reside. Thus, we can see that the motor neurons that innervate distal and proximal musculature are found mainly in the cervical and lumbar-sacral segments of the spinal cord, whereas those innervating axial musculature are found at all levels. The lower motor neurons are also distributed within the ventral horn at each

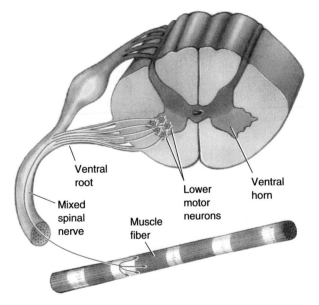

Figure 13.3
Innervation of muscle by lower motor neurons. The ventral horn of the spinal cord contains motor neurons that innervate skeletal muscle fibers.

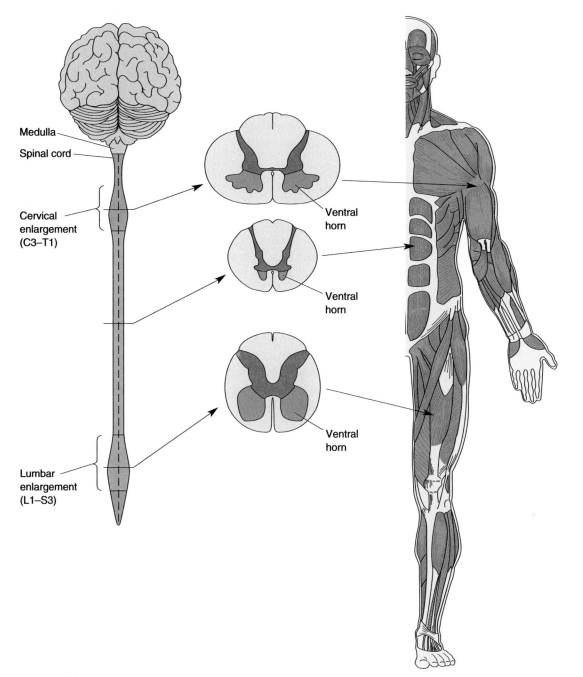

Medulla

Spinal cord

Cervical
enlargement
(C3–T1)

Ventral
horn

Ventral
horn

Ventral
horn

Lumbar
enlargement
(L1–S3)

Figure 13.4
Uneven distribution of motor neurons in the spinal cord. The cervical enlargement of the spinal cord contains the motor
neurons that innervate the arm muscles. The lumbar enlargement contains neurons that innervate the muscles of the leg.

spinal segment in a predictable way, depending on their function. The cells innervating the axial muscles are medial to those innervating the distal muscles, and the cells innervating flexors are dorsal to those innervating extensors (Figure 13.5).

Alpha Motor Neurons

Lower motor neurons of the spinal cord can be divided into two categories: alpha motor neurons and gamma motor neurons (discussed later in the chapter). **Alpha motor neurons** are directly responsible for the generation of force by muscle. An alpha motor neuron and all the muscle fibers it innervates collectively make up the elementary unit of motor control; Sherrington called this the **motor unit**. Muscle contraction results from the individual and combined actions of these motor units. The collection of alpha motor neurons that innervates a single muscle (e.g., the biceps brachii) is called a **motor neuron pool** (Figure 13.6).

Graded Control of Muscle Contraction by Alpha Motor Neurons. An alpha motor neuron communicates with a muscle fiber by releasing the neurotransmitter acetylcholine (ACh) at the neuromuscular junction, the specialized synapse between nerve and skeletal muscle (see Chapter 5). Because of the high reliability of neuromuscular transmission, the ACh released in response to one presynaptic action potential causes an EPSP in the muscle fiber that is large enough to trigger one postsynaptic action potential. By mechanisms we will discuss in a moment, a postsynaptic action potential causes a twitch—a rapid sequence of contraction and relaxation—in the muscle fiber. A sustained contraction requires a continual bar-

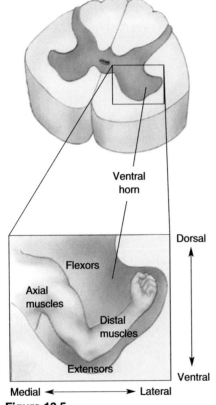

Figure 13.5
Distribution of lower motor neurons in the ventral horn. Motor neurons controlling flexors lie dorsal to those controlling extensors. Motor neurons controlling axial muscles lie medial to those controlling distal muscles.

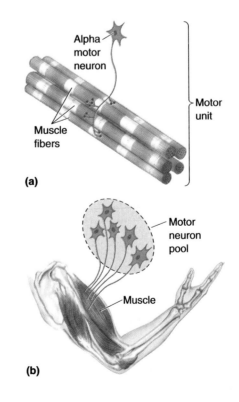

Figure 13.6
A motor unit and motor neuron pool. (a) A motor unit is an alpha motor neuron and all the muscle fibers it innervates. (b) A motor neuron pool is all the alpha motor neurons that innervate one muscle.

rage of action potentials. As for other types of synaptic transmission, high-frequency presynaptic activity causes temporal summation of the postsynaptic responses. Twitch summation increases the tension in the muscle fibers and smooths the contraction (Figure 13.7). The rate of firing of motor units is therefore one important way the CNS grades muscle contraction.

A second way the CNS grades muscle contraction is by recruiting additional synergistic motor units. The extra tension provided by the recruitment of an active motor unit depends on how many muscle fibers are in that unit. In the antigravity muscles of the leg (muscles that oppose the force of gravity when standing upright), each motor unit tends to be quite large, with an innervation ratio of over 1000 muscle fibers per single alpha motor neuron. In contrast, the smaller muscles that control the movement of the fingers and the rotation of the eyes are characterized by much smaller innervation ratios, as few as three muscle fibers per alpha motor neuron. In general, muscles with a large number of small motor units can be more finely controlled by the CNS.

Most muscles have a range of motor unit sizes, and these motor units are recruited in the order of smallest first, largest last. This orderly recruitment explains why finer control is possible when muscles are under light loads than when they are under greater loads. Small motor units have small alpha motor neurons, and large motor units have large alpha motor neurons. Thus, one way that orderly recruitment might occur is if small neurons, as a consequence of the geometry of their soma and dendrites, were more easily excited by signals descending from the brain. The idea that the orderly recruitment of motor neurons is due to variations in alpha motor neuron size, first proposed in the late 1950s by Harvard University neurophysiologist Elwood Henneman, is called the *size principle*.

Inputs to Alpha Motor Neurons. The lower motor neurons are controlled by their synaptic inputs in the ventral horn. *There are only three sources of input to an alpha motor neuron,* as shown in Figure 13.8. The first source is dorsal root ganglion cells with axons that innervate a specialized sensory apparatus embedded within the muscle called the *muscle spindle* (one of the proprioceptors introduced in Chapter 12). As we shall see, this input provides feedback about muscle length. The second source of input to an alpha motor neuron derives from upper motor neurons in the brain, notably those in the precentral gyrus of the cerebral cortex. These inputs terminate primarily on those alpha motor neurons that control distal muscles and play an important role in voluntary movement of the limbs and fingers. (This input will be discussed in more detail in Chapter 14.) The third and largest input to the spinal motor neuron derives from interneurons in the spinal cord. This input from interneurons may be excitatory or inhibitory and is part of the circuitry that generates the spinal motor programs.

Types of Motor Units

If you have ever dined on chicken, you know immediately that not all muscle is the same; there is the dark meat of the leg and the white meat of the breast and wings. The different appearance (and taste) of the different muscles is accounted for by distinctions in the biochemistry of the constituent muscle fibers. The red (dark) muscle fibers are characterized by a large number of mitochondria and enzymes specialized for oxidative energy metabolism. These fibers are relatively slow to contract but can sustain contraction for a long time without fatigue. They are typically found in the antigravity muscles of the leg and in the flight muscles of birds that fly (as opposed to chickens). In contrast, the pale (white) muscle fibers contain fewer mitochondria and rely mainly on anaerobic (without oxygen) metab-

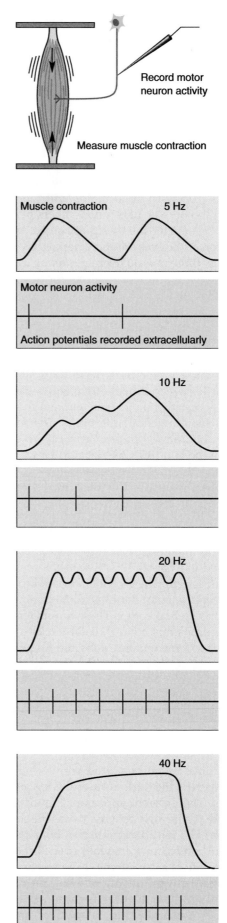

Figure 13.7
From muscle twitch to sustained contraction. A single action potential in an alpha motor neuron causes the muscle fiber to twitch. Summation of twitches causes a sustained contraction as the number and frequency of incoming action potentials increase.

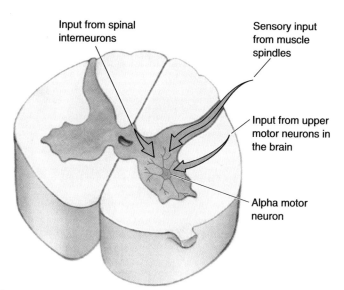

Input from spinal interneurons

Sensory input from muscle spindles

Input from upper motor neurons in the brain

Alpha motor neuron

Figure 13.8
The alpha motor neuron and its three sources of input.

olism. These fibers contract rapidly and powerfully, but they also fatigue rapidly. They are typical of muscles involved in escape reflexes; for example, the jumping muscles of frogs and rabbits. In humans, the arm muscles contain a large number of white fibers.

Even though both types of muscle fiber can (and usually do) coexist in a given muscle, each motor unit contains muscle fibers of only a single type. Thus, **fast motor units** contain rapidly fatiguing white fibers, and **slow motor units** contain slowly fatiguing red fibers. Just as the muscle fibers of the two types of units differ, so do many of the properties of the alpha motor neurons. For example, the motor neurons of fast units are generally bigger and have larger diameter, faster conducting axons; slow units have smaller diameter, more slowly conducting axons. The firing properties of the two types of motor neuron also differ. Fast motor neurons tend to generate occasional high-frequency bursts of action potentials (30–60 impulses per second), whereas slow motor neurons are characterized by relatively steady, low-frequency activity (10–20 impulses per second).

Neuromuscular Matchmaking. The precise matching of particular motor neurons to particular muscle fibers raises an interesting question. Since we've been talking about chickens, let's pose the question this way: Which came first, the muscle fiber or the motor neuron? Perhaps during early embryonic development there is a matching of the appropriate axons with the appropriate muscle fibers. Alternatively, we could imagine a situation in which the properties of the muscle are determined solely by the type of innervation it gets. If it receives a synaptic contact from a fast motor neuron, it becomes a fast fiber, and vice versa for slow units. This question was addressed in an experiment by John Eccles and his colleagues in which the normal innervation of a fast muscle was removed and replaced with a nerve that normally innervated a slow muscle (Figure 13.9). This procedure resulted in the muscle's taking on slow properties. These new properties included not only the type of contraction (slow, fatigue-resistant), but also a switch in much of the underlying biochemistry. This is referred to as a switch of muscle *phenotype*—its physical characteristics—because the types of proteins expressed by the muscle were altered by the new innervation. Work by Terje Lømo and his colleagues in Norway suggests that this switch in muscle phenotype can be induced simply by changing the activity in the motor

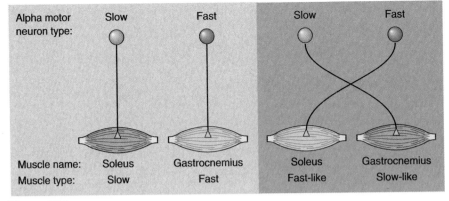

Figure 13.9
A crossed-innervation experiment. Forcing slow motor neurons to innervate a fast muscle causes the muscle to switch to assume slow properties.

neuron from a fast pattern (occasional bursts at 30–60 impulses per second) to a slow pattern (steady activity at 10–20 impulses per second). These findings are particularly interesting because they raise the possibility that *neurons* switch phenotype as a consequence of synaptic activity (experience) and that this may be a basis for learning and memory (discussed in detail in Chapters 19 and 20).

Besides the alterations imposed by *patterns* of motor neuron activity, muscle fibers are also changed simply by varying the *absolute amount* of activity. A long-term consequence of increased activity (especially due to isometric exercise) is hypertrophy, or exaggerated growth, of the muscle fibers. Conversely, prolonged inactivity leads to atrophy, or degeneration, of muscle fibers. Clearly, there is an intimate relationship between the lower motor neuron and the muscle fibers it innervates.

EXCITATION-CONTRACTION COUPLING

As we said, muscle contraction is initiated by the release of acetylcholine (ACh) from the axon terminals of the alpha motor neuron. ACh produces a large EPSP in the postsynaptic membrane due to the activation of nicotinic ACh receptors. Because the membrane of the muscle cell contains voltage-gated sodium channels, this EPSP is sufficient to evoke an action potential in the muscle fiber (but see Box 13.1). This action potential triggers the release of Ca^{2+} ions from an organelle inside the muscle fiber, which leads to contraction of the fiber. Relaxation occurs when the Ca^{2+} levels are lowered by reuptake into the organelle. To understand this process, we must take a closer look at the muscle fiber.

Muscle Fiber Structure

The structure of a muscle fiber is shown in Figure 13.10. Muscle fibers are formed early in fetal development by the fusion of muscle precursor cells, myoblasts, which are derived from the mesoderm (see Chapter 7). This fusion leaves each cell with more than one cell nucleus, so individual muscle cells are said to be *multinucleated*, and it makes them very long (hence the name "fiber"). Muscle fibers are delimited by an excitable cell membrane called the **sarcolemma**. Within the muscle fiber are a number of cylindrical structures called **myofibrils**, which contract in response to an action potential sweeping down the sarcolemma. The myofibrils are surrounded by the **sarcoplasmic reticulum**, an extensive intracellular sac that stores Ca^{2+} ions

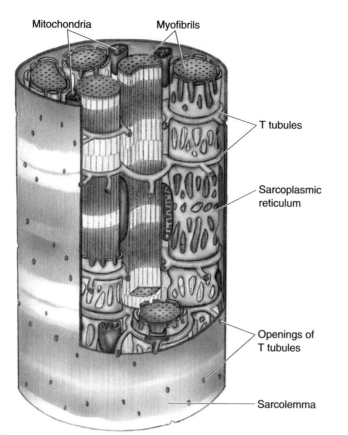

Mitochondria   Myofibrils

T tubules

Sarcoplasmic
reticulum

Openings of
T tubules

Sarcolemma

Figure 13.10
Structure of a muscle fiber. T tubules conduct electrical activity from the surface
membrane into the depths of the muscle fiber.

Box 13.1 | **OF SPECIAL INTEREST**

Myasthenia Gravis

The neuromuscular junction is a synapse seemingly designed to work every time. A presynaptic action potential causes the contents of hundreds of synaptic vesicles to be released into the synaptic cleft. The liberated ACh molecules act at densely packed nicotinic receptors in the postsynaptic membrane, and the resulting EPSP is many times larger than what is necessary to trigger an action potential, and twitch, in the muscle fiber—normally, that is. In a clinical condition called *myasthenia gravis*, released ACh is far less effective and neuromuscular transmission often fails. The name is derived from the Greek for "severe muscle weakness." The disorder is characterized by weakness and fatigability of voluntary muscles, typically including the muscles of facial expression, and it can be fatal if respiration is compromised. An unusual feature of myasthenia gravis is that the severity of the muscle weakness fluctuates, even over the course of a single day.

Myasthenia gravis is an *autoimmune disease*. For reasons we are only beginning to understand, the immune systems of afflicted individuals generate antibodies against their own nicotinic ACh receptors. The antibodies bind to the receptors, interfering with the normal actions of ACh at the neuromuscular junctions. In addition, the binding of antibodies to the receptors leads to secondary, degenerative changes in the structure of the neuromuscular junctions that also make transmission less efficient.

An effective treatment for myasthenia gravis is the administration of drugs that inhibit the enzyme acetylcholinesterase (AChE). Recall from Chapters 5 and 6 that AChE breaks down ACh in the synaptic cleft. In low doses, AChE inhibitors can strengthen neuromuscular transmission by prolonging the lifetime of released ACh. But the therapeutic window is narrow: as we saw in Box 5.4, too much ACh in the cleft leads to desensitization of the receptors and a block of neuromuscular transmission.

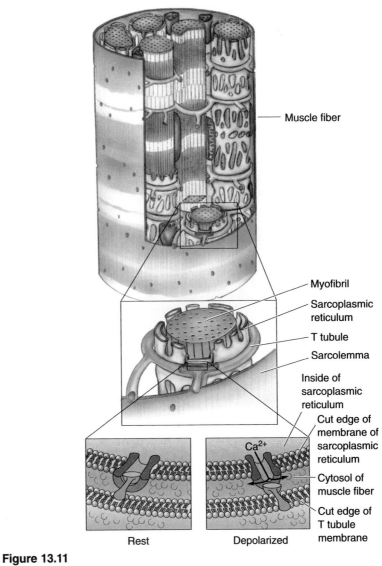

Figure 13.11
Release of Ca²⁺ from the sarcoplasmic reticulum. Depolarization of the T tubule membrane causes conformational changes in proteins that are linked to calcium channels in the sarcoplasmic reticulum, releasing stored Ca^{2+} into the cytosol of the muscle fiber.

(similar in appearance to the smooth ER of neurons; see Chapter 2). Action potentials sweeping along the sarcolemma gain access to sarcoplasmic reticulum deep inside the fiber by way of a network of tunnels called **T tubules** (T for transverse). These are like inside-out axons; the lumen of each T tubule is continuous with the extracellular fluid. Where the T tubule comes in close apposition to the sarcoplasmic reticulum, there is a specialized coupling of the proteins in the two membranes. A voltage-sensitive protein in the T tubule membrane is linked to a calcium channel protein in the sarcoplasmic reticulum. As illustrated in Figure 13.11, the arrival of an action potential in the T tubule membrane causes a conformational change in the voltage-sensitive membrane protein, which, in essence, "pulls the plug" on the calcium channel in the sarcoplasmic reticulum membrane. The resulting increase in intracellular free Ca^{2+} causes the myofibril to contract.

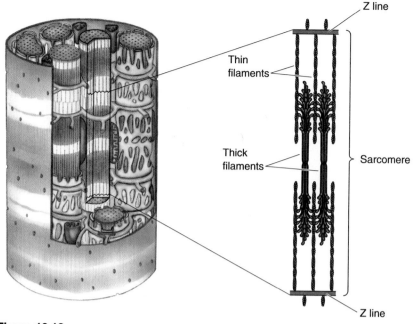

Figure 13.12
The myofibril: a closer look.

The Molecular Basis of Muscle Contraction

A closer look at the myofibril reveals how Ca^{2+} triggers contraction (Figure 13.12). The myofibril is divided into segments by disks called **Z lines**; two Z lines and the segment in between is called a **sarcomere**. Anchored to each side of the Z lines is a series of bristles called **thin filaments**. Bristles from adjacent Z lines face one another but do not come in contact. Between and among the two sets of bristles are a series of fibers called **thick filaments**. Muscle contraction occurs when the thin filaments slide along the thick filaments, bringing adjacent Z lines toward one another. In other words, the sarcomere becomes shorter in length. This sliding-filament model of sarcomere shortening is shown in Figure 13.13. The sliding of the filaments with respect to one another occurs because of the interaction between the major thick filament protein, **myosin**, and the major thin filament protein, **actin**. The exposed "heads" of the myosin molecules bind actin molecules and then undergo a conformational change that causes them to rotate (Figure 13.14). This rotation causes the thick filament to move with respect to the thin filament. At the expense of ATP, the myosin heads then disengage and "uncock" so that the process can repeat itself. Without Ca^{2+}, myosin cannot interact with actin because the myosin attachment sites on the actin molecule are covered by the protein **troponin**. The binding of Ca^{2+} to troponin exposes the sites where myosin binds to actin. Contraction continues as long as Ca^{2+} and ATP are available; relaxation occurs when the Ca^{2+} is sequestered by the sarcoplasmic reticulum. The reuptake of Ca^{2+} by the sarcoplasmic reticulum depends on the action of a calcium pump and hence also requires ATP.

We can summarize the steps of excitation-contraction coupling as follows:

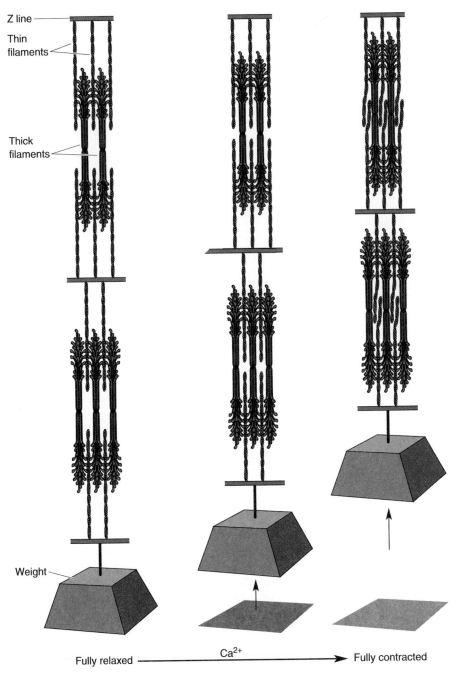

Z line

Thin filaments

Thick filaments

Weight

Fully relaxed ————————— Ca^{2+} —————————→ Fully contracted

Figure 13.13
The sliding-filament model of muscle contraction. Myofibrils shorten when the thin filaments slide toward one another on the thick filaments.

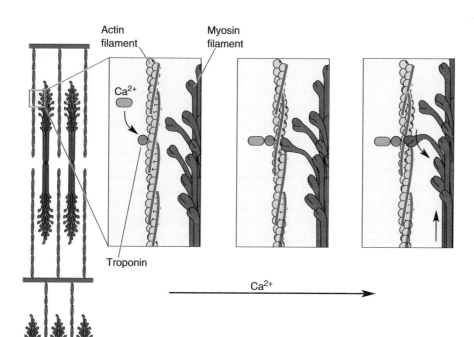

Figure 13.14
The molecular basis of muscle contraction. The binding of Ca²⁺ to troponin allows the myosin heads to bind to the actin filament. Then the myosin heads rotate, causing the filaments to slide with respect to one another.

Excitation
1. An action potential occurs in an alpha motor neuron axon.
2. ACh is released by the axon terminal of the alpha motor neuron at the neuromuscular junction.
3. Nicotinic receptor channels open, and the postsynaptic sarcolemma depolarizes (EPSP).
4. An action potential is generated in the muscle fiber and sweeps down the sarcolemma.
5. Depolarization of the T-tubules causes Ca²⁺ release from the sarcoplasmic reticulum.

Contraction
1. Ca²⁺ binds to troponin.
2. Myosin binding sites on actin are exposed.
3. Myosin heads bind actin.
4. Myosin heads rotate.
5. Myosin heads disengage at the expense of ATP.
6. The cycle continues as long as Ca²⁺ and ATP are present.

Relaxation
1. Ca²⁺ is sequestered by the sarcoplasmic reticulum by an ATP-driven pump.
2. Myosin binding sites on actin are covered by troponin.

You can now understand why death causes stiffening of the muscles, a condition known as *rigor mortis*. Starving the muscle cells of ATP both prevents detachment of the myosin heads and leaves the myosin attachment sites on the actin filaments exposed for binding. The end result is the formation of permanent attachments between the thick and thin filaments.

Since the proposal of the sliding-filament model in 1954 by English physiologists Hugh Huxley and Andrew Huxley, there has been a tremendous amount of progress in identifying the detailed molecular mechanisms of excitation-contraction coupling in muscle. This progress has resulted from a multidisciplinary approach to the problem, with critical contributions made by the use of electron microscopy, as well as biochemical and biophysical methods. The very recent application of molecular genetic techniques also promises to add important new information to our understanding of muscle function, in both health and disease (Box 13.2). This work is inspirational to neuroscientists investigating the molecular biology of neuron-neuron interactions. After all, neurons possess a cytoskeleton with actin and myosin, and they also have Ca^{2+} sequestering and releasing organelles.

SPINAL CONTROL OF MOTOR UNITS

We've traced the action potentials sweeping down the axon of the alpha motor neuron and seen how this causes contraction of the muscle fibers in the motor unit. Now let's explore how the activity of the motor neuron is itself controlled. We begin with a discussion of the first source of synaptic input to the alpha motor neuron introduced above—sensory feedback from the muscles themselves.

Proprioception from Muscle Spindles

As already mentioned, deep within most skeletal muscles are specialized structures called **muscle spindles** (Figure 13.15). A muscle spindle consists

---

Box 13.2 | **OF SPECIAL INTEREST**

Duchenne Muscular Dystrophy

Muscular dystrophy describes a group of inherited disorders, all of which are characterized by muscle weakness. One type, Duchenne muscular dystrophy, afflicts boys before adolescence. The disease is first detected as a weakness of the legs and usually puts its victims in wheelchairs by the time they reach age 12. The disease continues to progress, and afflicted males typically do not survive past the age of 30. The characteristic hereditary pattern of this disease, which afflicts only males but is passed on from their mothers, led to a search for a defective gene on the X chromosome. Major breakthroughs came in the late 1980s when the defective region of the X chromosome was identified. It was discovered that this region contains the gene for the cytoskeletal protein *dystrophin*. Boys with Duchenne muscular dystrophy lack the mRNA encoding this protein; a milder form of the disease, called Becker muscular dystrophy,

was found to be associated with an altered mRNA encoding only a portion of the dystrophin protein.

Dystrophin is a large protein that contributes to the muscle cytoskeleton lying just under the sarcolemma. It must not be an absolute requirement for muscle contraction, however, because movements in afflicted boys appear to be normal during the first few years of life. It is possible that the absence of dystrophin leads to secondary changes in the contractile apparatus, eventually resulting in muscle degeneration. Nonetheless, whatever the normal function of dystrophin ultimately proves to be, it is clear that our models of excitation-contraction coupling are based on knowledge of only a fraction of the proteins that are normally expressed by neurons and muscle fibers. As more is learned about the various proteins in the membrane and cytosol, we can anticipate significant revisions of these models. It is interesting to note that dystrophin is also concentrated in axon terminals in the brain, where it might contribute to excitation-secretion coupling.

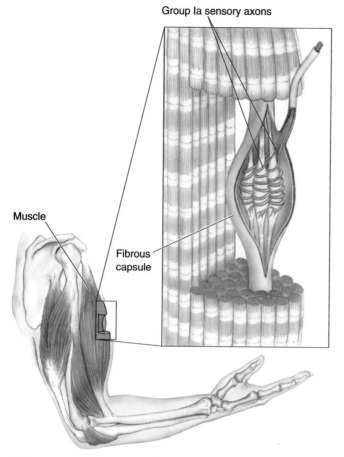

Group Ia sensory axons

Muscle

Fibrous
capsule

Figure 13.15
A muscle spindle and its sensory innervation.

of several types of specialized skeletal muscle fibers contained in a fibrous capsule. The middle third of the capsule is swollen, giving the structure the shape for which it is named. In this middle (equatorial) region, group Ia sensory axons wrap around the muscle fibers of the spindle. Recall from Chapter 12 that group I axons are the thickest myelinated axons in the body, meaning that they conduct action potentials very rapidly. Ia axons enter the spinal cord via the dorsal roots, branch repeatedly, and form excitatory synapses upon both interneurons and alpha motor neurons of the ventral horns. The Ia inputs are very powerful. Neurophysiologist Lorne Mendell, working at Harvard with Henneman, was able to show that a single Ia axon synapses on virtually every alpha motor neuron in the pool innervating the same muscle that contains the spindle. This discovery has raised interesting questions, not only about motor control, but also about how axons are able to establish connections with such precision (Box 13.3).

The Myotatic Reflex. The function of this sensory input to the spinal cord was first shown by Sherrington, who noted that when a muscle is pulled on, it tends to pull back (contract). The fact that this **myotatic reflex** (*myo* from the Greek for "muscle," *tatic* from the Greek for "stretch") involves sensory feedback from the muscle was shown by cutting the dorsal roots. This procedure eliminated the stretch reflex and also caused a loss of muscle tone. Sherrington deduced that the motor neurons must receive a continual synaptic input from the muscles. Later work showed that the discharge of Ia

Box 13.3 | **PATH OF DISCOVERY**

Synapse Search

by Lorne Mendell

In 1965, Elwood Henneman had just completed his seminal work on what is known today as the *size principle*, which states that motor neurons are recruited in the order of their size. He hypothesized that inputs to the individual neurons in a motor neuron pool are equivalent; therefore, differences in the properties of the motor neurons themselves determine their susceptibility to recruitment. Henneman proposed that a group Ia fiber from a single muscle spindle synapses on each of the motor neurons in the appropriate pool, and that size differences among the motor neurons determine which of them responds first to this input.

When I arrived in his laboratory in the Department of Physiology at Harvard Medical School in the summer of 1965 to begin my postdoctoral fellowship, Henneman and I initially explored the idea that more easily recruited small motor neurons reach action potential threshold more rapidly than large motor neurons for a given level of input. Neither of us had any experience with intracellular recording from motor neurons, so we set to work learning how to do it. These were long experiments: dissect the spinal cord most of the day and record most of the night. We carried out as many as 2–3 experiments per week. We worked together sharing the dissection; I dissected the spinal cord, and he exposed the leg nerves. Then we impaled motor neurons with microelectrodes and attempted to test our hypothesis. After several months of hard work, we opened the journal *Science* one day, and were disappointed to find that Daniel Kernell in Amsterdam had scooped us. He reported that small motor neurons are more easily excited than large motor neurons.

As interesting as these experimental results were to us, we felt that they were not definitive. The question we still really wanted to answer was whether each spindle afferent projected to each motor neuron in the pool, thus allowing equal inputs to motor neurons of different sizes. We began this phase by dissecting dorsal rootlets until one was found with the output of only a single muscle spindle. All the other dorsal roots carrying output from this muscle were

Lorne Mendell

then cut. Then when the nerve was stimulated, EPSPs produced in motor neurons were elicited by a *single* Ia afferent. Once such an afferent was isolated, as many motor neurons as possible were impaled in order to see the fraction to which the single Ia fiber projected. Our initial results were extremely encouraging because we found a very widespread projection. But, the EPSPs were very small and required signal averaging in order to be resolved. Fortunately, signal averagers had just been developed, and we were able to borrow one from David Hubel and Torsten Wiesel.

At this point several lucky accidents intervened. In several experiments, the electrode holding the last surviving dorsal rootlet was moved inadvertently, thereby blocking conduction. This ended the experiment, since all the other rootlets had been cut; but it inspired us to think of a better approach. We realized that cutting the remaining rootlets was unnecessary if, instead of triggering data collection from the motor neuron at a precise moment after stimulating the nerve, we used an action potential from a Ia afferent *itself* (recorded in the dorsal rootlet) as the trigger, and simply stretched the muscle to elicit activity. The precise timing of the incoming action potential and postsynaptic response ensured that the relationship between them was monosynaptic. Damaging the afferent then would not end the experiment since other dorsal rootlets would still be available. This procedure became known as *spike-triggered averaging* and has subsequently proved to be useful in many studies of functional connections in the nervous system.

We found that each Ia fiber does project to virtually every motor neuron innervating that muscle. The precision of these connections has made this synapse an important model system in the study of synaptic specificity and plasticity in the vertebrate CNS. (How does the Ia fiber come to innervate only the appropriate alpha motor neuron?) Henneman's size principle has survived decades of experimental scrutiny and remains the likely explanation for the orderly recruitment of motor units. Beyond that, however, this venerable idea stimulated research with wide-ranging and unforeseen consequences.

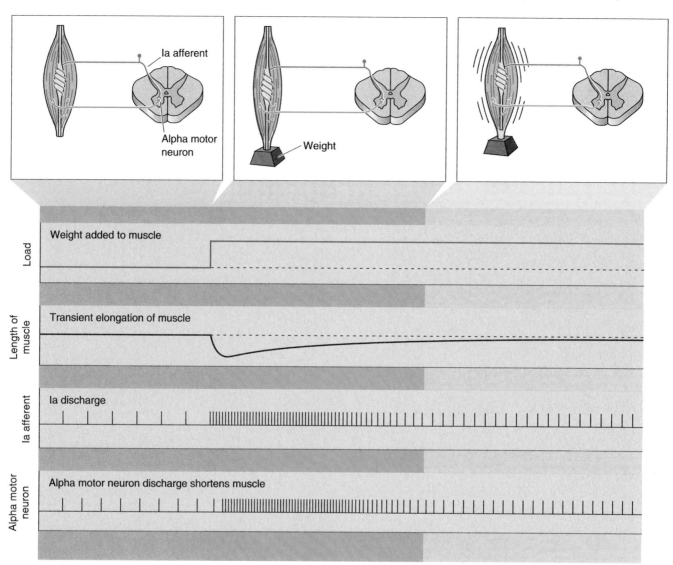

Figure 13.16
The myotatic reflex. This illustration shows the response of a Ia fiber and a motor neuron to the sudden addition of weight that stretches the muscle.

sensory axons is closely related to the length of the muscle. As the muscle is stretched, the discharge rate goes up; as the muscle is shortened and goes slack, the discharge rate goes down. The Ia fiber and the alpha motor neurons on which it synapses constitute the *monosynaptic myotatic reflex arc*: "monosynaptic" because only one synapse separates the primary sensory input from the motor neuron output.

Figure 13.16 shows how this reflex arc serves as an antigravity feedback loop. When a weight is placed on a muscle and the muscle starts to lengthen, the muscle spindles are stretched. The stretching of the equatorial region of the spindle leads to depolarization of the Ia axon endings due to the opening of mechanosensitive ion channels (see Chapter 12). The increased action potential discharge of the Ia axons synaptically depolarizes the alpha motor neurons, which respond by increasing their action potential frequency. This causes the muscle to contract, thereby shortening the muscle. The knee-jerk reflex that occurs when your doctor taps the tendon beneath your kneecap tests for the intactness of this reflex arc in the quadriceps muscle of your thigh (Figure 13.17).

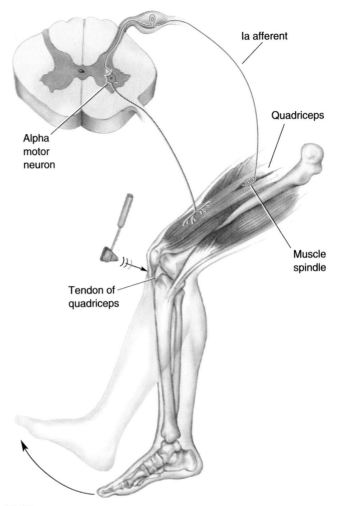

Figure 13.17
The knee-jerk reflex.

Gamma Motor Neurons

As we said, the muscle spindle contains modified skeletal muscle fibers within its fibrous capsule. These muscle fibers are called **intrafusal fibers**, to distinguish them from the more numerous **extrafusal fibers** that lie outside the spindle and form the bulk of the muscle. An important difference between the two types of muscle fibers is that only the extrafusal fibers are innervated by alpha motor neurons. The intrafusal fibers receive their motor innervation by another type of lower motor neuron called a **gamma motor neuron** (Figure 13.18).

Imagine a situation in which muscle contraction is commanded by an upper motor neuron. The alpha motor neurons respond, the extrafusal fibers contract, and the muscle shortens. The response of the muscle spindles is shown in Figure 13.19. If they went slack, the Ia fibers would become silent and the spindle would go "off the air," no longer providing information about muscle length. This does not happen, however, because the gamma motor neurons are also activated. Gamma motor neurons innervate the intrafusal muscle fibers at the two ends of the muscle spindle. Activation of these fibers causes a contraction of the two poles of the muscle spindle, thereby pulling on the noncontractile equatorial region and keeping the Ia afferents active. Notice that the activation of alpha and gamma motor neurons has opposite effects on Ia output; alpha activation alone decreases Ia activity, while gamma activation alone increases Ia activity.

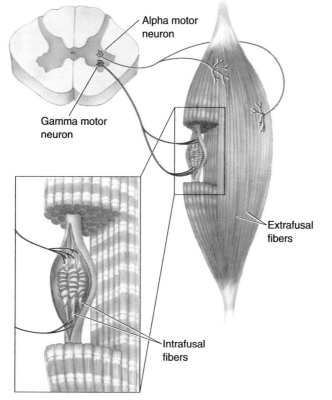

Figure 13.18
Alpha motor neurons, gamma motor neurons, and the muscle fibers they innervate.

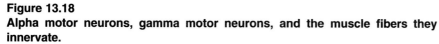

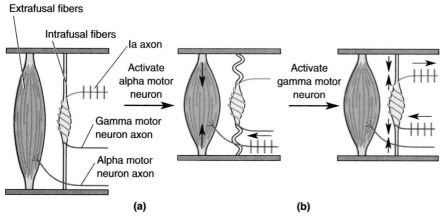

Figure 13.19
The function of gamma motor neurons. **(a)** Activation of alpha motor neurons causes the extrafusal muscle fibers to shorten. If the muscle spindle becomes slack, it goes "off the air" and no longer reports the length of the muscle. **(b)** Activation of gamma motor neurons causes the poles of the spindle to contract, keeping it "on-the-air."

Recall from our discussion above that the monosynaptic myotatic reflex arc can be viewed as a feedback loop. The principles of feedback control systems are that a set point is determined (in this case, the desired muscle length), deviations from the set point are detected by a sensor (the Ia axon endings), and deviations are compensated for by an effector system (the alpha motor neurons and extrafusal muscle fibers), returning the system to the set point. Changing the activity of the gamma motor neurons changes the set point of the myotatic feedback loop. This circuit, gamma motor neuron → intrafusal muscle fiber → Ia afferent → alpha motor neuron → extrafusal muscle fibers, is sometimes called the *gamma loop*.

Alpha and gamma motor neurons are simultaneously activated by descending commands from the brain. By regulating the set point of the myotatic feedback loop, the gamma loop provides additional control of alpha motor neurons and muscle contraction.

Proprioception from Golgi Tendon Organs

Muscle spindles are not the only source of proprioceptive inputs from the muscles. Another sensor in skeletal muscle is the **Golgi tendon organ**, which acts like a strain gauge; that is, it monitors muscle tension, or the force of contraction. Golgi tendon organs are located at the junction of the muscle and the tendon and are innervated by group Ib sensory axons.

It is important to note that while spindles are situated *in parallel* with the muscle fibers, Golgi tendon organs are situated *in series* (Figure 13.20). This different anatomical arrangement is what distinguishes the types of information these two sensors provide the spinal cord: Ia activity from the spindle encodes *muscle length* information, while Ib activity from the Golgi tendon organ encodes *muscle tension* information.

The Ib afferents enter the spinal cord, branch repeatedly, and synapse on interneurons in the ventral horn. Some of these interneurons form inhibitory connections with the alpha motor neurons innervating the same muscle. This is the basis for another spinal reflex called the *clasp-knife*, or *reverse*

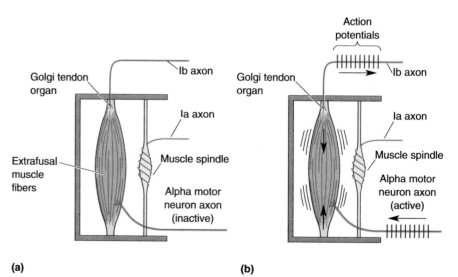

(a) **(b)**

Figure 13.20
Organization of muscle proprioceptors. (a) Muscle spindles are arranged parallel to the extrafusal fibers; Golgi tendon organs lie in series, between the muscle fibers and their points of attachment. (b) Golgi tendon organs respond to increased tension on the muscle and transmit this information to the spinal cord via type Ib sensory afferents. Because the activated muscle does not change length, the 1a afferents remain silent in this example.

myotatic, reflex (Figure 13.21). This reflex is demonstrated when a subject is asked to resist a force applied to bend the knee. At a certain level of tension on the contracting extensors, the muscles suddenly relax, and the knee collapses like the closing blade of a pocket knife. Thus, in extreme circumstances, this reflex arc protects the muscle from being overloaded. However, the normal function is to slow muscle contraction as the force of contraction increases. This type of proprioceptive feedback is also thought to be particularly important for the proper execution of fine motor acts, such as the manipulation of fragile objects with the hands, which require a steady, but not too powerful, grip.

Spinal Interneurons

The actions of Ib inputs from Golgi tendon organs on alpha motor neurons are entirely *polysynaptic*—they are all mediated by intervening spinal interneurons. Indeed, most of the input to the alpha motor neurons comes from interneurons of the spinal cord. Spinal interneurons receive synaptic input from primary sensory axons, descending axons from the brain, and collaterals of lower motor neuron axons. The interneurons are themselves networked together in a way that allows coordinated motor programs to be generated in response to their many inputs.

Inhibitory Input. Interneurons play a critical role in the proper execution of even the simplest reflexes. Consider the myotatic reflex, for example. Compensation for the lengthening of one set of muscles—the flexors of the elbow, for example—involves contraction of the flexors via the myotatic reflex, but also requires the relaxation of the antagonist muscles, the extensors. This process is called **reciprocal inhibition**, the contraction of one set of muscles accompanied by the relaxation of the antagonist muscles. The importance of this is obvious; imagine how hard it would be to lift something if your own antagonist muscles were constantly opposing you. In the case of the myotatic reflex, reciprocal inhibition occurs because collaterals of the Ia afferents synapse on inhibitory spinal interneurons that contact the alpha motor neurons supplying the antagonist muscles (Figure 13.22). Reciprocal inhibition is also used by descending pathways to overcome the

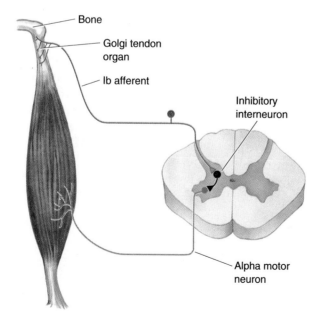

Figure 13.21
Circuitry of the clasp-knife reflex.

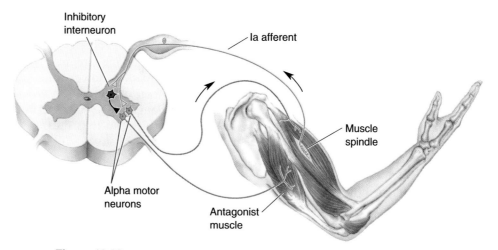

Inhibitory
interneuron

Ia afferent

Muscle
spindle

Alpha motor
neurons

Antagonist
muscle

Figure 13.22
Reciprocal inhibition of flexors and extensors of the same joint.

powerful myotatic reflex. Consider a situation in which the flexors of the elbow are voluntarily commanded to contract. The resulting stretch of the antagonist extensor muscles would activate their myotatic reflex arc, which would strongly resist flexion of the joint. However, the descending pathways that activate the alpha motor neurons controlling the flexors also activate interneurons, which inhibit the alpha motor neurons that supply the antagonist muscles.

Excitatory Input. Not all interneurons are inhibitory. An example of a reflex mediated in part by excitatory interneurons is the *flexor reflex* (Figure 13.23). This is a complex reflex arc used to withdraw a limb from an aversive stimulus (such as the withdrawal of your foot from the thumbtack in Chapter 3). This reflex is far slower than the myotatic reflex, indicating that a number of interneurons intervene between the sensory stimulus and the coordinated motor act. The pain fibers entering the spinal cord branch profusely and activate interneurons in several different spinal segments. These cells eventually excite the alpha motor neurons that control all the flexor muscles of the affected limb (and, needless to say, inhibitory interneurons are also recruited to inhibit the alphas that control the extensors).

You're walking along, and you step on a tack. Thanks to the flexor reflex, you reflexively yank your foot up. But where would that leave the rest of your body if nothing else happened? Falling to the floor, most likely. Luckily, an additional component of the reflex is recruited: the activation of extensor muscles and the inhibition of flexors *on the opposite side*. This is called the *crossed-extensor reflex*, and it is used to compensate for the extra load imposed by limb withdrawal on the antigravity extensor muscles of the opposite leg (Figure 13.24). Notice that this is another example of reciprocal inhibition, but in this case, activation of the flexors on one side of the spinal cord is accompanied by inhibition of the flexors on the opposite side.

Generation of Spinal Motor Programs for Walking

The crossed-extensor reflex, in which one side extends as the other side flexes, seems to provide a building block for locomotion. When you walk, you alternately withdraw and extend the two legs. All that is lacking is a mechanism to coordinate the timing. In principle, this could be a series of descending commands from upper motor neurons. However, as we already suspected from our consideration of headless chicken behavior, it seems

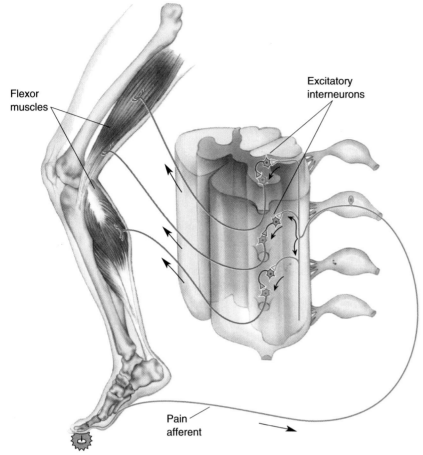

Flexor muscles

Excitatory interneurons

Pain afferent

Figure 13.23
Circuitry of the polysynaptic flexor reflex.

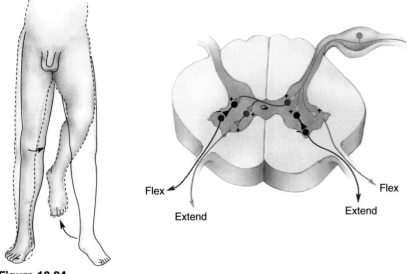

Flex

Extend

Flex

Extend

Figure 13.24
Circuitry of the crossed-extensor reflex.

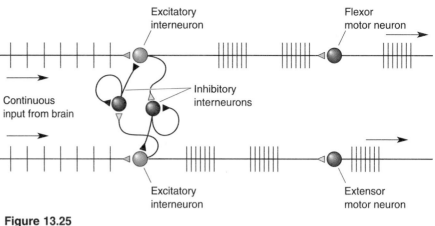

Figure 13.25
A possible circuit for rhythmic alternating activity.

likely that this control is exerted from within the spinal cord. Indeed, complete transection of the cat spinal cord at the mid-thoracic level leaves the hind limbs capable of generating coordinated walking movements. The circuit for the coordinated control of walking must reside, therefore, within the spinal cord. In general, circuits that give rise to rhythmic motor activity are called **central pattern generators**.

What does the central pattern generator for walking look like? We don't know the answer to this question, but it might look something like the circuit shown in Figure 13.25. According to this scheme, walking is initiated when a steady input excites two interneurons that connect to the motor neurons controlling the flexors and extensors, respectively. The interneurons respond to a continuous input by generating bursts of outputs. The activity of the two interneurons alternate because they inhibit each other via other, inhibitory, interneurons. Thus, a burst of activity in one interneuron strongly inhibits the other, and vice versa. Then, using the spinal cord circuitry of the crossed-extensor reflex, the movements of the opposite limb could be coordinated so that flexion on one side is accompanied by extension on the other. By adding further interneuronal connections between the lumbar and cervical spinal segments, we could account for the swinging of the arms that accompanies walking.

CONCLUDING REMARKS

We can draw several conclusions from the preceding discussion of the spinal control of movement. First, a great deal has been learned about movement and its spinal control by using a combination of experimental approaches, ranging from biochemistry to biophysics to behavior. Indeed, a complete understanding, whether of excitation-contraction coupling or central pattern generation, requires knowledge derived from every approach. Second, sensation and movement are inextricably linked even at the lowest levels of the neural motor system. The normal function of the alpha motor neuron depends on direct feedback from the muscles themselves and indirect information from the tendons, joints, and skin. Third, the spinal cord contains an intricate network of circuits for the control of movement; it is far more than a conduit for somatic sensory and motor information. Evidently, coordinated and complex patterns of activity in these spinal circuits can be driven by relatively crude descending signals. This leaves the question of precisely what the upper motor neurons contribute to motor control—the subject of the next chapter.

The Somatic Motor System
smooth muscle
striated muscle
cardiac muscle
skeletal muscle
muscle fiber
somatic motor system
flexion
extension
flexor
synergist muscle
extensor
antagonist muscle
axial muscle
proximal (girdle) muscle
distal muscle

The Lower Motor Neuron
alpha motor neuron
motor unit
motor neuron pool
fast motor unit
slow motor unit

Excitation-Contraction Coupling
sarcolemma
myofibril
sarcoplasmic reticulum
T tubules
Z line
sarcomere
thin filament
thick filament
myosin
actin
troponin

Spinal Control of Motor Units
muscle spindle
myotatic reflex
intrafusal fiber
extrafusal fiber
gamma motor neuron
Golgi tendon organ
reciprocal inhibition
central pattern generator

R E V I E W ☑
Q U E S T I O N S

1. What did Sherrington call the "final common pathway," and why?
2. Define, in one sentence, motor unit. How does it differ from motor neuron pool?
3. Which is recruited first, a fast motor unit or a slow motor unit? Why?
4. When and why does rigor mortis occur?
5. Your doctor taps the tendon beneath your kneecap and your leg extends. What is the neural basis of this reflex? What is it called?
6. What is the function of the gamma motor neurons?
7. Lenny, a character in Steinbeck's classic book *Of Mice and Men*, loved rabbits, but when he hugged them, they were crushed to death. Which type of proprioceptive input may have Lenny been lacking?

Brain Control of Movement

In Chapter 13, we discussed the organization of the peripheral somatic motor system: the joints, skeletal muscles, and their sensory and motor innervation. We saw that the final common pathway for behavior is the alpha motor neuron, that the activity of this cell is under the control of sensory feedback and spinal interneurons, and that reflex movements reveal the complexity of this spinal control system. In this chapter, we'll explore how the brain influences the activity of the spinal cord to command voluntary movements.

The central motor system is arranged as a hierarchy of control levels, with the forebrain at the top and the spinal cord at the bottom. It is useful to think of this motor control hierarchy as having three levels (Table 14.1). The highest level, represented by the association areas of neocortex and basal ganglia of the forebrain, is concerned with *strategy*: the goal of the movement and the movement strategy that best achieves the goal. The middle level, represented by the motor cortex and cerebellum, is concerned with *tactics*: the spatiotemporal sequences of muscle contractions required to smoothly and accurately achieve the strategic goal. The lowest level, represented by the brain stem and spinal cord, is concerned with *execution*: activation of the motor neuron and interneuron pools that generate the goal-directed movement and make any necessary adjustments of posture.

To appreciate the different contributions of the three hierarchical levels to movement, consider the actions of a baseball pitcher standing on the mound, preparing to pitch to a batter (Figure 14.1). The cerebral neocortex has information—based on vision, audition, somatic sensation, and proprioception—about precisely where the body is in space. Strategies must be devised to move the body from the current state to one in which a pitch is delivered and the desired outcome is attained (a swing and a miss). Several options are available—a curve ball, a fast ball, a knuckle ball, and so on—and these alternatives are filtered through the basal ganglia and back to the cortex until a final decision is made, based in large part on past experience (e.g., "This batter hit a home run last time I threw a fast ball"). The motor areas of cortex and the cerebellum then make the tactical decision (to throw the curve ball), and issue instructions to the brain stem and spinal cord. Activation of neurons in the brain stem and spinal cord then causes the movement to be executed. Properly timed activation of motor neurons in the cervical spinal cord generates a coordinated movement of the shoulder, elbow, wrist, and fingers. Simultaneously, brain stem input to the thoracic and lumbar spinal cord command the appropriate postural adjustments that keep the pitcher from falling over during the throw.

According to the laws of physics, the movement of a thrown baseball through space is *ballistic*, referring to a trajectory that cannot be altered. The movement of the pitcher's arm that throws the ball is also described as ballistic because it cannot be altered once initiated. This type of rapid voluntary movement is not under the same type of sensory feedback control that regulates antigravity postural reflexes (see Chapter 13). The reason is simple: The movement is too fast to be altered by sensory feedback. But the movement does not occur in the absence of sensory information. Sensory information *before* the movement was initiated was crucial in order to determine the start-

Table 14.1
The Motor Control Hierarchy

Level	Function	Structures
High	Strategy	Association areas of neocortex, basal ganglia
Middle	Tactics	Motor cortex, cerebellum
Low	Execution	Brain stem, spinal cord

ing positions of the limbs and body and to anticipate any changes in resistance during the throw. And sensory information *during* the movement is also important—not necessarily for the movement at hand, but for improving subsequent similar movements. The proper function of each level of the motor control hierarchy relies so heavily on sensory information that the motor system of the brain might properly be considered a *sensorimotor system*. At the highest level of the hierarchy, sensory information generates a mental image of the body and its relationship to the environment. At the middle level, tactical decisions are based on the memory of sensory information from past movements. At the lowest level, sensory feedback is used to maintain posture, muscle length, and tension before and after each voluntary movement.

In this chapter, we investigate this hierarchy of motor control and how each level contributes to the control of the peripheral somatic motor system. We start by exploring the pathways that bring information to the spinal motor neurons. From there we will ascend to the highest levels of the motor hierarchy, and then we'll fill in the pieces of the puzzle that bring the different levels together.

Figure 14.1
A baseball pitcher planning a pitch.

DESCENDING SPINAL TRACTS

How does the brain communicate with the motor neurons of the spinal cord? Axons from the brain descend through the spinal cord along two major pathways, shown in Figure 14.2. One is in the lateral column of the spinal cord, and the other is in the ventromedial column. Remember this rule of thumb: The **lateral pathway** is involved in voluntary movement of the distal musculature and is under direct cortical control, and the **ventromedial pathway** is involved in the control of posture and locomotion and is under brain stem control.

The Lateral Pathway

The most important component of the lateral pathway is the **corticospinal tract** (Figure 14.3a). Originating in the neocortex, it is the longest and one of the largest CNS tracts (10^6 axons). Two-thirds of the axons in the tract originate in areas 4 and 6 of the frontal lobe, collectively called **motor cortex**. Most of the remaining axons in the tract derive from the somatosensory areas of the parietal lobe (see Chapter 12). Axons from the cortex pass through the internal capsule bridging the telencephalon and thalamus, course through the midbrain and pons, and collect to form a tract at the base of the medulla. The tract forms a pyramid-shaped bulge running down the ventral surface of the medulla, explaining why it is called the **pyramidal tract**. At the junction of the medulla and spinal cord, the pyramidal tract decussates. This means that the *right* motor cortex directly commands the movement of the *left* side of the body and that the *left* motor cortex controls the muscles on the *right* side. As the axons cross, they collect in the lateral column of the spinal cord and form the lateral corticospinal tract. The corticospinal tract axons terminate in the dorsolateral region of the ventral horns and intermediate gray matter, the location of the motor neurons and interneurons that control the distal muscles, particularly the flexors (see Chapter 13).

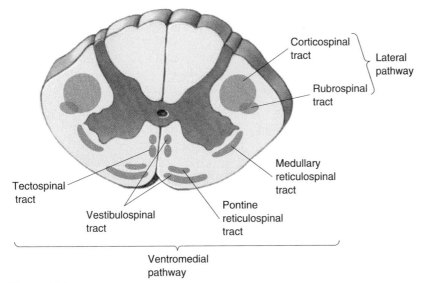

Figure 14.2
The descending tracts of the spinal cord. The lateral pathway, consisting of the corticospinal and rubrospinal tracts, controls voluntary movements of the distal musculature. The ventromedial pathway, consisting of the reticulospinal, vestibulospinal, and tectospinal tracts, controls postural muscles.

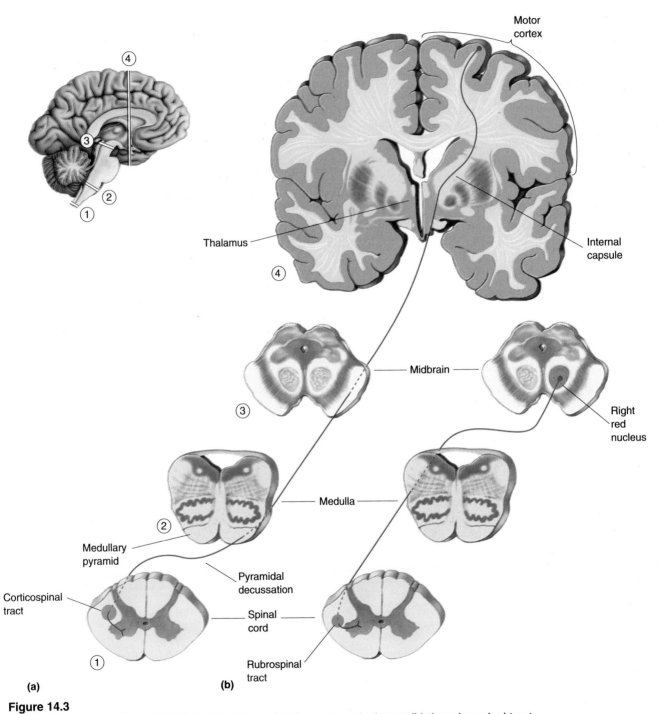

(a)   **(b)**

Figure 14.3
Origins and terminations of the lateral pathway. (**a**) The corticospinal tract; (**b**) the rubrospinal tract.

A much smaller component of the lateral pathway is the **rubrospinal tract**, which originates in the **red nucleus** of the midbrain, so named because of its distinctive pinkish hue in a freshly dissected brain (rubro is from the Latin for "red"). Axons from the red nucleus decussate in the pons and join those in the corticospinal tract in the lateral column of the spinal cord (Figure 14.3b). A major source of input to the red nucleus is the very region of frontal cortex that also contributes to the corticospinal tract. Indeed, it appears that in the course of primate evolution, this indirect corticorubrospinal pathway has largely been replaced by the direct corticospinal path. Thus, while the rubrospinal tract contributes importantly to motor control in many mammalian species, in humans it appears to be reduced, most of its functions subsumed by the corticospinal tract.

Effects of Lateral Pathway Lesions. The foundation for the modern view of the function of the lateral pathway was laid in the late 1960s by Donald Lawrence and Hans Kuypers. Experimental lesions in the lateral pathway (both corticospinal and rubrospinal tracts) in monkeys rendered them unable to make fractionated movements of the arms; that is, they could not move the shoulder, elbows, wrists, and fingers independently. Voluntary movements were also slower and less accurate. Despite this, the animals could sit upright and stand with normal posture. By analogy, a human with a lateral path lesion would be able to stand on the pitcher's mound, but unable to throw the ball.

Lesions in the corticospinal tracts alone caused a movement deficit as severe as that observed after lesions in the lateral columns. Interestingly, however, many functions gradually reappeared over the months following surgery. In fact, the only permanent deficit was some weakness of the distal flexors and an inability to move the fingers independently. A subsequent lesion in the rubrospinal tract completely reversed this recovery, however. These results suggest that the corticorubrospinal pathway was able, over time, to partially compensate for the loss of the corticospinal tract input.

The Ventromedial Pathway

The ventromedial pathway contains four descending tracts that originate in the brain stem and terminate among the spinal interneurons controlling proximal and axial muscles. These tracts are the vestibulospinal tract, the tectospinal tract, the pontine reticulospinal tract, and the medullary reticulospinal tract. Functionally, they may be divided into two groups: (1) the vestibulospinal and tectospinal tracts control the posture of the head and neck; (2) the pontine and medullary reticulospinal tracts control the posture of the trunk and the antigravity muscles of the limbs. We'll briefly discuss each one in turn.

Vestibulospinal and Tectospinal Tracts. The vestibulospinal and tectospinal tracts function to keep the head balanced on the shoulders as the body moves through space, and turn the head in response to new sensory stimuli. The **vestibulospinal tract** originates in the **vestibular nuclei** of the medulla, which relay sensory information from the vestibular apparatus in the inner ear (Figure 14.4a). The **vestibular apparatus** consists of a labyrinth of fluid-filled canals and cavities in the temporal bone that are closely associated with the cochlea. The motion of the fluid in this labyrinth, which accompanies movement of the head, activates hair cells (similar to the inner hair cells of the cochlea) that signal the vestibular nuclei via cranial nerve VIII. The vestibulospinal tracts project bilaterally down the spinal cord and activate the spinal circuits that compensate for head movement. Stability of the head is important because the head contains our eyes, and keeping the eyes stable, even as our body moves, ensures that our image of the world remains stable.

The **tectospinal tract** originates in the superior colliculus of the midbrain, which receives direct input from the retina (Figure 14.4b). Besides its retinal input, the superior colliculus receives projections from visual cortex, as well as afferents carrying somatosensory and auditory information. From this input, the superior colliculus constructs a map of the world around us; stimulation at one site in this map leads to an orienting response that directs the head and eyes to move so that the appropriate point of space is imaged on the fovea. Activation of the colliculus by the image of a runner sprinting toward second base, for example, would cause the pitcher to orient his head and eyes toward this important new stimulus.

Pontine and Medullary Reticulospinal Tracts. The reticulospinal tracts arise mainly from the **reticular formation** of the brain stem. The reticular formation runs the length of the brain stem at its core, just under the cerebral aqueduct and fourth ventricle. A complex meshwork of neurons and fibers, the reticular formation receives input from many sources and participates in many different functions. For the purposes of our discussion of motor control, the reticular formation may be divided into two parts that give rise to two different descending tracts: the pontine (medial) reticulospinal tract and the medullary (lateral) reticulospinal tract (Figure 14.5).

The **pontine reticulospinal tract** enhances the antigravity reflexes of the spinal cord. Activity in this pathway, by facilitating the extensors of the lower limb and (in humans) the flexors of the upper limb, helps maintain a standing posture by resisting the effects of gravity. This type of regulation is an important component of motor control: Keep in mind that most of the

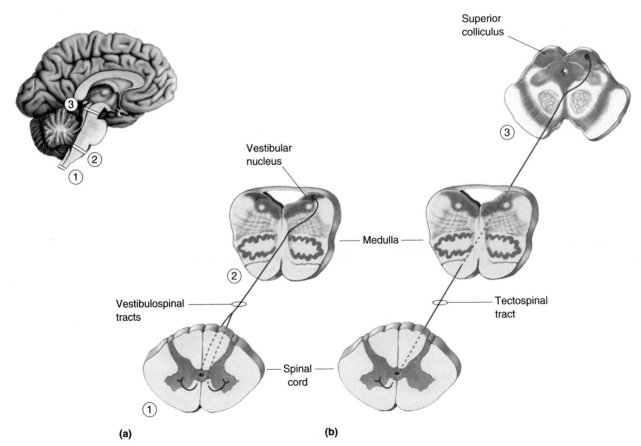

Figure 14.4
Origins and terminations of (a) the vestibulospinal tract and (b) the tectospinal tract. These components of the ventro-medial pathway control the posture of the head and neck.

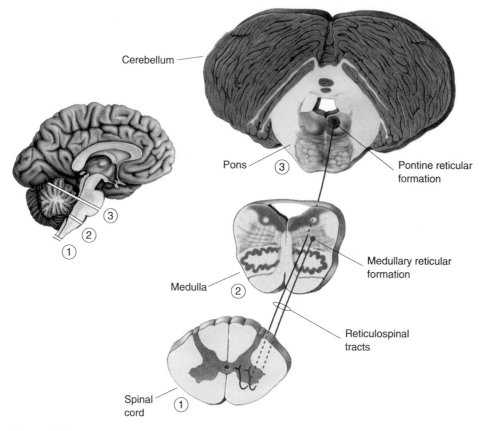

Figure 14.5
The pontine (medial) and medullary (lateral) reticulospinal tracts. These components of the ventromedial pathway control posture of the trunk and the antigravity muscles of the limbs.

time, the activity of ventral horn neurons maintains, rather than changes, muscle length and tension. The **medullary reticulospinal tract**, however, has the opposite effect; it liberates the antigravity muscles from reflex control. Activity in both reticulospinal tracts is controlled by descending signals from the cortex. A fine balance between them is required as the pitcher goes from standing on the mound to winding up and throwing the ball.

Figure 14.6 provides a simple summary of the major descending spinal tracts. The ventromedial pathway originates from several regions of the brain stem and participates mainly in the maintenance of posture and certain reflex movements. Initiation of a voluntary, ballistic movement, such as throwing a baseball, requires instructions that descend from the motor cortex along the lateral pathway. Motor cortex directly activates spinal motor neurons and also liberates them from reflex control by communicating with the nuclei of the ventromedial pathway. It is clear that the cortex is key for voluntary movement and behavior, so we will now focus our attention there.

PLANNING OF MOVEMENT BY THE CEREBRAL CORTEX

Although cortical areas 4 and 6 are called motor cortex, it is important to recognize that control of voluntary movement engages almost all of the neocortex. Goal-directed movement depends on knowledge of where the body is in space, where it intends to go, and on selection of a plan to get it there. Once a plan has been selected, it must be held in memory until the appro-

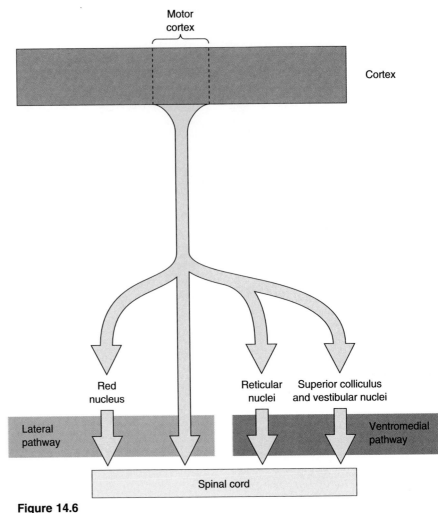

Figure 14.6
Summary of the major descending spinal tracts and where they originate.

priate time. Finally, instructions must be issued to implement the plan. To some extent, these different aspects of motor control are localized to different regions of the cerebral cortex. In this section, we explore some of the cortical areas implicated in motor planning. Later we'll look at how a plan is converted into action.

Motor Cortex

The motor cortex is a circumscribed region of the frontal lobe. Area 4 lies just anterior to the central sulcus on the precentral gyrus, and area 6 lies just anterior to area 4 (Figure 14.7). The definitive demonstration that these areas constitute motor cortex in humans came from the work of Canadian neurosurgeon Wilder Penfield. Recall from Chapter 12 that Penfield electrically stimulated the cortex in patients who were undergoing surgery to remove damaged bits of brain. The stimulation was used in an attempt to identify which regions of cortex should be spared from the knife. In the course of these operations, Penfield discovered that weak electrical stimulation of area 4 in the precentral gyrus would elicit a twitch of the muscles in a particular region of the body on the contralateral side. By systematically probing this region, it was established that there is a somatotopic organization in the human precentral gyrus much like that seen in the somatosensory areas of the postcentral gyrus (Figure 14.8). Area 4 is now often referred to as primary motor cortex or **M1.**

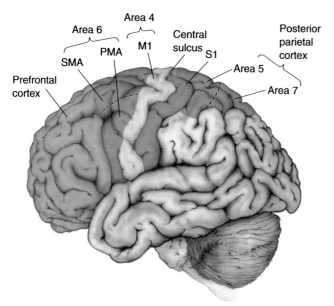

Figure 14.7
Areas of neocortex intimately involved in the planning and instruction of voluntary movement. Areas 4 and 6 constitute motor cortex.

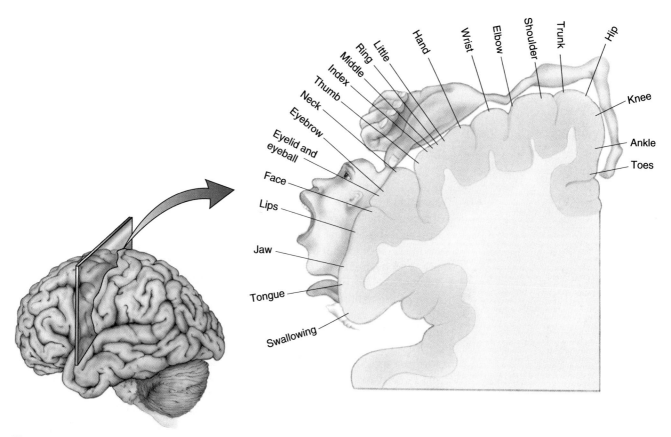

Figure 14.8
Somatotopic map of the human precentral gyrus.

The foundation for Penfield's discovery had been laid nearly a century before by Gustav Fritsch and Eduard Hitzig, who, in 1870, had shown that stimulation of the frontal cortex of anesthetized dogs would elicit movement of the contralateral side of the body (see Chapter 1). Then, around the turn of the century, David Ferrier and Charles Sherrington discovered that the motor area in primates was located in the precentral gyrus. By comparing the histology of this region in Sherrington's apes with that of the human brain, Australian neuroanatomist Alfred Walter Campbell concluded that cortical area 4 is motor cortex.

Campbell speculated that cortical area 6, just rostral to area 4, might be an area specialized for skilled voluntary movement. Penfield's studies 50 years later supported the conjecture that this was a "higher" motor area in humans by showing that electrical stimulation of area 6 could evoke complex movements of either side of the body. Penfield found two somatotopically organized motor maps in area 6: one in a lateral region he called the **premotor area**, or **PMA**, and one in a medial region called the **supplementary motor area**, or **SMA** (see Figure 14.7). These two areas appear to perform similar functions but on different groups of muscles. While SMA sends axons that innervate distal motor units directly, PMA connects primarily with reticulospinal neurons that innervate proximal motor units.

Contributions of Posterior Parietal and Prefrontal Cortex

Recall the baseball player, standing on the mound, preparing to pitch. It should be apparent that before the detailed sequence of muscle contractions for the desired pitch can be calculated, the pitcher must have information about the current position of his body in space and how it relates to the positions of the batter and the catcher. This mental body image depends on somatosensory, proprioceptive, and visual inputs to the posterior parietal cortex.

Two areas are of particular interest in the posterior parietal cortex: area 5, which is a target of inputs from the primary somatosensory cortical areas 3, 1, and 2 (see Chapter 12); and area 7, which is a target of higher-order visual cortical areas such as MT (see Chapter 10). Recall that human patients with lesions in these areas of the parietal lobes, as can occur after a stroke, show bizarre abnormalities of body image and the perception of spatial relations. In its most extreme manifestation, the patient will simply ignore the side of the body opposite the parietal lesion.

The parietal lobes are extensively interconnected with regions in the anterior frontal lobes that in humans are thought to be important for abstract thought, decision making, and the anticipation of the consequences of action. These "prefrontal" areas, along with the posterior parietal cortex, represent the highest levels of the motor control hierarchy, where decisions are made about what actions to take and their likely outcome (a curve ball followed by a strike). Prefrontal and parietal cortex both send axons that converge on cortical area 6. Recall that areas 6 and 4 together contribute most of the axons to the descending corticospinal tract. Thus, area 6 lies at the junction where signals encoding *what* actions are desired are converted into signals that specify *how* the actions will be carried out.

This general view of higher-order motor planning received dramatic support in a recent series of studies on humans carried out by Danish neurologist Per Roland and his colleagues. They used positron emission tomography (PET) to monitor changes in regional cortical blood flow as simple voluntary motor acts were planned and executed. Because there is a close link between local cerebral blood flow and neural activity, this method enables them to map the patterns of cortical activation that accompany voluntary movements. When the subjects were asked to perform a series of finger movements from memory, the following regions of cortex showed increased blood flow: the somatosensory and posterior parietal areas, parts of the prefrontal cortex (area 8), area 6, and area 4. These are the very regions of the cerebral cortex that, as discussed above, are thought to play a role in generating the intention to move and converting that intention to a plan of action. Interestingly, when the subjects were asked only to mentally rehearse the movement without actually moving the finger, area 6 remained active but area 4 did not.

Neuronal Correlates of Motor Planning

Recent experimental work on monkeys further supports the idea that area 6 (SMA and PMA) plays an important role in the planning of movement, particularly complex movement sequences of the distal musculature. Using a method developed in the late 1960s by Edward Evarts at the National Institutes of Health, it has been possible to record the activity of neurons in the motor areas of awake, behaving animals (Box 14.1). Cells in SMA typically increase their discharge rates about a second before the execution of a hand or wrist movement, consistent with their proposed role in planning movement (recall Roland's findings in humans). An important feature of this activity is that it occurs in advance of the movements of *either* hand, suggesting that the supplementary areas of the two hemispheres are closely linked via the corpus callosum. Indeed, movement deficits observed following an SMA lesion on one side, in both monkeys and humans, are

---

Box 14.1 | **OF SPECIAL INTEREST**

Behavioral Neurophysiology

Showing that a brain lesion impairs movement and that brain stimulation elicits movement does not tell us how the brain *controls* movement. To address this problem, we need to know how the discharge properties of neurons relate to different types of voluntary movement in the intact organism. PET scans have been extremely valuable in plotting out the distribution of activity in the brain as behaviors are performed, but they lack the resolution to track the millisecond-by-millisecond changes in the activity of individual neurons. The best method for this purpose is extracellular recording with metal microelectrodes (see Box 4.1). But how to do this in awake, behaving animals?

This problem was solved by Edward Evarts and his colleagues at the National Institutes of Health. Monkeys were trained to perform simple tasks; when the tasks were performed successfully, the monkeys were rewarded with a sip of fruit juice. For example, to study the brain's guidance of hand and arm movements, the monkey might be trained to

move its hand toward the brightest of several spots on a computer screen. Pointing to the correct spot earned it a juice reward. After training, the animals were anesthetized. In a simple surgical procedure, each monkey was fitted with a headpiece through a small opening in the skull, and a microelectrode was introduced into the brain. When the animals recovered from surgery, they showed no signs of discomfort from either the headpiece or the insertion of a microelectrode into the brain. Evarts and his colleagues then recorded the discharges of individual cells in the motor cortex as the animals made voluntary movements. In the example above, one could then see how the neuron's response changes when the animal points to different spots on the screen.

This is an example of what now is called behavioral neurophysiology, the recording of cellular activity in the brain of awake, behaving animals. By altering the simple task that the animal performs, the same method can be applied to the investigation of a wide range of neuroscientific problems, including attention, perception, learning, and movement (see Box 10.3).

particularly pronounced for tasks requiring the coordinated actions of the two hands, such as buttoning a shirt. In humans, a selective inability to perform complex (but not simple) motor acts is called *apraxia*.

You've heard the expression "Ready, set, go." The preceding discussion suggests that readiness depends on activity in the parietal and frontal lobes, along with important contributions from the brain centers that control levels of attention and alertness. "Set" may reside in the supplementary and premotor areas, where movement strategies are devised and held until they are executed. A good example is shown in Figure 14.9, based on the work of Michael Weinrich and Steven Wise at the National Institutes of Health. They monitored the discharge of a neuron in PMA as a monkey performed a task requiring a specific arm movement to a target. The monkey was first given an *instruction stimulus* informing him what the target would be ("Get set, monkey!"), followed after a variable delay by a *trigger stimulus* informing the monkey that it was OK to move ("Go, monkey!"). Successful performance of the task (i.e., waiting for the "Go" signal and then making the movement to the appropriate target) was rewarded by a sip of juice. The neuron in PMA began firing if the instruction was to move the arm to the left, and it continued to discharge until the trigger stimulus came on and the movement was initiated. If the instruction was to move to the right, this neuron did not fire (presumably another population of PMA cells became active under this condition). Thus, the activity of this PMA neuron reported the direction of the upcoming movement and continued to do so until the movement was made. Although we do not yet understand the details of the coding taking place in SMA and PMA, the fact that neurons in these areas are selectively active well before movements are initiated is consistent with a role in planning the movement.

Now, let's consider our baseball player standing on the mound. The decision has been made to throw a curve ball, but the batter walks away from the plate to clean his cleats. The pitcher stands motionless on the mound, muscles tensed, waiting for the batter to return. The pitcher is "set"; a select population of neurons in the premotor cortex (the cells that are planning the curve ball movement sequence) are bursting away in anticipation of the throw. Then the batter steps up to the plate, and an internally generated "Go" command is given. This command appears to be implemented with the participation of a major *subcortical* input to area 6, which is the subject of the next section. After that, we'll examine the origin of the "Go" command, the primary motor cortex.

THE BASAL GANGLIA

The major subcortical input to area 6 arises in a nucleus of the dorsal thalamus, called the **ventral lateral nucleus (VL)**. The input to this part of VL, called VLo, arises from the **basal ganglia** buried deep within the telencephalon. The basal ganglia, in turn, are targets of the cerebral cortex, particularly the frontal, prefrontal, and parietal cortex. Thus, we have a loop where information cycles from the cortex through the basal ganglia and thalamus and then back to the cortex, particularly the supplementary motor area (Figure 14.10). One of the functions of this loop appears to be the selection and initiation of willed movements.

Anatomy of the Basal Ganglia

The basal ganglia consist of the **caudate nucleus**, the **putamen**, the **globus pallidus**, and the **subthalamus**. In addition, we can add the **substantia nigra**, a midbrain structure that is reciprocally connected with the

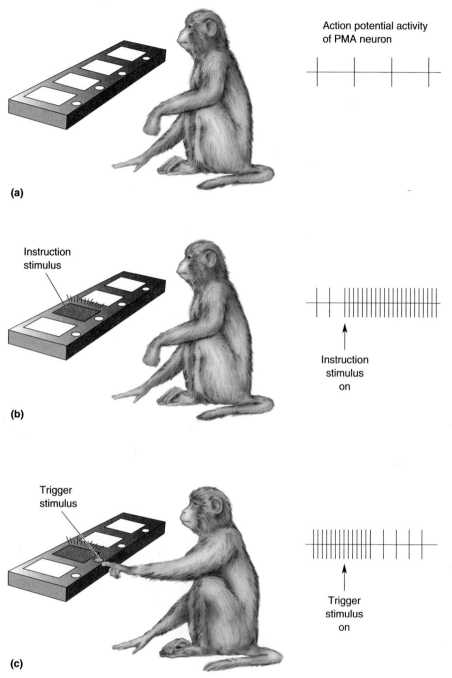

Figure 14.9
Discharge of a cell in the premotor area before a movement. (**a**) *Ready*. A monkey sits before a panel of lights. The task is to wait for an instruction stimulus that will inform him of the movement required to receive a juice reward, then perform the movement when a trigger stimulus goes on. Activity of a neuron in PMA is recorded during the task. (**b**) *Set*. The instruction stimulus occurs at the time indicated by the upward arrow, resulting in the discharge of the neuron in PMA. (**c**) *Go*: Shortly after the movement is initiated, the PMA cell ceases firing. (Source: Based on an experiment by Weinrich and Wise, 1982.)

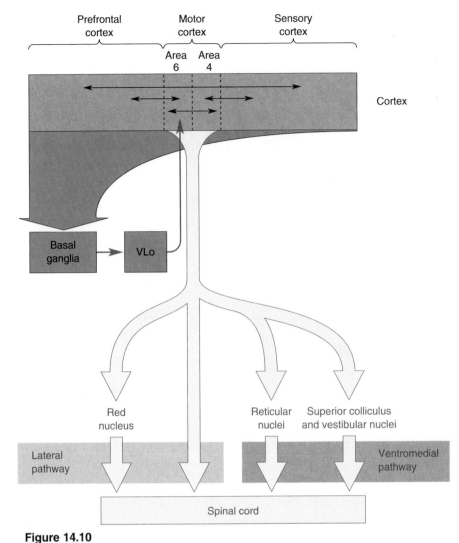

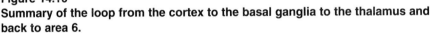

Figure 14.10
Summary of the loop from the cortex to the basal ganglia to the thalamus and back to area 6.

basal ganglia of the forebrain (Figure 14.11). The caudate and putamen together are called the **striatum**, which is the target of the cortical input to the basal ganglia. The globus pallidus is the source of the output to the thalamus. The other structures participate in various side loops that modulate the direct path:

Cortex → Striatum → Globus pallidus → VLo → Cortex (SMA).

Through the microscope, the neurons of the striatum appear randomly scattered, with no apparent order such as that seen in the layers of the cortex. But this bland appearance hides a degree of complexity in the organization of the basal ganglia that we are only now beginning to appreciate. It appears that the basal ganglia participate in a large number of parallel circuits, only a few of which are strictly motor. Other circuits are involved in certain aspects of memory and cognitive function. We will try to give a concise account of the motor function of the basal ganglia, simplifying what is a very complex and poorly understood part of the brain.

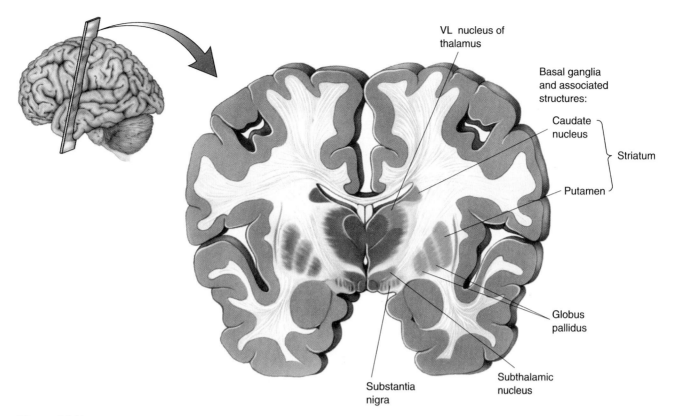

Figure 14.11
The basal ganglia and associated structures.

The Motor Loop

The most direct path in the motor loop through the basal ganglia originates with an excitatory connection from the cortex to cells in the putamen. The putamen cells make inhibitory synapses on neurons in the globus pallidus, which in turn make inhibitory connections with the cells in VLo. The thalamocortical connection (from VLo to SMA) is excitatory and facilitates the discharge of movement-related cells in SMA. This direct motor loop is summarized in Figure 14.12.

The functional consequence of cortical activation of the putamen is excitation of SMA by VL. Let's work out how this happens. At rest, neurons in the globus pallidus are spontaneously active and therefore inhibit VL. Cortical activation (1) excites putamen neurons, which (2) inhibit globus pallidus neurons, which (3) release the cells in VLo from inhibition, allowing them to become active. The activity in VLo boosts the activity of SMA. Thus, this part of the circuit acts as a positive feedback loop that may serve to focus, or funnel, the activation of widespread cortical areas onto the supplementary motor area of cortex. We can speculate that the "Go" signal for an internally generated movement occurs when activation of SMA is boosted beyond some threshold amount by the activity reaching it through this basal ganglia "funnel."

Basal Ganglia Disorders. Support for the view that the direct motor loop through the basal ganglia functions to facilitate initiation of willed movements has come from the study of several human diseases. One of these,

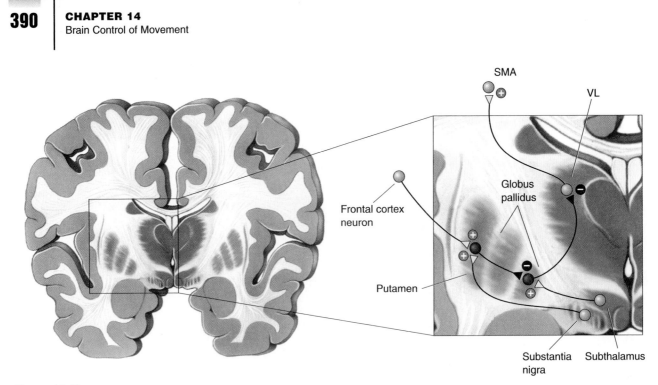

Figure 14.12
Wiring diagram of the basal ganglia motor loop. Synapses marked with a plus (+) are excitatory; those with a minus (−) are inhibitory.

Parkinson's disease, affects about 1% of all people over age 50. Parkinson's disease is characterized by *hypokinesia*, a paucity of movement. The symptoms include slowness of movement (bradykinesia), difficulty in initiating willed movements (akinesia), and increased muscle tone (rigidity). The organic basis of Parkinson's disease is degeneration of substantia nigra inputs to the striatum. These inputs use the neurotransmitter dopamine (DA), which normally facilitates the direct motor loop by activating cells in the putamen. In essence, the depletion of dopamine closes the funnel that feeds activity to SMA via the basal ganglia and VLo.

Parkinson's disease can be treated by administration of the compound dopa (L-dihydroxyphenylalanine, introduced in Chapter 6), which is a precursor to dopamine. Dopa crosses the blood-brain barrier and boosts DA synthesis in the cells that remain alive in the substantia nigra, thus alleviating some of the symptoms. However, dopa treatment does not alter the course of the disease, nor does it alter the rate at which substantia nigra neurons degenerate. (We will return to the topic of dopamine neurons in Chapter 15.)

Hyperkinesia, or excess movement, can also result from lesions that affect the basal ganglia. One example is a condition called **ballism**, which is characterized by violent, flinging movements of the extremities (somewhat like our baseball pitcher unintentionally winding up and throwing the ball while sitting in the dugout). As with Parkinson's disease, the pathology associated with ballism is known; it is caused by damage to the subthalamus (usually resulting from an interruption of its blood supply caused by stroke). The subthalamus, part of another side loop within the basal ganglia, excites neurons in the globus pallidus that project to VLo. Remember that excitation of the globus pallidus inhibits VLo. Thus, loss of excitatory drive to the globus pallidus facilitates VLo, in effect opening the funnel of activity to SMA.

In summary, the basal ganglia may facilitate movement by focusing activity from widespread regions of cortex onto SMA. Importantly, however, they also serve as a *filter* that keeps inappropriate movements from being expressed. We saw in Roland's PET studies that activity in SMA does not necessarily trigger movement, however. The initiation of voluntary movement also requires activation of area 4, the subject of the next section.

INITIATION OF MOVEMENT BY PRIMARY MOTOR CORTEX

SMA is heavily interconnected with M1, cortical area 4. The designation of area 4 on the precentral gyrus as primary motor cortex is somewhat arbitrary because this is not the only cortical area that contributes to the corticospinal tract or to movement. Nonetheless, it has been recognized since the time of Sherrington that this area has the lowest threshold for the elicitation of movement by electrical stimulation. In other words, stimulation intensities that are unable to evoke movement in other cortical areas are still effective in evoking movement from area 4, meaning that it has strong synaptic connections with the motor neurons. Focal electrical stimulation of area 4 evokes the movement of small groups of muscles and, as we discussed above, the somatic musculature is mapped systematically in this area. This ribbon of cortex that stretches the full length of the precentral gyrus is sometimes also called the **motor strip**.

Input-Output Organization of M1

The pathway by which motor cortex stimulation activates lower motor neurons originates in cortical layer V. Here there is a population of pyramidal neurons, some of which can be quite large (soma diameters approaching 0.1 mm). The largest cells were first described as a separate class by Russian anatomist Vladimir Betz in 1874 and are therefore called *Betz cells*. The layer V pyramidal cells in M1 receive their inputs primarily from two sources: other cortical areas and the thalamus. The major cortical inputs originate in the areas adjacent to area 4: area 6 immediately anterior; and areas 3, 1, and 2 immediately posterior (see Figure 14.7). The thalamic input to M1 arises mainly from another part of the ventral lateral nucleus, called VLc, which relays information from the cerebellum. Besides the direct projection to the spinal cord, layer V pyramidal cells also send collaterals to many subcortical sites involved in sensorimotor processing, especially the brain stem.

Coding of Movement in M1

For a time it was thought that the motor cortex consisted of a detailed mapping of the individual muscles such that activity of a single pyramidal cell would lead to activity in a single motor neuron pool. However, the view that has emerged from more recent work is that pyramidal cells can drive a group of muscles so that a limb moves toward a desired goal. Recordings from motor cortex neurons in behaving animals have revealed that a burst of activity occurs immediately before and during a voluntary movement, and that this activity appears to encode two aspects of the movement: force and direction.

Considering that cortical microstimulation studies had suggested the existence of a fine-grained movement map in M1, it came as something of a surprise when it was discovered that the movement direction tuning of individual M1 neurons is rather broad. This breadth of tuning is shown clearly in a type of experiment devised by Apostolos Georgopoulos and his colleagues, then working at Johns Hopkins University (Box 14.2). Monkeys were trained to move a joystick toward a small light whose position varied randomly around a circle. The M1 cells fired most vigorously during movement in one direction (180° in the example in Figure 14.13a), but also discharged during movements that varied ± 45° from the preferred direction.

Box 14.2   PATH OF DISCOVERY

Reaching a Goal
by Apostolos Georgopoulos

It was meant to be reaching and the brain. After I had finished medical school in Athens, Greece, and defended my doctoral dissertation on lipid metabolism, I thought that the most exciting research for the future would be to understand the brain mechanisms of behavior. I was indeed fortunate to be accepted by Dr. Vernon B. Mountcastle at the Johns Hopkins University for postdoctoral training in behavioral neurophysiology. It was good timing, for the technique of recording the activity of single cells in the brain of behaving monkeys had just been perfected by Ed Evarts at the National Institutes of Health and was already in use at the Mountcastle laboratory. When my family and I arrived in Baltimore on a Saturday afternoon in late October, we enjoyed the warm hospitality of the Mountcastles at their home in the beautiful Maryland countryside. It was that Sunday that Dr. Mountcastle said excitedly during a short walk in their garden, "You know, Pete, the cells in the posterior parietal cortex fire with reaching!" Incredible! One was then thinking of the posterior parietal cortex as a further processing stage of somatosensory information, and what a surprise to find there cells firing with active reaching movements! Although I did not collaborate with Dr. Mountcastle after that fellowship, I am still working on reaching and its brain correlates. This is almost "research imprinting!"

Reaching is a fascinating motor act. It is both fairly complicated and fairly simple. It is complicated because it is carried out by the coordinated action of several muscles and motion about the shoulder and elbow joints. It is simple because it is stereotyped and graceful and accurate, otherwise you wouldn't catch that ball so easily! The reaching movement is a vector in space with direction and amplitude. The first question, then, is: How does the brain specify these two parameters of reaching? One way to solve that problem would be to dedicate cells that are very specific for a particular direction. Then, activation of these cells would cause movement in that one direction. This idea was believable—even plausible—given the fact that cells in the visual cortex are rather

Apostolos Georgopoulos

specific (i.e., sharply tuned) with respect to the orientation of a luminous bar. Could a similar mechanism operate in the motor cortex with respect to the direction of movement?

We set out to test this hypothesis by training monkeys to reach to visual targets in various directions as we recorded activity in their M1 cells. The results were clear: individual cells in the motor cortex were not specific for a single direction of movement, but their activity was greatest during movement in a certain direction (the cell's "preferred direction") and decreased gradually with movements made increasingly farther away from it. The broad directional tuning indicates that a given cell participates in movements of various directions, and that, conversely, a movement in a particular direction will involve the activation of a whole population of cells. A vectorial neural code for the direction of reaching by the neuronal ensemble could be as follows: (1) A particular vector represents the contribution of a directionally tuned cell and points in the cell's preferred direction; (2) cell vectors are weighted by the change in cell activity during a particular movement; and (3) the sum of these vectors (i.e., the population vector) provides the unique outcome of the ensemble coding operation. We found that, indeed, the population vector points in the direction of the movement. Notice that we started with purely temporal spike trains (i.e., time series of action potentials) and ended up with a spatial measure (the population vector) that is a good predictor of what the movement direction is going to be.

Crucial experiments followed in which we used the population vector to decipher what the brain is working on while preparing the movement. The results showed that the population vector analysis was very useful in "reading out" local operations of neuronal circuits. This possibility was a long shot 12 years ago when we first applied this analysis, but one that came true against all odds!

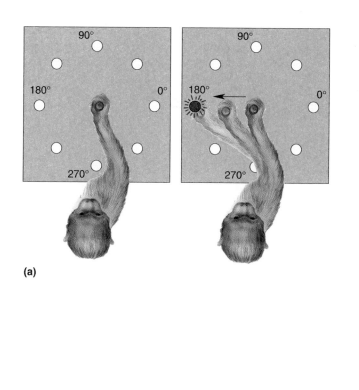

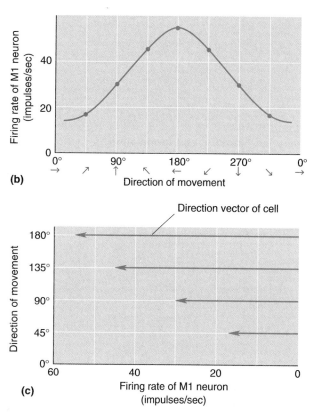

Figure 14.13
Responses of an M1 neuron during arm movements in different directions. (a) A monkey is instructed to move a handle towards a small light as responses of an M1 neuron are monitored. By having the monkey move in directions around the clock, the relationship between the cell's discharge rate and movement direction can be determined. **(b)** A tuning curve for an M1 neuron. This cell fires most during movements to the left. **(c)**. Because the cell in (b) responds best to leftward movement, it is represented by a vector pointing in that direction (the direction vector). The length of the vector is proportional to the firing rate of the cell. Notice that as the movement direction changes, the length of the direction vector changes. (Source: Based on an experiment by Georgopoulos et al., 1982.)

The coarseness in the directional tuning of the corticospinal neurons was certainly at odds with the high accuracy of the monkey's movements, suggesting that the direction of movement could not be encoded by the activity of individual cells that command movement in a single direction. Georgopoulos hypothesized that movement direction was encoded instead by the collective activity of a population of neurons.

To test the feasibility of this idea of population coding for movement direction, Georgopoulos and his colleagues recorded from over 200 different neurons in M1; for each cell, they constructed a directional tuning curve such as that shown in Figure 14.13b. From these data, the researchers knew how vigorously each of the cells in the population responded during movement in each direction. The activity of each cell was represented as a *direction vector* pointing in the direction that was best for that cell; the length of the vector represented how active that cell had been during a particular movement (Figure 14.14). The vectors representing each cell's activity could be plotted together for each direction of movement, then averaged to yield what the researchers called a *population vector*. They found excellent agreement between this average vector, representing the activity of the entire population of M1 cells, and the actual direction of movement (Figure 14.15).

This work suggests three important conclusions about how M1 commands voluntary movement: (1) that much of the motor cortex is active for every movement, (2) that the activity of each cell represents a single "vote"

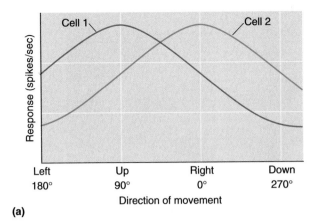

(a)

Figure 14.14

Direction vectors and population vectors. (a) The graph shows the tuning curves for two imaginary cells in the motor cortex (see Figure 14.13). Both cells fire during movement in a wide range of directions, but cell 1 fires best when movement is upward, while cell 2 responds best when the movement is from left to right. (b) The response of each cell can be represented as a direction vector. The direction vector always points in the preferred direction for the neuron, but the length varies depending on the number of action potentials fired by the cell during movement over a range of directions. For any given direction of movement, the direction vectors of the individual cells can be combined to yield a population vector, reflecting the strength of the response of both cells during this movement.

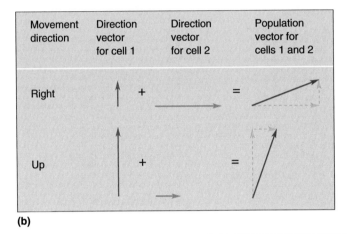

(b)

Figure 14.15

Prediction of the direction of movement by population vectors. Each cluster of lines reflects the direction vectors of many cells in M1. Line length reflects the discharge rate of each cell during a movement in eight different directions. For each direction, the individual direction vectors can be summed to yield the population vector (dashed arrow), which predicts the direction of movement. (Source: Georgopoulos et al., 1983.)

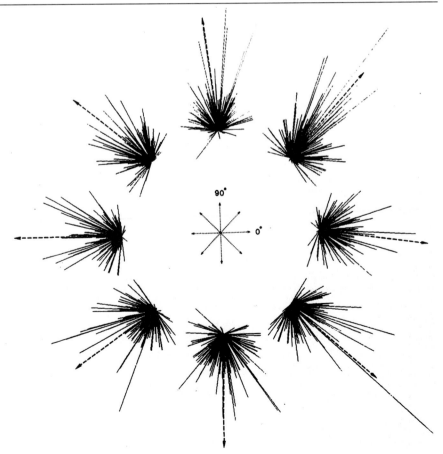

for a particular direction of movement, and (3) that the direction of movement is determined by a tally (and averaging) of the votes registered by each cell in the population. Although this population-coding scheme remains hypothetical in M1, experiments on the superior colliculus have shown conclusively that a population code is used by this structure to command precise eye movements (Box 14.3).

The Malleable Motor Map. An interesting prediction that derives from this scheme for motor control is that the larger the population representing a type of movement, the finer the possible control. From the motor map shown in Figure 14.8, we would predict that finer control should be possible for the hands and the muscles of facial expression, and indeed this is normally the case. Of course, fine movements of other muscles can be learned with experience; consider the finger, wrist, elbow, and shoulder movements of an accomplished cellist. Does this mean that cortical cells in M1 can switch allegiance from participation in one type of movement to another as skills are learned? The answer appears to be yes. John Donoghue, Jerome Sanes, and their students at Brown University have recently presented evidence indicating that such plasticity of the adult motor cortex is possible. For example, in one series of experiments using cortical microstimulation in rats, they found that the region of M1 that normally elicits

---

Box 14.3 **OF SPECIAL INTEREST**

Saccades and the Superior Colliculus

The word "saccade" is derived from the French word *saccader*, meaning "to jerk." Saccades are the ballistic movements of the eyes that we use to explore our visual surroundings. The eyes jump from one fixation point in space to another by saccadic movements. Saccades can be intentional, as when asked to "look to the left." Or they can be reflexive, as when you look toward something unexpected that pops into view out of the corner of your eye. In either case, saccades bring the image of a new visual target onto the foveae.

One of the structures that is important for the control of saccadic eye movements is the superior colliculus. Recall from Chapter 10 that the superior colliculus contains a retinotopic map, such that illumination of a point on the retina activates a circumscribed patch of cells in the superior colliculus. The superior colliculus also contains a *motor map* that is in register with the sensory map. Microstimulation of a given point in the colliculus will cause a saccade that brings the image of the corresponding point of visual space onto the central retina. (Recall that the superior colliculus contributes to the ventromedial spinal pathways that can also move the head.)

It had been recognized for some time that a point of light on the retina activates a large fraction of the cells in the superior colliculus, owing to their large receptive fields. The large population of active neurons in the motor map appeared to be at odds with the remarkable precision of the resulting visually driven saccades. This led James McIlwain at Brown University and David Sparks at the University of Pennsylvania, in the 1980s, to propose a population code similar to that suggested for the control of movement in the motor cortex. According to their idea, each cell in the patch of colliculus that is activated by a point of light registers a vote for a different saccade direction, depending on the precise location of the cell in the motor map. The actual saccade was proposed to be some average of these individual eye movement vectors.

David Sparks and his colleagues devised an ingenious test of this idea. They trained monkeys to look to a visual target for a juice reward (see Box 14.1). Then they inactivated neurons in a small region of the motor map with a microinjection of a local anesthetic. This treatment caused the saccades to become inaccurate in precisely the fashion that would be predicted by the loss of "registered voters" for the directions represented by the inactivated cells. This experiment indicates that the population-coding hypothesis is correct, at least for the control of eye movements by the superior colliculus.

whisker movements would evoke forelimb movements instead when the motor nerve that supplies the muscles of the snout was cut (Figure 14.16). These neuroscientists speculate that similar types of cortical reorganization might provide a basis for learning fine motor skills.

From the preceding discussion, we can imagine that when the time has come for our pitcher to wind up, the motor cortex generates a torrent of activity in the pyramidal tract. What could appear to be a discordant voice to the neurophysiologist recording from a single M1 neuron is part of a clear chorus of activity to the spinal motor neurons that generate the movement.

THE CEREBELLUM

It is not enough to simply command the muscles to contract. Throwing a curve ball requires a detailed *sequence* of muscle contractions, each one timed with great precision. Looking after this critical motor control function is the **cerebellum**, introduced in Chapter 7. Involvement of the cerebellum in this aspect of motor control is plainly revealed by cerebellar lesions; movements become uncoordinated and inaccurate. Take this simple test. Lay your arms in your lap for a moment, then touch your nose with one finger. Try it again with your eyes closed. No problem, right? Patients with cerebellar damage are incapable of performing this simple task. Instead of smoothly and simultaneously moving the shoulder, elbow, and wrist to bring the finger to rest on the nose, they will move each joint sequentially—first the shoulder, then the elbow, and finally the wrist. This is called *dysynergia*, decomposition of synergistic multijoint movement. Another characteristic deficit shown by these patients is that their finger movement will be *dysmetric*; they will either come up short of the nose or shoot past it, poking themselves in the face. You may recognize these symptoms as similar to those that accompany ethanol intoxication. Indeed, the clumsiness that accompanies alcohol abuse is a direct consequence of the depression of cerebellar circuits.

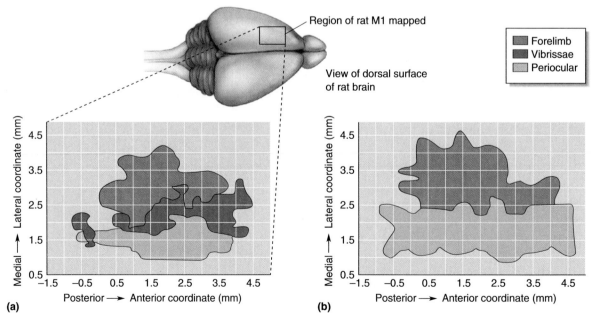

Figure 14.16

Malleable motor maps. (a) This map represents motor cortex from a normal rat. (b) This map represents a rat that had the motor nerve cut that supplies the vibrissae (whiskers). Notice that the regions of cortex that used to evoke movement of the vibrissae now cause muscle movement in the forelimb or around the eyes (periocular). (Source: Adapted from Sanes and Donoghue, 1990.)

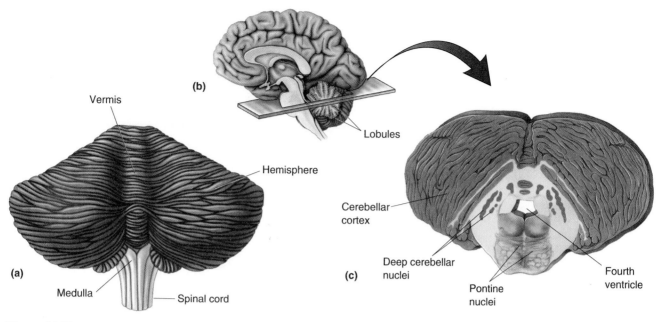

Figure 14.17
The cerebellum. (**a**) Dorsal view of the human cerebellum showing the vermis and hemispheres. (**b**) Midsagittal view of the brain showing the lobules of the cerebellum. (**c**) Cross section of the cerebellum showing the cortex and deep nuclei.

Anatomy of the Cerebellum

Cerebellar anatomy is shown in Figure 14.17. The cerebellum sits on stout stalks called peduncles that rise from the pons; the whole structure resembles a piece of broccoli or cauliflower. The visible part of the cerebellum is actually a thin sheet of cortex, which is repeatedly folded. The dorsal surface is characterized by a series of shallow ridges running transversely, called folia (singular: folium). In addition, there are deeper transverse fissures, revealed by making a sagittal slice through the cerebellum; these divide the cerebellum into 10 lobules. Together, folia and lobules serve to greatly increase the surface area of the cerebellar cortex, as the gyri of the cerebrum do for the cerebral cortex. Because of the high density of neurons in the cortex, the cerebellum, which constitutes only about one-tenth of the total volume of the brain, contains more than 50% of the total number of neurons in the CNS. There are also neurons embedded deep in the white matter of the cerebellum, forming the *deep cerebellar nuclei*, which relay most of the cortical output to various brain stem structures.

Unlike the cerebrum, the cerebellum is not obviously split down the middle. At the midline, the folia appear to run uninterrupted from one side to the other. The only distinguishing feature of the midline is a bump that runs like a backbone down the length of the cerebellum. This midline region is called the **vermis** (from the Latin word for "worm") and separates the two lateral **cerebellar hemispheres** from each other. The vermis and the hemispheres represent important functional divisions. The vermis sends output to the brain stem structures that contribute to the ventromedial descending spinal pathway, which, as we already discussed, controls the axial musculature. The hemispheres are related to the structures that contribute to the lateral pathway, particularly the cerebral cortex. For the purpose of illustration, we'll focus on the lateral cerebellum, which is particularly important for limb movements.

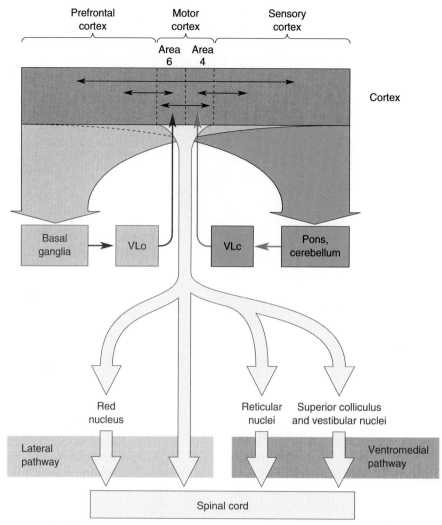

Figure 14.18
Summary of the motor loop through the cerebellum.

The Motor Loop Through the Lateral Cerebellum

The simplest circuit involving the lateral cerebellum constitutes yet another loop, shown schematically in Figure 14.18. Axons arising from layer V pyramidal cells in the sensorimotor cortex—frontal areas 4 and 6, somatosensory areas on the postcentral gyrus, and the posterior parietal areas—form a massive projection to clusters of cells in the pons, the **pontine nuclei**, which in turn feed the cerebellum. To appreciate the size of this pathway, consider that the corticopontocerebellar projection contains about 20 million axons; that's *20 times* more than in the pyramidal tract! The lateral cerebellum then projects back to the motor cortex via a relay in the ventral lateral nucleus of the thalamus (VLc).

From the effects of lesions in this pathway, we can deduce that it is critical for the proper execution of planned, voluntary, multijoint movements. Indeed, once the signal for movement intent has been received by the cerebellum, the activity of this structure appears to instruct the primary motor cortex with respect to movement direction, timing, and force. For ballistic movements, these instructions are based entirely on predictions about their outcome (because such movements are too fast for feedback to be of much immediate use). Such predictions are based on past experience; that is, they are learned. Therefore, the cerebellum is another important site for motor learning—it is a place where *what is intended* is compared with *what has happened*. When this comparison fails to meet expectations, compensatory modifications are made in certain cerebellar circuits.

Programming the Cerebellum. We will return to the details of cerebellar circuitry and how it is modified by experience in Chapter 20. But, for now, think about the process of learning a new skill (e.g., downhill skiing, tennis, juggling, throwing a curve ball). Early on, you must concentrate on new movements, which you will perform in a disjointed and uncoordinated fashion. Practice makes perfect, however. As you master the skill, the movements will become smooth, and eventually you will be able to perform them almost unconsciously. This process represents the creation of a new motor program that generates the appropriate movement sequences on demand without the need for conscious control.

We are reminded of a certain computer microchip manufacturer that advertises its product as the "computer inside." The cerebellum acts as the "brain inside" that unconsciously sees to it that the programs for skilled movement are executed properly and are adjusted whenever their execution fails to meet expectations.

CONCLUDING REMARKS

Let's return to the example of the baseball player one final time to put the different pieces of the motor control puzzle together. Imagine the pitcher walking to the mound. The spinal circuits of the crossed extensor reflex are engaged and coordinated by descending commands on the ventromedial pathway. Extensors contract, flexors relax; flexors contract, extensors relax.

Once on the mound, the pitcher is joined by the umpire. Into his outstretched hand the umpire drops a new baseball. The added weight stretches the flexors of the arm. Group Ia afferents become more active and cause monosynaptic excitation of the motor neurons innervating the flexors. The muscles contract to hold the ball up against gravity.

He is now ready to pitch. His neocortex is fully engaged and active as he looks at the catcher for the hand signal that instructs the pitch to be thrown. At the same time, the ventromedial pathways are working to maintain his standing posture. Although his body is still, the neurons of the ventral horns are firing madly under the influence of the ventromedial pathway, keeping the extensors of the lower leg flexed.

The catcher flashes the sign for the curve ball. The sensory information is communicated to the parietal and prefrontal cortex. These regions of cortex, and area 6, begin planning the movement strategy.

The batter steps up to the plate and is ready. Activity cycling through the basal ganglia increases, triggering the initiation of the pitch. In response to this input, SMA activity increases, followed immediately by activation of M1. Now instructions are sweeping down the axons of the lateral pathway. The cerebellum, activated by the corticopontocerebellar inputs, uses these instructions to coordinate the timing of the descending activity so the proper sequence of muscle contractions can occur. Cortical input to the reticular formation leads to the release of the antigravity muscles from reflex control. Finally, lateral pathway signals engage the motor neurons and interneurons of the spinal cord, which cause the muscles to contract.

The pitcher winds up and throws. The batter swings. The ball sails over the left-field fence. The crowd jeers; the manager curses; the owner frowns. Even as the cerebellum goes to work making adjustments for the next pitch, the player's body reacts. His face flushes; he sweats; he's angry and afraid. But these latter reactions are not the stuff of the somatic motor system. These are topics of Part III, Brain and Behavior.

KEY TERMS

Descending Spinal Tracts
lateral pathway
ventromedial pathway
corticospinal tract
motor cortex
pyramidal tract
rubrospinal tract
red nucleus
vestibulospinal tract
vestibular nuclei
vestibular apparatus
tectospinal tract
reticular formation
pontine reticulospinal tract
medullary reticulospinal tract

**Planning of Movement by the
   Cerebral Cortex**
M1
premotor area (PMA)
supplementary motor area (SMA)

The Basal Ganglia
ventral lateral nucleus (VL)

basal ganglia
caudate nucleus
putamen
globus pallidus
subthalamus
substantia nigra
striatum
Parkinson's disease
ballism

**Initiation of Movement by the
   Primary Motor Cortex**
motor strip

The Cerebellum
cerebellum
vermis
cerebellar hemispheres
pontine nuclei

✔ REVIEW QUESTIONS

1. List the components of the lateral and ventromedial descending spinal pathways. Which type of movement does each path control?
2. You are a neurologist presented with a patient who has the following symptom: an inability to independently wiggle the toes on the left foot, but with all other movements (walking, independent finger movement) apparently intact. You suspect a lesion in the spinal cord. Where?
3. PET scans can be used to measure blood flow in the cerebral cortex. What parts of the cortex show increased blood flow when a subject is asked to think about moving her right finger?
4. Why is dopa used to treat Parkinson's disease? How does it act to alleviate the symptoms?
5. Individual Betz cells fire during a fairly broad range of movement directions. How might they work together to command a precise movement?
6. Sketch the motor loop through the cerebellum. What movement disorders result from damage to the cerebellum?

PART III:
BRAIN & BEHAVIOR

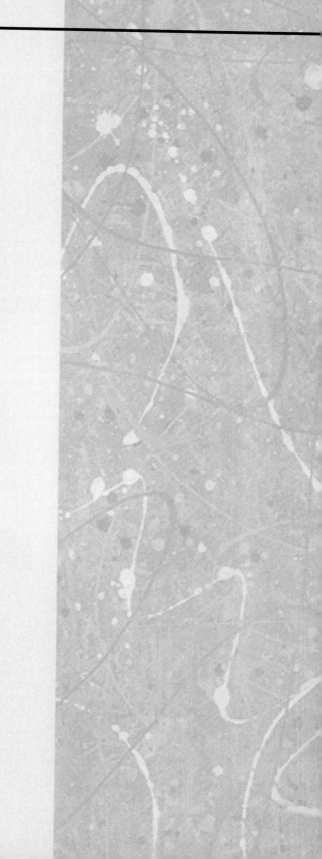

Chemical Control of Brain and Behavior

It should be obvious by now that knowing the organization of synaptic connections is essential to understanding how the brain works. It's not from a love of Greek and Latin that we belabor the neuroanatomy! Most of the connections we have described are precise and specific. For example, in order for you to read these words, there must be a very fine-grained neural mapping of the light falling on your retina—how else could you see the dot on this question mark? The information must be carried centrally and precisely dispersed to many parts of the brain for processing, coordinated with control of the motor neurons that closely regulate the six muscles of each eye as it scans the page.

In addition to anatomical precision, point-to-point communication in the sensory and motor systems requires mechanisms that restrict synaptic communication to the cleft between the axon terminal and its target. It just wouldn't do for glutamate released in somatosensory cortex to activate neurons in motor cortex! Furthermore, transmission must be brief enough to allow rapid responses to new sensory inputs. Thus, at these synapses, only minute quantities of neurotransmitter are released with each impulse, and these molecules are then quickly destroyed enzymatically or taken up by neighboring cells. The postsynaptic actions at transmitter-gated ion channels last only as long as the transmitter is in the cleft, a few milliseconds at most. Many axon terminals also possess presynaptic "autoreceptors" that detect the transmitter concentrations in the cleft and inhibit release if they get too high. These mechanisms ensure that this type of synaptic transmission is tightly constrained, both in space and in time.

The elaborate mechanisms that constrain point-to-point synaptic transmission bring to mind a telecommunications analogy. Telephone systems make possible very specific connections between one place and another; your mother in Tacoma can talk just to you in Providence, reminding you that her birthday was last week. The telephone lines can act like precise synaptic connections. The influence of one neuron (your mother) is targeted to a small number of other neurons (in this case, only you). The embarrassing message is limited to your ears only. For a real neuron in one of the sensory or motor systems discussed so far, its influence usually extends to the few dozen to hundred cells it synapses on—a conference call, to be sure, but still relatively specific.

Now imagine your mother being interviewed on a television talk show, which is broadcast via a cable network. In this case, the widespread cable connections may allow her to tell millions of people that you forgot her birthday, and the loudspeaker in each television set will announce the message to anyone within earshot. Likewise, certain neurons communicate with hundreds of thousands of other cells. These widespread systems also tend to act relatively slowly, over seconds to minutes. Because of their broad, protracted actions, such systems in the brain can orchestrate entire behaviors, ranging from falling asleep to falling in love. Indeed, many of the behavioral dysfunctions collectively known as mental disorders are believed to result specifically from imbalances of certain of these chemicals.

In this chapter, we take a look at three components of the nervous system that operate in expanded space and time (Figure 15.1). One component is the *secretory hypothalamus.* By secreting chemicals directly into the bloodstream, the secretory hypothalamus can influence functions throughout both brain and body. A second component is neurally controlled by the hypothalamus and exerts its effects outside the central nervous system (CNS), and was already introduced in Chapter 7 as the *autonomic nervous system (ANS).* Through extensive interconnections within the body, the ANS simultaneously controls the responses of many internal organs, blood vessels, and glands. The third component exists entirely within the CNS and consists of several related cell groups that differ with respect to the neurotransmitter they use. All of these cell groups extend their spatial reach with highly divergent axonal projections and prolong their actions by using metabotropic postsynaptic receptors. Members of this component of the nervous system are called the *diffuse modulatory systems of the brain.* The diffuse systems are believed to regulate, among other things, the level of arousal and mood.

THE SECRETORY HYPOTHALAMUS

Recall from Chapter 7 that the hypothalamus sits below the thalamus, along the walls of the third ventricle. It is connected by a stalk to the pituitary gland, which dangles below the base of the brain, just above the roof of your mouth (Figure 15.2). Although this tiny cluster of nuclei makes up less than 1% of the brain's mass, the influence of the hypothalamus on body physiology is enormous. Let's take a brief tour of the hypothalamus and then focus on some of the ways in which it exerts its powerful influence.

Overview of the Hypothalamus

The hypothalamus and dorsal thalamus are adjacent to one another, but their functions are very different. As we saw in the previous seven chapters, the dorsal thalamus lies in the path of all the point-to-point pathways whose destination is the neocortex. Accordingly, destruction of a small part of the dorsal thalamus can produce a discrete sensory or motor deficit: a little blind spot, or a lack of feeling on a portion of skin. In contrast, the *hypothalamus integrates somatic and visceral motor responses in accordance with the needs of the brain.* A tiny lesion in the hypothalamus can produce dramatic and often fatal disruptions of widely dispersed bodily functions.

Homeostasis. Life has certain requirements. For mammals, these include a narrow range of body temperatures and blood compositions. The hypothalamus regulates these levels in response to a changing external environment. This regulatory process is called **homeostasis**, the maintenance of a constant internal environment.

Consider temperature regulation. Biochemical reactions in many cells of the body are fine-tuned to occur at about 37°C. A variation of more than a few degrees in either direction can be catastrophic. Temperature-sensitive cells in the hypothalamus detect variations in brain temperature and orchestrate the appropriate responses. For example, when you stroll naked through the snow, the hypothalamus issues commands that cause you to shiver (generating heat in the muscles), develop goose pimples (a futile attempt to fluff up your nonexistent fur—a reflexive remnant from our hairier ancestors), and turn blue (shunting blood *away from* the cold surface

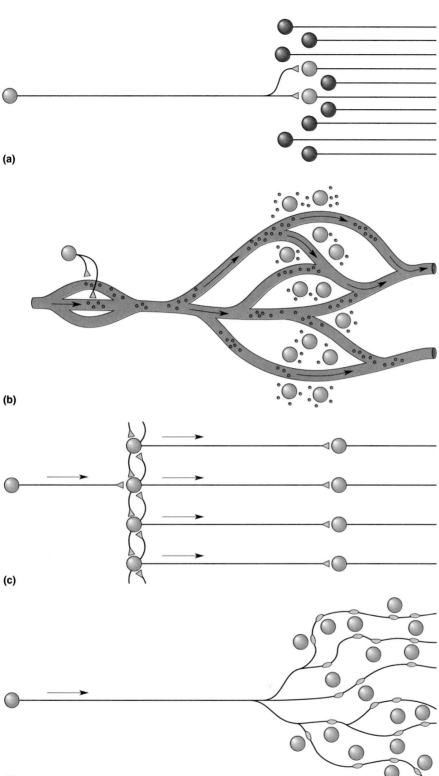

(a)

(b)

(c)

(d)

Figure 15.1
Patterns of communication in the nervous system. (a) Most of the systems
we have discussed in this book may be described as point-to-point. The proper
function of these systems requires restricted synaptic activation of target cells
and signals of brief duration. In contrast are three other components of the nervous system that act over great distances and for long periods of time. **(b)**
Neurons of the secretory hypothalamus affect their many targets by releasing
hormones directly into the bloodstream. **(c)** Networks of interconnected neurons
of the ANS can work together to activate tissues all over the body. **(d)** Diffuse
modulatory systems extend their reach with widely divergent axonal projections.

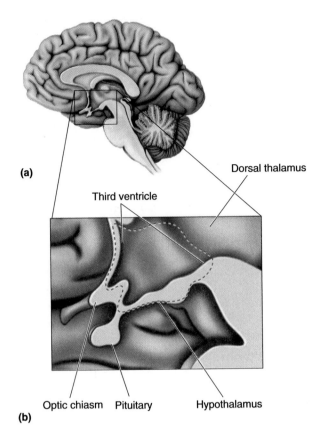

(a)

Dorsal thalamus

Third ventricle

Optic chiasm   Pituitary   Hypothalamus

(b)

Figure 15.2
Location of the hypothalamus and pituitary. (a) A midsagittal section. **(b)** Notice
that the hypothalamus forms the wall of the third ventricle and sits below the dorsal
thalamus. The dashed line indicates the approximate border of the hypothalamus.

tissues to keep the sensitive core of the body warmer). In contrast, when you
go for a jog in the tropics, the hypothalamus activates heat-loss mechanisms
that make you turn red (shunting blood *to* the surface tissues where heat can
be radiated away) and sweat (cooling the skin by evaporation).

Other examples of homeostasis are the tight regulation of blood volume,
pressure, salinity, acidity, and blood oxygen and glucose concentrations. The
means by which the hypothalamus achieves these different types of regulation are remarkably diverse.

Structure and Connections of the Hypothalamus. Each side of the hypothalamus can be divided into three zones: lateral, medial, and periventricular (Figure 15.3). The lateral and medial zones have extensive connections
with the brain stem and the telencephalon and regulate certain types of
behavior, as we will see in Chapter 16. Here we are concerned only with the
third zone, which actually receives much of its input from the other two.

The **periventricular zone** is so named because, with the exception of a
thin finger of neurons that are displaced laterally by the optic tract (called
the supraoptic nucleus), the cells of this region lie right next to the wall of
the third ventricle. Within this zone exists a complex mix of neurons with
different functions. One group of cells constitutes the *suprachiasmatic nucleus (SCN)*, which lies just above the optic chiasm. These cells receive direct
retinal innervation and function to synchronize circadian rhythms with the
daily light-dark cycle (see Chapter 17). Other cells in the periventricular
zone control the ANS, and regulate the outflow of the sympathetic and

parasympathetic innervation of the visceral organs. The cells in a third group, called *neurosecretory neurons*, extend axons down toward the stalk of the pituitary gland. These are the cells that now command our attention.

Pathways to the Pituitary

We have said that the pituitary dangles down below the base of the brain, which is true if the brain is lifted out of the head. In a living brain, the pituitary is gently held in a cradle of bone at the base of the skull. It deserves this special protection because it is the "mouthpiece" from which much of the hypothalamus "speaks" to the body. The pituitary has two lobes, posterior and anterior. The hypothalamus controls the two lobes in different ways.

Hypothalamic Control of the Posterior Pituitary. The largest of the neurosecretory cells, **magnocellular neurosecretory neurons**, extend axons around the optic chiasm, down the stalk of the pituitary, and into the posterior lobe (Figure 15.4). In the late 1930s, Ernst and Berta Scharrer, working at the University of Frankfurt in Germany, proposed that these neurons release chemical substances directly into the capillaries of the posterior lobe. At the time, this was quite a radical idea. It was well known that chemical messengers called hormones were released by glands into the bloodstream, but no one anticipated that a neuron could act like a gland or that a neurotransmitter could act like a hormone. The Scharrers were correct, however. The substances released into the blood by neurons are now called **neurohormones**.

The magnocellular neurosecretory cells release two neurohormones into the bloodstream, oxytocin and vasopressin. Both of these chemicals are peptides, each consisting of a chain of nine amino acids. **Oxytocin**, released during the final stages of childbirth, causes the uterus to contract and facilitates delivery of the newborn. It also stimulates the ejection of milk from the mammary glands. All lactating mothers know about the complex "let-down" reflex that involves the oxytocin neurons of the hypothalamus. Oxytocin release may be stimulated by the somatic sensations generated by a suckling baby. But the sight or cry of a baby (even someone else's) can also trigger the release of milk beyond the mother's conscious control. In each

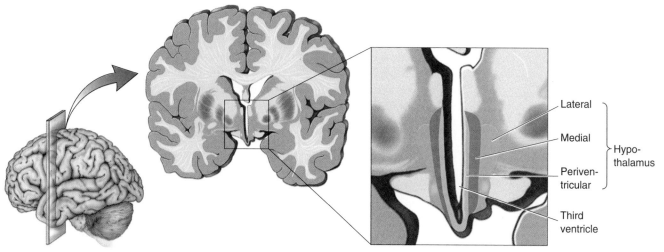

Figure 15.3
Zones of the hypothalamus. The hypothalamus is usually divided into three zones: lateral, medial, and periventricular. The periventricular zone receives inputs from the other zones, the brain stem, and the telencephalon. Neurosecretory cells in the periventricular zone secrete hormones into the bloodstream. Other periventricular cells control the autonomic nervous system.

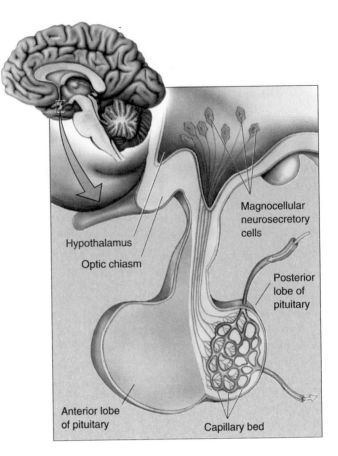

Figure 15.4
Magnocellular neurosecretory cells of the hypothalamus. Shown here is a midsagittal view of the hypothalamus and pituitary. Magnocellular neurosecretory cells secrete oxytocin and vasopressin directly into capillaries in the posterior lobe of the pituitary.

case, information about a sensory stimulus—somatic, visual, or auditory—reaches the cerebral cortex via the usual route, the thalamus, and the cortex ultimately stimulates the hypothalamus to trigger oxytocin release. The cortex can also suppress hypothalamic functions, as when anxiety inhibits the let-down of milk.

Vasopressin, also called *antidiuretic hormone (ADH)*, regulates proper blood volume and salt concentration. When one is deprived of water, blood volume decreases and blood salt concentration increases. These changes are detected by pressure receptors in the cardiovascular system and salt concentration-sensitive cells in the hypothalamus, respectively. Vasopressin-containing neurons receive information about these changes and respond by releasing vasopressin, which acts directly on the kidneys and leads to water retention and reduced urine production.

Under conditions of lowered blood volume and pressure, communication between the brain and the kidneys actually occurs in both directions (Figure 15.5). The kidneys secrete an enzyme into the blood called *renin*. Elevated renin sets off a sequence of biochemical reactions in the blood; *angiotensinogen*, a large protein released from the liver, is converted by renin to *angiotensin I*, which breaks down further to form another small peptide hormone, *angiotensin II*. Angiotensin II has direct effects on the kidney and blood vessels, which help increase blood pressure. But angiotensin II in the blood is also detected by the *subfornical organ*, a part of the brain that lacks a blood-brain barrier. Cells in the subfornical organ project axons into the hypothalamus where they activate, among other things, the vasopressin-containing neurosecretory cells. In addition, however, the subfornical organ activates cells in the lateral zone of the hypothalamus, somehow producing an overwhelming thirst. It may be difficult to accept, but it's true—to a limited extent, our brain is controlled by our kidneys! This example also illustrates that the means by which the hypothalamus maintains homeostasis goes beyond control of the visceral organs and can include the activation of entire behaviors.

Hypothalamic Control of the Anterior Pituitary. Unlike the posterior lobe, which really is a part of the brain, the anterior lobe of the pituitary is an actual gland. The cells of the anterior lobe synthesize and secrete a wide range of hormones that regulate secretions from other glands throughout the body (together constituting the endocrine system). The pituitary hormones act on the gonads, the thyroid glands, the adrenal glands, and the mammary glands (Table 15.1). For this reason, the anterior pituitary was traditionally described as the body's "master gland." But what controls the

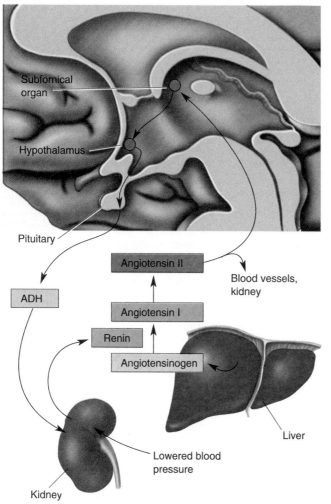

Figure 15.5
Communication from kidney to brain. Under conditions of lowered blood volume or pressure, the kidney secretes renin into the bloodstream. Renin in the blood promotes the synthesis of the peptide angiotensin II, which excites the neurons in the subfornical organ. The subfornical neurons stimulate the hypothalamus, causing an increase in vasopressin (ADH) production and a feeling of thirst.

Table 15.1
Hormones of the Anterior Pituitary

Hormone	Target	Action
Lutenizing hormone (LH)	Gonads	Ovarian, spermatic maturation
Follicle-stimulating hormone (FSH)	Gonads	Ovulation, spermatogenesis
Thyrotropin (TSH)	Thyroid	Thyroxin secretion (increases metabolic rate)
Adrenocorticotropin (ACTH)	Adrenal cortex	Cortisol secretion (mobilizes energy stores; inhibits immune system; other actions)
Growth hormone (GH)	All cells	Stimulation of protein synthesis
Prolactin	Mammary glands	Growth and milk secretion

master gland? The secretory hypothalamus. *The hypothalamus itself is the true master gland of the endocrine system.*

The anterior lobe is under the control of cells in the periventricular zone called **parvocellular neurosecretory neurons**. These hypothalamic neurons do not extend axons all the way into the anterior lobe; instead they communicate with their targets via the bloodstream (Figure 15.6). These neurons secrete what are called **hypophysiotropic hormones** into a uniquely specialized capillary bed at the floor of the third ventricle. These tiny blood vessels run down the stalk of the pituitary and branch again in the anterior lobe. This network of blood vessels is called the **hypothalamo-pituitary portal circulation.** Hypophysiotropic hormones secreted by hypothalamic neurons into the portal circulation travel downstream until they bind to specific receptors on the surface of pituitary cells. Activation of these receptors causes the pituitary cells to either secrete or stop secreting hormones into the general circulation.

Regulation of the adrenal glands illustrates how this system works. Located just above the kidneys, the adrenal glands consist of two parts, a shell called the **adrenal cortex** and a center called the **adrenal medulla**. The adrenal cortex produces the steroid hormone **cortisol**, which acts throughout the body to mobilize energy reserves and suppress the immune system, and it generally prepares us to carry on in the face of life's various stresses. In fact, a good stimulus for cortisol release is stress, ranging from physiological stress, such as a loss of blood; to positive emotional stimulation, such as falling in love; to psychological stress, such as anxiety over an upcoming exam. Parvocellular neurosecretory neurons that control the adrenal cortex determine whether a stimulus is stressful or not (as defined by the release of cortisol). These neurons lie in the periventricular hypothalamus and release a peptide called *corticotropin-releasing hormone* (CRH) into the blood of the portal circulation. CRH travels the short distance to the anterior pituitary, where, within about 15 seconds, it stimulates the release of *corticotropin*, or *adrenocorticotropic hormone* (ACTH). ACTH enters the general circulation and travels to the adrenal cortex where, within a few minutes, it stimulates cortisol release (Figure 15.7).

Blood levels of cortisol are, to some extent, self-regulated. Cortisol is a *steroid,* which is a class of biochemicals related to cholesterol. Thus, cortisol

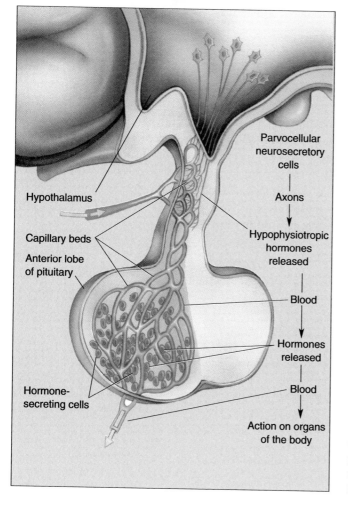

Hypothalamus

Capillary beds

Anterior lobe
of pituitary

Hormone-
secreting cells

Parvocellular
neurosecretory
cells

Axons

Hypophysiotropic
hormones
released

Blood

Hormones
released

Blood

Action on organs
of the body

Figure 15.6
Parvocellular neurosecretory cells of the hypothalamus.
Parvocellular neurosecretory cells secrete hypophysiotropic
hormones into specialized capillary beds of the hypothalamo-
pituitary portal circulation. These hormones travel to the anterior lobe of the pituitary, where they trigger or inhibit the
release of pituitary hormones from secretory cells.

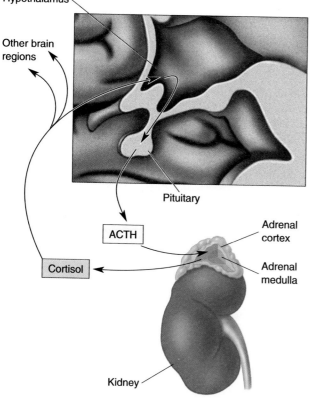

Hypothalamus

Other brain
regions

Pituitary

ACTH

Cortisol

Adrenal
cortex

Adrenal
medulla

Kidney

Figure 15.7
The stress response. Under conditions of physiological, emotional, or psychological stimulation or stress, the periventricular
hypothalamus secretes corticotropin-releasing hormone (CRH)
into the hypothalamo-pituitary portal circulation. This triggers the
release of adrenocorticotropic hormone (ACTH) into the general
circulation. ACTH stimulates the release of cortisol from the
adrenal cortex. Cortisol can act directly on hypothalamic neurons,
as well as on other neurons elsewhere in the brain.

is a lipophilic ("fat-loving") molecule, which will dissolve easily in lipid membranes and readily cross the blood-brain barrier. In the hypothalamus, cortisol interacts with specific receptors that lead to the inhibition of CRH release, thus ensuring that circulating cortisol levels do not get too high. Surprisingly, however, neurons with cortisol receptors are found widely distributed in the brain, not just in the hypothalamus. In these other CNS locations, cortisol has been shown to have significant effects on neuronal activity. Thus, we see that the release of hypophysiotropic hormones by cells in the secretory hypothalamus can produce widespread alterations in the physiology of both body and brain (Box 15.1).

Box 15.1 | **OF SPECIAL INTEREST**

Stress and the Brain

Biological stress is created by the brain, in response to real or imagined stimuli. The many physiological responses associated with stress help protect the body, and the brain, from the dangers that triggered the stress in the first place. But stress in chronic doses can have insidious harmful effects as well. Neuroscientists have only begun to sort out the relationship between stress, the brain, and brain damage.

Stress leads to the release of the steroid hormone cortisol from the adrenal cortex. Cortisol travels to the brain through the bloodstream and binds to receptors in the cytoplasm of many neurons. The activated receptors travel to the cell nucleus, where they stimulate gene transcription and ultimately protein synthesis. Steve Kerr, Philip Landfield, and their colleagues at the Bowman Gray School of Medicine in North Carolina have found that one consequence of cortisol's action is that neurons admit more Ca^{2+} through voltage-gated ion channels. This may be due to a direct change in the channels, or it may be indirectly caused by changes in the cell's energy metabolism. Whatever the mechanism, presumably in the short-term cortisol makes the brain better able to cope with the stress, perhaps by helping it figure out a way to avoid it!

But what about the effects of chronic, unavoidable stress? In Chapter 6 we learned that too much calcium can be a bad thing. If neurons become overloaded with calcium, they die (excitotoxicity). The question naturally arises: Can cortisol kill? Bruce McEwen and his colleagues at Rockefeller University, and Robert Sapolsky and his colleagues at Stanford University, have looked at this question in the rat brain. They found that daily injections of corticosterone (rat cortisol) for several weeks caused dendrites to wither on many neurons with corticosterone receptors. A few weeks later, these cells started to die. A similar result was found when, instead of daily hormone injections, the rats were stressed every day. Sapolsky's studies of baboons in Kenya further reveal the scourges of chronic stress. Baboons in the wild maintain a complex social hierarchy, and subordinate males steer clear of dominant males when they can. During one year when the baboon population boomed, local villagers caged many of the animals to prevent them from destroying their crops. Unable to escape the "top baboons" in the cages, many of the subordinate males subsequently died—not from wounds or malnutrition, but apparently from severe and sustained stress-induced effects. They had gastric ulcers, colitis, enlarged adrenal glands, and extensive degeneration of neurons in their hippocampus. Subsequent studies suggest that it is the direct effect of cortisol that damages the hippocampus. These effects of cortisol and stress resemble the effects of aging on the brain. Indeed, research has clearly shown that chronic stress causes premature aging of the brain.

It is not yet clear how the animal studies relate to humans, but we can consider them a cautionary tale. Modern life leads to tremendous, sustained stress for many people. Some athletes recklessly, and illegally, take large doses of various steroids for long periods in an effort to bulk-up their bodies. Do these damage the brain? More research is necessary to establish whether stress and steroids themselves can harm our brain, or if they facilitate the harmful effects of other disease processes. But in the mean time, it can't hurt to take it easy!

THE AUTONOMIC NERVOUS SYSTEM

Besides controlling the recipe of the hormonal soup that flows in our veins, the periventricular zone of the hypothalamus also controls the **autonomic nervous system** (**ANS**). The ANS is an extensive network of cells and fibers that are widely distributed inside the body cavity. "Autonomic" is from the Greek *autonomia*, roughly meaning "independence"; autonomic functions are usually carried out automatically, without conscious, voluntary control. They are also highly coordinated functions. Imagine a sudden crisis. In a morning class, as you are engrossed in a crossword puzzle, the instructor unexpectedly calls you to the blackboard to solve an impossible-looking equation. You are faced with a classic fight-or-flight situation, and your body reacts accordingly, even as your conscious mind frantically considers whether to blunder through it or beg off in humiliation. Your ANS triggers a host of physiological responses, including increasing heart rate and blood pressure, decreasing digestive functions, and mobilizing glucose reserves. These responses are all regulated by the **sympathetic division** of the ANS. Now imagine your relief as the class-ending bell suddenly rings, saving you from acute embarrassment and the instructor's anger. You settle back into your chair, breathe deeply, and read the clue for 24 DOWN. Within a few minutes, the activity of your sympathetic division decreases to low levels, and the functions of your **parasympathetic division** crank up again: your heart rate slows and blood pressure drops, digestive functions work harder on breakfast, and you stop sweating.

Notice that you may not have moved out of your chair throughout this unpleasant event. Maybe you didn't even move your pencil. But your body's internal workings reacted dramatically. Unlike the *somatic motor system*, whose alpha motor neurons can rapidly excite skeletal muscles with pinpoint accuracy, the actions of the ANS are always multiple, widespread, and relatively slow. Therefore, the ANS operates in expanded space and time. In addition, unlike the somatic motor system, which can only excite its peripheral targets, the ANS balances synaptic excitation and inhibition to achieve widely coordinated and graded control.

ANS Circuits

Together, the somatic motor system and the ANS constitute the total neural output of the CNS. The somatic motor system has a single job: it innervates and commands skeletal muscle fibers. The ANS has the complex task of commanding *every other* tissue and organ in the body that is innervated. Both systems have upper motor neurons in the brain that send commands to lower motor neurons, which actually innervate the peripheral target structures. However, they have some interesting differences (Figure 15.8). The cell bodies of all somatic lower motor neurons lie within the CNS, in either the ventral horn of the spinal cord or the brain stem. The cell bodies of all autonomic lower motor neurons lie outside the central nervous system, within cell clusters called **autonomic ganglia**. The neurons in these ganglia are called **postganglionic neurons**. Postganglionic neurons are driven by **preganglionic neurons**, whose cell bodies are in the spinal cord and brain stem. Thus, the somatic motor system controls its peripheral targets via a *monosynaptic* pathway, while the ANS uses a *disynaptic pathway.*

Sympathetic and Parasympathetic Divisions. The sympathetic and parasympathetic divisions operate in parallel, but they use pathways that are quite distinct structurally and in their transmitter systems. Preganglionic axons of the sympathetic division emerge only from the middle third of the spinal cord (thoracic and lumbar segments). In contrast, preganglionic axons of the parasympathetic division emerge only from the brain stem and the lowest (sacral) segments of the spinal cord, so the two systems complement each other anatomically (Figure 15.9). The preganglionic neurons of the sympathetic division lie within the *intermediolateral gray matter* of the spinal cord. They send their axons through the ventral roots to synapse in the ganglia of the **sympathetic chain,** which lies next to the spinal column. The preganglionic parasympathetic neurons, on the other hand, sit within a variety of brain stem nuclei and the lower (sacral) spinal cord, and their axons travel within several cranial nerves as well as the nerves of the sacral spinal cord. The parasympathetic axons travel much farther than the sympathetic axons, because the parasympathetic ganglia are typically located next to, on, or in their target organs (see Figures 15.8 and 15.9).

The targets of the autonomic nervous system include almost every part of the body, as shown in Figure 15.9. Both the sympathetic and the parasympathetic divisions:

- Innervate secretory glands (salivary, sweat, and various mucus-producing glands).
- Innervate the heart, blood vessels, and bronchi of the lungs to meet the energy demands of the body.
- Regulate the digestive and metabolic functions of the liver, gastrointestinal tract, and pancreas.
- Regulate the kidney, bladder, large intestine, and rectum.
- Are essential to the sexual responses of the genitals and reproductive organs.
- Interact with the body's immune system.

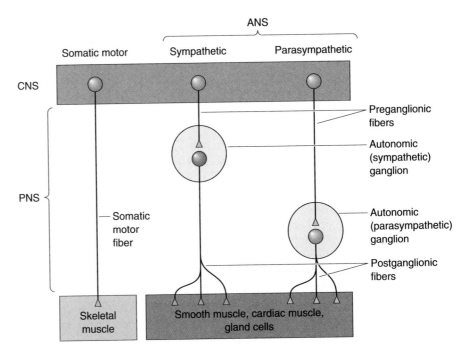

Figure 15.8
Organization of the three neural outputs of the CNS. The sole output of the somatic motor system is the lower motor neurons in the ventral horn of the spinal cord and the brain stem which Sherrington called the final common pathway for the generation of behavior. But some behaviors depend instead on the ANS, such as salivating, sweating, and genital stimulation. These visceral motor responses depend on the sympathetic and parasympathetic divisions of the ANS, whose lower motor neurons (i.e., postganglionic neurons) lie outside the CNS in autonomic ganglia.

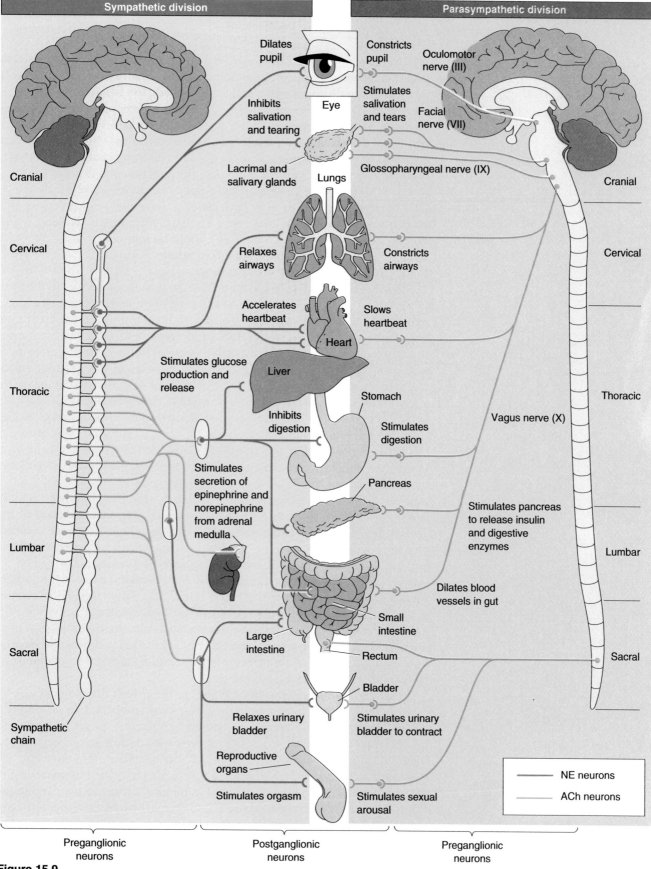

Figure 15.9

Chemical and anatomical organization of the sympathetic and parasympathetic divisions of the ANS. Notice that the preganglionic inputs of both divisions use ACh as a neurotransmitter. The postganglionic parasympathetic innervation of the visceral organs also uses ACh, but the postganglionic sympathetic innervation uses NE (with the exception of innervation of the sweat glands, which use ACh). The adrenal medulla receives preganglionic sympathetic innervation and secretes epinephrine into the bloodstream when activated.

The physiological influences of the sympathetic and parasympathetic divisions generally oppose one another. The sympathetic division tends to be most active during a crisis, real or perceived. The behaviors related to it are summarized in the puerile (but effective) mnemonic used by medical students, called the four-Fs: fight, flight, fright, and sex. The parasympathetic division facilitates various non-four-F processes, such as digestion, growth, immune responses, and energy storage. In most cases, the activity levels of the two ANS divisions are reciprocal; when one is high, the other tends to be low, and vice versa. The sympathetic division frenetically mobilizes the body for a short-term emergency at the expense of processes that keep it healthy over the long term. The parasympathetic division works calmly for the long-term good. Both cannot be stimulated strongly at the same time; their general goals are incompatible. Fortunately, neural circuits in the CNS inhibit activity in one division when the other is active.

Some examples will help illustrate how the balance of activity in the sympathetic and parasympathetic divisions control organ functions. The pacemaker region of the heart triggers each heartbeat without the help of neurons, but both divisions of the ANS innervate it and modulate it; sympathetic activity results in an increase in the rate of beating, while parasympathetic activity slows it down. The smooth muscles of the gastrointestinal tract are also dually innervated, but the effect of each division is the opposite of its effect on the heart. Intestinal motility, and thus digestion, is stimulated by parasympathetic axons and inhibited by sympathetic axons. Not all tissues receive innervation from both divisions of the ANS. For example, blood vessels of the skin, and the sweat glands, are innervated only by excitatory sympathetic axons. Lacrimal (tear-producing) glands are excited only by parasympathetic input.

Another example of the balance of parasympathetic-sympathetic activity is the curious neural control of the male sexual response. Erection of the human penis is a hydraulic process. It occurs when the penis becomes engorged with blood, which is triggered and sustained by parasympathetic activity. The curious part is that orgasm and ejaculation are triggered by *sympathetic* activity. You can imagine how complicated it must be for the nervous system to orchestrate the entire sexual act; parasympathetic activity gets it going (and keeps it going), but a shift to sympathetic activity is necessary to terminate it. Anxiety and worry—stress, in other words—and their attendant sympathetic activity, tend to inhibit erection and promote ejaculation. Not surprisingly, impotence and premature ejaculation are common complaints of the overstressed male.

The Enteric Division. The "little brain," as the **enteric division** of the ANS is sometimes called, is a unique neural system embedded in an unlikely place: the lining of the esophagus, stomach, intestines, pancreas, and gallbladder. It consists of two complicated networks, each with sensory nerves, interneurons, and autonomic motor neurons, called the *myenteric* (or Auerbach's) *plexus* and *submucous* (or Meissner's) *plexus* (Figure 15.10). These networks control many of the physiological processes involved in the transport and digestion of food, from oral to anal openings. The enteric system is not small; it contains about the same number of neurons as the entire spinal cord!

If the enteric division of the ANS qualifies as "brain" (which may be overstating the case), it is because it can operate with a great deal of independence. Enteric sensory neurons monitor tension and stretch of the gastrointestinal walls, the chemical status of stomach and intestinal contents, and hormone levels in the blood. This information is used by the enteric interneuronal circuits to control the activity levels of enteric output motor

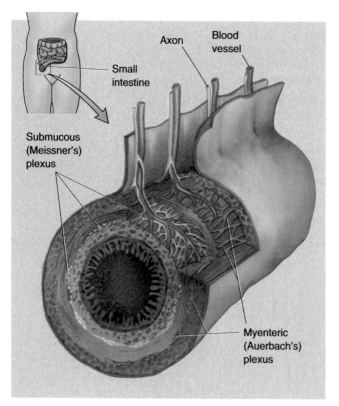

Figure 15.10
The enteric division of the ANS. This cross-sectional view of the small intestine shows the two networks of the enteric division: the myenteric plexus and the submucous plexus. They both contain visceral sensory and motor neurons that control the functions of the digestive organs.

neurons, which govern smooth muscle motility, production of mucous and digestive secretions, and the diameter of the local blood vessels. For example, consider a partially digested pizza making its way through the small intestine. The myenteric plexus ensures that lubricating mucus and digestive enzymes are delivered, that rhythmic (peristaltic) muscle action works to mix the pizza and enzymes thoroughly, and that intestinal blood flow increases to provide a sufficient fluid source and to transport newly acquired nutrients to the rest of the body.

The enteric division is not entirely autonomous. It receives input indirectly from the "real" brain via axons of the sympathetic and parasympathetic divisions. These provide supplementary control and can supersede the functions of the enteric division in some circumstances, such as acute stress.

Central Control of the ANS. As we have said, the hypothalamus is the main regulator of the autonomic preganglionic neurons. Somehow this diminutive structure integrates the diverse information it receives about the body's status, anticipates some of its needs, and provides a coordinated set of both neural and hormonal outputs. Essential to autonomic control are the connections of the periventricular zone to the brain stem and spinal cord nuclei that contain the preganglionic neurons of the sympathetic and parasympathetic divisions. The **nucleus of the solitary tract**, located in the medulla and connected with the hypothalamus, is another important center for autonomic control. In fact, some autonomic functions operate well even when the brain stem is disconnected from all structures above it, including the hypothalamus. The solitary nucleus integrates sensory information from the internal organs and coordinates output to the autonomic brain stem nuclei.

Neurotransmitters and the Pharmacology of Autonomic Function

Even people who never heard the word *neurotransmitter* know what it means to "get your adrenaline flowing." (In the United Kingdom the compound is called adrenaline, while in the U.S. we call it epinephrine.) Historically, the autonomic nervous system has probably taught us more than any other part of the body about how neurotransmitters work. Because the ANS is relatively simple compared to the CNS, we understand the ANS much better. In addition, neurons of the peripheral parts of the ANS are outside the blood-brain barrier, so all drugs that enter the bloodstream have direct access to them. The relative simplicity and accessibility of the ANS have led to a deeper understanding of the mechanisms of drugs that influence synaptic transmission.

Preganglionic Neurotransmitters. The primary transmitter of the peripheral autonomic neurons is acetylcholine (ACh), the same transmitter used at skeletal neuromuscular junctions. *The preganglionic neurons of both sympathetic and parasympathetic divisions release ACh.* The immediate effect is that the ACh binds to nicotinic ACh receptors (nAChR), which are ACh-gated channels, and evokes a fast EPSP that usually triggers an action potential in the postganglionic cell. This is very similar to the mechanisms of the skeletal neuromuscular junction, and drugs that block nAChRs in muscle, such as curare, also block autonomic output.

Ganglionic ACh does more than neuromuscular ACh, however. It also activates muscarinic ACh receptors (mAChR), which are metabotropic (G-protein-coupled) receptors that can cause both the opening and the closing of ion channels that lead to very slow EPSPs and IPSPs. These slow mAChR events are usually not evident unless the preganglionic nerve is activated repetitively. In addition to ACh, some preganglionic terminals release a variety of small, neuroactive peptides such as *NPY* (neuropeptide Y) and *VIP* (vasoactive intestinal polypeptide). These also interact with G-protein-coupled receptors and can trigger small EPSPs that last for several minutes. The effects of peptides are also modulatory; they do not usually bring the postsynaptic neurons to firing threshold, but they make them more responsive to the fast nicotinic effects when they do come along. Modulatory neurotransmitters of autonomic ganglia make the firing of postganglionic neurons very sensitive to the patterns of firing in preganglionic neurons.

Postganglionic Neurotransmitters. Postganglionic cells—the autonomic motor neurons that actually trigger glands to secrete, sphincters to contract or relax, and so on—use different neurotransmitters in the sympathetic and parasympathetic divisions of the ANS. Postganglionic parasympathetic neurons release ACh, but those of most parts of the sympathetic division use *norepinephrine* (*NE*). Parasympathetic ACh has a very local effect on its targets and acts entirely through mAChRs. In contrast, sympathetic NE often spreads far, even into the blood where it can circulate widely.

The autonomic effects of a variety of drugs that interact with cholinergic and noradrenergic systems can be confidently predicted, if you understand some of the autonomic circuitry and chemistry (see Figure 15.9). In general, drugs that promote the actions of norepinephrine *or* inhibit the muscarinic actions of acetylcholine are *sympathomimetic*; they cause effects that mimic activation of the sympathetic division of the ANS. For example, *atropine*, an antagonist of mAChRs, produces signs of sympathetic activation like dilation of the pupils. On the other hand, drugs that promote the muscarinic actions of ACh *or* inhibit the actions of NE are *parasympathomimetic*; they cause effects that mimic activation of the parasympathetic division of the ANS. For example, *propranolol*, an antagonist of the receptor for NE, slows the heart rate and lowers blood pressure. For this reason, propranolol is sometimes used to prevent the physiological consequences of stage fright.

But what about the familiar flow of adrenaline (i.e., epinephrine)? It is the compound released into the blood from the *adrenal medulla* when activated by preganglionic sympathetic innervation. Epinephrine is actually made from norepinephrine (noradrenaline in the U.K.), and it has effects on target tissues almost identical to those caused by sympathetic activation. Thus, the adrenal medulla is really nothing more than a modified sympathetic ganglion. You can imagine that as the adrenaline (i.e., the epinephrine) flows, a coordinated, bodywide set of sympathetic effects kicks in.

THE DIFFUSE MODULATORY SYSTEMS OF THE BRAIN

Consider what happens when you fall asleep. The internal commands "You are becoming drowsy" and "You are falling asleep" are messages that must be received by broad regions of the brain. Dispensing this information requires neurons with a particularly widespread pattern of axons. The brain has several such collections of neurons, each using a particular neurotransmitter and making widely dispersed, diffuse, almost meandering connections. Rather than carrying detailed sensory information, these cells often perform regulatory functions, modulating vast assemblies of postsynaptic neurons (such as the cerebral cortex, the thalamus, and the spinal cord) so that they become more or less excitable, more or less synchronously active, and so on. Collectively, they are a bit like the volume, treble, and bass controls on a radio, which do not change the lyrics or melody of a song but dramatically regulate the impact of both. In addition, different systems appear to be essential for aspects of motor control, memory, mood, motivation, and metabolic state. Many psychoactive drugs affect these modulatory systems, and the systems figure prominently in current theories about the biological basis of certain psychiatric disorders.

Anatomy and Functions of Diffuse Modulatory Systems

There are several **diffuse modulatory systems**, as we are calling them, and while they differ in structure and function, they have certain principles in common:

- Typically, the core of each system has a small set of neurons (several thousand).
- Neurons of the diffuse systems arise from the central core of the brain, most of them from the brain stem.
- Each neuron can influence many others, because each one has an axon that may contact more than 100,000 postsynaptic neurons spread widely across the brain.
- The synapses made by many of these systems seem designed to release transmitter molecules into the extracellular fluid so that they can diffuse to many neurons rather than be confined to the vicinity of the synaptic cleft.

We'll focus on the modulatory systems of the brain that use either norepinephrine (NE), serotonin (5-HT), dopamine (DA), or acetylcholine (ACh) as a neurotransmitter. Recall from Chapter 6 that all of these transmitters activate specific metabotropic (G-protein-coupled) receptors, and these receptors mediate most of their effects in the brain; for example, the brain has 10–100 times more metabotropic ACh receptors than ionotropic nicotinic ACh receptors. Because neuroscientists are still working hard to determine the exact functions of these systems in behavior, our explanations here will necessarily be vague.

The Noradrenergic Locus Coeruleus. Besides being a neurotransmitter in the peripheral ANS, NE is also used by neurons of the tiny **locus coeruleus** in the pons (from the Latin for "blue spot" because of the pigment in its cells). Each human locus coeruleus has about 12,000 neurons. We have two of them, one on each side. Axons leave the nucleus in several tracts but then fan out to innervate just about every part of the brain: all cerebral cortex, the thalamus and hypothalamus, the olfactory bulb, the cerebellum, the midbrain, and the spinal cord (Figure 15.11). The locus coeruleus must make some of the most diffuse connections in the brain, considering that just one of its neurons can make more than 250,000 synapses, and it can have one axon branch in the *cerebral* cortex and another in the *cerebellar* cortex!

Locus coeruleus cells seem to be involved in the regulation of attention, arousal, and sleep-wake cycles, as well as learning and memory, anxiety and pain, mood, and brain metabolism. This makes it sound as if the locus coeruleus may run the whole show. But the key word is "involved," which can mean almost anything. For example, our heart, liver, lungs, and kidneys are also involved in every brain function, for without them, all behavior would fail utterly. Because of its widespread connections, the locus coeruleus can influence virtually all parts of the brain. But to understand its actual functions, we start by determining what activates its neurons. Recordings from awake, behaving rats and monkeys show that locus coeruleus neurons are best activated by new, unexpected, nonpainful sensory stimuli in the animal's environment. They are least active when the animals are not vigilant, just sitting around quietly, digesting a meal. The locus coeruleus may participate in a general arousal of the brain during interesting events in the outside world. In a sense, the locus coeruleus acts like a sympathetic ganglion within the brain. Because NE can make neurons of the

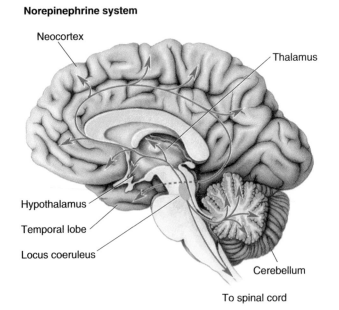

Norepinephrine system

Neocortex

Thalamus

Hypothalamus

Temporal lobe

Locus coeruleus

Cerebellum

To spinal cord

Figure 15.11
The noradrenergic diffuse modulatory system arising from the locus coeruleus. The small cluster of locus coeruleus neurons project axons that innervate vast areas of the CNS, including the spinal cord, cerebellum, thalamus, and cerebral cortex.

cerebral cortex more responsive to salient sensory stimuli, the locus coeruleus may function generally to increase brain responsiveness, speeding information processing by the point-to-point sensory and motor systems and making them more efficient.

The Serotonergic Raphe Nuclei. Serotonin-containing neurons are mostly clustered within the nine **raphe nuclei**. *Raphe* means "ridge" or "seam" in Greek, and indeed the raphe nuclei lie to either side of the midline of the brain stem. Each nucleus projects to different regions of the brain (Figure 15.12). Those more caudal, in the medulla, innervate the spinal cord, where they modulate pain-related sensory signals, as discussed in Chapter 12. Those more rostral, in the pons and midbrain, innervate most of the brain in much the same diffuse way as do the locus coeruleus neurons.

Similar to neurons of the locus coeruleus, raphe nuclei cells fire most rapidly during wakefulness, when an animal is aroused and active. Raphe neurons are quietest during sleep. The locus coeruleus and the raphe nuclei are part of a venerable concept called the *ascending reticular activating system*, which implicates the reticular "core" of the brain stem in processes that arouse and awaken the forebrain. This simple idea has been refined and redefined in countless ways since it was introduced in the 1950s, but its basic sense remains. Raphe neurons seem to be intimately involved in the control of sleep-wake cycles as well as the different stages of sleep. It is important to note that several other transmitter systems are involved in a coordinated way as well. We will discuss the involvement of the diffuse modulatory systems in sleep and wakefulness in Chapter 17.

Serotonergic raphe neurons have also been implicated in the control of mood and certain types of emotional behavior. We will return to serotonin and mood shortly when we discuss clinical depression, and in Chapter 16 we will discuss serotonin and the regulation of aggressive behavior.

The Dopaminergic Substantia Nigra and Ventral Tegmental Area. Although there are dopamine-containing neurons scattered throughout the

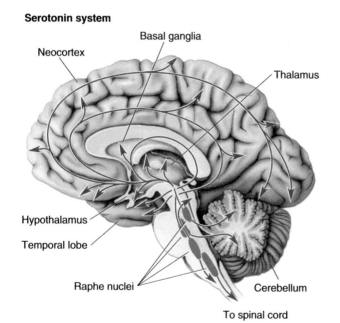

Figure 15.12
The serotonergic diffuse modulatory systems arising from the raphe nuclei.
The raphe nuclei are clustered along the midline of the brain stem and project extensively to all levels of the CNS.

CNS, including some in the retina, the olfactory bulb, and the periventricular hypothalamus, two closely related groups of dopaminergic cells have the characteristics of the diffuse modulatory systems (Figure 15.13). One of these arises in the *substantia nigra* in the midbrain. Recall from Chapter 14 that these cells project axons to the striatum (i.e., the caudate nucleus and the putamen), where they somehow facilitate the initiation of voluntary movements. Degeneration of the dopamine-containing cells in the substantia nigra is all that is necessary to produce the progressive, dreadful motor disorders of Parkinson's disease. This is not to say that we entirely understand the function of DA in motor control, but in general, it facilitates the initiation of motor responses by environmental stimuli.

The midbrain is also the origin of the other dopaminergic modulatory system, a group of cells that lie very close to the substantia nigra, in the *ventral tegmental area*. Axons from these neurons innervate a circumscribed region of the telencephalon that includes the frontal cortex and parts of the limbic system (see Chapter 16). This dopaminergic projection from the midbrain is sometimes called the *mesocorticolimbic dopamine system*. A number of different functions have been ascribed to this complicated projection. For example, there is evidence that it is involved in a "reward" system that somehow assigns value to, or *reinforces*, certain behaviors that are adaptive (like pursuing a prospective mate). The reward for activation of this system may be a pleasurable sensation. We will see in Chapter 16 that if rats (or humans) are given a chance to do so, they will work to electrically stimulate this pathway. In addition, this projection has been implicated in aspects of drug addiction and in psychiatric disorders, as we will discuss below.

The Cholinergic Basal Forebrain and Brain Stem Complexes. Acetylcholine is the familiar transmitter at the neuromuscular junction, at synapses in autonomic ganglia, and at postganglionic parasympathetic synapses. Cholinergic interneurons also exist within the brain, in the striatum and the cortex, for example. In addition, there are two major diffuse

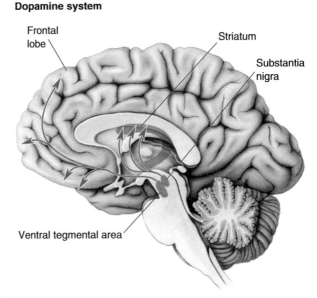

Dopamine system

Frontal lobe

Striatum

Substantia nigra

Ventral tegmental area

Figure 15.13
The dopaminergic diffuse modulatory systems arising from the substantia nigra and the ventral tegmental area. The substantia nigra and ventral tegmental area lie close together in the midbrain. They project to the striatum (caudate nucleus and putamen) and limbic and frontal cortical regions, respectively.

modulatory cholinergic systems in the brain, one of which is called the **basal forebrain complex**. It is a "complex" because the cholinergic neurons lie scattered among several related nuclei at the core of the telencephalon, medial and ventral to the basal ganglia. The best known of these are the *medial septal nuclei*, which provide the cholinergic innervation of the hippocampus, and the *basal nucleus of Meynert*, which provides most of the cholinergic innervation of the neocortex.

The function of the cells in the basal forebrain complex remains mostly unknown. But interest in this region has been fueled by the discovery that these are among the first cells to die during the course of Alzheimer's disease, which is characterized by a progressive and profound loss of cognitive functions. (However, there is widespread neuronal death in Alzheimer's disease, and no specific link between the disease and cholinergic neurons has been established.) Like the noradrenergic and serotonergic systems, the cholinergic system has been implicated in regulating general brain excitability during arousal and sleep-wake cycles. The basal forebrain complex may also play a special role in learning and memory formation.

The second diffuse cholinergic system has a long name: the *pontomesencephalotegmental cholinergic complex*. These are ACh-utilizing cells in the pons and midbrain tegmentum. This system acts mainly on the dorsal thalamus, where, together with the noradrenergic and serotonergic systems, it regulates the excitability of the sensory relay nuclei. These cells also project up to the telencephalon, providing a cholinergic link between the brain stem and basal forebrain complexes. Figure 15.14 shows the cholinergic systems.

Drugs and the Diffuse Modulatory Systems

Psychoactive drugs, those compounds with "mind-altering" effects, all act on the central nervous system, and most do so by interfering with chemical synaptic transmission. Many abused drugs act directly on the modulatory systems, particularly the noradrenergic, dopaminergic, and serotonergic systems.

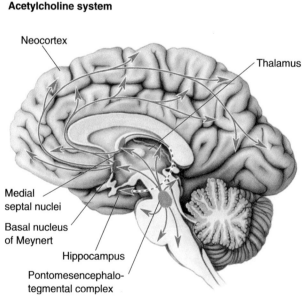

Acetylcholine system

Neocortex

Thalamus

Medial septal nuclei

Basal nucleus of Meynert

Hippocampus

Pontomesencephalo-tegmental complex

Figure 15.14
The cholinergic diffuse modulatory systems arising from the basal forebrain and brainstem. The medial septal nuclei and basal nucleus of Meynert project widely upon the cerebral cortex, including the hippocampus. The pontomesencephalotegmental complex projects to the thalamus and parts of the forebrain.

Hallucinogens. The use of *hallucinogens,* compounds that produce hallucinations, goes back thousands of years. Hallucinogenic compounds are contained in a number of plants consumed as part of religious ritual; for example, the *Psilocybe* mushroom by the Maya and the peyote cactus by the Aztec. The modern era of hallucinogenic drug use was unwittingly ushered in at the laboratory of Swiss chemist Albert Hofmann. In 1938, Hofmann chemically synthesized a new compound, *lysergic acid diethylamide,* abbreviated *LSD.* For five years, the LSD sat on the shelf. Then one day in 1943, Hofmann accidentally ingested some of the powder. His report of the effects attracted the immediate interest of the medical community. Psychiatrists began to use LSD in attempts to unlock the subconscious of mentally disturbed patients. Later the drug was discovered by intellectuals, artists, students, and the U.S. Defense Department, who investigated its "mind-expanding" effects. (A chief advocate of LSD use was former Harvard psychologist Timothy Leary.) In the 1960s LSD made its way to the street, where it remains today.

LSD is extremely potent. A dose sufficient to produce a full-blown hallucinogenic effect is only 25 *micrograms* (compared to a normal dose of aspirin at 650 *milligrams,* 25,000 times larger). Among the reported behavioral effects of LSD are a dreamlike state with heightened awareness of sensory stimuli, often with a mixing of perceptions such that sounds can evoke images, images can evoke smells, and so on.

The chemical structure of LSD (and the active ingredients of *Psilocybe* mushrooms and peyote) is very close to that of serotonin, suggesting that it acts on the serotonergic system. Indeed, LSD is a potent agonist at the serotonin receptors on the presynaptic terminals of neurons in the raphe nuclei. Activation of these receptors markedly inhibits the firing of raphe neurons. Thus, one known CNS effect of LSD is a reduction in the outflow of the brain's serotonergic diffuse modulatory system. It is interesting to note in this regard that decreased activity of the raphe nuclei is also characteristic of dream-sleep (see Chapter 17).

Can we conclude that LSD produces hallucinations by silencing the brain serotonin systems? If only drug effects on the brain were that simple! Unfortunately, there are problems with this hypothesis. For one, silencing neurons in the raphe nuclei by other means—by destroying them, for example—does not mimic the effects of LSD in experimental animals. Furthermore, animals still respond as expected to LSD after their raphe nuclei have been destroyed. Clearly, there is still some work to be done on this problem.

Thus, although a link between serotonin and the hallucinogenic effects of LSD is almost certain, we don't yet understand the precise nature of this relationship. Perhaps by inhibiting neurons of the raphe nuclei while at the same time activating 5-HT receptors, LSD causes hallucinations by superseding the naturally modulated release of serotonin.

Stimulants. In contrast to the uncertainties about hallucinogens and serotonin, it is clear that the powerful CNS stimulants *cocaine* and *amphetamine* both exert their effects at synapses made by dopaminergic and noradrenergic systems. Both drugs give users a feeling of increased alertness and self-confidence, a sense of exhilaration and euphoria, and a decreased appetite. Both are also *sympathomimetic,* which is to say that they cause peripheral effects that mimic activation of the sympathetic division of the ANS: increased heart rate, blood pressure, dilation of the pupils, and so on.

Cocaine is extracted from the leaves of the coca plant and has been used by Andean Indians for hundreds of years. In the mid-nineteenth century, cocaine turned up in Europe and North America as the magic ingredient in a wide range of concoctions touted by their salesmen as having medicinal

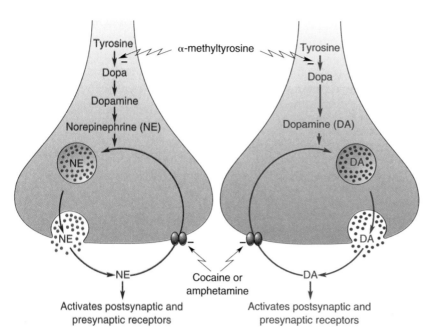

Figure 15.15
Stimulant drug action on the catecholamine axon terminal. On the left is a noradrenergic terminal and on the right is a dopaminergic terminal. Both neurotransmitters are catecholamines synthesized from the dietary amino acid tyrosine. Dopa
(3,4-dihydroxyphenylalanine) is an intermediate in the synthesis of both. The actions
of NE and DA are usually terminated by uptake back into the axon terminal.
Amphetamine and cocaine block this uptake, thus allowing NE and DA to remain in
the synaptic cleft longer.

value (an example is Coca-Cola, originally marketed in 1886 as a therapeutic agent, which contained both cocaine and caffeine). Cocaine use fell out
of favor early in the twentieth century, only to reemerge with a vengeance
in the late 1960s as a recreational drug. Ironically, one of the main reasons
for the rise in cocaine use during this period was the tightening of regulations against amphetamines. First chemically synthesized in 1887, amphetamines did not come into wide use until World War II, when they were
taken by soldiers of both sides (particularly aviators) to sustain them in
combat. Following the war, amphetamines became available as nonpre-
scription diet aids, as nasal decongestants, and as "pep-pills." Regulations
were finally tightened when it was recognized that amphetamines are, like
cocaine, highly addictive and dangerous in large doses.

The neurotransmitters dopamine and norepinephrine are *catecholamines*,
so named because of their chemical structure (see Chapter 6). The actions of
catecholamines released into the synaptic cleft are normally terminated by
specific uptake mechanisms. Cocaine and amphetamine both block this catecholamine uptake (Figure 15.15), although recent work suggests that
cocaine targets DA reuptake more selectively; amphetamine blocks NE and
DA reuptake *and* stimulates the release of DA. Thus, these drugs can prolong and intensify the effects of released DA or NE. Is this the means by
which cocaine and amphetamine cause their stimulant effects? There is
good reason for thinking so. For example, experimental depletion of brain
catecholamines by using synthesis inhibitors (such as α-methyltyrosine)
will abolish the stimulant effects of both cocaine and amphetamine.

Besides having a similar stimulant effect, cocaine and amphetamine
share another, more insidious behavioral action: Users often develop a

strong psychological dependence on the drugs. People will develop powerful cravings for prolonging and continuing drug-induced pleasurable feelings. These effects are believed to result specifically from the enhanced transmission in the mesocorticolimbic dopamine system during drug use. Remember, this system may normally function to reinforce adaptive behaviors. By short-circuiting the system, these drugs instead reinforce drug-seeking behavior. Indeed, just as rats will work to electrically stimulate the mesocorticolimbic projection, they will also work to receive an injection of cocaine.

The Diffuse Modulatory Systems and Mental Illness

Psychiatry has always been an inexact business because we simply do not know the root biological cause of most mental disorders. It is remarkable, therefore, that effective medical treatments have been developed for two of the most severe classes of mental illness: the affective disorders and the schizophrenias. The treatments in question are drugs that act on the diffuse modulatory systems of the brain. Because these drugs can often restore normal mental functions, it is believed that at least some of the psychoses may be caused by a malfunction of these modulatory systems.

Depression. *Affect* is the medical term for emotional state or mood; **affective disorders** are disorders of mood. An occasional, brief feeling of depression—getting "the blues"—is a common response to life's events, suffering a loss or disappointment, and we can hardly call this a disorder. However, the affective disorder psychiatrists and psychologists call **depression** is something more prolonged and much more severe, characterized by a feeling that one's emotional state is no longer under one's control. Other characteristics usually include insomnia, loss of appetite, feelings of worthlessness and guilt, and intrusive and morbid thoughts. An episode of depression can occur suddenly, often without obvious external cause, and if left untreated, it usually lasts 4–12 months. Another characteristic is that mood does not usually improve even when life's circumstances improve. Depression is a serious disease. It is a main precipitating cause of suicide, which claims more than 30,000 lives each year in the United States. Depression is also widespread. Perhaps as many as 20% of people will suffer a major, incapacitating bout of depression during their life. At the present time, an estimated 8 million Americans are afflicted.

The first real indication that depression might result from a problem with the central diffuse modulatory systems came in the 1960s. A drug called *reserpine*, introduced to control high blood pressure, caused psychotic depression in about 20% of cases. Reserpine depletes central catecholamines and serotonin by interfering with their loading into synaptic vesicles. Then it was found that another class of drugs introduced to treat tuberculosis caused a marked mood elevation. These drugs inhibit *monoamine oxidase (MAO)*, the enzyme that destroys catecholamines and serotonin. Another piece of the puzzle fell into place when it was recognized that the drug *imipramine*, introduced some years earlier as an antidepressant, inhibits the reuptake of released 5-HT and NE, thus promoting their action in the synaptic cleft. From these observations came the hypothesis that mood is closely tied to the levels of released norepinephrine and/or serotonin in the brain. According to this idea, depression is a consequence of a deficit in one of these diffuse modulatory systems.

Modern drug treatments for depression can be very effective, and most have in common enhanced neurotransmission at central serotonergic and/or noradrenergic synapses. The most popular treatments are (1) tricyclic compounds (so named because of their chemical structure) such as imipramine, which block the uptake of both NE and 5-HT; (2) selective uptake inhibitors like fluoxetine (trade name Prozac), which act only on serotonin terminals; and (3) MAO inhibitors such as phenelzine, which reduce the enzymatic degradation of 5-HT and NE (Figure 15.16).

Can we conclude from the effects of antidepressant drugs that depression is a consequence of reduced brain concentrations of serotonin or norepinephrine? If only the effects of drugs on the brain were that simple! Unfortunately, there are problems with a straightforward equation between mood and modulator. Perhaps most striking is the clinical finding that the antidepressant action of these drugs takes several weeks to develop, even though they have almost immediate effects on transmission at the modulatory synapses. Another concern is that other drugs that raise NE levels in the synaptic cleft, like cocaine, are not effective as antidepressants. A new hypothesis is that the effective drugs promote some sort of long-term adaptive change in the noradrenergic or serotonergic systems that alleviates the depression. Today's research focuses on the lasting consequences of antidepressant treatment on the brain, such as a slowly developing change in the number of receptors for NE or 5-HT.

Psychotherapies, without the use of drugs, can also be very effective for many people suffering from depression. In severe cases, electroconvulsive therapy (ECT) can be extremely helpful when drugs are not. ECT in its modern form is very controlled; it in no way resembles the wretched scene in the

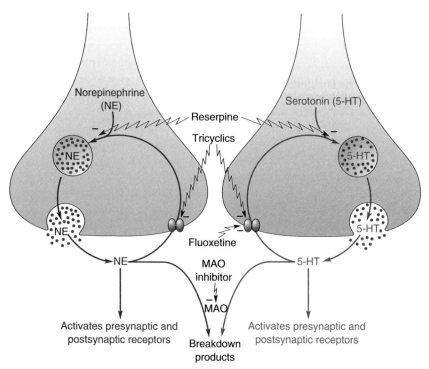

Figure 15.16
Drugs and the biochemical life cycles of norepinephrine and serotonin. The life cycle of a neurotransmitter includes its biosynthetic birth (see Figure 15.15), its packaging into vesicles and subsequent release from the nerve terminal, and either its reuptake or enzymatic destruction into breakdown products. Drugs can influence behavior by acting at any point in the cycle. Reserpine depletes stores of NE and 5-HT and in some people can cause severe depression. Some antidepressant drugs are MAO inhibitors, the tricyclics, and selective uptake inhibitors. MAO inhibitors enhance the actions of NE and 5-HT by preventing their enzymatic destruction. Tricyclics enhance NE and 5-HT action by blocking uptake. Fluoxetine acts the same way but is selective for serotonin.

movie *One Flew Over the Cuckoo's Nest*. Localized electrical stimulation triggers seizure discharge, but the patient is given anesthesia and muscle relaxants, and side effects are usually minimal. The reason that ECT works is entirely unknown.

Schizophrenia. Schizophrenia can be the most devastating of all mental illnesses. Striking during adolescence or early adulthood, the disease often persists for life. Like the affective disorders, schizophrenia is a major health problem in the United States; over 2 million people are afflicted.

The term *schizophrenia* was introduced around 1910 by Swiss psychiatrist Eugene Blueler, and it roughly means "divided mind." **Schizophrenia** is characterized by a loss of contact with reality, and fragmentation and disruption of thought, perception, mood, and movement. There are many variations in the manifestations of schizophrenia, and it is still not clear whether it is a single disease or several. Many psychiatrists recognize two types of schizophrenia based on their symptoms. Type I is characterized by psychotic episodes that often include bizarre delusions (such as "I am Jesus Christ"); hallucinations (hearing imaginary voices, for example); and disordered, often paranoid thoughts. Type II is characterized by a loss of appropriate emotional response (a "flat affect" or inappropriate laughter and/or crying), abnormal postures (called catatonia), and a lack of spontaneous speech. However, some patients exhibit both categories of symptoms, so it is not clear if there are truly different schizophrenic disease processes. MRI scans have revealed structural abnormalities in some of the brains of sufferers from both types of schizophrenia: The size of the ventricles is increased, probably because of a slight reduction in overall brain size; and there are particular reductions in the size of structures within the temporal lobes.

A link between DA and Type I schizophrenia comes from the effects of amphetamine. Remember from our discussion above that amphetamine enhances neurotransmission at catecholamine-utilizing synapses and causes the release of DA. Of course, amphetamine's normal stimulant action bears little resemblance to schizophrenia. However, because of its addictive properties, users of amphetamines often risk taking more and more to satisfy their cravings. The resulting overdose can lead to a psychotic episode that is virtually indistinguishable from Type I schizophrenia. This suggests that psychosis is somehow related to too much catecholamine in the brain.

A second reason to associate dopamine with schizophrenia relates to the CNS effects of antipsychotic drugs. In the 1950s it was discovered that the drug *chlorpromazine*, initially developed as an antihistamine, could prevent psychotic episodes in Type I schizophrenia. It was later found that chlorpromazine and related antipsychotic drugs, collectively called **neuroleptics**, are potent blockers of DA receptors, specifically the D_2 receptor. When a large number of neuroleptics are examined, the correlation between the dosage effective for controlling schizophrenia and their ability to bind to D_2 receptors is impressive (Figure 15.17). Indeed, these same drugs are effective in the treatment of amphetamine and cocaine psychoses. This suggests that psychotic episodes are triggered specifically by the activation of DA receptors.

Can we conclude from the effects of neuroleptics that schizophrenia results from an overactive dopamine system? If only the effects of drugs on the brain were that simple! We again run into the problem that although neuroleptics are fully effective in blocking DA receptors within hours after they are administered, their antipsychotic activity requires several weeks to become apparent. While some postmortem and PET studies have now shown that DA receptors are indeed more numerous in certain regions of the limbic forebrain in Type I schizophrenia, other studies, unfortunately, have failed to confirm these results. In addition, newly developed "atypi-

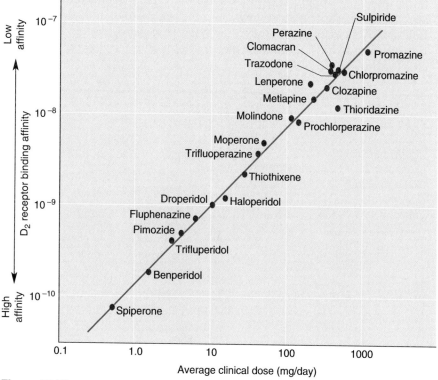

Figure 15.17
Neuroleptics and D₂ receptors. The neuroleptic dosages effective in controlling schizophrenia correlate well with the drugs' binding affinities for D_2 receptors. (Source: Adapted from Seeman, 1980.)

cal" neuroleptics, such as clozapine, are more effective on other neurotransmitter systems, such as 5-HT, than they are on the DA system. Thus, the link between dopamine and schizophrenia remains tantalizing, but enigmatic.

CONCLUDING REMARKS

In this chapter, we have examined three components of the nervous system that are characterized by the great reach of their influences. The secretory hypothalamus and autonomic nervous system communicate with cells all over the body, and the diffuse modulatory systems communicate with neurons in many different parts of the brain. They are also defined by the duration of their direct effects, which can range from minutes to hours. Finally, they are characterized by their chemical neurotransmitters. In many instances, the transmitter *defines* the system. For example, in the periphery, we can use the words "noradrenergic" and "sympathetic" interchangeably. The same thing goes for "raphe" and "serotonin" in the forebrain, and "substantia nigra" and "dopamine" in the basal ganglia. These chemical idiosyncrasies have permitted interpretations of drug effects on behavior that are not possible with most other neural systems. Thus, we have a good idea where in the brain amphetamine and cocaine exert their stimulant effects, and where in the periphery they act to raise blood pressure and heart rate.

At a detailed level, each of the systems discussed in this chapter performs different functions. But at a general level, they all *maintain brain homeostasis*: They regulate different processes within a certain physiological range. For example, the ANS regulates blood pressure within a range that is appropriate. Blood pressure variations optimize an animal's performance under different conditions. In a similar way, the noradrenergic locus coeruleus and serotonergic raphe nuclei regulate levels of consciousness and mood. These levels also vary within a range that is adaptive to the organism. Whether we speak of the control of blood pressure or the control of mood, however, malfunction of the regulatory system can have devastating effects. In the next two chapters, we will encounter these systems again in the context of specific functions.

KEY TERMS

The Secretory Hypothalamus
homeostasis
periventricular zone
magnocellular neurosecretory neuron
neurohormone
oxytocin
vasopressin
parvocellular neurosecretory neuron
hypophysiotropic hormone
hypothalamo-pituitary portal circulation
adrenal cortex
adrenal medulla
cortisol

The Autonomic Nervous System
autonomic nervous system (ANS)
sympathetic division
parasympathetic division

autonomic ganglia
postganglionic neuron
preganglionic neuron
sympathetic chain
enteric division
nucleus of the solitary tract

The Diffuse Modulatory Systems of the Brain
diffuse modulatory system
locus coeruleus
raphe nuclei
basal forebrain complex
affective disorder
depression
schizophrenia
neuroleptic

✔ REVIEW QUESTIONS

1. Battlefield trauma victims who have lost large volumes of blood often express a craving to drink water. Why?
2. You've stayed up all night trying to meet a term paper deadline. You now are typing frantically, keeping one eye on the paper and the other on the clock. How has the periventricular hypothalamus orchestrated your body's physiological response to this stressful situation? Describe in detail.
3. Why is the adrenal medulla often referred to as a modified sympathetic ganglion? Why isn't the adrenal cortex included in this description?
4. A number of famous athletes and entertainers have accidentally killed themselves by taking large quantities of cocaine. Usually the cause of death is heart failure. How would you explain the peripheral actions of cocaine?
5. What types of drugs are used to treat depression? What do they have in common?
6. Psychiatrists often refer to the dopamine theory of schizophrenia. Why do they believe dopamine is linked to schizophrenia? Why must we be cautious in accepting a simple equation between schizophrenia and too much dopamine?

Brain Mechanisms of Emotion

To appreciate the significance of emotions, one need only imagine life without them. Instead of the daily peaks and valleys we all experience, life would be a great empty plain. Without question, the expression of emotion is a large part of being human. Observing Mr. Spock in *Star Trek* and Arnold Schwarzenegger in *The Terminator*, we quickly decide that they are not "one of us" because they don't exhibit normal emotions.

In this chapter, we explore the neural basis of emotion. The subject is tricky because it is difficult to study emotions with many of the techniques used to study sensory and motor systems. If you are studying a sensory system, you can present a stimulus and seek neurons that respond to it. You can manipulate the stimulus to determine the stimulus attributes (orientation, sound frequency, etc.) that are best for evoking a response. But how can this technique be used to study emotion? It is not a straightforward matter to study emotions in animals that cannot tell us their subjective feelings. What we observe are the behavioral manifestations of the internal emotions. Therefore, we must carefully distinguish between emotional *experience* and emotional *expression*. What we know about the brain mechanisms of emotion has been derived from a synthesis of animal studies on emotional expression and clinical cases that give us insight into human feelings.

In the simplest sense, the study of emotion can be reduced to an input-output problem. The stimuli that evoke emotional responses usually come from our senses. The behavioral signs of emotion are controlled by the somatic motor system, the autonomic nervous system, and the secretory hypothalamus. Thus, one question we can ask is how incoming sensory stimuli lead to the behavioral and physiological responses indicative of emotional expression. The mechanisms of emotional experience are harder to grasp, but it is generally believed that the cerebral cortex plays a key role. The question is how sensory input or some internal signal ultimately leads to cortical activation characteristic of a specific emotion.

WHAT IS EMOTION?

Emotions—love, hate, disgust, joy, shame, envy, guilt, fear, anxiety, and so on—are feelings we all experience at one time or another. But what precisely defines those feelings? Are they sensory signals from our body, diffuse patterns of activity in our cortex, or something else? Questions of this sort have proven to be surprisingly difficult to answer and have led to the development of different theories about exactly what emotions are.

Theories of Emotion

In the nineteenth century, several highly regarded scientists, including Darwin and Freud, considered the role of the brain in the expression of emotion (Figure 16.1). Based on careful observations of emotional expression in animals and humans and emotional experience in humans, theories relating expression and experience were developed.

The James-Lange Theory. One of the first well-articulated theories of emotion was proposed in 1884 by the renowned American psychologist and philosopher William James. Related ideas were proposed by Danish psychologist Carl Lange. The theory that is now commonly known as the **James-Lange theory** of emotion proposed that we experience emotion *in response to* physiological changes in our body. For example, we feel sad because we cry rather than cry because we are sad. Our sensory systems send information to our brain about our current situation, and as a result our brain sends signals out to the body, changing muscle tone, heart rate, and so on. The sensory systems then react to the changes evoked by the brain, and it is this sensation that constitutes the emotion. According to James and Lange, the physiological changes *are* the emotion, and if they are removed, the emotion will go with them. This seems like a backward idea to many people today, as it did to many contemporaries of James and

Figure 16.1
Emotional expression of a cat terrified by a dog. This drawing is from Darwin's book, *The Expression of the Emotions in Man and Animals.* Darwin conducted one of the first extensive studies of emotional expression. (Source: Darwin, 1872/1955, p. 125.)

Lange. Until this theory was proposed, the commonly held conception was that an emotion is evoked by a situation and the body changes in response to the emotion. The James-Lange theory is the exact opposite.

Before you reject this theory as ridiculous, try one of the thought experiments suggested by James. Suppose you're boiling with anger about something that has just happened. Try to strip away all the physiological changes associated with the emotion. Your pounding heart is calmed, your tensed muscles are relaxed, and your flushed face is cooled. It is hard to imagine maintaining rage in the absence of any physiological signs. In fact, this little experiment isn't all that different from the technique used in meditation classes to relieve stress. Here is another example. You're on a first date with someone you are really attracted to, and you're swimming in a steaming soup of emotions that include happiness, love, lust, and anxiety. Then, poof! All at once the physiological signs of your infatuation are removed (like taking an imaginary cold shower). Are you still feeling the same emotional state? Probably not.

Even if it is true that emotion is intimately tied to physiological state, this doesn't mean that emotion cannot be felt in the absence of obvious physiological signs (a point even James and Lange would concede). But for strong emotions that are typically associated with physical change, there is a close relationship between emotion and its physiological manifestation, and it is not obvious which causes which.

The Cannon-Bard Theory. Although the James-Lange theory became popular in the early twentieth century, it wasn't long before it came under attack. In 1927, American physiologist Walter Cannon published a paper containing several compelling criticisms of the James-Lange theory and went on to propose a new theory. Cannon's theory was modified by Philip Bard, and the **Cannon-Bard theory** of emotion, as it came to be known, proposed that emotional experience can occur independently of emotional expression.

One of Cannon's arguments against the James-Lange theory was that emotions can be experienced even if physiological changes cannot be sensed. To support this claim, he offered the cases of animals he and others studied after transection of the spinal cord. Such surgery eliminated sensation of the body below the level of the cut, but it did not appear to abolish emotion. To the extent possible with muscular control of just the upper body or head, the animals still exhibited signs of experiencing emotions. Similarly, Cannon noted human cases in which a transected spinal cord did not diminish emotion. If emotional experience occurs when the brain senses physiological changes in the body, as the James-Lange theory proposed, then eliminating sensation should also eliminate emotions, and this did not appear to be the case.

A second observation of Cannon's that seems inconsistent with the James-Lange theory is that there is not a reliable correlation between the experience of emotion and the physiological state of the body. For example, fear is accompanied by increased heart rate, inhibited digestion, and increased sweating. However, these same physiological changes accompany other emotions, such as anger, and even nonemotional conditions of illness such as fever. How can fear be a consequence of the physiological changes, when these same changes are associated with states other than fear?

Cannon's new theory focused on the idea that the thalamus plays a special role in emotional sensations. In this theory, sensory input is received by the cerebral cortex, which in turn activates certain changes in the body. But according to Cannon, this stimulus-response neural loop is devoid of emotion. Emotions are produced when signals reach the thalamus either directly from the sensory receptors or by descending cortical input. In other

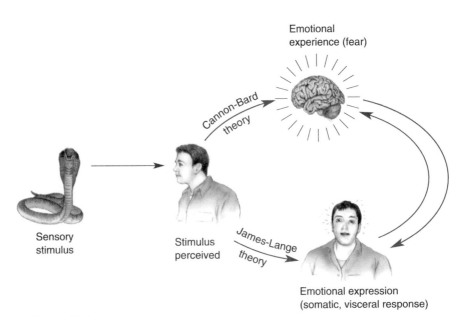

Figure 16.2
A comparison of the James-Lange and Cannon-Bard theories of emotion. In the James-Lange theory (red arrows), the man perceives the frightening animal and reacts. As a consequence of his body's response to the situation, he becomes afraid. In the Cannon-Bard theory (blue arrows), the frightening stimulus leads to the feeling of fear first, and then there is a reaction.

words, the character of the emotion is determined by the pattern of activation of the thalamus. An example may clarify the difference between this and the James-Lange theory. According to James and Lange, you feel sad because you cry; if you could prevent the crying, the sadness should go, too. In Cannon's theory, you don't have to cry to feel sad; there simply has to be the appropriate activation of your thalamus in response to the situation. The James-Lange and Cannon-Bard theories of emotion are compared in Figure 16.2.

From Theory to Experimental Studies

It would take too much space to discuss all the theories of emotion that have been proposed since the days of the James-Lange and Cannon-Bard theories. Subsequent work has demonstrated that each theory has merits as well as flaws. For instance, it has been shown, contrary to Cannon's statements, that fear and rage are associated with distinguishable physiological responses even though they both activate the sympathetic division of the ANS. Although this does not prove that these emotions are a result of distinct physiological responses, the responses are at least different.

Another interesting disagreement with the Cannon-Bard theory demonstrated by later studies is that emotion is sometimes affected by damage to the spinal cord. In one study of adult men with spinal injuries, there was a correlation between the extent of sensory loss and reported decreases in emotional experiences. This fascinating result might be used to revive the James-Lange theory that emotional experience depends on emotional expression, but other studies of people with spinal injuries have not always found a similar correlation. There is also some evidence that forcing oneself into the behavioral expression of an emotion—for instance, smiling—actually makes some people feel happier. Perhaps experiencing some emotions does depend on behavioral manifestations and experiencing others does not.

Ultimately, these questions can only be answered when we understand the neural basis for emotional experience. Although a complete understanding remains elusive, a valuable strategy has been to trace the pathways in the brain that link sensations (inputs) to the behavioral responses (outputs) that herald emotional experience. In the remainder of this chapter, we'll see that different emotions may depend on different neural circuits, but in many cases these circuits converge on the same parts of the brain.

THE LIMBIC SYSTEM CONCEPT

In previous chapters, we discussed how sensory information from peripheral receptors is processed along clearly defined, anatomically distinct pathways, to the neocortex. The components of a pathway collectively constitute a *system*. For example, neurons located in the retina, LGN, and striate cortex work together to serve vision, so we say they are part of the visual system. Is there a system, in this sense, responsible for experiencing emotions? During this century it has been proposed that the brain has such a system, known as the limbic system. Shortly, we will discuss the difficulties of trying to define a single system for emotion. But first, let's examine the origin of the limbic system concept.

Broca's Limbic Lobe

In a paper published in 1878, French neurologist Paul Broca noted that, on the medial surface of the cerebrum, all mammals possess a group of cortical areas that are distinctly different from the surrounding cortex. Using the Latin word for "border" (*limbus*), Broca named this collection of cortical areas the **limbic lobe** because they form a ring or border around the brain stem (Figure 16.3). According to this definition, the limbic lobe consists of the cortex around the corpus callosum, mainly in the cingulate gyrus, and the cortex on the medial surface of the temporal lobe, including a structure

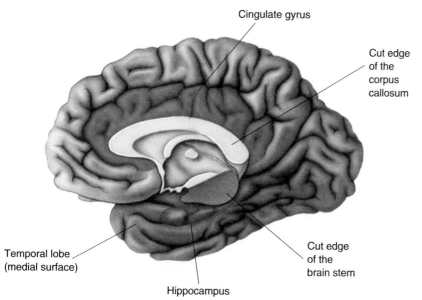

Figure 16.3
The limbic lobe. Broca defined the limbic lobe as those structures that form a ring around the brain stem and corpus callosum on the medial walls of the brain. The brain stem has been removed in the figure so that the medial surface of the temporal lobe is visible.

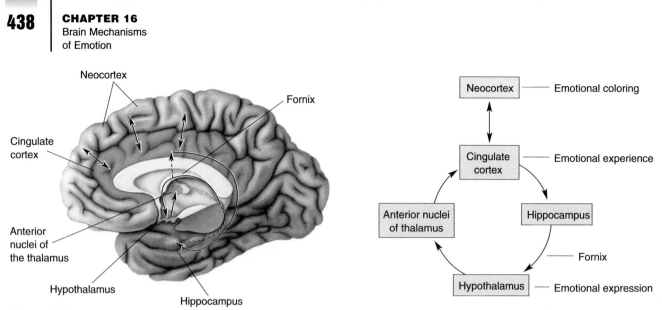

Figure 16.4
The Papez circuit. Papez believed that the experience of emotion was determined by activity in cingulate cortex and, less directly, other cortical areas. Emotional expression was thought to be governed by the hypothalamus. The cingulate cortex projects to the hippocampus, and the hippocampus projects to the hypothalamus by way of the bundle of axons called the fornix. Hypothalamic effects reach the cortex via a relay in the anterior thalamic nuclei.

called the hippocampus. Broca did not write about the importance of these structures for emotion, and for some time they were thought to be primarily involved in olfaction. However, the word *limbic*, and the structures in Broca's limbic lobe, subsequently came to be closely associated with emotion.

The Papez Circuit

By the 1930s, evidence suggested that a number of limbic structures are involved in emotion. Reflecting on the earlier work of Cannon, Bard, and others, American neurologist James Papez proposed that there is an "emotion system," lying on the medial wall of the brain, that links the cortex with the hypothalamus. Figure 16.4 shows the group of structures that has come to be called the **Papez circuit**. Each element is connected to another by a major fiber tract.

Papez believed, as do many scientists today, that the cortex is critically involved in the experience of emotion. Following damage to certain cortical areas, there are sometimes profound deficits in emotional behavior with little change in perception or intelligence (Box 16.1). Also, tumors located near cingulate cortex are associated with emotional disturbances including fear, irritability, and depression. Papez proposed that activity evoked in other neocortical areas by projections from cingulate cortex adds "emotional coloring." We saw in Chapter 15 that the hypothalamus integrates the actions of the ANS. In the Papez circuit, the hypothalamus governs the behavioral expression of emotion. The hypothalamus and neocortex are arranged so that each can influence the other, thus linking the expression and experience of emotion. In the circuit, cingulate cortex affects the hypothalamus via the hippocampus and fornix (the large bundle of axons leaving the hippocampus), whereas the hypothalamus affects cingulate cortex via the anterior thalamus. The fact that communication between the cortex and hypothalamus is bidirectional means that the Papez circuit could be compatible with both the James-Lange and the Cannon-Bard theories of emotion.

While anatomical studies demonstrated that the components of the Papez circuit were interconnected as Papez had indicated, there was only suggestive evidence that each was involved in emotion. One reason Papez

Box 16.1 | **OF SPECIAL INTEREST**

The Amazing Case of Phineas Gage

One of the most amazing studies ever conducted on the brain's influence on emotion was the result of an industrial accident. The unfortunate subject of this study was a 25-year-old foreman at a railroad construction site in Vermont named Phineas Gage. On September 13, 1848, he was tamping explosive powder into a hole in preparation for blasting at the construction site and made the mistake of looking away from what he was doing. The tamping iron he was holding contacted a rock, and the powder exploded. The consequences of this act are described by Dr. John Harlow in an 1848 article entitled "Passage of an Iron Rod Through the Head." When the charge went off, it sent the meter-long, 6 kg iron rod into Gage's head just below his left eye. After passing through his left frontal lobe, the rod exited the top of Gage's head.

Incredibly, after being carried to an ox cart, Gage sat upright on the ride to a nearby hotel and walked up a long flight of stairs to go inside. When Harlow first saw Gage at the hotel, he commented that "the picture presented was, to one unaccustomed to military surgery, truly terrific" (p. 390). As you might imagine, the projectile destroyed a considerable portion of the skull and left frontal lobe, and Gage lost a great deal of blood. The hole through his head was more than 9 cm in diameter. Harlow was able to stick the full length of his index finger into the hole from the top of Gage's head, and also upward from the hole in his cheek. Harlow dressed the wound as best he could. Over the following weeks, considerable infection developed. No one would have been surprised if Gage died. But incredibly, no more than a month after the accident, he was out of bed and walking around town.

Harlow corresponded with Gage's family for many years and in 1868 published a second article, "Recovery from the Passage of an Iron Bar Through the Head," describing Gage's life after the accident. After Gage recovered from his wounds, he was seemingly normal except for one thing: His personality was drastically and permanently changed. When he tried to return to his old job as construction foreman, the company found he had changed so much for the worse that they wouldn't rehire him. According to Harlow, before the accident Gage was considered "the most efficient and capable foreman. . . . He possessed a well-balanced mind, and was looked upon by those who knew him as a shrewd, smart business man, very persistent in executing all

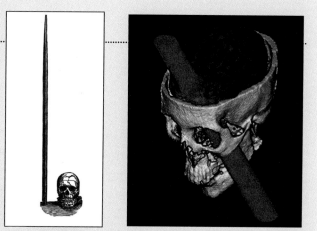

Figure A. (Sources: left, Harlow, 1868, p. 347; right, Damasio et al., 1994, p. 1104.)

his plans of operation" (pp. 339–340). After the accident, Harlow describes him as follows:

> He is fitful, irreverent, indulging at times in the grossest profanity (which was not previously his custom), manifesting but little deference for his fellows, impatient of restraint or advice when it conflicts with his desires, at times pertinaciously obstinate, yet capricious and vacillating, devising many plans of future operation, which are no sooner arranged than they are abandoned in turn for others appearing more feasible . . . His mind was radically changed, so decidedly that his friends and acquaintances said he was "no longer Gage." (pp. 339–340)

There are no psychological test results to tell us what had happened to Phineas Gage's intellect. However, from Harlow's account of the man's life after the accident, it appears that Gage's personality was altered far more than his intelligence. Phineas went on to live for another 12 years, seemingly without direction.

No autopsy was performed after Gage's death, but his skull and the tamping iron have been preserved in a museum at Harvard Medical School. Figure A (left) shows a sketch made by Harlow indicating the size of the tamping iron relative to Gage's skull. Recently, Hanna and Antonio Damasio and their colleagues at the University of Iowa made new measurements of the skull and used modern imaging techniques to assess the damage to Gage's brain. Their reconstruction of the tamping iron's path is shown in Figure A (right). The iron rod severely damaged cortex in the frontal lobes of both hemispheres, and it was this damage that caused Gage to behave like an ill-tempered child, giving expression to an endless succession of strong emotions. His unfortunate accident was one of the first demonstrations of the important role that the frontal lobes play in emotion and its expression.

thought the hippocampus is involved in emotion is that it is affected by the virus responsible for rabies. An indication of rabies infection, and an aid in its diagnosis, is the presence of abnormal cytoplasmic bodies in neurons, especially in the hippocampus. Because rabies is characterized by hyperemotional responses such as exaggerated fear and aggressiveness, Papez reasoned that the hippocampus must be involved in normal emotional experience. Although there was little evidence related to the role of the anterior thalamus, clinical reports stated that lesions in this area led to apparent emotional disturbances such as spontaneous laughing and crying.

You may have noticed the correlation between the elements composing both the Papez circuit and Broca's limbic lobe. Because of their similarity, the group of structures hypothetically responsible for the sensation and expression of emotion are often referred to as the **limbic system**, even though Broca's limbic lobe originally had nothing to do with emotion. The term *limbic system* was popularized in 1952 by American physiologist Paul MacLean, who argued that the limbic structures form one of three primary functional divisions of the brain. These three divisions, according to MacLean's speculative theory, are the reptilian brain, the old mammalian brain, and the new mammalian brain, and they evolved in that order. According to this theory, the evolution of a limbic system (the old mammalian brain) enabled animals to experience and express emotions and emancipated animals from the stereotyped behavior dictated by their brain stem (the reptilian brain). In this scheme, the evolution of neocortex gave higher animals, and especially humans, the capacity for problem solving and rational thought (Box 16.2).

Difficulties with the Single Emotion System Concept

We have defined a group of interconnected anatomical structures roughly encircling the brain stem as a limbic system. To the extent that many of these structures are involved in emotion, we can validly refer to a system of emotion as the limbic system. Experimental work supports the hypothesis that some of the structures in Broca's limbic lobe and the Papez circuit play a role in emotion. On the other hand, some of the components of the Papez circuit are no longer thought to be important for the expression of emotion, such as the hippocampus.

The critical point seems to be a conceptual one concerning the definition of an emotion system. What should we include in the system? For example, if a nucleus of cells in the brain plays a role in both emotion and some other function such as olfaction or learning and memory, should it be included? Should we include every structure that contributes in any way to the experience or expression of any type of emotion? Part of the problem is the word *system*. It doesn't seem logical to list every structure somehow related to emotion and call it a system, implying that the parts work together to perform a common function. Given the diversity of emotions we experience, there is no compelling reason to think that only one system rather than several is involved. Conversely, there is good evidence that some structures involved in emotion are also involved in other functions; there is not a one-to-one structure-function relationship. We are only beginning to learn how the experience and expression of emotion arise in the brain. Although the term *limbic system* is still commonly used, there are those who question the utility of trying to define a single, discrete emotion system. Our approach will be to focus on a few specific emotions for which there is strong evidence of the involvement of certain neural circuits.

FEAR AND ANXIETY

Suppose one night you're walking over to a friend's house and you decide to take a shortcut. You become anxious when you find yourself on

Box 16.2 | **PATH OF DISCOVERY**

Origins of the Limbic System

by Paul MacLean

Paul MacLean

Triggered by an experience in early childhood, a question that has long puzzled me is why human beings with their unrivaled intelligence are often driven to behave in a way that seems irrational and conducive to trouble. After college, this and more pervasive questions pertaining to the origin and meaning of life persuaded me that going into medicine and focusing on functions of the brain would be more likely to throw light on my questions than a career in philosophy.

In 1947 (after World War II service in the Medical Corps, and then a short civilian practice), a U.S. Public Health fellowship gave me the opportunity to engage in research under the aegis of Dr. Stanley Cobb, neuropathologist, neuropsychiatrist, and founder of the Department of Psychiatry at Massachusetts General Hospital. My research required my devising improved electrodes for recording the electrical activity of structures near the base of the brain, particularly that of the hypothalamus, which was regarded as playing a central role in basic drives and emotional reactions. I was interested in a condition called psychomotor epilepsy, in which the patient commonly has no convulsive seizure but experiences one or more of a wide variety of emotional and other symptoms, followed by simple or complicated forms of unremembered behavior. Gibbs et al. reported that the epileptic disturbance appeared to arise in the anterior temporal region. Our recordings indicated that the spike discharge reached greatest amplitude near the medial temporal lobe where the hippocampal cortex is located.

Why should an irritative lesion in this region result not only in emotional manifestations, but also in symptoms involving one or more of the sensory systems? In seeking an answer, I stumbled upon a paper by Papez entitled "A Proposed Mechanism of Emotion" (1937) in which he emphasized that the cortex of the so-called "rhinencephalon" enveloping Broca's great limbic lobe was the only cerebral cortex with strong hypothalamic connections. He proceeded to point out a number of clinical and experimental conditions associated with emotional or autonomic manifestations.

On a visit to Papez in 1948, he showed me by dissecting a human brain how pathways from the somatic, auditory, and visual cortex might reach the hippocampal formation. Thereafter, I wrote a paper on recent developments supporting the Papez theory of emotion, giving emphasis to neurosurgical findings of Penfield and others that (both then and since) have provided the best evidence that the structures in question derive information in terms of affective feelings that guide behavior involved in self-preservation and procreation. To lessen the accent on olfactory function, I referred to the rhinencephalon as the "visceral brain," using the term in the sense of "strong inward feelings." I later found that the expression *visceral brain* led to misunderstanding because in physiological parlance, "visceral" applies only to the visceral organs. Consequently, I reverted to Broca's descriptive word "limbic" and used the expression *limbic system* when referring in a modular sense to the limbic cortex and its "primary" connections with the brain stem. This explains how the term *limbic system* entered the literature in 1952.

Continuing research would indicate that the limbic system provides a neural substrate for the evolution of mammals and their distinctive family way of life. The limbic cingulate cortex in particular seems to be identified with forms of behavior that distinguish the evolutionary transition from the reptilian therapsids to mammals—namely, nursing conjoined with maternal care; the mammalian separation cry; and play.

In my paper on the Papez theory (and in my writings since then), I have used the phenomenology of psychomotor epilepsy and knowledge of its underlying mechanisms to emphasize the importance of the limbic integration of internal and external experience for an affective sense of self and reality; feelings of belief and conviction; and the memory of ongoing events. I also suggested that because of the relatively simple structure of the evolutionarily older limbic cortex, it might be unable to communicate in verbal terms with the most highly evolved neocortex. This, I noted, might help to explain the difference between what we "feel" and what we "know."

dark streets that are not particularly safe. You see a group of threatening-looking guys approaching you from the other end of the block, and now you are more than just anxious—you're frightened. Thinking about the anxiety and fear you would experience in this situation will remind you of the dramatic response the sympathetic division of your autonomic nervous system can evoke. As discussed in Chapter 15, even before you have a behavioral reaction, your hypothalamus orchestrates a response in your ANS that affects virtually every part of your body, from increased heart rate and respiration to sweating.

From a neurobiological standpoint, the question is how incoming sensory information leads to the behavioral and physiological responses associated with fear and anxiety. While we don't have the complete answer yet, there is strong evidence that a structure called the *amygdala*, which is buried in the temporal lobe, plays a key role. Before discussing recent studies of the amygdala however, let's first look at a very influential experiment conducted in the 1930s that focused attention on the involvement of the temporal lobes in fear.

The Klüver-Bucy Syndrome

Shortly after Papez' proposal of an emotion circuit in the brain, neuroscientists Heinrich Klüver and Paul Bucy, at the University of Chicago, found that bilateral removal of the temporal lobes, or temporal *lobectomy*, in rhesus monkeys has a dramatic effect on the animals' responses to fearful situations. The surgery produces numerous bizarre behavioral abnormalities that Klüver and Bucy placed into five categories: psychic blindness, oral tendencies, hypermetamorphosis, altered sexual behavior, and emotional changes. This constellation of symptoms is called the **Klüver-Bucy syndrome**. Although our focus is the emotional disturbances, we will briefly discuss the other symptoms to fully appreciate the syndrome.

Klüver and Bucy observed that the monkeys appeared to have *psychic blindness* because they did not seem to recognize common objects or understand their meaning despite the fact that they could see. They would pick up and examine every object within view and, as the term *oral tendencies* implies, place each object into their mouth. They seemed to use their mouth instead of their eyes to identify each object. Klüver and Bucy put a variety of objects into the room with each monkey. Inedible items, such as a live mouse, a glass, and a nail, were examined by mouth and discarded. Pieces of food were examined the same way and then eaten. If a hungry monkey was shown a group of objects it had seen before, intermixed with food, the monkey would still go through the process of picking up each object for study. A normal hungry monkey in the same situation would make a bee-line for the food. One thing that this behavior seemed to show, in addition to psychic blindness and oral tendencies, was an irresistible compulsion to examine things. The animals appeared to be obsessed by the urge to run around and touch everything and place each found object into the mouth. This behavior is what Klüver and Bucy meant by the term *hypermetamorphosis*. There were also changes in sexual behavior. Some of the monkeys showed a strikingly increased interest in sex, including masturbation as well as heterosexual and homosexual acts. (The experimental notes of Klüver and Bucy read something like a script for a XXX monkey movie.) These behaviors were not observed in normal monkeys.

The emotional changes in monkeys with Klüver-Bucy syndrome were most dramatically represented by an apparent decrease in fear. For example, a normal wild monkey will avoid humans and other animals. In the presence of an experimenter, it will usually crouch in a corner and remain still;

if approached, it will dash off to a safer corner. This sort of fear and excitement was not seen in the monkeys with bilateral temporal lobectomies. Experimental monkeys would not only approach and touch the human, they would even let the human stroke them and pick them up. Remember, this is an otherwise wild monkey. The same fearlessness was exhibited in the presence of other animals that monkeys normally fear. Even after approaching and being attacked by a natural enemy such as a snake, the monkey would go back and try to examine it again. You might think this behavior indicates stupidity or memory loss rather than fearlessness, but other evidence suggested decreased emotion. For example, there was a significant decrease in the vocalizations and facial expressions usually associated with emotion. It appeared that both the normal experience and the normal expression of emotion were severely decreased by the temporal lobectomy. Virtually all the symptoms of the Klüver-Bucy syndrome reported in monkeys have also been seen in humans with temporal lobe lesions. In addition to visual recognition problems, oral tendencies, and hypersexuality, people appear to have "flattened" emotions.

The Amygdala

In interpreting the findings of Klüver and Bucy, we must keep in mind that a large amount of brain tissue was removed. Removal of the temporal lobes involves not only temporal cortex but also all of the subcortical structures in that area including the amygdala and hippocampus. Some of the symptoms, especially psychic blindness, probably resulted from the removal of visual cortical areas in the temporal lobes (see Chapter 10). However, the emotional disturbances probably resulted from destruction of the amygdala. Indeed, there is considerable evidence that the amygdala is involved in numerous aspects of emotion, not just fear.

Anatomy of the Amygdala. The **amygdala** is situated in the pole of the temporal lobe, just below the cortex on the medial side. Because it is almond-shaped, its name derives from the Greek word for "almond." The amygdala is a complex of nuclei that are commonly divided into three groups: the **basolateral nuclei**, the **corticomedial nuclei**, and the central nucleus (Figure 16.5). Afferents to the amygdala come from a large variety of sources, including neocortex in all lobes of the brain, as well as the hippocampal and cingulate gyri. Of particular interest here is the fact that information from all the sensory systems feeds into the amygdala, particularly the basolateral nuclei. Each sensory system has a different projection pattern to the amygdala nuclei, and interconnections within the amygdala allow the integration of information from different sensory systems. Two major pathways connect the amygdala with the hypothalamus: the *ventral amygdalofugal pathway* and the *stria terminalis*.

Effects of Amygdala Destruction and Stimulation. It has been demonstrated in several species that bilateral ablation of the amygdala has the effect of flattening emotion in a manner similar to the Klüver-Bucy syndrome. Bilateral amygdalectomy can profoundly reduce fear as well as having effects on aggression and memory. There are reports that rats so treated will approach a sedated cat and nibble its ear, and that a wild lynx will become as docile as a house cat. The fearlessness is believed to result from the destruction of nuclei in the basolateral region of the amygdala. We will have more to say about amygdala involvement in aggression and memory a little later.

There are very few cases of humans with selective damage to the amygdala, but Ralph Adolphs and his colleagues at the University of Iowa recently studied a 30-year old woman, known as S. M., with bilateral destruction

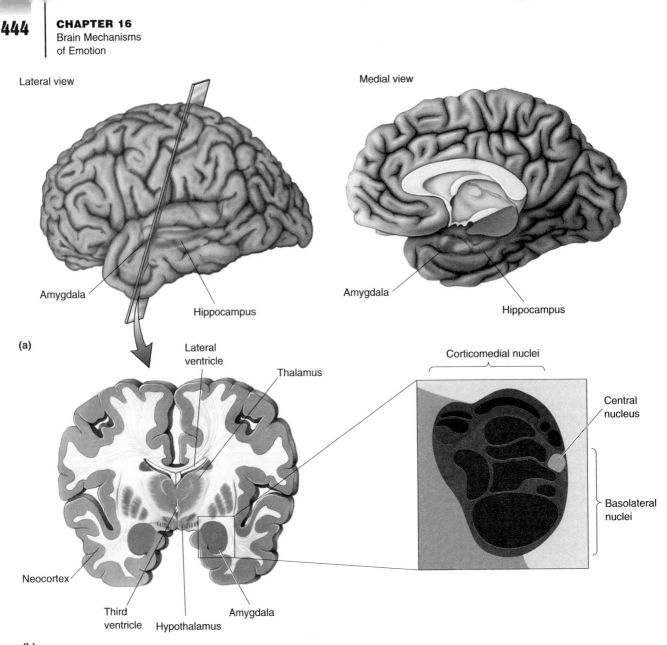

Figure 16.5
Cross section of the amygdala. (a) Lateral and medial views of the temporal lobe showing the location of the amygdala in relation to the hippocampus. **(b)** The brain is sectioned coronally to show the amygdala. The basolateral nuclei (red) receive visual, auditory, gustatory, and tactile afferents. The corticomedial nuclei (purple) receive olfactory afferents.

of the amygdala resulting from Urbach-Wiethe disease. S. M. has normal intelligence and is perfectly able to identify people from photographs. However, she has difficulty recognizing certain emotions expressed by people in the photographs. When asked to categorize the emotion expressed in a person's face, she is normally able to recognize happiness, sadness, and disgust. She is somewhat less likely to describe an angry expression as angry and much less likely to describe a fearful expression as afraid. It appears that the amygdala lesion selectively decreases her ability to recognize fear.

If removing the amygdala reduces the expression and recognition of fear, what happens if the intact amygdala is electrically stimulated? Depending on the site, amygdala stimulation can lead to a state of increased vigilance or attention. Stimulation of the lateral portion of the amygdala in cats can elicit a combination of fear and violent aggression. Electrical stimulation of the amygdala in humans has been reported to lead to anxiety and fear.

A Neural Circuit for Learned Fear

Through socialization or painful experience, we all learn to avoid certain behaviors for fear of being hurt. If you once received a painful shock as a child by pushing a paper clip into an electrical outlet, you probably never did it again. Memories associated with fear can be quickly formed and long-lasting. Although the amygdala is not thought to be a primary location for memory storage, it does seem to be involved in giving emotional content to memories.

A number of different experiments suggest that neurons in the amygdala can "learn" to respond to stimuli associated with pain, and after such learning these stimuli evoke a fearful response. In an experiment performed by Bruce Kapp and his colleagues at the University of Vermont, rabbits were trained to associate the sound of a tone with mild pain. The researchers made use of the fact that a normal sign of fear in rabbits is a change in heart rate. An animal was placed in a cage, and at various times it would hear one of two tones. One tone was followed by a mild electrical shock to the feet through the metal floor of the cage, the other tone was benign. After training, Kapp's group found that the rabbit's heart rate developed a fearful response to the tone associated with pain, but not to the benign tone. Prior to training, neurons in the central nucleus of the amygdala failed to respond to the tones used in the experiment. However, after training, neurons in the central nucleus responded to the shock-related tone but not to the benign one. Joseph LeDoux of New York University has shown that after this type of fear conditioning, amygdala lesions eliminate the learned visceral responses, such as the changes in heart rate and blood pressure.

LeDoux has proposed a circuit to account for learned fear. Auditory information is sent to the basolateral region of the amygdala, where cells in turn send axons to the central nucleus. Efferents from the central nucleus project to the hypothalamus, which can alter the state of the ANS, and to the periaqueductal gray matter in the brain stem, which can evoke behavioral reactions via the somatic motor system (Figure 16.6).

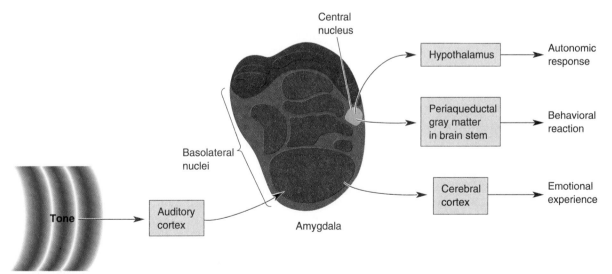

Figure 16.6
A neural circuit for learned fear. Through training, a sound tone becomes associated with pain. The presumed fear response is mediated by the amygdala. The emotional stimulus reaches the basolateral nuclei of the amygdala by way of auditory cortex, and the signal is relayed to the central nucleus. Efferents from the amygdala project to the brain stem periaqueductal gray matter, causing the behavioral reaction to the stimulus, and to the hypothalamus, resulting in the autonomic response. The experience of emotion presumably involves projections to cerebral cortex.

ANGER AND AGGRESSION

In American culture, we are somewhat ambivalent about aggressiveness as a personality trait and about acts of violent aggression. We chastise some people who behave overly aggressively but praise others for aggressively pursuing their goals. Murder is considered a capital offense, but wartime killing is considered not only acceptable but honorable. Clearly, we distinguish different forms of aggression in humans. Likewise, we can categorize different types in animals. One animal may act aggressively toward another for many reasons: to kill for food, to defend offspring, to win a mate, to scare off a potential adversary. Though it has not been proven, there is some evidence that different types of aggression are regulated differently by the nervous system.

Aggression is a multifaceted behavior that is not a product of a single isolated system in the brain. One factor that has an influence on aggression is the level of male sex hormones, or androgens. In animals, there is a correlation between seasonal androgen levels and aggressive behavior. Consistent with one of the roles of androgens, injections of testosterone (one of the androgens) can make an immature animal more aggressive, and castration can reduce aggressiveness. In humans the relationship is less clear, although some have claimed that there is a connection between testosterone levels and aggressive behavior in violent criminals. In any case, there is strong evidence for a neurobiological component to aggression, which is our focus here.

A useful distinction can be made between predatory aggression and affective aggression. **Predatory aggression** involves attacks made against a member of a different species for the purpose of obtaining food, such as a lion hunting a zebra. Attacks of this type are typically accompanied by relatively few vocalizations, and they are aimed at the head and neck of the prey. Predatory aggression is not associated with high levels of activity in the sympathetic division of the ANS. **Affective aggression** is for show rather than to kill for food, and it involves high levels of activity in the sympathetic division of the ANS. An animal in this state will typically make vocalizations while adopting a threatening or defensive posture. A cat hissing and arching its back at the approach of a dog is a good example. The behavioral and physiological manifestations of both types of aggression must be mediated by the somatic motor system and the ANS, but the pathways must diverge at some point to account for the dramatic differences in the behavioral responses.

The Hypothalamus and Aggression

One of the first structures linked to aggressive behavior was the hypothalamus. Although some of the early experiments were crude by today's standards, they pointed the way for later studies.

Sham Rage. Experiments performed in the 1920s showed that a remarkable behavioral transformation took place in cats or dogs whose cerebral hemispheres had been removed. Animals that were not easy to provoke prior to the surgery would go into a state of violent rage with the least provocation after the surgery. For instance, a violent response might be produced by an act as mild as scratching a dog's back. This state was called **sham rage** because the animal demonstrated all the behavioral manifestations of rage, but in a situation that normally would not cause anger. It was also a sham in the sense that the animals would not actually attack as they normally might.

Perhaps these animals had every reason to be angry after such destructive surgery, but there is something to be learned from these experiments. While

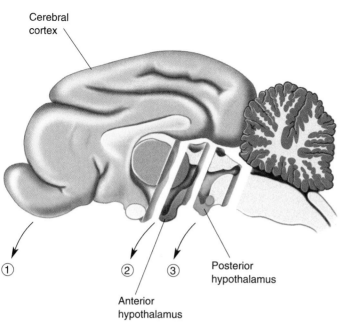

Cerebral
cortex

① ② ③ Posterior
hypothalamus

Anterior
hypothalamus

Figure 16.7
Brain transections and sham rage. If the cerebral hemispheres are removed and
the hypothalamus is left intact (1), sham rage results. A similar result is obtained if
the anterior hypothalamus is removed in addition to the cerebral cortex (1 + 2). If the
posterior hypothalamus is removed in addition to the anterior hypothalamus (1 + 2 +
3), sham rage does not result.

this extreme behavioral condition resulted from removing all of both cerebral hemispheres (telencephalon), the behavioral effect could be reversed by
making the lesion just a little bit larger to include portions of the diencephalon, particularly the hypothalamus. Sham rage was observed if the
anterior hypothalamus was destroyed along with the cortex, but it was not
seen if the lesion was extended to include the posterior half of the hypothalamus (Figure 16.7). The implication is that the posterior hypothalamus
may be particularly important for the expression of anger and aggression
and that normally it is inhibited by the telencephalon. But we must bear in
mind that the lesions were large, and something other than the posterior
hypothalamus may have been destroyed with the larger lesion.

Electrical Stimulation of the Hypothalamus. In a series of pioneering studies begun in the 1920s, W. R. Hess at the University of Zurich investigated the
behavioral effects of electrically stimulating the diencephalon. Hess made
small holes in the skulls of anesthetized cats and implanted electrodes into the
brain. After the animal awoke from the anesthesia, a small electrical current
was passed through the electrodes, and behavioral effects were noted. Various
structures were stimulated, but here we will focus on the effects of stimulating
different regions of the hypothalamus. The variety of responses from stimulating slightly different portions of the hypothalamus is amazing, considering
that it is such a small part of the brain. Depending on where the electrode is
placed, stimulation may cause the animal to sniff, pant, eat, or express behaviors characteristic of fear or anger. These behavioral reactions illustrate the two
primary functions of the hypothalamus discussed in Chapter 15: homeostasis
and emotional expression. There can be changes in heart rate, pupillary dilation, and gastrointestinal motility, to name a few. Because stimulation of some
parts of the hypothalamus also elicits behavior characteristic of fear and rage,
we hypothesize that the hypothalamus is an important component of the system normally involved in expressing these emotions.

The expression of rage Hess evoked by hypothalamic stimulation was similar to the sham rage seen in animals whose cerebral hemispheres had been removed. With a small application of electrical current, a cat would spit, growl, fold its ears back, and its hair would stand on end. This complex set of behaviors would normally occur when the cat felt threatened by an enemy. Sometimes the cat would suddenly run as if fleeing an imaginary attacker. If the intensity of the stimulation was increased, the animal might make an actual attack, swatting with a paw or leaping onto an imaginary adversary. When the stimulation was stopped, the rage disappeared as quickly as it started, and the cat might even curl up and go to sleep.

In a series of studies conducted at the Yale University Medical School in the 1960s, John Flynn found that affective aggression and predatory aggression could be elicited by stimulating different areas of the hypothalamus (Figure 16.8). Affective aggression (also known as a threat attack) was observed with stimulation of the medial hypothalamus. Similar to the rage response reported by Hess, the animal would arch its back, hiss, and spit but usually not actually attack a victim such as a nearby rat. Predatory aggression (called by Flynn a silent-biting attack) was evoked by stimulation of the lateral hypothalamus. While the back might be somewhat arched and the hair slightly on end, predatory aggression was not accompanied by the dramatic threatening gestures of affective aggression. Nonetheless, in this "quiet attack," the cat would move swiftly toward a rat and viciously bite its neck.

The Midbrain and Aggression

There are two major pathways by which the hypothalamus sends signals involving autonomic function to the brain stem: the **medial forebrain bundle** and the **dorsal longitudinal fasciculus**. Axons from the lateral hypothalamus make up part of the medial forebrain bundle, and these project to the *ventral tegmental area* in the midbrain. Stimulation of the ventral tegmental area can elicit behaviors characteristic of predatory aggression, just as stimulation of the lateral hypothalamus does. Conversely, lesions in the ventral tegmental area can disrupt offensive aggressive behaviors. One finding suggesting that the hypothalamus influences aggressive behavior via its effect on the ventral tegmental area is that hypothalamic stimulation will not evoke aggression if the medial forebrain bundle is cut. Interestingly, aggressive behavior is not entirely eliminated by this surgery, suggesting that this route is important when the hypothalamus is involved but that the hypothalamus need not always be involved.

The medial hypothalamus sends axons to the **periaqueductal gray matter** of the midbrain by way of the dorsal longitudinal fasciculus. Electrical stimulation of the periaqueductal gray matter can produce affective aggression, and lesions located there can disrupt this behavior.

The Amygdala and Aggression

In our discussion of the neurobiology of anxiety and fear, we saw that the amygdala plays a key role. The amygdala is also involved in aggressive behavior. In an experiment performed by American scientist Karl Pribram and his colleagues in 1954, amygdala lesions were shown to have a major effect on social interactions in a colony of eight male rhesus monkeys. Having lived together for some time, the animals had established a social hierarchy. The first intervention made by the investigators was to make bilateral amygdala lesions in the brain of the most dominant monkey. After this animal returned to the colony, it fell to the bottom of the hierarchy, and the previously next subordinate monkey became dominant. Presumably, the sec-

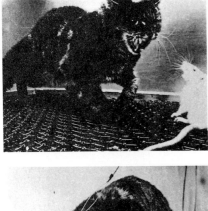

Figure 16.8
Rage reactions in cats with hypothalamic stimulation. (a) Stimulation of the medial hypothalamus produces affective aggression (threat attack). **(b)** Stimulation of the lateral hypothalamus evokes predatory aggression (silent-biting attack). (Source: Flynn, 1967, p. 45.)

ond monkey in the hierarchy discovered that the "top banana" had become more placid and less difficult to challenge. After an amygdalectomy was performed on the new dominant monkey, it likewise fell to the bottom of the hierarchy. This suggests that the amygdala is important for the aggression normally involved in maintaining a position in the social hierarchy.

Experiments in which amygdala subdivisions were electrically stimulated or destroyed suggest that it has multiple effects on aggressive behavior by way of its connections to the hypothalamus and other structures. Electrical stimulation of the basolateral nuclei produces affective aggression presumably through the effects of the efferents in the ventral amygdalofugal pathway on the hypothalamus and brain stem nuclei. Lesions in the basolateral nuclei reduce affective aggression.

The corticomedial nuclei have an inhibitory influence on aggression. Neurons in the corticomedial nuclei send axons through the stria terminalis to the hypothalamus. Lesions in either the corticomedial nuclei or the stria terminalis markedly increase predatory aggression. Therefore, we can infer that this portion of the amygdala may normally have an inhibitory effect on the hypothalamus, suppressing predatory aggression.

Surgery to Reduce Human Aggression. Recognizing that aggression is reduced by amygdalectomy in animals, some neurosurgeons thought it might be possible to similarly modify aggressive behavior in violent humans. It was thought by some that violent behavior frequently resulted from seizures in the temporal lobe. In a human amygdalectomy, electrodes are passed through the brain and down into the temporal lobe. By making neural recordings along the way and imaging the electrodes using X-rays, it is possible to get the tip of the electrode into the amygdala. Electrical current is passed through the electrode, or a solution is injected, to destroy all or part of the amygdala. Clinical reports claim considerable success in reducing aggressive asocial behavior, increasing the ability to concentrate, decreasing hyperactivity, and reducing seizures.

This type of brain surgery, a method of treating behavioral disorders, is called **psychosurgery**. Earlier in this century it was common to treat severe disorders involving anxiety, aggression, or neuroses with psychosurgical techniques, including the frontal lobotomy (Box 16.3). However, destroying a significant portion of the brain is a drastic measure. Removing the amygdala and/or other structures is irreversible, and one never knows until after recovery whether there was inadvertent damage to cognitive or behavioral function.

Serotonin and Aggression

Studies indicate that the neurotransmitter serotonin may be involved in aggression. Serotonin-containing neurons are located in the raphe nuclei of the brain stem, and they ascend in the medial forebrain bundle and project to the hypothalamus and various limbic structures involved in emotion (see Figure 15.12).

One link between serotonin and aggression comes from studies of induced aggression in rodents. Mice and rats that are isolated in a small cage for four weeks will frequently become hyperactive and extremely aggressive toward other mice. Although the isolation has no effect on the *level* of serotonin in the brain, there is a decrease in the *turnover rate* (the rate of synthesis, release, and resynthesis) of this neurotransmitter. Moreover, this decrease is only found in the mice that later become unusually aggressive, not in those relatively unaffected by the isolation. Also, female mice typically do not become aggressive following isolation, and they show no decrease in serotonin turnover. Evidence indicates that drugs that block the synthesis or release of serotonin increase aggressive behavior. In one study, the drug

Box 16.3 | OF SPECIAL INTEREST

The Frontal Lobotomy

Ever since the discoveries by Klüver, Bucy, and others that brain lesions can alter emotional behavior, clinicians have attempted surgery as a means to treat severe behavioral disorders in humans. Probably no other brain operation has received as much media coverage as the frontal lobotomy. Whether it's in a science fiction book or a punk rock song, stories are told of mind-altering surgery. Today it is difficult for many people to imagine that destruction of a large portion of the brain was once thought to be therapeutic. Indeed, in 1949 the Nobel Prize in Medicine was awarded to Dr. Egas Moniz for his development of the frontal lobotomy technique. Even stranger is the fact that Moniz was shot in the spine and partially paralyzed by a patient—either a tragedy or poetic justice, depending on your point of view. One doesn't hear about lobotomies being performed anymore, but tens of thousands were performed following World War II.

Little theory supported the development of the lobotomy. In the 1930s, John Fulton and Carlyle Jacobsen of Yale University reported that frontal lobe lesions had a calming effect in chimpanzees. It has been suggested that frontal lesions have this effect because of the destruction of limbic structures and, in particular, connections with frontal and cingulate cortex. Sad to say, this drastic surgery was done with barely more of a guiding principle than this: The limbic system controls emotion; therefore, people with emotional problems might be helped by destroying part of the system. In other words, a little emotion is a good thing but too much is debilitating and this can be surgically corrected.

A frightening variety of techniques were used to produce lesions in the frontal lobes. In the technique known as transorbital lobotomy, shown in Figure A, a knife was inserted through the thin bone at the top of the eye's orbit. The handle was then swung medially and laterally to destroy cells and interconnecting pathways. Thousands of people were lobotomized with this technique because it was so simple that it could even be performed in the physician's office. With this technique, called "ice pick psychosurgery"

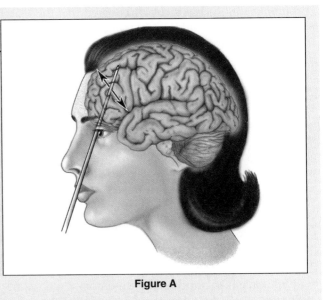

Figure A

by some, the physician could not see what was being destroyed. However, it left no outward scars.

Frontal lobotomy reportedly had beneficial effects on people with a number of disorders, including psychosis, depression, and various neuroses. The effect of the surgery was described as a relief from anxiety and escape from thoughts that were unendurable. Only later did a pattern of less pleasant side effects emerge. While frontal lobotomy can be performed with little decrease in IQ or loss of memory, it does have other profound effects. The changes that appear to be related to the limbic system are a blunting of emotional responses and a loss of the emotional component of thoughts. In addition, it is commonly found that lobotomized patients develop "inappropriate behavior" or an apparent lowering of moral standards. Similar to Phineas Gage, patients have considerable difficulty planning and working towards goals. Lobotomized patients have trouble concentrating and are easily distracted.

Perhaps more to the point, with our modest understanding of the neural circuitry underlying emotion and other brain functions, it is hard to justify destroying a large portion of the brain. Fortunately, treatment with lobotomy decreased fairly rapidly, and today drug therapy is primarily used instead for serious emotional disorders.

PCPA (parachlorophenylalanine), which blocks serotonin synthesis, was administered to rats, and the rats increased their attacks on the mice.

Recombinant DNA techniques have recently been used to produce mice lacking a certain serotonin receptor (the 5-HT$_{1B}$ receptor). Activation of this receptor is known to decrease aggression. The mutant mice lacking the receptor behaved normally and were not abnormally aggressive living together as a group. But when they were put in a stressful situation, such as having a new mouse placed in their house, the mutants were far more aggressive. An important question is where these serotonin receptors are located. Anatomical labeling techniques indicate that they may be found in the amygdala and periaqueductal gray matter, as well as the raphe nuclei and basal ganglia. The localization of this receptor in structures thought to be involved in aggressive behavior offers the potential of linking pharmacological studies of serotonin with the behavioral research we've examined.

REINFORCEMENT AND REWARD

In the early 1950s, James Olds and Peter Milner at the California Institute of Technology conducted an experiment in which a rat had an electrode implanted in its brain such that the brain could be locally stimulated at any time. The rat freely roamed about in a box about 3 feet square. Each time the rat went to a certain corner of the box, its brain was stimulated. After the first stimulation, the rat walked away but quickly returned and was again stimulated. Soon the rat was spending all of its time in the corner, apparently seeking the electrical stimulation. In a brilliant twist on this experiment, Olds and Milner set up a new box for the rat containing a lever that would deliver a brief stimulus to the brain when stepped on (Figure 16.9).

Figure 16.9
Electrical self-stimulation by a rat. When the rat presses on the lever, it receives a brief electrical current to an electrode in its brain.

At first the rat wandered about the box and stepped on the lever by accident, but before long it was pressing the lever repeatedly. This behavior is called **electrical self-stimulation**. Sometimes the rats would become so involved in pressing the lever that they would shun food and water, stopping only after collapsing from exhaustion.

Thus, reinforcing brain stimulation was discovered, and it raises many questions. What structures must be stimulated to get this kind of reinforcement? Why does the rat repeatedly stimulate itself? Is the effect pleasurable? Does the rat feel something akin to the satisfaction it might receive from food or sex? Olds and Milner's unexpected finding led to many subsequent studies. Deciphering the neural mechanisms underlying this artificial form of reinforcement will hopefully shed light on the physiological basis for normal reinforcing behaviors such as eating, drinking, and sex, and abnormal ones such as addiction.

Electrical Self-Stimulation and Reinforcement

It is not clear why the rats repeatedly pressed the lever in Olds and Milner's electrical self-stimulation experiments. One interpretation is that the rat got a positive feeling from the stimulation and therefore returned for more. Thus, the brain sites leading to reinforcing stimulation were called **pleasure centers**. But, that term misrepresents the results of the experiments in two significant ways. First, we don't know that something like pleasure was experienced. Perhaps the self-stimulation makes the rat want more stimulation without its being pleasurable. We know that humans will sometimes feel compelled to eat or drink alcohol even when these behaviors do not evoke a pleasurable feeling. Second, even if the stimulation is pleasurable, there may not be a particular "center" responsible for the reinforcement. Perhaps stimulation at a large number of diffusely scattered sites will produce self-stimulation, or maybe the stimulation affects bundles of axons rather than a nucleus of cells.

In the decades following Olds and Milner's discovery, many self-stimulation sites have been found in limbic structures and elsewhere in the rat's brain. These include the *septal area*, lateral hypothalamus, the medial forebrain bundle, the ventral tegmental area, and the dorsal pons (Figure 16.10). Stimulation of the medial forebrain bundle produces powerful reinforcement, and it has been studied more than any other structure. A smaller number of sites produce aversive behavior when electrically stimulated. For example, animals will actively avoid behavior that results in stimulation of certain parts of their brain. They also learn to perform tasks that will terminate stimulation at these sites. These "displeasure centers," or negative-reinforcement sites, are located in more medial portions of the hypothalamus and lateral parts of the midbrain tegmental area. Stimulation at these locations may evoke a negative feeling such as fear, or it may excite a pathway that is usually active in a negative-reinforcement situation such as fleeing from a predator.

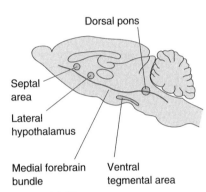

Dorsal pons

Septal area

Lateral hypothalamus

Medial forebrain bundle

Ventral tegmental area

Figure 16.10
Self-stimulation sites in the rat brain.
Rats will self-stimulate when electrodes are placed in the areas colored red.

Brain Stimulation in Humans

To determine the sensations evoked by brain stimulation, it would be desirable to stimulate a person's brain and ask how it feels. Obviously, this is not normally feasible or ethical. However, certain surgical procedures performed on humans require that they be alert during surgery and that their brain be electrically stimulated. As we have mentioned elsewhere in this book, the most common example is surgical treatment of severe epilepsy. When the brain is electrically stimulated, pleasurable feelings are sometimes evoked, so the term *pleasure center* may not be entirely inappropriate. However, in some cases where a patient can electrically self-stimulate, the brain site they choose to repeatedly stimulate does not cause a pleasurable feeling. Why do they do it? To answer this question, let's focus on two patients studied by Robert Heath at the Tulane University School of Medicine in the 1960s.

The first patient had severe narcolepsy; he would abruptly go from being awake into a deep sleep. (Narcolepsy and sleep will be discussed in Chapter 17.) The condition significantly interfered with his life and obviously made it difficult to hold a job. He was fitted with 14 electrodes in different areas of the brain in the hope of finding a self-stimulation site that might keep him alert. When he stimulated his hippocampus, he reported feeling mild pleasure. Stimulation of his midbrain tegmentum made him feel alert but unpleasant. The site he chose to frequently self-stimulate was the septal area of the forebrain (Figure 16.11). Stimulating this area made him more alert and gave him a good feeling, which he described as building up to orgasm. He reported that he would sometimes push the button over and over, trying unsuccessfully to achieve orgasm, ultimately ending in frustration. Despite the frustration, the most frequently self-stimulated site was associated with a positive feeling.

The second patient's case is a bit more complex. This person had electrodes implanted at 17 brain sites in the hope of learning something about the location of his severe epilepsy. He reported pleasurable feelings with stimulation of the septal area and the midbrain tegmentum. Consistent with the first case above, septal stimulation was associated with sexual feelings. The midbrain stimulation gave him a "happy drunk" feeling. Other mildly positive feelings were produced by stimulation of the amygdala and caudate nucleus. Interestingly, the site that was most frequently stimulated was in the medial thalamus, even though stimulation here induced an irritable feeling, one that was less pleasurable than stimulation at other locations. The patient stated that the reason he stimulated this area the most was that it gave him the feeling he was about to recall a memory. He repeated the stimulation in a futile attempt to fully bring the memory into his mind, even though, in the end, this process proved to be frustrating.

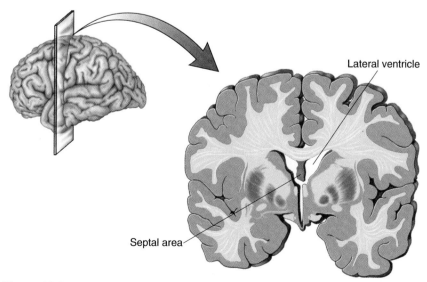

Figure 16.11
The septal area. The septal area is located in the rostral forebrain below the lateral ventricle.

The two specific cases above, and a host of others, lead us to the conclusion that self-stimulation sites in the human brain are not always synonymous with pleasure. There is often some reward or anticipated reward associated with the stimulation, but the experience is not always pleasant.

Dopamine and Reinforcement

One proposed explanation of the large number of scattered self-stimulation sites is that they are all interconnected by a common pathway involved in normal reward behavior. It is interesting to note that high rates of self-stimulation are obtained with electrodes placed in the medial forebrain bundle and ventral tegmental area. As discussed in Chapter 15, cell bodies of dopaminergic neurons are located in the ventral tegmental area (as well as the substantia nigra) of the midbrain, and they send axons through the medial forebrain bundle to many areas of the brain (see Figure 15.14). The medial forebrain bundle also contains descending fibers that could conceivably carry reward signals to the ventral tegmental area.

There is also pharmacological evidence linking dopamine to reinforcement behavior. DA agonists such as amphetamine increase the rate of self-stimulation. Injecting a rat with a drug that blocks DA receptors (e.g., haloperidol) decreases self-stimulation. But we must be cautious in interpreting this finding because dopamine is also important for the brain's control of movement. We saw in Chapter 14 that degeneration of the substantia nigra causes a depletion of dopamine leading to the movement disorder known as Parkinson's disease. Therefore, it is conceivable that rats given DA antagonists simply cannot press a lever, although that does not appear to be the case.

While much evidence suggests that dopamine is involved in reinforcement and reward, the nature of this association is not clear. Indeed, certain experimental data are difficult to reconcile with the hypothesis that dopaminergic neurons in the medial forebrain bundle play a critical role. For instance, several studies have shown that cutting the medial forebrain bundle does not have the devastating effect on self-stimulation one might expect. Also, it is important to remember that self-stimulation sites are not identical to the brain sites that receive dopaminergic input from the midbrain. Thus, we have the somewhat unsettled situation in which pharmacological experiments suggest an important role for dopamine, but lesion studies, among others, raise concerns. Dopamine may be important in some situations, but it cannot simply be the reward transmitter for the brain.

CONCLUDING REMARKS

We have examined several neural pathways that appear to be involved in the experience and expression of emotion. Our approach has been somewhat the opposite of that taken by Papez in the 1930s. With good insight, Papez proposed a large emotion system comprised of numerous structures along the midline of the brain. Although the role of each structure was far from clear, he knew that they were anatomically connected. Our discussion of the brain mechanisms of emotion has focused on a smaller number of structures. Rather than evaluating every component of the Papez circuit, we built up small circuits that experiments suggest are most clearly involved in emotion. Some structures, such as the amygdala, appear to be important for several different emotions.

We spent a lot of time discussing fear and aggression but no time discussing the physiology underlying many other emotions, such as hate and joy. The reason is simply the limitations of experimental technique. To study emotion, we must be able to reliably evoke that emotion in an animal and then find a way to describe the responses. This is why we are closer to understanding the neural circuitry involved in emotional expression than emotional experience. Even the studies of electrical self-stimulation in humans must be critically viewed because the effect of the stimulation is not simply a pleasant feeling, nor were these brains completely normal to begin with.

Keep in mind for later chapters that the structures apparently involved in emotion also have other functions. For a considerable time after the limbic lobe was defined by Broca, it was thought to be primarily an olfactory system. And even though our perspective has changed a lot since Broca's time, parts of the brain involved in olfaction have been included in the definition of the limbic system. We will see in Chapter 19 that some of the limbic structures are also important for learning and memory. How is it that these structures evolved to perform so many functions? We don't really know. There are interesting relationships between emotion, memory, and olfaction. Emotional events usually result in particularly enduring memories. Most people also feel that olfaction—a whiff of a certain perfume, for example—powerfully evokes memories.

KEY TERMS

What Is Emotion?
James-Lange theory
Cannon-Bard theory

The Limbic System Concept
limbic lobe
Papez circuit
limbic system

Fear and Anxiety
Klüver-Bucy syndrome
amygdala
basolateral nuclei
corticomedial nuclei

Anger and Aggression
predatory aggression
affective aggression
sham rage
medial forebrain bundle
dorsal longitudinal fasciculus
periaqueductal gray matter
psychosurgery

Reinforcement and Reward
electrical self-stimulation
pleasure center

REVIEW ✔ QUESTIONS

1. According to the James-Lange and Cannon-Bard theories of emotion, what is the relationship between the anxiety you would feel after oversleeping for an exam and your physical responses to the situation?
2. How have the definition of the limbic system and thoughts about its function changed since the time of Broca?
3. What procedures will produce an abnormal rage reaction in an experimental animal? How do we know that the animals *feel* angry?
4. What changes in emotion were observed following temporal lobectomy by Klüver and Bucy? Of the numerous anatomical structures they removed, which is thought to be closely related to changes in temperament?
5. Why might performing bilateral amygdalectomy on a dominant monkey in a colony result in that monkey's becoming a subordinate?
6. Does stimulation of "pleasure centers" always evoke pleasure? Why might a rat repeatedly self-stimulate a portion of its brain even though the experience is not pleasurable?
7. What evidence is there that dopamine plays an important role in reinforcement? What evidence is there to the contrary? What behavioral abnormality was linked to dopamine in Chapter 15?
8. What assumptions about limbic structures underlie the surgical treatment of emotional disorders?

Rhythms of the Brain

Earth is a rhythmic environment. Temperature, precipitation, and daylight vary with the seasons; light and dark trade places each day; tides ebb and flow. To compete effectively and therefore survive, an animal's behavior must oscillate with the cadences of its environment. Brains have evolved a variety of systems for rhythmic control. Sleeping and waking are the most striking periodic behavior. But some rhythms controlled by the brain have much longer periods, as in hibernating animals, and many have shorter periods, such as the cycles of breathing, the steps of walking, the repetitive stages of one night's sleep, and the electrical rhythms of the cerebral cortex. The functions of some rhythms are obvious, while others are obscure, and some rhythms indicate pathology.

In this chapter, we explore selected brain rhythms, beginning with the fast and proceeding to the slow. The forebrain, especially the cerebral cortex, produces a range of rapid electrical rhythms that are easily measured and that closely correlate with interesting behaviors, including sleep. We discuss the electroencephalogram, or EEG, because it is the classical method of recording brain rhythms and is essential for studying sleep. Sleep is treated at length because it is complex, ubiquitous, and so dear to our hearts. Finally, we summarize what is known about the timers that regulate the ups and downs of our hormones, temperature, alertness, and metabolism. Almost all physiological functions of the body change according to daily cycles known as circadian rhythms. The clocks that time circadian rhythms are in the brain, calibrated by the sun via the visual system, and they profoundly influence our health and well-being.

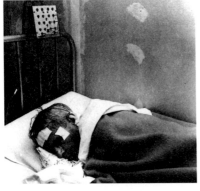

Figure 17.1
A subject in a sleep research study. The subject shown here is American sleep researcher Nathaniel Kleitman, codiscoverer of REM sleep. The white patches on his head are pieces of tape holding EEG electrodes, and those next to his eyes hold electrodes that monitor his eye movements. (Source: Carskadon, 1993.)

THE ELECTROENCEPHALOGRAM

Very often the forest is more interesting than the trees. Similarly, we are often less concerned with the activities of single neurons than with understanding the activity of a large population of neurons. The **electroencephalogram** (EEG) is a measurement that enables us to glimpse the generalized activity of the cerebral cortex. The roots of the EEG lie in work done by English physiologist Richard Caton in 1875. Caton made electrical recordings from the surface of dog and rabbit brains using wire electrodes and long strings that cast moving shadows on the wall to indicate electrical events. The human EEG was first described by Austrian psychiatrist Hans Berger in 1929. Berger observed that the waking and sleeping EEGs are distinctly different. Today, the EEG is mostly used to help diagnose certain neurological conditions, especially the seizures of epilepsy, and for research purposes, notably to study sleep.

Recording Brain Waves

Recording an EEG is relatively simple. The method is usually noninvasive, and it is painless. Countless people have slept through entire nights wearing EEG electrodes in the comfort of sleep research laboratories (Figure 17.1). The electrodes are wires taped to the scalp, along with conductive paste to ensure a low-resistance connection. As shown in Figure 17.2, some two dozen electrodes are fixed to standard positions on the head and connected to banks of amplifiers and recording devices. Small voltage fluctuations, usually a few tens of microvolts (µV) in amplitude, are measured between selected pairs of electrodes. Different regions of the brain—anteri-

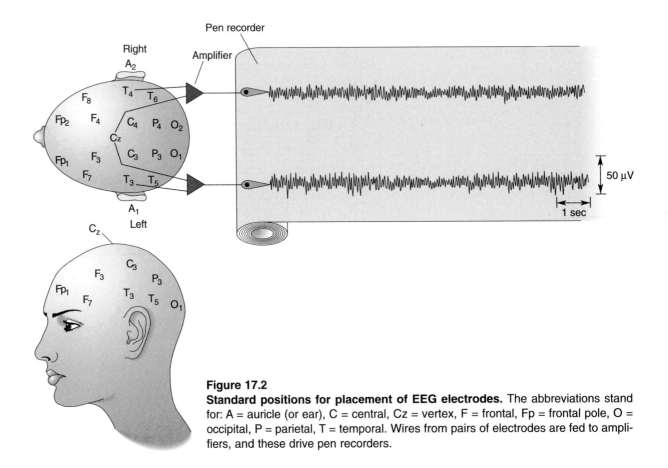

Figure 17.2
Standard positions for placement of EEG electrodes. The abbreviations stand for: A = auricle (or ear), C = central, Cz = vertex, F = frontal, Fp = frontal pole, O = occipital, P = parietal, T = temporal. Wires from pairs of electrodes are fed to amplifiers, and these drive pen recorders.

or and posterior, left and right—can be examined by selecting the appropriate electrode pairs. The typical EEG record is a set of many simultaneous squiggles, indicating voltage changes between pairs of electrodes.

What part of the nervous system generates the endless squiggles of an EEG? For the most part, an EEG measures the currents that flow during synaptic excitation of the dendrites of many pyramidal neurons in the cerebral cortex, which lies right under the skull and makes up 80% of the brain's mass. But the electrical contribution of any single cortical neuron is exceedingly small, and the signal must penetrate several layers of non-neural tissue, including the meninges, fluid, bones of the skull, and skin, to reach the electrodes (Figure 17.3). Therefore, it takes many thousands of underlying neurons, activated together, to generate an EEG signal big enough to see at all. This has an interesting consequence: The amplitude of the EEG signal strongly depends on how *synchronous* is the activity of the underlying neu-

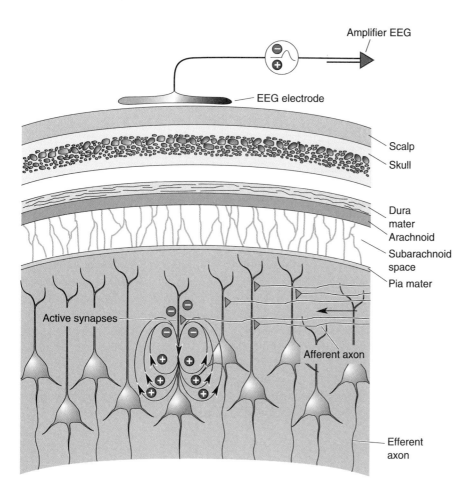

Figure 17.3
Generation of very small electrical fields by synaptic currents in pyramidal cells. In this case, the synapse is on the upper part of the dendrite. When the afferent axon fires, the presynaptic terminal releases glutamate, which opens numerous cation channels. Positive current flows into the dendrite, leaving a slight negativity in the extracellular fluid. Current spreads down the dendrite and escapes out of its deeper parts, leaving the extracellular fluid slightly positive at those sites. The EEG electrode (referred to a second electrode at some distance away) measures this signal through thick layers of tissue. Only if thousands of cells contribute their small voltage does the signal become large enough to see at the surface of the scalp. (Notice that the convention in EEG work is to plot the signals with negativity upward.)

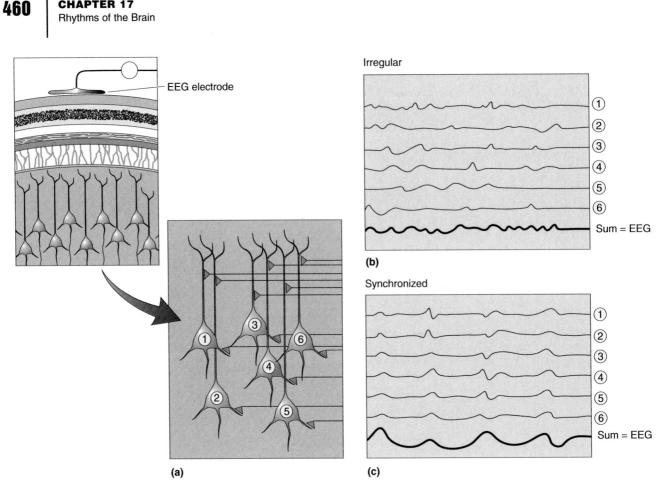

Figure 17.4
Generation of large EEG signals by synchronous activity. (a) In a population of pyramidal cells located under an EEG electrode, each neuron receives many synaptic inputs. **(b)** If the inputs fire at irregular intervals, the summed activity detected by the electrode will be of small amplitude. **(c)** If the same number of inputs fire within a narrow time window, the resulting EEG will be much larger.

rons. When a group of cells is excited simultaneously, the tiny signals sum to generate one large surface signal. However, when each cell receives the same amount of excitation, but spread out in time, the summed signals are meager and irregular (Figure 17.4). Notice that in this case the *number* of activated cells and the *total amount of excitation* may not have changed, only the timing of activity. If synchronous excitation of this group of cells is repeated again and again, the resulting EEG will consist of large, rhythmic waves. We often describe rhythmic EEG signals in terms of their relative amplitude, suggesting how synchronous the underlying activity is (although other factors may contribute to amplitude as well).

EEG Rhythms

EEG rhythms vary dramatically and often correlate with particular states of behavior (such as level of attentiveness, sleeping, or waking) and pathology (seizures or coma). Figure 17.5 shows part of a normal EEG. The rhythms are categorized by their frequency range, and each range is named after a Greek letter. *Beta* rhythms are fastest, anything greater than about 14 Hz, and signal an activated cortex. *Alpha* rhythms are about 8–13 Hz and are associated with quiet, waking states. *Theta* rhythms are 4–7 Hz and occur during some sleep states. *Delta* rhythms are quite slow, less than 4 Hz, often large in amplitude, and are a hallmark of deep sleep.

While analysis of an EEG will never tell us *what* a person is thinking, it can help us know *if* a person is thinking. In general, high-frequency/low-amplitude rhythms are associated with alertness and waking, or the dreaming stages of sleep. Low-frequency/high-amplitude rhythms are associated with nondreaming sleep states, or the pathological state of coma. This is logical because when the cortex is most actively engaged in processing information, whether generated by sensory input or by some internal process, the activity level of cortical neurons is relatively high but also relatively unsynchronized. In other words, each neuron, or very small group of neurons, is vigorously involved in a slightly different aspect of a complex cognitive task; it fires rapidly, but not quite simultaneously with most of its neighbors. This leads to low synchrony, so EEG amplitude is low, and beta waves dominate. By contrast, during deep sleep, cortical neurons are not engaged in information processing, and large numbers of them are phasically excited by a common, slow, rhythmic input. In this case synchrony is high, so EEG amplitude is high.

Mechanisms and Meanings of Brain Rhythms

Electrical rhythms abound in the cerebral cortex. But how are they generated, and what functions do they perform? Let's take a look at each of these questions in turn.

Generation of Synchronous Rhythms. There are two fundamental ways in which the activity of a large set of neurons will produce synchronized oscillations: (1) They may all take their cues from a central clock, or *pacemaker*; or (2) they may share or distribute the timing function among themselves by mutually exciting or inhibiting one another. The first mechanism is analogous to a leader and a band, with each musician playing in strict time to the baton's changing beat (Figure 17.6). The second mechanism is more subtle, because the timing arises from the collective behavior of the cortical neurons themselves. Musically, it is more like a jam session. The concept is easily demonstrated by a group of friends, even nonmusical ones. Simply tell them to clap together, but give them no instructions about how fast to clap or whose beat to follow. Within one or two claps, they will all be clapping in synchrony! How? By listening and watching each other, they adjust their clapping rates to match that of the others. A key factor is person-to-person interaction; in a network of neurons, these interactions occur via synaptic connections. People tend to clap within a narrow range of fre-

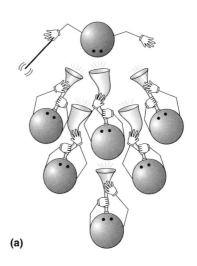

(a)

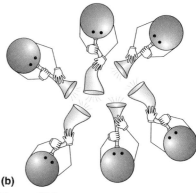

(b)

Figure 17.6
Two mechanisms of synchronous rhythms. (a) Synchronous rhythms may be led by a pacemaker or **(b)** arise from the collective behavior of all participants.

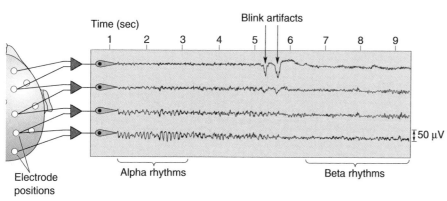

Figure 17.5
A normal EEG. The subject is awake and quiet, and recording sites are indicated at the left. The first few seconds show normal alpha activity, which has frequencies of 8–13 Hz and is largest in the occipital regions. About halfway through the recording, the subject opened his eyes, signaled by the large blink artifacts on the top traces (arrows), and alpha rhythms were suppressed.

quencies, so they don't have to adjust their timing very far in order to clap in synchrony. Likewise, some neurons may fire at certain frequencies much more than others. This kind of collective, organized behavior can generate rhythms of impressive dimensions, which can move in space and time. Have you ever been part of a human wave in the stands of a sold-out football stadium?

Many different circuits of neurons can generate rhythmic activity. A very simple one, consisting of only one excitatory and one inhibitory neuron, is shown in Figure 17.7.

Within the mammalian brain, rhythmic, synchronous activity is usually coordinated by a combination of the pacemaker and the collective methods. The thalamus, with its massive input to all of the cortex, can act as a powerful pacemaker. Under certain conditions, thalamic neurons can generate very rhythmic action potential discharges. But how do thalamic neurons oscillate? Thalamic cells have a particular set of voltage-gated ion channels that allow each cell to generate very rhythmic, self-sustaining discharge patterns even when there is no external input to the cell (Figure 17.8). The rhythmic activity of each thalamic pacemaker neuron then becomes synchronized with many other thalamic cells via a hand-clapping kind of collective interaction. Synaptic connections between excitatory and inhibitory thalamic neurons force each individual neuron to conform to the rhythm of the group. These coordinated rhythms are then passed to the cortex by the

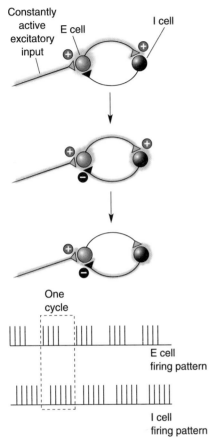

Figure 17.7
A two-neuron oscillator. One excitatory cell (E cell) and one inhibitory cell (I cell) synapse upon each other. As long as there is a constant excitatory drive (which does not have to be rhythmic) onto the E cell, activity will tend to trade back and forth between the two neurons. One activity cycle through the network will generate the pattern of firing shown in the dashed box.

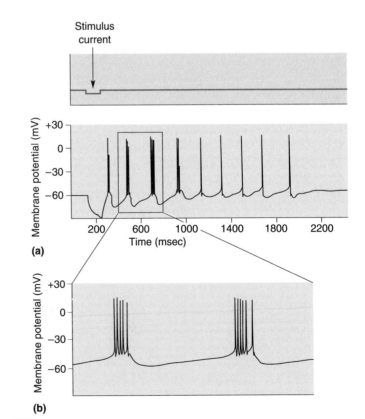

Figure 17.8
A one-neuron oscillator. At times during sleep states, thalamic neurons fire in rhythmic patterns that do not reflect their input. Shown here are intracellular recordings of membrane voltage in such a case. **(a)** A short pulse (less than 0.1 sec) of stimulus current was applied, and the cell responded with almost 2 sec of rhythmic firing, first with bursts at about 5 Hz, and then with single spikes. **(b)** Two of the bursts expanded in time; each burst is a cluster of 6 or 7 action potentials. (Source: Adapted from Bal and McCormick, 1993, Fig. 2.)

thalamocortical axons, which excite cortical neurons. In this way, a relatively small group of centralized thalamic cells (acting as band-leader) can compel a much larger group of cortical cells (acting as the band) to march to the thalamic beat (Figure 17.9).

Some rhythms of the cerebral cortex do not depend on the thalamic pacemaker but rely instead on the collective, cooperative interactions of cortical neurons themselves. In this case the excitatory and inhibitory interconnections of the neurons result in a coordinated, synchronous pattern of activity that may remain localized or can spread to encompass larger regions of cortex.

Functions of Brain Rhythms. Cortical rhythms are fascinating to watch, and they parallel so many interesting human behaviors that we are compelled to ask: Why so many rhythms? More importantly, do they serve a purpose? There are no satisfactory answers yet. Ideas abound, but relevant data are scarce. One hypothesis for sleep-related rhythms is that they are the brain's way of disconnecting the cortex from sensory input. When you are awake, the thalamus allows sensory information to be relayed up to the cortex, or to pass through it. When you are asleep, thalamic neurons enter a self-generated rhythmic state that prevents organized sensory information from penetrating upward to the cortex. While this idea has intuitive appeal (most people do prefer to sleep in a dark, quiet environment), it does not explain why *rhythms* are necessary. Why not just steadily inhibit the thalamus and allow the cortex to rest quietly?

A function for fast rhythms in the awake cortex has recently been proposed. Recall from Chapter 10 that one scheme for understanding visual perception takes advantage of the fact that cortical modules responding to the same object are synchronously active. Walter Freeman, a neurobiologist at the University of California, Berkeley, has pioneered hypotheses suggesting that neural rhythms are used to coordinate activity between regions of the nervous system. Both sensory and motor systems of the awake brain often generate bursts of synchronously active neurons that give rise to EEG oscillations in the 30–80 Hz range. It may be that by momentarily synchronizing the fast oscillations generated by different regions of cortex, the brain binds together various neural components into a single perceptual construction. For example, when you are trying to catch a basketball, different groups of neurons that simultaneously respond to the specific shape, color, movement, distance, and even identity and importance of the basketball will tend to oscillate synchronously. The fact that the oscillations of these scattered groups of cells (those that together encode "basketballness") are highly *synchronous* would somehow tag them as a meaningful group, distinct from other nearby neurons, thereby unifying the disjointed neural pieces of the "basketball puzzle." The evidence for this idea is indirect, far from proven, and understandably controversial.

For now, the functions of rhythms in the cerebral cortex are largely a mystery. One plausible hypothesis is that most rhythms have no direct function. Instead, they may be intriguing but unimportant by-products of the tendency for brain circuits to be strongly interconnected, with various forms of excitatory feedback. When something excites itself, whether it is an audio amplifier or the human stadium wave, it often leads to instability or oscillation. Feedback circuits are essential for the cortex to do all the marvelous things it does for us. Oscillations may be the unfortunate but unavoidable consequence, unwanted but tolerated by necessity. Even without a function, EEG rhythms provide us with a convenient window on the functional states of the brain.

The Seizures of Epilepsy

Seizures, the most extreme form of synchronous brain activity, are always a sign of pathology. A **generalized seizure** involves the entire cerebral cor-

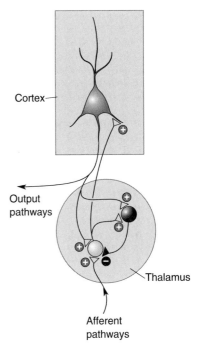

Cortex

Output
pathways

Thalamus

Afferent
pathways

Figure 17.9
Rhythms in the thalamus drive rhythms in the cerebral cortex. The thalamus can generate rhythmic activity because of the intrinsic properties of its neurons and because of its synaptic interconnections. Yellow represents populations of excitatory neurons, and black represents a population of inhibitory neurons.

tex of both hemispheres. A **partial seizure** involves only a circumscribed area of the cortex. In both cases, the neurons within the affected areas fire with a synchrony that never occurs during normal behavior. As a consequence, seizures are usually accompanied by very large EEG patterns. The cerebral cortex, probably because of its extensive feedback circuitry, is never far from the runaway excitation we know as a seizure. Isolated seizures are not uncommon during a lifetime, and 7–10% of the general population have at least one. When a person experiences repeated seizures, the condition is known as **epilepsy**. About 1% of the U.S. population (25,000,000 people) has epilepsy.

Epilepsy is not so much a disease itself as it is a symptom of disease. Its causes can sometimes be identified, and they include tumors, trauma, metabolic dysfunction, infection, vascular disease, and genetic predisposition, but in many cases the cause of epilepsy is not known. It is highly unlikely that a single mechanism accounts for all seizures. Research suggests that some seizures may reflect an upset of the delicate balance of synaptic excitation and inhibition in the brain. Other seizures may be due to excessively strong or dense excitatory interconnections. Drugs that block GABA receptors are very potent *convulsants* (seizure-promoting agents). The withdrawal of chronic depressant drugs, such as alcohol or barbiturates, may also trigger seizures. A variety of drugs are useful in the suppression of seizures, and these *anticonvulsants* tend to counter excitability in various ways. For example, some act by prolonging the inhibitory actions of GABA (e.g., barbiturates and benzodiazepines), while others decrease the tendency for certain neurons to fire high-frequency action potentials (e.g., phenytoin and carbamazepine).

The behavioral features of a seizure depend on the neurons involved and the patterns of their activity. During most forms of generalized seizures, virtually all cortical neurons participate, so behavior is completely disrupted for many minutes. Consciousness is lost, while all muscle groups may be driven by tonic (i.e., ongoing) activity or by clonic (i.e., rhythmic) patterns, or by both in sequence. *Absence seizures* occur during childhood and consist of less than 30 sec of generalized, 3 Hz EEG waves accompanied by loss of consciousness. Strangely, the motor signs of an absence seizure are subtle—a fluttering of the eyelids or a twitching of the mouth.

Partial seizures can be fascinating to study and instructive anatomically. If they begin in a small area of motor cortex, they can cause clonic movement of part of a limb. In the late 1800s, American neurologist John Hughlings Jackson observed the progression of seizure-related movements across the body, looked for the scars in his patients' brains after they died, and correctly inferred the basic somatotopic map of the motor cortex (see Chapter 14). If seizures begin in a sensory area, they may trigger an abnormal sensation, or *aura*, such as an odd smell or sparkling lights. Most bizarre are the partial seizures that elicit more well-formed auras such as *déjà vu* (the feeling that something has happened before) or hallucinations. These will sometimes involve the cortex of the temporal lobes, including the hippocampus and amygdala, and will impair memory, thought, and consciousness. Finally, partial seizures may spread uncontrollably and become generalized seizures.

SLEEP

Sleep and dreams—they are mysterious, even mystical to some people, and a favorite subject for art and literature, philosophy and science. Sleep is a powerful master. Each night we abandon our companions, our work and our play, and enter the cloister of sleep. We have only limited control over the decision; we can postpone sleep for a while, but eventually it overwhelms us. We spend about one-third of our lives sleeping, and one-quarter of that time in a state of active dreaming. Sleep is universal among higher vertebrates, perhaps among all animals. Prolonged sleep deprivation is devastating to proper functioning, at least temporarily; and in some animals (though not humans), it may even cause death. Sleep is essential to our lives—almost as important as eating and breathing. But why do we sleep? What purpose does it serve? To quote from a book by James Horne entitled *Why We Sleep*, "Many people feel that, despite 50 years of research, all we can conclude about the function of sleep is that it overcomes sleepiness" (1988, p. 1). But one of the wonderful things about science is that the lack of consensus inspires a flourishing of theories, and sleep research is no exception.

We can still describe what we cannot explain, and sleep has been richly studied. Let's begin with a definition: *Sleep is a readily reversible state of reduced responsiveness to, and interaction with, the environment.* (Coma and general anesthesia are not readily reversible and do not qualify as sleep.) In the following sections, we discuss the phenomenology and neural mechanisms of sleep and dreaming.

The Functional States of the Brain

During a normal day, you experience two very different and noticeable types of behavior: waking and sleeping. It is much less obvious that your sleep also has very distinct phases. Several times during a night you enter a state called **rapid eye movement sleep**, or **REM sleep**, when your EEG looks more awake than asleep, your body (except for your eye muscles) is immobilized, and you conjure up the vivid, detailed illusions we call dreams. The rest of the time you spend in a state called **non-REM sleep** (or NREM sleep), in which the brain does not usually generate complex dreams. (Non-REM sleep is also sometimes called slow-wave sleep because of its domination by large, slow EEG rhythms.) These fundamental behavioral states—awake, non-REM sleep, and REM sleep—are produced by three distinct states of brain function (Table 17.1). Each state is also accompanied by large shifts in body function.

Non-REM sleep seems to be a state designed for rest. Muscle tension throughout the body is reduced, and movement is minimal. It is important to realize that the body is *capable* of movement during non-REM sleep, but only rarely does the brain command it to move, usually to briefly adjust the body's position. The temperature and energy consumption of the body are lowered. Because of an increase in activity of the parasympathetic division of the ANS, heart rate, respiration, and kidney function all slow down, and digestive processes increase. The brain also seems to rest. Its rate of energy use, and the general firing rates of its neurons, are at their lowest point of

Table 17.1
Characteristics of the Three Functional States of the Brain

	Awake	Non-REM sleep	REM sleep
EEG	Low voltage, fast	High voltage, slow	Low voltage, fast
Sensation	Vivid, externally generated	Dull or absent	Vivid, internally generated
Thought	Logical, progressive	Logical, repetitive	Vivid, illogical, bizarre
Movement	Continuous, voluntary	Occasional, involuntary	Muscle paralysis; movement commanded by brain but not carried out
Rapid eye movement	Often	Rare	Often

the day. The slow, large-amplitude EEG rhythms indicate that the neurons of the cortex are oscillating in relatively high synchrony, and experiments suggest that most sensory input cannot even reach the cortex. While there is no way to know for certain what people are thinking when they are asleep, indications are that mental processes also hit their daily low during the non-REM state. When awakened, people usually recall nothing, or only very vague thoughts. Detailed, entertaining types of dreams are rare, although not absent, during non-REM sleep. William Dement, a prominent sleep researcher at Stanford University, characterizes non-REM sleep as *an idling brain in a movable body.*

In contrast, Dement calls REM sleep *an active, hallucinating brain in a paralyzed body.* REM sleep is dreaming sleep. Although the REM period accounts for only a small part of our sleep, it is the part most researchers get excited about (and this is the state that most excites the brain), probably because dreams are so intriguing and enigmatic. If you awaken someone during REM sleep, as Dement, Eugene Aserinsky, and Nathaniel Kleitman first did in the mid-1950s, he will likely report visually detailed, lifelike episodes, often with bizarre story lines—the kinds of dreams we love to talk about and try to interpret. The physiology of REM sleep is also bizarre. The EEG looks almost indistinguishable from that of an active, waking brain, with fast, low-voltage fluctuations. In fact, the oxygen consumption of the brain (a measure of its energy use) is higher in REM sleep than when the brain is awake and concentrating on difficult mathematical problems. The paralysis that occurs during the REM state is almost a total loss of skeletal muscle tone, or **atonia**. Most of the body is actually *incapable* of moving! Respiratory muscles do continue to function, but just barely. The muscles controlling eye movement and the tiny muscles of the inner ear are the exceptions; these are strikingly active. With lids closed, the eyes occasionally dart rapidly back and forth. These bursts of rapid eye movement are the best predictors of vivid dreaming, and 90–95% of people awakened during or after them report dreams. Physiological control systems are dominated by sympathetic activity during REM sleep. Inexplicably, the body's temperature control system simply quits, and core temperature begins to drift downward. Heart and respiration rates increase but become irregular. In healthy people, during REM sleep the clitoris and penis become engorged with blood and erect, although this usually has nothing to do with the sexual content of dreams. Overall, during REM sleep the brain seems to be doing everything except resting.

The Sleep Cycle

Even a good night's sleep is not a steady, unbroken journey. Sleep takes the brain through a repetitive roller-coaster ride of activity (Box 17.1). Roughly 75% of total sleep time is spent in non-REM and 25% in REM, with periodic cycles between these states throughout the night (Figure 17.10a). Non-REM sleep is generally divided into four distinct stages. During a normal night, we slide through the stages of non-REM, then into REM, then back through the non-REM stages again, repeating the cycle about every 90 minutes. These cycles are examples of **ultradian rhythms**, which have faster periods than circadian rhythms.

EEG rhythms during the stages of sleep are shown in Figure 17.10b. As an average, healthy adult becomes drowsy and begins to sleep, she first enters non-REM stage 1. Stage 1 is transitional sleep, when the EEG alpha rhythms of relaxed waking become less regular and wane, and the eyes make slow, rolling movements. Stage 1 is fleeting, usually lasting only a few minutes. It is also the lightest stage of sleep, meaning that we are most eas-

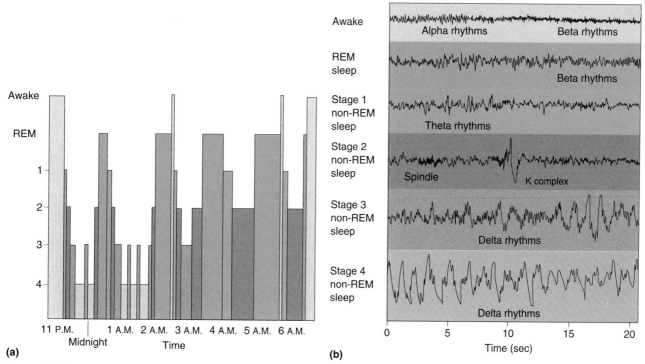

(a)

(b)

Figure 17.10

The stages of sleep through one night. (a) This graph represents falling asleep at 11 P.M., and entering non-REM sleep stage 1. The night's sleep progresses through the deeper stages of non-REM, then up and into REM. The cycle is repeated several times. However, each cycle tends to have shorter and shallower non-REM periods and longer REM periods. **(b)** EEG rhythms during the stages of sleep. (Source: Adapted from Horne, 1988, Fig. 1.1.)

Box 17.1 | **OF SPECIAL INTEREST**

Walking, Talking, and Screaming in Your Sleep

Sleep is not always serene and stationary. Talking, walking, and screaming are common, and they usually occur during non-REM sleep. If this seems surprising, remember that REM sleep is accompanied by almost total body paralysis. You would therefore be incapable of walking or talking during REM sleep, even if your dream "urged" you to do it.

Sleepwalking, or *somnambulism*, peaks at about age 11. Although 40% of us were sleepwalkers as children, few of us sleepwalk as adults. Sleepwalking usually occurs during the first stage 4 non-REM period of the night. A full-blown sleepwalking incident involves open eyes and movement around the room, the house, or even outside, with enough awareness to avoid objects and climb stairs. Cognitive functioning and judgment are severely impaired. It is often difficult to awaken sleepwalkers because they are in deep, slow-wave sleep. The best treatment is a guiding hand back to bed. Sleepwalkers usually have no memory of the incident the next morning.

Almost everyone practices sleep talking, *somniloquy*, now and then. Unfortunately, sleeping speech is usually so garbled or nonsensical that a curious listener is disappointed by its emptiness. More dramatic are *sleep terrors*, which are most common in children 5–7 years old. A girl screams in the middle of the night. Her parents rush to her bedside, frantic to know what has terrified her. The girl cries inconsolably, unable to explain her horrifying experience. After 10 agonizing minutes of shrieking and flailing, she finally sleeps quietly, leaving the parents shaken and baffled. The next morning she is bright and cheerful, with no recollection of the night's misadventure. Sleep terrors are distinctly different from nightmares, which are vivid, complex dreams, outwardly quiet, that occur during REM sleep. By contrast, sleep terrors begin in stage 3 or 4 of non-REM sleep, and the experience is not dreamlike but a feeling of uncontrollable panic, accompanied by greatly increased heart rate and blood pressure. They usually pass with age and are not a symptom of a psychiatric disorder.

ily awakened. Stage 2 is slightly deeper and may last 5–15 min. Its characteristics include the occasional 8–14 Hz oscillation of the EEG called the sleep spindle, which is known to be generated by a thalamic pacemaker. In addition, a high-amplitude sharp wave called the K complex is observed. Eye movements almost cease. Next follows stage 3, and the EEG begins large-amplitude, slow delta rhythms. Eye and body movements are absent. Stage 4 is the deepest stage of sleep, with large EEG rhythms of 2 Hz or less. During the first cycle of sleep, stage 4 may persist for 20–40 min. Then sleep begins to lighten again, ascends to stage 2 for 10–15 minutes, and suddenly enters a brief period of REM sleep, with its fast EEG beta rhythms and sharp, frequent eye movements.

As the night progresses, there is a general reduction in the duration of non-REM sleep, particularly in stages 3 and 4, and an increase in the REM periods. Half of the night's REM sleep occurs during its last third, and the longest REM cycles may last 30–50 min. Still, there seems to be an obligatory refractory period of about 30 min between periods of REM; in other words, each REM cycle is followed by at least 30 min of non-REM sleep before the next REM period can begin.

What is a normal night's sleep? Your mother may have insisted that you need a "good 8 hours" of sleep each night. Research suggests that normal requirements vary widely among adults, from about 5–10 hr per night. The average length is about 7.5 hr, and the sleep duration of about 68% of young adults is between 6.5 and 8.5 hr. What is the proper length of sleep time for you? The best measure of successful sleep is the quality of your time awake. You need a certain amount of sleep in order to maintain a reasonable level of alertness. Too much daytime sleepiness can be more than annoying; it can be dangerous if it interferes with driving, for example. Because of the wide variations among individuals, you must decide for yourself how much sleep you need.

Why Do We Sleep?

All mammals, birds, and reptiles appear to sleep, although only mammals and some birds have a REM phase. Sleep time varies widely, from about 18 hr in bats and opossums to about 3 hr in horses and giraffes. Some people argue that a behavior this pervasive must have a critical function; otherwise some species would have lost the need to sleep through evolution.

Some animals apparently have more reason *not* to sleep than others. Imagine living your entire life in deep or turbulent water, yet needing to breathe air every minute or so. Even a quick nap would be awkward, at best. This is precisely the situation with dolphins, yet they sleep about as much as humans do. Remarkably, bottlenose dolphins sleep with only one cerebral hemisphere at a time: About 2 hr asleep on one side, then 1 hr awake on both sides, 2 hr asleep on the other side, and so on, for a total of 12 hr per night (Figure 17.11). (This gives new meaning to being "half asleep.") Bottlenose dolphins seem to have no REM sleep. The blind Indus dolphin of Pakistan lives in an even more arduous environment. It uses sonar to navigate through muddy, turbid, sweeping currents, and during monsoon season it must never stop swimming or it will come to grief on the rocks and debris of the flooded estuary it calls home. Still, the Indus dolphin manages to sleep, snatching "microsleeps" 4–6 sec long while continuing to swim slowly. Its many microsleeps add up to about 7 hr in a 24 hr-day. Dolphins have evolved extraordinary sleep mechanisms that adapt them to a demanding environment. But the fact that dolphins are not sleepless reinforces our question: What is so important about sleeping?

No single theory of sleep function is widely accepted, but the most reasonable ideas fall into two categories: theories of *restoration* and theories of *adaptation*. The first category is a commonsense explanation: We sleep in order to rest and recover, and to prepare to be awake again. The second category is less obvious: We sleep to keep us out of trouble, to hide from predators when we are most vulnerable or from other harmful features of the environment, or to conserve energy.

If sleep is restorative, what is it restoring? Quiet rest is certainly not a substitute for sleep. Sleeping does something more than simple resting. Prolonged sleep deprivation may lead to serious physical and behavioral problems (Box 17.2). Unfortunately, no one has yet identified a particular physiological process that is clearly restored by sleep, an essential substance that is made or a toxin that is destroyed while sleeping. Sleep does prepare us to be effectively awake again. But does sleep renew us in the same way that eating and drinking do, by replacing essential substances, or the way the healing of a wound repairs damaged tissues? For the most part, evidence indicates that sleep is not a time of increased tissue repair for the body. However, it is possible that brain regions such as the cerebral cortex can achieve some form of essential "rest" only during non-REM sleep.

Adaptation theories of sleep take many forms. Some large animals eat small animals; a stroll in the moonlight is far too risky for a squirrel living in owl and fox territory. The squirrel's best strategy may be to stay safely tucked away in an underground burrow during the night, and sleep is a good way to enforce such isolation. At the same time, sleep may be an adaptation for conserving energy. While sleeping, the body does only just enough work to stay alive, core temperature drops, temperature regulation is depressed, and the rate of calories burned is kept low.

Functions of Dreaming and REM Sleep

People of many ancient cultures believed that dreams were a window on some higher world and a source of information, guidance, power, or enlightenment. Perhaps they were right, but the collective wisdom of the past does not agree on exactly how to interpret the meaning of dreams. Today we must take a step backward and first ask whether dreams even *have* meaning. Dreams are difficult to study. Obviously, we can't directly observe the dreams of someone else, and even the dreamer has access to them only after she has awakened, and perhaps forgotten or distorted the

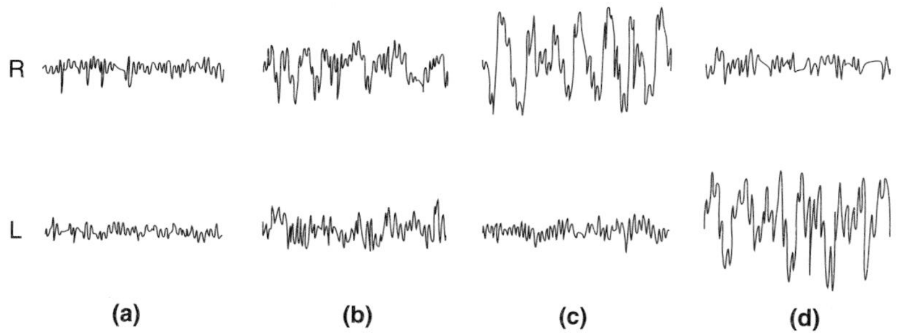

Figure 17.11
Sleep in the bottlenose dolphin. These are EEG patterns recorded from the right (R) and left (L) hemispheres. **(a)** High-frequency activity on both sides during alert wakefulness. **(b)** Low-amplitude delta rhythms of an intermediate sleep stage on the right side, and to some extent on the left. **(c)** Large delta rhythms of deep sleep only on the right side, with fast activation on the left. **(d)** The patterns shift to opposite hemispheres some time later. (Source: Mukhametov, 1984, Fig. 1.)

experience. Modern explanations of dreaming lean heavily on studies of REM sleep rather than dreaming because the phenomena of REM can be objectively measured. But it is important to remember that the two are not synonymous. Some dreams can occur outside of REM sleep, and REM sleep has many peculiar features that have nothing to do with dreaming.

Do we need to dream? No one knows, but the body does seem to crave REM sleep. It is possible to deprive sleepers of REM sleep specifically, by waking them every time they enter the REM state; when they fall asleep a minute or two later, it is inevitably into a non-REM state, and they can accumulate an entire night of relatively pure non-REM sleep. As Dement first observed, after several days of this annoying treatment, sleepers attempt to enter the REM state much more frequently than normal. When they are finally allowed to sleep undisturbed, they experience *REM rebound* and spend more time in REM proportional to the duration of their deprivation. Most studies have not found that REM deprivation causes any psychological harm during the daytime. Again, it is important not to interpret REM deprivation as dream deprivation, since during REM deprivation, dreams may continue to occur during sleep onset and during non-REM periods.

Sigmund Freud suggested many functions for dreams. For Freud, dreams were disguised wish-fulfillment, an unconscious way for us to express our sexual and aggressive fantasies, which are forbidden while we are awake. Bad dreams might help us conquer the anxiety-provoking events of life. Recent theories of dreaming are more biologically based. Allan Hobson and Robert McCarley of Harvard University propose an "activa-

---

Box 17.2 | **OF SPECIAL INTEREST**

The Longest All-Nighter

In 1963, Randy Gardner was a 17-year-old high school student with an ambitious idea for a San Diego Science Fair project. On December 28 he awoke at 6 A.M. to begin. When he finished 11 days (264 hours) later, he had broken the world's record for continuous wakefulness, under the continuous scrutiny of two friends and, during the last 5 days, fascinated sleep researchers. He had used no drugs, not even caffeine. The experience was not pleasant. Randy quickly became irritable, nauseated, had trouble remembering, and could not even watch television after 2 days. By the fourth day he had mild delusions and overwhelming fatigue, and by the seventh day he had tremors, his speech was slurred, and his EEG no longer showed alpha rhythms. Fortunately, he did not become psychotic, despite the predictions of some "experts." On the contrary, on his last awake night he beat one of his better-rested observers at an arcade baseball game, and he gave a coherent account of himself at a national press conference. When he finally went to bed, he slept for almost 15

hours straight, then stayed awake 23 hours to wait for nightfall, and slept for another 10.5 hours. After the first sleep his symptoms had mostly disappeared, and within a week he was sleeping and behaving normally.

One of the most interesting things about Randy's ordeal is that there were no lasting harmful effects. The same is not true for some animals deprived of sleep. If rats are kept awake for long periods, they progressively lose weight while consuming much more food, they become weak, accumulate stomach ulcers and internal hemorrhages, and even die. They seem to suffer from an impairment of their ability to regulate body temperature and metabolic needs. Total sleep deprivation is not necessary. Prolonged REM sleep loss alone is detrimental. These results may imply that sleep does provide something physiologically essential.

tion-synthesis" hypothesis, which explicitly rejects the Freudian, psychological interpretations (Box 17.3). Instead, dreams, or at least some of their bizarre features, are seen as the associations and memories of the cerebral cortex that are elicited by the random discharges of the pons during REM sleep. Thus the pontine neurons, via the thalamus, *activate* various areas of the cerebral cortex, elicit well-known images or emotions, and the cortex then tries to *synthesize* the disparate images into a sensible whole. Not sur-

---

Box 17.3 | **PATH OF DISCOVERY**

Dreams and the Brain

by J. Allan Hobson

J. Allan Hobson

My story has three parts: how I discovered REM sleep, how I discovered its chemical control by the brain stem, and how I discovered that dreaming was understandable at the level of brain chemistry.

Discovering REM sleep was easy. In 1961 I simply accepted Fred Snyder's invitation to spend a night in his NIMH (National Institutes of Mental Health) sleep lab. Of course, REM sleep had already been described 8 years earlier, in 1953, by Aserinsky and Kleitman. In 1961 I discovered REM *for myself*. I saw it; I saw the brain waves flatten out. I saw the eyes dancing. And I heard the experimental subject give a detailed dream report when awakened. For me, that night was the scientific equivalent of a religious conversion experience because REM sleep was the scientific problem that I had been looking for. Although I was groggy and sleep-deprived the next morning, I sang all the way home because I knew that by unraveling the mechanisms of REM I could understand how and why the brain dreams. To a young psychiatrist, that seemed like a marvelous way to study the physiological basis of our nightly madness.

Unraveling the chemical mechanism of REM sleep generation took a little longer. If the two 1975 *Science* papers in which Bob McCarley and I described the aminergic-cholinergic reciprocal interaction model can be taken as an end point, it took 14 years. Of course this story is incomplete and is still developing today. Along the way, the most surprising observation that we made was that both the noradrenergic locus coeruleus and the serotonergic raphe neurons turned *off* in REM sleep. This finding was a surprise because everyone had predicted the opposite. It was suddenly clear that these two aminergic systems supported waking and not sleep, as had been theorized, so this was also a second powerful "Ah, ha!" experience. It opened the door to our 1973 discovery that REM sleep could be triggered by cholinergic stimulation of the paramedian pontine reticular for-

mation. And that in turn led to our 1991 description of a REM sleep regulatory region in the far lateral peribrachial area where cholinergic stimulation produces increases in REM that last for a week to 10 days.

Knowing that the brain's neuromodulatory balance shifted from aminergic dominance in waking to cholinergic dominance in REM also emboldened us to propose, in our two 1977 papers outlining our activation-synthesis dream theory, that all of the distinctive cognitive features of dreaming were electrically and chemically determined. By challenging the reigning Freudian theory of psychoanalysis, these heretical articles elicited more letters to the editor than the *American Journal of Psychiatry* had ever before received. Naturally and understandably, most of the letters attacked us as insensitive materialists and unenlightened philistines. We proposed, for example, that dream amnesia could be ascribed to aminergic demodulation of the forebrain (rather than to Freudian repression), while bizarre imagery could be ascribed to self-stimulation of the brain by the disinhibited cholinergic system (rather than the Freudian disguise of unacceptable unconscious wishes). Building upon the foundation of this activation-synthesis hypothesis, we have recently documented and analyzed many other emotional and cognitive features of dreaming in terms of the underlying neurophysiology.

Whether any of these contributions really count as "discoveries" is not for me to say. A more modest claim is simply that we have helped create a solid neurocognitive approach to dreaming, a fascinating and informative human mental state, with many formal similarities to major mental illness. It is clear to me that we have just scratched the surface of the dreaming brain and that the opportunities for future scientific progress in this exciting field are limitless.

prisingly, the "synthesized" dream product may be quite bizarre and even nonsensical because it is triggered by the semirandom activity of the pons. Evidence for the activation-synthesis hypothesis is mixed. On the one hand, it predicts the weirdness of dreams and their correlation with REM. However, still unexplained is how random activity can trigger the complex and fluid stories that many dreams contain, nor how it can evoke dreams that recur night after night.

There have been many suggestions that REM sleep, and perhaps dreams themselves, have an important role in memory. None of the evidence is definitive, but there are intriguing hints that REM sleep somehow aids the integration or consolidation of memories. Depriving humans or rats of REM sleep can impair their ability to learn a variety of tasks. Some studies show an increase in the duration of REM sleep after an intense learning experience. In an intriguing recent study, Israeli neuroscientist Avi Karni and his colleagues trained people to do a simple visual task, asking them to identify the orientation of lines in their peripheral visual field. With repeated practice over days, people got much better at this task; surprisingly, their performance also improved between evening and morning, after a night's sleep. Karni found that if people were deprived of REM sleep, their learning of the task did not improve overnight. Depriving them of non-REM sleep, on the other hand, actually enhanced their performance. Karni hypothesizes that this kind of memory requires a period of time to strengthen, and that REM sleep is particularly effective for this purpose.

You may have heard about sleep-learning—the notion that you can study for an exam by simply listening to a tape of the material while you blissfully snooze away. Sounds like a student's fantasy, right? Unfortunately, it is exactly that, and no more. There is no scientific evidence for sleep-learning, and careful studies have shown that the very few things recalled the next morning were heard when the subjects briefly woke up. In fact, sleep is a profoundly amnesic state. Most of our dreams, for example, seem to be lost forever. Although we dream profusely during each of the four or five REM periods each night, usually only the last dream before waking is remembered. Also, when we wake up to do something in the middle of the night, we have often forgotten the incident by morning.

At this point you are probably confused about the functions of dreaming and REM sleep. So are we. Unfortunately, there is not enough evidence to support or dismiss any of the theories we have discussed. There are also many other creative and plausible ideas that we have not had the space to present.

Neural Mechanisms of Sleep

Until the 1940s, it was generally believed that sleep was a passive process: Deprive the brain of sensory input, and it will fall asleep. However, when the sensory afferents to an animal's brain are blocked, the animal continues to have cycles of waking and sleeping. Sleep is an active process and requires the participation of a variety of brain regions. As we saw in Chapter 15, wide expanses of the cortex are actually controlled by very small collections of neurons much deeper in the brain. These cells act like the switches or tuners of the forebrain, altering cortical excitability and gating the flow of sensory information into it. The full details of these control systems are complex and not fully understood. But we can summarize a few basic principles:

1. The neurons most critical to the control of sleeping and waking are part of the diffuse modulatory neurotransmitter systems.
2. The brain stem modulatory neurons using norepinephrine and serotonin fire during waking and enhance the awake state; some neurons using acetylcholine enhance critical REM events, and other cholinergic neurons are active during waking.

3. The diffuse modulatory systems control the rhythmic behaviors of the thalamus, which in turn controls many EEG rhythms of the cerebral cortex; slow, sleep-related rhythms of the thalamus apparently block the flow of sensory information up into the cortex.
4. Sleep also involves activity in descending modulatory systems, for example, to actively inhibit motor neurons during dreaming.

There are three basic kinds of evidence for the localization of sleep mechanisms in the brain. Lesion data reveal changes in function after a part of the brain is removed, results of stimulation experiments identify changes following activation of a brain region, and recordings of neural activity determine the relationship between that activity and different brain states.

Wakefulness and the Ascending Reticular Activating System. Lesions in the brain stem of humans can cause sleep and coma, suggesting that it has neurons whose activity is essential to keep us awake. Italian neurophysiologist Giuseppe Moruzzi and his colleagues, working in the 1940s and 1950s, began to sort out the neurobiology of the brain stem's control of waking and arousal. They found that lesions in the midline structures of the brain stem caused a state similar to non-REM sleep, but lesions in the lateral tegmentum, which interrupted ascending sensory inputs, did not. Conversely, electrical stimulation of the midline tegmentum of the midbrain, within the reticular formation, transformed the cortex from the slow, rhythmic EEGs of non-REM sleep to a more alert and aroused state with an EEG similar to that of waking. Moruzzi called this ill-defined region of stimulation the *ascending reticular activating system*, mentioned in Chapter 15. This area is now much better defined, anatomically and physiologically, and it is clear that Moruzzi's stimulation was affecting many different sets of ascending modulatory systems.

Several sets of neurons increase their firing rates in anticipation of awakening and during various forms of arousal. They include cells of the locus coeruleus, which contain norepinephrine, serotonin-containing cells of the raphe nuclei, acetylcholine-containing cells of the brain stem and basal forebrain, and neurons of the midbrain that use histamine as a neurotransmitter. Collectively, these neurons synapse directly on the entire thalamus, cerebral cortex, and other brain regions. The general effects of their transmitters are a depolarization of neurons, an increase in their excitability, and a suppression of rhythmic forms of firing. These effects are most clearly seen in the relay neurons of the thalamus (Figure 17.12).

Falling Asleep and the Non-REM State. Falling asleep involves a progression of changes over several minutes, culminating in the non-REM state. It is not clear what initiates non-REM sleep, but there is a general decrease in the firing rates of most brain stem modulatory neurons (those using NE, 5-HT, and ACh). Although most regions of the basal forebrain seem to promote alertness and arousal, a subset of its cholinergic neurons increases their firing rate with the onset of non-REM sleep and are silent during wakefulness.

Early stages of non-REM sleep include the EEG sleep spindles, described earlier, which are generated by the inherent rhythmicity of thalamic neurons. As non-REM sleep progresses, spindles disappear and are replaced by slow delta rhythms (less than 4 Hz). Delta rhythms also seem to be a product of thalamic cells, occurring when their membrane potentials become even more negative than during spindle rhythms (and much more negative than they are during waking). Synchronization of activity during spindle or delta rhythms is due to neural interconnections within the thalamus, and possibly between the thalamus and cortex. Because of the strong, two-way excitatory connections between the thalamus and cortex, rhythmic activity in one is often strongly and widely projected upon the other.

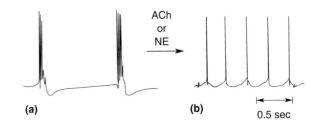

(a) (b) 0.5 sec

Figure 17.12
Acetylcholine (ACh) and norepinephrine (NE) modulate the rhythmicity of the thalamus and cortex. In this case, either ACh *or* NE shifts cells from **(a)** an intrinsic burst-firing mode to **(b)** a single-spiking mode. This may be what happens during transitions from non-REM sleep to the waking state. (Source: Adapted from Steriade, McCormick, and Sejnowski, 1993, Fig. 5D.)

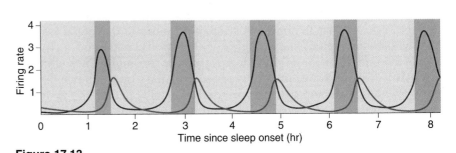

Figure 17.13
Control of the onset and offset of REM periods by brain stem neurons. This graph shows the relative firing rates of REM sleep-associated neurons during a single night. Periods of REM are colored green. "REM-on" cells are cholinergic neurons of the pons, and they increase their firing rates just before the onset of REM (red line). "REM-off" cells are noradrenergic and serotonergic neurons of the locus coeruleus and raphe nuclei, respectively, and their firing rates increase just before the end of REM (blue line). (Source: Adapted from McCarley and Massaquoi, 1986, Fig. 4B.)

Mechanisms of REM Sleep. REM is such a different state from non-REM that we would expect some clear neural distinctions. The cortex is at least as active during REM sleep as it is during waking. For example, visual cortex neurons that respond vigorously and specifically to light when an animal is awake are often very active (spontaneously) during REM sleep as well. Neurons of the motor cortex also fire rapidly and generate organized motor patterns that attempt to command the entire body but succeed only with a few muscles of the eye, the inner ear, and those essential for respiration. The elaborate dreams of REM sleep certainly require the cerebral cortex. However, the cortex is not necessary for the *production* of REM sleep.

The control of REM sleep, as with the other functional brain states, derives from diffuse modulatory systems in the core of the brain stem, particularly the pons. The firing rates of the two major systems of the upper brain stem, the locus coeruleus and the raphe nuclei, decrease to almost nothing with the onset of REM (Figure 17.13). However, there is a concurrent sharp increase in the firing rates of ACh-containing neurons in the pons, and some evidence suggests that cholinergic neurons induce REM sleep. It is probably the action of ACh during REM sleep that causes the thalamus and cortex to behave so much like they do in the waking state. REM periods are terminated when NE and 5-HT-containing neurons begin firing again.

Why don't we act out our dreams? The same core brain stem systems that control the sleep processes of the forebrain also actively inhibit our spinal motor neurons, preventing the descending motor activity from expressing itself as actual movement. This is clearly an adaptive mechanism, protecting us from ourselves. There are rare people, usually elderly men, who do seem to act out their dreams; they have a hazardous condition known as *REM sleep behavior disorder*. These people often sustain repeated injuries, and even their spouses have fallen victim to their nocturnal flailings. One man dreamed he was in a football game and tackled his bedroom bureau. Another imagined he was defending his wife from attack, when in fact he was beating her in her bed. The basis for this REM disorder seems to be disruption of the brain stem systems that normally mediate REM atonia. Experimental lesions in certain parts of the pons can cause a similar condition in cats. During REM periods, they may seem to chase imaginary mice or investigate invisible intruders. Disorders of REM control mechanisms may also underlie the problems of people with narcolepsy (Box 17.4).

Sleep-Promoting Factors. Sleepiness is one of the most familiar consequences of infectious diseases such as the common cold and the flu. There may be direct links between the immune response to infection and the regulation of sleep. Sleep researchers have searched intensively for a chemical in the blood or cerebrospinal fluid that promotes or even causes sleep. Many sleep-promoting substances have been identified in sleep-deprived

---

| Box 17.4 | **OF SPECIAL INTEREST** |

Narcolepsy

Narcolepsy is a bizarre and disabling mayhem of sleeping and waking. Despite the sound of the name, it is not a form of epilepsy. It can include some or all of the following manifestations. *Excessive daytime sleepiness* can be severe and often leads to unwanted "sleep attacks." *Cataplexy* is a sudden muscular paralysis, while consciousness is maintained. It is often brought on by strong emotional expression, such as laughter or tears, or by surprise or sexual arousal, and it usually lasts less than one minute. *Sleep paralysis*, a similar loss of muscle control, occurs during the transition between sleeping and waking. It sometimes occurs in the absence of narcolepsy, and it can be very disconcerting; although conscious, a person may be unable to move or speak for several minutes. *Hypnagogic hallucinations* are graphic dreams, often frightening, that also accompany sleep onset and may occur following sleep paralysis. Sometimes such dreams flow smoothly with real events that occurred just prior to falling asleep.

EEG monitoring reveals a distinct difference between narcoleptic and normal sleep. A narcoleptic person goes directly from waking into a REM phase, whereas normal adult sleepers always enter a long period of non-REM first. Most of the symptoms of narcolepsy may be interpreted as an abnormal intrusion of the characteristics of REM sleep into waking.

The prevalence of narcolepsy varies greatly. For example, there are about 250,000 patients in the U.S. (about 1 in 1,000 people), but the condition is very rare in Israel (1 in 500,000 people). The biological basis for narcolepsy is not understood, although there is evidence for some change in the cholinergic systems of the brain. In some cases, but not all, narcolepsy runs in families. But the story is complex. Several genes, as well as environmental and behavioral factors, have been implicated. Narcolepsy also occurs in goats, donkeys, ponies, and more than a dozen breeds of dogs. There is no cure for narcolepsy, and treatment consists of trying to relieve the symptoms. Frequent naps and CNS stimulants may help daytime sleepiness, while tricyclic antidepressant drugs (which have REM-suppressant effects) may reduce cataplexy.

animals. Most interact with the body's immune system. In the 1970s, physiologist John Pappenheimer, of Harvard University, identified a muramyl dipeptide in the spinal fluid of sleep-deprived goats that facilitated non-REM sleep. Muramyl peptides are usually produced only by the cell walls of bacteria, not brain cells, and they also cause fever and stimulate immune cells of the blood. It is not clear how they appear in spinal fluid, but they may be synthesized by bacteria in the intestines. Another sleep-promoting factor, interleukin-1, is synthesized by the brain, in glia, and in macrophages, cells throughout the body that scavenge foreign material. Interleukin-1 is also a peptide that stimulates the immune system.

The precise function of sleep-promoting substances is still unclear. They are not likely to be primary controllers of sleep. For one thing, they are not present in effective concentrations in a normally sleepy, as opposed to sleep-deprived, animal. They also tend to suppress the time spent in REM sleep, so their "sleep promotion" is quite selective. Nevertheless, the common linkage between sleep promotion and immune stimulation might mean that sleep is one way to help fight infection.

BRAIN CLOCKS

Almost all land animals coordinate their behavior with **circadian rhythms**, matching their functions to the daily cycles of lightness and darkness that result from the spin of the Earth (from the Latin *circa*, "approximately," and *dies*, "day"). The precise schedules of circadian rhythms vary among species. Some animals are active during daylight hours, others only at night, and others mainly at the transitional periods of dawn and dusk. Most physiological and biochemical systems in the body also rise and fall with daily rhythms; body temperature, blood flow, urine production, hormone levels, hair growth, and metabolic rate all fluctuate (Figure 17.14).

When the cycles of daylight and darkness are removed from an animal's environment, circadian rhythms continue on more or less the same schedule because the primary clocks for circadian rhythms are not astronomical (the Sun and Earth) but biological, in the brain. Brain clocks, like all clocks, are imperfect and require occasional resetting. Now and then you readjust your watch to keep it in synch with the rest of the world (or at least the time given on the radio). Similarly, external stimuli, such as light and dark, or daily temperature changes, help adjust the brain's clocks to keep them synchronized with the coming and going of the sunlight. Circadian rhythms have been well studied at both behavioral and cellular levels. Brain clocks are especially useful for studying the link between the activity of specific neurons and behavior.

Circadian Rhythms

The first evidence for a biological clock came from a brainless organism, the mimosa plant. The mimosa raises its leaves during the day and lowers them at night. It seemed obvious to many people that the plant simply reacts to sunlight, in some kind of reflex movement. In 1729, French physicist Jean Jacques d'Ortous de Mairan tested the obvious; he put some mimosa plants in a dark closet and found that they continued to raise and lower their leaves. But a surprising new observation can still lead to a wrong conclusion. It was de Mairan's opinion that the plant was still somehow sensing the sun's movements, even in the darkness. More than a century later, Swiss botanist Augustin de Candolle showed that a similar plant in the dark moved its leaves up and down every 22 hr, rather than 24 hr according to the sun's movement. This implied that the plant was not responding to the sun, and very likely had an internal biological clock.

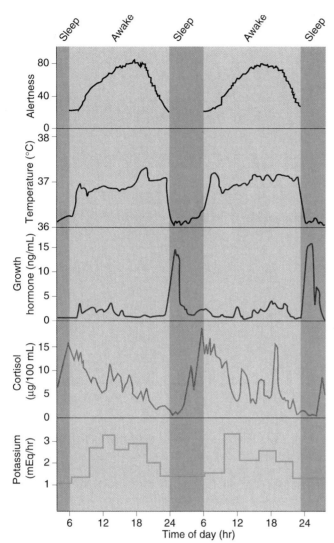

Figure 17.14
Circadian rhythms of physiological functions. Fluctuations over two consecutive days are shown here. Alertness and core body temperature vary similarly. However, growth hormone and cortisol levels in the blood are highest during sleep, although at different times. The bottom graph shows the excretion of potassium by the kidneys, which is highest during the day. (Source: Adapted from Coleman, 1986, Fig. 2.1.)

Environmental time cues (light/dark, temperature and humidity variations) are collectively termed **zeitgebers** (German for "time-givers"). In the presence of zeitgebers, animals become *entrained* to the day-night rhythm and maintain an activity cycle of exactly 24 hr. Obviously, even small, consistent errors of timing could not be tolerated for long. A 24.5-hr cycle would, within 3 weeks, completely shift an animal from daytime to nighttime activity. When mammals are completely deprived of zeitgebers, they settle into a rhythm of activity and rest that often has a period more or less than 24 hr, in which case their rhythms are said to *free-run*. In mice, the natural free-running period is about 23 hr, in hamsters it is close to 24 hr, and in humans it tends to be 24.5–25.5 hr (Figure 17.15).

It is quite difficult to separate a human from all possible zeitgebers. Even inside a laboratory, society provides many subtle time cues, such as the sounds of machinery, the comings and goings of people, and the on-off cycling of heating and air conditioning. Some of the most secluded

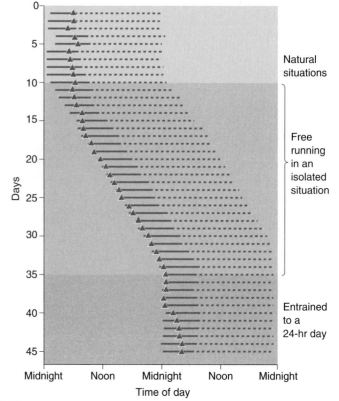

Figure 17.15
Circadian rhythms of sleep and wakefulness. Shown here is a daily plot of one person's sleep-wake cycles. Each horizontal line is a day; solid lines indicate sleep, and dashed lines indicate waking. A triangle (▲) indicates the point of the day's lowest body temperature. The subject was first exposed to 9 days of natural 24-hr cycles of light and dark, noise and quiet, and air temperature. During the middle 25 days, all time cues were removed, but the subject was free to set his own schedule. Notice that the sleep-wake cycles remained stable, but each lengthened to about 25 hr. The subject was now *free-running*. Notice also that the low point of body temperature shifts from the end of the sleep period to the beginning. During the last 11 days, a 24-hr cycle of light and meals was reintroduced, the subject again entrained to a day-long rhythm, and body temperature gradually shifted back to its normal point in the sleep cycle. (Source: Adapted from Dement, 1976, Fig. 2.)

environments are deep caves, which have been the sites for several isolation studies. When people in caves are allowed to set their own schedules of activity for months on end—waking and sleeping, turning lights on and off, and eating when they choose—they initially settle into roughly a 25-hr rhythm. But after days to weeks, their activity may begin to free-run with a surprisingly long period of 30–36 hr. They stay awake for about 20 hr straight, then sleep for about 12 hr, and this pattern seems perfectly normal to them at the time.

In isolation experiments, behavior and physiology do not always continue to cycle together. Body temperature sometimes continues to rise and fall over the relatively short 25-hr periods, while sleep-wake cycles are much longer. This means that the rhythms of temperature and sleeping-waking, which are normally locked to a 24-hr period, become desynchronized. Normally our lowest body temperature occurs shortly before we awaken in the morning, but when desynchronized, this temperature nadir can drift, first moving earlier into the sleep period, and then into waking time. The consequences of desynchronization are unpleasant, but the implications are interesting. Sleep quality and waking comfort are impaired when cycles are desynchronized. A similar desynchronization may occur temporarily when we travel and force our bodies suddenly into a new sleep-wake cycle. This is the familiar experience known as jet-lag, and the best cure is bright light, which helps resynchronize our biological clocks. One implication of this desynchronization is that the body has more than one biological clock, because sleeping-waking and temperature can cycle at their own pace, uncoupled from one another.

The primary zeitgeber for mammals is the light-dark cycle. However, the mother's hormone levels may be the first zeitgeber for some mammals, already entraining their activity levels in the womb. In studies of various adult animals, effective zeitgebers have also included the periodic availability of food or water, social contact, environmental temperature cycles, and noise-quiet cycles. Although many of these are much less effective than light-dark cycles, they may be important for particular species, in certain circumstances.

A Clock in the Brain: The Suprachiasmatic Nucleus

A biological clock that produces circadian rhythms consists of several components:

Light sensor → clock → output pathway

One or more input pathways are sensitive to light and dark, and entrain the clock and keep its rhythm coordinated with the circadian rhythms of the environment. The clock itself continues to run and keep its basic rhythm even when the input pathway is removed. Output pathways from the clock allow it to control certain brain and body functions according to the timing of the clock.

Mammals have a tiny pair of neuron clusters in the hypothalamus that serves as a biological clock: the **suprachiasmatic nuclei (SCN)**, introduced in Chapter 15. Each SCN has a volume of less than $0.3 \, mm^3$, and its neurons are among the smallest in the brain. They are located on either side of the midline, bordering the third ventricle (Figure 17.16). When the SCN is stimulated electrically, circadian rhythms can be shifted in a predictable way. Removal of both nuclei abolishes circadian rhythmicity of sleeping and waking, feeding and drinking, and activity (Figure 17.17). The brain's internal rhythms never return without an SCN. Lesions in the SCN do not abolish sleeping, however, and animals will continue to coordinate their sleeping and waking with light-dark cycles if they are present. Therefore, there must be other control systems that regulate the sleep process itself.

Because behavior is normally synchronized with light-dark cycles, there must also be a photosensitive mechanism for resetting the brain clock. The SCN accomplishes this via the retinohypothalamic tract: Axons from ganglion cells in the retina synapse directly on the dendrites of SCN neurons. This input from the retina is necessary and sufficient to entrain sleeping and waking cycles to night and day. When recordings are made from neurons of the SCN, many are indeed sensitive to light. Unlike the more familiar neurons of the visual pathways discussed in Chapter 10, SCN neurons have very large, nonselective receptive fields and respond to the luminance of light stimuli rather than their orientation or motion.

Output axons of the SCN mainly innervate nearby parts of the hypothalamus, but some also go to the midbrain and other parts of the diencephalon. Because almost all SCN neurons use GABA as their primary neurotrans-

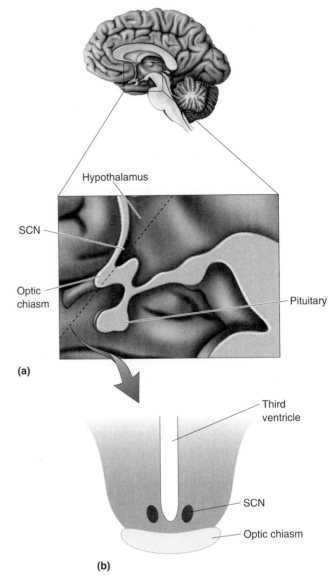

Figure 17.16
The human suprachiasmatic nuclei. There are two suprachiasmatic nuclei (SCN) within the hypothalamus, just above the optic chiasm and next to the third ventricle. **(a)** Sagittal view. **(b)** Frontal view, sectioned at the dashed line in part a.

mitter, presumably they inhibit the neurons they innervate. It is not yet clear how the SCN sets the timing of so many important behaviors. Extensive lesions in the efferent SCN pathways disrupt circadian rhythms. In addition to the axonal output pathways, SCN neurons may rhythmically secrete the peptide neuromodulator vasopressin (see Chapter 15). When the two SCNs are ablated in hamsters, the animals lose their circadian rhythmicity. However, transplantation of a new SCN can restore rhythms within 2–4 weeks (Box 17.5).

SCN Mechanisms

How do neurons of the SCN keep time? We don't have the answer at the molecular level. But it's clear that each SCN cell is a minuscule clock. The ultimate isolation experiment has been simply to remove neurons from the SCN of a rat and grow them alone in a tissue culture dish, segregating them from the rest of the brain and from each other. Nevertheless, their rates of action potential firing, glucose utilization, vasopressin production, and protein synthesis continue to vary with rhythms of about 24 hr, just as they do

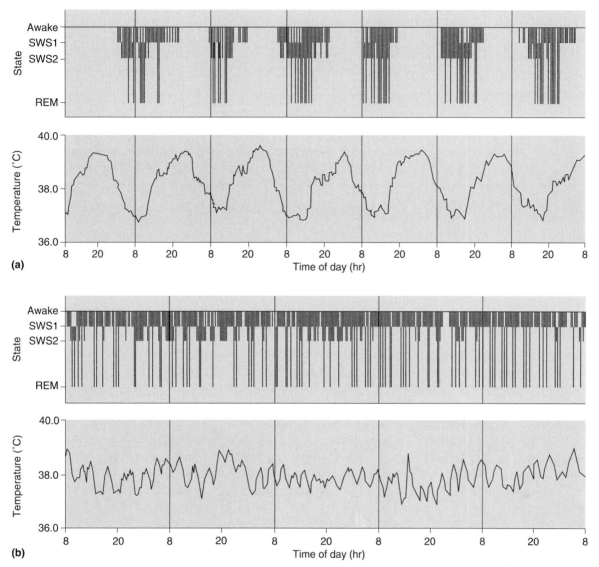

Figure 17.17
The SCN and circadian rhythms. (a) Normal squirrel monkeys kept in a constantly lit environment display circadian rhythms of about 25.5 hr. The graph shows the stages of waking-sleeping and concurrent variations in body temperature. The animals' activity states were defined as awake, two levels of non-REM sleep (SWS1 or SWS2), or REM sleep. **(b)** Circadian rhythms are abolished in monkeys with lesions in both SCN, kept in the same constant-light environment. Notice that persistent high-frequency rhythms of both activity and temperature result from SCN lesions. (Source: Adapted from Edgar, Dement, and Fuller, 1993, Figs. 1, 3.)

in the intact brain (Figure 17.18). Other isolated mammalian neurons do not have circadian rhythmicity. SCN cells in culture can no longer be entrained to light-dark cycles (input from the eyes is necessary for this), but their basic rhythmicity remains intact and expresses itself just as it does when an animal is deprived of zeitgebers.

SCN cells communicate their rhythmic message to the rest of the brain through efferent axons, using action potentials in the usual way, and rates of SCN cell firing vary with a circadian rhythm. However, action potentials are not necessary for SCN neurons to maintain their rhythm. When tetrodotoxin (TTX), a blocker of sodium channels, is applied to SCN cells, it blocks their action potentials but has no effect on the rhythmicity of their metabolism and biochemical functions. When the TTX is removed, action potentials resume firing with the same phase and frequency they had originally, before the TTX, implying that the SCN clock keeps running even without action potentials. SCN action potentials are like the hands of a

---

Box 17.5 **OF SPECIAL INTEREST**

Mutant Hamster Clocks

Golden hamsters are perfectionists of circadian timing. When placed in constant darkness, they continue sleeping and waking, running on their wheels, and eating and drinking over an average period of 24.1 hr, for weeks on end. It was this dependability that made neuroscientists Martin Ralph and Michael Menaker, then working at the University of Oregon, take heed when one of the hamsters in their laboratory began punching in with 22.0-hr cycles during a period of 3 weeks in the dark. This maverick male was bred with three females of unimpeachable circadian character (their free-running periods were 24.01, 24.03, and 24.04 hr—quite normal). When 20 pups from the three resulting litters were tested in the dark, their free-running periods were evenly split into two narrow groups. Half had periods of 24.0 hr and half had periods of 22.3 hr. Further cross-breeding showed that the hamsters with the shorter circadian periods had one mutant copy of a gene (called *tau*) that was dominant over their normal gene. After further breeding, Ralph and Menaker also found that animals with two copies of the mutant *tau* gene had free-running periods of only 20 hr!

Hamsters with mutant circadian rhythms provided a convincing way to answer a fundamental question: Is the SCN the brain's circadian clock? Ralph, Menaker, and their colleagues found that when both SCNs of a hamster were ablated, rhythms were entirely lost. But rhythms could be restored to these ablated animals by simply transplanting a new SCN into their hypothalamus and waiting about 1 week. The key finding was that hamsters receiving transplants adopted the circadian rhythm of the *transplanted* SCN, not the rhythm they were born with. In other words, if a genetically normal hamster with SCN lesions received an SCN from a donor with one copy of the mutant *tau* gene, it subsequently cycled at about 22 hr. If its transplanted SCN came from an animal with two mutant *tau*s, it cycled at 20 hr. This is very compelling evidence that the SCN is the master circadian clock in the hamster brain, and probably in our own brain as well.

Short circadian periods were often devastating to a mutant hamster's lifestyle when it was placed into normal, 24-hr light and dark cycles. A hamster's normal preference is to be active at night, but most *tau* animals could not completely entrain to the 24-hr rhythm. Instead, they found their activity periods continually shifting through various parts of the light-dark cycle. A similar problem sometimes occurs in people, most often in the elderly. Due to an age-dependent shortening of the circadian rhythm, overwhelming sleepiness begins in early evening, and awakening comes at 3:00 or 4:00 in the morning. Some people are unable to entrain their sleep-wake cycles to a daily rhythm and, like the mutant hamsters, find their activity cycles constantly shifting with respect to daylight.

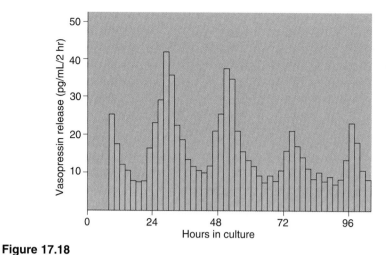

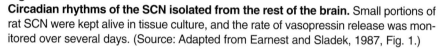

Figure 17.18
Circadian rhythms of the SCN isolated from the rest of the brain. Small portions of rat SCN were kept alive in tissue culture, and the rate of vasopressin release was monitored over several days. (Source: Adapted from Earnest and Sladek, 1987, Fig. 1.)

clock; removing the clock's hands does not stop the clock from working, but it does make it very difficult to read the time.

If each SCN cell is a clock, there must be some mechanism to coordinate those thousands of cellular clocks so that the SCN as a whole gives a single, clear message about time to the rest of the brain. Light information from the retina certainly helps do this, but SCN cells also communicate directly with each other. However, even the coordination of rhythms *between* SCN cells seems to be independent of action potentials and normal synaptic transmission because TTX does not block them. Also, the SCN of the very young rat brain coordinates circadian rhythms perfectly well, even *before* it has developed any synapses. The nature of neuron-to-neuron communication within the SCN is not understood, but in addition to classical synapses, it may include other chemical signals, direct electrical interactions, or the participation of glia.

CONCLUDING REMARKS

Rhythms are ubiquitous in the mammalian central nervous system. They also span a broad range of frequencies, from nearly 100 Hz in the cortical EEG to once per year (0.00000003 Hz) for many seasonal behaviors, such as the autumn mating of deer, the winter hibernation of chipmunks, and the instinct that drives migrating swallows to return to Capistrano, California, every March 19—in 200 years they have only missed the date twice! While the purpose of some rhythms is obvious, the functions of many neural rhythms are unknown. Indeed, some rhythms may have no function at all but arise as a secondary consequence of neural interconnections that are essential for other, nonrhythmic, purposes. Among the most conspicuous yet inexplicable of brain rhythms is sleep.

Sleep offers a fascinating set of problems for neuroscience. Unlike most studies of single ion channels, single neurons, or the systems mediating perception and movement, sleep research begins with profound ignorance about a most basic question: Why? We still admit ignorance about why we spend one-third of our lives sleeping, most of that time languid and vegetative and the rest of it paralyzed and hallucinating. Sleep and dreams may have no vital function, but they can be studied and enjoyed nevertheless. Ignoring the functional question will not be a satisfying approach for long, however. For most neuroscientists, asking "Why?" remains the deepest and most challenging problem of all.

KEY TERMS

The Electroencephalogram
electroencephalogram (EEG)
generalized seizure
partial seizure
epilepsy

Sleep
rapid eye movement sleep (REM sleep)

non-REM sleep
atonia
ultradian rhythm

Brain Clocks
circadian rhythm
zeitgeber
suprachiasmatic nucleus (SCN)

✔ REVIEW QUESTIONS

1. Why do EEGs with relatively fast frequencies tend to have smaller amplitudes than EEGs with slower frequencies?
2. The human cerebral cortex is very large and must be folded extensively to fit within the skull. What do the foldings of the cortical surface do to the brain signals that are recorded by an EEG electrode at the scalp?
3. Sleep seems to be a behavior of every species of mammal, bird, and reptile. Does this mean that sleep performs a function essential for the life of these higher vertebrates? If you do not think so, what might be an explanation for the abundance of sleep?
4. An EEG during REM sleep is very similar to an EEG when awake. How do the brain and body in REM sleep *differ* from the brain and body when awake?
5. What is a likely explanation for the brain's relative insensitivity to sensory input during REM sleep as compared to the waking state?
6. The SCN receives direct input from the retinas, via the retinohypothalamic tract, and this is how light-dark cycles can entrain circadian rhythms. If the retinal axons were somehow disrupted, what would be the likely effect on a person's circadian rhythms of sleeping and waking?

Wiring the Brain

We have seen that most of the operations of the brain depend on remarkably precise interconnections among its 100 billion neurons. As an example, consider the precision in the wiring of the visual system, from retina to lateral geniculate nucleus (LGN) to cortex, shown in Figure 18.1. All retinal ganglion cells extend axons into the optic nerve, but only ganglion cell axons from the nasal retinas cross at the optic chiasm. Axons from the two eyes are mixed in the optic tract, but in the LGN they are sorted out again (1) by ganglion cell type, (2) by eye of origin (ipsilateral or contralateral), and (3) by retinotopic position. LGN neurons project axons into the optic radiations that travel via the internal capsule to the primary visual (striate) cortex. Here, they terminate (1) only in cortical area 17, (2) only in specific cortical layers (mainly layer IV), and (3) again according to cell type and retinotopic position. Finally, the neurons in layer IV make very specific connections with cells in other cortical layers that are appropriate for binocular vision and are specialized to enable the detection of contrast borders. How did such precise wiring arise?

Back in Chapter 7, we looked at the embryological and fetal development of the nervous system to understand how it changed from a simple tube in the early embryo into the structures that we recognize in the adult as the brain and spinal cord. Here we'll take another look at brain development, this time to see how connections are formed and modified as the brain matures. We will discover that most of the wiring in the brain is specified by genetic instructions that enable axons to detect the correct pathways and the correct targets. However, a small but important component of the final wiring depends on sensory information about the world around us during early childhood. In this way, nurture and nature both contribute to the final structure and function of the nervous system. We will be using the central visual system as an example whenever possible, so you may want to quickly review Chapter 10 before continuing.

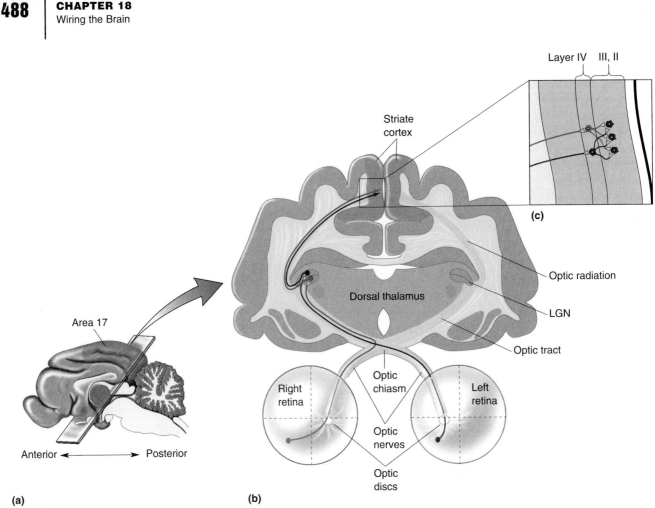

Figure 18.1
Organization of the mature mammalian retinogeniculocortical pathway. (a) A midsagittal view of a cat brain showing the location of primary visual cortex (striate cortex, area 17). The sectional plane, illustrated in part b, reveals all the components of the ascending visual pathway. (b) Notice that the temporal retina of the right eye and the nasal retina of the left eye project axons via the optic nerve and optic tract to the LGN of the right dorsal thalamus. Inputs from the two eyes remain segregated in separate layers at the level of this synaptic relay. LGN neurons project to striate cortex via the optic radiations. These axons terminate mainly in layer IV, where inputs serving the two eyes continue to be segregated. (c) The first site of major convergence of inputs from the two eyes is in the projection of layer IV cells onto cells in layer III.

THE GENESIS OF NEURONS

The first step in wiring the nervous system together is generating neurons. Consider as an example the striate cortex. In the adult, there are six cortical layers, and the neurons in each of these layers have characteristic appearances and connections. Neuronal structure develops in three major steps: cell proliferation, cell migration, and cell differentiation.

Cell Proliferation

Recall from Chapter 7 that the brain develops from the walls of the five fluid-filled vesicles. These fluid-filled spaces remain in the adult and constitute the ventricular system. Very early in development, the walls of the vesicles consist of only two layers: the ventricular zone and the marginal zone. The *ventricular zone* lines the inside of each vesicle, and the *marginal zone* faces the overlying pia. Within these layers of the telencephalic vesicle, a cellular ballet is performed that gives rise to all the neurons and glia of the visual cortex. The choreography of cell proliferation is described below, and the five "positions" correspond to the circled numbers in Figure 18.2:

1. *First position:* A cell in the ventricular zone extends a process that reaches upward toward the pia.
2. *Second position:* The nucleus of the cell migrates upward from the ventricular surface toward the pial surface; the cell's DNA is copied.
3. *Third position:* The nucleus, containing two complete copies of the genetic instructions, settles back to the ventricular surface.
4. *Fourth position:* The cell retracts its arm from the pial surface.
5. *Fifth position:* The cell divides in two.

The newly formed *daughter cell* then migrates away from the ventricular zone to take up its position in the cortex, leaving the *stem cell* in the ventricular zone to undergo more divisions. Ventricular zone stem cells repeat this pattern until all the neurons of the cortex have been generated. In humans, all neocortical neurons are born between the fifth week and the fifth month of gestation (pregnancy). Think about it: *At birth, you have all the neocortical neurons you're ever going to have.* Take good care of them.

Cortical cells can be divided into glia and neurons, and the neurons can be further divided depending on what layer they reside in, what dendritic morphology they possess, and what neurotransmitter they use. Conceivably, this diversity could arise from different types of stem cell. In other words, there could be one class of stem cell that gives rise only to layer

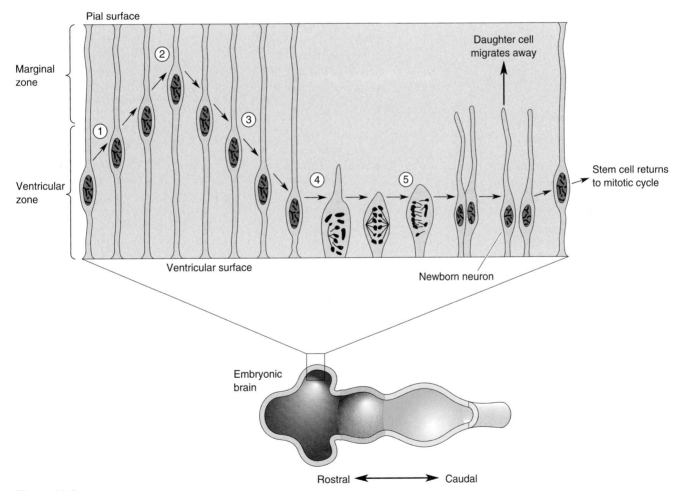

Figure 18.2
The choreography of cell proliferation. The wall of the brain vesicles initially consists of only two layers, the marginal zone and the ventricular zone. Each cell performs a characteristic "dance" as it divides, shown here from left to right. The circled numbers correspond to the five "positions" described in the text.

VI pyramidal cells, another that gives rise to layer V cells, and so on. However, this is not the case. Multiple cell types, including neurons and glia, can arise from the same stem cell. Cell fate appears to be specified instead by the environment surrounding the daughter cells shortly after division. Because this environment changes systematically as gestation proceeds, neuronal fate depends on the time each cell is born. Thus, neurons born early in the genesis of the cortex become layer VI cells. Later-dividing cells become layer V neurons, followed by the neurons of layers IV, III, and II.

Cell Migration

Daughter cells migrate by slithering along thin fibers that radiate from the ventricular zone toward the pia (Figure 18.3). These fibers are derived from specialized **radial glial cells**, providing the scaffold on which the cortex is built. For the most part, the movement of the immature neurons, called **neuroblasts**, follows this radial path from the ventricular zone toward the surface of the brain (Box 18.1). When cortical assembly is complete, the radial glia withdraw their radial processes.

The neuroblasts destined to become layer VI cells are among the first to migrate away from the ventricular zone, and they accumulate below the marginal zone and form a cell layer called the **cortical plate**. Next come the layer V cells, followed by layer IV cells, and so on. Notice that each new wave of neuroblasts migrates radially right past those in the existing cortical plate. In this way, the cortex is said to be assembled *inside-out* (Figure 18.4).

Cell Differentiation

The process in which the neuroblast takes on the appearance and characteristics of a neuron is called *cell differentiation* (Figure 18.5). Differentiation begins almost as soon as the neuroblast arrives in the cortical plate. Thus, layer V and VI neurons have differentiated even before layer II cells have migrated into the cortical plate.

Differentiation begins with the appearance of neurites sprouting off the cell body. At first these neurites all appear about the same, but soon one becomes recognizable as the axon and the others as dendrites. Differentiation will occur even if the neuroblast is removed from the brain and placed in tissue culture. For example, cells destined to become neocortical pyramidal cells will often assume the same characteristic dendritic architecture in tissue culture. This means that differentiation is programmed well before the neuroblast arrives at its final resting place. However, there is evidence that the complexity of the dendritic trees also depends on "environmental" factors in the cortex, such as the amount of presynaptic innervation. Indeed, there is even evidence that dendrites might be wriggling around inside our heads all the time, depending on levels of circulating hormones and the amount of synaptic stimulation the dendrites are receiving.

THE GENESIS OF CONNECTIONS

As neurons differentiate, they extend axons that must find their appropriate targets. Think of this development of long-range connections, or pathway formation, in the CNS as occurring in three phases: pathway selection, target selection, and address selection. Let's understand the meaning of these terms in the context of the development of the visual pathway from the retina to the LGN, as shown in Figure 18.6.

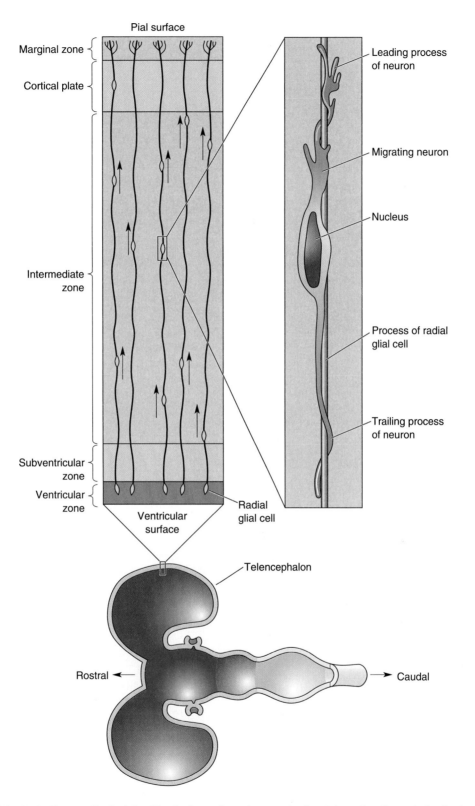

Figure 18.3
Migration of neuroblasts to the cortical plate. The bottom view shows a horizontal section through the brain early in development. Expanded views at the top show neuroblasts crawling along the thin processes of the radial glia en route to the cortical plate, which forms just under the marginal zone.

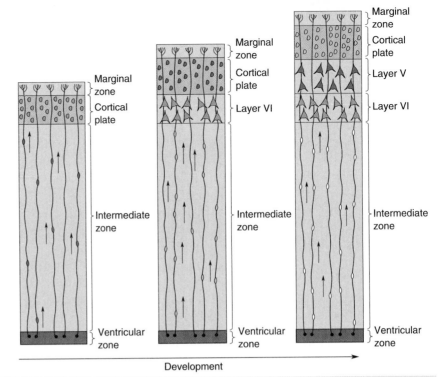

Figure 18.4
Inside-out development of the cortex. The first cells to migrate to the cortical plate are those that will form layer VI. As these differentiate into neurons, the neuroblasts destined to become layer V cells migrate past and collect in the cortical plate. This process repeats again and again until all layers of the cortex have differentiated.

Development

Box 18.1 | **OF SPECIAL INTEREST**

Dividing Up the Neocortex

The neocortex is often described as a sheet of tissue. In reality, however, cortex is much more like a patchwork quilt, with many different structurally distinct areas stitched together. One of the triumphs of human evolution is the creation of increasingly distinct neocortical areas that are specialized for increasingly sophisticated analysis. It is natural to wonder exactly how all these areas arise during development.

Cortical neurons are born in a zone that lines the ventricle and then migrate along radial glia to take up their final position in one of the cortical layers. Thus, it seems reasonable to conclude that cortical areas in the adult brain simply reflect an organization that is already present in the ventricular zone of the fetal telencephalon. According to this idea, the ventricular zone contains something like a microfilm record of the future cortex, which is projected onto the wall of the telencephalon as development proceeds. For this scheme to work, the migrating neuroblast must climb up a single radial glial fiber and not swing randomly from vine to vine, so to speak, on its way to the cortical plate.

If migration is strictly radial, then all the offspring of a stem cell should migrate to exactly the same neighborhood of the cortex. Researchers Christopher Walsh and Constance Cepko of Harvard University devised an ingenious way to test this idea. They inserted a specially engineered segment of DNA into the chromosomes of a single stem cell. Every time the stem cell divided, a copy of this "reporter" DNA was smuggled into the daughter cell. The reporter DNA would announce its presence by turning the cell blue.

Contrary to the prevailing belief at the time, Walsh and Cepko found that blue cells spread out over a large area of the cortex. The wanderlust of sibling neuroblasts was later confirmed by using time-lapse video microscopy. It now appears that at least one-third of all neuroblasts stray considerable distances as they migrate radially toward the cortical plate.

Although these new findings have challenged older notions, there is still no consensus on how different cortical areas emerge during development. Indeed, the drama is being played out monthly in scientific journals, with neuroscientists on all sides of the issue. In the meantime, however, we can humbly ask: If not the ventricular zone, what else can divide up the neocortex during development?

Recall that each cortical area has different patterns of connections, particularly with the dorsal thalamus. Area 17 receives input from the LGN, area 3 receives input from the ventral posterior nucleus, and so on. What if the thalamic inputs instruct the cortex to differentiate into different areas? Researchers Brad Schlaggar and Dennis O'Leary of the Salk Institute addressed this question in a novel way. In rats, the thalamic fibers wait in the cortical white matter and do not enter the cortex until a few days after birth. Schlaggar and O'Leary peeled off the cortex in newborn rats, rotated it 180°, and stuck it back on! This created a situation in which the thalamic fibers from the ventral posterior nucleus were waiting under the cortex of the *occipital lobe* instead of the parietal lobe, as is normally the case. Remarkably, the fibers invaded the new territory and claimed it in the name of the somatosensory system. These results suggest the thalamus may specify the pattern of the cortical quilt.

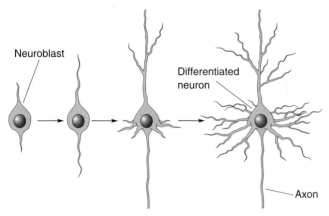

Figure 18.5
Differentiation of a neuroblast into a neuron.

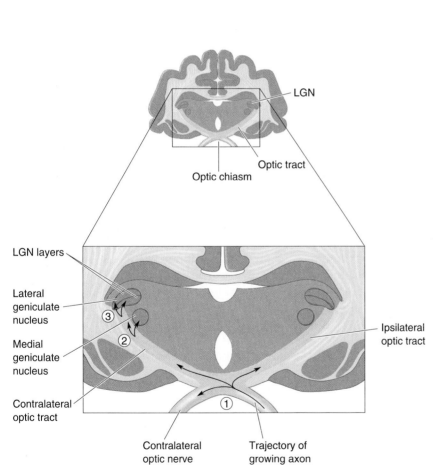

Figure 18.6
The three phases of pathway formation. The growing retinal axon must make several choices to find its correct target in the LGN. (1) During pathway selection, the axon must choose the correct path. (2) During target selection, the axon must choose the correct structure to innervate. (3) During address selection, the axon must choose the correct cells to synapse with in the target structure.

Imagine for a moment that you must lead a growing retinal ganglion cell axon to the correct location in the LGN. First you travel down the optic stalk toward the brain. But soon you reach the optic chiasm at the base of the brain and must decide which fork in the road to take. You have three choices: You can enter the optic tract on the same side, you can enter the optic tract on the opposite side, or you can dive into the other optic nerve. The correct path depends on the location in the retina of your ganglion cell, and on the cell type. If you came from the nasal retina, you would cross over at the chiasm into the contralateral optic tract; but if you came from the temporal retina, you would stay in the tract on the same side. And in no case would you enter the other optic nerve. These are examples of the decisions that must be made by the growing axon during *pathway selection*.

Having forged your way into the dorsal thalamus, you are now confronted with a dozen possible targets; for example, the medial geniculate nucleus. The correct choice, of course, is the *lateral* geniculate nucleus. This decision is called *target selection*.

But finding the correct target still isn't enough. You must now find the correct layer of the LGN. You also must make sure that you sort yourself out with respect to other invading retinal axons so that retinotopy in the LGN is established. These are examples of the decisions that must be made by the growing axon during *address selection*.

We will see that each of the three phases of pathway formation depends critically on communication between cells. This communication occurs in several ways: direct cell-cell contact, contact between cells and the extracellular secretions of other cells, and communication between cells over a distance via diffusible chemicals. As the pathways develop, the neurons also begin to communicate via action potentials and synaptic transmission.

The Growing Axon

Once the neuroblast has migrated to take up its appropriate position in the nervous system, the neuron differentiates and extends the processes that will ultimately become the axon and dendrites. At this early stage, however, the axonal and dendritic processes appear quite similar and collectively are still called neurites. The growing tip of a neurite is called a **growth cone** (Figure 18.7).

The growth cone is specialized to identify an appropriate path for neurite elongation. The leading edge of the growth cone consists of flat sheets of membrane called *lamellipodia* that undulate in rhythmic waves like the wings of a stingray swimming along the ocean bottom. Extending from the lamellipodia are thin spikes called *filopodia*, which constantly probe the environment, moving in and out of the lamellipodia. Growth of the neurite occurs when a filopodium, instead of retracting, takes hold of the substrate (the surface on which it is growing) and pulls the advancing growth cone forward.

The extracellular environment that confronts the growth cone is often a dense thicket of fibrous proteins that are secreted by other cells (the extracellular matrix; see below). The growth cone appears to clear its path by secreting proteases, enzymes that digest protein. This is one example of the specialized biochemical mechanisms that are expressed by growing axons. The unique biochemistry of the growth cone can be used by neuroscientists to detect sites of axonal growth in the brain. For example, growth-associated protein (GAP43) is found in high concentrations only in neurons that are undergoing axonal elongation, and only in the axon, not the dendrites.

Pathway Formation

Retinal axons grow along the substrate provided by the extracellular matrix of the ventral wall of the optic stalk. **Extracellular matrix** is the net-

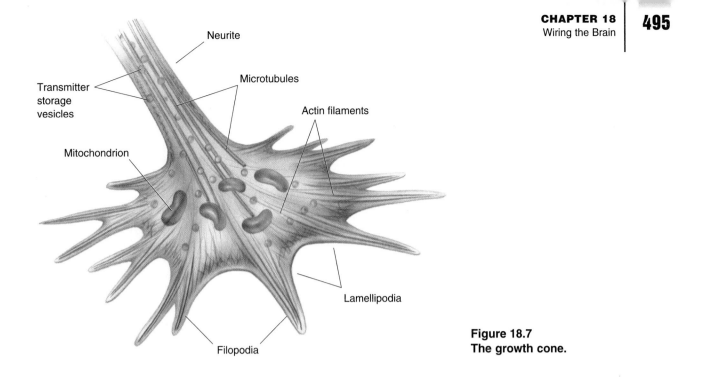

Neurite

Transmitter storage vesicles

Microtubules

Actin filaments

Mitochondrion

Lamellipodia

Filopodia

Figure 18.7
The growth cone.

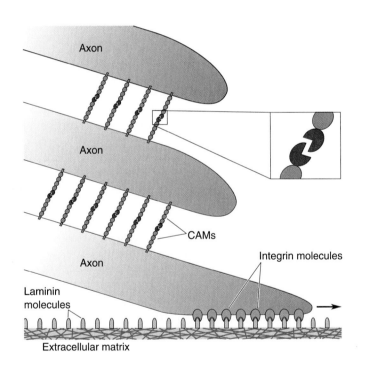

Axon

Axon

Axon

CAMs

Integrin molecules

Laminin molecules

Extracellular matrix

Figure 18.8
Fasciculation. The bottom axon grows along the molecular "highway" of the extracellular matrix. The other axons ride piggyback, sticking to one another by the interaction of cell-adhesion molecules (CAMs) on their surfaces.

work of fibrous glycoproteins that are deposited in the space between cells. One glycoprotein in this matrix is *laminin*. The growing axons express special surface molecules called *integrins* that bind laminin, and this interaction promotes axonal elongation. Thus, the extracellular matrix along the optic stalk forms a molecular "highway" on which retinal axons grow.

Travel down the molecular highway is aided by **fasciculation**, a mechanism that causes axons that are growing together to stick together (Figure 18.8). Fasciculation is due to the expression of specific surface molecules called **cell-adhesion molecules** (**CAM**s). The CAMs in the membrane of neighboring axons bind tightly to one another, causing the axons to grow in unison.

Extracellular matrix can be repellent as well as attractive to growing axons, depending on the cell surface molecules the axons express. Thus, axons from the temporal retina grow toward the midline at the chiasm but encounter a signal there that causes them to veer sharply away. In contrast, nasal axons continue right across the midline and into the contralateral optic tract. These differences must be explained by differential expression of cell surface molecules based on cell position in the retina. Indeed, there is now evidence that the concentration of certain ganglion cell surface molecules varies progressively from the nasal to the temporal retina. Such a gradient in axon surface markers is believed to be matched to complementary gradients on the surfaces of cells in target structures, and this match gives rise to retinotopy. This idea, that chemical markers on growing axons are matched with complementary chemical markers on their targets, is called the **chemoaffinity hypothesis** (Box 18.2).

Besides the markers that are deposited in the extracellular matrix, diffusible molecules act over a distance to attract growing axons toward their targets, like the aroma of freshly brewed java might attract the coffee lover. Although the existence of such chemoattractants was proposed over a century ago by Cajal and was inferred by many experimental studies since then, only very recently has an attractant molecule been identified. In 1994 Marc Tessier-Lavigne and his colleagues, working at the University of California, San Francisco, isolated and characterized the *netrins*, proteins that are released by specific cells in the embryonic CNS and used to guide growing axons toward them.

When axons reach their target, they often encounter a new extracellular environment that retards further growth. These environmental signals can be the absence of specific glycoproteins in the extracellular matrix; for example, the absence of laminin. In addition, axonal growth can be inhibited by

---

The Frog Retinotectal Projection and the Chemoaffinity Hypothesis

One of the favorite model systems of developmental neurobiologists is the frog retinotectal projection. Recall from Chapter 10 that the tectum is the amphibian homologue of the mammalian superior colliculus. The tectum receives retinotopically ordered input from the contralateral eye and uses this information to organize movements in response to visual stimulation, such as lunging after a fly passing overhead. Thus, this system can be used to investigate the mechanisms that generate orderly maps in the CNS.

Another advantage of amphibians is that their CNS axons will regenerate after being cut, which is not true for mammals. In an important series of experiments beginning in the 1940s, Roger Sperry, at California Institute of Technology, took advantage of this property to investigate how the retinotopic map was established in the tectum. In one experiment, Sperry cut the optic nerve, rotated the eye 180° in the orbit, and then allowed the upside-down nerve to regenerate. Despite the fact that the axons in the optic nerve were now scrambled from where they would occur naturally, the axons grew into the tectum to the same sites that they occupied originally. It was as if every retinal axon carried a ticket for a reserved seat in the tectum, and it didn't matter what entrance to the tectum each one used. These sorts of observations led Sperry to propose the *chemoaffinity hypothesis*, which is still with us today, and which partly earned him the Nobel Prize in 1981.

What about those frogs with the upside-down eyes? When a fly passed overhead, these frogs still lunged. But they jumped down instead of up! Can you see why?

diffusible signals released from target structures. For example, in some systems it has been shown that the application of neurotransmitters can inhibit axonal growth, probably by raising calcium concentrations in the growth cone.

The number and nature of the chemical signals that guide activity-independent pathway formation are currently under intense investigation. It should be clear even from this brief description that considerable order can develop in the visual system solely under the influence of these mechanisms.

Synapse Formation

When the growth cone comes in contact with its target, it collapses and forms a synapse. The details of the mechanisms of synapse formation in the CNS are still sketchy; most of what is known comes from studies of the neuromuscular junction. At the neuromuscular junction, the first step appears to be the induction of a cluster of postsynaptic receptors under the site of contact. This is triggered by an interaction between proteins secreted by the growth cone and the target membrane. At the neuromuscular junction, one of these proteins is called *agrin*, which is deposited in the extracellular space at the site of contact (Figure 18.9). The layer of proteins in this space is called the **basal lamina**. The interaction between agrin and receptors in the muscle cell membrane causes the aggregation of postsynaptic acetylcholine receptors. Thus, agrin in the basal lamina acts as a sort of molecular liaison between the axon and its target.

The interaction between axon and target occurs in both directions, and the induction of a presynaptic terminal also appears to involve proteins in the basal lamina. Basal lamina factors provided by the target cell evidently can stimulate Ca^{2+} entry into the growth cone, which triggers neurotransmitter release. Thus, although the final maturation of synaptic structure may

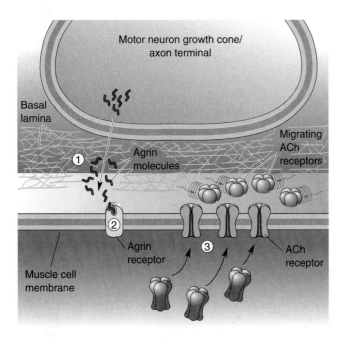

Figure 18.9
Steps in the formation of a neuromuscular synapse. (1) The growing motor neuron secretes the protein agrin into the basal lamina. (2) Agrin interacts with specific receptors in the muscle cell membrane. This interaction leads to (3) the clustering of ACh receptors in the postsynaptic membrane.

take a matter of weeks, rudimentary synaptic transmission appears very rapidly after contact is made. Besides mobilizing transmitter, Ca^{2+} entry into the presynaptic terminal also triggers changes in the cytoskeleton that cause the growth cone to collapse and assume the appearance of a terminal bouton.

ELIMINATION OF CELLS AND SYNAPSES

The mechanisms of pathway formation we've discussed are sufficient to establish considerable order in the connections of the fetal brain. For example, in the visual system these mechanisms ensure (1) that retinal axons reach the LGN, (2) that geniculate axons reach layer IV of striate cortex, and (3) that both of these sets of axons form synapses in their target structures in proper retinotopic order. But the job of wiring together the nervous system isn't finished yet. A prolonged period of development follows, from before birth all the way through adolescence, in which these connections are refined. It may come as a surprise that one of the most significant refinements is a large-scale *reduction* in the numbers of all those newly formed neurons and synapses.

Cell Death

Entire populations of neurons are eliminated during pathway formation, a process known as *naturally occurring cell death*. After axons have reached their targets and synapse formation has begun, there is a progressive decline in the number of presynaptic axons and neurons. Cell death reflects competition for **trophic factors**, life-sustaining nutrients, that are provided in limited quantities by the target cells. This process is believed to produce the proper match in the number of pre- and postsynaptic neurons (Figure 18.10).

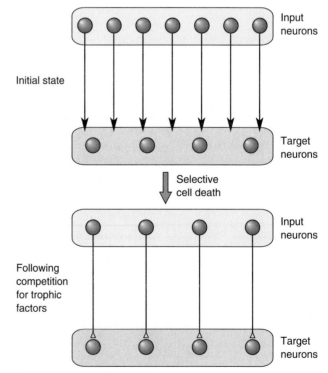

Figure 18.10
Matching inputs with targets by selective cell death. The input neurons are believed to compete with one another for limited quantities of trophic factors produced by the target neurons.

A peptide called **nerve-growth factor** (**NGF**) was the first trophic factor to be identified in the 1940s by Italian biologist Rita Levi-Montalcini. NGF is produced by the targets of axons in the sympathetic division of the ANS. Levi-Montalcini and Stanley Cohen found that the injection of antibodies to NGF into newborn mice resulted in the total degeneration of the sympathetic ganglia. NGF, produced and released by the target tissue, is taken up by the sympathetic axons and transported retrogradely, where it acts to promote neuronal survival. Indeed, if axoplasmic transport is disrupted, the neurons will die despite the release of NGF by the target tissue. Their pioneering work earned Levi-Montalcini and Cohen the 1986 Nobel Prize.

NGF is one of a family of related trophic proteins collectively called the **neurotrophins**. One neurotrophin important for the survival of visual cortical neurons is called **brain-derived neurotrophic factor**, or **BDNF**. Recently it was discovered that BDNF and the other neurotrophins can act at specific cell surface receptors. Most of these receptors are neurotrophin-activated protein kinases that phosphorylate tyrosine residues on their substrate proteins (recall phosphorylation from Chapter 6). Research is under way to understand the downstream consequences of activating these receptors, fueled in part by the prospect of new treatments to rescue dying neurons.

Changes in Synaptic Capacity

Each neuron can receive on its dendrites and soma a finite number of synapses. This number is the *synaptic capacity* of the neuron. Throughout the nervous system, synaptic capacity peaks early in development and then declines as the neurons mature. For example, in the striate cortex of all species examined so far, the synaptic capacity of immature neurons exceeds that of adult cells by about 50%. In other words, visual cortical neurons in the infant brain receive one-and-a-half times as many synapses as do the neurons in adults.

When do cortical neurons lose all those synapses? Yale University scientists Jean-Pierre Bourgeois and Pasko Rakic recently conducted a detailed study to address this question in the striate cortex of the macaque monkey. They discovered that synaptic capacity was remarkably constant in the striate cortex from infancy until the time of puberty. However, during the subsequent adolescent period, synaptic capacity declined sharply—by almost 50% in just over 2 years. A quick calculation revealed the following startling fact: The loss of synapses in the primary visual cortex during adolescence occurs at an average rate of *5000 per second*. (No wonder adolescence is such a trying time!)

While changes in neuronal synaptic capacity during development are well documented, relatively little is known about the mechanisms that govern this change in the CNS. But, once again, the neuromuscular junction has provided a useful model system for the study of synaptic elimination. Initially, a muscle fiber may receive input from several different motor neurons. Eventually, however, this polyneuronal innervation is lost, and each muscle fiber receives synaptic input from a single alpha motor neuron (Figure 18.11). This process is regulated by electrical activity in the muscle. Silencing the activity of the muscle fiber leads to a retention of polyneuronal innervation, while stimulation of the muscle accelerates the elimination of all but one input. Exactly what that activity regulates, and how this regulation leads to the retention of one and only one input, are problems that are currently being investigated.

ACTIVITY-DEPENDENT SYNAPTIC REARRANGEMENT

Imagine a neuron that has a synaptic capacity of 12 synapses and receives inputs from 3 presynaptic neurons, A, B, and C (Figure 18.12). One

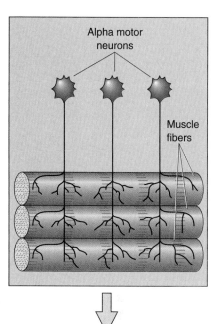

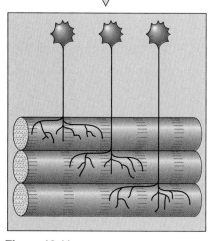

Figure 18.11
Synapse elimination. Initially each muscle fiber receives inputs from several alpha motor neurons. Over the course of development, all inputs but one are lost.

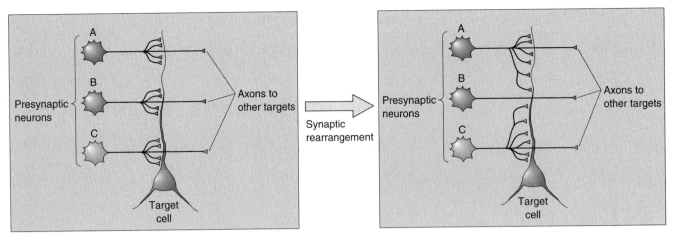

Figure 18.12
Synaptic rearrangement. The target cell receives the same number of synapses in both cases, but the innervation pattern has changed.

arrangement would be that each of the 3 presynaptic neurons provides 4 synapses apiece. Another arrangement is that neurons A and C provide 6 synapses apiece and neuron B provides none. A change from one such pattern of synapses to another is called *synaptic rearrangement*. There is abundant evidence for widespread synaptic rearrangement in the immature brain.

Synaptic rearrangement is the final step in the process of address selection. Unlike most of the earlier steps of pathway formation, *synaptic rearrangement occurs as a consequence of neural activity and synaptic transmission*. In the visual system, some of this activity-dependent shaping of connections occurs prior to birth in response to spontaneous neuronal discharges. However, significant activity-dependent development occurs after birth and is influenced profoundly by sensory experience during childhood. Thus, we will find that the ultimate performance of the adult visual system is determined to a significant extent by the quality of the visual environment during the early postnatal period. In a very real sense, *we learn to see during a critical period of postnatal development.*

Macaque monkeys and cats have typically been used as models for studies of activity-dependent visual system development because, like humans, both of these species have good binocular vision. The neuroscientists who pioneered this field were none other than David Hubel and Torsten Wiesel who, as you will recall from Chapter 10, also laid the foundation for our current understanding of the central visual system in the adult brain. In 1981, they shared the Nobel Prize with Roger Sperry (see Box 18.2).

Segregation of Synaptic Inputs

A feature of the development of both the LGN and the striate cortex, first discovered in the macaque monkey by Pasko Rakic, is that the inputs that carry information from the two eyes initially converge onto the same target cells. Then, over a period of several weeks, these inputs segregate out into the eye-specific LGN layers and ocular dominance columns we learned about in Chapter 10. We will discuss segregation of synaptic inputs in the LGN and striate cortex in turn.

Segregation of Retinal Axons in the LGN. The first axons to reach the LGN are usually those from the contralateral retina, and they spread out to occupy the entire nucleus. Somewhat later, the ipsilateral projection arrives and intermingles with the axons of the contralateral eye. Then the axons from the two eyes segregate into the eye-specific domains that are characteristic of the adult nucleus. The injection of tetrodotoxin (TTX) into the eyeball prevents this process of segregation, showing that it is dependent on activity generated in the retina (recall that TTX blocks action potentials). What is the source of the activity, and how does it orchestrate segregation?

Since segregation occurs in the fetus, prior to the development of photoreceptors, the activity cannot be driven by light stimulation. Rather, it appears that ganglion cells are spontaneously active during this period of fetal development. This activity is not random, however. Recent studies by Carla Shatz and her colleagues at Stanford University indicate that ganglion cells fire in quasisynchronous "waves" that spread across the retina (Box 18.3). The origin of the wave and its direction of propagation may be random, but during each wave the activity in a ganglion cell is highly correlated with the activity of its nearest neighbors. And because these waves are generated independently in the two retinas, the activity patterns arising in the two eyes are not correlated with respect to each other.

Segregation is thought to depend on a process of synaptic stabilization whereby only retinal terminals that are active at the same time as their postsynaptic LGN target neuron are retained. This hypothetical mechanism of synaptic plasticity was first articulated by Canadian psychologist Donald Hebb in the 1940s. Consequently, synapses that can be modified in this way are called **Hebb synapses**, and synaptic rearrangements of this sort are called **Hebbian modifications**. According to this hypothesis, whenever a wave of retinal activity drives a postsynaptic LGN neuron to fire action potentials, the synapses between them are stabilized. In other words, *neurons that fire together, wire together* (Figure 18.13). Because the activity from the two eyes does not occur at the same time, the inputs will compete on a "winner-takes-all" basis until one input is retained and the other is eliminated, thus leading to complete segregation of the two inputs. (We'll look at the role of activity in synaptic elimination in a moment.)

Segregation of LGN Inputs in the Striate Cortex. Similar to the situation in the LGN, the afferents serving the two eyes are initially intermingled in cortical layer IV and then segregate under the influence of activity (Figure 18.14). In the macaque monkey, this segregation of ocular dominance columns begins before birth and is normally completed by the third postnatal week. In the cat, the process occurs entirely after birth.

To the extent that segregation occurs postnatally, the formation of ocular dominance columns can be affected by the deprivation of normal vision. This is most dramatically demonstrated by an experimental manipulation used by Hubel and Wiesel, called **monocular deprivation**, in which one eyelid is sealed closed. If monocular deprivation is begun shortly after birth, for the duration of the period of natural segregation, the striking result is that the "open-eye" columns develop to be much wider than the "closed-eye" columns (Figure 18.15). If the deprivation is begun later, after the period of natural segregation, then anatomical effects are not observed in layer IV. Thus, a **critical period** exists for this type of plasticity which in the macaque monkey lasts until about 6 weeks of age.

Within this critical period, the anatomical effects of monocular deprivation can be reversed simply by closing the previously open eye and opening the previously closed eye. The result of this "reversed-occlusion" manipulation is that the shrunken ocular dominance columns of the formerly closed eye expand, and the expanded columns of the formerly open eye shrink. These and other experiments suggest that within this 6-week critical period, even after ocular dominance column segregation appears to be anatomically complete (after 3 weeks), the afferents serving the two eyes

Box 18.3 | **PATH OF DISCOVERY**

Waves in the Eye

by Carla Shatz

Carla Shatz

When Hubel and Wiesel discovered the system of ocular dominance columns in the visual cortex, everyone wanted to know how such an incredibly beautiful pattern of neural connections could form during development. It seemed impossible that this almost crystalline array could form in any way other than by prespecification, with the LGN axons from each eye growing from the outset directly to the correct part of layer IV. The idea that connections might form in an imprecise pattern first and later become refined as a result of correlations in the firing of sets of neurons was not yet well formulated. I can remember the amazement I felt when I heard Pasko Rakic tell us in a seminar that initially in development the ocular dominance columns in layer IV were not present, but rather formed gradually from initial patterns in which the LGN axons representing both eyes were intermixed in layer IV. Not only that, but in monkeys, Rakic found that the columns had begun to form in the womb, long before neural activity driven by pattern vision is even possible.

How could the ocular dominance columns form without vision? Even more intriguing, Rakic and others also found that the LGN layers were not present initially. Retinal ganglion cell axons from the two eyes are intermixed with each other and then sort to form the eye-specific layers even earlier in development than column formation in cortex. At those times in development, the rods and cones are not even present in the retina! I wondered how an activity-dependent mechanism could work at all so early in development. Perhaps the retinal ganglion cells themselves are generating spontaneous activity. But if so, according to mathematical models, the activity had to have special properties; it had to be highly correlated so that neighboring ganglion cells fired in near synchrony while distant cells were silent. It would be a real challenge to study such an idea because you would have to record simultaneously from many retinal ganglion cells in the developing retina and see when each cell fired and where it was located.

Sometimes exciting discoveries can happen because you are in the right place at the right time and you are not afraid to try things that seem unlikely to work. Just when Rachel Wong, a postdoctoral fellow in my lab, and I began thinking about how to approach this problem, my colleague at Stanford University, Denis Baylor, and his postdoctoral fellow Markus Meister, had begun wonderful experiments using a special multielectrode array to record simultaneously from many neurons in the salamander retina. The design of this unique electrode array was such that it could not be put into the eye, but rather small pieces of the retina had to be placed on the array *in vitro*. While this approach worked well for cold-blooded vertebrates like the salamander, it seemed very unlikely to work for the mammalian retina. Still, we thought that just maybe we could take the developing retina and keep it alive *in vitro* long enough to see if ganglion cells were firing spontaneously and synchronously.

I will never forget the afternoon when Rachel and Markus had the first successful experiment. The room was darkened to prevent any possibility of a contribution of light-driven activity from developing photoreceptors. Each electrode on the array was connected to an amplifier and it was possible, by turning switches, to monitor the action potential activity recorded at each electrode by listening to it on a loudspeaker. It sounded like waves crashing on the beach. At each recording site, ganglion cells would fire in a burst of action potentials and then take a long pause and then resume firing in the same rhythmic pattern over and over again for many hours. It was mesmerizing to discover a small piece of the brain so busy so early in development! What's more, the spatial pattern of activity resembled a wave that periodically swept across the retina in random directions.

The glimpse of this small piece of the brain generating such elegant patterns of correlated activity on its own was an incredible experience. And it was a special treat to be able to conduct these experiments with such talented and generous colleagues. Although much of scientific life can be spent working alone, these highly interactive collaborations bring delight and can result in new and exciting discoveries not possible from a single individual.

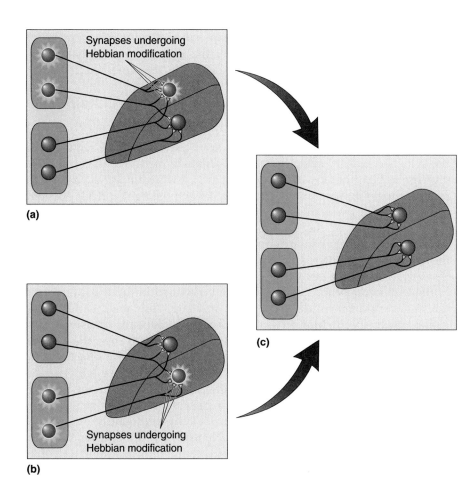

(a)

(b)

(c)

Synapses undergoing Hebbian modification

Synapses undergoing Hebbian modification

Figure 18.13
Plasticity at Hebb synapses. Illustrated here are two target neurons in the LGN with inputs from different eyes. Inputs from the two eyes initially overlap and then segregate under the influence of activity. **(a)** The two input neurons in one eye, at the top, fire at the same time. This is sufficient to cause the top LGN target neuron to fire, but not the bottom target neuron. The active inputs onto the active target undergo Hebbian modification and become more effective. **(b)** This is the same situation as in part a except now the two input neurons in the other eye, at the bottom, are active simultaneously, causing the bottom target neuron to fire. **(c)** Over time, neurons that fire together, wire together. Notice also that input cells that fire out of synch with the target, lose their link.

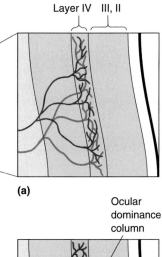

Striate cortex

Layer IV   III, II

(a)

Ocular dominance column

(b)

Figure 18.14
Segregation of ocular dominance columns in cat striate cortex. (a) Initially the inputs from the LGN serving the two eyes (different colors) are intermingled in layer IV. **(b)** Over the course of fetal and early postnatal development, the inputs from the two eyes segregate into ocular dominance columns in layer IV.

exist in a sort of *dynamic equilibrium* that can still be disrupted by deprivation. At the end of the critical period, however, the afferents apparently lose their capacity for growth and retraction and, in a sense, are cemented in place.

Convergence of Synaptic Inputs on Neurons Outside of Layer IV

Although the streams of information from the two eyes are segregated in the LGN and layer IV of striate cortex, eventually they must be combined to give rise to binocular vision. The anatomical basis for binocular vision is the convergence of inputs from layer IV cells serving the right and left eyes onto cells in layer III. These are among the last connections to be specified during the development of the retinogeniculocortical pathway. Again, activity-dependent synaptic rearrangement plays a major role in this process. Because the neural activity patterns depend mainly on the visual environment, synaptic rearrangements of this sort are also said to be *experience-dependent*.

Binocular connections are formed and modified under the influence of the visual environment during infancy and early childhood. Unlike segregation of eye-specific domains, which evidently depends on *asynchronous* patterns of activity spontaneously generated in the two eyes, the *establishment of binocular receptive fields depends on correlated patterns of activity that arise from the two eyes as a consequence of vision*. This has been demonstrated clearly by experiments that bring the patterns of activity from the two eyes out of register. For example, monocular deprivation, which replaces patterned activity in one eye with random activity, profoundly disrupts the binocular connections in striate cortex. Neurons outside of layer IV, which normally have binocular receptive fields, respond only to stimulation of the nondeprived eye after even a brief period of monocular deprivation. This change in the binocular organization of the cortex is called an **ocular dominance shift** (Figure 18.16).

These effects of monocular deprivation are not merely a passive reflection of the anatomical changes in layer IV discussed earlier. An ocular dominance shift, reflecting changes in intracortical connections, will occur in response to monocular deprivation initiated in the macaque monkey at 1 year of age, well beyond the period of susceptibility of LGN axonal arbors. However, plasticity of these intracortical connections is also limited to a critical period of postnatal life. For example, in the cat, where it has been studied in greatest detail, ocular dominance plasticity peaks at about 1 month of age and then progressively declines to a very low level by 3–4 months (Figure 18.17). The critical period for modification of binocular connections in the macaque monkey extends to about age 2 years, and estimates are that this plasticity ends in humans at about age 10. These critical periods coincide with the times of greatest growth of the head and eyes in these species. Thus, it is believed that plasticity of binocular connections is normally required to maintain good binocular vision throughout this period of rapid growth. The hazard associated with such activity-dependent fine-tuning is that these connections are also highly susceptible to deprivation.

Synaptic Competition

As you well know, a muscle that is not used regularly will atrophy and lose strength; hence the saying, "Use it or lose it." Is the disconnection of activity-deprived synapses simply a consequence of disuse? This does not appear to be the case in striate cortex, because the disconnection of a deprived eye input requires that the open-eye inputs be active. Rather, a process of **binocular competition** evidently occurs, where the inputs from

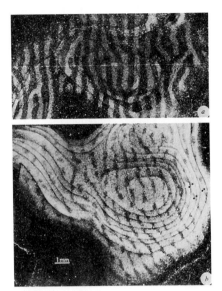

Figure 18.15
Modification of ocular dominance stripes after monocular deprivation. Tangential sections through layer IV of macaque monkey striate cortex illuminated to show the distribution of radioactive LGN terminals serving one eye. **(a)** A normal monkey. **(b)** A monkey that had been monocularly deprived for 18 months, starting at 2 weeks of age. The nondeprived eye had been injected with radioactive tracer (see Figure 10.17), revealing expanded ocular dominance columns in layer IV. (Source: Wiesel, 1982, p. 585.)

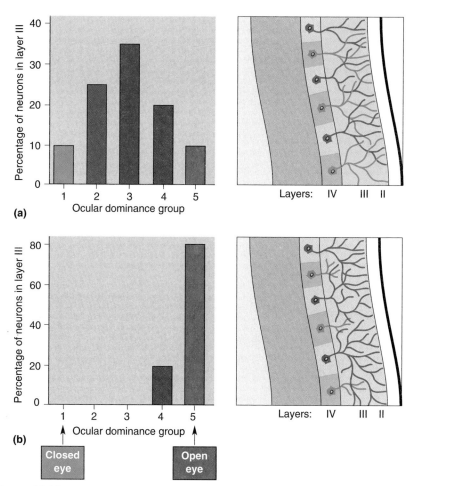

(a)

(b)

Closed eye

Open eye

Figure 18.16
The ocular dominance shift. Shown here are ocular dominance histograms constructed after electrophysiological recording from neurons in the striate cortex of **(a)** normal cats and **(b)** cats that had been monocularly deprived early in life. The bars show the percentage of neurons in layer III in each of five ocular dominance categories. Cells in groups 1 and 5 are activated by stimulation of either the left or the right eye, respectively, but not both. Cells in group 3 are activated equally well by either eye. Cells in groups 2 and 4 are binocularly activated but show a preference for either the left or the right eye, respectively. The histogram in part a reveals that the majority of neurons in the visual cortex of a normal animal are driven binocularly. The histogram in part b shows that a period of monocular deprivation leaves few neurons responsive to the deprived eye. The illustrations to the right of the histograms show the likely anatomical basis for this change in physiology.

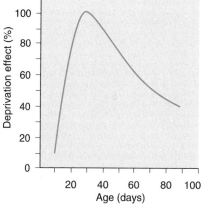

Figure 18.17
The critical period for plasticity of binocular connections in the kitten. Plotted here is the sensitivity of binocular connections in cat striate cortex to monocular deprivation at different postnatal ages. The deprivation effect is the percentage of neurons in area 17 whose responses are dominated by stimulation of the nondeprived eye. The critical period in cats begins at about 3 weeks and ends at about 3 months. (Source: Adapted from Olson and Freeman, 1980.)

the two eyes actively compete for synaptic territory. If the activity of the two eyes is correlated and equal in strength, the two inputs will be retained on the same cortical cell. However, if this balance is disrupted by depriving one eye, the more active input will somehow displace the deprived synapses or cause them to be less effective.

Competition in visual cortex is demonstrated by the effects of **strabismus**, a condition in which the eyes are not perfectly aligned (i.e., are "cross-eyed" or "wall-eyed"). This is a common visual disorder in humans that can result in the permanent loss of stereoscopic vision. Experimental strabismus is produced by surgically or optically misaligning the two eyes, thereby causing visually evoked patterns of activity from the two eyes to arrive out of synch in the cortex. If you press gently with a finger alongside one eye, you can see the consequences of misalignment of the two eyes. As a result of this manipulation of visual experience, there is a total loss of binocular receptive fields even though the two eyes retain equal representation in the cortex (Figure 18.18). This is a clear demonstration that the disconnection of inputs from one eye occurs as the result of competition rather than disuse (the two eyes are equally active, but for each cell, a "winner takes all"). If produced early enough, strabismus can also sharpen the segregation of ocular dominance columns in layer IV.

The changes in ocular dominance and binocularity after deprivation have clear behavioral consequences. An ocular dominance shift after monocular deprivation leaves the animal visually impaired in the deprived eye, and the loss of binocularity associated with strabismus completely eliminates stereoscopic depth perception. However, neither of these effects is irreversible if corrected early enough in the critical period. The clinical lesson is clear, that congenital cataracts or ocular misalignment must be corrected in early childhood, as soon as surgically feasible, to avoid permanent visual disability.

Modulatory Influences

With increasing age, there appear to be additional constraints on the forms of activity that will cause modifications of cortical circuits. Before birth, spontaneously occurring bursts of retinal activity are sufficient to orchestrate aspects of address selection in the LGN and cortex. After birth, an interaction with the visual environment is of critical importance. However, even visually driven retinal activity may be insufficient for modifications of binocularity during this critical period. There is increasing experimental evidence that such modifications also require that the animals pay attention to visual stimuli and use vision to guide behavior. For example, modifications of binocularity following monocular stimulation do not occur when animals are kept anesthetized, even though it is known that cortical neurons respond briskly to visual stimulation under this condition. These and related observations have led to the proposal that synaptic plasticity in the cortex requires the release of extraretinal "enabling factors" that are linked to behavioral state (level of alertness, for example).

Some progress has been made in identifying the physical basis of these enabling factors. Recall that a number of diffuse modulatory systems innervate the cortex (see Chapter 15). These include the noradrenergic inputs from the locus coeruleus and the cholinergic inputs from the basal forebrain. The effects of monocular deprivation have been studied in animals in which these modulatory inputs to striate cortex were eliminated. This was found to cause a substantial impairment of ocular dominance plasticity outside of cortical layer IV, even though transmission in the retinogeniculocortical pathway was apparently normal (Figure 18.19).

The mechanism for modulation of synaptic plasticity by acetylcholine and norepinephrine remains to be determined. However, it is known that these neurotransmitters increase the excitability of neurons, thus making it more likely that cortical neurons will respond with action potentials to visual stimulation.

Elementary Mechanisms of Synaptic Plasticity

Much of the activity-dependent development of the visual system can be explained using Hebb synapses. This is particularly obvious for the development of binocular connections, in which axons from layer IV cells that converge onto the same layer III cell are consolidated when (and only when)

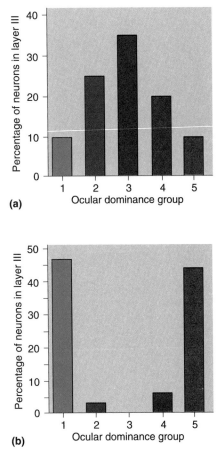

Figure 18.18
Effects of strabismus on cortical binocularity. (a) An ocular dominance histogram from a normal animal like that in Figure 18.16a. **(b)** In this case, the eyes have been brought out of alignment by cutting one of the eye muscles. After a brief period of strabismus, binocular cells are almost completely absent. The cells in visual cortex are driven by either the right or the left eye, but not by both.

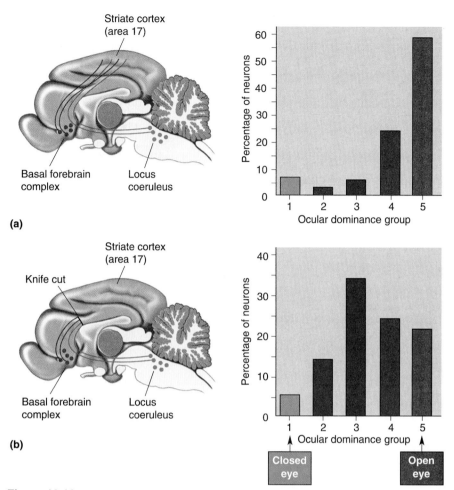

Figure 18.19
Dependence of plasticity of binocular connections on modulatory inputs. (a) A midsagittal view of a cat brain showing the trajectory of two modulatory inputs to striate cortex. One arises in the locus coeruleus and uses NE as a transmitter, and the other arises in the basal forebrain complex and uses ACh as a transmitter. The activity of both of these inputs is related to levels of attention and alertness. If these systems are intact, monocular deprivation will produce the expected ocular dominance shift, shown in the histogram at the right. **(b)** The result of depleting the cortex of these modulatory inputs. Now, monocular deprivation has little effect on the binocular connections in striate cortex. (Source: Adapted from Bear and Singer, 1986.)

they carry synchronous patterns of activity. Why is it that neurons that fire together, wire together? The answer requires knowledge of the mechanisms of excitatory synaptic transmission in the visual system.

Excitatory Synaptic Transmission in the Immature Visual System. Glutamate is the transmitter at all of the modifiable synapses we have discussed (retinogeniculate, geniculocortical, and corticocortical), and it activates several subtypes of postsynaptic receptors. Recall from Chapters 5 and 6 that neurotransmitter receptors may be divided into two broad categories: G-protein-coupled, or metabotropic, receptors; and transmitter-gated ion channels. Postsynaptic glutamate-gated ion channels allow the passage of positively charged ions into the postsynaptic cell and may be further divided into AMPA receptors and NMDA receptors. AMPA and NMDA receptors are colocalized at many synapses.

An NMDA receptor has two unusual features that distinguish it from an AMPA receptor (Figure 18.20). First, NMDA receptor conductance is volt-

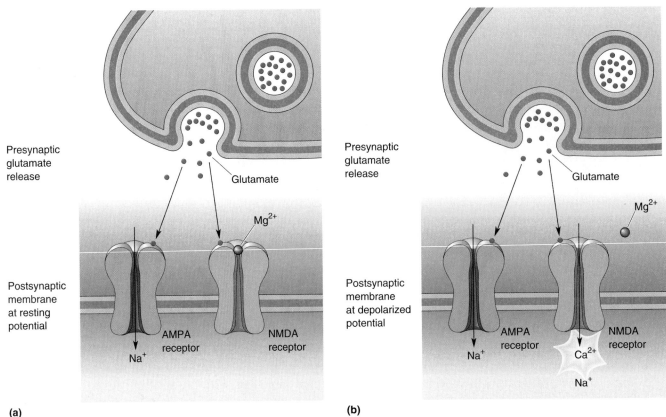

Figure 18.20
NMDA receptors signal coincident pre- and postsynaptic activity. (a) Presynaptic activation causes the release of glutamate, which acts on postsynaptic AMPA receptors and NMDA receptors. At the negative resting membrane potential, the NMDA receptors pass little ionic current because they are blocked with Mg^{2+} ions. **(b)** If glutamate release coincides with depolarization sufficient to displace the Mg^{2+} ions, then Ca^{2+} will enter the postsynaptic neuron via the NMDA receptor. Hebbian synaptic plasticity could be explained if the Ca^{2+} admitted by the NMDA receptor were to trigger an enhancement of synaptic effectiveness.

age-gated, owing to the action of Mg^{2+} at the channel. At the resting membrane potential, the inward current through the NMDA receptor is interrupted by the movement of Mg^{2+} ions into the channel, where they become lodged. As the membrane is depolarized, however, the Mg^{2+} is displaced from the channel, and current is free to pass into the cell. Thus, substantial current through the NMDA receptor channel requires the concurrent release of glutamate by the presynaptic terminal *and* depolarization of the postsynaptic membrane. The other distinguishing feature of an NMDA receptor is that its channel conducts Ca^{2+} ions. Therefore, *the magnitude of the Ca^{2+} flux passing through the NMDA receptor channel specifically signals the level of pre- and postsynaptic coactivation.*

Long-Term Potentiation. Perhaps NMDA receptors in the cortex and LGN serve as Hebbian detectors of coincident pre- and post-synaptic activity, and Ca^{2+} entry through the NMDA receptor channel triggers the biochemical mechanisms that modify synaptic effectiveness. This idea can be tested by performing an experiment in which layer IV neurons are stimulated electrically to fire action potentials while their postsynaptic targets, the layer III neurons, are strongly depolarized. Such a coincidence of pre- and postsynaptic activation should strongly activate postsynaptic NMDA receptors. If the hypothesis is correct, it should also lead to a lasting increase in the effectiveness of the stimulated synapses.

Figure 18.21 shows the results of such an experiment. As predicted, pairing low-frequency stimulation of layer IV with intracellular depolarization of a cell in layer III results in a long-lasting increase in the effectiveness of the stimulated synapses. This form of synaptic modification, called **long-term potentiation (LTP)**, is prevented by application of drugs that block NMDA receptors. Thus, the strong activation of NMDA receptors that occurs when pre- and postsynaptic neurons fire together may partially account for why they wire together during visual system development. (We will discuss LTP and its molecular basis further in Chapter 20.)

Long-Term Depression. Hebb synapses can explain a lot, but there is another side to the story. In the case of strabismus, for example, synapses whose activity fails to correlate with that of the postsynaptic cell are weakened and then eliminated. In other words, *neurons that fire out of synch, lose their link*. What is the mechanism responsible for this form of synaptic plasticity?

Weak coincidences could be signaled by lower levels of NMDA receptor activation and less Ca^{2+} influx. Recent experiments suggest that the lower level of Ca^{2+} admitted under these conditions triggers an opposite form of synaptic plasticity, **long-term depression (LTD)**, whereby the active synapses are decreased in effectiveness. (We will also discuss LTD further in Chapter 20.) Thus, one current hypothesis is that the maintenance of some connections formed during development may depend on their success in evoking an NMDA receptor-mediated response beyond some threshold level. Failure to achieve this threshold leads to disconnection. Both processes depend on activity originating in the retina, NMDA receptor activation, and postsynaptic Ca^{2+} entry.

This role for NMDA receptors in activity-dependent synaptic rearrangement during development is still hypothetical. Nonetheless, it has been shown that drugs that block NMDA receptors injected into the brain disrupt (1) the natural segregation of eye-specific inputs in the LGN, (2) binocular competition of geniculate terminals in layer IV, and (3) the modification of binocular connections in the superficial cortical layers. Thus, we can conclude that NMDA receptor activation is crucial for some aspect of activity-dependent synaptic plasticity.

Why Do Critical Periods End?

Although plasticity of visual connections persists in the adult brain, it is well-established that the range over which this plasticity occurs constricts with increasing age. Early in development, gross rearrangements of axonal arbors are possible, while in the adult, plasticity appears to be restricted to local changes in synaptic efficacy. In addition, the adequate stimulus for evoking a change also appears to be increasingly constrained as the brain matures. An obvious example is the fact that simply patching one eye will cause a profound alteration in the binocular connections of the superficial layers during infancy, but by adolescence this type of experience fails to cause a lasting alteration in cortical circuitry.

We still have no satisfactory answer to the question of why critical periods end. As we learn more about the elementary mechanisms of axonal pathfinding and synaptic plasticity, we'll gain insight about how these processes are regulated. Meanwhile, here are three current hypotheses for you to think about.

Plasticity diminishes when axon growth ceases. We've seen that there is a period of several weeks when the geniculate arbors can contract and expand within layer IV under the influence of visual experience. Thus, a factor that limits the critical period in layer IV could be a loss of the capability for changes in axonal length, which in turn may be due to changes in

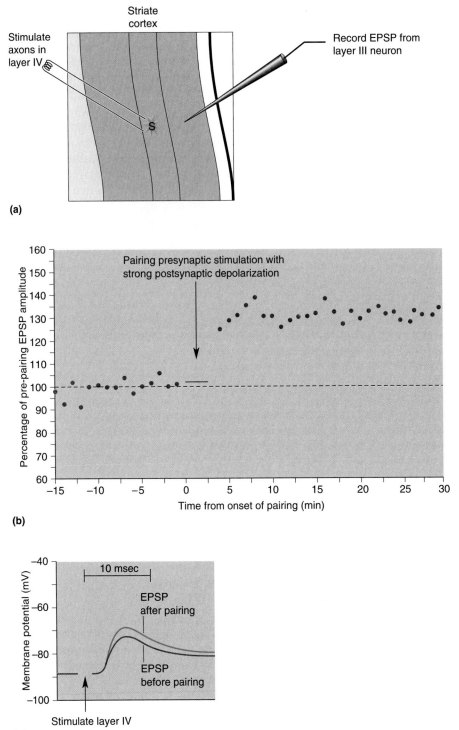

(a)

(b)

(c)

Figure 18.21

Hebb synapses in layer III of striate cortex. Shown here is the record of an experiment demonstrating Hebbian synaptic modification of the connections between cortical layers IV and III in the rat. **(a)** The site, S, was electrically stimulated every 30 sec, and the intracellular EPSP was recorded in a layer III neuron. **(b)** A plot of the EPSP amplitude versus time during the experiment. At the time indicated, the electrical stimuli to layer IV were paired with strong depolarization of the intracellularly recorded layer III neuron. As a consequence of this pairing, the intracellular response of the layer III neuron to the test stimulation of layer IV was potentiated. **(c)** Representative EPSPs measured before and after pairing. (Source: Adapted from Kirkwood and Bear, 1994a.)

the extracellular matrix or surface glycoproteins expressed by the geniculate axons.

Plasticity diminishes when synaptic transmission matures. The end of a critical period may reflect changes in the elementary mechanisms of synaptic plasticity. There is now evidence that some excitatory amino acid receptors change during postnatal development. For example, it has been shown that activation of metabotropic glutamate receptors stimulates very different postsynaptic responses in striate cortex during the critical period when binocular connections are most susceptible to monocular deprivation. In addition, the properties of NMDA receptors change in layer IV at the same time as the end of the critical period for ocular dominance column segregation. This change can be detected at the single channel level and may result from a developmental switch in NMDA receptor subunit composition.

Plasticity diminishes when cortical activation is constrained. As development proceeds, certain types of activity may be filtered by successive synaptic relays to the point where they no longer activate NMDA receptors, or other elementary mechanisms, sufficiently to trigger plasticity. As mentioned previously, ACh and NE facilitate synaptic plasticity in the superficial cortical layers, perhaps simply by enhancing polysynaptic intracortical transmission. A decline in the effectiveness of these neurotransmitters, or a change in the conditions under which they are released, may contribute to the decline in plasticity. Indeed, there is evidence that supplementing the adult cortex with NE can restore some degree of modifiability. There is also evidence that intrinsic inhibitory circuitry is late to mature in the striate cortex. Consequently, patterns of activity that may have gained access to modifiable synapses in superficial layers early in postnatal development may be tempered by inhibition in the adult.

The question of why critical periods end is important. Synaptic modification and rewiring of circuitry provide the capacity for some recovery of function after CNS damage. However, such recovery is disappointingly limited in the adult brain. On the other hand, recovery of function after brain injury can be nearly 100% in the immature nervous system when synaptic rearrangements are widespread. Thus, knowledge of how plasticity is regulated during normal development may suggest ways to promote recovery from damage later in life.

CONCLUDING REMARKS

We have seen that the generation of circuitry during brain development occurs mostly prior to birth and is guided by cell-cell communication through physical contact and by diffusible chemical signals. Nonetheless, while most of the "wires" find their proper place before birth, the final refinement of synaptic connections, particularly in the cortex, occurs during infancy and is influenced by the sensory environment. Although we have focused on the visual system, other sensory and motor systems are also readily modified by the environment during critical periods of early childhood. In this way, our brain is a product not only of our genes, but also of the world we grow up in.

The end of developmental critical periods does not signify an end to experience-dependent synaptic plasticity in the brain. It is possible that some higher cortical areas in the frontal and temporal lobes exhibit plasticity in adults as robust as that seen in area 17 in juveniles. In any case, the environment must modify the brain throughout life at some level, or there would be no basis for memory formation. In the next two chapters, we will take a close look at the neurobiology of learning and memory. We will see that the mechanisms of synaptic plasticity proposed to account for learning bear a close resemblance to those believed to play a role in synaptic rearrangement during development.

KEY TERMS

The Genesis of Neurons
radial glial cell
neuroblast
cortical plate

The Genesis of Connections
growth cone
extracellular matrix
fasciculation
cell-adhesion molecule (CAM)
chemoaffinity hypothesis
basal lamina

Elimination of Cells and Synapses
trophic factor
nerve-growth factor (NGF)

neurotrophin
brain-derived neurotrophic factor
    (BDNF)

**Activity-Dependent Synaptic
    Rearrangement**
Hebb synapse
Hebbian modification
monocular deprivation
critical period
ocular dominance shift
binocular competition
strabismus
long-term potentiation (LTP)
long-term depression (LTD)

REVIEW QUESTIONS

1. What does it mean that the cortex develops "inside-out"?
2. Describe the three phases of pathway formation. In which phase (or phases) does neural activity play a role?
3. What are three ways that Ca^{2+} ions are thought to contribute to the processes of synapse formation and rearrangement?
4. How are the elimination of polyneuronal innervation of a muscle fiber and the segregation of retinal terminals in the LGN similar? How do these processes differ?
5. Not long ago, when a child was born with strabismus, the defect usually was not corrected until after adolescence. Today, surgical correction is always attempted during early childhood. Why? How does strabismus affect the connections in the brain, and how does it affect vision?
6. Children are often able to learn several languages apparently without effort, while most adults must struggle to master a second language. From what you know about brain development, why might this be true?

Memory Systems

It is difficult to study the brain without developing a sense of awe about how well it works. The different sensory, motor, limbic, and modulatory systems we have explored contain billions of individual neural elements with enormous numbers of interconnections. As we saw in Chapter 18, the development of these connections is an orderly process that follows certain rules. Wiring the brain isn't magic—at least not entirely! But as impressive and orderly as prenatal development is, we are by no means complete creatures when we are born. From the moment we take our first breath, and possibly before, the sensory stimuli we experience modify our brain and influence our behavior. Indeed, one of the main goals of the first 20 years of our lives is to learn the skills we need in order to survive in the world. We learn an enormous number of things, some straightforward (snow is cold), and others more abstract (an isosceles triangle has two sides of equal length).

There is a close relationship between what we called experience-dependent brain development in Chapter 18 and what we call learning in this chapter. Visual experience during infancy is essential for the normal development of visual cortex, but it also allows us to recognize an image of our mother's face. Visual development and learning probably employ similar mechanisms, but at different times and in different cortical areas. Viewed in this way, learning and memory are the lifelong adaptations of brain circuitry to the environment. They enable us to respond appropriately to situations we have experienced before. This example also illustrates that a particular form of information is likely to be stored in those parts of the brain that normally process that type of information. Thus, we can anticipate that different parts of the brain will participate in different types of memory.

In this chapter, we discuss the anatomy of memory: how different parts of the brain fit together to store particular types of information. In Chapter 20, we will focus on the elementary synaptic mechanisms that can store information in the brain.

TYPES OF MEMORY AND AMNESIA

By a simple definition, **learning** is the acquisition of new information or knowledge. **Memory** is the retention of learned information. We learn and remember lots of different things, and it is important to appreciate that these various things might not be processed and stored by the same neural "hardware." There is no single brain structure or cellular mechanism that accounts for all learning. Moreover, the way in which information of a particular type is stored may change over time.

Declarative and Procedural Memory

Psychologists have studied learning and memory extensively. As a result, they have distinguished what appear to be different types. Several schemes have been proposed distinguishing what appear to be different categories of memory. A useful distinction for our purposes is between declarative memory and procedural memory. During the course of our lives, we learn many facts—the capital of Thailand is Bangkok; the coyote will never catch the roadrunner. We also store memories of life's events— "I had Cheerios for breakfast"; "I heard a boring chemistry lecture yesterday." Memory for facts and events is called **declarative memory**. Memory for skills or behavior is called **procedural memory**. We learn to play the piano, throw a Frisbee, or tie our shoes, and somewhere that information is stored in our brain. Generally, declarative memories can be accessed for conscious recollection and procedural memories cannot. However, the procedures we learn can be performed without conscious recollection. As the old saying goes, you never forget how to ride a bicycle. You may not explicitly remember the day you first rode a two-wheeler on your own (the declarative part of the memory), but your brain remembers what to do when you're on one (the procedural part of the memory).

Another distinction is that declarative memories are often easy to form and are easily forgotten, whereas procedural memories tend to require repetition and practice over a longer time to form but are less likely to be forgotten. Think of the difference between memorizing the capitals of foreign countries versus learning to ski. Declarative memory is usually what we mean by the word "memory" in everyday usage, whereas procedural memory is more like "habit." In this chapter, we'll focus on declarative memory. While there is no clear limit to the number of declarative memories the brain can store, there can be great diversity in the ease and speed with which new information is acquired. Studies of humans with abnormally good memories suggest that the limit on the storage of declarative information is remarkably high (Box 19.1).

Short-Term, Long-Term, and Working Memory

Within the category of declarative memory, we can make a further distinction between short-term and long-term memory. When someone tells you his or her phone number, you can retain it for a limited period of time by repeating the number to yourself. If the number is too long (e.g., a phone number with extra numbers for a foreign country), you may have trouble remembering the number at all. Memory that is temporary, limited in capacity, and requires continual rehearsal is called **short-term memory**. If someone interrupts your thoughts while you are trying to hold the phone number in short-term memory, you'll probably forget it. But through some mysterious and imperceptible process, at some point you will remember the phone number even if you are interrupted. The information has been

Box 19.1 | OF SPECIAL INTEREST

An Extraordinary Memory

In the 1920s, a man named Sherashevsky came to the office of the Russian psychologist Aleksandr Luria. Thus began a 30-year study of the uncommon memory of this man Luria referred to simply as S. Luria's fascinating description of this study is contained in a little book entitled *The Mind of a Mnemonist*. When S. came to Luria's lab, Luria initially studied him by giving conventional tests such as memorizing lists of words, numbers, or nonsense syllables. He'd read the list once and then ask S. to repeat it. Much to Luria's surprise, he couldn't come up with a test S. could not pass. Even when 70 words were read in a row, S. could repeat them forward, backward, and in any other order. During the many years they worked together, Luria never found a limit to S.'s memory. In tests of his retention, S. demonstrated that he remembered lists he had previously seen even 15 years earlier!

How did he do it? S. described several factors that may have contributed to his great memory. One was his unusual sensory response to stimuli—he retained a vivid image of things he saw. When shown a table of 50 numbers, he claimed that it was easy to later read off numbers in one row or along the diagonal because he simply had to call up a visual image of the entire table. Interestingly, when he occasionally made errors in recalling tables of numbers written on a chalkboard, they appeared to be "reading" errors rather than memory errors. For instance, if the handwriting was sloppy, he would mistake a 3 for an 8 or a 4 for a 9. It was as if he was seeing the chalkboard and numbers all over again when he was recalling the information.

Another interesting aspect of S.'s sensory response to stimuli was a powerful form of synesthesia. *Synesthesia* is a phenomenon in which sensory stimuli evoke sensations usually associated with different stimuli. For example, when S. heard a sound, in addition to hearing he would see splashes of colored light and perhaps have a certain taste in his mouth.

After learning that his memory was unusual, S. left his job as a reporter and became a professional stage performer—a mnemonist. In order to remember huge lists of numbers or tables of words given by members of the audience trying to stump him, he complemented his lasting sensory responses to stimuli and his synesthesia with memory "tricks." To remember a long list of items, he made use of the fact that each item evoked some sort of visual image. As the list was read or written, S. imagined himself walking through his hometown; as each item was given, he placed its evoked image along his walk— the image evoked by item 1 by the mailbox, the image for item 2 by this bush, and so on. To later recall the items, he walked the same route and picked up the items he had put down. Though we may not have the complex synesthetic sensations of S., this technique of associating things with something we know well, such as our neighborhood, is one we all can use.

But not everything about S.'s memory was to his advantage. While the complex sensations evoked by stimuli helped him remember lists of numbers and words, they interfered with his ability to integrate and remember more complex things. He had trouble recognizing faces because each time a person's expression changed, he would also "see" changing patterns of light and shade, which would confuse things. He also wasn't very good at following a story read to him. Rather than ignoring the exact words and focusing on the important ideas, as most of us would do, S. was overwhelmed by an explosion of sensory responses. Just imagine how bewildering it would be to be bombarded by constant visual images evoked by each word, plus sounds and images evoked by the tone of voice of the person reading the story. No wonder S. had trouble!

Another problem S. experienced, which is hard for us to appreciate, was the inability to forget. This became a particular problem when he was a professional mnemonist and gave numerous performances in which he was asked to remember things written on a chalkboard. He would look at the chalkboard and see things written there on many different occasions. He tried various tricks to try and forget old information, such as mentally erasing the board, but these didn't work. Only by the strength of his attention and by actively telling himself to let information slip away could he forget. It was as if the effort most of us use for remembering and the ease with which we forget were reversed for S.

We don't know the neural basis for S.'s remarkable memory. Perhaps he lacked the same sort of segregation most of us have between sensations in different sensory systems. This may have contributed to an uncommonly strong multimodal coding of memories. Maybe his synapses were more malleable than normal. Unfortunately, we'll never know.

put into **long-term memory**, which is more permanent, has a much greater capacity, and does not require continual rehearsal. We take this amazing memory phenomenon completely for granted every minute of our lives. We just assume that when we experience things, our brain will change appropriately to preserve a record of the experience. The process of storing new information in long-term memory is called **consolidation.** Note that consolidation does not necessarily require short-term memory as an intermediary (Figure 19.1).

An elaboration of the short-term memory concept that has been made in recent years is working memory. **Working memory** is a more general term for temporary information storage that allows for the possibility that several types of information are held at the same time. From a neuroscience perspective, a key feature of working memory that distinguishes it from short-term memory is that there may be multiple sites in the brain where temporary storage occurs, rather than a single short-term memory system. This implies that we could not be consciously aware of all the information held in working memory at the same time, in different parts of the brain.

Let's take the example of riding a bicycle. This is a complex task requiring several types of information to be simultaneously processed. Some of this is sensory information: Is that a large pothole I see up ahead; is that a car I hear coming up behind me? Other information is cognitive or motor: Do I turn left here to get home; should I pedal faster to prepare for the hill ahead? It seems unlikely that these various types of information can all be handled by a single short-term memory system. Instead, multiple sensory and motor cortical areas may retain information and computations for short periods of time, and we pay attention to them as the need arises. Also, we may call up previously stored facts, such as the path leading home, that are currently needed. These facts may be temporarily placed into a more readily accessible form. Note from the bicycle example that information held in working memory may or may not be newly acquired sensory information. Also, the incoming sensory information is not necessarily retained in long-term memory; it may be used and then discarded.

Interestingly, there are reports of humans with cortical lesions who have normal short-term memory for information coming from one sensory system (e.g., they can remember the same number of visually seen numbers as other people) but a profound deficit in short-term memory when informa-

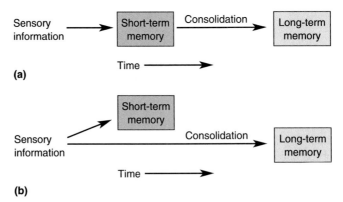

Figure 19.1
Short-term and long-term memory. Sensory information can be temporarily held in short-term memory, but permanent storage into long-term memory requires consolidation. **(a)** Information may be consolidated from short-term memory. **(b)** Alternatively, information processing necessary for consolidation may occur separately from short-term memory.

tion comes from a different sense (e.g., they cannot remember more than one number spoken to them). These cases are consistent with the idea that there are multiple temporary storage areas in the brain.

Amnesia

As we all know, forgetting is nearly as common an occurrence as learning in our daily lives. That's normal and unavoidable. However, certain diseases and injuries to the brain cause a serious loss of memory and/or the ability to learn called **amnesia**. Concussion, chronic alcoholism, encephalitis, brain tumor, and stroke can all disrupt memory. You've probably seen a movie or television show in which a person experiences a great trauma and wakes up the next day not knowing who he or she is and not remembering the past. That kind of absolute amnesia for past events and information is actually quite rare. It is more common for trauma to cause limited amnesia along with other nonmemory deficits. If amnesia is not accompanied by any other cognitive deficit, it is called *dissociated* (i.e., the memory problems are dissociated from any other problems). We will focus on cases of dissociated amnesia because a clear relationship can be drawn between memory deficits and brain injury.

Following trauma to the brain, memory loss may manifest itself in two different ways: retrograde amnesia and anterograde amnesia (Figure 19.2). In **retrograde amnesia**, there is memory loss for events prior to the trauma. In other words, you forget things you already knew. In severe cases, there might be complete amnesia for all declarative information learned before the trauma. More often, retrograde amnesia follows a pattern in which events in the months or years preceding the trauma are forgotten, but memory is increasingly good for older memories. This is quite different from **anterograde amnesia**, which is an inability to form new memories follow-

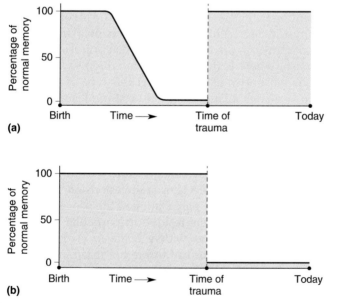

(a)

(b)

Figure 19.2
Amnesia produced by trauma to the brain. (a) In retrograde amnesia, events for a period of time prior to the trauma are forgotten, but memories from the distant past and the period following the trauma are intact. **(b)** In anterograde amnesia, events prior to the trauma can be remembered, but there are no memories for the period following the trauma.

ing brain trauma. If the anterograde amnesia is severe, a person may be completely incapable of learning anything new. In milder cases, learning may be slower and require more repetition than normal. In clinical cases, there is often a mixture of retrograde and anterograde amnesias of different degrees of severity.

To make this more concrete, consider the case of a 45-year-old man who experienced some brain trauma at the age of 40. If he has severe retrograde amnesia, he will be unable to remember many things that occurred prior to age 40; if he has severe anterograde amnesia, he will remember nothing from age 40 until the present time.

A form of amnesia that involves a much shorter period of time is called transient global amnesia. In **transient global amnesia**, there is a sudden onset of anterograde amnesia that only lasts for a period of minutes to days, accompanied by retrograde amnesia for recent events preceding the attack. This type of amnesia can result from brief cerebral ischemia, in which the blood supply to the brain is temporarily reduced, or concussion to the head from trauma such as a car accident or a hard blow while playing football. There have been reports of transient global amnesia brought on by physical stress, drugs, cold showers, and even sex, presumably because all of these affect cerebral blood flow. Many cases have been linked to use of the antidiarrheal drug clioquinol (which has been taken off the market). In one case, a college student visiting Paris took clioquinol on the fifth day of a six-day holiday, during which she traveled with friends. When she got home, she discovered she couldn't remember her vacation at all! The memories for this short period were absent years later, but otherwise there was no lasting effect of the clioquinol. While we don't know exactly what causes transient global amnesia, it is probably a consequence of temporary blood deprivation to structures essential for learning and memory.

THE SEARCH FOR THE ENGRAM

Now that we've discussed several types of memory and amnesia, let's turn our attention to the parts of the brain that are involved in memory storage. The physical representation or location of a memory is called an **engram**, also known as a *memory trace*. When you learn the meaning of a word in French, where is this information stored—where is the engram? The technique most often used to answer this type of question is the time-honored experimental ablation method of Marie-Jean-Pierre Flourens (see Chapter 1).

Lashley's Studies of Maze Learning in Rats

In the 1920s, American psychologist Karl Lashley conducted experiments in order to study the effects of brain lesions on learning in rats. Well aware of the cytoarchitectural parcellation of the neocortex, Lashley set out to determine whether engrams resided in particular association areas of cortex (see Chapter 7), as was widely believed at the time. In a typical experiment, he trained a rat to run through a maze in order to get a food reward (Figure 19.3a). On the first trial, the rat was slow getting to the food because it would enter blind alleys and have to turn around. After running through the same maze a number of times, the rat learned to avoid blind alleys and go straight to the food. The question Lashley investigated was how performance on this task was affected by making a lesion in some part of the rat's cortex (Figure 19.3b). He found that rats given brain lesions before learning took more trials before they could run the maze without going down blind alleys. The lesions seemed to interfere with their ability to

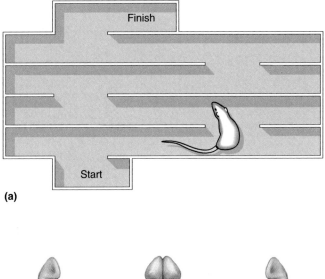

(a)

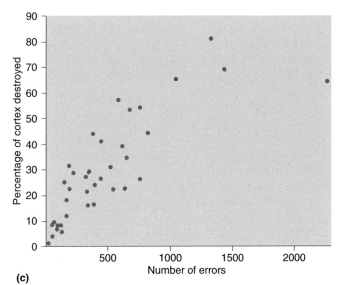

Lateral view of   Dorsal view   Lateral view of
left hemisphere   right hemisphere

(b)

(c)

Figure 19.3
Effects of cortical lesions on maze performance. (a) The rat is trained to run
through a maze from start to finish without entering blind alleys. **(b)** Three different
cortical lesions are shown in blue, yellow, and red. **(c)** The greater the percentage of
cortex destroyed, the more errors are made while the rats learn to run the maze. The
number of errors shown is cumulative across trials, suggesting that rats with larger
lesions had difficulty remembering which arms of the maze were blind alleys.
(Source: Adapted from Lashley, 1929, Figs. 2, 16, and 23.)

learn. In another group, a lesion was made after a rat had learned to run the maze without making mistakes. After the lesion, the rat made mistakes and went down blind alleys it had previously learned to avoid. It appeared that the lesion damaged or destroyed the memory for how to reach the food. What was the effect of the size and location of the lesion? Interestingly, Lashley found that the severity of the deficits caused by the lesions (both learning and remembering) correlated with the *size* of the lesions (Figure 19.3c) but was apparently unrelated to the *location* of the lesion within the cortex.

Based on these findings, Lashley speculated that all cortical areas contribute *equally* to learning and memory; it is simply a matter of getting poorer performance on the maze task as the lesion gets bigger and the ability to remember the maze worsens. If true, this would be a very important finding because it implies that engrams are based on neural changes spread throughout the cortex rather than being localized to one area. The main problem with this interpretation is that it goes beyond what can safely be concluded from the data. Notice how large the lesions are in Figure 19.3b. Perhaps one of the reasons Lashley did not find a difference in the effects of lesions at various locations is that he made lesions so large that they damaged several cortical areas involved in learning the maze task. Another problem was that the rats might solve the maze in several different ways—by sight, feel, and smell—and the loss of one memory might be compensated for by another.

Subsequent research has proven Lashley's conclusions to be incorrect. All cortical areas do not contribute equally to memory. Nonetheless, he was correct that memories are distributed. Lashley had an important and lasting impact on the study of learning and memory because he led other scientists to consider ways in which memories might be distributed among the many neurons of the cerebral cortex.

Hebb and the Cell Assembly

Lashley's most famous student was Donald Hebb, introduced in Chapter 18. Hebb reasoned that it was crucial to understand how external events are represented in the activity of the brain before one can hope to understand how and where these representations are stored. In a remarkable book published in 1949 entitled *The Organization of Behavior*, Hebb proposed that the internal representation of an object (let's say a circle drawn on a piece of paper) consists of all the cortical cells that are activated by this stimulus. This group of simultaneously active neurons Hebb called a **cell assembly** (Figure 19.4). Hebb imagined that all these cells were connected to one another by reciprocal connections. The internal representation of the object was held in short-term memory as long as activity reverberated through the connections of the cell assembly. Hebb further hypothesized that if activation of the cell assembly persisted long enough, consolidation would occur by a "growth process" that made these reciprocal connections more effective (i.e., neurons that fired together would wire together; recall Hebb synapses from Chapter 18). Subsequently, if only a fraction of the cells of the assembly were activated by a later stimulus, as in Figure 19.4, the now-powerful reciprocal connections would cause the whole assembly to become active again, thus recalling the entire internal representation of "circle."

Hebb's important message was that the engram (1) could be widely distributed among the connections that link the cells of the assembly and (2) could involve the same neurons that are involved in sensation and perception. Destruction of only a fraction of the cells of the assembly would not be expected to eliminate the memory, thus possibly explaining Lashley's results. Hebb's ideas stimulated the development of neural network com-

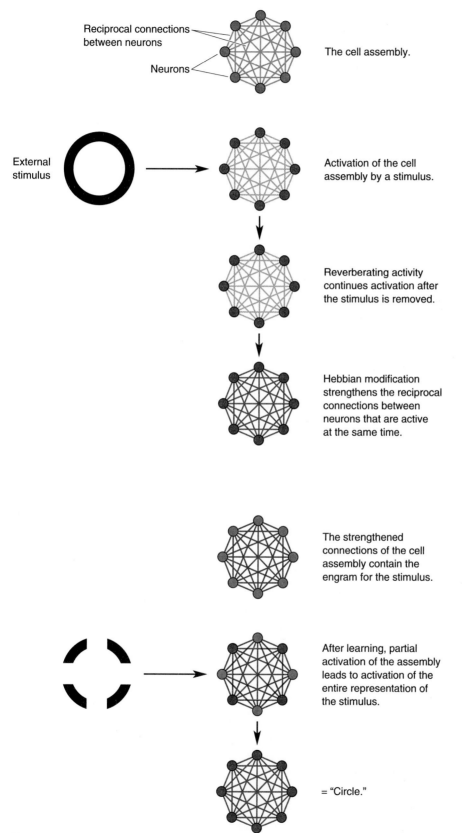

Reciprocal connections between neurons

Neurons

The cell assembly.

External stimulus

Activation of the cell assembly by a stimulus.

Reverberating activity continues activation after the stimulus is removed.

Hebbian modification strengthens the reciprocal connections between neurons that are active at the same time.

The strengthened connections of the cell assembly contain the engram for the stimulus.

After learning, partial activation of the assembly leads to activation of the entire representation of the stimulus.

= "Circle."

Figure 19.4
Hebb's cell assembly and memory storage.

puter models. Although his original assumptions had to be modified slightly, these models have successfully reproduced many features of human memory (Box 19.2).

Localization of Declarative Memories in the Neocortex

According to Hebb, if an engram is based on information from only one sensory modality, it should be possible to localize it within the regions of cortex that serve this modality. For example, if the engram relies only on visual information, then we would expect it to reside within the visual cortex. Studies of visual discrimination in monkeys are consistent with this proposal.

Macaque monkeys can be trained to perform visual discriminations (e.g., they can tell the difference between pairs of objects based on their shapes). After being trained on this task and becoming proficient at it, a lesion is made in inferotemporal cortex (area IT), a high-order visual area in the inferior temporal lobe (Figure 19.5a). With this brain lesion, the animal can no longer perform the discrimination task even though its basic visual capacities remain intact. It appears that the previously learned task has been forgotten or information about it can no longer be accessed. As in Lashley's experiments, the implication is that the memory for the task is stored in the cortex. However, in this case of a task specific to vision, the memory appears to be stored in a high-order visual area. Stated another way, inferotemporal cortex is both a visual area and an area involved in memory storage.

Further evidence that inferotemporal cortex is involved in certain types of memory storage comes from physiological experiments in which the response properties of individual neurons are examined. For example, recordings made from IT neurons suggest that they may encode memories of faces (Figure 19.5b). In a typical experiment, the monkey is alert, and an elec-

---

Box 19.2 BRAIN FOOD

A Model of a Distributed Memory

In recent years, models of neural systems have become increasingly widespread. In some cases, a model can provide insight into the workings of a system that are otherwise hard to gain. One area in which models have been helpful is the study of systems for learning and memory. Let's look at a simple model for distributed memory storage.

To keep things simple, we'll examine a nervous system consisting of 3 sensory neurons (the inputs) and 15 postsynaptic neurons (the outputs). In our tiny system, we'll take the inputs to represent red, green, and blue retinal cones. We're going to ignore the complexities of visual processing and assume that these three cones project to output neurons whose responses we use to name color. Suppose that we initially don't know anything about naming colors. In the model we can represent this by a low

response in all 15 of the output neurons, regardless of what cone is activated (Figure Aa).

As we learn to name colors, the synapses between the inputs and outputs change in one of two ways. The simplest way is that most of the synapses weaken and a few strengthen, such that each input ends up being connected to just one output. If you stimulate each cone, there is one output neuron that responds, and this response corresponds to the word "red," "green," or "blue" (Figure Ab). The most important thing to recognize is that the synapses between the inputs and outputs store the memory that associates a particular input with a particular output pattern. In this simple scheme, the memory of the appropriate color name is entirely stored in one synapse.

Another way that the neural network might evolve as we learn to name colors is for each input cone to remain connected to many of the output neurons (7 in Figure Ab). In this case when we stimulate one cone we get a response pattern spanning 7 output cells. There is a distinct response pattern corre-

continued on page 525

continued from page 524

sponding to each color name, but now that pattern is distributed across 7 output neurons. Likewise, the memory relating inputs to color names is spread across multiple synapses. This is precisely what Hebb meant by a distributed memory.

Our neural system for color naming will work with either scheme for connecting inputs and outputs. However, there are beneficial features to the distributed memory. One useful feature of the distributed system is that it naturally incorporates interpolation. This means that if more than one of the inputs is activated at one time, the response pattern is in between the patterns obtained with stimulation of an individual input. For example, if a greenish-blue light partially activates both the green and the blue cones, the response pattern has parts of the separate patterns for blue and green and a larger patch in between (Figure Ac). This seems like an

intuitively reasonable way to represent this intermediate color.

Another potential advantage is that the distributed system is more immune from catastrophic memory loss if output neurons die. In the nondistributed scheme, the color name red is totally dependent on the response of a single output neuron. If this cell dies, say goodbye to red forever. Contrast this with the distributed system. While it is true that there is a greater number of output neurons that would affect each color name if they died, the effect would be rather small. If cells are lost in the distributed system, there is a graceful degradation of the response patterns associated with stimulating each cone, rather than a potential catastrophic loss (Figure Ad). This relative immunity from the effects of cell loss is a great advantage. Neurons in the human brain die every day, and it is probably because of the distributed nature of memory that we don't suddenly lose memories for people and events.

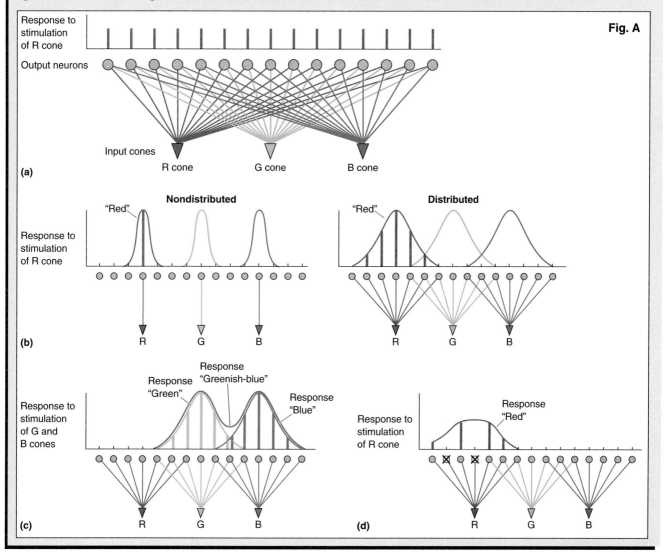

Fig. A

(a) Response to stimulation of R cone — Output neurons — Input cones: R cone, G cone, B cone

(b) **Nondistributed** / **Distributed** — Response to stimulation of R cone — "Red" — R, G, B

(c) Response to stimulation of G and B cones — Response "Green", Response "Greenish-blue", Response "Blue" — R, G, B

(d) Response to stimulation of R cone — Response "Red" — R, G, B

ron to multiple presentations of familiar faces (other monkeys the subject frequently sees) is recorded. The face cell responds more to some faces than to others. (This is analogous to a neuron in striate cortex, which responds more to a slit of light at some orientations than others.) When new faces unfamiliar to the monkey are introduced, interesting neuronal responses result, as shown in Figure 19.5c. The first time new faces are seen, the cell responds at about the same moderate level to all of them. However, with a couple of additional exposures, the response increasingly changes such that some faces evoke a significantly greater response than others. The cell is becoming selective in its response to these new stimuli as we watch it! With continued presentation of the same group of faces, the response of the neuron to each pattern becomes more stable. We can speculate that the response selectivity of this neuron and others is part of a distributed code for the representation (i.e., memory) of many different faces. This dynamic aspect of responses in area IT supports Hebb's view that the brain can use cortical areas for both the processing of sensory information and the storage of memories.

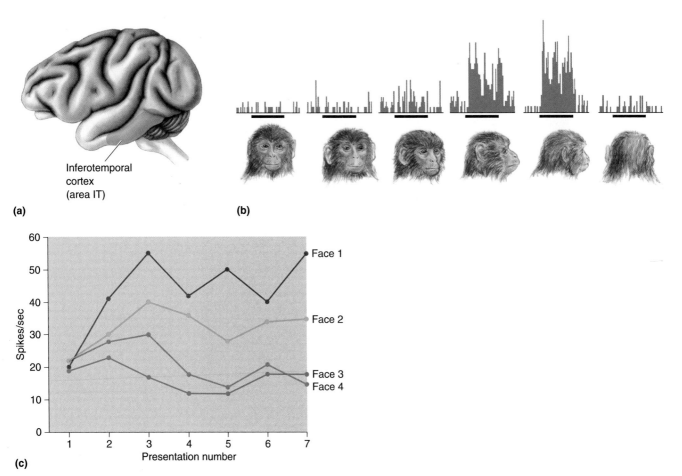

Figure 19.5
Responses to faces in inferotemporal cortex. (a) The location of area IT in the inferior temporal lobe. **(b)** Responses of a face cell. The histograms show the response of a neuron (spikes/sec) in monkey inferotemporal cortex to different views of a monkey's head. The horizontal bar under each histogram indicates when the stimulus was presented. **(c)** Changing responses of a cell to unfamiliar faces. When the four faces are presented for the first time, there is a moderate response to each. With subsequent presentations, the cell becomes more responsive to faces 1 and 2 and less responsive to faces 3 and 4. (Sources: Part b, Desimone et al., 1984, Fig. 6; part c, adapted from Rolls et al., 1989, Fig. 1.)

Electrical Stimulation of the Human Temporal Lobes

One of the most intriguing and controversial studies implicating the neocortex of the temporal lobe in the storage of declarative memory traces was performed on living humans. In Chapters 12 and 14 we discussed the work of Wilder Penfield, in which patients, as part of surgical treatment for severe epilepsy, had their brain electrically stimulated at numerous locations prior to ablation of the seizure-prone region. Stimulation of somatic sensory cortex caused the patient to experience tingling sensations in regions of skin, whereas stimulation of motor cortex caused a certain muscle to twitch.

Electrical stimulation of the temporal lobe, however, occasionally produced more complex sensations than those obtained with stimulation in other brain areas. In a number of cases, Penfield's patients described sensations that sounded like hallucinations or recollections of past experiences. This is consistent with reports that epileptic seizures of the temporal lobes can evoke complex sensations, behaviors, and memories. Here is part of Penfield's account of one operation:

> At the time of operation, stimulation of a point on the anterior part of the first temporal convolution on the right caused him [the patient] to say, "I feel as though I were in the bathroom at school." Five minutes later, after negative stimulations elsewhere, the electrode was reapplied near the same point. The patient then said something about "street corner." The surgeon asked him, "where" and he replied, "South Bend, Indiana, corner of Jacob and Washington." When asked to explain, he said he seemed to be looking at himself—at a younger age. (Penfield, 1958, p. 25)

Another patient reported a similar sense of experiencing flashbacks. When her temporal cortex was stimulated, she said, "I think I heard a mother calling her little boy somewhere. It seemed to be something that happened years ago." With stimulation at another location, she said, "Yes, I hear voices. It is late at night around the carnival somewhere—some sort of traveling circus. . . . I just saw lots of big wagons that they use to haul animals in."

Are these people reexperiencing events from earlier in their life because memories are evoked by the electrical stimulation? Does this mean that memories are stored in the neocortex of the temporal lobe? These are tough questions! One interpretation of these findings is that the sensations are recollections of past experiences. The fact that these elaborate sensations were only obtained when the stimulation was in the temporal lobe suggests that the temporal lobe may play a special role in memory storage. However, there are other aspects to the findings that don't clearly support the hypothesis that engrams are being electrically activated. For instance, in some of the cases in which the stimulated part of the temporal lobe was removed, the memories that had been evoked by stimulation in that area could be evoked by stimulation somewhere else. In other words, the memory had not been "cut out." Also, it is important to appreciate that complex sensations were only reported by a minority of the patients, and all of these patients had abnormal cortex associated with their epilepsy.

There is no way to prove with certainty whether the complex sensations evoked by temporal lobe stimulation are recalled memories. However, there is no question that the consequences of temporal lobe stimulation and temporal lobe seizures can be qualitatively different from stimulation of other areas of the neocortex. We will now have a closer look at the structure of the temporal lobes and the elements within them that are strongly implicated in learning and memory.

THE TEMPORAL LOBES AND DECLARATIVE MEMORY

The temporal lobe is located under the temporal bone, so named because the hair of the temples is often the first to go gray with the passage of time (*tempus* is Latin for "time"). The association of the temporal lobe with time was fortuitous because considerable evidence points to this region of the brain as being particularly important for the recording of past events. The temporal lobes contain the temporal neocortex, which may be a site of long-term memory storage. Also within the temporal lobe are the hippocampus and other structures, which, as we will see, are critical for the formation of declarative memories.

The Effects of Temporal Lobectomy

If the temporal lobe is particularly important for learning and memory, one would expect that removing both temporal lobes would have a profound effect on these functions. Recall from the discussion of the Klüver-Bucy syndrome in Chapter 16 that bilateral temporal lobectomy does seem to have some effect on memory. The monkeys studied by Klüver and Bucy following temporal lobectomy had a peculiar way of interacting with their environment, in addition to a host of other abnormalities. The monkeys explored their room by placing objects in their mouth. If an object was edible, they would eat it; if it wasn't, they would drop it. However, their behavior suggested that they did not have basic perceptual deficits. In the words of Klüver and Bucy, they exhibited "psychic blindness"—even though they could see things, they did not appear to understand with their eyes what the objects were. They would repeatedly go back to the same inedible objects, put them in their mouth, then toss them aside. This problem with object recognition is probably related to memory function in the temporal lobe.

Human Case Study: H. M. A renowned case of amnesia resulting from temporal lobe damage provides further evidence for the importance of this region in memory. A man known by his initials, H. M., has been studied more than any other amnesic. H. M. experienced minor epileptic seizures beginning around age 10, and, as he aged, they became more serious generalized seizures involving convulsions, tongue biting, and loss of consciousness. It is not clear what led to the onset of the seizures, but they may have resulted from damage sustained in an accident in which he was knocked off his bicycle at the age of 9 and left unconscious for 5 minutes. After graduating from high school he got a job, but despite heavy medication with anticonvulsants, his seizures increased in frequency and severity to the point that he was unable to work. In 1953, at the age of 27, H. M. had an operation in which an 8-cm length of medial temporal lobe was bilaterally excised, including cortex, the underlying amygdala, and the anterior two-thirds of the hippocampus, in a last-ditch attempt to assuage the seizures (Figure 19.6). The surgery was successful in alleviating the seizures.

The removal of much of the temporal lobes had little effect on H. M.'s perception, intelligence, or personality. H. M. is still alive, and in most every way, he appears to be normal. However, this normal appearance belies the reality that the surgery left him with an amnesia so profound that he is incapable of performing basic human activities. H. M. has partial retrograde amnesia for the years preceding the operation. However, much more serious is his extreme anterograde amnesia. While he can remember a great deal about his childhood, he is unable to remember someone he met 5 minutes earlier. Dr. Brenda Milner of the Montreal Neurological Institute has worked with H. M. for 40 years, but, incredibly, she has to introduce herself to him every time they meet. In describing H. M.'s amnesia, Milner has said

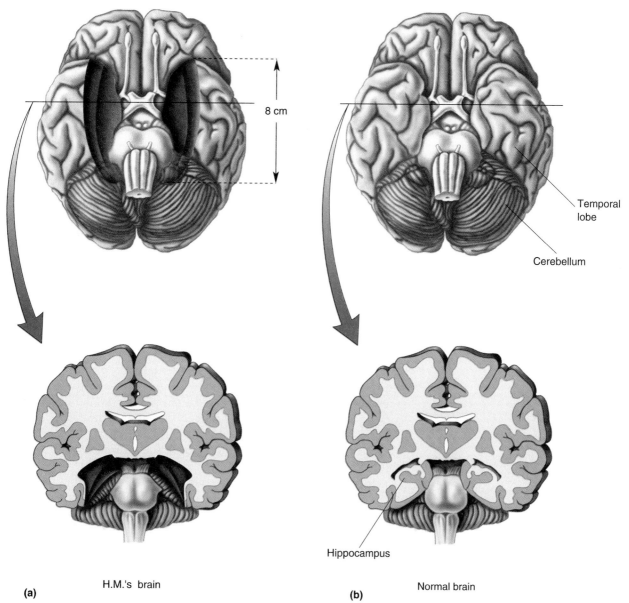

Figure 19.6
H. M.'s brain lesion. (a) The medial temporal lobe was removed from both hemispheres in H. M.'s brain. **(b)** A normal brain showing the location of the hippocampus and cortex, removed from H. M.'s brain. (Source: Adapted from Scoville and Milner, 1957, Fig. 2.)

it appears that he forgets events as quickly as they occur. If he is told to remember a number and is then distracted, he will not only forget the number, he will also forget that he had even been asked to remember one. H. M. has moved since 1953, and consequently he cannot find his way around his own neighborhood. He always underestimates his age, and he is unable to recognize a current picture of himself.

To be clear about the nature of H. M.'s amnesia, we must contrast what was lost with what is retained. He remembers his childhood, so long-term memories formed before the surgery and his ability to recollect past events were not destroyed. His short-term memory is also normal. For instance, with constant rehearsal he can remember a list of six numbers, although any interruption will cause him to forget. He simply has an extreme inability to form new declarative memories. Importantly, he is able to learn new tasks (i.e., form new *procedural* memories). For example, he was taught to draw by looking at his hand in a mirror, a task that takes a good deal of practice for anyone. The odd thing is that he has learned to perform new tasks (procedural learning) despite the fact that he has no recollection of the specific experiences in which he was taught to do them (the declarative component of the learning). In other words, he became very good at mirror writing, but he had no memory of ever having practiced. The characteristics of H. M.'s amnesia reinforce the idea that the neuroanatomy and neural mechanisms underlying procedural and declarative memory, as well as short-term and long-term memory, are not identical. In our search to understand the role of the medial temporal lobe in learning and memory, we are led to focus on the processing and consolidation of new declarative memories.

The Medial Temporal Lobes and Memory Processing

In the medial temporal lobe, there is a group of interconnected structures that appears to be of great importance for declarative memory consolidation. The key structures are the hippocampus, the nearby cortical areas, and the pathways that connect these structures with other parts of the brain (Figure 19.7). As we saw in Chapter 7, the **hippocampus** is a folded structure situated medial to the lateral ventricle. Lateral to the hippocampus are three important cortical regions that surround the rhinal sulcus: the **entorhinal cortex**, which occupies the medial bank of the rhinal sulcus; the **perirhinal cortex**, which occupies the lateral bank; and the **parahippocampal cortex**, which lies lateral to the rhinal sulcus. (We'll refer to entorhinal cortex and perirhinal cortex collectively as rhinal cortex.)

Inputs to the medial temporal lobe come from the association areas of the cerebral cortex containing highly processed information from all sensory modalities (Figure 19.8). For instance, inferior temporal visual cortex projects to the medial temporal lobe, but low-order visual areas such as striate cortex do not. This means that the input contains complex representations, perhaps of behaviorally important sensory information, rather than responses to simple features such as light-dark borders. Input first reaches the rhinal and parahippocampal cortex before being passed to the hippocampus. A major output pathway from the hippocampus is the **fornix**, which loops around the thalamus before terminating in the hypothalamus.

An Animal Model of Human Amnesia. The effects of temporal lobectomy, and particularly the amnesia of H. M., make a strong case that one or more structures in the medial temporal lobe are essential for the formation of declarative memories. If these structures are damaged, severe anterograde amnesia results. There has been intensive investigation into the particular structures in the medial temporal lobe that are essential for memory formation. For the most part, these studies use the experimental ablation technique to assess whether the removal of some part of the temporal lobe affects memory.

Because the macaque monkey brain is similar in many ways to the human brain, macaques are frequently studied to further our understanding of human amnesia. The monkeys are most often trained to perform a task called **delayed non-match to sample (DNMS)** (Figure 19.9). In this type of experiment, a monkey faces a table that has several small wells in its surface. It first sees the table with one object on it covering a well. The object might be a square block or an eraser (the sample stimulus). The monkey is trained to displace the object so that it can grab a food reward in the well under the object. After getting the food, a screen is put down to prevent the monkey from seeing the table for some period of time (the delay interval).

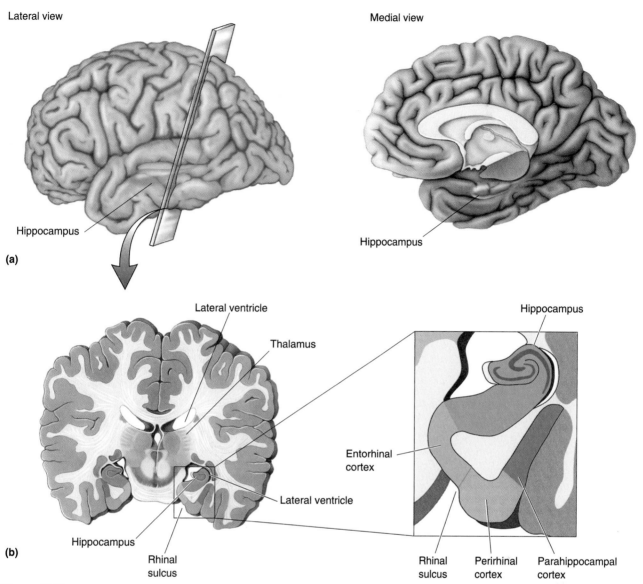

Lateral view

Medial view

Hippocampus

(a)

Hippocampus

Lateral ventricle

Thalamus

Hippocampus

Entorhinal
cortex

Lateral ventricle

(b)

Hippocampus

Rhinal
sulcus

Rhinal
sulcus

Perirhinal
cortex

Parahippocampal
cortex

Figure 19.7
Structures in the medial temporal lobe. (a) Lateral and medial views show the location of the hippocampus in the temporal lobe. **(b)** The brain is sectioned coronally to show the hippocampus and cortex of the medial temporal lobe.

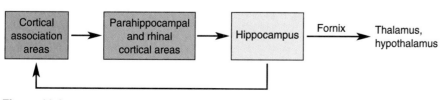

Cortical
association
areas

Parahippocampal
and rhinal
cortical areas

Hippocampus

Fornix

Thalamus,
hypothalamus

Figure 19.8
Information flow through the medial temporal lobe.

Finally, the animal gets to see the table again, but now there are two objects on it: One is the same as before, and another is new. The monkey's task is to displace the new object (the non-matching object) in order to get another food reward in a well below it. Normal monkeys are relatively easy to train on the non-matching task and get very good at it, probably because it exploits their natural curiosity for novel objects. With delays between the two stimulus presentations of anywhere from a few seconds to 10 minutes, the monkey correctly displaces the non-matching stimulus on about 90% of the trials. Memory required in the DNMS task has been called **recognition memory** because it involves the ability to judge whether a stimulus has been seen before. It is probably reasonable to assume that if the animal can perform this task with a delay several minutes long, it must have formed a long-term memory for the sample object.

In the early 1980s, experiments performed by Mortimer Mishkin and his colleagues at the National Institute of Mental Health and Larry Squire and his coworkers at the University of California, San Diego, demonstrated that severe deficits on the DNMS task resulted from bilateral medial temporal lesions in macaque monkeys. Performance was close to normal if the delay between the sample stimulus and the two stimuli was short (a few seconds). This is very important because it indicates that the monkey's perception was still intact after the ablation and it remembered the DNMS procedure. But when the delay was increased from a few seconds to a few minutes, the monkey made increasingly more errors choosing the non-matching stimulus (Figure 19.10). With the lesion, the animal was no longer as good at remembering what the sample stimulus was in order to choose the other object. Its behavior suggests that it forgot the sample stimulus if the delay was too long. The deficit in recognition memory produced by the lesion was not modality-specific, since this deficit was also observed if the monkey was allowed to touch but not see the objects.

The monkeys with medial temporal lesions appeared to provide a good model of human amnesia. As with H. M., the amnesia involved declarative rather than procedural memory, it seemed more anterograde than retrograde, and long-term memory was more impaired than short-term memo-

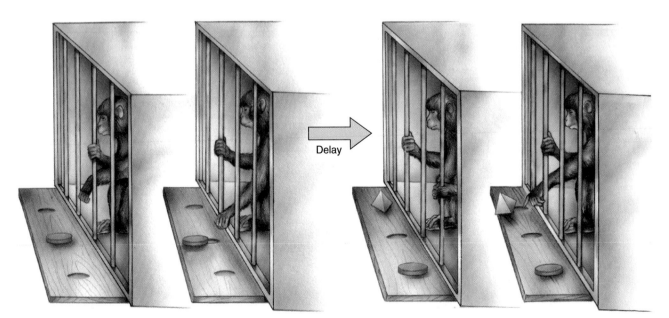

Figure 19.9
The delayed non-match to sample task. (Source: Adapted from Mishkin and Appenzeller, 1987, p. 84.)

ry. This animal model provided a window into the neural basis for human memory. As Squire describes in Box 19.3, there was a great deal of interest in knowing the particular structures responsible for amnesia. Note that the lesions that were originally found to produce recognition memory deficits in monkeys were quite large. They included the hippocampus, amygdala, and rhinal cortex. At one time it was thought that the key structures damaged in the lesions were the hippocampus and amygdala. However, recent work has shown that selective amygdala lesions have no effect on this form of memory, and lesions of the hippocampus alone produce only relatively mild amnesia. For example, Squire studied a man known by the initials R. B. who had bilateral hippocampal damage as a result of oxygen deprivation during surgery. Although R. B. clearly had difficulty forming new memories, this anterograde amnesia was not nearly as severe as that seen in H. M. The latest research indicates that the most severe memory deficits actually result from damage to the perirhinal cortex.

Together with the hippocampus, the cortex in and around the rhinal sulcus evidently performs a critical transformation on the information coming from association cortex. One hypothesis is that these medial temporal structures consolidate the memory into cortex. However, it is also possible that they serve as an essential intermediate processing stage that involves something other than consolidation. There is a hint that these cortical regions may also do something else. In H. M. and possibly in R. B., there was some retrograde amnesia. It might be that the cortex of the medial temporal lobes stores memories temporarily before they are ultimately transferred in a more permanent form to the neocortex elsewhere.

The Diencephalon and Memory Processing

As we have seen, lesions in the medial temporal lobe, especially the perirhinal cortex, produce profound amnesia. However, other lesions can also disrupt memory. Outside the temporal lobe, one of the brain regions most associated with memory and amnesia is the diencephalon. This isn't surprising, considering that the diencephalon is heavily connected with the temporal lobes.

The three regions of the diencephalon implicated in the processing of recognition memory are the anterior and dorsomedial nuclei in the thalamus and the mammillary bodies in the hypothalamus. Recall that a major output of the hippocampal formation is a bundle of axons making up the fornix. Most of these axons project to the mammillary bodies (Figure 19.11). Neurons in the mammillary bodies then project to the anterior nucleus of the thalamus. This circuit from hippocampus to hypothalamus to anterior nucleus and then to cingulate cortex should sound familiar; it is half of the Papez circuit we discussed in Chapter 16. The dorsomedial nucleus of the thalamus also receives input from temporal lobe structures, including the amygdala and inferotemporal neocortex, and it projects to virtually all of frontal cortex.

Large midline thalamic lesions in monkeys produce relatively severe deficits on the delayed non-match to sample task. These lesions damage the anterior and dorsomedial nuclei of the thalamus, producing retrograde degeneration in the mammillary bodies. Bilateral lesions limited to either the dorsomedial nuclei or the anterior nuclei produce significant, but milder deficits. The limited data available suggest that rather mild deficits are seen after lesions in only the mammillary bodies.

Human Case Study: N. A. There have been reports for many years suggesting a relationship between human amnesia and damage to the diencephalon. A particularly dramatic example is the case of a man known as

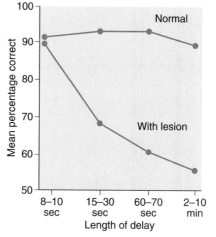

Figure 19.10
Performance on a delayed non-match to sample task. The Y-axis shows the percentage of correct choices made by monkeys as a function of the length of the delay interval. Performance of normal monkeys is compared with performance of monkeys with large bilateral medial temporal lesions. (Source: Adapted from Squire, 1987, Fig. 49.)

Box 19.3 | PATH OF DISCOVERY

Larry Squire

Memory Recollections
by Larry Squire

The year 1978 was a turning point for my thinking about the anatomy of memory. In that year J. A. Horel published a paper that challenged conventional wisdom about the anatomical basis of memory impairment in amnesic patients, including the famous patient H. M. H. M.'s amnesia was profound, making a strong case that critical memory structures are located in the medial temporal lobe. Most people thought that H. M.'s severe amnesia was, at least in part, a consequence of damage to his hippocampus. So it was notable in 1978 when Horel, after a review of the available data, suggested that amnesia might instead be the result of damage to a band of white matter, the temporal stem, and that the hippocampus was not relevant.

My research up to that time involved studies of learning and memory in amnesic patients, but it seemed likely that animal experiments would be needed to clarify these matters. In the 1960s and 1970s there were attempts to develop an animal model in both rats and monkeys and to produce memory deficits similar to H. M.'s, but the results were varied and confusing. I thought it would be best to use monkeys, so I looked for a scientist experienced with monkeys to collaborate with. I had the great fortune to persuade Stuart Zola-Morgan to come join me in San Diego. Stuart set up the lab, and we began experiments making different lesions in the medial temporal lobe to determine what structures were most important for memory.

Mortimer Mishkin at the National Institutes of Health had shown that large medial temporal lobe lesions including both the hippocampus and the amygdala produced a severe deficit in recognition memory, as measured by the delayed non-match to sample task. Separate lesions of just the amygdala or hippocampus produced milder deficits, so Mishkin proposed that both structures played an important role in recognition memory. Other experiments by Mishkin, and some by us, subsequently showed that the large lesion produced deficits analogous to those in H. M.—for example, long-term but not short-term memory was impaired and the amnesia was multimodal. It gradually became clear that we had a good animal model for human amnesia.

What remained was to make more selective lesions to determine what critical structures were damaged in these large lesions. We found evidence against the temporal stem idea proposed by Horel in our first study, which Stuart and I did in collaboration with Mort Mishkin. I was also skeptical from the beginning that in the monkey the amygdala was critical for the kind of memory impaired in H. M. In part this was because I thought it was involved in other things such as emotion and emotional learning. Also, studies of humans had suggested that amnesia was more associated with posterior temporal lobe lesions rather than anterior temporal lobe lesions (that would include the amygdala), and rat studies even at that time seemed to show that hippocampal damage could cause memorylike deficits without damaging the amygdala. By the late 1980s with the help of David Amaral, a fine neuroanatomist, we had found that the amygdala was not actually involved in amnesia. We also had studied the amnesic known as R. B. who taught us that selective hippocampal lesions can produce amnesia, but not nearly as severe as that of H. M. So we were left with this question: If hippocampal lesions produce only mild deficits and amygdala lesions no deficits at all, what else is there in H. M.'s lesion that is responsible for his devastating amnesia?

On March 14, 1987, it dawned on Stuart and me how we might solve this puzzle. We came to the realization that we really didn't understand the anatomy of the anterior temporal lobe sufficiently to know what the original large lesions in the monkey were destroying in addition to the amygdala. We decided to postpone surgery. We took the old brain sections from our experiments from several years earlier to David Amaral. David had been studying the neuroanatomy of this region for some time, and one of the important things he had done was to show that the boundaries of perirhinal cortex needed to be expanded. Perirhinal cortex, situated beneath the amygdala in the anterior medial temporal lobe, had been damaged in previous studies that targeted the amygdala. After looking at the sections from each of the monkeys, David said that all the lesions (in these animals who had had severe memory deficits) had damage to the perirhinal cortex. I remember throwing my hands up and saying, "That's it!"

We went on to perform new experiments with lesions that explicitly included perirhinal cortex. Mishkin's laboratory has since made similar lesions, and there is now a consensus that, of all the medial temporal lobe structures involved in visual recognition memory, the perirhinal cortex is the single most important. The important structures seem to be the hippocampus and the adjacent cortex—parahippocampal, entorhinal, and perirhinal cortex. Now, after all those years of work, we think we understand in broad outline the specific brain structures and connections that when damaged give rise to amnesia in humans.

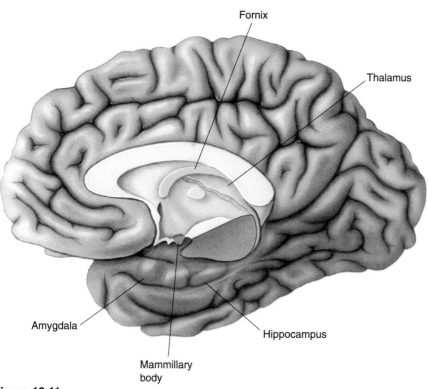

Fornix

Thalamus

Amygdala

Hippocampus

Mammillary
body

Figure 19.11
Components of the diencephalon involved in memory. The thalamus and mammillary bodies receive afferents from structures in the medial temporal lobe.

N. A. In 1959 at the age of 21, N. A. was a radar technician in the U. S. Air Force. One day he was sitting down assembling a model in his barracks, while behind him a roommate played with a miniature fencing foil. N. A. turned at the wrong moment and was stabbed. The foil went through his right nostril, taking a leftward course into his brain. Many years later when a CT scan was performed, the only obvious damage was a lesion in his left dorsomedial thalamus, though there may have been other damage.

After his recovery, N. A.'s cognitive ability was normal but his memory was impaired. He had relatively severe anterograde amnesia and retrograde amnesia for a period of about 2 years preceding the accident. While he could remember some faces and events from the years following his accident, even these memories were sketchy. He had difficulty watching television because during commercials, he'd forget what had previously happened. In a sense, he lived in the past by preferring to wear old familiar clothes and keeping his hair in a crewcut.

Although N. A.'s amnesia was less severe than H. M.'s, the quality was strikingly similar. There was preservation of short-term memory, recollection of old memories, and general intelligence. Along with an inability to form new declarative memories, there was retrograde amnesia for a period of years preceding the accident that produced the amnesia. The similarities in the effects of medial temporal and diencephalic lesions suggest that these interconnected areas are part of a system serving the common function of forming new declarative long-term memories.

Korsakoff's Syndrome. Further support for a role of the diencephalon in memory comes from Korsakoff's syndrome. **Korsakoff's syndrome** commonly results from chronic alcoholism and is characterized by confusion, confabulations, severe memory impairment, and apathy. As a result of poor nutrition, alcoholics may develop a thiamin deficiency, which can lead to a variety of symptoms including abnormal eye movements, loss of coordination, and tremors. This condition can be treated with supplemental thiamin. If left untreated, thiamin deficiency can lead to structural brain damage, which does not respond to thiamin treatment. The structural damage produces Korsakoff's syndrome. Although all cases of Korsakoff's syndrome are not associated with damage to the same parts of the brain, there are usually lesions in the dorsomedial thalamus and the mammillary bodies.

In addition to anterograde amnesia, Korsakoff's syndrome involves more severe retrograde amnesia than the other cases we've discussed. However, there does not seem to be a good correlation between the severity of anterograde amnesia and retrograde amnesia. This is consistent with the other studies of amnesia we've discussed, suggesting that the mechanisms involved in consolidation (disrupted in anterograde amnesia) are largely distinct from processes used to recall memories (disrupted in retrograde amnesia). Based on a small number of cases such as that of N. A., it is suspected that anterograde amnesia results from damage to the thalamus and mammillary bodies. It is not clear which damage is responsible for the retrograde amnesia, but in addition to diencephalic lesions, Korsakoff's patients sometimes have damage to the cerebellum, brain stem, and neocortex.

MEMORY FUNCTIONS OF THE HIPPOCAMPUS

The experiments we've discussed indicate that one important role of the medial temporal lobes is in declarative memory processing or consolidation. However, this is not the only mnemonic function of the structures in this region. Research on the rat hippocampus suggests that it is involved in a type of memory used in performing various spatial tasks and tasks involving sensory discriminations. This type of memory has been categorized by different researchers as spatial memory, working memory, or relational memory. We will see shortly where the controversy about naming comes from. But first, let's look at what is known about the physiology of the hippocampus and the effects of lesions.

Effects of Hippocampal Lesions in Rats

A role for the hippocampus in memory has been demonstrated by experiments in which rats are trained to get food in a radial arm maze. Devised by David Olton and his colleagues at Johns Hopkins University, the *radial arm maze* consists of arms, or passageways, radiating from a central platform (Figure 19.12a). If a normal rat is put in such a maze, it will explore until it finds the food at the end of each arm. With practice, the rat becomes efficient at finding all the food, going down each arm of the maze just once (Figure 19.12b). In order to run through the maze without going twice into one of the arms, the rat uses visual or other cues around the maze to remember where it has already been.

If the hippocampus is destroyed before the rat is put in the maze, performance will differ from normal in an interesting way. In one sense, rats with lesions seem normal; they learn to go down the arms of the maze and eat the food placed at the end of each arm. But unlike normal rats, they never learn to do this efficiently. Rats with hippocampal lesions go down the same arms more than once, only to find no food after the first trip, and they leave other arms containing food unexplored for an abnormally long time. It appears that the rats can learn the task in the sense that they go down the arms in search of food. But they cannot seem to remember which arms they've already been down.

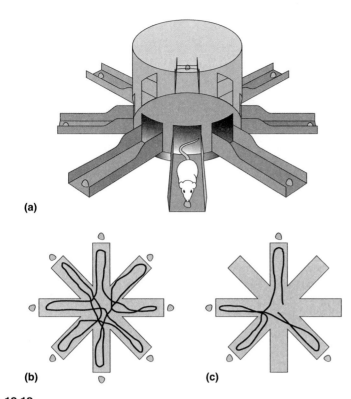

Figure 19.12
Following a rat through a radial arm maze. (a) An 8-arm radial maze. **(b)** The path of a rat through a maze in which all the arms contain food. **(c)** If a rat learns that 4 of the 8 arms never contain food, it will ignore these and follow a path to only the baited arms. (Source: Parts b and c, adapted from Cohen and Eichenbaum, 1993, Fig. 7.4.)

A variation on the radial arm experiment illustrates an important subtlety in the deficit produced by destroying the hippocampus. Instead of placing food at the end of all the arms of the maze, food is placed only at the ends of certain arms and never in the other ones. After a bit of practice, a normal rat learns to avoid going down the arms that never contain food (Figure 19.12c). At the same time, the rat learns to efficiently get the food in the other arms, entering each arm just once. How do rats with hippocampal lesions do on this task? Interestingly, just like normal rats, they are able to learn to avoid the arms that never contain food. But they still are not able to get the food from the other arms without wasting time going down the same arms more than once. Isn't this a little odd? How can we argue that the lesion disrupts the ability to learn the locations of arms that have already been entered, even though the rat can learn to avoid arms that never contain food? Evidently, the key to making sense of these findings is that the information about the no-food arms is always the same each time the rat goes in the maze, whereas the information about which arms the rat has already entered varies from one trial to the next.

Place Cells

In a fascinating series of experiments begun in the early 1970s, John O'Keefe and his colleagues at University College London showed that many neurons in the hippocampus selectively respond when a rat is in a particular location in its environment. Suppose we have a microelectrode implanted in the hippocampus of a rat while it scurries about inside a large box. At first the cell is quiet, but when the rat moves into the northwest corner of the box, the cell starts firing. When it moves out of the corner, the firing stops; when it returns, the cell starts firing again. The cell only responds when the rat is in that one portion of the box (Figure 19.13a). This location that evokes the greatest response is called the neuron's *place field*. We try recording from another hippocampal cell and it too has a place field, but this one only fires when the rat goes to the center of the box. For obvious reasons, these neurons are called **place cells**.

In some ways, place fields are similar to the receptive fields of neurons in the sensory systems. For instance, the location of the place field is related to sensory input such as visual stimuli in the environment. In our experiment with the rat in the box, we could paint images above the four corners, such as a star above the northwest corner, a happy face above the southeast corner, and so on. Consider a cell that responds only when the rat is in the northwest corner of the box, near the painted star. Suppose we take the rat out of the box and blindfold it. We then secretly go back and rotate the box 180° so that now the northwest corner has the happy face and the southeast corner has the star. Will the cell we were previously studying respond when the animal is in the northwest corner, or will it respond in the corner where the star is now located (the southeast corner)? We put the rat back in the box and take off the blindfold. It starts exploring, and the neuron becomes active when the rat goes to the corner near the star. This demonstrates that, at least under some conditions, the response is based on visual stimuli.

While place cells are similar to receptive fields in some ways, there are also major differences. For instance, once the animal has become familiar with the box with the images painted in each corner, a neuron will continue to fire when the rat goes into the northwest corner even if we turn off the lights so that the animal cannot see the location markers. Evidently, the responses of place cells are related to where the animal *thinks* it is. If there are obvious visual cues (such as the star and happy face), the place fields will be based on these cues. But if there are no cues (e.g., because the lights are out), the place cells will still be location-specific as long as the animal has had enough time to explore the environment and develop a sense of where it is.

Performance in the radial arm maze, discussed above, may utilize these place cells, which code for location. Of particular importance in this regard is the finding that the place fields are dynamic. For instance, let's say we first let a rat explore a small box, and we determine the place fields of several cells. Then we cut a hole in a side of the box so the animal can explore a larger area. Initially, there will not be any place fields outside the small box. But after the rat has explored its new expanded environment, some cells will develop place fields outside the small box (Figure 19.13b). These cells seem to *learn* in the sense that they alter their place fields to suit the current environment. It's easy to imagine how these sorts of cells could be involved in remembering arms already visited in the radial arm maze. And, if they are involved in running the maze, it certainly makes sense that performance is degraded by destroying the hippocampus.

Spatial Memory, Working Memory, and Relational Memory

From our discussion of the hippocampus to this point, it may seem that its role is easily defined. First, we saw that performance in a radial arm maze, which requires memory for the locations of arms already explored, is disrupted by hippocampal lesions. Second, the responses of neurons in the hippocampus, the place cells, suggest that these neurons are specialized for location memory. This is consistent with a hypothesis originally put for-

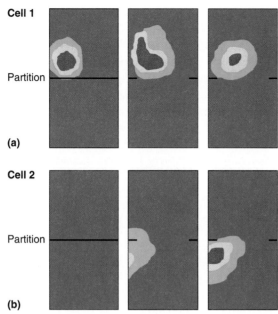

Cell 1

Partition

(a)

Cell 2

Partition

(b)

Figure 19.13
Place cells in the hippocampus. A rat explores a small box for 10 minutes (left panels); then a partition is removed so that it can explore a larger area (center and right panels). **(a)** Color coding indicates the area in the box where one place cell in the hippocampus responds: red = large response, yellow = moderate response, light blue = weak response, dark blue = no response. This cell has a place field in the small upper box; when the partition is removed, it stays in the same location. **(b)** In this case, an electrode is next to a cell in the hippocampus that does not respond when the animal is in the small upper box (left). Over the next 20 minutes, a place field develops in the new larger box (center and right). (Source: Adapted from Wilson and McNaughton, 1993, Fig. 2.)

ward by O'Keefe, stating that the hippocampus is specialized for creating a spatial map of the environment. In one sense it is undeniable that the hippocampus, at least in rats, plays an important role in spatial memory. However, it has been argued by others that this is not the best description of what the hippocampus does. In Olton's original studies using the radial arm maze, he described the result of hippocampal lesions as a deficit in working memory. The rats were not able to retain recently acquired information concerning arms already explored. Thus working memory might be one aspect of hippocampal function. This would explain why the rats given lesions could avoid arms that never contained food but still not remember which arms they had recently visited. Presumably after training, the information about no-food arms was saved in long-term memory, but working memory was still required to avoid the arms where food had already been retrieved.

A more recent hypothesis attempting to integrate a range of experimental findings has been put forward by behavioral neuroscientists Neal Cohen at the University of Illinois, Howard Eichenbaum at the State University of New York at Stony Brook, and their colleagues. Cohen and Eichenbaum describe the function of the hippocampus, in conjunction with other structures of the medial temporal lobe, as involving relational memory. The basic idea of **relational memory** is that highly processed sensory information comes into the hippocampus and nearby cortex, and processing occurs leading to the storage of memories in a manner that ties together or relates all the things happening at the time the memory was stored. For example, as you read this book you may form memories relating multiple things: specific facts, illustrations that catch your eye, interesting passages, the arrangement of material on the page, and information about the sounds or events going on around you as you read. You may have had the experience of searching for a particular passage in a book by searching for a page that looks a certain way that you remember. It is also a common event that remembering one thing (such as the theme song to an old television show) brings back a flood of related facts (the characters in the show, your living room at home, the friends you watched with, and so on). Interconnectedness is a key feature of declarative memory storage.

Whether or not relational memory is the principal function of the hippocampus has not been settled. However, it is a useful concept. In the maze experiments, a case can be made that the rat performs the task by storing memories of the arms it has been down in terms of the sensory cues in the room and the time that an arm was entered. There is also evidence suggesting that place cells may encode relational information of types other than spatial location. A simple example is that the responses of neurons that have place fields are sometimes also affected by other factors such as the speed or direction the rat is moving. The responses of hippocampal neurons are also sometimes determined by nonspatial factors. This was demonstrated in an experiment in which Eichenbaum, Cohen, and their colleagues trained rats to discriminate odors. At one end of the rat's cage were two ports putting out two different odors for the animal to sniff (Figure 19.14). For each pair of odors, the animal was trained to go toward the port releasing one odor and avoid the other port. The researchers found that some neurons in the hippocampus became selectively responsive for particular pairs of odors. Moreover, the neurons were particular about which odor was at which port—they would respond strongly with odor 1 at port A and odor 2 at port B, but not with the odors switched to the opposite ports. This indicates that the response of the hippocampal neurons relates the specific odors, their spatial locations, and the fact that they are presented separately or together. It was also shown that hippocampal lesions produce deficits on this discrimination task.

THE NEOCORTEX AND WORKING MEMORY

Earlier we said that working memory was likely to reside in multiple locations in the brain. As we've seen, there may be a working memory function of the hippocampus demonstrated by rats retrieving food in a radial arm maze. Here we'll take a closer look at two locations in the neocortex

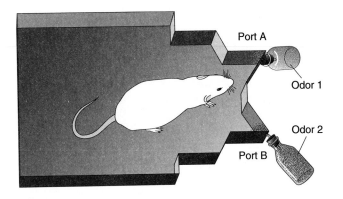

Figure 19.14
Cohen and Eichenbaum's odor discrimination experiment. (Source: Adapted from Eichenbaum, Fagan, Mathews, and Cohen, 1988, Fig. 1.)

that appear to be involved in working memory: the prefrontal cortex and the lateral intraparietal area (area LIP).

Prefrontal Cortex and Working Memory

One of the most obvious anatomical differences between primates (especially humans) and other mammals is that primates have a large frontal lobe. The rostral end of the frontal lobe, the **prefrontal cortex**, is particularly highly developed (Figure 19.15). Compared to the sensory and motor cortical areas, the function of prefrontal cortex is relatively poorly understood. But because it is so well developed in humans, it is often assumed that prefrontal cortex is responsible for those characteristics, such as self-awareness and the capacity for complex planning and problem solving, that distinguish us from other animals. One reason for thinking that prefrontal cortex may be involved in learning and memory is that it is interconnected with the medial temporal lobe and diencephalic structures previously discussed (Figure 19.16).

Some of the first evidence suggesting that the frontal lobe is important for learning and memory came from experiments performed in the 1930s employing a *delayed response task*. In this task, a monkey first sees food being placed in a well below one of two identical covers in a table. Then there is a delay period in which the animal cannot see the table. Finally, the animal is allowed to see the table again and receives the food as a reward if it chooses the correct well. Large prefrontal lesions seriously degrade performance on this delayed response task, and other tasks including a delay period. Moreover, the monkeys perform increasingly poorly as the delay period is lengthened. These results imply that prefrontal cortex plays some important role in memory.

Experiments conducted more recently suggest that prefrontal cortex is involved with working memory for problem solving and the planning of behavior. One piece of evidence comes from the behavior of humans with lesions in prefrontal cortex. These people usually perform better than those with medial temporal lesions on simple memory tasks such as recalling information after a delay period. However, in other more complex tasks, humans with prefrontal damage show marked deficits. Recall the case of Phineas Gage, discussed in Chapter 16. Having sustained severe frontal lobe damage (an iron bar passing through the head qualifies as severe!), Gage had a difficult time maintaining a course of behavior after his injury. Although he could carry out behaviors appropriate for different situations, he had difficulty planning and organizing these behaviors, probably because of the damage to his frontal lobe.

A task that brings out the problems associated with prefrontal cortical damage is the Wisconsin card-sorting test. A person is asked to sort a deck of cards having a variable number of colored geometric shapes on them

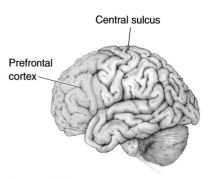

Figure 19.15
Prefrontal cortex. The prefrontal cortex at the rostral end of the frontal lobe receives afferents from the medial dorsal nucleus of the thalamus.

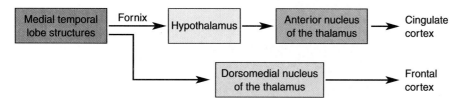

Figure 19.16
Connections between cortical association areas and the medial temporal lobe. Prefrontal cortex as well as cingulate cortex receive afferents from structures of the medial temporal lobe.

(Figure 19.17). The cards can be sorted by color, shape, or number of symbols, but at the beginning of the test, the subject isn't told which category to use. However, by putting cards into stacks and being informed when errors occur, the subject learns what the current sorting category is. After 10 correct card placements are made, the sorting category is changed, and the routine starts over again. In order to perform well on this test, the person must use memory for previous cards and errors made, in order to plan the next card placement. People with prefrontal lesions have great difficulty on this task when the sorting category is changed; they continue to sort according to a rule that no longer applies. It appears that they have difficulty using recent information (i.e., data in working memory) to change their behavior. The same sort of deficit is seen if they are asked to trace a path through a maze on a table. They will repeatedly make the same mistake of going down the same blind alley. In other words, they don't learn from their recent experience in the same way that a normal person does. And when they make mistakes, they may have difficulty returning to an earlier point in the maze and instead start over from the beginning.

The neurons in prefrontal cortex represent a variety of response types, some of which may reflect a role in working memory. Figure 19.18 shows two response patterns obtained while a monkey performed a delayed response task. The neuron in the top trace responded while the animal saw the location of the food, was unresponsive during the delay interval, and responded again when the animal made a choice. This pattern may be directly related to the presentation of the stimuli. Perhaps more interesting is the response pattern of the other neuron, which responded only during the delay interval. This cell was not directly activated by the stimuli in the first or second interval in which it saw the food wells. The increased activity during the delay period may be related to the retention of information needed to make the correct choice after the delay (i.e., working memory).

Area LIP and Working Memory

In recent years, several other cortical areas have been found to contain neurons that appear to retain working memory information. In Chapter 14 we saw an example in area 6 (see Figure 14.10). Another example is provided

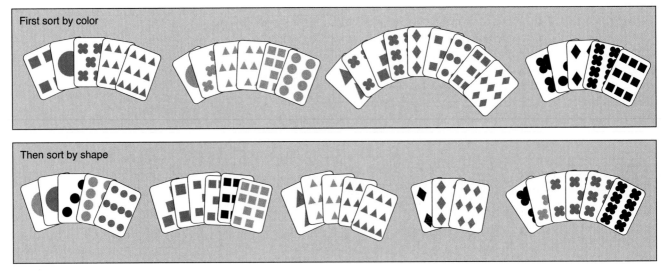

Figure 19.17
The Wisconsin card-sorting test. Cards containing various numbers of colored symbols must first be sorted by color. After a string of correct responses is made, the sorting category is changed to shape.

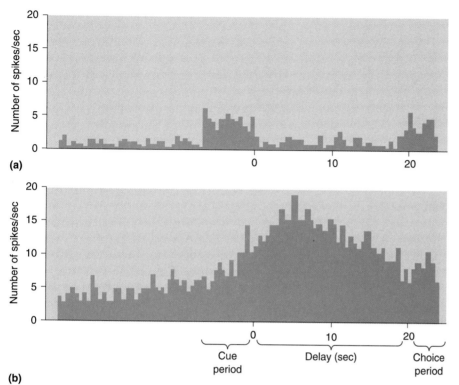

(a)

(b)

Figure 19.18
Neural responses in monkey prefrontal cortex. The two histograms show the activity of cells in prefrontal cortex recorded while the animal performed a delayed response task. During a 7-sec cue period, within the view of the monkey, food is placed into one of two food wells. During the delay period, the animal cannot see the food wells; after the delay, it is allowed to choose a well to receive a food reward (the choice period). **(a)** This cell responds when the well is first baited with food and again when the animal exposes the well to get the food. **(b)** This cell responds strongest during the delay period when there is no visual stimulus. (Source: Adapted from Fuster, 1973, Fig. 2.)

by **lateral intraparietal cortex (area LIP)**, buried in the intraparietal sulcus (Figure 19.19). Saccadic eye movements can be elicited by electrically stimulating area LIP. The responses of many neurons in area LIP have patterns that suggest a working memory function. This is seen in a *delayed saccade task*, in which a monkey fixates on a spot on a computer screen and a target is briefly flashed at a peripheral location (Figure 19.20a). After the target disappears, there is a variable length delay. At the end of the delay period, the fixation spot goes off, and the animal's eyes make a saccadic movement to the remembered location of the target. The response of an LIP neuron while a monkey performs this task is shown in Figure 19.20b. The neuron begins firing shortly after the peripheral target is presented; this seems like a normal stimulus-evoked response. But the cell keeps firing throughout the delay period in which there is no stimulus, until saccadic movements finally occur. Further experiments using this saccade task suggest that the response of the LIP neuron is temporarily holding information that will be used to produce the saccades.

Other areas in parietal and temporal cortex have been shown to have analogous working memory responses. These areas seem to be modality-specific, just as the responses in area LIP are specific to vision. This is consistent with the clinical observation that there are specific auditory and visual working memory deficits in humans produced by cortical lesions.

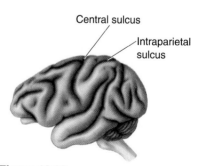

Central sulcus

Intraparietal sulcus

Figure 19.19
Lateral intraparietal cortex (LIP) buried in the intraparietal sulcus.

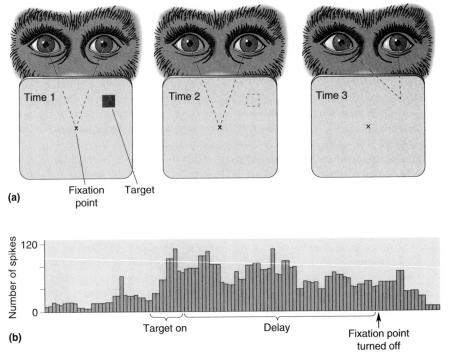

(a)

Fixation   Target
point

(b)

Target on   Delay   Fixation point
turned off

Figure 19.20
The delayed saccade task. (a) After fixating on a central point, a target goes on and off at a peripheral location. There is a delay period after the target goes off in which the monkey continues to fixate on the central point. At the end of the delay period, the fixation point goes off, and the animal moves its eyes to the remembered location of the target. **(b)** The histogram shows the response of an LIP neuron. The neuron begins firing when the target is presented and continues firing until after the fixation point is gone and the saccadic eye movement begins. (Source: Adapted from Goldman-Rakic, 1992, p. 113, and Gnadt and Anderson, 1988, Fig. 2.)

CONCLUDING REMARKS

A clear message from the topics discussed in this chapter is that learning and memory are not confined to a single place in the brain. It is not the case that a small number of specialized "memory cells" act to store our life experiences independent from the rest of brain function. In the formation of declarative memories, it appears that highly processed sensory information coming from the cortical association areas is fed into the medial temporal cortex and processing goes on there and in the associated structures of the diencephalon, before memories are finally stored in a more permanent form in neocortex.

Our discussion leaves several large questions unanswered. One of these is the basis for procedural memory. The case of H. M. makes a dramatic point that procedural learning and memory can be preserved even in the most devastating cases of human amnesia. A person can learn to perform new tasks, yet retain absolutely no memory of having learned. A second unanswered question that is even more general is: How are memories stored? We've said that engrams may exist in temporal lobe neocortex, among other places. But what is the physiological basis for the memory storage? When we try to remember a phone number, an interruption can make us forget, suggesting that memories are initially held in a particularly fragile form. However, long-term memory is much more robust, being able to survive interruption, anesthesia, and the normal bumps and traumas that life involves. Partly because of this robustness, it is thought that memories are ultimately stored in structural changes in neocortex. Our brain is constantly undergoing rewiring, to a certain degree, to adapt to life's experiences. The nature of these structural changes in the brain, which underlie learning and memory, are the subject of Chapter 20.

KEY TERMS

Types of Memory and Amnesia
learning
memory
declarative memory
procedural memory
short-term memory
long-term memory
consolidation
working memory
amnesia
retrograde amnesia
anterograde amnesia
transient global amnesia

The Search for the Engram
engram
cell assembly

The Temporal Lobes and Declarative Memory
hippocampus
entorhinal cortex

perirhinal cortex
parahippocampal cortex
fornix
delayed non-match to sample (DNMS)
recognition memory
Korsakoff's syndrome

Memory Functions of the Hippocampus
place cell
relational memory

The Neocortex and Working Memory
prefrontal cortex
lateral intraparietal cortex (area LIP)

REVIEW QUESTIONS ✔

1. If you try to recall how many windows there are in your house by mentally walking from room to room, are you using declarative memory, procedural memory, or both?
2. What abilities and disabilities do you think a person completely lacking short-term memory would have?
3. Why did Lashley conclude that all cortical areas contribute equally to learning and memory? Why was this conclusion later called into question?
4. What evidence is there that long-term memories are stored in neocortex?
5. If you were using a microelectrode to record from the brain and you suspected that a neuron you encountered was involved in storing long-term memories, how would you test that hypothesis?
6. What are place cells, and where are they found? In what ways are the response characteristics of place cells different from the receptive fields of sensory neurons?
7. What role does the hippocampus play in spatial memory, working memory, and relational memory?
8. What is working memory, and in what brain areas have neural correlates of working memory been observed?

CHAPTER
20

Synaptic Mechanisms of Memory

INTRODUCTION

PROCEDURAL LEARNING

SIMPLE SYSTEMS: INVERTEBRATE MODELS OF LEARNING

VERTEBRATE MODELS OF LEARNING

OF MOLECULES AND MEMORIES

CONCLUDING REMARKS

KEY TERMS

REVIEW QUESTIONS ✔

An important first step in understanding the neurobiology of memory is identifying *where* different types of information are stored. As we saw in the last chapter, answers to this question are beginning to yield to basic neuroscientific research. However, an equally important question concerns *how* information is stored. As Hebb pointed out (and as research using computer models has amply confirmed), memories can result from subtle alterations in synapses, and these alterations can be widely distributed in the brain. This helps narrow the search for a physical basis of memory—synaptic modifications—but it also raises a dilemma. The synaptic modifications that underlie memory may be too small and too widely distributed to be observed and studied experimentally.

These considerations inspired some researchers to look to the nervous systems of simple invertebrate animals for insights into the molecular mechanisms of memory. Although such animals obviously aren't going to remember the definition of "isosceles triangle," they do display certain types of simple procedural memory. Invertebrate research has clearly shown that Hebb was right: Memories *can* reside in synaptic alterations. Moreover, it has been possible to identify some of the molecular mechanisms that lead to this synaptic plasticity. Although nonsynaptic changes have also been found that may account for some types of memory, invertebrate research leaves little doubt that the synapse is an important site of information storage.

It would be disappointing if the best we could do was understand how a sea slug learns. We really want to know how our own brain forms memories. Hope for this understanding now comes from the study of synaptic plasticity in regions of the mammalian brain associated with different types of memory. Theoretical work and invertebrate studies point to lasting alterations in the effectiveness of synapses in the brain as a basis for memory. Therefore, researchers have used electrical stimulation of the brain to produce measurable synaptic alterations whose mechanisms can be studied. One of the interesting conclusions of this research is that the mechanisms of activity-dependent synaptic plasticity in the adult brain have much in common with those operating during development to ensure that the brain is wired together properly. Another conclusion is that the basic molecular machinery operating in the mammalian brain to alter synaptic effectiveness is quite similar to that which is responsible for memory formation in invertebrates.

There is a growing sense of optimism among neuroscientists that we may soon know the physical basis of memory. This investigation has benefited from combined approaches of researchers in disciplines ranging from psychology to molecular biology. In this chapter, we'll take a look at some of their discoveries.

PROCEDURAL LEARNING

Recall that *declarative memories* have a certain ethereal quality. They are easily formed and easily forgotten, and they may result from small modifications of synapses that are widely distributed in the brain. These characteristics make this type of memory particularly challenging to study at the synaptic level. However, *procedural memories* have characteristics that make them more amenable to investigation. Besides the fact that these memories are particularly robust, they can be formed along simple reflex pathways that link sensations to movements. Procedural learning involves learning a motor response (procedure) in reaction to a sensory input. It is typically divided into two types: nonassociative learning and associative learning.

Nonassociative Learning

Nonassociative learning describes the change in the behavioral response that occurs over time in response to a single type of stimulus. It is divided into two types: habituation and sensitization.

Habituation. Suppose you live in a dorm where there is a single telephone in the hall. When the phone rings, you run to answer it. But every time you do, the call is for a dormmate. Over time, you stop reacting to the ringing of the phone and eventually no longer even hear it. This type of learning, **habituation**, is learning to ignore a stimulus that lacks meaning (Figure 20.1a). You are habituated to a lot of stimuli. Perhaps as you read this sentence cars and trucks are passing by outside, a dog is barking, your roommate is playing the Grateful Dead for the hundredth time—and all this goes on without your really noticing. You have habituated to these stimuli.

Sensitization. Suppose you're walking down the sidewalk on a well-lit city street at night and suddenly there is a blackout. You hear footsteps behind you, and though normally this wouldn't disturb you, now you nearly jump out of your skin. Car headlights appear, and you react by side-stepping away from the street. The strong sensory stimulus (the blackout) caused **sensitization** that intensifies your response to all stimuli, even ones that previously evoked little or no reaction (Figure 20.1b).

Associative Learning

During **associative learning**, we form associations between events. Two types are usually distinguished: classical conditioning and instrumental conditioning.

Classical Conditioning. This type of learning was discovered and characterized in dogs by the famous Russian physiologist Ivan Pavlov around the turn of the last century. **Classical conditioning** involves associating a stimulus that evokes a measurable response with a second stimulus that normally does not evoke this response. The first type of stimulus, the one that normally evokes the response, is called the *unconditional stimulus* (*US*) because no training (conditioning) is required for it to yield a response. In Pavlov's experiments, the US was the sight of a piece of meat, and the response was salivation by the dog. The second type of stimulus, the one that normally does not evoke this same response, is called the *conditional stimulus* (*CS*) because this one requires training (conditioning) before it will yield a response. In Pavlov's experiments, the CS was an auditory stimulus, for example, the sound of a bell. Training consisted of repeatedly *pairing* the sound with the presentation of the meat. After many of these pairings the meat was withheld, and the animal salivated to the sound alone. The dog had learned an association between the sound (CS) and the presentation of

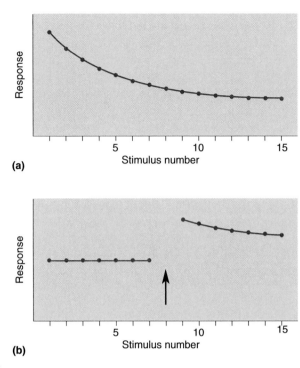

(a)

(b)

Figure 20.1
Types of nonassociative learning. (a) In habituation, repeated presentation of the same stimulus produces a progressively smaller response. (b) In sensitization, a strong stimulus (arrow) results in an exaggerated response to all subsequent stimuli.

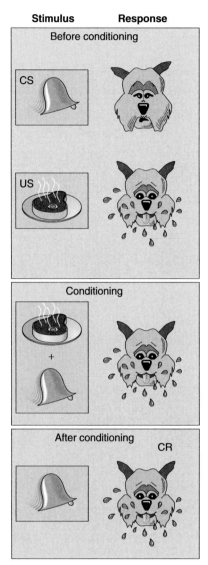

Figure 20.2
Classical conditioning. Prior to conditioning, the sound of a bell (the CS) elicits no response, in sharp contrast to the response elicited by the sight of a piece of meat (the US). Conditioning entails pairing the sound of the bell with the sight of the meat. The dog learns that the bell predicts the meat.

meat (US); the dog learned that the sound *predicted* the meat. The learned response to the conditioned stimulus is called the *conditioned response* (CR) (Figure 20.2).

It is important to understand that there are certain timing requirements in successful classical conditioning. Conditioning will occur if the US and CS are presented simultaneously or if the CS precedes the US by a short interval. However, if the CS precedes the US by too much, the conditioning will be much weaker or absent. Conditioning typically does not occur if the CS follows the US. The synaptic mechanism of classical conditioning will have to account for these stringent timing requirements.

Instrumental Conditioning. This type of associative learning was discovered and studied by Columbia University psychologist Edward Thorndike early in this century. In **instrumental conditioning**, an individual learns to associate a response, a motor act, with a meaningful stimulus, typically a reward such as food. As an example, consider what happens when a hungry rat is placed in a box with a lever that dispenses food. In the course of exploring the box, the rat bumps the lever and out pops a piece of food. After this happy accident occurs a few more times, the rat learns that pressing the lever leads to a food reward. The rat will then work the lever (and eat the food) until it is no longer hungry. This should sound familiar. Recall from previous chapters experiments in which monkeys were trained to make responses as subtle as saccadic eye movements in order to receive a juice reward. Behavioral neurophysiology makes use of instrumental learning. The reward need not be food or drink, either. Remember that rats will lever-press for a cocaine reward or for electrical stimulation of the medial forebrain bundle. Instrumental conditioning will also occur if a response, instead of evoking a rewarding stimulus, prevents the occurrence of an aversive stimulus such as a foot shock.

As in classical conditioning, a predictive relationship is learned during instrumental conditioning. In classical conditioning, it is learned that one stimulus (CS) predicts another stimulus (US). In instrumental conditioning, it is learned that a particular behavior is associated with a particular consequence. Also like classical conditioning, timing is important. Successful instrumental conditioning requires the stimulus to occur shortly after the response.

However, because motivation plays such a large part in instrumental conditioning (after all, only a hungry rat will lever-press for a food reward), the underlying neural circuits are believed to be considerably more complex than those involved in simple classical conditioning. Therefore, we'll leave instrumental conditioning. Instead, we'll focus on the simpler forms of learning and memory for which mechanisms have been identified in the nervous system of an invertebrate animal.

SIMPLE SYSTEMS: INVERTEBRATE MODELS OF LEARNING

Over the course of neuroscience history, a large menagerie of invertebrate creatures has been used for experimentation. You are already familiar with the squid and the contribution of its giant axon and giant synapse to our understanding of cellular neurophysiology (see Chapters 4 and 5). Other invertebrate species that have been used include cockroaches, flies, bees, leeches, and nematodes. The reason is that invertebrate nervous systems offer some important experimental advantages:

- *Small nervous systems.* Neurons in the invertebrate nervous systems number as few as a thousand, roughly 10 million times less than in the human brain.
- *Large neurons.* Many invertebrate neurons are very large, making them easy to study electrophysiologically.
- *Identifiable neurons.* Neurons in invertebrates can be catalogued based on size, location, and electrophysiological properties and can therefore be identified from individual to individual.
- *Identifiable circuits.* Identifiable neurons make the same connections with one another from individual to individual.
- *Simple genetics.* The small genomes and rapid life cycles of some invertebrates (such as flies and nematodes) make them ideal for studies of the genetic and molecular biological basis of learning.

Another advantage—if one is careful in the selection of the species—is that the subject of the experiment can be eaten at the end of the day. For example, the inhibitory neurotransmitter GABA was first characterized in and chemically isolated from New England lobsters.

Invertebrates can be particularly useful for analysis of the neural basis of behavior. Granted, the behavioral repertoire of the average invertebrate is rather limited. Nonetheless, many invertebrate species exhibit the simple forms of learning we discussed above: habituation, sensitization, and classical conditioning. One species in particular has been used to study the neurobiology of learning, the sea slug *Aplysia californica* (Figure 20.3).

Nonassociative Learning in *Aplysia*

If someone blows gently at your eye, you blink. Eventually, however, you will habituate (assuming that the air puff is not painful). Similarly, if a jet of water is squirted onto a fleshy region of *Aplysia* called the siphon, the siphon and the gill will retract (Figure 20.4). This is called the *gill-withdrawal reflex*. Like the eye blink, the gill-withdrawal reflex displays habituation after repeated presentation of the water jet. In a pioneering series of experiments

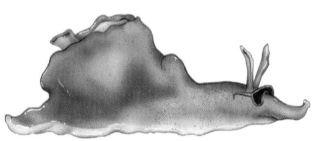

Figure 20.3
Aplysia californica. One cool sea slug, used for neurobiological studies of learning and memory.

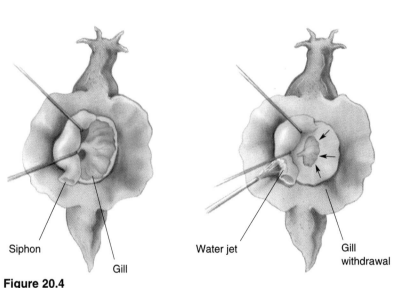

Siphon

Gill

Water jet

Gill withdrawal

Figure 20.4
The gill-withdrawal reflex in Aplysia.

beginning in the 1960s, Eric Kandel and his colleagues, now at Columbia University, set out to determine where this procedural memory resides and how it is formed.

Habituation of the Gill-Withdrawal Reflex. Sensory information from the siphon travels along a nerve until it enters a region of the *Aplysia* nervous system called the abdominal ganglion (Figure 20.5). Here the information is distributed to motor neurons and to interneurons. One of the motor neurons that receives direct monosynaptic sensory input from the siphon is identified as L7, and this cell innervates the muscles that produce gill withdrawal. Therefore, research has focused on understanding how this simple monosynaptic reflex arc changes during habituation.

The first important question concerns where the habituation occurs. Figure 20.6 shows a simplified wiring diagram of the gill-withdrawal reflex. Repeated stimulation of the siphon skin leads to progressively less contraction of the gill-withdrawal muscles. The change underlying this habituation could occur (1) at the sensory nerve endings in the skin, making them less sensitive to the squirt of water; (2) at the muscle, making it less responsive to synaptic stimulation by the motor neuron; or (3) at the synapse between the sensory neuron and the motor neuron. The first possibility was ruled out by making microelectrode recordings from the sensory neuron as habituation occurred. The sensory neuron continued to fire action potentials in

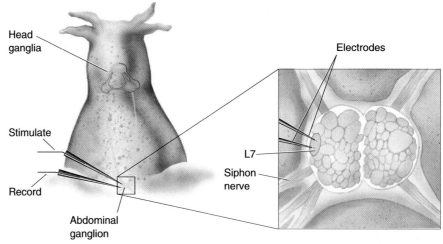

Figure 20.5
The abdominal ganglion of *Aplysia*. The gill withdrawal reflex involves neurons within the abdominal ganglion that can be dissected and studied electrophysiologically.

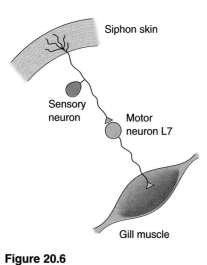

Figure 20.6
A simple wiring diagram for the gill-withdrawal reflex. The sensory neuron which detects stimuli applied to the skin of the siphon synapses directly on the motor neuron that causes the gill to withdraw.

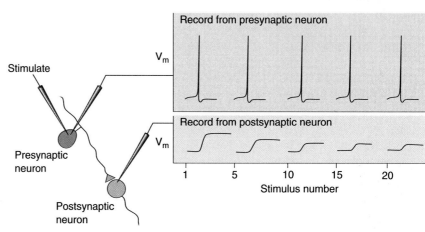

Figure 20.7
Habituation at the cellular level. Repeated electrical stimulation of a sensory neuron leads to a progressively smaller EPSP in the postsynaptic motor neuron.

response to stimulation of the skin, even as the motor response decreased. Similarly, the third possibility was ruled out by electrically stimulating the motor neuron (L7) and showing that it always evoked the same amount of muscle contraction. This left the second possibility: that habituation occurs at the synapse joining the sensory input to the motor neuron. Indeed, simply giving repetitive electrical stimulation to the sensory neuron was sufficient to cause a progressive decrease in the size of the postsynaptic EPSP (Figure 20.7).

Identifying a synaptic modification as a possible basis for habituation still leaves unanswered the question of where the synapse is modified. Let's consider the possibilities. After habituation, there could be (1) less neurotransmitter release by the presynaptic axon or (2) decreased postsynaptic responsiveness to the transmitter (perhaps caused by fewer receptors). Recall from Chapter 5 that transmitter molecules are released in discrete packets called *quanta*, believed to correspond to the contents of individual synaptic vesicles. Vincent Castelluci and Kandel performed a sophisticated quantal analysis of transmission at this synapse and concluded that *after habituation there are fewer quanta released per action potential.* The sensitivity of the postsynaptic cell to neurotransmitter did not change. In other words, habituation of the gill-withdrawal reflex is associated with a *presynaptic* modification.

Why is neurotransmitter release reduced after repeated stimulation of the sensory nerve terminal? Recall the sequence of events that couples the arrival of the action potential in the terminal with the release of neurotransmitter. A critical step is the entry of Ca^{2+} into the terminal, which is mediated by voltage-gated calcium channels. In the nerve terminal of the sensory neuron, it appears that these channels become progressively and persistently less effective when they are repeatedly opened. It is still unknown exactly why this is so, but the consequence of less presynaptic Ca^{2+} entry per action potential is less transmitter released.

Sensitization of the Gill-Withdrawal Reflex. To cause sensitization of the gill-withdrawal reflex, Kandel and his colleagues applied a brief electrical shock to the head of *Aplysia.* This resulted in exaggerated gill withdrawal in response to stimulation of the siphon. Using the strategy outlined above, they identified a site of plasticity showing changes that correlate with the behavior. Once again, it turned out to be a modification of transmitter release in the sensory nerve terminal.

How might stimulation of the head lead to sensitization of the gill-withdrawal reflex? Understanding the answer to this question requires that we add a third neuron to the wiring diagram (Figure 20.8). This third cell, L29, is activated by the head shock and makes a synapse *on the axon terminal of the sensory neuron.* The neurotransmitter released by L29 is believed to be serotonin (5-HT). Serotonin sets in motion a molecular cascade that sensitizes the sensory axon terminal so that it lets in more Ca^{2+} per action potential.

The serotonin receptor on the sensory axon terminal is a G-protein-coupled metabotropic receptor. Stimulation of this receptor leads to the production of intracellular second messengers. In the case of the *Aplysia* sensory

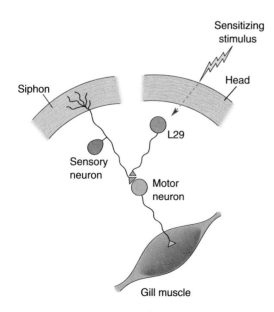

Figure 20.8
Wiring diagram for sensitization of the gill-withdrawal reflex. A sensitizing stimulus to the head indirectly activates an interneuron, L29, which makes an axoaxonic synapse on the terminal of the sensory neuron.

nerve terminal, this second messenger is *cyclic AMP* produced from ATP by the enzyme *adenylyl cyclase* (Figure 20.9). As we learned in Chapters 5 and 6, cAMP activates *protein kinase A*, and this enzyme phosphorylates (attaches phosphate groups to) various proteins. In the sensory nerve terminal, one of these proteins is a potassium channel, and phosphorylation of this channel causes it to close. The closure of potassium channels in the axon terminal leads to a prolongation of the presynaptic action potential. This results in more Ca^{2+} entry through voltage-gated Ca^{2+} channels during the action potential and therefore more quanta of transmitter are released (Figure 20.10).

Associative Learning in *Aplysia*

In the 1980s, researchers discovered that *Aplysia* also could be classically conditioned. Again, the response measured was the withdrawal of the gill. The unconditional stimulus (US) was a strong shock to the tail. The conditional stimulus (CS) was stimulation of the siphon that was so gentle that it did not cause much of a response. It was discovered that if stimulation of the tail (US) was *paired* with stimulation of the siphon (CS), the subsequent response to the siphon alone (CR) was much greater than what could be accounted for by sensitization (Figure 20.11). As in Pavlov's experiments,

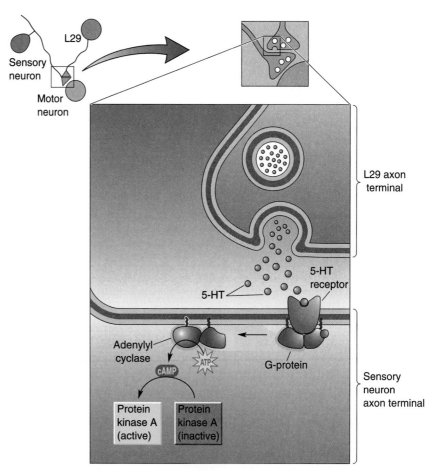

Figure 20.9
A mechanism for sensitization of the gill-withdrawal reflex. Serotonin released by L29 in response to the head shock leads to G-protein-coupled activation of adenylyl cyclase in the sensory axon terminal. Activation of this enzyme leads to the production of cyclic AMP, which in turn activates protein kinase A. Protein kinase A attaches phosphate groups to a potassium channel, causing it to close.

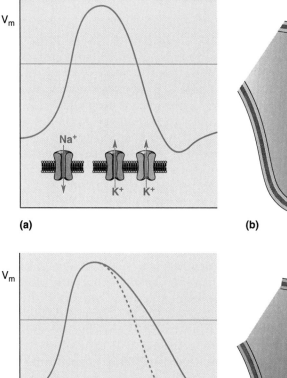

(a)

(b)

(c)

(d)

Figure 20.10
The effect of decreased potassium conductance in the sensory axon terminal. (a) The trace shows membrane voltage changes during an action potential. The rising phase is caused by the opening of voltage-gated sodium channels, and the falling phase is caused by the closing of the sodium channels and the opening of potassium channels. In the axon terminal, voltage-gated Ca^{2+} channels stay open as long as the membrane voltage exceeds a threshold value, indicated by the red line. (b) The resulting entry of Ca^{2+} stimulates the release of neurotransmitter. (c) A decrease in K^+ conductance after sensitization prolongs the action potential. (d) The voltage-gated Ca^{2+} channels stay open longer, thereby admitting more Ca^{2+} into the terminal. This causes more transmitter to be released per action potential.

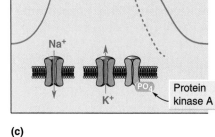

(a)

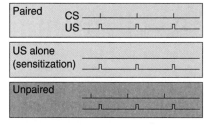

(b)

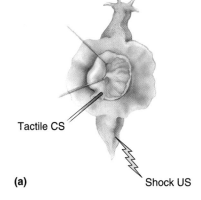

(c)

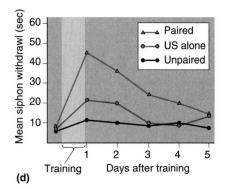

(d)

Figure 20.11
Classical conditioning in *Aplysia*. (a) A gentle water jet to the siphon is the CS. A shock to the tail is the US. The response measured is the withdrawal of the gill. (b) The wiring diagram for classical conditioning. The US activates the same serotonergic cell (L29) that is activated during sensitization. (c) Timing of the CS and US during three different types of training. (d) Plotted here is the magnitude of the gill withdrawal in response to the CS. After pairing (classical conditioning), the animal withdraws the gill in response to the CS, which before training was ineffective in eliciting a response. (Source: Adapted from Dudai, 1989, Fig. 4.3.)

timing was critical. Conditioning occurred only if the CS (siphon stimulus) preceded the US (tail stimulus) by ≤ 0.5 seconds.

Once again, a critical modification occurs at the synapse between the sensory and motor neuron and involves increased neurotransmitter release. To understand how this comes about, we need to focus on what is happening at that synapse during the CS-US pairing. At the *cellular* level, the CS is represented by the arrival of an action potential in the sensory axon terminal, and the US is represented by the release of serotonin by L29 (activated by the tail shock). At the *molecular* level, the CS is represented by the influx of Ca^{2+}, and the US is represented by the G-protein-coupled activation of the enzyme adenylyl cyclase in the terminal (Table 20.1). Remember that adenylyl cyclase generates cAMP. In 1991, a key discovery was made by Thomas Abrams and his colleagues at Columbia University that *in the presence of elevated [Ca^{2+}], adenylyl cyclase churns out more cAMP* (Figure 20.12). More cAMP means more activation of protein kinase A, more phosphorylation of potassium channels, and therefore the release of more transmitter molecules. Therefore, in the case of classical conditioning of the gill-withdrawal reflex, adenylyl cyclase may serve as a detector of CS-US coincidence. According to this hypothesis, *learning* occurs when a presynaptic Ca^{2+} pulse coincides with (or just precedes) the G-protein-coupled activation of adenylyl cyclase, which stimulates the production of a lot of cAMP. *Memory* occurs when potassium channels are phosphorylated, and transmitter release is enhanced.

Pavlov would have been surprised by these results. He had assumed that this type of associative memory required the sophisticated neural hardware of the cerebral cortex. Now we see that even a sea slug can be classically conditioned, and that many of the defining characteristics of the conditioning can be accounted for by the properties of a single type of enzyme.

We've seen that a great deal can be learned about the neurobiology of behavior in invertebrates. It is important to realize, however, that none of the *Aplysia* studies have uncovered the cellular basis of learning, only some *cellular correlates* of learning. Although the studied synaptic changes can accompany behavioral learning, they may not be essential for learning to occur. In other words, there may be (and, in some instances, definitely are) additional mechanisms that contribute to habituation, sensitization, and classical conditioning. Thousands of neurons in the *Aplysia* nervous system are active during even the simplest reflexes, and synaptic changes related to learning are likely to be widely distributed among them. Nonetheless, the invertebrate studies have been invaluable for identifying candidate molecular mechanisms for learning and memory.

Table 20.1

Cellular and Molecular Correlates of Classical Conditioning in *Aplysia*

Stimulus	Level of Analysis		
	Behavioral	Cellular	Molecular
CS	Touch to siphon	Presynaptic action potential	Presynaptic influx of Ca^{2+}
US	Shock to tail	Release of serotonin	G-protein activation of adenylyl cyclase in presynaptic terminal

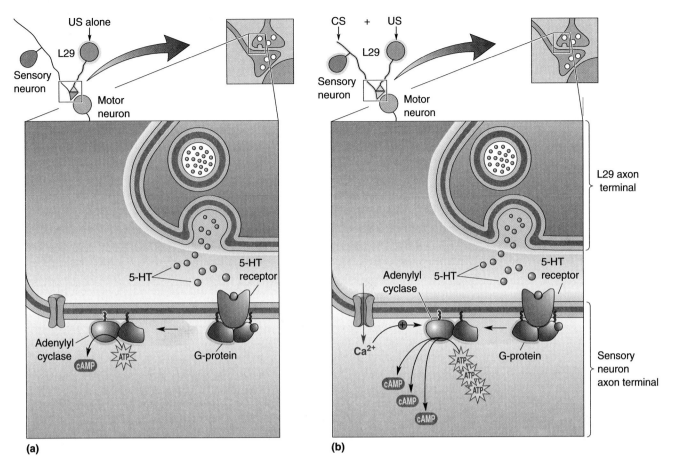

Figure 20.12
The molecular basis for classical conditioning in *Aplysia*. (a) The US alone leads to activation of the motor neuron (via an interneuron, not shown) and to sensitization of the sensory input by the same mechanism illustrated in Figures 20.9 and 20.10. **(b)** Pairing the CS and the US causes greater activation of adenylyl cyclase than either stimulus does by itself because the CS admits Ca^{2+} into the presynaptic terminal. The Ca^{2+} (by interacting with a protein called calmodulin, not shown) increases the response of adenylyl cyclase to G-proteins.

VERTEBRATE MODELS OF LEARNING

Let's summarize what we've learned from the invertebrate studies about the possible neural basis of memory:

- Learning and memory can result from modifications of synaptic transmission.
- Synaptic modifications can be triggered by the conversion of neural activity into intracellular second messengers.
- Memories can result from alterations in existing synaptic proteins.

Keep these points in mind as we explore different types of activity-dependent synaptic plasticity in the mammalian brain.

Synaptic Plasticity in the Cerebellar Cortex

In Chapter 14, we discussed the cerebellum in the context of motor control. We concluded that the cerebellum was an important site for motor learning, a place where corrections are made when the outcome of movements fails to meet expectations. Because these corrections are believed to be made by modifications of synaptic connections, the cerebellum has become a model system for the study of the synaptic basis of learning in the mammalian brain.

Anatomy of the Cerebellar Cortex. Of special interest is the **cerebellar cortex**, the sheet of tissue that lies just under the surface of the cerebellum (Figure 20.13a). The cortex consists of two layers of neuronal cell bodies, the *Purkinje cell layer* and the *granule cell layer*, and these are separated from the pial surface by a *molecular layer* that is largely devoid of somata. **Purkinje cells**, named for the Czech neuroanatomist who first described them in 1837, have a number of interesting features. First, their dendrites extend only into the molecular layer, where they branch like a fan, flattened in one plane. Second, Purkinje cell axons synapse on neurons in the deep cerebellar nuclei, which are the major output cells of the cerebellum. Thus, Purkinje cells are in a powerful position to modify the output of the cerebellum. Third, Purkinje cells use GABA as a neurotransmitter, so their influence on cerebellar output is inhibitory.

Purkinje cell dendrites are directly contacted by one of the two major inputs to the cerebellum. This input arises in a nucleus of the medulla called the **inferior olive**, which integrates information from muscle proprioceptors. Axons from the inferior olive are called **climbing fibers** because they

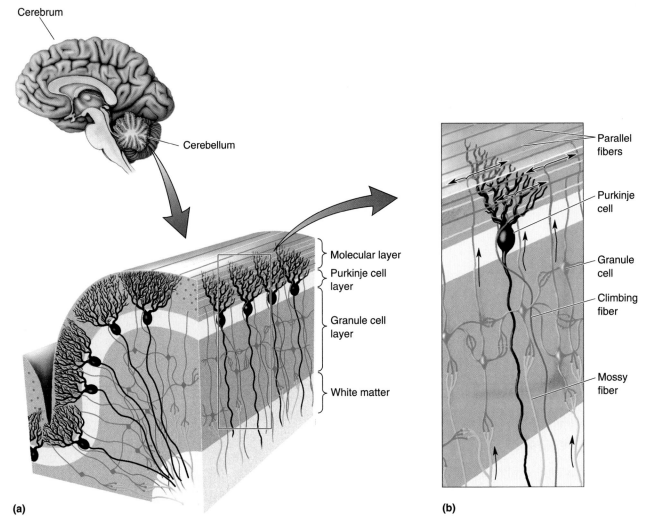

(a) **(b)**

Figure 20.13
Structure of the cerebellar cortex. (a) A view of the cortex showing the organization of the granule cell, Purkinje cell, and molecular layers. **(b)** The major inputs to the Purkinje cells are parallel fibers arising from granule cells and climbing fibers arising from the inferior olive. The major input to the granule cells is the mossy fibers, arising from neurons in the pontine nuclei.

twist around the Purkinje cell dendrites like a vine on the branches of a tree. Each Purkinje cell receives input from only one inferior olive cell, but this input is very powerful. A single climbing fiber axon makes hundreds of excitatory synapses on the dendritic tree of its target Purkinje neuron. An action potential in a climbing fiber generates an exceptionally large EPSP that is always sufficient to strongly activate the postsynaptic Purkinje cell.

The second major input to the cerebellum arises from a variety of brain stem cell groups, notably the pontine nuclei that relay information from the cerebral neocortex. These inputs, called **mossy fibers**, synapse on **cerebellar granule cells,** which form a layer just below the Purkinje cells (Figure 20.13b). Granule cells are very small, very tightly packed, and very numerous. (Estimates are that granule cells may make up half the total number of neurons in the brain!) The granule cells give rise to axons that ascend into the molecular layer, where they branch like a T. Each branch, called a **parallel fiber**, runs straight for several millimeters in a direction that intersects the plane of the Purkinje cell dendrites at a right angle, like wires passing a telephone pole. Therefore, a single parallel fiber has only a brief encounter with any one Purkinje cell, but along its length it encounters many. However, although a Purkinje cell receives only one synapse from each passing parallel fiber, it does so from as many as 100,000 different fibers.

This unusual convergence of parallel and climbing fiber inputs onto Purkinje cell dendrites has fascinated neuroscientists ever since it was described originally by Cajal. In the early 1970s, James Albus at the Goddard Space Flight Center in Greenbelt, Maryland, suggested a way that this arrangement could serve motor learning. He proposed (1) that the climbing fiber input carries error signals indicating that a movement has failed to meet expectations, and (2) that corrections are made by adjusting the effectiveness of the parallel fiber inputs to the Purkinje cell. The idea that motor learning is served by plasticity of the parallel fiber–Purkinje cell synapse actually was proposed a few years earlier by David Marr, then at Cambridge University in England. What is now called the **Marr-Albus theory of motor learning** specifically predicts plasticity of the parallel fiber synapse *if it is active at the same time as the climbing fiber input to the postsynaptic Purkinje cell.*

Long-Term Depression in the Cerebellar Cortex. Masao Ito and his colleagues at the University of Tokyo tested the prediction of the Marr-Albus theory directly. To monitor the effectiveness of parallel fiber synapses on Purkinje cells, they applied brief electrical stimulation to the parallel fibers and measured the size of the EPSP in the Purkinje cell. Then, to induce synaptic plasticity, they paired stimulation of the climbing fibers with stimulation of the parallel fibers. Remarkably, they found that after this pairing procedure, activation of the parallel fibers alone resulted in a smaller postsynaptic response in the Purkinje cell. This type of modification could last at least an hour and therefore was termed **long-term depression**, or simply **LTD** (Figure 20.14).

An important property of LTD is that it only occurs in those parallel fiber synapses that are active at the same time as the climbing fibers. Other parallel fiber synapses that were not stimulated in conjunction with the climbing fibers did not exhibit the plasticity. This property, that only the active inputs show the synaptic plasticity, is called **input specificity**.

Immediately we can see similarities between cerebellar LTD and classical conditioning in *Aplysia*. In *Aplysia* there is an input-specific synaptic modification when presynaptic activity in the sensory axon (event 1) occurs at the same time as stimulation of the sensory axon terminal with serotonin (event 2). In the cerebellar cortex, there is an input-specific modification

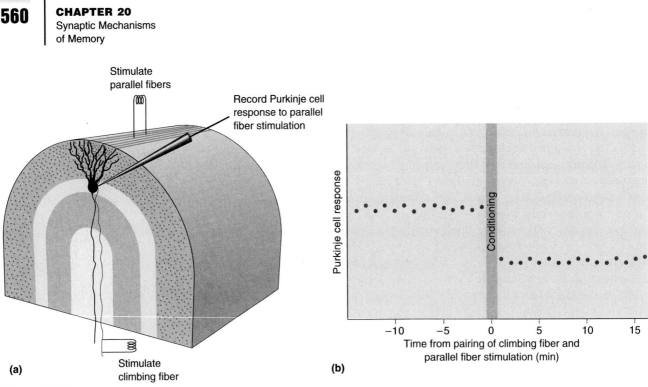

Figure 20.14

Cerebellar long-term depression. (a) The experimental arrangement for demonstrating LTD. The magnitude of the Purkinje cell response to stimulation of a "beam" of parallel fibers is monitored. Conditioning involves pairing parallel fiber stimulation with climbing fiber stimulation. **(b)** A graph of an experiment performed in this way. After the pairing, there is a long-term depression of the response to parallel fiber stimulation.

when activation of the parallel fiber–Purkinje cell synapse (event 1) occurs at the same time as depolarization of the postsynaptic Purkinje cell by the climbing fiber (event 2). However, note that the site of convergence of events 1 and 2 is different in the two systems. In *Aplysia*, convergence occurs in the presynaptic axon terminal, while in the cerebellar cortex, convergence apparently occurs in the Purkinje cell dendrite. Another difference, obviously, is the nature of the synaptic change. After classical conditioning, the sensory-motor synapse becomes *more* effective, while after pairing parallel and climbing fiber activity, the parallel fiber–Purkinje synapse becomes *less* effective.

As with *Aplysia*, now we must ask what side of the parallel fiber–Purkinje synapse is modified. Recall that in *Aplysia*, the modifications were always presynaptic alterations in transmitter release. But in the cerebellum, the synaptic modification was discovered to be postsynaptic. Specifically, LTD was found to result from a decrease in the effectiveness of the glutamate receptor that mediates transmission at this synapse, called the *AMPA receptor*. After LTD, the AMPA receptor is desensitized, meaning that a change in the protein has occurred so the channel does not stay open as long as normal in the presence of glutamate.

Mechanisms of Cerebellar LTD. To understand how paired stimulation of the climbing and parallel fibers leads to LTD, we need to focus on the Purkinje cell dendrite where these signals converge. First, we can ask what is so important about activating the climbing fiber synapses on the Purkinje cell. Recall that this is a very powerful input, causing a large postsynaptic EPSP that always stimulates the Purkinje cell to fire an action potential. However, in addition to activating voltage-gated sodium channels (causing the action potential), this depolarization is powerful enough to activate voltage-gated calcium channels that exist in the membrane of the Purkinje cell dendrites. Thus, climbing fiber activation is associated with a surge of Ca^{2+} into the Purkinje cell dendrite. To assess the importance of this Ca^{2+}

signal in the induction of LTD, a substance that binds Ca^{2+} and prevents rises in $[Ca^{2+}]$, called a *Ca^{2+} chelator*, was injected into the Purkinje cell. This treatment blocked LTD. This and related experiments have led to the conclusion that the critical signal provided by climbing fiber activation is a surge in $[Ca^{2+}]$ in the Purkinje cell dendrite.

What is special about activation of the parallel fibers? The transmitter released after parallel fiber activation is glutamate, and, as we said above, one of the postsynaptic receptors is the AMPA receptor. The AMPA receptor is the channel that mediates the EPSP by allowing Na^+ ions into the Purkinje cell dendrite. However, there is a second type of glutamate receptor postsynaptic to the parallel fibers. This is a *metabotropic glutamate receptor* that is coupled via a G-protein to the enzyme *phospholipase C*. Activation of this enzyme leads to the production of a second messenger (diacylglycerol), which then activates *protein kinase C* (Figure 20.15).

The evidence to date indicates that LTD is caused when *three* intracellular signals occur at the same time: (1) a rise in $[Ca^{2+}]_i$ due to climbing fiber activation, (2) a rise in $[Na^+]_i$ due to AMPA receptor activation, and (3) activation of protein kinase C due to metabotropic receptor activation. Precisely what happens next is still under investigation, but it certainly involves the phosphorylation of proteins by protein kinase C, and the end result is a decrease in the conductance of the AMPA receptor channel. In this model, *learning* occurs when rises in $[Ca^{2+}]_i$ and $[Na^+]_i$ coincide with the activation of protein kinase C. *Memory* occurs when AMPA channels are modified and excitatory postsynaptic currents are depressed.

If we make a leap of faith and assume that LTD plays a role in motor learning (not yet proven), then we see that (1) learning and memory can result from modifications of synaptic transmission, (2) synaptic modifications can be triggered by the conversion of neural activity into intracellular second messengers, and (3) memories can result from alterations in existing synaptic proteins. Does this sound familiar? Figure 20.16 compares the mechanisms of classical conditioning in *Aplysia* and LTD in the cerebellum.

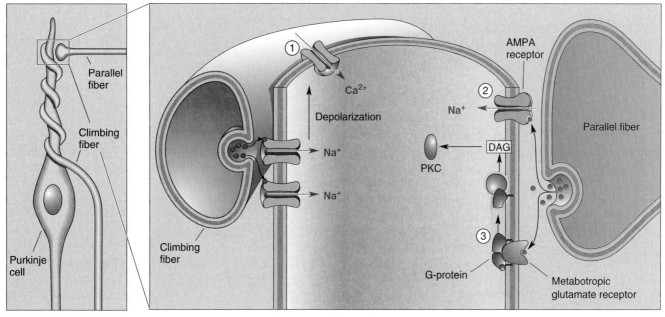

Figure 20.15
A mechanism of LTD induction in the cerebellum. (1) Climbing fiber activation strongly depolarizes the Purkinje cell dendrite, which leads to activation of voltage-gated Ca^{2+} channels. (2) Parallel fiber activation leads to Na^+ entry through AMPA receptors, and (3) the generation of diacylglycerol (DAG) via stimulation of the metabotropic receptor. DAG activates protein kinase C (PKC).

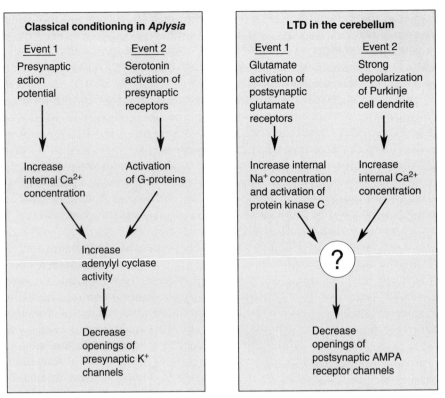

Figure 20.16
A comparison of the events leading to classical conditioning in *Aplysia* and to LTD in the cerebellum.

Synaptic Plasticity in the Hippocampus and Neocortex

Recall from Chapter 19 that declarative memory, the type of memory you'll rely on to pass your next exam, involves the neocortex and structures in the medial temporal lobe, including the hippocampus. In 1973, an important discovery was made in the hippocampus by Timothy Bliss and Terje Lømo, working together in Norway (Box 20.1). They found that brief, high-frequency electrical stimulation of an excitatory pathway to the hippocampus produced a long-lasting enhancement in the strength of the stimulated synapses. This effect is now known as **long-term potentiation**, or **LTP** (first introduced in Chapter 18). LTP is considered by many to hold the secret for how declarative memories are formed in the brain. Let's explore LTP where it is usually studied, in the hippocampus.

Anatomy of the Hippocampus. The hippocampus, illustrated in Figure 20.17, consists of two thin sheets of neurons folded onto each other. One sheet is called the **dentate gyrus** and the other sheet is called **Ammon's horn**. Ammon's horn has four divisions, of which we need concern ourselves with only two: **CA3** and **CA1** (CA stands for *cornu Ammonis*, Latin for "Ammon's horn").

Recall from Chapter 19 that a major input to the hippocampus is the *entorhinal cortex*. Entorhinal cortex sends information to the hippocampus by way of a bundle of axons called the **perforant path**. Perforant path axons synapse on neurons of the dentate gyrus. Dentate gyrus neurons give rise to axons (also called mossy fibers) that synapse on cells in CA3. The CA3 cells give rise to axons that branch. One branch leaves the hippocampus via the fornix. The other branch, called the **Schaffer collateral**, forms synapses on the neurons of CA1. These connections are summarized in Figure 20.17.

Box 20.1 PATH OF DISCOVERY

Discovering LTP
by Tim Bliss and Terje Lømo

Tim Bliss

Terje Lømo

By the late summer of 1966, Terje Lømo had been a Ph.D. student in Per Andersen's laboratory at the University of Oslo for nearly a year. Together, they had been studying synaptic transmission in corticothalamic pathways in the rabbit, and Lømo was now ready to move onto an independent project. Andersen, who had published a series of pioneering electrophysiological papers on the hippocampus a few years earlier, proposed that Lømo examine the phenomenon of frequency potentiation in a monosynaptic cortical pathway, the perforant path input from entorhinal cortex to the granule cells of the hippocampus.

Frequency potentiation refers to the growth in the size of a synaptic response during repetitive stimulation. After characterizing the properties of the evoked synaptic response to single stimuli, Lømo tried repetitive high-frequency stimulation. During stimulation at 10–15 Hz, he found that the slope of the EPSP became steeper, and that more and more granule cells responded with action potentials. Stimulation at this rate could only be continued for a few seconds without the danger of causing seizures, so it was necessary to allow a rest of several minutes between successive trains of stimuli. To Lømo's surprise, subsequent trains always elicited larger responses. When he tested with single shocks, these were also enhanced after a train and could remain so for several hours. Lømo reported these observations at a meeting of the Scandinavian Physiological Society in Turku, Finland in September 1966, and the abstract of that meeting contains the first published reference to what later came to be known as long-term potentiation (LTP). Scientists in the international neuroscience community of that time apparently were not avid readers of the proceedings section of *Acta Physiologica Scandinavica*, so there for a time the matter rested. Lømo returned to the detailed characterization and distribution of evoked synaptic potentials in the dentate gyrus, work which was to form the bulk of the thesis which he completed in 1969.

Meanwhile, in Montreal in the summer of 1967, Tim Bliss was writing his thesis on activity-dependent plasticity in the isolated cortical slab, an ingenious preparation developed by his supervisor, Ben Delisle Burns, in which cortical tissue was separated from long-range cortical and subcortical inputs by surgical undercutting. Bliss recorded evoked action potentials from neurons with weak test shocks and showed that brief changes in synaptic activity could induce long-term changes in the probability of firing. The problem he faced was that the underlying cortical circuitry was so complex that he could not interpret these changes at the synaptic level. What was needed was a monosynaptic cortical pathway. His thoughts turned toward the hippocampus, even as they turned toward home. Back in London as a postdoctoral scientist at the National Institute for Medical Research in Mill Hill, he started experiments on the hippocampus. Before long, he was offered the opportunity to work with Per Andersen in Oslo, whose pioneering descriptions of synaptic activity in hippocampal circuitry he had so admired in Montreal. "By the way," Andersen had told him when they first talked in London, "there's a fellow in Oslo called Lømo who has some results which will interest you."

During that year in Oslo, starting in the fall of 1968, Bliss worked mostly with Andersen and Skrede, charting the so-called trisynaptic circuit which carries information through the hippocampus from dentate granule cells to CA3 pyramidal cells and thence to CA1 pyramidal cells. But one day a week, he and Lømo, whose results Bliss had indeed found curiously interesting, set out to repeat and extend Lømo's original observations. Neither will forget the excitement and elation of those days—and nights—when they confirmed the existence of LTP in the perforant path; they established that it could be induced by brief trains which did not cause frequency potentiation and documented the independent potentiation of the synaptic and spike components of the evoked response. By the end of the year they knew they had identified a set of excitatory synapses in the brain that could retain a trace of past activity for many hours, and was therefore a potential synaptic candidate for information storage.

continued on page 564

Owing to its very simple architecture and organization, the hippocampus is an ideal place to study synaptic transmission in the mammalian brain. In the late 1960s, it was discovered that the hippocampus actually could be removed from the brain (in experimental animals) and cut up like a loaf of bread, and that the resulting slices could be kept alive *in vitro* for many hours. In such a *brain slice preparation*, fiber tracts can be stimulated electrically and synaptic responses recorded. Because cells in the slice can be observed, stimulating and recording electrodes can be positioned with the precision that was previously reserved for invertebrate preparations. This preparation has greatly facilitated the study of LTP.

Properties of LTP in CA1. Although LTP was first demonstrated at the perforant path synapses on the neurons of the dentate gyrus, today most of the experiments on the mechanism of LTP are performed on the Schaffer collateral synapses on the CA1 pyramidal neurons in brain slice preparations.

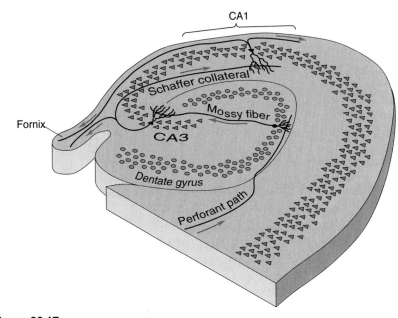

Figure 20.17
Some microcircuits of the hippocampus.

Box 20.1 | **PATH OF DISCOVERY**

continued from page 563

At the end of 1969, Lømo, his thesis completed (and with LTP almost as an afterthought), traveled to London to take up a postdoctoral fellowship with Bernard Katz and Ricardo Miledi at University College London. Bliss returned to Mill Hill soon afterwards and set about repeating and extending on English rabbits the observations which he and Lømo had so readily seen in Norwegian rabbits. He failed.

Lømo came up to Mill Hill to help solve the crisis. Together they failed. The situation was not looking good for LTP, when Tony Gardner-Medwin salvaged the newborn field, not to mention the pride of English rabbits, when he and Bliss showed that, in awake animals with implanted electrodes, tetanic stimulation induced potentiation which could last for days. The crisis was over, even if the mystery of the anaesthetized English rabbit and its indifference to LTP remains.

In a typical experiment, the effectiveness of the Schaffer collateral synapse is monitored by giving a bundle of presynaptic axons a brief electrical stimulus and then measuring the size of the resulting EPSP in a postsynaptic CA1 neuron (Figure 20.18). This method is similar to the way parallel fiber responses are monitored in the cerebellar cortex. Usually such a test stimulation is given every minute or so for 15–30 minutes to ensure that

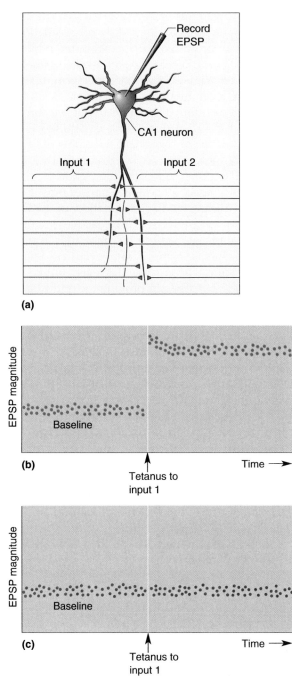

Figure 20.18
Long-term potentiation in CA1. (a) The response of a CA1 neuron is monitored as two inputs are alternately stimulated. LTP is induced in input 1 by giving this input a tetanus. **(b)** The graph shows a record of the experiment. The tetanus to input 1 (arrow) yields a potentiated response to stimulation of this input. **(c)** LTP is input-specific, so there is no change in the response to input 2 after a tetanus to input 1.

the baseline response is stable. Then, to induce LTP, the same axons are given a **tetanus**, which is a brief burst of high-frequency stimulation (typically 50–100 stimuli at a rate of 100/sec). Usually this tetanus induces LTP, and subsequent test stimulation evokes an EPSP that is much greater than it was during the initial baseline period. In other words, the tetanus has caused a modification of the stimulated synapses so that they are more effective. Other synaptic inputs onto the same neuron that did not receive tetanic stimulation do not show LTP. Therefore, like cerebellar LTD, hippocampal LTP is *input-specific*.

One remarkable feature of this plasticity is that it can be induced by a brief tetanus, lasting less than a second, consisting of stimulation at frequencies well within the range of normal axon firing. A second remarkable feature of LTP is its longevity. LTP induced in CA1 of awake animals can last many weeks, possibly even a lifetime. No wonder this form of synaptic plasticity has attracted interest as a candidate mechanism for declarative memory.

Subsequent research has shown that high-frequency stimulation is not an absolute requirement for LTP. Rather, what is required is that *synapses be active at the same time that the postsynaptic CA1 neuron is strongly depolarized.* In order to achieve the necessary depolarization with a tetanus, (1) synapses must be stimulated at frequencies high enough to cause temporal summation of the EPSPs, and (2) enough synapses must be active simultaneously to cause significant spatial summation of EPSPs. This second requirement is called **cooperativity** because coactive synapses must cooperate to produce enough depolarization to cause LTP. Thus, like cerebellar LTD, LTP results when synaptic stimulation (event 1) coincides with strong postsynaptic depolarization (event 2). However, unlike the cerebellum, where a single powerful synapse can provide the critical depolarization, in the hippocampus, adequate depolarization requires that many excitatory synapses be active at the same time.

Consider for a moment how the cooperativity property of hippocampal LTP could be used to form associations. Imagine a hippocampal neuron receiving synaptic inputs from three sources: A, B, and C. Initially, no single input is strong enough to evoke an action potential in the postsynaptic neuron. Now imagine that inputs A and B repeatedly fire at the same time. Because of spatial summation, inputs A and B are now capable of firing the postsynaptic neuron and, further, of causing LTP. Only the active synapses will be potentiated, and these, of course, are those belonging to inputs A and B. Now, because of potentiation of their synapses, *either* input A *or* input B can fire the postsynaptic neuron (but not input C). Thus, LTP has caused an association of inputs A and B. In this way, the sight of a rose could be associated with the smell of a rose (they often occur at the same time), but never with the smell of an onion (Figure 20.19).

Speaking of associations, remember the idea of a Hebb synapse introduced in Chapter 18 to account for aspects of visual development? LTP in CA1 is Hebbian: Inputs that fire together, wire together.

Mechanisms of LTP in CA1. Excitatory synaptic transmission in the hippocampus is mediated by glutamate receptors. As is the case at the parallel fiber–Purkinje cell synapse in the cerebellum, Na^+ ions passing through the AMPA subclass of glutamate receptor are responsible for the EPSP at the Schaffer collateral–CA1 pyramidal cell synapse. However, unlike the cerebellum, CA1 neurons also have postsynaptic *NMDA receptors.* Recall that these glutamate receptors have the unusual property that they conduct Ca^{2+} ions, but only when glutamate binds *and* the postsynaptic membrane is depolarized. Thus, Ca^{2+} entry through the NMDA receptor specifically signals when pre- and postsynaptic elements are active at the same time (recall Figure 18.20).

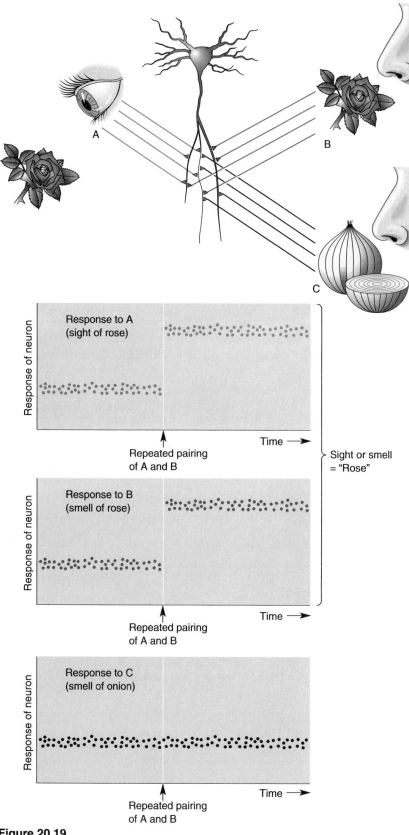

Figure 20.19
A rose is a rose, but it is not an onion. Because the sight and smell of the rose
occur at the same time, the inputs carrying this information to a neuron may undergo LTP, thus forming an association between the two stimuli.

Considerable evidence now links this rise in postsynaptic $[Ca^{2+}]_i$ to the induction of LTP. For example, LTP induction is prevented if NMDA receptors are pharmacologically inhibited, or if rises in postsynaptic $[Ca^{2+}]_i$ are prevented by the injection of a Ca^{2+} chelator into the postsynaptic neuron. The rise in $[Ca^{2+}]_i$ is believed to activate two protein kinases: *protein kinase C* and *calcium-calmodulin-dependent protein kinase II*. Pharmacological inhibition of either kinase blocks the induction of long-term potentiation.

Following the rise in postsynaptic $[Ca^{2+}]_i$ and the activation of the kinases, however, the molecular trail that leads to a potentiated synapse gets harder to follow. Current research suggests that this trail may actually branch (Figure 20.20). One path appears to lead toward an increased effectiveness of the postsynaptic AMPA receptors, perhaps by way of phosphorylation. The other branch may lead toward an increased presynaptic release of neurotransmitter. To make matters worse for students of LTP, in different labs using slightly different experimental conditions, the mechanisms for the expression of LTP may be predominantly postsynaptic, presynaptic, or a combination of both.

Even if LTP is only partly explained by a modification of presynaptic neurotransmitter release, this still raises a problem. Remember that the initial step in LTP induction is the detection by the NMDA receptor of (1) glutamate released into the synaptic cleft and (2) postsynaptic membrane depolarization. When these conditions are satisfied, then Ca^{2+} enters the *postsynaptic* cell. If this ultimately results in an increase in *presynaptic* transmitter release, then *there must be a signal that travels from the postsynaptic neuron to the presynaptic axon terminal*. Such a signal is called a **retrograde messenger**.

A number of potential retrograde messengers have been studied. However, to date there is still no agreement on the identity of the elusive retrograde messenger involved in LTP. Some really interesting ideas are being entertained, however. For example, one of the current hypotheses is that a *gas* is released by the postsynaptic neuron, which travels back to the presynaptic terminal. The gases in question are *nitric oxide* and *carbon monoxide*. Both can be produced in response to NMDA receptor activation, will easily pass through cell membranes, and have a very short chemical lifespan so they won't spread far from the stimulated synapse (a necessary requirement for explaining input specificity).

Long-Term Depression in CA1. We have seen that information can be stored either in decreases in synaptic effectiveness (cerebellar LTD) or in increases in synaptic effectiveness (hippocampal LTP). However, there is no reason these two types of synaptic plasticity cannot reside in one location. In fact, we have already learned that hippocampal responses can be modified in both directions. Remember the place cells in Chapter 19? When the spatial environment changed, *two* changes occurred in the place cells: They became *more* responsive to a new location and *less* responsive to an old location. These observations are supported by computer models indicating that memories are more efficiently stored by synaptic modifications that occur in both directions.

These considerations have led researchers to search for long-term depression in the hippocampus, and several forms have now been described. Of particular interest is a form of LTD that appears to represent a reversal of LTP in CA1. It was recently discovered that if the Schaffer collaterals are stimulated at low frequencies, input-specific LTD will occur. Surprisingly, LTD, like LTP, is triggered by postsynaptic Ca^{2+} entry through the NMDA receptor.

But how can the same signal, Ca^{2+} entry through the NMDA receptor, trigger *both* LTP *and* LTD? Recent research indicates that the lower elevations in $[Ca^{2+}]_i$ that occur during low-frequency stimulation may activate

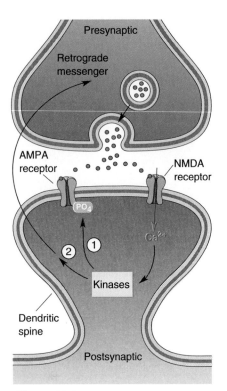

Figure 20.20
Routes for the expression of LTP in CA1. Ca^{2+} entering through the NMDA receptor activates protein kinases. This can cause LTP (1) by changing the effectiveness of postsynaptic AMPA receptors or (2) by generating a retrograde messenger that leads to a lasting increase in neurotransmitter release.

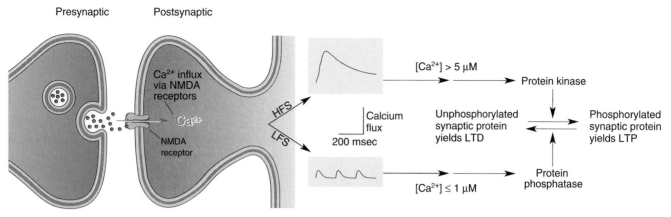

Figure 20.21
A model for how Ca²⁺ can trigger both LTP and LTD in the hippocampus. High-frequency stimulation (HFS) yields LTP by causing a large elevation of [Ca²⁺]. Low-frequency stimulation (LFS) yields LTD by causing a smaller elevation of [Ca²⁺]. (Source: Adapted from Bear and Malenka, 1994, Fig. 1.)

different enzymes from those activated by high $[Ca^{2+}]_i$. Specifically, while high $[Ca^{2+}]_i$ activates protein kinases, it appears that lower elevations in $[Ca^{2+}]_i$ may activate *protein phosphatases*. Recall that these enzymes pluck phosphate groups off phosphoproteins. Therefore, if LTP is putting them on (phosphate groups), LTD may be taking them off. Figure 20.21 summarizes how LTP and LTD might interact in CA1 neurons.

LTP, LTD, and Memory. LTP and LTD have attracted a lot of interest because it is believed that these mechanisms of synaptic plasticity contribute to the formation of declarative memories. For this to be true, however, these forms of plasticity must be demonstrated in the neocortex, not just the hippocampus, because neocortex is the likely site of declarative long-term memory. Fortunately, recent research indicates that the types of NMDA-receptor-dependent synaptic plasticity that have been characterized in the hippocampus also occur in the neocortex (Figure 20.22). It appears that synaptic plasticity throughout the cerebral cortex may be governed according to the same rules and may use the same mechanisms. And it's interesting that these mechanisms closely resemble those that have been identified to play a role in the development of cortical connections (see Chapter 18).

But what evidence exists to link LTP and LTD to memory? So far, all that we've described is a possible neural basis for a memory of having one's brain electrically stimulated! The most useful approach has been to see whether methods that block LTP also block learning and memory. A test of spatial memory in rats, called the **Morris water maze**, is normally used because proper performance on this task is known to depend on the hippocampus (Figure 20.23). In this test, a rat is placed in a pool filled with milky water. Submerged just below the surface in one location is a small platform that allows the rat to escape. A naive rat placed in the water will swim around until it bumps into the hidden platform, and then it will climb onto it. Normal rats quickly learn the spatial location of the platform and on subsequent trials waste no time swimming straight to it. But rats with bilateral hippocampal damage never seem to remember the location of the platform.

To assess the possible role of hippocampal LTP in the learning of spatial locations, University of Edinburgh psychologist Richard Morris (inventor of the maze) and his colleagues injected an NMDA-receptor blocker into the hippocampus of rats that were being trained in a water maze. Unlike normal animals, these rats never remembered the location of the submerged

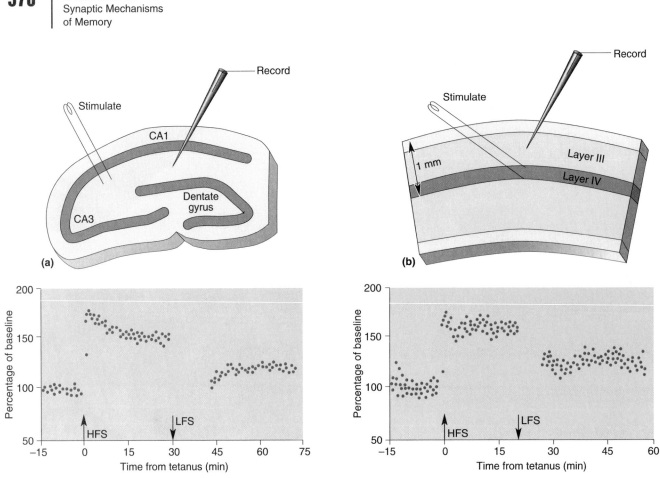

Figure 20.22
A comparison of LTP and LTD in (a) the hippocampus and (b) the neocortex. In both structures, the same types of conditioning stimulation yield the same types of synaptic plasticity. (Source: Adapted from Kirkwood et al., 1993, Fig. 1.)

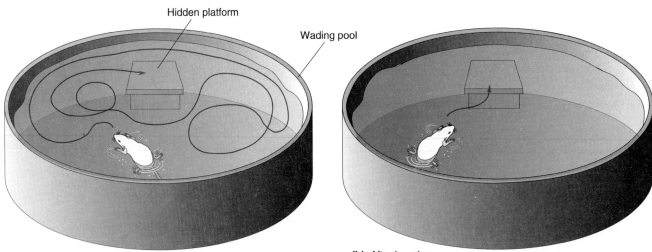

Figure 20.23
The Morris water maze. (a) The trajectory a rat might take to find a hidden platform the first time the rat is placed in the pool. **(b)** After training, the rat knows where the platform is and swims straight to it. Learning this task requires the activation of NMDA receptors in the hippocampus.

platform. Therefore, like LTP and LTD, spatial learning seems to require the activation of NMDA receptors in the hippocampus. Recent work has focused on isolating and genetically deleting the enzymes and receptors that are critical for synaptic plasticity to see if such deletions also affect learning and memory (Box 20.2). So far, it appears that learning and LTP do have many similar requirements.

OF MOLECULES AND MEMORIES

We have now seen in several different model systems that memory can result from experience-dependent alterations in synaptic transmission. In almost every case, synaptic transmission is modified as a result of changing the number of phosphate groups that are attached to proteins in the synaptic membrane. In the case of sensitization and classical conditioning in *Aplysia*, this occurs at certain potassium channels in the presynaptic membrane. In the case of both LTD in the cerebellum and LTP in the hippocampus, it is very likely that this occurs at the postsynaptic AMPA receptor.

Adding phosphate groups to a protein can change synaptic effectiveness and form a memory, but only as long as the phosphate groups remain attached to that protein. Phosphorylation as a long-term memory mechanism is problematic for two reasons:

1. Phosphorylation of a protein is not permanent. Over time, the phosphate groups are removed, thereby erasing the memory.
2. Protein molecules themselves are not permanent. Most proteins in the brain have a lifespan of less than 2 weeks and are undergoing a continual process of replacement. Memories tied to changes in individual protein molecules would not be expected to survive this rate of molecular turnover.

Thus, we need to consider the mechanisms that might convert what initially is a change in synaptic protein phosphorylation to a form that can last a lifetime.

Persistently Active Protein Kinases

Phosphorylation of synaptic proteins, and memory, could be maintained if the kinases—the enzymes that attach phosphate groups to proteins—were made to stay "on" all the time. Normally the kinases are tightly regulated and are "on" only in the presence of a second messenger. But what if learning changed these kinases so they no longer required the second messenger? The relevant synaptic proteins would remain phosphorylated all the time.

There is now evidence that some kinases can become independent of their second messengers. Let's consider as an example the changes that occur in one protein kinase during LTP in the hippocampus.

Protein Kinase C and LTP. Recall that Ca^{2+} entry into the postsynaptic cell and activation of protein kinase C are required for the induction of LTP in CA1. Recent research has shown that protein kinase C stays "on" long after $[Ca^{2+}]_i$ has fallen back to a low level. Furthermore, this persistent kinase activity is important for the maintenance of the potentiated synaptic response.

How might protein kinase C be switched permanently on? The answer requires some knowledge of how this enzyme is normally regulated (Figure 20.24). The molecule is built like a pocket knife, with two parts connected by a hinge. One part, the *catalytic* region, performs the phosphorylation reaction. The other part is called the *regulatory* region. Normally, in the absence of the appropriate second messenger, the knife is closed and the catalytic region is covered by the regulatory region. This keeps the enzyme "off." The normal action of the second messenger is to pry the knife open, but only as

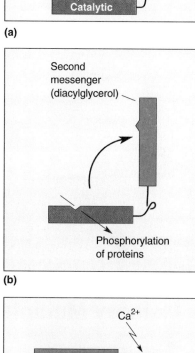

(a)

(b)

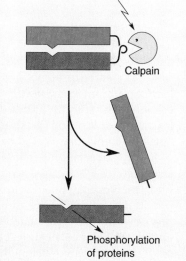

(c)

Figure 20.24
Regulation of protein kinase C. (a) The hingelike molecule is normally "off" when the catalytic region is covered by the regulatory region. **(b)** The hinge opens upon activation of the molecule by the appropriate second messenger, freeing the catalytic region to phosphorylate other proteins. **(c)** A large elevation of Ca^{2+} can activate the protease calpain, which breaks the hinge of the protein kinase C molecule and enables the catalytic region to stay "on" permanently.

long as the messenger is present. When the messenger is removed, the molecule usually snaps shut and the kinase turns "off" again. After LTP, however, it appears that the knife is actually broken at the hinge. This frees the catalytic region to float away into the cytoplasm and stay permanently "on."

How is the hinge of the protein kinase molecule broken? One hypothesis is that the large increase in $[Ca^{2+}]_i$ that occurs during an LTP-producing tetanus activates an enzyme called *calpain*. Calpain is a *protease*, an enzyme that breaks the peptide bonds that hold proteins together. Activation of calpain is known to break the hinge on the protein kinase C molecule. Interestingly, drugs that inhibit calpain can block LTP induction and some forms of spatial learning.

Protein Synthesis and Structural Plasticity

Persistent kinases are likely to contribute to the maintenance of a synaptic modification, but apparently only for a limited time (minutes to hours). After

---

Box 20.2 **OF SPECIAL INTEREST**

Memory Mutants

Of the several hundred thousand proteins manufactured by a neuron, some may be more important than others when it comes to learning. It is even possible that some proteins are *uniquely* involved in learning and memory. Needless to say, we could gain considerable insight into the molecular basis of learning and learning disorders, if such hypothetical "memory molecules" could be identified.

Recall that each protein molecule is the readout of a segment of DNA called a gene. One way to identify a "memory protein" is to delete genes one at a time and see if specific learning deficits result. This is precisely the strategy that Seymour Benzer, Yadin Dudai, and their colleagues at the California Institute of Technology tried using the fruit fly *Drosophila melanogaster*. *Drosophila* has long been a favorite species of geneticists, but one might reasonably question to what extent a fruit fly learns. Fortunately, *Drosophila* can perform the same tricks that *Aplysia* has mastered: habituation, sensitization, and classical conditioning. For example, fruit flies can learn that a particular odor predicts a shock. They demonstrate this memory after training by flying away when the odor is presented. The strategy is to produce mutant flies by exposing them to chemicals or X-rays. They are then bred and screened for behavioral deficits. The first mutant displaying a fairly specific learning deficit was described in 1976 and called *Dunce*. Other memory-deficient mutants were later described and given vegetable names such as *Rutabaga* and *Cabbage*. The next challenge was to

identify exactly which proteins had been deleted. It turned out that all three of these memory mutants lacked particular enzymes in cAMP-associated signaling pathways. Perhaps this isn't too surprising, considering the key role this second messenger system plays in procedural memory in *Aplysia*.

In *Drosophila* studies, the mutations were made at random, followed by extensive screening, first to find a learning deficit and then to determine exactly which gene was missing. Very recently, however, genetic engineering techniques have made it possible to make very specific deletions of known genes in mammals. Thus, for example, in 1992 Susumu Tonegawa, Alcino Silva, and their colleagues at MIT were able to isolate and delete one form (α) of the calcium-calmodulin-dependent protein kinase II in mice. Recall that experiments had already suggested that this enzyme is critical for the induction of long-term potentiation. Sure enough, these mice have a clear deficit in LTP in the hippocampus and the neocortex. And, when tested in the Morris water maze, they were found to have a severe memory deficit. Thus, these mice are memory mutants, just like their distant cousins *Dunce*, *Rutabaga*, and *Cabbage*.

Are we to conclude that the missing proteins in these mutants are the elusive "memory molecules"? No. All of these mutants show other behavioral deficits besides memory. We can only conclude at present that animals growing up without these proteins are unusually stupid. However, these studies do underscore the critical importance of specific second messenger pathways in translating a fleeting experience into a lasting memory.

that, a requirement for long-term memory is the synthesis of new protein. This protein is used to assemble new synapses.

Protein Synthesis and Memory. The possible role of new protein synthesis in memory has been investigated extensively since the introduction of drugs in the 1960s that selectively inhibit the assembly of protein from messenger RNA. Protein synthesis inhibitors can be injected into the brains of experimental animals as they are trained to perform a task, and deficits in learning and memory can be assessed. These studies reveal that if brain protein synthesis is inhibited at the time of training, the animals learn normally but fail to remember when tested days later (Figure 20.25). A deficit in long-term memory is also often observed if the inhibitors are injected shortly after training. However, the memories become increasingly resistant to inhibition of protein synthesis as the interval between the training and the injection of inhibitor is increased.

Very similar effects of protein synthesis inhibitors have been observed in the model systems we've been discussing. In *Aplysia*, repeated application of sensitizing stimuli causes a form of long-term memory that can last many days. The application of protein synthesis inhibitors at the time of training has no effect on sensitization measured hours later but completely blocks development of the long-term memory. Likewise, inhibition of protein synthesis at the time of a tetanus has no effect on the induction of LTP in the hippocampus. However, instead of lasting days to weeks, the synaptic potentiation gradually disappears over a few hours.

How are we to interpret these results? From what we have learned so far, memory formation appears initially to involve the rapid modification of existing synaptic proteins. These modifications, perhaps with the help of persistently active kinases, work against the factors that would erase our memory (such as molecular turnover). It's a losing battle, unless a new protein arrives at the synapse and converts the temporary change in the synapse to a more permanent one.

Structural Plasticity and Memory. How does the synapse make use of the timely arrival of a new protein? In the case of synapses that are strengthened after learning, it appears that this protein is used to construct brand new synapses. This has been best demonstrated in *Aplysia*, where it is possible to examine identified synapses made by the sensory neuron in the gill-withdrawal reflex. Long-term (but not short-term) sensitization causes a *doubling* of the number of synapses made by this neuron! Furthermore, this increase in synapse number slowly decays at the same rate as the long-term memory. Thus, in this system, long-term memory is associated with the formation of new synapses and forgetting is associated with a loss of these synapses.

Do similar structural changes occur in the mammalian nervous system after learning? This is a difficult problem to solve experimentally because of the complexity of the mammalian brain and the distributed nature of memory. One approach has been to compare brain structure in animals that have had ample opportunity to learn with that of animals that have had little chance to learn. Thus, putting a laboratory rat in a "complex" environment filled with toys and playmates (other rats) has been shown to increase the number of synapses per neuron in the occipital cortex by about 25%. Another approach has been to investigate structural changes after long-term potentiation in the hippocampus. It appears that LTP is also accompanied by an increased number of excitatory synapses.

Structural changes after learning need not be confined to *increases* in synapse number, however. For example, long-term habituation of the gill-withdrawal reflex in *Aplysia* is associated with a *decrease* (by one-third) in the number of synapses made by the sensory neuron. And recent research

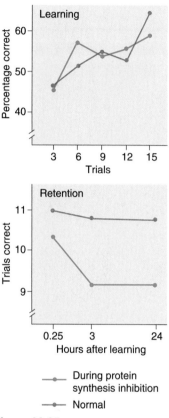

Figure 20.25
The effect of protein synthesis inhibitors on learning and memory. Typically, inhibitors of protein synthesis have no effect on the rate of learning a new task. However, long-term memory will fail to form if new protein synthesis is prevented during training. (Source: Adapted from Davis and Squire, 1984, Fig. 1.)

suggests that the conditioning stimulation that yields long-term depression in the cerebellar cortex also decreases the number of parallel fiber–Purkinje cell synapses.

It is important to recognize that there are limits to structural plasticity in the adult brain. As we discussed in Chapter 18, large changes in brain circuitry are generally confined to critical periods of early life. The growth and retraction of most axons in the adult CNS are restricted to no more than a few tens of micrometers. But it is now very clear that the end of a critical period does not necessarily signify an end to changes in the structure of axon terminals or the effectiveness of their synapses.

CONCLUDING REMARKS

Learning and memory can occur at synapses. Regardless of the species, brain location, and memory type, many of the underlying mechanisms appear to be universal. Events are represented first as changes in the electrical activity of the brain, then as second messenger molecules, and next as modifications of existing synaptic proteins. These temporary changes are converted to permanent ones—and long-term memory—by altering the structure of the synapse. In many forms of memory, this involves new protein synthesis and the assembly of new microcircuits. In other forms of memory, existing circuits may be disassembled. In either case, learning requires many of the same mechanisms that were used to refine brain circuitry during development.

One universal feature is the involvement of Ca^{2+} ions. Clearly, calcium does far more than build strong bones and teeth. Not only is it critical for neurotransmitter secretion and muscle contraction, it is involved in nearly every form of synaptic plasticity. Because it is a charge-carrying ion on the one hand and a potent second messenger substance on the other, Ca^{2+} has a unique ability to directly couple electrical activity with long-term changes in the brain.

Can basic neuroscience research take us from ions to intelligence? From calcium to cognition? If your performance on the next exam is a fair measure of intelligence and cognition, and if synaptic plasticity truly is the basis of declarative memory, it would appear that the answer is yes.

KEY TERMS

Procedural Learning
nonassociative learning
habituation
sensitization
associative learning
classical conditioning
instrumental conditioning

Vertebrate Models of Learning
cerebellar cortex
Purkinje cell
inferior olive
climbing fiber
mossy fiber
cerebellar granule cell

parallel fiber
Marr-Albus theory of motor learning
long-term depression (LTD)
input specificity
long-term potentiation (LTP)
dentate gyrus
Ammon's horn
CA3
CA1
perforant path
Schaffer collateral
tetanus
cooperativity
retrograde messenger
Morris water maze

1. Outline the steps involved in the presynaptic release of neurotransmitter. Why would the closure of a potassium channel in the presynaptic axon terminal change the amount of Ca^{2+} entering and change the amount of neurotransmitter released?

2. Rabbits can be classically conditioned to blink in response to a tone. This is accomplished by repeatedly pairing the tone with an air puff to the eye. Richard Thompson and his colleagues at Stanford University have made the following observations: Learning fails to occur, and the memory is wiped out, if the cerebellum is surgically removed; the air puff activates cells in the inferior olive; the tone activates cerebellar mossy fibers. Using your knowledge of synaptic plasticity in the cerebellum, propose a mechanism for classical conditioning in the rabbit.

3. In Figure 20.16, the mechanisms of classical conditioning in *Aplysia* and LTD in the cerebellar cortex are compared. Expand this comparison to include LTP in the hippocampus. What would events 1 and 2 be? How do these signals converge to affect a common intracellular process? How is the synaptic change expressed?

4. What property of the NMDA receptor makes it well suited to detect coincident pre- and postsynaptic activity? How could Ca^{2+} entering through the NMDA receptor possibly trigger both LTP and LTD in CA1 and neocortex?

5. People given electroconvulsive shocks to the head forget events that occurred shortly before the shock. One interpretation of this retrograde amnesia is that short-term memory and/or the consolidation into long-term memory of these events was disrupted by the shock. How might protein synthesis be involved in the conversion of short-term into long-term memory?

6. In H. M. and R. B. (Chapter 19), destruction of the hippocampus appears to have impaired the mechanism that "fixes" new memories in the neocortex. Propose a mechanism explaining why this might be true.

CHAPTER
21

Language and Attention

One of the exciting aspects of current neuroscience research is that complex functions of the brain are being studied that would have seemed out of reach not long ago. In this chapter, we'll look at two complex facets of human behavior, attention and the use of language. Language and attention have been studied for many years by linguists and psychologists, and now the underlying brain processes are being examined by neuroscientists.

The use of language is a uniquely human behavior that obviously has had a great effect on human society. Much of what we know about language processing is derived from studies of language deficits resulting from brain damage. Numerous different aspects of language can be selectively disrupted, including speech, comprehension, and naming, suggesting that language is processed in multiple, anatomically distinct, stages. We will use studies of brain lesions, brain stimulation, and brain imaging to investigate components of language that appear to occur at different locations in the brain.

Humans are constantly shifting attention between different stimuli in the environment, which appears to be a crucial aspect of the ability to exert extra mental energy on the performance of selected tasks. Attention has significant beneficial effects on behavioral performance, and we will outline some of these. Recent physiological experiments on attention give us a dramatic new view of brain function in which the receptive field properties of neurons change to suit the needs of ongoing behavioral tasks.

LANGUAGE AND THE BRAIN

Our use of language—the fact that we have a brain sophisticated enough for language—is one of the key features that distinguishes humans from other animals (Box 21.1). More than just sounds, language is a system by which sounds, symbols, and gestures are used for communication.

It is estimated that there are about 10,000 languages and dialects throughout the world. Languages differ in many ways, such as the order in which nouns and verbs are arranged. But despite differences in syntax, from the English spoken in the House of Lords to the dialects of aboriginal tribes around the globe, all languages convey the subtleties of human experience and emotion. Consider the fact that no mute tribe of people has ever been found, not even in the remotest corner of the world. Many believe that this is a consequence of the fact that the human brain has evolved special language-processing systems. These systems are present in newborn babies; if a child is in a normal language environment, he or she will inevitably learn to speak and understand language. Consistent with this idea, it has been found that children acquire language in a similar manner in all cultures.

The Discovery of Specialized Language Areas in the Brain

As in so many other areas of neuroscience, it has only been in the last century that our understanding of a clear relationship between language and the brain has emerged. Much of what we know about the importance of certain brain areas is derived from studies of aphasia. **Aphasia** is the partial or complete loss of language abilities following brain damage, often without the loss of cognitive faculties or the ability to move the muscles used in speech. During the Greek and Roman Empires, it was commonly thought that speech is controlled by the tongue and that speech disorders originate there rather than in the brain. If a head injury resulted in a loss of speech, the treatment involved special gargles or massage of the tongue. By the sixteenth century, it had been noted that a person could suffer speech impairment without paralysis of the tongue. However, in spite of this step forward, treatment still included such procedures as cutting the tongue, drawing blood, and applying leeches. Around 1770, Johann Gesner published a relatively modern theory of aphasia, describing it as the inability to associate images or abstract ideas with their expressive verbal symbols. He attributed this loss to brain damage resulting from disease. Gesner's definition makes the important observation that in aphasia, cognitive ability may remain intact, but some function specific to verbal expression is lost. Despite the incorrect association Franz Joseph Gall and later phrenologists made between skull shape and brain function (see Chapter 1), they, too, made an important observation about aphasia. They reasoned that cases of brain lesions, in which speech is lost but other mental faculties are retained, suggest that there is a specific region of the brain used for speech.

In 1825, based on many case studies, French physician Jean-Baptiste Bouillaud proposed that speech is specifically controlled by the frontal lobes. But it took another four decades before this idea was generally accepted. In 1861, Simon Alexandre Ernest Aubertin, Bouillaud's son-in-law, described the case of a man who had shot away his frontal bone in a failed suicide attempt. In treating this man, Aubertin discovered that if a spatula was pressed against the exposed frontal lobe while the man was speaking, his speech was immediately halted and did not resume until the pressure was released. He inferred that the pressure on the brain interfered with the normal function of a cortical area in the frontal lobe.

OF SPECIAL INTEREST

Is Language Unique to Humans?

To determine whether animals other than humans use language, one must be specific about what language is. If it is defined simply as communication, then most animals certainly do use language. But this definition really misses the point. Human language is a remarkably complex, flexible, and powerful *system* for communication that involves the creative use of words according to the rules of a systematic grammar. Is there anything similar in other animals? Actually, there are two questions: Do animals naturally use language? Can animals be taught human language?

Concerning the natural communication of animals, there is no evidence that they use a system even close to the complexity of humans. Chimpanzees and monkeys use a variety of sounds in social interactions to communicate alarm, warn adversaries, claim territory, and so on. However, compared to humans, nonhuman primates have a very limited system that uses sounds in stereotypical situations. Human language is a much more creative system. Limited only by the rules of grammar, it is effectively infinite. New word combinations and sentences are constantly being made, and the combinations have clear meaning according to the meaning of the individual words plus the rules used in arranging them. It is possible that further observations of nonhuman primates or dolphins might reveal that they combine their vocalizations in endlessly creative ways, but at present there is no evidence to support this.

The second question is whether animals can be taught human language. In the 1940s, several psychologists tried raising baby chimpanzees just like human children, including teaching them to speak. Despite extensive training, the chimps never learned to say more than a few words. This wasn't really a fair test though, because the position of the larynx in chimpanzees makes the animal unable to make the sounds of human speech. In more recent studies, animals have been trained to use either American Sign Language gestures or objects such as plastic pieces in various shapes to indicate words. Well-known cases include the chimpanzee named Washoe trained by Allen and Beatrice Gardner and the gorilla named Koko trained by Francine Patterson. Without question

THE FAR SIDE   By GARY LARSON

What we say to dogs

Okay, Ginger! I've had it! You stay out of the garbage! Understand, Ginger? Stay out of the garbage, or else!

what they hear

blah blah GINGER blah blah blah blah blah blah blah blah GINGER blah blah blah blah blah blah...

Figure A

the animals learn the meanings of gestures, but this proves nothing more than that the animals have good memories. The big question is whether they combine the symbols in original ways according to some grammatical system in order to express new things. The answer is mired in controversy. Proponents of animal language report instances in which animals use combinations of symbols to describe a novel situation. For instance, after learning the sign for water and the sign for bird, one chimpanzee put the signs together as "water bird" when it was near a swimming swan. This may or may not indicate language usage. Opponents of animal language have argued that animals never really learn sign language. The chimpanzee may simply have made separate gestures for water and bird. Part of the problem is that the animals frequently combine symbols in illogical or unintelligible ways and only occasionally come up with something like "water bird." Do the seemingly logical combinations of symbols reflect moments of language lucidity, or are they just coincidences that the experimenters choose to emphasize? It is not easy to know, and more controlled experiments are ongoing. For now, it is probably prudent to take claims of animal language with a grain of salt.

continued on page 580

Broca's Area and Wernicke's Area. Also in 1861, French neurologist Paul Broca had a patient who was almost entirely unable to speak. Broca invited Aubertin to examine the brain, and they concluded that this man also had a lesion in the frontal lobes. Perhaps because of a change in the scientific climate, Broca's case study appears to have swung popular opinion around to the idea that there is a language center in the brain. In 1863, Broca published a paper describing eight cases in which language was disturbed by damage to the frontal lobe in the left hemisphere. Additional similar cases, along with reports that speech is not disturbed by right hemispheric lesions, led Broca, in 1864, to propose that language expression is controlled by only one hemisphere, almost always the left. This view is supported by results of a more modern procedure for assessing the role of the two hemispheres in language, called the **Wada procedure**, in which a single hemisphere of the brain is anesthetized (Box 21.2). In most cases, anesthesia of the left hemisphere, but not the right, disrupts speech. If one hemisphere is thought to be more heavily involved in a particular task, it is said to be *dominant*. The region of the dominant left frontal lobe that Broca identified as being critical for articulate speech has come to be called **Broca's area** (Figure 21.1). Broca's work is of considerable significance because it was the first clear demonstration that brain functions can be anatomically localized.

In 1874, German neurologist Karl Wernicke reported that lesions in the left hemisphere, in a region distinct from Broca's area, also disrupt normal speech. Located on the superior surface of the temporal lobe between auditory cortex and the angular gyrus, this region is now commonly called **Wernicke's area** (see Figure 21.1). The nature of the aphasia Wernicke observed is different from that associated with damage to Broca's area. Having established that there are two language areas in the left hemisphere, Wernicke and others proceeded to construct maps of language processing in the brain. Interconnections between auditory cortex, Wernicke's area, Broca's area, and the muscles required for speech were hypothesized, and different types of language disabilities were attributed to damage in different parts of this system.

Box 21.1 | **OF SPECIAL INTEREST**

continued from page 579

In response to the question that titles this box, the current answer appears to be that language *is* unique to humans. To some people this conclusion may seem unlikely or even distasteful—how could *we* be the only species with language? But don't confuse language with intelligence. Language is not necessary for thought. Monkeys, as well as humans raised without any language, can do many things requiring abstract reasoning. Many creative people say they do some of their best thinking without words. Albert Einstein claimed that many of his ideas about relativity came from visually thinking of himself riding on a beam of light looking around at clocks and other objects. In any event, Fido probably does think, but he doesn't need language (Figure A). The other thing people find hard to believe is that language could have evolved in humans without more primitive languages in other species. In his wonderful book *The Language Instinct*, Steven Pinker compares the evolution argument for language to a similar argument about the elephant's trunk. The trunk is obviously a prominent feature and is essential to the elephant's lifestyle. Nonetheless, there simply aren't living animals with lesser forms of the trunk. It seems to be unique to the elephant, just as language seems to be unique to humans.

Box 21.2 | OF SPECIAL INTEREST

The Wada Procedure

A simple procedure used to study the function of a single cerebral hemisphere is the *Wada procedure*, developed by John Wada at the Montreal Neurological Institute. A fast-acting barbiturate, such as sodium amytal, is injected into the carotid artery on one side of the neck (Figure A). The drug is preferentially carried in the bloodstream to the hemisphere ipsilateral to the injection, where it acts as an anesthetic for about 10 minutes. The effects are sudden and dramatic. Within a matter of seconds the limbs on the side of the body contralateral to the injection become paralyzed, and somatic sensation is lost. By asking the patient to answer questions, one can assess his or her ability to speak. If the injected hemisphere is dominant for speech, the patient will be completely unable to talk until the anesthesia wears off. If the injected hemisphere is not dominant, the person will continue to speak throughout the procedure. Table A shows that in 96% of right-handed people and 70% of left-handed people, the left hemisphere is dominant for speech. Because 90% of all people are right-handed, this means that the left hemisphere is dominant for language in roughly 93% of people. While small but significant numbers of people with either handedness have a dominant right hemisphere, only in left-handers are bilateral representations of speech seen. In the Wada procedure this is indicated when an injection into either hemisphere has some disruptive effect on speech, though specifics of the disruption may be different for the two hemispheres.

Table A
Hemispheric Control of Speech in Relation to Handedness

| Handedness | Number of Cases | Speech Representation (%) | | |
		Left	Bilateral	Right
Right	140	96	0	4
Left	122	70	15	15

(Source: Rasmussen and Milner, 1977, Table 1.)

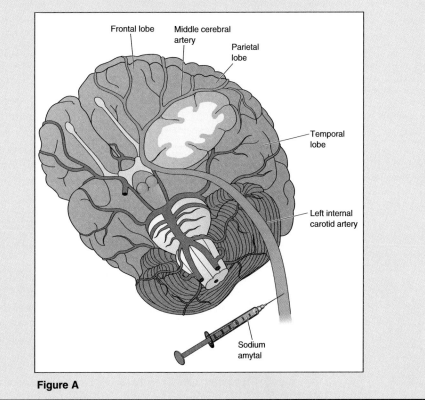

Figure A

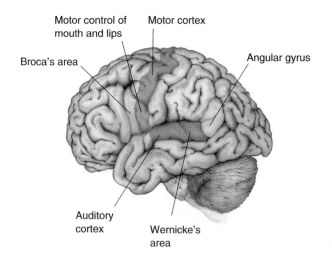

Figure 21.1
Key components of the language system in the left hemisphere. Broca's area in the frontal lobe lies next to the area that controls the mouth and lips in motor cortex. Wernicke's area on the superior surface of the temporal lobe is situated between auditory cortex and the angular gyrus.

Although the terms *Broca's area* and *Wernicke's area* are still commonly used, be aware that these terms are not without problems. The borders of these areas are not clearly defined, and they appear to be quite variable from one person to the next. We will also see that each area may be involved in more than one language function. However, that is a more recent finding that will only make sense after we look at the aphasias produced by damage to Broca's and Wernicke's areas.

Types of Aphasia and Their Causes

As anyone who has ever tried to learn a foreign language knows, language is complicated. Trying to understand how such a complex process works in the brain is a challenge. The oldest technique for studying the relationship between language and the brain involves correlating functional deficits with lesions in particular brain areas. The occurrence of distinct types of aphasia, as shown in Table 21.1, suggests that language is processed in several stages, at several locations in the brain.

Broca's Aphasia. The syndrome called **Broca's aphasia** is also known as motor or nonfluent aphasia because the person has difficulty speaking even though he or she can understand language heard or read. The case of David Ford is typical. Ford was a radio operator in the Coast Guard when, at age 39, he suffered a stroke. He remained an intelligent man, but he had little control over his right arm and leg (demonstrating that his lesion was in the left hemisphere). His speech was also abnormal, as the following discussion with psychologist Howard Gardner illustrates:

> I asked Mr. Ford about his work before he entered the hospital.
> "I'm a sig . . . no . . . man . . . uh, well, . . . again." These words were emitted slowly, and with great effort. The sounds were not clearly articulated; each syllable was uttered harshly, explosively, in a throaty voice. With practice, it was possible to understand him, but at first I encountered considerable difficulty in this.
> "Let me help you," I interjected. "You were a signal . . . "
> "A sig-nal man . . . right," Ford completed my phrase triumphantly.
> "Were you in the Coast Guard?"

"No, er, yes, yes . . . ship . . . Massachu . . . chusetts . . . Coastguard . . . years."
He raised his hands twice, indicating the number "nineteen."

. . .

"Could you tell me Mr. Ford, what you've been doing in the hospital?"
"Yes, sure. Me go, er, uh, P.T. nine o'cot, speech . . . two times . . . read . . . wr
. . . ripe, er, rike, er, write . . . practice . . . get-ting better."
"And have you been going home on weekends?"
"Why, yes . . . Thursday, er, er, er, no, er, Friday . . . Bar-ba-ra . . . wife . . . and,
oh, car . . . drive . . . purnpike . . . you know . . . rest and . . . tee-vee."
"Are you able to understand everything on television?"
"Oh, yes, yes . . . well . . . al-most." Ford grinned a bit. (Gardner, 1974, pp. 60–61.)

People with Broca's aphasia have difficulty saying anything, often paus-
ing to search for the right word. The inability to find words is called **anomia**
(literally meaning "no name"). Interestingly, there are certain "overlearned"
things Broca aphasics can say without much hesitation, such as the days of
the week and the Pledge of Allegiance. The hallmark of Broca's aphasia is a
telegraphic style of speech, in which mainly *content words* (nouns, verbs, and
adjectives carrying content specific to the sentence) are used. For instance,
when Mr. Ford was asked about being in the Coast Guard, his answer contained the words ship, Massachusetts, Coast Guard, and years, but little
else. Many *function words* (articles, pronouns, and conjunctions connecting
the parts of the sentence grammatically) are left out (there are no ifs, ands,
or buts). Also, verbs are frequently not conjugated. In the jargon of aphasia
deficits, the inability to construct grammatically correct sentences is called
agrammatism. There are some peculiar nuances to the agrammatical tendencies in Broca's aphasia. In the case of Ford, he could read and use the words
"bee" and "oar" but had difficulty with the more common words "be" and
"or." The problem isn't related to the sound but to whether or not the word
is a noun. In a similar vein, Broca aphasics have difficulty repeating things
spoken to them, though they tend to be better at familiar nouns such as
"book" and "nose." Sometimes they substitute incorrect sounds or words.
Ford says "purnpike" for "turnpike." These are called *paraphasic errors*.

Table 21.1
Characteristics of Different Types of Aphasia

Type of Aphasia	Site of Brain Damage	Comprehension	Speech	Impaired Repetition	Paraphasic Errors
Broca's	Motor association cortex of frontal lobe	Good	Nonfluent, agrammatical	Yes	Yes
Wernicke's	Posterior temporal lobe	Poor	Fluent, grammatical, meaningless	Yes	Yes
Conduction	Arcuate fasciculus	Good	Fluent, grammatical	Yes	Yes
Global	Portions of temporal and frontal lobes	Poor	Very little	Yes	—
Transcortical motor	Frontal lobe, anterior to Broca's area	Good	Nonfluent, agrammatical	No	Yes
Transcortical sensory	Cortex near the junction of temporal, parietal, and occipital lobes	Poor	Fluent, grammatical, meaningless	No	Yes
Anomic	Inferior temporal lobe	Good	Fluent, grammatical	No	No

In contrast to the speech difficulties in Broca's aphasia, comprehension is generally quite good. In the dialogue above, Ford seems to understand the questions asked of him, and, for the most part, he says he understands what he sees on television. In Gardner's study, Ford was able to answer simple questions such as, "Does a stone float on water?" However, more difficult questions demonstrated that he did not have completely normal comprehension abilities. If he was told, "The lion was killed by the tiger; which animal is dead?" or "Put the cup on top of the fork and place the knife inside the cup," he had difficulty understanding. This is probably related to the fact that he generally had trouble with the function words "by" in the first example and "on top of" in the second example.

Because the most obvious difficulty is in producing speech, Broca's aphasia is thought of as a language disturbance toward the motor end of the language system. Language is understood but not easily produced. While it is true that Broca aphasics are worse at speech than other types of aphasics, several things suggest that there is more to the syndrome. As pointed out above, comprehension is generally good, but comprehension deficits can be demonstrated by tricky questions. Also, patients sometimes have considerable anomia, suggesting that they have problems "finding" words as well as making the appropriate sounds.

Wernicke suggested that the area damaged in Broca's aphasia contains memories for the fine series of motor commands required for articulating word sounds. Because Broca's area is near the part of motor cortex that controls the mouth and lips, there is an appealing logic to this idea. Wernicke's theory is still held by some, but there are other ways of looking at the problem. For instance, the difference in the aphasic's ability to use content words and function words suggests that Broca's area and nearby cortex may be specifically involved in making grammatical sentences out of words. This might explain why Gardner's Mr. Ford could produce sounds such as "bee" and "oar" when they represent content words, but not when the sounds represent the function words "be" and "or."

Wernicke's Aphasia. When Wernicke noted that superior temporal lesions can lead to aphasia, the syndrome he observed was quite distinct from Broca's aphasia. Indeed, Wernicke suggested that aphasia is of two general types. In Broca's aphasia, speech is disturbed but comprehension is relatively intact. In **Wernicke's aphasia**, speech is fluent but comprehension is poor. (Although these definitions are oversimplified, they are useful for remembering the syndromes.)

Let's consider the case of Philip Gorgan, another patient studied by Gardner.

> "What brings you to the hospital?" I asked the 72-year-old retired butcher four weeks after his admission to the hospital.
>
> "Boy, I'm sweating, I'm awful nervous, you know, once in a while I get caught up, I can't mention the tarripoi, a month ago, quite a little, I've done a lot well, I impose a lot, while, on the other hand, you know what I mean, I have to run around, look it over, trebbin and all that sort of stuff."
>
> I attempted several times to break in, but was unable to do so against this relentlessly steady and rapid outflow. Finally, I put up my hand, rested it on Gorgan's shoulder, and was able to gain a moment's reprieve.
>
> "Thank you, Mr. Gorgan. I want to ask you a few—"
>
> "Oh sure, go ahead, any old think you want. If I could I would. Oh, I'm taking the word the wrong way to say, all of the barbers here whenever they stop you it's going around and around, if you know what I mean, that is tying and tying for repucer, repuceration, well, we were trying the best that we could while another time it was with the beds over there the same thing . . . " (Gardner, 1974, pp. 67–68.)

Clearly, Mr. Gorgan's speech is altogether different from that of Mr. Ford. Gorgan's speech is fluent, and he has no trouble using function words as well as content words. If you didn't understand English, the speech would probably sound normal because of its fluency. However, the content does not make much sense. It is a strange mixture of clarity and gibberish. Along with their far greater output of speech compared to Broca aphasics, Wernicke aphasics also make far more paraphasic errors. Gorgan would sometimes use the correct sounds but in an incorrect sequence, such as "plick" instead of "clip." Occasionally he would stumble around the correct sound or word, as when, in another conversation, he called a piece of paper "piece of handkerchief, pauper, hand pepper, piece of hand paper." Interestingly, he would sometimes use an incorrect word but one categorically similar to the correct word, such as "knee" instead of "elbow."

Because of the stream of unintelligible speech, it is difficult to assess with speech alone whether the Wernicke aphasic comprehends what he hears or reads. Indeed, one of the intriguing things about Wernicke aphasics is that they frequently appear undisturbed by the sound of their own speech and the speech of others, even though they probably don't understand either. Comprehension is usually assessed by asking the patient to respond in a nonverbal manner. For instance, the patient could be asked to put object A on top of object B. Questions and commands of this sort quickly lead to the conclusion that Wernicke aphasics do not understand most instructions. They are completely unable to comprehend questions of the sort understood by Broca aphasics. When Gorgan was presented with commands written on cards ("Wave goodbye," "Pretend to brush your teeth"), he was often able to read the words but never acted as if he understood what the words meant.

Gorgan's strange speech was mirrored in his writing and his ability to play music. When Gardner gave him a pencil, he spontaneously took it and wrote "Philip Gorgan. This is a very good beautifyl day is a good day, when the wether has been for a very long time in this part of the campaning. Then we want on a ride and over to for it culd be first time . . . " (p. 71). Likewise, when he sang or played the piano, pieces of the appropriate song were intermixed with musical gibberish, and he had a difficult time ending, just as in his speech.

Insight about the possible function of Wernicke's area is provided by its location on the superior temporal gyrus near primary auditory cortex. Wernicke's area may play a critical role in relating incoming sounds to their meaning. In other words, it is an area specialized for storing memories of the sounds that make up words. It has been suggested that Wernicke's area is a high-order area for sound recognition in the same sense that inferior temporal cortex is thought to be a high-order area for visual recognition. (Recall the psychic blindness observed in monkeys with temporal lesions in Chapter 16.) A sound recognition deficit would explain why Wernicke aphasics don't comprehend speech well. However, there must be more to Wernicke's area to account for the odd speech patterns. Speech in Wernicke's aphasia suggests that Broca's area and the system responsible for speech production are running without control over content. The speech zooms along, swerving in every direction like a car with a sleepy driver at the wheel.

Aphasia and the Wernicke-Geschwind Model. Shortly after making his observations on what came to be called Wernicke's aphasia, Wernicke proposed a model for language processing in the brain. Later extended by Norman Geschwind at Boston University, this model is known as the **Wernicke-Geschwind model.** The key elements in the system are Broca's area, Wernicke's area, the *arcuate fasciculus*, a bundle of axons connecting the two cortical areas, and the *angular gyrus*. The model also includes sensory and motor areas involved in receiving and producing language. To understand what the model entails, we'll consider the performance of two tasks.

The first task is the repetition of spoken words (Figure 21.2). When sounds of incoming speech reach the ear, the auditory system processes the sounds, and neural signals eventually reach auditory cortex. According to the model, the sounds are not understood as meaningful words until they are processed in Wernicke's area. In order to repeat the words, word-based signals are passed to Broca's area from Wernicke's area via the arcuate fasciculus. In Broca's area, the words are converted to a code for the muscular movements required for speech. Output from Broca's area is sent to the nearby motor cortical areas responsible for moving the lips, tongue, larynx, and so on.

The second task we'll consider is reading written text aloud (Figure 21.3). In this case, the incoming information is processed by the visual system through striate cortex and higher-order visual cortical areas. The visual signals are then passed to the angular gyrus at the junction of the occipital, pari-

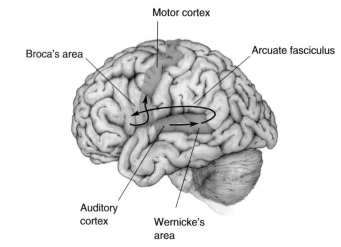

Figure 21.2
Repeating a spoken word according to the Wernicke-Geschwind model.

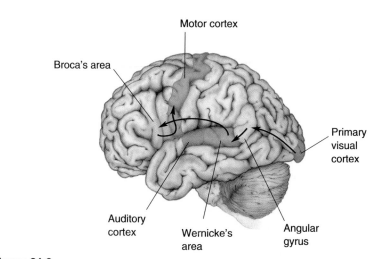

Figure 21.3
Repeating a written word according to the Wernicke-Geschwind model.

etal, and temporal lobes. In the cortex of the angular gyrus it is assumed that a transformation occurs so that the output evokes the same pattern of activity in Wernicke's area as would be evoked if the words were spoken rather than written. From this point, the processing follows the same progression as in the first example: Wernicke's area to Broca's area to motor cortex.

This model offers simple explanations for key elements of Broca's and Wernicke's aphasia. A lesion in Broca's area seriously interferes with speech production because the proper signals can no longer be sent to motor cortex. On the other hand, comprehension is relatively intact because Wernicke's area is undisturbed. A lesion in Wernicke's area produces great comprehension problems because this is the site of the transformation of sounds into words. The ability to speak is unaffected because Broca's area is still able to drive the muscles required for speech.

The Wernicke-Geschwind model has several errors and oversimplifications. For example, words read do not have to be transformed into a pseudo-auditory response, as suggested in the reading task described above. In fact, visual information can reach Broca's area from visual cortex without making a stop at the angular gyrus. One of the dangers inherent in any model is overstating the significance of a given cortical area for a particular function. It has recently been found that the severity of Broca's and Wernicke's aphasias depends on how much cortex is damaged beyond the limits of Broca's and Wernicke's areas. Also, aphasia is influenced by damage to subcortical structures such as the thalamus and caudate nucleus, which are not in the model. In surgical cases where parts of cortex are removed, the resulting language deficits are usually milder than the deficits resulting from stroke, which affects both cortical and subcortical structures.

Another important factor is that there is often significant recovery of language function after a stroke, and it appears that other cortical areas can sometimes compensate for what is lost. As in many neurological syndromes, young children recover extremely well, but even adults, especially left-handers, can show good recovery of function.

A final problem with the Wernicke-Geschwind model is that most aphasias involve both comprehension and speech deficits. Mr. Ford, with Broca's aphasia, had good comprehension but was confused by complex questions. Conversely, Mr. Gorgan, with Wernicke's aphasia, had several speech abnormalities in addition to a severe lack of comprehension. Therefore, in cortical processing, the sharp functional distinctions between regions as implied by the model do not exist. In response to these shortcomings, more complex models have been proposed, but they retain some of the features of the Wernicke-Geschwind model (Figure 21.4). Despite the problems with the Wernicke-Geschwind model, it continues to be of clinical use because of its simplicity and approximate validity.

Conduction Aphasia. The value of a model is not only its ability to account for previous observations but also its ability to predict. Wernicke showed that his language-processing model could predict that a form of aphasia would result from a lesion that disconnects Wernicke's area from Broca's area but leaves both areas themselves intact. In the Wernicke-Geschwind model, this could be accomplished by a lesion in the fibers composing the arcuate fasciculus. In reality, such disconnection lesions usually involve damage to parietal cortex in addition to the arcuate fasciculus, but Broca's and Wernicke's areas are spared.

Sure enough, aphasia from such lesions was demonstrated and is now known as **conduction aphasia**. As the model predicts, based on the preservation of Broca's and Wernicke's areas, comprehension is good and speech is fluent. The patient is typically able to express himself through speech without difficulty. The deficit that chiefly characterizes conduction aphasia is difficulty in repeating words. In response to hearing a few words, the patient may attempt to repeat what was said, but the repetition will substitute words, omit words, and include paraphasic errors. Repetition is usually best with nouns and short common expressions, but it may fail entirely if the spoken words are function words, polysyllabic words, or nonsense sounds. Interestingly, a person with conduction aphasia will comprehend sentences he reads aloud, even though what is said aloud contains many paraphasic errors. This is consistent with the idea that comprehension is good and the deficit occurs between regions involved in comprehension and speech.

One of the sad yet fascinating things about aphasia is the diversity of syndromes that occur following strokes. While the syndromes challenge any language model, each one offers a clue to our understanding of language processing. Characteristics of a few other aphasias are listed in Table 21.1.

Aphasia in Bilinguals and the Deaf. Cases of aphasia in bilingual people and deaf people provide fascinating insight about language processing in the brain. Suppose a person knows two languages before she has a stroke. Does the stroke produce aphasia in one language and not the other, or do both languages suffer equally? The answer depends on several factors, including the order in which the languages were learned, the fluency achieved in each language, and how recently the language was used. The consequences of a stroke are not always predictable, but language tends to be relatively more preserved in the language learned more fluently and earlier in life. If the person learned two languages at the same time to equiva-

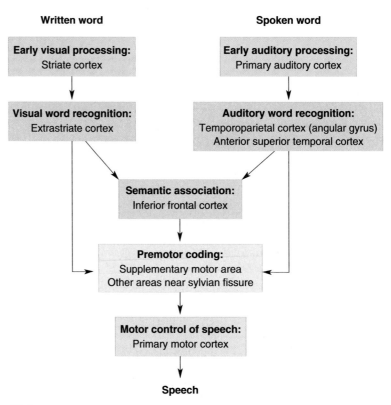

Figure 21.4
A model of language processing. Stages in the processes of repeating a written or spoken word are indicated. Below each stage is the cortical location where task-specific activity was observed with PET imaging. (Source: Adapted from Petersen et al., 1988, Fig. 3.)

Me

Index finger
points to and
touches chest.

Cat

Draw out two
whiskers with
thumb and
index finger.

Figure 21.5
"Speaking" in American Sign Language.

lent levels of fluency, a lesion will probably produce similar deficits in both
languages. If the languages were learned at different times in life, it is likely that one language will be affected more than the other. The implication is
that the second language may make use of different, though overlapping,
populations of neurons than the first.

The study of language deficits in those who are deaf and/or know sign
language suggests that there is some universality to language processing in
the brain. American Sign Language uses hand gestures to express all the
ideas and emotions most of us convey with spoken language (Figure 21.5).
Left hemispheric lesions in people who use sign language appear to cause
language deficits similar to those seen in verbal aphasics. There are cases
analogous to Broca's aphasia in which comprehension is good but the ability
to "speak" through sign language is severely impaired. Importantly, the ability to move the hands is not impaired (i.e., the problem is not with motor
control). Rather, the deficit is specific to the use of hand movements for the
expression of language. There are also sign language versions of Wernicke's
aphasia in which the patient signs fluently but with many mistakes, while
having difficulty comprehending the signing of others. In one unusual case,
a hearing man who was the child of deaf parents learned both sign language
and verbal language. A left hemispheric stroke initially gave him global
aphasia, but his condition significantly improved with time. The important
observation is that his verbal and sign languages recovered together, as if
overlapping brain areas were used. While there do appear to be aphasias in
those using sign language analogous to speech aphasias, there is also evidence that signing aphasia and speaking aphasia can be produced by left
hemispheric lesions in somewhat different locations.

Lessons Learned from Split-Brain Studies

We have seen that damage to certain parts of the brain leads to a variety
of aphasias. As the early work of Broca indicated, language is usually not
handled equally by the two cerebral hemispheres. Some of the most valuable and fascinating findings on the language differences of the two hemispheres come from **split-brain studies** in which the hemispheres are surgically disconnected. Communication between the cerebral hemispheres is

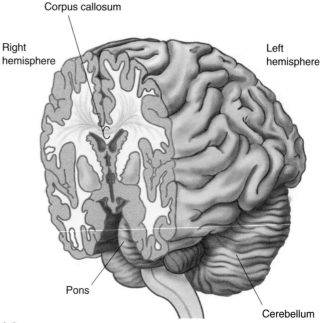

Corpus callosum

Right
hemisphere

Left
hemisphere

Pons

Cerebellum

Figure 21.6
The corpus callosum. The corpus callosum is the largest bundle of axons providing communication between the cerebral hemispheres.

served by several bundles of axons known as commissures. Recall from Chapter 7 that the largest of these is the great cerebral commissure, also called the **corpus callosum** (Figure 21.6). The corpus callosum consists of about 200 million axons crossing between the hemispheres. Surely such a huge bundle of fibers must be of great importance. Surprisingly, until about 1950, it had proven difficult to demonstrate any important role of the corpus callosum.

In split-brain procedures, the skull is opened and the axons making up the corpus callosum are severed (Figure 21.7). The hemispheres may retain some communication via the brain stem or smaller commissures (if they aren't also severed), but most of the intercerebral communication is lost. In the 1950s, Roger Sperry and his colleagues at the University of Chicago and later at the California Institute of Technology performed a series of experiments, using split-brain animals, to explore the function of the corpus callosum and the separated cerebral hemispheres. (You may recall Sperry from Box 18.2, describing his studies of the formation of retinotopic maps in the frog's tectum.) Sperry's group confirmed earlier reports that cutting the corpus callosum in a cat or monkey has no noticeable effect on the animal's behavior. Temperament is unchanged, and the animal appears to be normal in coordination, reaction to stimuli, and ability to learn. However, in cleverly devised experiments, Sperry's group showed that the animals sometimes acted as if they had two separate brains.

Language Processing in Split-Brain Humans. Because split-brain monkeys did not appear to have any major deficits, surgeons felt they were justified in cutting the corpus callosum as a last resort in treating certain types of severe epilepsy in humans. They hoped to prevent the spread of epileptic activity from one hemisphere to the other. It may seem questionable to cut 200 million axons on the assumption that they are not very important, but the surgery is often beneficial in restoring a seizure-free life. Michael Gazzaniga at New York University, who initially worked with Sperry, studied a number of these people. His techniques were modifications of those used with experimental animals.

One key methodological feature of studying split-brain humans involves careful control that visual stimuli are presented to only one cerebral hemisphere. Gazzaniga did this by taking advantage of the fact that objects to the left of the point of fixation are seen by only the right hemisphere, and objects to the right are seen by only the left hemisphere, as long as the eyes can't move to bring the image onto the fovea (Figure 21.8). Pictures or words were flashed on for a fraction of a second using a device with a cameralike shutter. Because the images were presented for a shorter time than that required to move the eyes, the images were seen only by one hemisphere.

Although split-brain humans are normal in most every way, there is a striking asymmetry in their ability to verbalize answers to questions posed separately to the two hemispheres. For instance, numbers, words, and pictures visually presented only in the right visual field are repeated or described with no difficulty because the left hemisphere is dominant for language. Likewise, objects that can be manipulated only by the right hand (out of view of both eyes) can be described. These findings would be entirely unremarkable except for the fact that such simple verbal descriptions of sensory input are impossible for the right hemisphere.

If an image is shown only in the left visual field or an object is felt only by the left hand, the person will be unable to describe it and will usually say

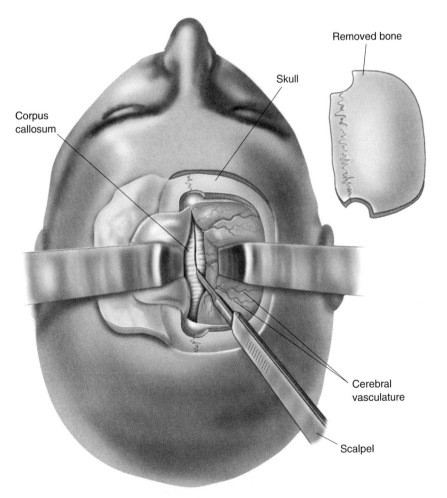

Figure 21.7
Split-brain surgery in a human. To reach the corpus callosum, a portion of the skull is removed and the cerebral hemispheres are retracted.

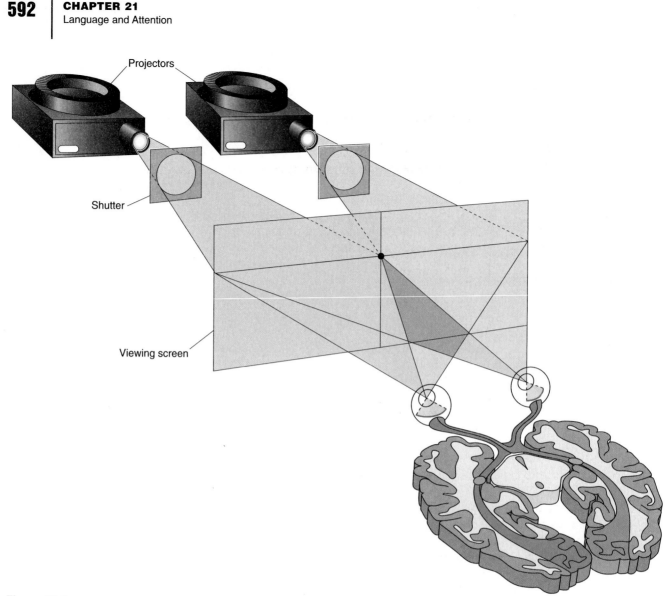

Projectors

Shutter

Viewing screen

Figure 21.8
Visual stimulation of one hemisphere in humans. A visual stimulus is briefly flashed to the left or right visual field by means of a shutter. The display time is shorter than the time needed to generate a saccadic eye movement, thus assuring that only one hemisphere sees the stimulus.

that nothing is there (Figure 21.9). An object could be covertly placed in a patient's left hand and there would be no verbal indication of even noticing. This absence of response is a consequence (and demonstration) of the fact that speech in most people is controlled by the left hemisphere. If you think about the implications for split-brain people, you'll realize that they have an unusual existence. Following their surgery, they are unable to describe anything to the left of their visual fixation point: the left side of a person's face, the left side of the room, and so on. What is startling is that this doesn't seem to disturb the patients.

While there is a dramatic inability of the right hemisphere to speak, this does not mean it knows nothing of language. It can be demonstrated that the right hemisphere can read and understand numbers, letters, and short words as long as the required response is nonverbal. In one experiment, the right hemisphere is shown a word that is a noun. As already mentioned, the person will say he sees nothing. Of course, that's the talkative left hemisphere speaking, and it *didn't* see anything. But if he is urged to use his left

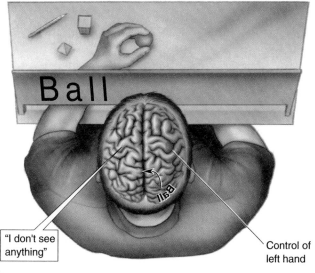

Figure 21.9
Demonstrating language comprehension in the right hemisphere. If a split-brain person sees a word in the left visual field, he will say he sees nothing. This is because the left hemisphere, which usually controls speech, did not see the word, and the right hemisphere, which saw the word, cannot speak. However, the left hand, which is controlled by the right hemisphere, can pick out the object corresponding to the word by touch alone.

hand to select a card containing a picture corresponding to the word he saw, or pick out an object by touch, he can do it (see Figure 21.9). The right hemisphere isn't able to do this with more complex words or sentences, but the results clearly imply that the right hemisphere does have language comprehension.

There is also evidence suggesting that the right hemisphere understands complex pictures despite its inability to say so. In one experiment, a subject was shown a series of pictures in her left visual field, and at one point a nude photo appeared in the series. When the experimenter asked what she saw, she said nothing, but then she began to laugh. She told the experimenter that she didn't know what was funny, but that perhaps it was the machine used in the experiment.

As an aside, there were some things that the right hemisphere appeared to be better at than the left hemisphere. For instance, even though the split-brain patients were right-handed and thus their left hemispheres were much more practiced at drawing, the left hand controlled by the right hemisphere was better at drawing or copying figures containing three-dimensional perspective. The patients were also better at solving complex puzzles with their left hand. It has also been reported that the right hemisphere is somewhat better at perceiving nuances in sound.

In some of the split-brain studies, conflicting behaviors were initiated by the two hemispheres, apparently because they were thinking differently. In one task a patient was asked to arrange a group of blocks to match a pattern on a small card. He was told to do this using only his right hand (left hemisphere), which is not generally good at this type of task. As the right hand struggled to arrange the blocks, the left hand (right hemisphere), which knew how to do it, reached in to take over. Only the restraint of the experimenter kept the left hand from pushing the right one out of the way to solve the puzzle. Another patient studied by Gazzaniga would sometimes find himself pulling his pants down with one hand and pulling them up with the other. These bizarre behaviors make a strong case that there are two independent brains controlling the two sides of the body.

The results of these split-brain studies demonstrate that the two hemispheres can function as independent brains and that they have different language abilities. Although the left hemisphere is usually dominant for language, the right hemisphere has significant skills in comprehending language. It is important to keep in mind that the split-brain studies test the ability of each hemisphere to perform on its own. Presumably in the intact brain, the callosum allows for synergistic interactions between the hemispheres for language and other functions.

Anatomical Asymmetry and Language

In the nineteenth century there were reports of anatomical differences between the two hemispheres. For example, it was noted that the left sylvian fissure is longer and less steep than the right (Figure 21.10). However, as recently as the 1960s, there was considerable doubt about whether there are significant cortical asymmetries. Because of the strikingly asymmetrical control of speech demonstrated by the Wada procedure, it would be interesting to know if there is an anatomical difference between the hemispheres. Some of the first good quantitative data demonstrating hemispheric differences came from the work of Geschwind and his colleague Walter Levitsky. Initial observations were made on postmortem brains, but more recently, **magnetic resonance imaging (MRI)** has allowed examination of the living brain (Box 21.3).

The most significant difference seen was in a region called the **planum temporale** on the superior surface of the temporal lobe (Figure 21.11). Based on measurements of 100 brains, Geschwind and Levitsky found that in about 65% of the brains, the left planum temporale was larger than the right, whereas in only about 10%, the right was larger. In some instances, the left area was more than five times larger than the right. It has been suggested that the left planum temporale is larger because this hemisphere is dominant for speech. However, this hypothesis has not been proven. Interestingly, the asymmetry in this area is seen even in the human fetus, suggesting that it is not a developmental consequence of the use of the left hemisphere for speech. Indeed, apes also tend to have a larger left planum temporale. This suggests the possibility that, if the planum temporale is related to language, speech became dominant in the left hemisphere because of a preexisting size difference.

It may have occurred to you that a functional human asymmetry more obvious than language is handedness. More than 90% of humans are right-

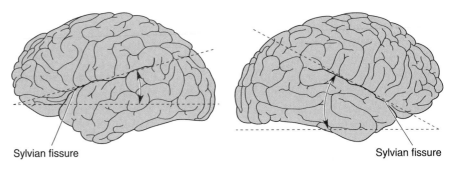

Sylvian fissure

Left hemisphere

Sylvian fissure

Right hemisphere

Figure 21.10

Asymmetry of the sylvian fissure. In most right-handed people, the sylvian fissure in the left hemisphere is longer and runs at a shallower angle than the fissure in the right hemisphere. (Adapted from Geschwind, 1979, p. 192.)

Magnetic Resonance Imaging

Magnetic resonance imaging (MRI) is a general technique used to determine the amount of certain atoms at different locations in the body. It has become an important tool in neuroscience because it can be used noninvasively to obtain a detailed picture of the nervous system, particularly the brain.

In the most common form of MRI, the presence of hydrogen atoms is quantified, for instance in water or fat in the brain. An important fact from physics is that when a hydrogen atom is put in a magnetic field, its nucleus (which consists of a single proton) can exist in either of two states: a high-energy state or a low-energy state. In the brain there are many protons in each of these states, but there are more in the low energy state.

The key trick used in MRI is to make the protons jump from one state to the other. To do this, energy is added to the protons by passing an electromagnetic wave (i.e., a radio signal) through the head while it is positioned between the poles of a large magnet. When the radio signal is set at just the right frequency, the protons in the low-energy state will absorb energy from the signal and hop to the higher-energy state. The frequency at which the protons absorb energy is called the *resonant frequency* (hence the name magnetic resonance). When the radio signal is turned off, some of the protons fall back down to the lower state, and in doing so, they emit a radio signal of their own at a particular frequency. The signal emitted by the protons can be picked up with a radio receiver. The stronger the signal, the more hydrogen there must have been between the poles of the magnet.

If we used the procedure described above, we'd simply get a measurement of the total amount of hydrogen in the head. However, it's possible to measure the amount of hydrogen at a fine spatial scale.

To do this we take advantage of the fact that the frequency at which the protons emit energy is proportional to the size of the magnetic field. In the MRI machines used in hospitals, the magnetic fields vary from one side of the magnet to the other. This gives a spatial code to the radio waves emitted by the protons; high-frequency signals come from hydrogen atoms near the strong side of the magnet, and low-frequency signals come from the weak side of the magnet. The last step in the MRI process is to orient the gradient of the magnet at many different angles relative to the head and measure the amount of hydrogen. A sophisticated computer program is then used to make a single image from all of the measurements, showing a picture of the distribution of hydrogen atoms in the head. Figure A shows a lateral view of the brain in a living human, obtained by MRI. In Figure B a horizontal slice through the brain reveals the planum temporale. Notice that in the images the white and gray matter can easily be distinguished. This difference makes it possible to see the effects of demyelinating diseases on white matter in the brain. MRI images also reveal lesions in the brain because tumors and inflammation generally increase the amount of extracellular water.

In the past decade, methods have been developed for monitoring blood flow and oxygenation in different portions of the brain using MRI. These methods provide a means of noninvasively measuring neural activity in the brain because increased brain activity is correlated with increases in blood flow and oxygenation. This technique is called functional MRI and it is an important tool for examining normal and abnormal processing in the brain.

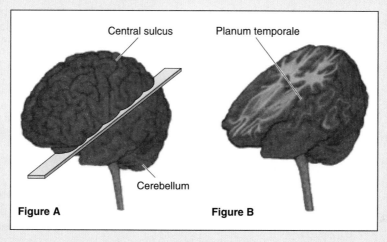

Figure A

Central sulcus

Cerebellum

Figure B

Planum temporale

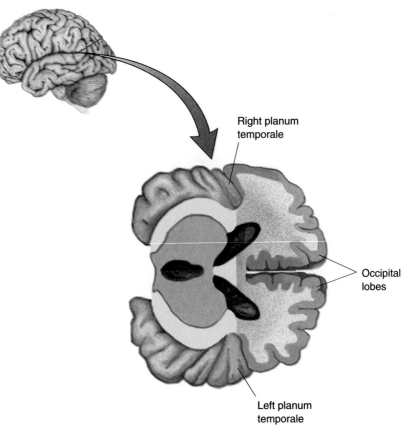

Figure 21.11
Asymmetry of the planum temporale. This region on the superior temporal lobe is usually significantly larger in the left hemisphere. (Source: Adapted from Geschwind and Levitsky, 1968, Fig. 1.)

handed and usually relatively clumsy with their left hand. This implies that in some way the left hemisphere is specialized for fine motor control. Is this related to the left hemispheric dominance for language? The answer is not known, but it is interesting that humans are different from nonhuman primates in regard to handedness as well as language. While animals of many species show a consistent preference for using one hand over the other, there are typically equal numbers of left-handers and right-handers.

Language Studies Using Brain Stimulation and PET Imaging

Until recently, the main way language processing in the brain could be examined was by correlating language deficits with postmortem analysis of brain damage. However, there are other techniques that allow language function to be studied in the brains of living humans. Electrical brain stimulation and PET imaging are two of these techniques.

Effects of Brain Stimulation on Language. At several points in this book, we have discussed the electrical brain stimulation studies of Wilder Penfield. Without general anesthesia, patients were able to report the effects of stimulation at different cortical sites. In the course of these experiments, Penfield noted that stimulation at certain locations affected speech. These effects fell into three main categories: vocalizations, speech arrest, and speech difficulties similar to aphasia.

Stimulation of motor cortex in the area that controls the mouth and lips caused immediate speech arrest (Figure 21.12). Such a response is logical because the activated muscles sometimes pulled the mouth to one side or clenched the jaw shut. Stimulation of motor cortex occasionally evoked cries or rhythmic vocalizations. It is important to note that these effects were found with electrical stimulation of motor cortex on either side of the brain. Penfield found three other areas where electrical stimulation interfered with speech, but these were only in the dominant left hemisphere. One of these areas appeared to correspond to Broca's area. If this area was stimulated while a person was speaking, speech was either stopped entirely (with strong stimulation), or there was hesitation in the speech (with weaker stimulation). Some patients were unable to name objects they could name before and after the brain stimulation. Occasionally, they would substitute an incorrect word. They apparently experienced a mild transient form of anomia. Confusion about words and speech arrest also resulted in some patients by stimulation at two other sites, one in the posterior parietal lobe near the sylvian fissure and the other in the temporal lobe. These latter two areas were in the vicinity of the arcuate fasciculus and Wernicke's area, although they did not align perfectly with those areas.

It is somewhat reassuring to see that electrical stimulation selectively affects speech in the brain areas roughly corresponding to those responsible for aphasia. However, the consequences of stimulation are surprisingly variable between nearby cortical sites and between subjects. In studies similar to those of Penfield, neurosurgeon George Ojemann, at the University of Washington, has found that the effects of stimulation can sometimes be quite specific. For example, stimulation of small parts of cortex at different locations can interfere with naming, reading, or repetition of facial movements (Figure 21.13). What makes these findings all the more intriguing is that different results are sometimes obtained at nearby stimulation sites, and similar results can be obtained at distant sites. These results may indicate that the language areas in the brain are much more complex than is implied by the Wernicke-Geschwind model. Areas used in language are also

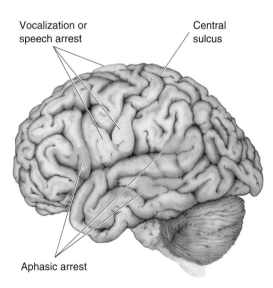

Figure 21.12
Sites where electrical brain stimulation affects language. Stimulation of motor cortex causes vocalizations or speech arrest by activating facial muscles. At other sites, stimulation causes an aphasic arrest in which language is agrammatical or anomia is observed. (Source: Adapted from Penfield and Rasmussen, 1950, Fig. 56.)

more extensive than simply Broca's and Wernicke's areas, as they have been found to include other cortical areas, as well as parts of the thalamus and striatum. Within Broca's area and Wernicke's area there may be specialized regions, possibly on the scale of functional columns in somatosensory cortex or ocular dominance columns in visual cortex. It appears that the large language areas identified on the basis of aphasic syndromes may well encompass a good deal of finer structure.

PET Imaging of Language Processing. With the advent of modern imaging techniques, it has become possible to observe normal language processing. With **positron emission tomography (PET)**, the level of neural activity in different parts of the brain is inferred from regional blood flow (Box 21.4). In one study of language processing, researchers used PET imaging to observe the differences in brain activity between the sensory responses to words and the production of speech. They began by measuring cerebral blood flow with the subject at rest. They then had the person either listen to

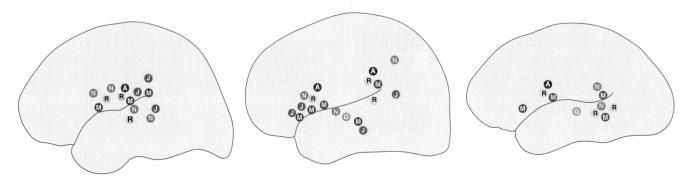

Figure 21.13
Effects of brain stimulation in three patients being treated for epilepsy. Patients were awake, and difficulties in speaking or reading were noted. N = naming difficulty with intact speech (anomia), A = arrest of speech, G = grammatical errors, J = jargon (fluent speech with frequent errors), R = failure to read, M = facial movement errors. (Source: Adapted from Ojemann and Mateer, 1979, Fig. 1.)

Box 21.4 **BRAIN FOOD**

PET Imaging of Brain Activity

If one had the ideal neurophysiological recording technique, it would be possible to record the activity of individual neurons in the brain completely noninvasively. Unfortunately, there isn't such a method—at least not yet. Nonetheless, positron emission tomography (PET) provides beautiful images of brain activity with a spatial resolution of about 5–10 mm.

The basic procedure is very simple. A radioactive solution is introduced into the bloodstream, and the radioactivity is located in the brain where the blood takes it. The solution is radioactive because it contains atoms that emit positrons (positively charged electrons) when the atoms disintegrate. Wherever

the blood goes, positrons will be emitted. The positrons interact with electrons to produce photons of light. The locations of the positron emitting atoms are found by detectors that pick up the light.

One powerful application of PET is measuring metabolic activity in the brain. In a technique developed by Louis Sokoloff and his colleagues at the National Institute of Mental Health, 2-deoxyglucose (2-DG) is bound to a positron-emitting isotope of fluorine or oxygen. This radioactive 2-DG is injected into the bloodstream, and it travels to the brain. Metabolically active neurons normally use glucose, and they'll also take up the 2-DG. The amount they take up is proportional to their activity. However, unlike glucose, once the 2-DG is phosphorylated in the neuron, it cannot leave through the cell membrane. Thus, the amount of radioactive 2-DG accu-

continued on page 599

Box 21.4 | **BRAIN FOOD**

continued from page 598

mulated in a neuron and the number of positron emissions indicate the neuron's level of activity.

In a typical PET application, a person's head is placed in an apparatus surrounded by detectors (Figure A). Using computer algorithms, the light (resulting from positron emissions) reaching each of the detectors is recorded. With this information, levels of activity for populations of neurons at various sites in the brain can be calculated. Putting together these measurements from different sites creates an image of the brain activity pattern. The researcher monitors brain activity while the subject performs a task, such as moving a finger or reading aloud. Different tasks "light up" different brain areas. In order to obtain a picture of the activity evoked by a particular behavioral or thought task, a subtraction technique is used. Even in the absence of any sensory stimulation, the PET image will contain a great deal of brain activity. To create an image of brain activity resulting from a specific task, such as a person looking at a picture, this background activity is subtracted out (Figure B).

Although PET imaging has proven to be a valuable technique, it has significant limitations. Because the spatial resolution is only 5–10 mm, the images show the activity of many thousands of cells. Electrodes are still needed to study individual neurons. Also, one PET brain scan may take one to many minutes to obtain. This, plus concerns about radiation exposure, limit the number of scans that can be obtained from any one person in a reasonable period of time. By comparison, MRI scans can be obtained faster and with greater resolution, and there is no radiation risk. With the development of functional MRI (see Box 21.3), it has become possible to monitor brain activity without the drawbacks of PET imaging.

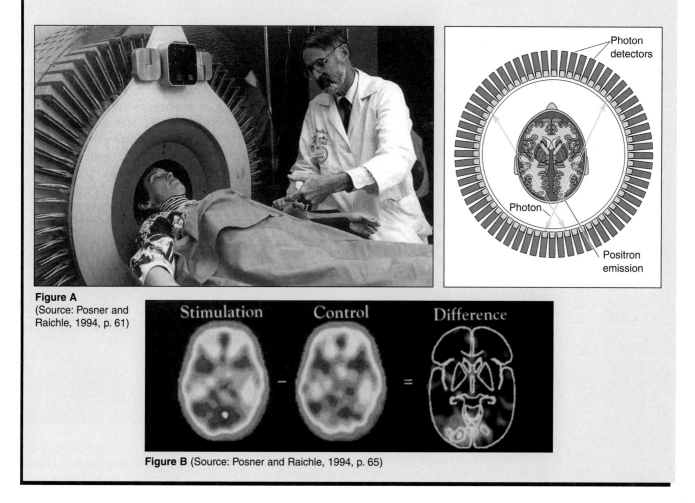

Figure A
(Source: Posner and Raichle, 1994, p. 61)

Stimulation — Control = Difference

Figure B (Source: Posner and Raichle, 1994, p. 65)

words being read or look at words presented on a monitor. By subtracting the levels of blood flow at rest from the levels during listening or seeing, blood flow levels were obtained specifically corresponding to the activity evoked by the sensory input. The results are shown in the top half of Figure 21.14. Not surprisingly, the visual stimuli evoked increased brain activity in striate cortex and extrastriate cortex and the auditory stimuli elicited activity in primary and secondary auditory cortex. However, the areas activated in extrastriate cortex and secondary auditory cortex did not respond to visual and auditory stimuli that were not words. These areas may be specialized for the coding of words either seen or heard. The visual stimuli did not evoke noticeably increased activity in the area of the angular gyrus and Wernicke's area, as one would expect based on the Wernicke-Geschwind model.

Another task studied with PET imaging was repeating words. In order to know what words to repeat, subjects must perceive and process the words by either the visual or the auditory system. Thus, the brain activity seen in the repetition task should include a component associated with the basic perceptual process as well as a component associated with speech. To isolate the speech component, the response pattern previously obtained in the simple sensory task was subtracted out. In other words, the image for "speaking words" equals an image corresponding to "repeating spoken words" minus an image corresponding to "listening to words." After the subtraction, the blood flow pattern indicated high levels of activity in primary motor cortex and the supplementary motor area (Figure 21.14, bottom left). There was also increased blood flow around the sylvian fissure near Broca's area. However, the PET images showed such activity *bilaterally*, and it was even observed when the subjects were instructed to move their mouth and tongue without speaking. Since there is good evidence that Broca's area is unilateral, it may not be showing up in these images, for reasons unknown.

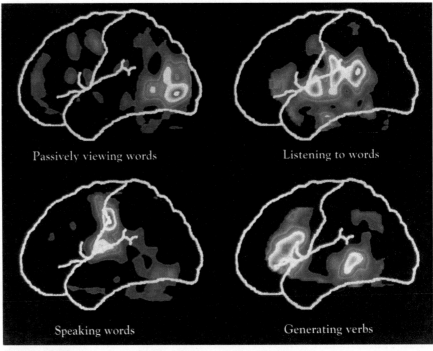

Passively viewing words   Listening to words

Speaking words   Generating verbs

Figure 21.14

PET imaging of sensation and speech. Relative levels of cerebral blood flow are color coded. Red indicates the highest levels and progressively lower levels are represented by orange, yellow, green, and blue. (Source: Posner and Raichle, 1994, p. 115.)

The final task required the subjects to think a little. For each word presented, the subject had to state a use for the word (e.g., if "cake" was presented, the subject said "eat"). To isolate activity specific to this verb-noun association task, the blood flow pattern obtained previously for speaking words was subtracted. Areas that were activated in the association task are located in the left inferior frontal area, the anterior cingulate gyrus, and the posterior temporal lobe (Figure 21.14, bottom right). The activity in frontal and temporal cortex is believed to be related to the performance of the word-association task, whereas the activity in cingulate cortex is probably related to attention.

These results from PET imaging are consistent with the overall placement of language areas inferred from studies of aphasia. However, as in the brain stimulation studies we discussed, the PET results suggest that language processing involves more complex mechanisms than a simple interaction between two major language areas (Broca's and Wernicke's). Because language involves many different skills, such as naming, articulation, grammar usage, and comprehension, this conclusion seems reasonable. An important question is whether there are separate systems involved in these various language functions. The model in Figure 21.4 may be a more realistic description of language processing than the Wernicke-Geschwind model, but there is clearly much to be learned. The hope is that further brain imaging studies will clarify language systems in the brain.

ATTENTION

Picture yourself at a crowded party where you are talking with a friend, numerous other conversations are going on around you, and loud music is playing. You are bombarded by sound from all directions. Yet somehow you're able to concentrate on the conversation you're having and ignore most of the other noise and conversations. You are *paying attention* to the one conversation. Behind you, you hear someone mention your name, and you decide to eavesdrop. Without turning around, you start focusing your attention on this other conversation to find out what's being said about you.

The ability to select one conversation to listen to out of many going on at the same time is an example of attention. The act of differentially processing simultaneous sources of information is called *selective attention*. Attention can also be exerted in other sensory modalities. In the visual system, attention allows us to concentrate on one object over many others in our visual field. Interactions between modalities also occur. For example, if you are performing an attention-demanding visual task, such as reading, you will be less sensitive to incoming sounds. Clearly, attention has to do with preferential processing of sensory information. Amidst all the sights, sounds, and tastes coming into our brain, we are able to selectively attend to some information and ignore the rest. Why, you might ask, do we do that? After all, if all this sensory information successfully makes its way into the brain, why not process all of it?

One possibility is that the brain simply cannot process all the incoming sensory information simultaneously. For example, striate cortex in a macaque monkey occupies about 10% of all cerebral cortex. By current estimates, there are more than 30 other visual areas, but many are much smaller than striate cortex. It is likely that these other areas cannot process as much information as V1. If this is true, then attention plays a key role in selecting what information should receive the limited processing resources of the brain. Another possibility is that even if the brain could process all the sensory information at once, there might be a performance advantage to focusing on one thing at a time. Maybe the situation is analogous to juggling—it's a lot easier to keep one ball in the air than five.

Only recently have neuroscientists begun to explore the neural consequences of attention. We will look at some of these fascinating physiological studies. But first, let's examine a few key findings concerning the behavioral manifestations of attention. We'll focus on visual attention because that is where behavioral studies can best be related to physiological research.

Behavioral Consequences of Attention

Under most conditions, if we want to visually scrutinize something, we move our eyes so that the object of interest is imaged on the fovea in each eye. Implicit in this behavior is the fact that most of the time we are paying attention to something imaged on our fovea. However, it is possible to shift attention to objects imaged on parts of the retina outside the fovea. Shifting attention to some location on the retina enhances visual processing in several ways. Two ways that we'll look at are enhanced detection and faster reaction times.

Attention Enhances Detection. Figure 21.15 shows an experiment used to study the effects of directing visual attention to different locations. During the entire course of an experiment, the observer fixated on a central point. The observer's task was to say whether a target stimulus was flashed on at a location to the left of the fixation point, to the right, or not at all. Sounds easy, right? Actually, the task was difficult because the target presented to either side of the fixation point was a small circle and was flashed on for only about 15 msec. The experiment had several special procedures for identifying the effects of attention. Each trial began with the presentation of a cue stimulus at the fixation point. The cue was either a plus sign, an arrow pointing left, or an arrow pointing right. After the cue was extinguished, there was a variable delay period during which the fixation point was seen. In half the trials there was no further stimulus, and in the other half a small circle was flashed on for 15 msec at either the left or right position. A key element of the experiment is that the cue was used to direct attention. If the central cue was a plus sign, it was equally likely that a little circle would appear at either the left or the right. If the cue was a left arrow, it was four times more likely that a target would appear on the left than the right. If the cue was a right arrow, it was four times more likely that a target would appear on the right than the left. The observer had to keep her eyes pointing straight ahead, but in order to make the most correct responses on the difficult task of detecting the flashed circles, it would be advantageous to make use of the cue. For instance, if the cue was a left arrow, it would be beneficial to try and pay attention to the left target location more than the right.

For each of the subjects used in this experiment, the data collected consisted of the percentage of the time a circle was correctly detected. Because there was no target circle on half the trials, the observers could not get a high percentage correct by "cheating" (i.e., by saying there was always a target at the side where the arrow pointed). In trials where the central cue was a plus sign, the observers detected a target stimulus on about 60% of the trials in which one was presented. When the cue was a right arrow, the observers detected a target stimulus at the right on about 80% of the trials in which one was presented there. However, when the cue pointed right, the observers detected a target stimulus at the left on only about 50% of the trials in which one was presented there. With the appropriate left-right reversal, the results were about the same with left arrows.

What do these data mean? To answer, we must imagine what one of the observers was doing. Evidently the expectation of the observer based on the cues influenced her ability to detect the subsequent targets. It appears that the arrow cues caused the observer to shift her attention to the side where the arrow pointed, even though her eyes didn't move. Presumably this shift of

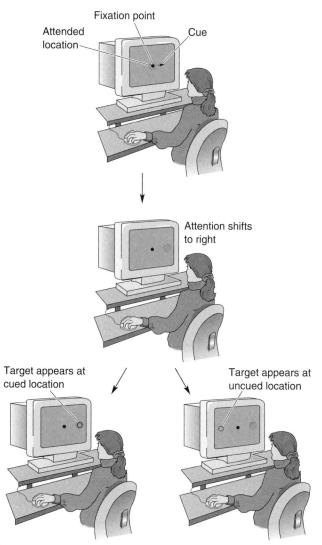

Figure 21.15
An experiment to measure the effect of attention on detection. While an observer maintains steady fixation, a cue directs her to shift her attention to one side of the computer screen. In each trial, the observer indicates whether a circular target is seen on either side of the screen.

attention made it easier to detect the flashed targets compared to the trials when the central cue was a plus sign. Conversely, the observer was less sensitive to the targets on the side of fixation opposite to where the arrow cue pointed. Based on these results and those from many other similar experiments, our first conclusion about the behavioral effects of attention is that it makes things easier to detect. This is probably one of the reasons that we can listen in on one conversation among many when we give it our attention.

Attention Speeds Reaction Times. Using an experimental model similar to the one above, it has been demonstrated that attention increases the speed of reactions in perceptual studies. This experiment was performed by cognitive psychologist Michael Posner at the University of Oregon, who has conducted extensive studies of attention using behavioral and imaging techniques (Box 21.5). In a typical experiment, an observer fixated on a central point on a computer screen, and target stimuli were presented to either the left or the right of the fixation point. However, in this experiment the

Localizing Cognitive Operations

by Michael I. Posner

Michael I. Posner

For the last 30 years, graduate students at the University of Oregon have gathered at my house each Tuesday evening to discuss research. One night during the late 1970s, we read papers by Vernon Mountcastle and Robert Wurtz discussing the recent discovery of single neurons in the parietal lobe of monkeys that were involved in visual attention. In our research we found that people could shift attention to cues within 100 milliseconds entirely covertly without any eye movement or other overt adjustment, and when they did so they were faster and more accurate in response to a stimulus that occurred at the attended location. Someone in the group made the reasonable suggestion that the two findings should be related. This suggestion was more radical than it might now seem, because, at the time, most cognitive psychologists supported the view that mental activity was about software and the physical system, whether brain or computer, that ran the software was not important. Nonetheless, being a rather concrete person, I launched what proved to be a 10-year odyssey to discover how brain systems represent the mental operations involved in higher cognition.

The first step involved patients with strokes in the parietal lobe. We set up a laboratory in a hospital in Portland. There, with the help of Oscar Marin, we recruited patients. We found that patients who had suffered strokes in the parietal lobe, even when the neurologist claimed they were recovered, lacked the ability to shift attention toward events in the visual field opposite the lesion. We discovered a very specific abnormality in their ability to disengage visual attention. What was even more fascinating was that patients with lesions in the midbrain and thalamus also had problems with visual attention, but each patient group had difficulty with different mental operations involved in attention shifts. This led to a hypothesis about how the brain localizes mental operations.

The next step involved an even longer move, this time to St. Louis, Missouri. There, Marcus Raichle had refined the use of positron emission tomography to trace local blood flow in normal people. The St. Louis group was anxious to find a psychologist interested in using PET to study cognition. It was my view that PET could test the proposition that mental operations were localized in specific neural areas. We first examined reading and listening to English words. That choice was dictated by the potential utility of the findings in helping neurosurgeons avoid areas involved in language. We found very specific brain areas that processed the visual, phonological, and semantic codes of these words. These findings provided confirmation of the general view of localization that arose from the clinical work. Later, the same methods showed that areas of the superior parietal lobe were involved in disengaging and shifting attention, again supporting the patient and monkey work.

Would it be possible to observe the orchestration of these brain areas in real time? Together with Don Tucker, we were able to record event-related potentials (ERP) from large numbers of scalp electrodes (Figure A). This could be done while people performed various tasks, such as reading words. We found scalp signatures of the brain areas active in PET and were able to show their recruitment in real time. This PET, ERP, and lesion work suggesting the localization of mental operations is reported in more detail in the book *Images of Mind* that Raichle and I published in 1994.

We still gather each Tuesday night to discuss research. Sometimes it's a struggle to stay awake, but you never know who or exactly which question will launch the next inquiry.

Figure A

observer was told to wait until he perceived a stimulus at either location and to press a button when he did. A measurement was made of how long it took the observer to react to the presentation of a stimulus and press a button. Preceding the target was a cue stimulus, either a plus sign or an arrow pointing left or right. The arrows indicated the side to which a stimulus was more likely to appear, whereas the plus sign meant that either side was equally likely.

Results from this experiment demonstrated that an observer's reaction times were influenced by where the central cue told him to direct his attention. When the central cue was a plus sign, it took about 250–300 msec to press the button. When an arrow cue correctly indicated where a target would appear (e.g., right arrow and right target), reaction times were 20–30 msec faster. Conversely, when the arrow pointed in one direction and the target appeared at the opposite location, it took 20–30 msec longer to react to the target and press the button. The reaction time included time for transduction in the visual system, time for visual processing, time to make a decision, time to code for the finger movement, and time to press the button. Nonetheless, there was a small but reliable effect based on which direction the arrows directed the observer's attention (Figure 21.16). If we assume that attention to visual objects does not have a direct effect on visual transduction or motor coding, we are left with the hypothesis that attention can alter the speed of visual processing or the time to make a decision about pressing the button.

Neglect Syndrome as an Attentional Disorder. In Chapter 12, we briefly discussed the **neglect syndrome**, in which a person appears to ignore objects, people, and sometimes their own body to one side of their center of gaze. Some have argued that this syndrome is a unilateral deficit in attention. The manifestations of the neglect syndrome can be so bizarre that they're hard to believe if not directly observed. In mild cases, the behavior may not be apparent from casual observation. But in severe cases, the patient acts as if half of the universe no longer exists. He may shave only one side of his face, brush the teeth on only one side of his mouth, dress only one side of his body, and eat food from only one side of his plate. Because neglect syndrome is less common following left hemispheric damage, it has primarily been studied in regard to neglect of the left half of space as a result of damage to right cerebral cortex. In addition to neglecting objects to the left side, some patients exhibit denial. For instance, they may say that their left hand isn't really paralyzed or in extreme cases refuse to believe that a limb on their left side is part of their body. Refer to Figure 12.24 as a typical example of the distorted sense of space these patients have. If asked to make a drawing, they may crowd all the features into the right half, leaving the left half blank. A particularly dramatic example of this is in the paintings shown in Figure 21.17, which were made by an artist as he recovered from a stroke. If a person with a neglect syndrome is asked to close her eyes and point toward the midline of her body, she typically will point too far to the right, as if there has been a shrinkage of the left half. If blindfolded and asked to explore objects placed on a table before them, patients behave normally in exploring objects to the right but are haphazard about probing to the left. All of these examples point toward a problem in relating to the space around them.

Neglect syndrome is most commonly associated with lesions in posterior parietal cortex in the right hemisphere, but it has also been reported to occur following damage to right hemisphere prefrontal cortex, cingulate cortex, and other areas. It has been proposed that posterior parietal cortex is involved in attending to objects at different positions in extrapersonal

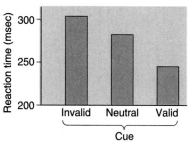

Figure 21.16
The effect of cueing on reaction time. In neutral cue trials, the cue was a plus sign, which gave no indication of the likely location of the following target. In valid cue trials, the arrow-shaped cue pointed to the location where the target appeared, speeding reactions to the targets. When the cue was invalid, pointing in a direction opposite to where the target later appeared, reaction times were slower. (Source: Adapted from Posner, Snyder, and Davidson, 1980, Fig. 1.)

Figure 21.17
Self-portraits during recovery from a stroke that caused a neglect syndrome.
Two months after suffering a stroke affecting parietal cortex on the right side, the person made the upper-left portrait. There is virtually no left side to the face in the painting. Three-and-a-half months after the stroke, there is some detail in the left side but not nearly as much as the right side (upper right). At 6 months (lower left) and 9 months (lower right) after the stroke, there is increasingly more treatment of the left side of the painting. (Source: Posner and Raichle, 1994, p. 152.)

space. If this is true, then neglect syndrome might be a disruption of the ability to shift attention. One piece of evidence supporting this hypothesis is that objects in the right visual field of patients with neglect syndrome are sometimes abnormally effective in capturing attention, and patients may experience difficulty disengaging their attention from an object on this side. It is not clear why the syndrome is more often seen with right hemispheric damage than left hemispheric damage. The right hemisphere appears to be dominant for understanding spatial relationships, and in split-brain studies it has been shown to be superior at solving complex puzzles. This seems consistent with the greater loss of spatial sense after right hemispheric lesions. One hypothesis is that the left hemisphere is involved in attending to objects in the right visual field, whereas the right hemisphere is involved in attending to objects in the left and right visual fields. This would account for the asymmetric effects of left and right hemispheric lesions, but at present there is only suggestive evidence in support of this hypothesis.

Physiological Effects of Attention

What is happening in the brain when we shift our attention to something? While it's possible that attention is a property of cognition, recent experiments demonstrate that the effects of attention can be observed in sensory areas such as visual cortex. But it is important to recognize a critical distinction. We will primarily discuss experiments demonstrating that brain activity changes when attention is devoted to a stimulus. While this activity change may *result* from attention, it may not have anything to do with the *control* of attention. The question of how attention is allocated to particular objects is quite a different matter.

PET Imaging of Attention. What we would like to know is what happens in the brain when we pay attention to something. For example, if we attend to a particular location or object, is there increased neural activity associated with it? If so, does attention affect all visual cortical areas or only a select few? These questions have been addressed by using PET imaging in humans. As we have seen, PET imaging can provide a picture of brain activity associated with mental processes. Steven Petersen and his colleagues at Washington University have used PET imaging while humans performed a same-different discrimination task (Figure 21.18). An image was flashed on a computer screen for about half a second; after a delay period, a second image was flashed. Each image was composed of small elements that could vary in shape, color, and speed of motion. The task of the observer was to indicate whether the two successive images were the same or different. While the speed, shape, and color of the elements could all change between the two images, the observer was instructed to base his same-different judgments on any one or all of the attributes. Accordingly, brain activity in response to identical stimuli could be measured when the person was attending to different features. Of course, the brain activity would simply indicate where neurons were activated by the visual patterns, unless something were added to the experiment. To isolate the effect of attention, two different versions of the experiment were conducted. In *selective-attention* experiments, subjects viewed the stimuli and performed the same-different task after being told to pay attention to just one of the features (shape, color, or speed). In *divided-attention* experiments, subjects simultaneously monitored all features and based their same-different judgments on changes in any feature. The researchers then subtracted the divided-attention responses from the selective-attention responses to obtain an image of changes in brain activity associated with attention to one feature.

Figure 21.19 illustrates the results. Different areas of cortex had higher activity when different attributes of the stimuli were being discriminated. For instance, whereas ventromedial occipital cortex was affected by attention in color and shape discrimination tasks, it was not affected in the speed discrimination task. Conversely, areas in parietal cortex were influenced by

Figure 21.18
Same-different stimuli used for PET imaging. The observer sees two frames, each containing moving elements that can change in shape, color, and speed of motion. The observer responds by indicating whether the stimuli are the same or different.

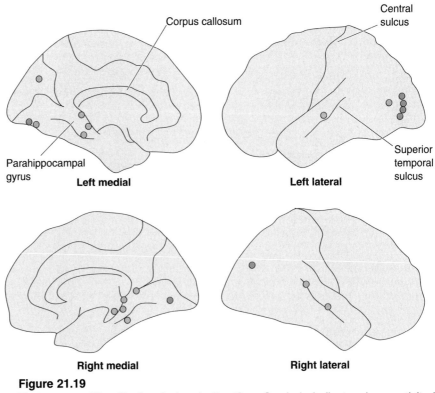

Figure 21.19
Feature-specific effects of visual attention. Symbols indicate where activity in PET images was higher in selective-attention experiments relative to divided-attention experiments. In selective-attention experiments, the same-different judgments were based on speed (green), color (blue), or shape (orange). (Source: Adapted from Corbetta et al., 1990, Fig. 2.)

attention to the motion task but not the other tasks. While it is not possible to know with certainty which cortical areas were highlighted in these experiments, several visual areas, including V1, V2, and MT (V5), did not appear to be influenced by attention. Areas of heightened activity in the color and shape tasks may have corresponded to areas V4 and IT and other visual cortical areas in the temporal lobe. The area most affected by performing the motion task was next to area MT.

The important point to learn from these and other PET experiments is that numerous cortical areas appear to be affected by attention, and that the greatest attentional effects are seen in "late" rather than "early" areas in the visual system. Attention selectively increases brain activity, but the particular areas affected depend on the nature of the behavioral task performed. We'll now examine a couple of these areas in detail and see how studies in behaving monkeys have clarified the role of attention.

Enhanced Neuronal Responses in Parietal Cortex. The perceptual studies discussed earlier show that in carefully constructed experiments, attention can be moved independently of eye position. But what happens normally when you move your eyes? Let's assume that when you are scrutinizing an object focused on your fovea, your attention is also directed to the fovea. If a bright flash of light appeared in the periphery of your visual field, you normally would make a saccade toward the sudden flash so that you could look at it with your fovea. But what happened to your attention? Did it move before, during, or after the eye movement? Behavioral studies have shown that shifts in attention can occur in about 50 msec, whereas saccades take about 200 msec. Therefore, it is quite likely that attention is drawn to the flash before your eyes move.

The assumption that attention changes location prior to an eye movement underlies an experiment performed by neurophysiologists Robert Wurtz, Michael Goldberg, and David Robinson at the National Institutes of Health. They recorded neural activity at several locations in the brains of monkeys to determine whether there was increased activity that might be related to attention, prior to eye movements. Assuming there is a relationship between attentional shifts and eye movements, it is logical to examine parts of the brain involved in generating saccades.

The researchers recorded from neurons in the posterior parietal cortex of monkeys while the animals performed a simple behavioral task (Figure 21.20). This cortical area is thought to be involved in directing eye movements, in part because electrical stimulation here will evoke saccades. Parietal neurons have rather large visual receptive fields, perhaps encompassing 25% of the entire visual field. In the task, a monkey fixated on a spot on a computer screen; when a target appeared at a different location, it made a saccade to that location. In each trial the receptive field of a cortical neuron was located, and the target used in the behavioral task was positioned so that it appeared within the receptive field. As expected, the neuron was excited when the target flashed in its receptive field (Figure 21.21a). However, the key observation Wurtz and his colleagues made was that the response of many parietal cortex neurons was significantly enhanced (a rapid burst of action potentials) when the animal subsequently made a saccade to the target (Figure 21.21b). This effect was spatially selective because the enhancement was not seen if a saccade was made to a location outside the receptive field (Figure 21.21c). This is an important point because it indicates that the brain was not simply generally more excitable.

The enhancement that occurred before the saccades is curious if the parietal neurons were simply responding to the visual stimuli. Why should the response to a target stimulus depend on an eye movement that happens long *after* the target is turned on? One explanation is that the increased activity prior to the saccades was a consequence of attention shifting to the

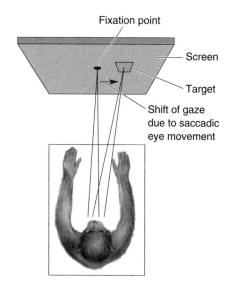

Figure 21.20
A behavioral task for directing a monkey's attention. While recordings are made from the posterior parietal cortex, the monkey fixates on a point on a computer screen. When a peripheral target appears (usually in a cell's receptive field), the animal makes a saccade to the target. (Source: Adapted from Wurtz, Goldberg, and Robinson, 1982, p. 128.)

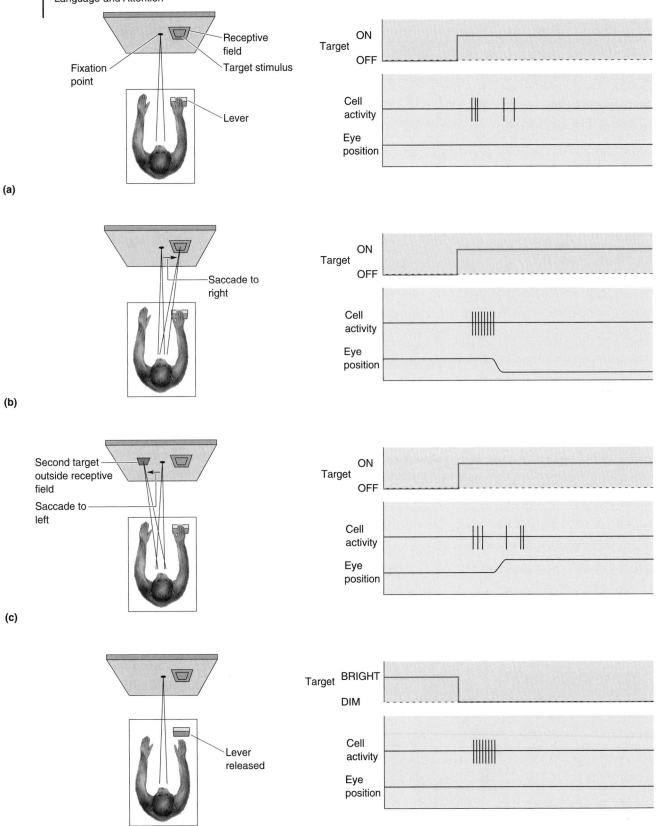

(a)

(b)

(c)

(d)

Figure 21.21
The effect of attention on the response of a neuron in posterior parietal cortex. (a) A neuron in posterior parietal cortex responds to a target stimulus in its receptive field. **(b)** The response is enhanced if the target presentation is followed by a saccade to the target. **(c)** The enhancement effect is spatially selective, as it is not seen if a saccade is made to a stimulus not in the receptive field. **(d)** Enhancement is also seen when the task requires the animal to release a hand lever when the peripheral spot dims. (Source: Adapted from Wurtz, Goldberg, and Robinson, 1982, p. 128.)

location inside the receptive field of the neuron. Another possibility is that the enhanced response was a premotor signal related to coding for the subsequent eye movement, just as neurons in motor cortex fire before hand movements. In order to address this possibility, they performed a variation on the experiment in which the behavioral response was changed from a saccade to a hand movement (Figure 21.21d). Again, there was an enhanced response to the target in the receptive field. This shows that the response enhancement was not a premotor signal for a saccade, but rather was related to attention.

It isn't difficult to see how response enhancement of the kind observed in posterior parietal cortex could be involved in the behavioral effects of attention discussed earlier. If attention drawn to one location in the visual field by a cueing stimulus increases the response to other stimuli near that location, this could account for the spatially selective enhancement in the ability to detect a target. Likewise, it is conceivable that an increased response could lead to more rapid visual processing and ultimately faster reaction times, as seen in the perceptual experiments.

Receptive Field Changes in Area V4. In a fascinating series of experiments, Robert Desimone and his colleagues at the National Institute of Mental Health have revealed surprisingly specific effects of attention on the response properties of neurons in visual cortical area V4. In one experiment, monkeys performed a same-different task with pairs of stimuli within the receptive fields of V4 neurons. The animal compared two successive stimuli and pushed a lever one way with its hand if the stimuli were the same and the opposite way if the stimuli were different. As an example, suppose that a particular V4 cell responded strongly to vertical and horizontal red bars of light in its receptive field but did not respond to vertical or horizontal green bars. The red bars were "effective" stimuli and the green bars were "ineffective" stimuli. While the monkey fixated, two stimuli (each either effective or ineffective) were briefly presented in the receptive field, and after a delay period, two more stimuli were presented. In an experimental session, the animal was instructed to base its same-different judgments on the successive stimuli at one of the two locations within the receptive field. In other words, the animal had to pay attention to one location in the receptive field, but not the other, in order to perform the task.

Consider what happened in a trial when effective stimuli appeared at the attended location and ineffective stimuli appeared at the other location (Figure 21.22a). Not surprisingly the V4 neuron responded strongly in this situation because there were perfectly good "effective" stimuli in the receptive field. Suppose the monkey was then instructed to base its same-different judgments on the stimuli at the other location in the receptive field (Figure 21.22b). At this location, only green ineffective stimuli were presented. The response of the neuron should have been the same as before because exactly the same stimuli were in the receptive field, right? Wrong! Even though the stimuli were identical, on average the responses of V4 neurons were less than half as great when the animal attended to the area in the neuron's receptive field containing the ineffective stimuli. It's as if the receptive field contracted around the attended area, decreasing the response to the effective stimuli at the unattended location. The location specificity of attention observed in this experiment may be directly related to that discussed earlier in the human perceptual experiments. Perceptually, detection is enhanced at attended locations compared to unattended locations. It's not too great a conceptual leap to imagine that the difference in ease of detection at the attended and unattended locations is based on the higher activity evoked by effective stimuli at the attended location.

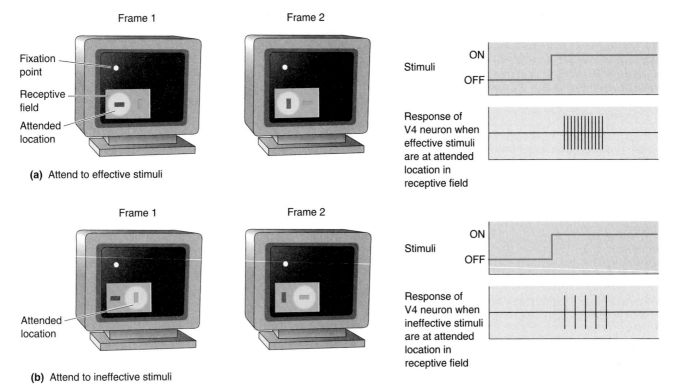

Frame 1 Frame 2

Fixation point

Receptive field

Attended location

(a) Attend to effective stimuli

Stimuli — ON / OFF

Response of V4 neuron when effective stimuli are at attended location in receptive field

Frame 1 Frame 2

Attended location

(b) Attend to ineffective stimuli

Stimuli — ON / OFF

Response of V4 neuron when ineffective stimuli are at attended location in receptive field

Figure 21.22
Stimuli used to study attentional effects on neurons in visual cortical area V4. The yellow circle indicates whether the monkey is attending to **(a)** the left or **(b)** the right location in the receptive field. For this neuron, red bars of light are effective in producing a response and green bars are ineffective.

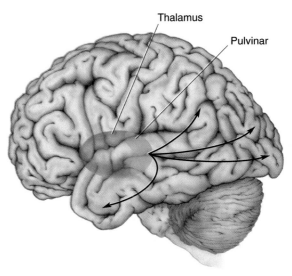

Thalamus

Pulvinar

Figure 21.23
Pulvinar projections to the cortex. The pulvinar nucleus is in the posterior thalamus. It sends widespread efferents to areas of cerebral cortex, including areas V1, V2, MT, parietal cortex, and inferior temporal cortex.

How Is Attention Directed?

We have discussed the effects of attention on the responses of neurons in several cortical areas. But what is controlling attention? At present, there is no clear answer to this question. Certainly, the specificity observed in the physiological studies puts considerable demand on the neural mechanisms guiding attention.

One structure that has been studied for its possible role in guiding attention is the *pulvinar nucleus* of the thalamus. Several properties of the pulvinar make it interesting. For example, it has reciprocal connections with most visual cortical areas of the occipital, parietal, and temporal lobes, giving it the potential to modulate widespread cortical activity (Figure 21.23). Also, humans with pulvinar lesions appear to have difficulty focusing their attention on objects on the side contralateral to the lesion. They respond abnormally slowly to stimuli on the contralateral side, particularly when there are competing stimuli on the ipsilateral side. It has been proposed that such a deficit is a reduced ability to focus attention on objects in the contralateral visual field. A similar phenomenon has been observed in monkeys. When muscimol, an agonist of the inhibitory neurotransmitter GABA, is unilaterally injected into the pulvinar, the activity of the neurons is suppressed. Behaviorally, the injection produces difficulty in shifting attention to contralateral stimuli, which seems similar to the effect of pulvinar lesions in humans. Interestingly, injection of the GABA antagonist bicuculline appears to facilitate shifting attention to the contralateral side.

There are several problems with the hypothesis that the pulvinar directs attention. It has been found that unilateral deactivation of the superior colliculus or posterior parietal cortex has behavioral effects similar to deactivation of the pulvinar. Also, bilateral pulvinar lesions do not have a large effect on attention, as would be expected if it were a critical part of an attention system.

Although there is interesting evidence from human and animal studies suggesting that the pulvinar is involved in attention, it is not any more clearly involved than the superior colliculus or the posterior parietal cortex. Perhaps these structures work together to guide attention. However, the possibility must also be considered that there is no central control for attention, but that attention is a property of the interactions between many areas.

CONCLUDING REMARKS

After studying the brain systems responsible for sensation, memory, motor control, and behavior, it may be easy to forget that the human brain is much more than a passive information processor. In this chapter, we've explored several aspects of higher brain function that demonstrate the opposite. For example, through the process of selective attention, we dynamically allocate our mental energies to perform chosen tasks while ignoring other stimuli that might interfere. The effects of attention even include changes in receptive field properties that not long ago were considered static. A different sort of adaptability is seen in human language. Specialized language areas have evolved in the human brain that endow us with an incredibly flexible and creative system for communication.

As the techniques of neuroscience research expand, it is interesting to ponder what aspects of higher brain function might one day be understood. For example, will neuroscience ever explain consciousness? The nature of consciousness has interested scholars for ages. The British physicist, biochemist, and codiscoverer of the helical structure of DNA, Francis Crick, recently published an interesting book called *The Astonishing Hypothesis*. In it he proposes that it isn't too soon to start thinking about experiments that might teach us about consciousness, and indeed, some neuroscientists have already started.

Think about some of the things we've discussed in this and the preceding chapters. Among others, neuroscience research is exploring such topics as memory, recognition, attention, language, and emotion. These are the key components of the human mind. And what is consciousness? We really don't know the answer, but a reasonable guess might include interactions between memory systems and an attentional mechanism. Perhaps our short-term memory allows us to pay attention to our own ongoing status—we are aware, and we're aware that we're aware. Speculations aside, the point is that neuroscience is approaching problems that would have seemed entirely beyond the grasp of science not long ago. You've learned a lot about how the brain works and what it does. Even more exciting is the realization that neuroscience techniques will allow us to address new and grander questions in the future.

KEY TERMS

Language and the Brain
aphasia
Wada procedure
Broca's area
Wernicke's area
Broca's aphasia
anomia
Wernicke's aphasia
Wernicke-Geschwind model

conduction aphasia
split-brain study
corpus callosum
magnetic resonance imaging (MRI)
planum temporale
positron emission tomography (PET)

Attention
neglect syndrome

✔ REVIEW QUESTIONS

1. How is it possible for a split-brain human to speak intelligibly if the left hemisphere controls speech? Isn't this inconsistent with the fact that the left hemisphere must direct motor cortex in both hemispheres to coordinate movements of the mouth?
2. What can you conclude about the normal function of Broca's area from the observation that there are usually some comprehension deficits in Broca's aphasia? Must Broca's area itself be directly involved in comprehension?
3. Pigeons can be trained to press one button when they want food and to press other buttons when they see particular visual stimuli. This means the bird can look around and "name" things it sees. How would you determine whether or not the pigeon is using a new language—"button-ese"?
4. What does the Wernicke-Geschwind language processing model explain? What data are inconsistent with this model?
5. What differences do you think there are between the conscious states of a human with neglect syndrome and a split-brain human who can only describe things in the right visual field?
6. The following headline appeared in the newspaper: "Drunk Gets Nine Months in Violin Case." This phrase has two reasonable interpretations, and somewhere your brain activity must change when you switch interpretations. What kind of experiment would you use to look for changing activity correlated with the cognitive change? Where would you look?

GLOSSARY

A1. Primary auditory cortex.

absolute refractory period. The period of time, measured from the onset of an action potential, during which another action potential cannot be triggered.

accommodation. The focusing of light by changing the shape of the eye's lens.

ACh (acetylcholine). An amine that serves as a neurotransmitter at many synapses in the PNS and CNS, including at the neuromuscular junction.

actin. A cytoskeletal protein in all cells, and the major thin filament protein in a muscle fiber; causes muscle contraction by specific chemical interactions with myosin.

action potential. A brief fluctuation in membrane potential caused by the rapid opening and closing of voltage-gated ion channels. Action potentials sweep like a wave along axons to transfer information from one place to another in the nervous system.

active zone. A presynaptic membrane differentiation that is the site of neurotransmitter release.

adenylyl cyclase. An enzyme that catalyzes the conversion of ATP to cAMP, a second messenger.

adrenal cortex. The outer segment of the adrenal gland; releases cortisol when stimulated by the pituitary hormone ACTH.

adrenal medulla. The inner segment of the adrenal gland, innervated by preganglionic sympathetic fibers; releases epinephrine.

adrenaline. A catecholamine neurotransmitter synthesized from norepinephrine; also called **epinephrine**.

affective aggression. A threatening or defensive form of aggression accompanied by vocalizations and a high level of activity in the ANS.

affective disorder. A mental disorder of mood and affect, such as depression.

afferent. An axon coursing toward and innervating a given structure.

after-hyperpolarization. The hyperpolarization that follows strong depolarization of the membrane. The after-hyperpolarization following an action potential is also called the **undershoot**.

agnosia. The inability to recognize objects, even though simple sensory skills appear to be normal; most commonly caused by damage to posterior parietal areas of the brain.

alpha motor neuron. The neuron that innervates the extrafusal fibers of skeletal muscle.

amacrine cell. A neuron in the retina that projects neurites laterally in the inner plexiform layer.

amino acid. A chemical building block of protein molecules containing a central carbon atom, an amino group, a carboxyl group, and a variable R group.

Ammon's horn. A layer of neurons in the hippocampus that sends axons into the fornix.

amnesia. A severe loss of memory or the ability to learn. See also **anterograde amnesia**, **retrograde amnesia**, **transient global amnesia**.

AMPA receptor. A subtype of glutamate receptor; a glutamate-gated ion channel that is permeable to Na^+ and K^+.

amygdala. An almond-shaped nucleus in the anterior temporal lobe, thought to be involved in emotion and certain types of learning and memory.

analgesia. The absence of normal sensations of pain.

anion. A negatively charged ion.

anomia. A difficulty or an inability finding words.

ANS. See **autonomic nervous system**.

antagonist muscle. A muscle that acts against another at the same joint.

anterior. A direction meaning toward the nose, or rostral.

anterograde amnesia. The inability to form new memories.

anterograde transport. Axoplasmic transport from the soma to the axon terminal.

aphasia. A partial or complete loss of language abilities following brain damage. See also **Broca's aphasia**, **conduction aphasia**, **Wernicke's aphasia**.

aqueous humor. The fluid between the cornea and the lens of the eye.

arachnoid membrane. The middle of the three meninges that cover the surface of the CNS.

area 17. Primary visual cortex.

area MT. An area of the neocortex, at the junction of the parietal and temporal lobes, that receives input from striate cortex and appears to be specialized for the detection of stimulus movement; also called V5.

aspinous neuron. A neuron lacking dendritic spines.

associative learning. The learning of associations between events; two types are usually distinguished: **classical conditioning** and **instrumental conditioning**.

astrocyte. A glial cell in the brain that supports neurons and regulates the extracellular ionic and chemical environment.

atonia. The absence of muscle tone.

ATP (adenosine triphosphate). The molecule that is the cell's energy source. The hydrolysis of ATP to produce adenosine diphosphate (ADP) releases energy that fuels most of the biochemical reactions of the neuron. ADP is converted back to ATP in the mitochondria.

attenuation reflex. The contraction of muscles in the middle ear, resulting in a reduction in auditory sensitivity.

audition. The sense of hearing.

auditory canal. A channel leading from the pinna to the tympanic membrane.

auditory nerve. Cranial nerve VIII, consisting of axons projecting from the spiral ganglion to the cochlear nuclei.

autonomic ganglia. Peripheral ganglia of the sympathetic and parasympathetic divisions of the ANS.

autonomic nervous system (ANS). The part of the PNS that innervates the internal organs, blood vessels, and glands; also called the visceral PNS; consists of sympathetic, parasympathetic, and enteric divisions.

autoradiography. A method for visualizing sites of radioactive emissions in tissue sections.

autoreceptor. A receptor in the membrane of a presynaptic axon terminal that is sensitive to the same neurotransmitter that is released by that terminal.

axial muscle. A muscle that controls movements of the trunk of the body.

axon. A neurite specialized to conduct nerve impulses, or action potentials, normally in a direction away from the soma.

axon collateral. A branch of an axon.

axon hillock. Swelling of the axon where it joins the soma.

axon terminal. The end region of the axon, usually a site of synaptic contact with another cell; also called the **terminal bouton** or the presynaptic terminal.

axoplasmic transport. The process of transporting materials down an axon.

ballism. A movement disorder caused by damage to the subthalamus, characterized by excess and uncontrolled ballistic movement.

barbiturate. A depressant drug that makes inhibition more effective. Barbiturates bind to the $GABA_A$ receptor and increase the duration of channel openings in the presence of GABA.

basal forebrain complex. Several cholinergic nuclei of the telencephalon, including the medial septal nuclei and basal nucleus of Meynert.

basal ganglia. A collection of associated cell groups in the basal forebrain, including the caudate nucleus, putamen, globus pallidus, and subthalamus.

basal lamina. A layer of proteins in the space between a nerve terminal and the muscle cell it innervates.

basal telencephalon. The region of the telencephalon lying deep in the cerebral hemispheres.

basilar membrane. A membrane separating the scala tympani and scala media in the cochlea.

basolateral nuclei. A group of nuclei in the basolateral amygdala whose axons constitute the ventral amygdalofugal pathway.

benzodiazepine. A depressant drug that makes inhibition more effective. Benzodiazepines bind to the $GABA_A$ receptor and increase the frequency of channel openings in the presence of GABA.

binocular competition. A process believed to occur during the development of the visual system whereby the inputs from the two eyes actively compete to innervate the same cells.

binocular receptive field. The receptive field of a neuron that responds to stimulation of either eye.

binocular visual field. The portion of the visual field viewed by both eyes.

bipolar cell. Any cell having only two neurites; in the retina, the cell that connects photoreceptors to ganglion cells.

blob. A collection of cells, mainly in striate cortical layers II and III, that is characterized by a high level of the enzyme cytochrome oxidase.

blob channel. The visual information processing channel that passes through the parvocellular and koniocellular layers of the LGN and converges on the blobs of striate cortical layer III; believed to process information about color.

blood-brain barrier. A specialization of the walls of brain capillaries that limits the movement of bloodborne substances into the extracellular fluid of the brain.

bouton en passant. A swollen region where an axon makes a synaptic contact as it passes by another neuron.

brain stem. The diencephalon, midbrain, pons, and medulla.

brain. The part of the CNS contained in the skull, consisting of the cerebrum, cerebellum, brain stem, and retinas.

brain-derived neurotrophic factor (BDNF). A neurotrophin found in the brain that promotes the survival of cortical neurons.

Broca's aphasia. A language disturbance, also known as motor or nonfluent aphasia, in which a person has difficulty speaking or repeating words but can understand language.

GLOSSARY

Broca's area. A region of the frontal lobe that is associated with Broca's (motor) aphasia when damaged.

bundle. A collection of axons that run together but that do not necessarily have the same origin and destination.

CA1. A region of Ammon's horn in the hippocampus that receives input from the neurons of CA3.

CA3. A region of Ammon's horn in the hippocampus that receives input from the neurons of the dentate gyrus.

calcium-calmodulin-dependent protein kinase (CaMK). A protein kinase activated by elevations of internal Ca^{2+} concentration.

calcium-pump. An ion pump that removes cytosolic Ca^{2+} ions.

cAMP. See **cyclic adenosine monophosphate**.

Cannon-Bard theory. A theory of emotion proposing that emotional experience is independent of emotional expression and is determined by the pattern of thalamic activation.

capsule. A collection of axons that connect the cerebrum with the brain stem.

cardiac muscle. The muscle of the heart.

catecholamines. The neurotransmitters dopamine, norepinephrine, and epinephrine.

cation. A positively charged ion.

caudal. A direction meaning toward the tail, or posterior.

caudate nucleus. A part of the basal ganglia in the basal forebrain, involved in motor control.

cell assembly. A group of simultaneously active neurons that may be the internal representation of an external event or object.

cell body. The central region of the neuron containing the nucleus; also called the **soma** or **perikaryon**.

cell-adhesion molecule (CAM). A molecule on the cell surface that causes cells to adhere to one another.

center-surround receptive field. A visual receptive field with a circular center region and a surround region forming a ring around the center. Stimulation of the center produces a response opposite to the response generated by stimulation of the surround.

central nervous system (CNS). The brain (including the retinas) and spinal cord.

central pattern generator. A neural circuit that gives rise to rhythmic motor activity.

central sulcus. The sulcus that divides the frontal lobe from the parietal lobe.

cerebellar cortex. The sheet of gray matter lying just under the pial surface of the cerebellum.

cerebellar granule cell. A neuron in the cerebellar cortex that receives input from mossy fibers and gives rise to parallel fibers that innervate Purkinje cells.

cerebellar hemispheres. The lateral regions of the cerebellum.

cerebellum. A structure derived from the rhombencephalon, attached to the brain stem at the pons; an important movement control center.

cerebral aqueduct. The canal filled with cerebrospinal fluid within the midbrain.

cerebral cortex. The layer of gray matter that lies just under the surface of the cerebrum.

cerebral hemispheres. The two sides of the cerebrum, derived from the paired telencephalic vesicles.

cerebrospinal fluid (CSF). The fluid produced by the choroid plexus that flows through the ventricular system to the subarachnoid space.

cerebrum. The largest part of the forebrain; also called the telencephalon.

cGMP. See **cyclic guanosine monophosphate**.

characteristic frequency. The sound frequency to which a neuron in the auditory system gives its greatest response.

chemical synapse. A synapse in which presynaptic activity stimulates the release of neurotransmitter, which activates receptors in the postsynaptic membrane.

chemoaffinity hypothesis. The hypothesis that chemical markers on growing axons are matched with complementary chemical markers on their targets.

chemoreceptor. Any sensory receptor selective for chemicals.

chromosome. A structure in the cell nucleus containing a single linear thread of DNA.

ciliary muscle. A muscle that controls the shape of the eye's lens.

circadian rhythm. Any rhythm with a period of about one day.

classical conditioning. A learning procedure used to associate a stimulus that evokes a measurable response with another stimulus that normally does not evoke this response.

climbing fiber. The axon of an inferior olive neuron that innervates a Purkinje cell of the cerebellum.

CNS. See **central nervous system**.

cochlea. The spiral-shaped bony structure in the inner ear that contains the hair cells that transduce sound.

cochlear amplifier. Outer hair cells, including the motor proteins in the outer hair cell membrane, responsible for amplifying the displacements of the basilar membrane in the cochlea.

cochlear nucleus. See **dorsal cochlear nucleus, ventral cochlear nucleus**.

color-opponent cell. A cell in the visual system with an excitatory response to wavelengths of light having one color and an inhibitory response to wavelengths of another color. The color pairs that cancel each other are red-green and blue-yellow.

commissure. Any collection of axons that connect one side of the brain with the other side.

complex cell. A type of visual cortical neuron that has an orientation selective receptive field without distinct ON and OFF subregions.

concentration gradient. A difference in concentration from one region to another.

conductance. See **electrical conductance**.

conduction aphasia. A type of aphasia associated with damage to the arcuate fasciculus, characterized by good comprehension and speech but difficulty repeating words.

cone photoreceptor. A retinal photoreceptor containing one of three different photopigments that are maximally sensitive to different wavelengths of light. Cones are concentrated in the fovea, specialized for daytime vision, and responsible for all color vision.

conjunctiva. The membrane that folds back from the eyelids and attaches to the sclera of the eye.

consolidation. The process of storing new information into long-term memory.

contralateral. On the opposite side of the midline.

cooperativity. The property of LTP reflecting the requirement that many inputs be active at the same time during a tetanus to induce LTP. See also **long-term potentiation (LTP)**.

cornea. The transparent external surface of the eye.

coronal plane. An anatomical plane of section that divides the nervous system into anterior and posterior parts.

corpus callosum. The great cerebral commissure consisting of axons connecting the cortex of the two cerebral hemispheres.

cortex. Any collection of neurons that forms a thin sheet, usually at the brain's surface.

cortical module. The chunk of cerebral cortex that is necessary and sufficient to analyze one discrete point in a sensory surface.

cortical plate. A cell layer of the immature cerebral cortex containing undifferentiated neurons.

cortical white matter. The collection of axons lying just below the cerebral cortex.

corticomedial nuclei. A group of nuclei in the medial amygdala whose axons constitute the stria terminalis.

corticospinal tract. The tract that originates in the neocortex and terminates in the spinal cord, involved in the control of voluntary movement.

cortisol. A steroid hormone released by the adrenal cortex that mobilizes energy stores, inhibits the immune system, and has direct actions on some CNS neurons.

cranial nerves. The nerves that exit from the brain.

critical period. A limited period of time when a particular aspect of brain development is sensitive to a change in the external environment.

current. See **electrical current**.

cyclic adenosine monophosphate (cAMP). A second messenger formed from ATP by the action of the enzyme adenylyl cyclase.

cyclic guanosine monophosphate (cGMP). A second messenger formed from GTP by the action of the enzyme guanylate cyclase.

cytoarchitectural map. A map, usually of cerebral cortex, based on cytoarchitectural differences.

cytoarchitecture. The arrangement of neuronal cell bodies in parts of the brain.

cytochrome oxidase. A mitochondrial enzyme concentrated in cells that form the blobs in striate cortex.

cytoplasm. Everything contained in the confines of the cell membrane, excluding the nucleus.

cytoskeleton. The internal scaffolding—consisting of microtubules, neurofilaments, and microfilaments—that gives a cell its characteristic shape.

cytosol. The watery fluid inside the cell.

Dale's principle. The idea that a neuron has a unique identity with respect to neurotransmitter.

dark adaptation. The process by which the retina becomes more sensitive to light in dim lighting conditions.

dark current. The inward Na^+ current that occurs in photoreceptors in the dark.

declarative memory. Memory for facts and events.

delayed non-match to sample (DNMS). A behavioral task in which animals are trained to displace one of two alternative objects that does not match a previously seen sample object.

dendrite. A neurite specialized to receive synaptic inputs from other neurons.

GLOSSARY

dendritic spine. A small bag of membrane that protrudes from the dendrites of some cells and receives synaptic input.

dendritic tree. All the dendrites of a single neuron.

dentate gyrus. A layer of neurons in the hippocampus that receives input from the entorhinal cortex.

depolarize. To make the membrane potential less negative.

depression. An affective disorder characterized by prolonged, severe impairment of mood; may include anxiety, sleep disturbances, and other physiological disturbances.

dermatome. A region of skin innervated by the pair of dorsal roots from one spinal segment.

diacylglycerol (DAG). A second messenger molecule formed by the action of phospholipase C on the membrane phospholipid phosphatidylinositol-4,5-bisphosphate. DAG activates the enzyme protein kinase C.

diencephalon. A region of the brain stem derived from the prosencephalon. Diencephalic structures include the thalamus and the hypothalamus.

differentiation. The process whereby structures become more elaborate and specialized.

diffuse modulatory system. One of several systems of CNS neurons that project widely and diffusely onto large areas of the brain and use modulatory neurotransmitters.

diffusion. The temperature-dependent movement of molecules from regions of high concentration to regions of low concentration, causing them to become more evenly distributed.

diopter. A unit of measurement for refractive power of the eye; the reciprocal of the focal distance.

direction selectivity. The property of cells in the visual system that respond only when stimuli move within a limited range of directions.

distal muscle. A muscle that controls the hands, feet, or digits.

DNA (deoxyribonucleic acid). A double-stranded molecule constructed from four nucleic acids that contains the genetic instructions for a cell.

dopa. A chemical precursor of dopamine and the other catecholamines.

dopamine (DA). A catecholamine neurotransmitter synthesized from dopa.

dorsal. A direction meaning toward the back.

dorsal cochlear nucleus. A nucleus in the medulla that receives afferents from the spiral ganglion in the cochlea.

dorsal column. A white matter tract on the dorsal side of the spinal cord, carrying touch and proprioceptive axons to the brain stem.

dorsal column nuclei. A pair of nuclei located in the posterior medulla; target of dorsal column axons, mediating touch and proprioceptive input from the limbs and trunk.

dorsal column-medial lemniscal pathway. An ascending somatic sensory pathway that mediates information about touch, pressure, vibration, and limb proprioception.

dorsal horn. The dorsal region of the spinal cord containing neuronal cell bodies.

dorsal longitudinal fasciculus. A bundle of axons reciprocally connecting the hypothalamus and midbrain periaqueductal gray matter.

dorsal root. A bundle of sensory axons that emerges from a spinal nerve and attaches to the dorsal side of the spinal cord.

dorsal root ganglion. A collection of cell bodies of the sensory neurons that are part of the somatic PNS. There is one dorsal root ganglion for each spinal nerve.

duplex theory of sound localization. The principle that two schemes function in sound localization: interaural time delay at low frequencies and interaural intensity difference at high frequencies.

dura mater. The outermost of the three meninges that cover the surface of the CNS.

EEG. See **electroencephalogram**.

efferent. An axon originating in, and coursing away from, a given structure.

electrical conductance. A measure of the ease with which charged particles can migrate from one point to another, represented by the symbol g and measured in units called siemens. Conductance is the inverse of the resistance and is related to electrical current and voltage by Ohm's law.

electrical current. The rate of movement of electrical charge, represented by the symbol I and measured in units called amperes (amp).

electrical potential. The force exerted on an electrically charged particle, represented by the symbol V and measured in units called volts; also called **voltage** or potential difference.

electrical resistance. A measure of the difficulty with which charged particles can migrate from one point to another, represented by the symbol R and measured in units called ohms. Resistance is the inverse of the conductance and is related to electrical current and voltage by Ohm' law.

electrical self-stimulation. Electrical stimulation that an animal can voluntarily deliver to a portion of its brain.

electrical synapse. A synapse in which electrical current can flow directly from one cell to another via a gap junction.

electroencephalogram (EEG). Electrical activity generated by the brain and recorded from the scalp.

endolymph. The fluid that fills the scala media, containing high K$^+$ and low Na$^+$ concentrations.

endorphin. One of many endogenous opioid peptides with actions similar to morphine; present in many brain structures, particularly those related to pain.

engram. The physical representation or location of a memory.

enteric division. A division of the ANS that innervates the digestive organs, and consists of the myenteric and submucous plexuses.

entorhinal cortex. A cortical region in the medial temporal lobe that occupies the medial bank of the rhinal sulcus; provides input to the hippocampus.

ependymal cell. A type of glial cell that provides the lining of the brain ventricular system.

epilepsy. A chronic brain disorder characterized by recurrent seizures.

epinephrine. A catecholamine neurotransmitter synthesized from norepinephrine; also called **adrenaline**.

EPSP. See **excitatory postsynaptic potential**.

EPSP summation. A simple form of synaptic integration whereby EPSPs add together to produce a larger postsynaptic depolarization.

equilibrium potential. See **ionic equilibrium potential**.

Eustachian tube. An air-filled tube connecting the middle ear to the mouth.

excitable membrane. Any membrane capable of generating action potentials. The membrane of axons and muscle cells is excitable.

excitatory postsynaptic potential (EPSP). Depolarization of the postsynaptic membrane potential by the action of a synaptically released neurotransmitter.

exocytosis. The process whereby material is released from an intracellular vesicle into the extracellular space by fusion of the vesicle membrane with the cell membrane.

extension. The direction of movement that opens a joint.

extensor. A muscle that causes extension when it contracts.

extracellular matrix. The network of fibrous proteins deposited in the space between cells.

extrafusal fiber. The muscle fiber in skeletal muscle that lies outside muscle spindles and receives innervation from alpha motor neurons.

extraocular muscle. A muscle that moves the eye in the orbit.

falling phase. The part of an action potential characterized by the rapid fall of membrane potential from positive to negative.

fasciculation. The process whereby axons growing together stick to one another.

fast motor unit. A motor unit with a large alpha motor neuron innervating rapidly contracting and rapidly fatiguing white muscle fibers.

flexion. The direction of movement that closes a joint.

flexor. A muscle that causes flexion when it contracts.

forebrain. The region of the brain derived from the rostral primary embryonic brain vesicle; also called the prosencephalon. Forebrain structures include the telencephalon and the diencephalon.

fornix. A bundle of axons that originates in the hippocampal formation, loops around the thalamus, and terminates in the diencephalon.

fourth ventricle. The space filled with cerebrospinal fluid within the hindbrain.

fovea. The pit or depression in the retina at the center of the macula. The fovea in humans contains only cone photoreceptors and is specialized for high-acuity vision.

frequency. The number of waves (or other discrete events) per second, expressed in Hz.

frontal lobe. The region of the cerebrum lying anterior to the central sulcus, under the frontal bone.

G-protein. A membrane-bound protein that binds GTP when activated by a membrane receptor. Active G-proteins can stimulate or inhibit other membrane-bound proteins.

G-protein-coupled receptor. A membrane protein that activates G-proteins when it binds neurotransmitter.

GABA. See **gamma-aminobutyric acid**.

gamma motor neuron. A motor neuron that innervates the intrafusal muscle fibers.

gamma-aminobutyric acid (GABA). An amino acid synthesized from glutamate. GABA is the major inhibitory neurotransmitter in the CNS.

ganglion. A collection of neurons in the PNS.

ganglion cell layer. The layer of the retina closest to the center of the eye, containing ganglion cells.

GLOSSARY

ganglion cell. The cell in the retina that receives input from the bipolar cells and sends an axon into the optic nerve.

gap junction. A specialized junction where a narrow gap between two cells is spanned by protein channels (connexons) that allow ions to pass directly from one cell to another.

gating. A property of many ion channels, making them open or closed in response to specific signals such as membrane voltage or the presence of neurotransmitters.

gene. A segment of DNA carrying the instructions for a single protein.

gene expression. The process of transcribing the information from a gene into mRNA.

generalized seizure. Pathologically large and synchronous neural activity that spreads to encompass the entire cerebral hemispheres.

globus pallidus. A part of the basal ganglia in the basal forebrain, involved in motor control.

glomerulus. A cluster of neurons in the olfactory bulb that receives input from olfactory receptor neurons.

glutamate (Glu). An amino acid; the major excitatory neurotransmitter in the CNS.

glycine (Gly). An amino acid; an inhibitory neurotransmitter at some locations in the CNS.

Goldman equation. A mathematical relationship used to predict membrane potential from the concentrations and membrane permeabilities of ions.

Golgi apparatus. An organelle that sorts and chemically modifies proteins that are destined for delivery to different parts of the cell.

Golgi stain. A method of staining brain tissue that shows neurons and all of their neurites; named for its discoverer, Italian histologist Camillio Golgi (1843–1926).

Golgi tendon organ. A specialized structure within the tendons of skeletal muscle that senses muscle tension.

Golgi Type I neuron. A neuron in the brain with a long axon that transfers information from one region of the brain to another.

Golgi Type II neuron. A neuron in the brain with a short axon that does not extend beyond the vicinity of the cell body.

gray matter. A generic term for a collection of neuronal cell bodies in the CNS.

Gray's type I synapse. A chemical synapse in the CNS with asymmetrical membrane differentiations.

Gray's type II synapse. A chemical synapse in the CNS with symmetrical membrane differentiations.

growth cone. The specialized tip of a growing neurite.

GTP. Guanosine triphosphate.

gustation. The sense of taste.

gustatory nucleus. A nucleus in the brain stem that receives primary taste input.

gyrus. A bump or bulge lying between the sulci of the cerebrum. Plural: gyri.

habituation. A form of nonassociative learning leading to decreased behavioral responses to repeated stimulation.

hair cell. An auditory cell that transduces sound into a change in membrane potential.

Hebb synapse. A synapse that exhibits Hebbian modifications.

Hebbian modification. An increase in the effectiveness of a synapse caused by the simultaneous activation of pre- and postsynaptic neurons.

helicotrema. A hole at the apex of the cochlea that connects the scala tympani to the scala vestibuli.

hertz (Hz). The unit of frequency equivalent to cycles per second.

hindbrain. The region of the brain derived from the caudal primary embryonic brain vesicle; also called the rhombencephalon. Hindbrain structures include the cerebellum, pons, and medulla.

hippocampus. A region of the cerebral cortex lying adjacent and medial to the olfactory cortex. In humans, the hippocampus is in the temporal lobe and may play a special role in learning and memory.

histology. The microscopic study of the structure of tissues.

homeostasis. The balanced functioning of physiological processes and maintenance of an organism's constant internal environment.

horizontal cell. A cell in the retina that projects neurites laterally in the outer plexiform layer.

horizontal plane. An anatomical plane of section that divides the nervous system into dorsal and ventral parts.

hyperalgesia. A reduced threshold for pain, an increased response to painful stimuli, or a spontaneous pain that follows localized injury.

hypophysiotropic hormone. A peptide hormone, such as CRH or GRH, released into the blood by the parvocellular neurosecretory neurons of the hypothalamus, which stimulates or inhibits the secretion of hormones from the anterior pituitary.

hypothalamo-pituitary portal circulation. A system of blood vessels that carries hypophysiotropic hormones from the hypothalamus to the anterior pituitary.

hypothalamus. The ventral part of the diencephalon, involved in the control of the ANS and the pituitary gland.

immunocytochemistry. An anatomical method that uses antibodies to study the location of molecules within cells.

in situ **hybridization.** A method for localizing strands of mRNA within cells.

incus. An ossicle in the middle ear whose shape somewhat resembles an anvil.

inferior colliculus. A nucleus in the midbrain from which all ascending auditory signals project to the MGN.

inferior olive. A nucleus of the medulla that gives rise to climbing fiber input to the cerebellar cortex.

inhibitor. A drug or toxin that blocks the normal action of a protein or a biochemical process.

inhibitory postsynaptic potential (IPSP). Change of the postsynaptic membrane potential by the action of a synaptically released neurotransmitter making the postsynaptic neuron less likely to fire action potentials.

inner ear. The cochlea, which is part of the auditory system, plus the labyrinth, which is part of the vestibular apparatus.

inner hair cell. An auditory cell located between the modiolus and the rods of Corti; the primary transducer of sound into an electrochemical signal.

inner nuclear layer. The layer of the retina containing the cell bodies of bipolar, horizontal, and amacrine cells.

inner plexiform layer. The layer of the retina, lying between the ganglion cell layer and the inner nuclear layer, that contains the neurites and synapses between bipolar cells, amacrine cells, and ganglion cells.

innervate. To form synapses with.

inositol-1,4,5-triphosphate (IP$_3$). A second messenger molecule formed by the action of phospholipase C on the membrane phospholipid phosphatidylinositol-4,5-bisphosphate. IP$_3$ causes the release of Ca^{2+} from intracellular stores.

input specificity. A property of some forms of synaptic plasticity such that only the active synapses onto a neuron are modified.

instrumental conditioning. A learning procedure used to associate a response, such as a motor act, with a meaningful stimulus, such as food.

intensity. The amplitude of a wave. Sound intensity is the amplitude of the sound wave that perceptually determines loudness.

internal capsule. A large collection of axons that connects the telencephalon with the diencephalon.

internal resistance. The resistance to electrical current flow longitudinally down a cable or neurite, represented by the symbol r_i.

interneuron. Any neuron that is not a sensory or motor neuron. The term is also used to describe a CNS neuron whose axon does not leave the structure in which it resides.

intrafusal fiber. The specialized muscle fiber contained within a muscle spindle that receives motor innervation from gamma motor neurons.

ion. An atom or molecule that has a net electrical charge due to a difference in the number of electrons and protons.

ion channel. A membrane-spanning protein that forms a pore allowing passage of ions from one side of the membrane to the other.

ion pump. An enzyme that transports ions across a membrane at the expense of metabolic energy.

ion selectivity. A property of ion channels that are selectively permeable to some ions and not to others.

ionic driving force. The difference between the real membrane potential, V_m, and the ionic equilibrium potential, E_{ion}.

ionic equilibrium potential. The electrical potential difference that exactly balances an ionic concentration gradient, represented by the symbol E_{ion}.

IP$_3$. See **inositol-1,4,5-triphosphate**.

ipsilateral. On the same side of the midline.

IPSP. See **inhibitory postsynaptic potential**.

iris. The circular, pigmented muscle that controls the size of the pupil.

James-Lange theory. A theory of emotion proposing that the subjective experience of emotion is a consequence of physiological changes in the body.

kainate receptor. A subtype of glutamate receptor; a glutamate-gated ion channel that is permeable to Na$^+$ and K$^+$.

Klüver-Bucy syndrome. The constellation of symptoms, including psychic blindness, oral tendencies, hypermetamorphosis, altered sexual behavior, and emotional changes, that are observed in humans and monkeys after bilateral temporal lobectomy.

koniocellular LGN layer. A layer of the LGN containing very small cells, lying just ventral to each magnocellular and parvocellular layer.

Korsakoff's syndrome. A neurological syndrome resulting from chronic alcoholism, characterized by confusion, confabulations, apathy, and amnesia.

GLOSSARY

large, dense-core vesicle. A spherical membrane-enclosed vesicle, about 100 nm in diameter, containing peptides intended for secretion by exocytosis; also called **secretory granule**.

lateral. A direction meaning away from the midline.

lateral geniculate nucleus (LGN). The thalamic nucleus that relays information from the retina to the primary visual cortex.

lateral intraparietal cortex (area LIP). A cortical area buried in the intraparietal sulcus involved in vision and visually guided behavior.

lateral pathway. The axons in the lateral column of the spinal cord that are involved in the control of voluntary movements of the distal musculature and are under direct cortical control.

lateral ventricle. The space filled with cerebrospinal fluid within each cerebral hemisphere.

layer of photoreceptor outer segments. The layer of the retina farthest from the center of the eye, containing the light-sensitive elements of the photoreceptors.

learning. The acquisition of new information or knowledge.

lemniscus. A tract that meanders through the brain like a ribbon.

length constant. A parameter used to describe how far changes in membrane potential can passively spread down a cable such as an axon or a dendrite, represented by the symbol λ. The length constant λ is the distance at which the voltage change falls to 37% of its original value; λ depends on the ratio of r_m to r_i.

lens. The transparent structure lying between the aqueous humor and the vitreous humor that enables the eye to adjust its focus to different viewing distances.

LGN. See **lateral geniculate nucleus**.

ligand-binding method. A method that uses radioactive receptor ligands (agonists or antagonists) to identify the location of neurotransmitter receptors.

light adaptation. The process by which the retina becomes less sensitive to light in bright lighting conditions.

limbic lobe. The hippocampus and cortical areas bordering the mammalian brain stem, which Broca proposed as a distinct lobe of the brain.

limbic system. A group of structures, including those in the limbic lobe and Papez circuit, that are anatomically interconnected and are involved in emotion, learning, and memory.

locus. A small, well-defined group of cells.

locus coeruleus. A small nucleus located bilaterally in the pons; its neurons use norepinephrine as their neurotransmitter and project widely upon all levels of the CNS.

long-term depression (LTD). A long-lasting decrement of the effectiveness of synaptic transmission that follows certain types of conditioning stimulation.

long-term memory. Information storage that is relatively permanent and does not require continual rehearsal.

long-term potentiation (LTP). A long-lasting enhancement of the effectiveness of synaptic transmission that follows certain types of conditioning stimulation.

M-type ganglion cell. A type of retinal ganglion cell characterized by a large cell body and dendritic arbor, a transient response to light, and no sensitivity to different wavelengths of light.

M1. Primary motor cortex, area 4.

macula. A yellowish spot in the middle of the retina with relatively few large blood vessels, containing the fovea.

magnetic resonance imaging (MRI). A technique for imaging the brain in which hydrogen atoms are localized by changing their atomic state in a magnetic field.

magnocellular channel (M channel). The visual information processing channel that begins with the M-type retinal ganglion cells and leads to layer IVB of striate cortex; believed to process information about visual movement.

magnocellular LGN layer. A layer of the LGN receiving synaptic input from the M-type retinal ganglion cells.

magnocellular neurosecretory neuron. A large neuron of the periventricular and supraoptic nuclei of the hypothalamus that projects to the posterior pituitary and secretes oxytocin or vasopressin into the blood.

malleus. An ossicle in the middle ear attached to the tympanic membrane whose shape somewhat resembles a hammer.

Marr-Albus theory of motor learning. The theory that parallel fiber synapses on Purkinje cells are modified when their activity coincides with climbing fiber activity.

mechanoreceptor. Any sensory receptor selective for mechanical stimuli, such as hair cells of the inner ear, various receptors of the skin, or stretch receptors of skeletal muscle.

medial. A direction meaning toward the midline.

medial forebrain bundle. A large bundle of axons coursing through the hypothalamus carrying efferents from the dopaminergic, noradrenergic, and serotonergic neurons in the brain stem as well as fibers interconnecting the hypothalamus, limbic structures, and midbrain tegmental area.

medial geniculate nucleus (MGN). A relay nucleus in the thalamus through which all auditory information passes on its way from the inferior colliculus to auditory cortex.

medial lemniscus. A white matter tract of the somatic sensory system, carrying axons from dorsal column nuclei to the thalamus.

medulla. The part of the hindbrain caudal to the pons and cerebellum.

medullary reticulospinal tract. A tract originating in the medullary reticular formation and terminating in the spinal cord, involved in the control of movement.

membrane differentiation. Dense accumulations of protein in the membranes on either side of a synaptic cleft.

membrane potential. The voltage across a cell membrane, represented by the symbol V_m.

membrane resistance. The resistance to electrical current flow across the membrane, represented by the symbol r_m.

memory. The retention of learned information.

meninges. Three membranes that cover the surface of the CNS: the dura mater, arachnoid membrane, and pia mater.

metabotropic receptor. A G-protein-coupled receptor whose primary action is to stimulate or inhibit an intracellular biochemical response.

MGN. See **medial geniculate nucleus**.

microelectrode. A probe used to measure the electrical activity of cells. Microelectrodes have a very fine tip and can be fashioned from etched metal or glass pipettes filled with electrically conductive solutions.

microfilament. A polymer of the protein actin, forming a braided strand 5 nm in diameter. Microfilaments contribute to the cytoskeleton.

microglial cell. A type of cell that functions as a phagocyte in the nervous system to remove debris left by dead or dying neurons and glia.

microionophoresis. A method of applying drugs and neurotransmitters in very small quantities to cells.

microtubule. A polymer of the protein tubulin, forming a straight, hollow tube, 20 nm in diameter. Microtubules contribute to the cytoskeleton and play an important role in axoplasmic transport.

midbrain. The region of the brain derived from the middle primary embryonic brain vesicle, also called the mesencephalon. Midbrain structures include the tectum and the tegmentum.

middle ear. The tympanic membrane plus the ossicles.

midline. The line that bisects the nervous system into right and left halves.

midsagittal plane. An anatomical plane of section through the midline that is perpendicular to the ground. A section in the midsagittal plane divides the nervous system into right and left halves.

miniature postsynaptic potential. The change in postsynaptic membrane potential caused by the action of neurotransmitter released from a single synaptic vesicle.

mitochondrion. An organelle responsible for cellular respiration. Mitochondria generate ATP using the energy produced by the oxidation of food.

modiolus. A conical bony structure around which the cochlea spirals.

modulation. A term used to describe the actions of neurotransmitters that do not directly evoke postsynaptic potentials but modify the cellular response to EPSPs and IPSPs generated by other synapses.

monocular deprivation. An experimental manipulation that deprives one eye of normal vision.

Morris water maze. A task used to assess spatial memory in which a rodent must swim to the location of a hidden platform sitting below the surface of a pool of water.

mossy fiber. The axon of a pontine neuron that innervates cerebellar granule cells.

motor cortex. Cortical areas 4 and 6, which are directly involved in the control of voluntary movement.

motor end-plate. The postsynaptic membrane at the neuromuscular junction.

motor neuron. A neuron that synapses on a muscle cell and causes muscle contraction.

motor neuron pool. All the alpha motor neurons innervating the fibers of a single skeletal muscle.

motor strip. A name for area 4 on the precentral gyrus.

motor unit. One alpha motor neuron and all the muscle fibers it innervates.

mRNA (messenger ribonucleic acid). A molecule constructed from four nucleic acids that carries the genetic instructions for the assembly of a protein from the nucleus to the cytoplasm.

multipolar neuron. A neuron with three or more neurites.

GLOSSARY

muscarinic receptor. A subtype of ACh receptor that is G-protein-coupled.

muscle fiber. A multinucleated skeletal muscle cell.

muscle spindle. A specialized structure within skeletal muscles that senses muscle length; also called a stretch receptor. Muscle spindles provide sensory information to neurons in the spinal cord via group Ia afferents.

myelin. A membranous wrapping around axons provided by oligodendroglia in the CNS and Schwann cells in the PNS.

myofibril. A cylindrical structure within a muscle fiber that contracts in response to an action potential.

myosin. A cytoskeletal protein in all cells, and the major thick filament protein in a muscle fiber; causes muscle contraction by chemical interactions with actin.

myotatic reflex. The reflex that leads to muscle contraction in response to muscle stretch. The myotatic reflex is mediated by the monosynaptic connection of a group Ia afferent from a muscle spindle with an alpha motor neuron innervating the same muscle.

neglect syndrome. A neurological disorder in which a part of the body or a part of the visual field is ignored or suppressed; most commonly associated with damage to posterior parietal areas of the brain.

neocortex. The cerebral cortex with six or more layers, found only in mammals.

Nernst equation. A mathematical relationship used to calculate an ionic equilibrium potential.

nerve-growth factor (NGF). A neurotrophin required for the survival of the cells of the sympathetic division of the ANS; also important for aspects of CNS development.

neural crest. The primitive embryonic PNS, consisting of neural ectoderm that pinches off laterally as the neural tube forms.

neural tube. The primitive embryonic CNS, consisting of a tube of neural ectoderm.

neurite. Any thin tube extending from a neuronal cell body. Neurites can be further divided into axons and dendrites.

neuroblast. An immature neuron, prior to cell differentiation.

neurofilament. A type of intermediate filament found in neurons. Neurofilaments measure 10 nm in diameter, and are an important component of the neuronal cytoskeleton.

neurohormone. A hormone released by neurons into the bloodstream.

neuroleptic. A class of drugs, also known as the major tranquilizers, helpful in treating certain psychoses such as schizophrenia.

neuromuscular junction. A chemical synapse between a spinal motor neuron axon and a skeletal muscle fiber.

neuron doctrine. The concept that the neuron is the elementary functional unit of the brain, and that neurons communicate with each other by contact, not continuity.

neuronal membrane. The barrier, about 5 nm thick, that separates the inside of the cell from the outside; consists of a phospholipid bilayer with proteins embedded in it; encloses the intracellular organelles and vesicles.

neuropharmacology. The study of the effects of drugs on nervous system tissue.

neurotransmitter. A chemical that is released by a presynaptic element upon stimulation and activates postsynaptic receptors.

neurotrophin. A member of a family of related neuronal trophic factors, including nerve growth factor (NGF) and brain-derived neurotrophic factor (BDNF).

nicotinic ACh receptor. A class of ACh-gated ion channel found in various locations, notably at the neuromuscular junction.

Nissl stain. A class of basic dyes that stain the somata of neurons. Named for its discoverer, German histologist Franz Nissl (1860–1919).

nitric oxide (NO). A gas produced from the amino acid arginine that serves as an intercellular messenger.

NMDA receptor. A subtype of glutamate receptor; a glutamate-gated ion channel that is permeable to Na^+, K^+, and Ca^{2+}. Inward ionic current through the NMDA receptor is voltage-dependent due to a magnesium block at negative membrane potentials.

nociceptor. Any receptor selective for potentially harmful stimuli; may induce sensations of pain.

node of Ranvier. A space between two consecutive myelin sheaths where an axon comes in contact with the extracellular fluid.

non-REM sleep. A stage of sleep characterized by large, slow EEG waves; a paucity of dreams; and some muscle tone; also known as NREM sleep, or slow-wave sleep.

nonassociative learning. The change in the behavioral response that occurs over time in response to a single type of stimulus; divided into two types: habituation and sensitization.

norepinephrine (NE). A catecholamine neurotransmitter synthesized from dopamine; also called noradrenaline.

nucleus. 1. The roughly spherical organelle in the cell body containing the chromosomes. 2. A general term used to describe a clearly distinguishable mass of neurons, usually deep in the brain.

nucleus of the solitary tract. A brain stem nucleus that receives sensory input and uses it to coordinate autonomic function via its outputs to other brain stem and forebrain nuclei, and the hypothalamus.

occipital lobe. The region of the cerebrum lying under the occipital bone.

ocular dominance column. A region of striate cortex receiving information predominantly from one eye.

ocular dominance shift. A change in visual cortex interconnections that makes more neurons responsive to one eye or the other.

OFF bipolar cell. A bipolar cell of the retina that depolarizes in response to dark (light OFF) in the center of its receptive field.

Ohm's law. The relationship between electrical current (I), voltage (V), and conductance (g): I = gV. Because electrical conductance is the inverse of resistance (R), Ohm's law may also be written V = IR.

olfaction. The sense of smell.

olfactory bulb. A bulb-shaped brain structure derived from the telencephalon that receives input from olfactory receptor neurons.

olfactory cortex. The region of the cerebral cortex connected to the olfactory bulb and separated from the neocortex by the rhinal fissure.

olfactory epithelium. A sheet of cells lining part of the nasal passages that contains olfactory receptors neurons.

oligodendroglial cell. A glial cell that provides myelin in the CNS.

ON bipolar cell. A bipolar cell of the retina that depolarizes in response to light (light ON) in the center of its receptive field.

opioid receptor. A membrane protein that selectively binds natural (e.g., endorphin) and unnatural (e.g., morphine) opiate substances.

optic chiasm. The structure in which the right and left optic nerves converge and partially decussate to form the optic tracts.

optic disk. The location on the retina where optic nerve axons leave the eye.

optic nerve. The bundle of ganglion cell axons that passes from the eye to the optic chiasm.

optic radiation. A collection of axons coursing from the LGN to the visual cortex.

optic tectum. A term used to describe the superior colliculus, particularly in nonmammalian vertebrates.

optic tract. A collection of retinal ganglion cell axons stretching from the optic chiasm to the brain stem. Important targets of the optic tract are the LGN and superior colliculus.

organ of Corti. The auditory receptor organ that contains hair cells, rods of Corti, and supporting cells.

organelle. A membrane-enclosed structure inside a cell. Examples are the nucleus, mitochondrion, endoplasmic reticulum, and Golgi apparatus.

orientation column. A column of visual cortical neurons stretching from layer II to layer VI that responds best to the same stimulus orientation.

orientation selectivity. The property of a cell in the visual system that responds to a limited range of stimulus orientations.

ossicle. One of three small bones in the middle ear.

outer ear. The pinna plus the auditory canal.

outer hair cell. An auditory receptor cell located farther from the modiolus than the rods of Corti.

outer nuclear layer. The layer of the retina containing the cell bodies of the photoreceptors.

outer plexiform layer. The layer of the retina, lying between the inner nuclear layer and the outer nuclear layer, that contains the neurites and synapses between photoreceptors, horizontal cells, and bipolar cells.

oval window. A hole in the bony cochlea at which movement of the ossicles is transferred to movement of the fluids in the cochlea.

overshoot. The part of an action potential when the membrane potential is more positive than zero millivolts.

oxytocin. A small peptide hormone released from the posterior pituitary by magnocellular neurons; stimulates uterine contractions and the let-down of milk from mammary glands.

P-type ganglion cell. A type of retinal ganglion cell characterized by a small cell body and dendritic arbor, a sustained response to light, and sensitivity to different wavelengths of light.

Pacinian corpuscle. A mechanoreceptor of the deep skin, selective for high-frequency vibrations.

Papez circuit. A circuit of anatomical structures interconnecting the hypothalamus and cortex, which Papez proposed to be an emotion system.

GLOSSARY

papilla. A small protuberance on the surface of the tongue that contains taste buds.

parahippocampal cortex. A cortical region in the medial temporal lobe that lies lateral to the perirhinal cortex.

parallel fiber. The axon of a cerebellar granule cell that innervates Purkinje cells.

parallel processing. The idea that different stimulus attributes are processed by the brain in parallel, using distinct pathways.

parasympathetic division. A division of the ANS; its peripheral axons emerge from the brain stem and sacral spinal cord; it maintains heart rate and respiratory, metabolic, and digestive functions under normal conditions.

parietal lobe. The region of the cerebrum posterior to the central sulcus lying under the parietal bone.

Parkinson's disease. A movement disorder caused by damage to the substantia nigra, characterized by paucity of movement, difficulty in initiating willed movement, and resting tremor.

partial decussation. A partial crossing of an axonal pathway from one side of the CNS to the other.

partial seizure. Pathologically large and synchronous neural activity that remains localized to a relatively small region of the brain.

parvocellular LGN layer. A layer of the LGN receiving synaptic input from the P-type retinal ganglion cells.

parvocellular neurosecretory neuron. A small neuron of the medial and periventricular hypothalamus that secretes hypophysiotropic peptide hormones into the hypothalamo-pituitary portal circulation to stimulate or inhibit the release of hormones from the anterior pituitary.

parvocellular-interblob channel (P-IB channel). The visual information processing channel that begins with the P-type retinal ganglion cells and leads to the interblob regions of layer III. It is believed to process information about object shape.

patch clamp. A method that enables an investigator to hold constant the membrane potential of a patch of membrane while current through a small number of membrane channels is measured.

peptide bond. The covalent bond between the amino group of one amino acid and the carboxyl group of another.

perforant path. The axonal pathway from the entorhinal cortex to the dentate gyrus of the hippocampus.

periaqueductal gray matter (PAG). A region surrounding the cerebral aqueduct in the core of the mid-brain with descending pathways that can inhibit the transmission of pain-causing signals.

perikaryon. The central region of the neuron containing the nucleus; also called the **soma** or **cell body**.

perilymph. The fluid that fills the scala vestibuli and scala tympani, containing low K^+ and high Na^+ concentrations.

peripheral nervous system (PNS). The parts of the nervous system other than the brain and spinal cord. The PNS includes all the spinal ganglia and nerves, cranial nerves III–XII, and the ANS.

perirhinal cortex. A cortical region in the medial temporal lobe that occupies the lateral bank of the rhinal sulcus.

periventricular zone. A hypothalamic region that lies most medially, bordering the third ventricle.

phase locking. Consistent firing of an auditory neuron at the same phase of a sound wave.

pheromone. An olfactory stimulus used for chemical communication between individuals.

phosphodiesterase (PDE). The enzyme that breaks down the cyclic nucleotide second messengers cAMP and cGMP.

phospholipase C (PLC). An enzyme that cleaves the membrane phospholipid phosphatidylinositol-4,5-bisphosphate to form the second messengers diacylglycerol (DAG) and inositol triphosphate (IP_3).

phospholipid bilayer. The arrangement of phospholipid molecules that forms the basic structure of the cell membrane. The core of the bilayer is lipid, creating a barrier to water and to water-soluble ions and molecules.

phosphorylation. A biochemical reaction in which a phosphate group (PO_4^{2-}) is transferred from ATP to another molecule. Phosphorylation of proteins by protein kinases changes their biological activity.

photoreceptor. A specialized cell in the retina that transduces light energy into changes in membrane potential.

pia mater. The innermost of the three meninges that cover the surface of the CNS.

pinna. The funnel-shaped outer ear consisting of skin covering cartilage.

place cell. A neuron in the rat hippocampus that only responds when the animal is in a certain region of space.

planum temporale. An area on the superior surface of the human temporal lobe that is frequently larger in the left hemisphere than the right.

pleasure center. A name originally given to any self-reinforcing sites in the brain, regardless of whether the electrical stimulation evoked pleasure.

PNS. See **peripheral nervous system**.

point image. All the neurons activated by stimulation of one point in the visual field.

polypeptide. A string of amino acids held together with peptide bonds.

polyribosome. A collection of several ribosomes, floating freely in the cytoplasm.

pons. The part of the rostral hindbrain that lies ventral to the cerebellum and the fourth ventricle.

pontine nuclei. The clusters of neurons that relay information from the cerebral cortex to the cerebellar cortex.

pontine reticulospinal tract. A tract originating in the pontine reticular formation and terminating in the spinal cord, involved in the control of movement.

positron emission tomography (PET). A technique used for imaging brain activity by measuring the flow of blood containing radioactive atoms that emit positrons.

posterior. A direction meaning toward the tail, or caudal.

posterior parietal cortex. The posterior region of the parietal lobe, mainly Brodmann's areas 5 and 7, involved in visual and somatosensory integration and attention.

postganglionic neuron. A peripheral neuron of the sympathetic and parasympathetic divisions of the **ANS**; its cell body lies in autonomic ganglia and its axons terminate on peripheral organs and tissues.

postsynaptic density. A postsynaptic membrane differentiation that is the site of neurotransmitter receptors.

potential. See **electrical potential**.

predatory aggression. Attack behavior, often with the goal of obtaining food, accompanied by few vocalizations and low activity of the ANS.

prefrontal cortex. The cortical area at the rostral end of the frontal lobe that receives input from the dorsomedial nucleus of the thalamus.

preganglionic neuron. A neuron of the sympathetic or parasympathetic divisions of the ANS; its cell body lies in the CNS (spinal cord or brain stem) and its axons extend peripherally to synapse on postganglionic neurons in the autonomic ganglia.

premotor area (PMA). The lateral part of cortical area 6, involved in the control of voluntary movement.

primary auditory cortex. Brodmann's area 41, located on the superior surface of the temporal lobe; also called A1.

primary gustatory cortex. The area of neocortex that receives taste information from the ventral posteri-

or medial (VPM) nucleus; corresponding mainly to Brodmann's area 43.

primary sensory neuron. A neuron specialized to detect environmental signals at the body's sensory surfaces.

primary somatosensory cortex. The Brodmann's areas 3a, 3b, 1, and 2, located in the postcentral gyrus; also called **S1**.

procedural memory. Memory for skills.

proprioception. Sensation of body position and movement using sensory signals from muscle, joints, and skin.

proprioceptor. A sensory receptor from the muscles, joints, and skin that contributes to proprioception.

protein. A molecule constructed from amino acids according to genetic instructions.

protein kinase A (PKA). A protein kinase that is activated by the second messenger cAMP.

protein kinase. A class of enzyme that phosphorylates proteins. This reaction changes the conformation of the protein and its biological activity.

protein kinase C (PKC). A protein kinase activated by the second messenger diacylglycerol.

protein phosphatase. An enzyme that removes phosphate groups from proteins.

protein synthesis. The assembly of protein molecules in the cell's cytoplasm according to genetic instructions.

proximal (girdle) muscle. A muscle that controls the shoulder or pelvis.

psychosurgery. Brain surgery used to treat behavioral disorders.

pupil. The opening that allows light to enter the eye and strike the retina.

pupillary light reflex. The reflex, involving retinal inputs to brain stem neurons that control the iris, that causes the diameter of the pupil to become larger in dim light and smaller in bright light.

Purkinje cell. A cell in the cerebellar cortex that projects an axon to the deep cerebellar nuclei.

putamen. A part of the basal ganglia in the basal forebrain, involved in motor control.

pyramidal cell. A neuron characterized by a pyramidal-shaped cell body and elongated dendritic tree, found in the cerebral cortex.

pyramidal tract. A tract running along the ventral medulla that carries corticospinal axons.

quantal analysis. A method for determining how many vesicles release neurotransmitter during normal synaptic transmission.

GLOSSARY

radial glial cell. A glial cell in the embryonic brain extending a process from the ventricular zone to the surface of the brain, along which immature neurons and glia migrate.

raphe nuclei. Clusters of serotonergic neurons that lie along the midline of the brain stem, from the midbrain to the medulla, and project diffusely upon all levels of the CNS.

rapid eye movement sleep (REM sleep). A stage of sleep characterized by low-amplitude, high-frequency EEG waves, vivid dreams, rapid eye movements, and atonia.

rate-limiting step. In the series of biochemical reactions that leads to the production of a chemical, the one step that limits the rate of synthesis.

receptive field. The region of a sensory surface (retina, skin) that, when stimulated, changes the membrane potential of a neuron.

receptor. 1. A specialized protein that detects chemical signals, such as neurotransmitters, and initiates a cellular response. 2. A specialized cell that detects environmental stimuli and generates neural responses.

receptor agonist. A drug that binds to a receptor and activates it.

receptor antagonist. A drug that binds to a receptor and inhibits its function.

receptor potential. A stimulus-induced change in the membrane potential of a sensory receptor.

receptor subtype. One of several different receptors to which a neurotransmitter binds.

reciprocal inhibition. The process whereby the contraction of one set of muscles is accompanied by the relaxation of the antagonist muscles.

recognition memory. Memory required to perform the delayed non-match to sample task.

red nucleus. A cell group in the midbrain involved in the control of movement.

refraction. The bending of light rays that can occur when they travel from one transparent medium to another.

Reissner's membrane. The cochlear membrane that separates the scala vestibuli from the scala media.

relational memory. A type of memory in which all the events occurring at a given time are stored in a manner linking them together.

relative refractory period. The period of time following an action potential during which more depolarizing current is required to achieve threshold than normal.

resistance. See **electrical resistance**.

resting membrane potential. The membrane potential, or membrane voltage, maintained by a cell when it is not generating action potentials. Neurons have a resting membrane potential of about –65 mV.

reticular formation. A region of the brain stem ventral to the cerebral aqueduct and fourth ventricle, involved in many functions including the control of posture and locomotion.

reticular lamina. A thin sheet of tissue that holds the tops of hair cells in the organ of Corti.

retina. The thin layer of cells at the back of the eye that transduces light energy into neural activity.

retinofugal projection. The pathway that carries information from the retina to the visual cortex.

retinotectal projection. A collection of axons coursing from the retina to the superior colliculus.

retinotopy. The topographic organization of visual pathways where neighboring cells on the retina feed information to neighboring cells in a target structure.

retrograde amnesia. Memory loss for events prior to an illness or brain trauma.

retrograde messenger. Any chemical messenger that communicates information from the postsynaptic side of the synapse to the presynaptic side.

retrograde transport. Axoplasmic transport from the axon terminal to the soma.

rhodopsin. The rod photoreceptor photopigment.

ribosome. The structure in the cell that assembles new proteins from amino acids according to the instructions carried by mRNA.

rising phase. The first part of an action potential characterized by the rapid depolarization of the membrane.

rod photoreceptor. A retinal photoreceptor containing rhodopsin and specialized for low light levels.

rostral. A direction meaning toward the nose, or anterior.

rough endoplasmic reticulum (rough ER). A membrane-enclosed organelle with ribosomes attached to its outer surface. Rough ER is a site of synthesis for proteins that are destined to be inserted into membrane or to be enclosed by membrane.

round window. A membrane-covered hole in the bony cochlea that is continuous with the scala tympani in the cochlea.

rubrospinal tract. A tract originating in the red nucleus and terminating in the spinal cord, involved in the control of movement.

S1. Primary somatosensory cortex.

sagittal plane. An anatomical plane parallel to the midsagittal plane.

saltatory conduction. Action potential propagation down a myelinated axon.

sarcolemma. The outer cell membrane of a muscle fiber.

sarcomere. The contractile element between Z lines in a myofibril. The sarcomere contains the thick and thin filaments that slide along one another to cause muscle contraction.

sarcoplasmic reticulum. An organelle within a muscle fiber that stores Ca^{2+} and releases it when stimulated by an action potential in the T tubules.

scala media. A channel in the cochlea that lies between the scala vestibuli and the scala tympani.

scala tympani. A channel in the cochlea that runs from the helicotrema to the round window.

scala vestibuli. A channel in the cochlea that runs from the oval window to the helicotrema.

Schaffer collateral. An axon of a CA3 neuron that innervates neurons in CA1 of the hippocampus.

schizophrenia. A mental disorder characterized by a loss of contact with reality; fragmentation and disruption of thought, perception, mood, and movement; delusions, hallucinations, and disordered memory.

Schwann cell. A glial cell that provides myelin in the **PNS**.

sclera. The tough outer wall of the eyeball; the "white of the eye."

second messenger. A short-lived chemical signal in the cytosol that can trigger a biochemical response. Second messenger formation is usually stimulated by a first messenger (a neurotransmitter or hormone) acting at a G-protein-coupled cell surface receptor. Examples of second messengers are cAMP, cGMP, IP_3.

second messenger cascade. A multistep process that couples activation of a neurotransmitter receptor to activation of intracellular enzymes.

secretory granule. A spherical membrane-enclosed vesicle, about 100 nm in diameter, containing peptides intended for secretion by exocytosis; also called large, dense-core vesicle.

sensitization. A form of nonassociative learning leading to an intensified response to all stimuli.

serotonin (5-HT). An amine neurotransmitter, 5-hydroxytryptamine.

sham rage. Behavior produced by brain lesions; a display of great anger, but in a situation that would not normally cause anger.

short-term memory. Information storage that is temporary, limited in capacity, and requires continual rehearsal.

shunting inhibition. A form of synaptic inhibition in which the main effect is to reduce r_m, thereby shunting depolarizing current generated at excitatory synapses.

simple cell. A cell found in striate cortex having an elongated orientation selective receptive field with distinct ON and OFF subregions.

skeletal muscle. The type of muscle, derived from the mesodermal somites, that is under voluntary control.

slow motor unit. A motor unit with a small alpha motor neuron innervating slowly contracting and slowly fatiguing red muscle fibers.

smooth endoplasmic reticulum (smooth ER). A membrane-enclosed organelle that is heterogeneous and performs different functions in different locations.

smooth muscle. A type of muscle in the digestive tract, arteries, and related structures; innervated by the ANS and not under voluntary control.

sodium-potassium pump. An ion pump that removes intracellular Na^+ ions and concentrates intracellular K^+ ions, using ATP as its energy source.

soma. The central region of the neuron containing the nucleus; also called the **cell body** or **perikaryon**.

somatic motor system. The skeletal muscles and the parts of the nervous system that control them.

somatic PNS. The part of the PNS that innervates the skin, joints, and skeletal muscles.

somatic sensation. The senses of touch, temperature, body position, and pain.

somatotopy. The topographic organization of somatic sensory pathways where neighboring receptors in the skin feed information to neighboring cells in a target structure.

spatial summation. Summation of EPSPs generated at more than one synapse on the same cell.

spike-initiation zone. The region of neuronal membrane, characterized by a high density of voltage-gated sodium channels, where action potentials are normally initiated.

spinal canal. The space filled with cerebrospinal fluid within the spinal cord.

spinal cord. The part of the CNS in the vertebral column.

spinal nerve. A nerve attached to the spinal cord that innervates the body.

spinal segment. One set of dorsal and ventral roots plus the portion of spinal cord related to them.

spinothalamic pathway. An ascending somatic sensory pathway, traveling from spinal cord to the thalamus via the lateral spinothalamic columns, that mediates information about pain, temperature, and some touch.

GLOSSARY

spiny neuron. A neuron with dendritic spines.

spiral ganglion. A collection of cells located in the modiolus of the cochlea that receives input from hair cells and sends output to the cochlear nuclei in the medulla via the auditory nerve.

split-brain study. An examination of behavior in animals or humans that have had the cerebral hemispheres disconnected by cutting the corpus callosum.

stapes. An ossicle in the middle ear attached to the oval window that somewhat resembles a stirrup.

stellate cell. A neuron characterized by a radial, starlike distribution of dendrites.

stereocilium. A hairlike cilium attached to the top of a hair cell in the organ of Corti.

strabismus. A condition in which the eyes are not perfectly aligned.

striate cortex. Primary visual cortex, Brodmann's area 17; also called **V1**.

striated muscle. A type of muscle with a striated, or striped, appearance. Striated muscle is divisible into two categories: skeletal and cardiac.

striatum. A collective term for the caudate nucleus and putamen.

substantia. A group of related neurons deep within the brain, but usually with less distinct borders than those of nuclei.

substantia gelatinosa. A thin, dorsal part of the dorsal horn of the spinal cord that receives input from unmyelinated C fibers; important in the transmission of nociceptive signals.

substantia nigra. A cell group in the midbrain that uses dopamine as a neurotransmitter and innervates the striatum.

subthalamus. A part of the basal ganglia in the basal forebrain, involved in motor control.

sulcus. A groove in the surface of the cerebrum running between neighboring gyri. Plural: sulci.

superior colliculus. A structure in the tectum of the midbrain that receives direct retinal input and directs saccadic eye movements.

superior olive. A nucleus in the caudal pons that receives afferents from the cochlear nuclei and sends efferents to the inferior colliculus.

supplementary motor area (SMA). The medial part of cortical area 6, involved in the control of voluntary movement.

suprachiasmatic nucleus (SCN). A small nucleus of the hypothalamus, just above the optic chiasm, that receives retinal innervation and synchronizes circadian rhythms with the daily light-dark cycle.

sympathetic chain. A series of interconnected sympathetic ganglia of the ANS that lie adjacent to the vertebral column; they receive input from preganglionic sympathetic fibers and project postganglionic fibers to target organs and tissues.

sympathetic division. A division of the ANS; its peripheral axons emerge from the thoracic and lumbar spinal cord; it activates a variety of physiological responses in fight-or-flight situations, including increased heart rate, respiration, blood pressure, energy mobilization, and decreased digestive and reproductive functions.

synapse. The region of contact where a neuron transfers information to another cell.

synaptic cleft. The region separating the presynaptic and postsynaptic membranes.

synaptic transmission. The process of transferring information from one cell to another at a synapse.

synaptic vesicle. A membrane-enclosed structure, about 50 nm in diameter, containing neurotransmitter and found at a site of synaptic contact.

synergist muscle. A muscle that contracts together with other muscles to produce movement in one direction.

T tubule. A membrane-enclosed tunnel running within a skeletal muscle fiber that links excitation of the sarcolemma with the release of Ca^{2+} from the sarcoplasmic reticulum.

taste bud. A cluster of cells, including taste receptor cells, in papillae of the tongue.

taste receptor cell. A modified epithelial cell that transduces taste stimuli.

tectorial membrane. A sheet of tissue that hangs over the organ of Corti in the cochlea.

tectospinal tract. A tract originating in the superior colliculus and terminating in the spinal cord, involved in the control of head and neck movement.

tectum. The part of the midbrain lying dorsal to the cerebral aqueduct.

tegmentum. The part of the midbrain lying ventral to the cerebral aqueduct.

telencephalon. A region of the brain derived from the prosencephalon. Telencephalic structures include the paired cerebral hemispheres, containing cerebral cortex and the basal telencephalon.

temporal lobe. The region of the cerebrum lying under the temporal bone.

temporal summation. Summation of EPSPs generated in rapid succession at the same synapse.

terminal arbor. Branches at the end of an axon terminating in the same region of the nervous system.

terminal bouton. The end region of the axon, usually a site of synaptic contact with another cell; also called **axon terminal**.

tetanus. A term used by neuroscientists to describe repetitive stimulation.

tetrodotoxin (TTX). A toxin that blocks Na^+ permeation through voltage-gated sodium channels, thereby blocking action potentials.

thalamus. The dorsal part of the diencephalon, highly interconnected with the cerebral neocortex.

thermoreceptor. A sensory receptor selective for temperature changes.

thick filament. A part of the cytoskeleton of a muscle cell, containing myosin. The thick filaments lie between and among the thin filaments, and slide along them to cause muscle contraction.

thin filament. A part of the cytoskeleton of a muscle cell, containing actin. The thin filaments are anchored to the Z-bands, and slide along the thick filaments to cause muscle contraction.

third ventricle. The space filled with cerebrospinal fluid within the diencephalon.

threshold. The level of depolarization that is just enough to trigger an action potential.

tonotopy. The systematic organization within an auditory structure on the basis of characteristic frequency.

tract. A collection of CNS axons having a common site of origin and a common destination.

transcription. The process of synthesizing an mRNA molecule according to genetic instructions encoded in DNA.

transducin. The G-protein that couples rhodopsin to the enzyme phosphodiesterase in photoreceptors.

transduction. The transformation of sensory stimulus energy into a cellular signal, such as a receptor potential.

transient global amnesia. A bout of anterograde and retrograde amnesia that lasts for a period of minutes to days.

translation. The process of synthesizing a protein molecule according to genetic instructions carried by an mRNA molecule.

transmitter-gated ion channel. A membrane protein forming a pore that is permeable to ions and gated by a neurotransmitter.

transporter. A protein involved in the transport of neurotransmitters across membranes.

trigeminal nerve. Cranial nerve V; attaches to the pons and carries primarily sensory axons from the head, mouth, and dura mater, and motor axons of mastication.

trophic factor. Any molecule that promotes cell survival.

troponin. A protein that binds Ca^{2+} in a muscle cell and thereby regulates the interaction of myosin with actin.

tympanic membrane. A membrane at the internal end of the auditory canal that moves in response to variations in air pressure; also called eardrum.

ultradian rhythm. Any rhythm with a period significantly less than one day.

undershoot. The part of an action potential when the membrane potential is more negative than at rest; also called an **after-hyperpolarization**.

unipolar neuron. A neuron with a single neurite.

V1. Primary visual cortex, or striate cortex.

vasopressin. A small peptide hormone released from the posterior pituitary by magnocellular neurons; promotes water retention and decreased urine production by the kidney; also called antidiuretic hormone (ADH).

ventral. A direction meaning toward the belly.

ventral cochlear nucleus. A nucleus in the medulla that receives afferents from the spiral ganglion in the cochlea.

ventral horn. The ventral region of the spinal cord containing neuronal cell bodies.

ventral lateral nucleus (VL). A nucleus of the thalamus that relays information from the basal ganglia and cerebellum to the motor cortex.

ventral posterior (VP) nucleus. The main thalamic relay nucleus of the somatic sensory system.

ventral posterior medial (VPM) nucleus. The part of the ventral posterior nucleus of the thalamus that receives somatosensory input from the face, including afferents from the tongue.

ventral root. A bundle of motor neuron axons that emerges from the ventral spinal cord and joins sensory fibers to form a spinal nerve.

ventricular system. The spaces filled with cerebrospinal fluid inside the brain, consisting of the lateral ventricles, the third ventricle, the cerebral aqueduct, and the fourth ventricle.

ventromedial pathway. The axons in the ventromedial column of the spinal cord that are involved in the control of posture and locomotion and are under brain stem control.

vermis. The midline region of the cerebellum.

vestibular apparatus. A part of the inner ear specialized for the detection of head motion.

GLOSSARY

vestibular nuclei. The nuclei in the medulla that receive input from the vestibular apparatus of the inner ear.

vestibulospinal tract. A tract originating in the vestibular nuclei of the medulla and terminating in the spinal cord, involved in the control of movement and posture.

visceral PNS. The part of the PNS that innervates the internal organs, blood vessels, glands; also called the autonomic nervous system.

vision. The sense of sight.

visual acuity. The ability of the visual system to resolve two nearby points.

visual angle. A way to describe distance across the retina; an object that subtends an angle of 3.5° will form an image on the retina that is 1 mm across.

visual field. The total region of space that is viewed by both eyes, when the eyes are fixated on a point.

visual hemifield. The half of the visual field to one side of the fixation point.

vitreous humor. The jellylike substance filling the eye between the lens and the retina.

volley principle. The idea that high sound frequencies are represented in the pooled activity of a number of neurons, each of which fires in a phase-locked manner.

voltage clamp. A method that enables an investigator to hold the membrane potential constant while transmembrane currents are measured.

voltage. The force exerted on an electrically charged particle, represented by the symbol V and measured in units called volts; also called **electrical potential** or potential difference.

voltage-gated potassium channel. A membrane protein forming a pore that is permeable to K^+ and gated by depolarization of the membrane.

voltage-gated sodium channel. A membrane protein forming a pore that is permeable to Na^+ and gated by depolarization of the membrane.

Wada procedure. A procedure in which one cerebral hemisphere is anesthetized in order to test the function of the other hemisphere.

Wernicke's aphasia. A language disturbance in which speech is fluent but meaningless, and comprehension is poor.

Wernicke's area. An area on the superior surface of the temporal lobe between auditory cortex and the angular gyrus that is associated with Wernicke's aphasia when damaged.

Wernicke-Geschwind model. A model for language processing involving interactions between Broca's area and Wernicke's area with sensory and motor areas.

white matter. A generic term for a collection of CNS axons. When a freshly dissected brain is cut open, axons appear white.

working memory. The temporary recollection or retention of information required for ongoing behavior.

Young-Helmholtz trichromacy theory. The theory that the brain assigns colors based on the relative activation of the three types of photoreceptors.

Z line. A band delineating sarcomeres in a myofibril.

zeitgeber. Any environmental cue, such as the light-dark cycle, that signals the passage of time.

zone of Lissauer. The outermost region of the dorsal horn of the spinal cord, carrying unmyelinated pain and temperature axons.

REFERENCES AND SUGGESTED READINGS

Chapter 1

Clarke E, Dewhurst K. An illustrated history of brain function. Los Angeles: University of California Press, 1972.

Clarke E, O'Malley C. The human brain and spinal cord. Los Angeles: University of California Press, 1968.

Corsi P, ed. The enchanted loom. New York: Oxford University Press, 1991.

Crick F. The astonishing hypothesis: the scientific search for the soul. New York: Macmillan, 1994.

Finger S. Origins of neuroscience. New York: Oxford University Press, 1994.

National Academy of Sciences Institute of Medicine. Science, medicine, and animals. Washington DC: National Academy Press, 1991.

United States Office of Science and Technology Policy, Subcommittee on Brain and Behavioral Sciences. Report to Congress: Maximizing human potential, April, 1991.

Worden F, Swazey J, Adelman G, eds. The neurosciences: paths of discovery. Cambridge, MA: MIT Press, 1975.

Chapter 2

Alberts B, Bray D, Lewis J, Raff M, Roberts K, Watson JD. Molecular biology of the cell. 3rd ed. New York: Garland, 1994.

DeFilipe J, Jones EG. Cajal on the cerebral cortex. New York: Oxford University Press, 1988.

Finger S. Origins of neuroscience. New York: Oxford University Press, 1994.

Geren BB. The formation from the Schwann cell surface of myelin in the peripheral nerves of chick embryos. Experimental Cell Research 1954;7:558–562.

Grafstein B, Forman DS. Intracellular transport in neurons. Physiological Reviews 1980;60:1167–1283.

Hall Z. An introduction to molecular neurobiology. Sunderland, MA: Sinauer, 1992.

Harris KM, Stevens JK. Dendritic spines of CA1 pyramidal cells in the rat hippocampus: serial electron microscopy with reference to their biophysical characteristics. Journal of Neuroscience 1989;9:2982–2997.

Hammersen F. Histology. Baltimore: Urban & Schwarzenberg, 1980.

Hubel DH. Eye, brain and vision. New York: Scientific American Library, 1988.

Levitan I, Kaczmarek L. The neuron: cell and molecular biology. New York: Oxford University Press, 1991.

Peters A, Palay SL, Webster HdeF. The fine structure of the nervous system. 3rd ed. New York: Oxford University Press, 1991.

Robertson JD. New observations on the ultrastructure of the membranes of frog peripheral nerve fibers. Journal of Biophysical and Biochemical Cytology 1957;3:1043–1048.

Widnell CC, Pfenninger KH, eds. Essential cell biology. Baltimore: Williams & Wilkins, 1990.

Chapter 3

Hille B. Ionic channels of excitable membranes. 2nd ed. Sunderland, MA: Sinauer, 1992.

Levine J, Miller K. Biology: discovering life. 2nd ed. Lexington, MA: DC Heath, 1994.

Nicholls J, Martin A, Wallace B. From neuron to brain. 3rd ed. Sunderland, MA: Sinauer, 1993.

Ransom BR, Goldring S. Slow depolarization in cells presumed to be glia in cerebral cortex of cat. Journal of Neurophysiology 1973;36:869–878.

Shepherd G. Neurobiology. 3rd ed. New York: Oxford, 1994.

Stryer L. Biochemistry. 3rd ed. New York: WH Freeman, 1988.

Chapter 4

Agmon A, Connors BW. Correlation between intrinsic firing patterns and thalamocortical synaptic responses of neurons in mouse barrel cortex. Journal of Neuroscience 1992; 12:319–329.

Connors B, Gutnick M. Intrinsic firing patterns of diverse neocortical neurons. Trends in Neurosciences 1991;13:99–104.

Hille B. Ionic channels of excitable membranes. 2nd ed. Sunderland, MA: Sinauer, 1992.

Hodgkin A. Chance and design in electrophysiology: an informal account of certain experiments on nerve carried out between 1942 and 1952. Journal of Physiology (London) 1976;263:1–21.

Huguenard J, McCormick D. Electrophysiology of the neuron. New York: Oxford, 1994.

Llinás R. The intrinsic electrophysiological properties of mammalian neurons: insights into central nervous system function. Science 1988;242:1654–1664.

Neher E, Sakmann B. The patch clamp technique. Scientific American 1992;266: 28–35.

Nicholls J, Martin A, Wallace B. From neuron to brain. 3rd ed. Sunderland, MA: Sinauer, 1993.

Noda M, Ikeda T, Suzuki H, et al. Expression of functional sodium channels from cloned cDNA. Nature 1986;322:826–828.

Shepherd G. Neurobiology. 3rd ed. New York: Oxford University Press, 1994.

Unwin N. The structure of ion channels in membranes of excitable cells. Neuron 1989;3:665–676.

Chapter 5

Bloedel JR, Gage PW, Llinás R, Quastel DM. Transmitter release at the squid giant synapse in the presence of tetrodotoxin. Nature 1966;212:49–50.

Eccles J. Developing concepts of the synapses. Journal of Neuroscience 1990;10: 3769–3781.

Fatt P, Katz B. An analysis of the end-plate potential recorded with an intracellular electrode. Journal of Physiology (London) 1951;115:320–370.

Furshpan E, Potter D. Transmission at the giant motor synapses of the crayfish. Journal of Physiology (London) 1959;145:289–325.

Hall Z. An introduction to molecular neurobiology. Sunderland, MA: Sinauer, 1992.

REFERENCES AND SUGGESTED READINGS

Heuser J, Reese T. Evidence for recycling of synaptic vesicle membrane during transmitter release at the frog neuromuscular junction. Journal of Cell Biology 1973; 57:315–344.

Jessel T, Kandel E. Synaptic transmission: a bidirectional and self-modifiable form of cell-cell communication. Cell 1993;72: 1–30.

Kelly R. Storage and release of neurotransmitters. Cell 1993;72:43–53.

Levitan I, Kaczmarek L. The neuron: cell and molecular biology. New York: Oxford University Press, 1991.

Llinás R, Steinberg IZ, Walton K. Presynaptic calcium currents and their relation to synaptic transmission: voltage clamp study in squid giant synapse and theoretical model for the calcium gate. Proceedings of the National Academy of Sciences of the USA. 1976;73:2913–2922.

Llinás R, Sugimori M, Silver RB. Microdomains of high calcium concentration in a presynaptic terminal. Science 1992; 256:677–679.

Loewi O. From the workshop of discoveries. Lawrence: University of Kansas Press, 1953.

Neher E, Sakmann B. The patch clamp technique. Scientific American 1992;266: 44–51.

Shepherd G. Neurobiology. 3rd ed. New York: Oxford University Press, 1994.

Sherrington C. Integrative action of the nervous system. New Haven: Yale University Press, 1906.

Siegel G, Albers R, Agranoff B, Katzman R, eds. Basic neurochemistry: molecular, cellular and medical aspects. 4th ed. Boston: Little, Brown & Co, 1989.

Unwin N. Neurotransmitter action: opening of ligand-gated ion channels. Cell 1993;72:31–41.

Chapter 6

Cooper J, Bloom F, Roth R. The biochemical basis of neuropharmacology. 6th ed. New York: Oxford University Press, 1991:454.

Gilman A. G proteins: Transducers of receptor-generated signals. Annual Review of Biochemistry 1987;56:615–649.

Hall Z. An introduction to molecular neurobiology. Sunderland, MA: Sinauer, 1992.

Hille B. Ionic channels of excitable membranes. 2nd ed. Sunderland, MA: Sinauer, 1992.

McGeer P, Eccles J, McGeer E. Molecular neurobiology of the mammalian brain. 2nd ed. New York: Plenum, 1987.

Meldrum B, Moroni F, Simon R, Woods J, eds. Excitatory amino acids. New York: Raven Press, 1991.

Nicholls J, Martin A, Wallace B. From neuron to brain. 3rd ed. Sunderland, MA: Sinauer, 1993.

Nicoll R, Malenka R, Kauer J. Functional comparison of neurotransmitter receptor subtypes in the mammalian nervous system. Physiological Reviews 1990; 70:513–565.

Siegel G, Agranoff B, Albers R, Molinoff P, eds. Basic neurochemistry: molecular, cellular and medical aspects. 4th ed. New York: Raven Press, 1989.

Snyder S. Drugs and the brain. New York: WH Freeman, 1986.

Snyder S. Drug and neurotransmitter receptors in the brain. Science 1984;224:22–31.

Chapter 7

Creslin E. Development of the nervous system: a logical approach to neuroanatomy. CIBA Clinical Symposium 1974;26: 1–32.

Diamond M. Enriching heredity: the impact of the environment on the anatomy of the brain. New York: Macmillan, 1988.

Diamond M, Law F, Rhodes H, Linder B, Rosenweig MR, Krech D, Bennett EL. Increases in cortical depth and glial numbers in rats subjected to enriched environment. Journal of Comparative Neurology 1966;128:117–126.

Diamond M, Scheibel A, Murphy G, Harvey T. On the brain of a scientist: Albert Einstein. Experimental Neurology 1985;88:198–204.

Martin J. Neuroanatomy: text and atlas. New York: Elsevier, 1989.

Nauta W, Feirtag M. Fundamental neuroanatomy. New York: WH Freeman, 1986.

Watson C. Basic human neuroanatomy: an introductory atlas. 5th ed. New York: Little, Brown & Co, 1995.

Chapter 8

Buck LB, Axel R. A novel multigene family may encode odorant receptors: a molecular basis for odor recognition. Cell 1991; 65:175–187.

Corey DP, Roper SD. Sensory transduction. New York: Rockefeller University Press, 1992.

Engen T. Odor sensation and memory. New York: Praeger, 1991.

Getchell TV, Doty RL, Bartoshuk LM, Snow JB. Smell and taste in health and disease. New York: Raven Press, 1991.

Gilbertson TA. The physiology of vertebrate taste reception. Current Opinion in Neurobiology 1993;3:532–539.

Kauer JS. Contributions of topography and parallel processing to odor coding in the vertebrate olfactory pathway. Trends in Neurosciences 1991;14:79–85.

Lancet D. Exclusive receptors. Nature 1994;372:321–322.

Lowe G, Gold GH. Nonlinear amplification by calcium-dependent chloride channels in olfactory receptor cells. Nature 1993; 366:283–286.

McLaughlin S, Margolskee RF. The sense of taste. American Scientist 1994;82: 538–545.

Nakamura T, Gold GH. A cyclic nucleotide-gated conductance in olfactory receptor cilia. Nature 1987;325:442–444.

Reed RR. Signaling pathways in odor detection. Neuron 1992;8:205–209.

Ressler J, Sullivan SL, Buck LB. A zonal organization of odorant receptor gene expression in the olfactory epithelium. Cell 1993;73:597–609.

Ressler J, Sullivan SL, Buck LB. A molecular dissection of spatial patterning in the olfactory system. Current Opinion in Neurobiology 1994;4:588–596.

REFERENCES AND SUGGESTED READINGS

Roper SD. The microphysiology of peripheral taste organs. Journal of Neuroscience 1992;12:1127–1134.

Sato T. Recent advances in the physiology of taste cells. Progress in Neurobiology 1980;14:25–67.

Chapter 9

Barlow H. Summation and inhibition in the frog's retina. Journal of Physiology (London) 1953;119:69–78.

Baylor DA. Photoreceptor signals and vision. Investigative Ophthalmology and Visual Science 1987;28:34–49.

Daw NW. Color-coded cells in goldfish, cat, and rhesus monkey. Investigative Ophthalmology 1972;11:411–417.

Dowling JE, Werblin FS. Synaptic organization of the vertebrate retina. Vision Research 1971;Suppl 3:1–15.

Dowling JE. The retina: an approachable part of the brain. Cambridge, MA: Belknap Press, 1987.

Enroth-Cugell C, Robson JG. Functional characteristics and diversity of cat retinal ganglion cells. Investigative Ophthalmology and Visual Science 1984;25:250–257.

Fesenko EE, Kolesnikov SS, Lyubarsky AL. Induction by cyclic GMP of cationic conductance in plasma membrane of retinal rod outer segment. Nature 1985;313:310–313.

Kuffler S. Discharge patterns and functional organization of the mammalian retina. Journal of Neurophysiology 1953;16:37–68.

Moses RA, Hart WM, eds. Adler's physiology of the eye. 3rd ed. St. Louis: Mosby, 1987.

Schnapf JL, Baylor DA. How photoreceptor cells respond to light. Scientific American 1987;256:40–47.

Wässle H, Boycott B. Functional architecture of the mammalian retina. Physiological Reviews 1991;71:447–480.

Chapter 10

Barlow H. Single units and sensation: a neuron doctrine for perceptual psychology? Perception 1972;1:371–394.

Daw NW. The psychology and physiology of colour vision. Trends in Neurosciences 1984;7:330–335.

Grafstein B, Laureno R. Transport of radioactivity from eye to visual cortex in the mouse. Experimental Neurology 1973; 39:44–57.

Gray C, Singer W. Stimulus-specific neuronal oscillations in orientation columns of cat visual cortex. Proceedings of the National Academy of Sciences USA 1989;86:1698–1702.

Hendry S, Yoshioka T. A neurochemically distinct third channel in the macaque dorsal lateral geniculate nucleus. Science 1994;264:575–577.

Hubel D. Explorations of the primary visual cortex, 1955–78 (Nobel lecture). Nature 1982;299:515–524.

Hubel D. Eye, brain and vision. New York: WH Freeman, 1988.

Hubel D, Wiesel T. Receptive fields, binocular interaction and functional architecture in the cat's visual cortex. Journal of Physiology (London) 1962;160:106–154.

Hubel D, Wiesel T. Receptive fields and functional architecture of monkey striate cortex. Journal of Physiology (London) 1968;195:215–243.

Hubel D, Wiesel T. Functional architecture of the macaque monkey visual cortex (Ferrier lecture). Proceedings of the Royal Society of London. Series B 1977;198:1–59.

LeVay S, Wiesel TN, Hubel DH. The development of ocular dominance columns in normal and visually deprived monkeys. Journal of Comparative Neurology 1980;191:1–51.

Livingstone M, Hubel D. Anatomy and physiology of a color system in the primate visual cortex. Journal of Neuroscience 1984;4:309–356.

Martin K. A brief history of the "feature detector." Cerebral Cortex 1994;4:1–7.

Salzman C, Britten K, Newsome W. Cortical microstimulation influences perceptual judgments of motion detection. Nature 1990;346:174–177.

Zeki S. A vision of the brain. London: Blackwell Scientific Publications, 1993.

Chapter 11

Ashmore JF, Kolston PJ. Hair cell based amplification in the cochlea. Current Opinion in Neurobiology 1994;4:503–508.

Hudspeth AJ. The hair cells of the inner ear. Scientific American 1983;248:54–64.

Hudspeth AJ, Markin V. The ear's gears. Physics Today 1994;47:22–28.

Middlebrooks JC, Green DM. Sound localization by human listeners. Annual Review of Psychology 1991;42:135–159.

Moore BCJ. An introduction to the psychology of hearing. San Diego: Academic Press, 1989.

Pickles JO. Early events in auditory processing. Current Opinion in Neurobiology 1993;3:558–562.

Rose JE, Hind JE, Anderson DJ, Brugge JF. Some effects of stimulus intensity on response of auditory nerve fibers in the squirrel monkey. Journal of Neurophysiology 1971;24:685–699.

Ruggero MA, Rich NC. Furosemide alters organ of Corti mechanics: evidence for feedback of outer hair cells upon the basilar membrane. Journal of Neuroscience 1991;11:1057–1067.

Russel IJ, Cody AR, Richardson GP. The responses of inner and outer hair cells in the basal turn of the guinea-pig cochlea and in the mouse cochlea grown in vitro. Hearing Research 1986;22:199–216.

Simmons JA. A view of the world through the bat's ear: the formation of acoustic images in echolocation. Cognition 1989;33:155–199.

von Békésy G. Experiments in hearing. New York: McGraw-Hill, 1960. (edited and translated by EG Wever)

Chapter 12

Brown AC. Somatic sensation: peripheral aspects. In: Patton HD, Fuchs AF, Hille B, Scher AM, Steiner R, eds. Textbook of physiology. Philadelphia: WB Saunders, 1989;1:298–313.

Darian-Smith I, ed. Handbook of physiology—Section 1. The nervous system, vol. III. Sensory processes. Bethesda, MD: American Physiological Society, 1984.

REFERENCES AND SUGGESTED READINGS ■

Fields HL. Pain. New York: McGraw-Hill, 1987.

Johnson KO, Hsiao SS. Neural mechanisms of tactual form and texture perception. Annual Review of Neuroscience 1992; 15:227–250.

Kaas JH. Somatosensory system. In: Paxinos G, ed. The human nervous system. San Diego: Academic Press, 1990:813–844.

Kaas SH, Nelson RH, Sur M, Merzenich MM. Organization of somatosensory cortex in primates. In: Schmitt FO, Worden FG, Adelman G, Dennis SG, eds. The organization of the cerebral cortex. Cambridge, MA: MIT Press,1981: 237–262.

Loewenstein WR. Mechano-electric transduction in the Pacinian corpuscle: initiation of sensory impulses in a mechanoreceptor. In: Loewenstein WR, ed. Handbook of sensory physiology. New York: Springer-Verlag, 1971;1:269–290.

McMahon SB, Koltzenburg M. Itching for an explanation. Trends in Neurosciences, 1992;15:497–501.

Melzack R, Wall PD. The challenge of pain. New York: Basic Books, 1983.

Penfield W, Rasmussen T. The cerebral cortex of man. New York: Macmillan, 1952.

Sacks O. The man who mistook his wife for a hat and other clinical tales. New York: Summit, 1985.

Schmidt RF, ed. Fundamentals of sensory physiology. New York: Springer-Verlag, 1978.

Springer SP, Deutsch G. Left brain, right brain. New York: WH Freeman, 1989.

Vallbo ÅB. Single-afferent neurons and somatic sensation in humans. In: Gazzaniga MS, ed. The cognitive neurosciences. Cambridge, MA: MIT Press, 1995:237–251.

Vallbo ÅB, Johansson RS. Properties of cutaneous mechanoreceptors in the human hand related to touch sensation. Human Neurobiology 1984;3:3–14.

Woolsey TA, Van der Loos H. The structural organization of layer IV in the somatosensory region (SI) of mouse cerebral cortex: the description of a cortical field composed of discrete cytoarchitectonic units. Brain Research 1970;17: 205–242.

Chapter 13

Brown T. The intrinsic factors in the act of progression in the mammal. Proceedings of the Royal Society of London. Series B 1911;84:308–319.

Buller A, Eccles J, Eccles R. Interactions between motoneurons and muscles in respect to the characteristic speeds of their responses. Journal of Physiology (London) 1960;150:417–439.

Drachman DB. Myasthenia gravis. New England Journal of Medicine 1994; 330:1797–1810.

Evarts E, Wise S, Bousfield D, ed. The motor system in neurobiology. New York: Elsevier, 1985.

Ghez C. Muscles: effectors of the motor system. In: Kandel E, Schwartz J, Jessel T, eds. Principles of neural science. New York: Elsevier, 1991:548–563.

Henneman E, Somjen G, Carpenter D. Functional significance of cell size in spinal motoneurons. Journal of Neurophysiology 1965;28:560–580.

Huxley A, Niedergerke R. Structural changes in muscle during contraction. Interference microscopy of living muscle fibres. Nature 1954;173:971–973.

Huxley H, Hanson J. Changes in cross-striations of muscle during contraction and stretch and their structural interpretation. Nature 1954;173:973–976.

Lømo T, Westgaard R, Dahl H. Contractile properties of muscle: control by pattern of muscle activity in the rat. Proceedings of the Royal Society of London. Series B 1974;187:99–103.

Mendell L, Henneman E. Terminals of single Ia fibers: distribution within a pool of 300 homonymous motor neurons. Science 1968;160:96–98.

Rowland L. Clinical concepts of Duchenne muscular dystrophy. The impact of molecular genetics. Brain 1988;111:479–495.

Sherrington C. The integrative action of the nervous system. 2nd ed. New Haven: Yale University Press, 1947.

Chapter 14

Brooks V. The neural basis of motor control. New York: Oxford University Press, 1986.

Campbell A. Histological studies on the localization of cerebral function. Cambridge: Cambridge University Press, 1905.

Donoghue J, Sanes J. Motor areas of the cerebral cortex. Journal of Clinical Neurophysiology 1994;11:382–396.

Georgopoulos A, Kalaska J, Caminiti R, Massey J. On the relations between the direction of two-dimensional arm movements and cell discharge in primate motor cortex. Journal of Neuroscience 1982; 2:1527–1537.

Georgopolous A, Caminiti R, Kalaska R, Massey J. Spatial coding of movement: a hypothesis concerning the coding of movement direction by motor control populations. In: Massion J, Paillard J, Shultz W, Wiesendanger M, eds. Neural coding of motor performance. Experimental Brain Research Suppl. 1983;7:327–336.

Ghez C. The control of movement. In: Kandel E, Schwartz J, Jessel T, eds. Principles of neural science. New York: Elsevier, 1991:533–547.

Lawrence D, Kuypers H. The functional organization of the motor system in the monkey. II. The effects of lesions of the descending brain-stem pathways. Brain 1968;91: 15–36.

Lawrence D, Kuypers H. The functional organization of the motor system in the monkey: I. The effects of bilateral pyramidal lesions. Brain 1968;91:1–14.

Lee C, Rohrer W, Sparks D. Population coding of saccadic eye movements by neurons in the superior colliculus. Nature 1988;332:357–360.

McIlwain JT. Distributed spatial coding in the superior colliculus: a review. Visual Neuroscience 1991;6:3–13.

Porter R, Lemon R. Corticospinal function and voluntary movement. Oxford: Clarendon Press, 1993:428.

Roland P, Larsen B, Lassen N, Skinhøf E. Supplementary motor area and other cortical areas in organization of voluntary movements in man. Journal of Neurophysiology 1980;43:118–136.

■ REFERENCES AND SUGGESTED READINGS

Sanes J, Donoghue J. Organization and adaptability of muscle representations in primary motor cortex. In: Caminiti R, Johnson R, Burnod Y, eds. Control of arm movements in space: neurophysiological and computational approaches. Berlin: Springer-Verlag, 1990:103–127.

Weinrich M, Wise S. The premotor cortex of the monkey. Journal of Neuroscience 1982;2:1329–1345.

Chapter 15

Andreasen NC. The mechanisms of schizophrenia. Current Opinion in Neurobiology 1994;4:245–251.

Appenzeller O. The autonomic nervous system: an introduction to basic and clinical concepts. New York: Elsevier, 1990.

Barajas-López C, Huizinga JD. New transmitters and new targets in the autonomic nervous system. Current Opinion in Neurobiology 1993;3:1020–1027.

Bloom FE, Lazerson A, Hofstader L. Brain, mind, and behavior. New York: WH Freeman, 1985:323.

Caroll CR. Drugs in modern society. Dubuque, IA: WC Brown, 1985.

Cooper JR, Bloom FE, Roth RH. The biochemical basis of neuropharmacology. 6th ed. New York: Oxford University Press, 1991.

Jänig W, McLachlan EM. Characteristics of function-specific pathways in the sympathetic nervous system. Trends in Neurosciences 1992;15:475–481.

Kerr DS, Campbell LW, Hao S-Y, Landsfield PW. Corticosteroid modulation of hippocampal potentials: increased effect with aging. Science 1989;245:1505–1509.

Kerr DS, Campbell LW, Applegate MD, Brodish A, Landsfield PW. Chronic stress-induced acceleration of electrophysiologic and morphometric biomarkers of hippocampal aging. Journal of Neuroscience 1991;11:1316–1324.

Koob GF. Drugs of abuse: anatomy, pharmacology and function of reward pathways. Trends in Pharmacological Sciences 1992;13:177–184.

McEwen BS, Schmeck HM. The hostage brain. New York: Rockefeller University Press, 1994.

Sapolsky RM. Why zebras don't get ulcers: a guide to stress, stress-related diseases, and coping. New York: WH Freeman, 1994.

Sapolsky RM, Krey LC, McEwen BS. The neuroendocrinology of stress and aging: the glucocorticoid cascade hypothesis. Endocrine Reviews 1986;7:284–301.

Seeman P. Brain dopamine receptors. Pharmacological Reviews 1980;32:229–313.

Snyder SH. Drugs and the brain. New York: Scientific American Books, 1986.

Strange PG. Brain biochemistry and brain disorders. New York: Oxford University Press, 1992.

Swanson LW. The hypothalamus. In: Björkland A, Hökfelt T, Swanson LW, eds. Handbook of chemical neuroanatomy. New York: Elsevier, 1987;5(I);1–124.

Trimble MR. Biological psychiatry. New York: John Wiley & Sons, 1988.

Watanabe Y, Gould E, McEwen BS. Stress induces atrophy of apical dendrites of hippocampal CA3 pyramidal neurons. Brain Research 1992;588:341–345.

Chapter 16

Adolphs R, Tranel D, Damasio H, Damasio A. Impaired recognition of emotion in facial expressions following bilateral damage to the human amygdala. Nature 1994; 372:669–672.

Aggleton JP. The contribution of the amygdala to normal and abnormal emotional states. Trends in Neurosciences 1993; 16:328–333.

Bard P. On emotional expression after decortication with some remarks on certain theoretical views. Psychological Reviews 1934;41:309–329.

Broca P. Anatomie comparé de circonvolutions cérébrales. Le grand lobe limbique et la scissure limbique dans le série des mammifères. Revue d'Anthropologie 1878; 1:385–498.

Cannon WB. The James-Lange theory of emotion. American Journal of Psychology 1927;39:106–124.

Davis M. The role of the amygdala in fear and anxiety. Annual Review of Neuroscience 1992:353–375.

Damasio H, Grabowski T, Frank R, Galaburda AM, Damasio AR. The return of Phineas Gage: clues about the brain from the skull of a famous patient. Science 1994; 264:1102–1105.

Darwin C. The expression of the emotions in man and animals. New York: The Philosophical Library, 1955 (1872 original edition).

Flynn JP. The neural basis of aggression in cats. In: Glass DC, ed. Neurophysiology and emotion. New York: Rockefeller University Press, 1967.

Fulton JF. Frontal lobotomy and affective behavior. A neurophysiological analysis. New York: WW Norton, 1951.

Harlow JM. Passage of an iron rod through the head. Boston Medical and Surgical Journal 1848;39:389–393.

Harlow JM. Recovery from the passage of an iron bar through the head. Publication of the Massachusetts Medical Society 1868; 2:329–347.

Heath RG. Electrical self-stimulation of the brain in man. American Journal of Psychiatry 1963;120:571–577.

Hess WR. Diencephalon: autonomic and extrapyramidal functions. New York: Grune & Stratton, 1954.

Jacobsen CF, Wolf JB, Jackson TA. An experimental analysis of the functions of the frontal association areas in primates. Journal of Nervous and Mental Disease 1935;82:1–14.

James W. What is an emotion? Mind 1884;9:188–205.

Kalin NH. The neurobiology of fear. Scientific American 1993; 268:94–101.

Kapp BS, Pascoe JP, Bixler MA. The amygdala: a neuroanatomical systems approach to its contributions to aversive conditioning. In: Butler N, Squire LR. Neuropsychology of memory. New York: Guilford, 1984.

Klüver H, Bucy PC. Preliminary analysis of functions of the temporal lobes in monkeys. Archives of Neurology and Psychiatry 1939;42:979–1000.

Lange CG. Uber Gemuthsbewegungen. Liepzig: T. Thomas, 1887.

LeDoux JE. Emotion, memory and the brain. Scientific American 1994;270:50–57.

REFERENCES AND SUGGESTED READINGS ∎

Lowe J, Carrol D. The effects of spinal injury on the intensity of emotional experience. British Journal of Clinical Psychology 1985;24:135–136.

MacLean PD. The limbic system ("visceral brain") and emotional behavior. Archives of Neurology and Psychiatry 1955;73:130–134.

Olds J, Milner P. Positive reinforcement produced by electrical stimulation of the septal area and other regions of the rat brain. Journal of Comparative Physiological Psychology 1954;47:419–427.

Panksepp J. Toward a general psychobiological theory of emotions. Behavioral and Brain Sciences 1982;5:407–467.

Papez JW. A proposed mechanism of emotion. Archives of Neurology and Psychiatry 1937;38:725–743.

Pribram KH. Towards a science of neuropsychology (method and data). In: Patton RA, ed. Current trends in psychology and the behavioral sciences. Pittsburgh: University of Pittsburgh Press, 1954.

Routtenberg A. The reward system of the brain. Scientific American 1978;239:154–164.

Saudou F, Amara DA, Dierich A, LeMeur M, Ramboz S, Segu L, Buhot M, Hen R. Enhanced aggressive behavior in mice lacking 5-HT$_{1B}$ receptor. Science 1994;265:1875–1878.

Sheard MH. Aggressive behavior: effects of neural modulation by serotonin. In: Simmel EC, Hahn ME, Walters JK, eds. Aggressive behavior: genetic and neural approaches. Hillsdale, NJ: Lawrence Erlbaum Associates, 1983.

Thompson JG. The psychobiology of emotions. New York: Plenum Press, 1988.

Valenstein ES. Brain control. New York: John Wiley & Sons, 1973.

Wise RA, Rompre P. Brain dopamine and reward. Annual Review of Psychology 1989;40:191–225.

Woodworth RS, Sherrington CS. A pseudoaffective reflex and its spinal path. Journal of Physiology (London) 1904; 31:234–243.

Chapter 17

Aldrich MS. The neurobiology of narcolepsy-cataplexy. Progress in Neurobiology 1986;41:533–541.

Aréchiga H. Circadian rhythms. Current Opinion in Neurobiology 1993;3:1005–1010.

Bal T, McCormick DA. Mechanisms of oscillatory activity in guinea-pig nucleus reticularis thalami *in vitro*: a mammalian pacemaker. Journal of Physiology (London) 1993;468:669–691.

Carskadon MA, ed. Encyclopedia of sleep and dreaming. New York: Macmillan, 1993.

Coleman RM. Wide awake at 3:00 A.M. by choice or by chance? New York: WH Freeman, 1986.

Dement WC. Some must watch while some must sleep. San Francisco: San Francisco Book Company, 1976.

Earnest DJ, Sladek CD. Circadian vasopressin release from perfused rat suprachiasmatic explants in vitro: effects of acute stimulation. Brain Research 1987; 422:398–402.

Edgar DM, Dement WC, Fuller CA. Effect of SCN lesions on sleep in squirrel monkeys: evidence for opponent processes in sleep-wake regulation. Journal of Neuroscience 1993;13:1065–1079.

Freeman W. The physiology of perception. Scientific American 1991;264:78–85.

Gray CM. Synchronous oscillations in neuronal systems: mechanisms and functions. Journal of Comparative Neuroscience 1994;1:11–38.

Hobson JA. The dreaming brain. New York: Basic Books, 1986.

Hobson JA. Sleep and dreaming. Current Opinion in Neurobiology 1993;10:371–382.

Horne J. Why we sleep. Oxford: Oxford University Press, 1988.

Karni A, Tanne D, Rubenstein BS, Akenasy JJM, Sagi D. Dependence on REM sleep of overnight performance of a perceptual skill. Science 1994;265:679–682.

Klein DC, Moore RY, Reppert SM, eds. Suprachiasmatic nucleus: the mind's clock. New York: Oxford University Press, 1991.

Lamberg L. Body rhythms: chronobiology and peak performance. New York: William Morrow & Co., 1994.

McCarley RW, Massaquoi SG. A limit cycle reciprocal interaction model of the REM sleep oscillator system. American Journal of Physiology 1986;251:R1011.

Moruzzi G. Reticular influences on the EEG. Electroencephalography and Clinical Neurophysiology 1964;16:1.

Mukhametov LM. Sleep in marine mammals. In: Borbély AA, Valatx JL, eds. Sleep mechanisms. Munich: Springer-Verlag, 1984:227–238.

Pappenheimer JR, Koski G, Fencl V, Karnovsky ML, Krueger J. Extraction of sleep-promoting factor S from cerebrospinal fluid and from brains of sleep-deprived animals. Journal of Neurophysiology 1975; 38:1299–1311.

Ralph MR, Foster RG, Davis FC, Menaker M. Transplanted suprachiasmatic nucleus determines circadian period. Science 1990;247:975–978.

Ralph MR, Menaker M. A mutation of the circadian system in golden hamsters. Science 1988;241:1225–1227.

Steriade M, McCormick DA, Sejnowski TJ. Thalamocortical oscillations in the sleeping and aroused brain. Science 1993;262:679–685.

Takahashi JS, Kornhauser JM, Koumenis C, Eskin A. Molecular approaches to understanding circadian oscillations. Annual Review of Physiology 1993;55:729–753.

Vitaterna MH, King DP, Chang A, Kornhauser JM, Lowrey PL, McDonald JD, Dove WF, Pinto LH, Turek FW, Takahashi JS. Mutagenesis and mapping of a mouse gene, Clock, essential for circadian behavior. Science 1994;264:719–725.

Winson J. The biology and function of rapid eye movement sleep. Current Opinion in Neurobiology 1993;3:243–248.

Chapter 18

Bear MF, Kleinschmidt A, Gu Q, Singer W. Disruption of experience-dependent synaptic modifications in striate cortex by infusion of an NMDA receptor antagonist. Journal of Neuroscience 1990;10:909–925.

REFERENCES AND SUGGESTED READINGS

Bear MF, Singer W. Modulation of visual cortical plasticity by acetylcholine and noradrenaline. Nature 1986;320:172–176.

Bourgeois J, Rakic P. Changes of synaptic density in the primary visual cortex of the macaque monkey from fetal to adult stage. Journal of Neuroscience 1993;13: 2801–2820.

Brown M, Hopkins W, Keynes R. Essentials of neural development. Cambridge: Cambridge University Press, 1991.

Carmignoto G, Vicini S. Activity-dependent decrease in NMDA receptor responses during development of the visual cortex. Science 1992;258:1007–1011.

Dudek SM, Bear MF. A biochemical correlate of the critical period for synaptic modification in the visual cortex. Science 1989;246:673–675.

Ghosh A, Carnahan J, Greenberg M. Requirement for BDNF in activity-dependent survival of cortical neurons. Science 1994;263:1618–1623.

Goodman C, Shatz C. Developmental mechanisms that generate precise patterns of neuronal connectivity. Cell 1993;72: 77–98.

Ip N, Yancopoulos G. Neurotrophic factor receptors: just like other growth factor receptors. Current Opinion in Neurobiology 1994;4:400–405.

Kennedy T, Serafini T, de la Torre JR, Tessier-Lavigne M. Netrins are diffusible chemotropic factors for commissural axons in the embryonic spinal cord. Cell 1994;78:425–435.

Kirkwood A, Bear MF. Hebbian synapses in visual cortex. Journal of Neuroscience 1994a;14:1634–1645.

Kirkwood A, Bear MF. Homosynaptic long-term depression in the visual cortex. Journal of Neuroscience 1994b;14: 3404–3412.

LeVay S, Stryker MP, Shatz CJ. Ocular dominance columns and their development in layer IV of the cat's visual cortex: a quantitative study. Journal of Comparative Neurology 1978;179: 223–244.

Meister M, Wong R, Baylor D, Shatz C. Synchronous bursts of action potentials in ganglion cells of the developing mammalian retina. Science 1991;252:939–943.

Olson CR, Freeman RD. Profile of the sensitive period for monocular deprivation in kittens. Experimental Brain Research 1980;39:17–21.

Rakic P. Development of visual centers in the primate brain depends on binocular competition before birth. Science 1981;214:928–931.

Schlagger BL, O'Leary DD. Potential of visual cortex to develop an array of functional units unique to somatosensory cortex. Science 1991;252:1556–1560.

Sperry R. Chemoaffinity in the orderly growth of nerve fiber patterns and connections. Proceedings of the National Academy of Sciences USA 1963;4: 703–710.

Walsh C, Cepko C. Widespread dispersion of neuronal clones across functional regions of the cerebral cortex. Science 1992;255:434.

Wiesel T. Postnatal development of the visual cortex and the influence of the environment. Nature 1982;299:583–592.

Chapter 19

Chorover SL, Schiller PH. Short-term retrograde amnesia in rats. Journal of Comparative and Physiological Psychology 1965;59:73–78.

Cohen NJ, Eichenbaum H. Memory, amnesia, and the hippocampal system. Cambridge, MA: MIT Press, 1993.

Desimone R, Albright TD, Gross CG, Bruce C. Stimulus-selective properties of inferior temporal neurons in the macaque. Journal of Neuroscience 1984;4: 2051–2062.

Dudai Y. The neurobiology of memory. New York: Oxford University Press, 1989.

Eichenbaum H, Fagan H, Mathews P, Cohen NJ. Hippocampal system dysfunction and odor discrimination learning in rats: impairment or facilitation depending on representational demands. Behavioral Neuroscience 1988;102:331–339.

Funahashi S, Chafee MV, Goldman-Rakic PS. Prefrontal neuronal activity in rhesus monkeys performing a delayed anti-saccade task. Nature 1993;365:753–756.

Fuster JM. Unit activity in prefrontal cortex during delayed-response performance: neuronal correlates of transient memory. Journal of Neurophysiology 1973;36: 61–78.

Fuster JM. Memory in the cerebral cortex. Cambridge, MA: MIT Press, 1995.

Goldman-Rakic P. Working memory and the mind. Scientific American 1992;267: 111–117.

Gnadt JW, Andersen RA. Memory related motor planning activity in posterior parietal cortex of macaque. Experimental Brain Research 1988;70: 216–220.

Hebb DO. The organization of behavior: a neuropsychological theory. New York: John Wiley & Sons, 1949.

Klüver H, Bucy PC. Preliminary analysis of functions of the temporal lobes in monkeys. Archives of Neurology and Psychiatry 1939;42:979–1000.

Lashley KS. Brain mechanisms and intelligence. Chicago: University of Chicago Press, 1929.

Luria A. The mind of a mnemonist. Cambridge, MA: Harvard University Press, 1968.

Meunier M, Bachevalier J, Mishkin M, Murray EA. Effects on visual recognition of combined and separate ablations of the entorhinal and perirhinal cortex in rhesus monkeys. Journal of Neuroscience 1993;13:5418–5432.

Mishkin M, Appenzeller T. The anatomy of memory. Scientific American 1987; 256:80–89.

O'Keefe JA. Place units in the hippocampus of the freely moving rat. Experimental Neurology 1979;51:78–109.

O'Keefe JA, Nadel L. The hippocampus as a cognitive map. London: Oxford University Press, 1978.

Olton DS, Samuelson RJ. Remembrance of places passed: spatial memory in rats. Journal of Experimental Psychology 1976;2:97–116.

Olton DS, Becker JT, Handelman GE. Hippocampus, space, and memory. Behavioral and Brain Sciences 1979; 2:313–365.

REFERENCES AND SUGGESTED READINGS ▮

Penfield W. The excitable cortex in conscious man. Liverpool: Liverpool University Press, 1958.

Penfield W, Rasmussen T. The cerebral cortex of man. New York: Macmillan, 1950.

Rolls ET, Baylis GC, Hasselmo ME, Nalwa V. The effect of learning on the face selective responses of neurons in the cortex in the superior temporal sulcus of the monkey. Experimental Brain Research 1989; 76:153–164.

Scoville WB, Milner B. Loss of recent memory after bilateral hippocampal lesions. Journal of Neurology, Neurosurgery and Psychiatry 1957;20:11–21.

Squire LR. Memory and brain. New York: Oxford University Press, 1987.

Wilson MA, McNaughton BL. Dynamics of the hippocampal ensemble code for space. Science 1993;261:1055–1058.

Zola-Morgan S, Squire LR, Amaral DG, Suzuki WA. Lesions of perirhinal and parahippocampal cortex that spare the amygdala and hippocampal formation produce severe memory impairment. Journal of Neuroscience 1989;9: 4355–4370.

Zola-Morgan S, Squire LR, Clower RP, Rempel NL. Damage to the perirhinal cortex exacerbates memory impairment following lesions to the hippocampal formation. Journal of Neuroscience 1993;13:251–265.

Chapter 20

Bailey CH, Kandel ER. Structural changes accompanying memory storage. Annual Review of Neuroscience 1993;55: 397–426.

Bear MF, Malenka RC. Synaptic plasticity: LTP and LTD. Current Opinion in Neurobiology 1994;4:389–400.

Bliss TVP, Lømo T. Long-lasting potentiation of synaptic transmission in the dentate area of the anaesthetized rabbit following stimulation of the perforant path. Journal of Physiology (London) 1973;232:331–356.

Bliss TVP, Collingridge GL. A synaptic model of memory: long-term potentiation in the hippocampus. Nature 1993;361:31–39.

Bourne HR, Nicoll R. Molecular machines integrate coincident synaptic signals. Neuron 1993;10:65–75.

Carew TJ, Hawkins RD, Kandel ER. Differential classical conditioning of a defensive withdrawal reflex in Aplysia californica. Science 1983;219:397–400.

Carew TJ, Sahley CL. Invertebrate learning and memory: from behavior to molecules. Annual Review of Neuroscience 1986; 9:435–487.

Castellucci VF, Kandel ER. A quantal analysis of the synaptic depression underlying habituation of the gill-withdrawal reflex in Aplysia. Proceedings of the National Academy of Sciences USA 1974; 77:7492–7496.

Christie BR, Kerr DS, Abraham WC. The flip side of synaptic plasticity: Long-term depression mechanisms in the hippocampus. Hippocampus 1994;4: 127–135.

Davis HP, Squire LR. Protein synthesis and memory. Psychological Bulletin 1984; 96:518–559.

Dudai Y. The neurobiology of memory. New York: Oxford University Press, 1989:340.

Dudek SM, Bear MF. Homosynaptic long-term depression in area CA1 of hippocampus and the effects of NMDA receptor blockade. Proceedings of the National Academy of Sciences USA 1992; 89:4363–4367.

Grant SGN, Silva AJ. Targeting learning. Trends in Neurosciences 1994;17:71–75.

Greenough WT, Bailey CH. The anatomy of memory: convergence of results across a diversity of tests. Trends in Neurosciences 1988;11:142–147.

Ito M. The cerebellum and neural control. New York: Raven Press, 1984.

Kandel ER, Schwartz JH. Molecular biology of learning: modulation of transmitter release. Science 1982;218:433–443.

Kirkwood A, Dudek SD, Gold JT, Aizenman CD, Bear MF. Common forms of synaptic plasticity in hippocampus and neocortex in vitro. Science 1993;260: 1518–1521.

Linden DJ, Connor JA. Cellular mechanisms of long-term depression in the cerebellum. Current Opinion in Neurobiology 1993; 3:401–406.

Linden DJ. Long-term synaptic depression in the mammalian brain. Neuron 1994; 12:457–472.

Morris RGM, Anderson E, Lynch GS, Baudry M. Selective impairment of learning and blockade of long-term potentiation by an N-methyl-D-aspartate receptor antagonist, AP5. Nature 1986;319:774–776.

Morris RGM, Kandel ER, Squire LR. The neuroscience of learning and memory: cells, neural circuits and behavior. Trends in Neurosciences 1988;11: 125–127.

Mulkey RM, Herron CH, Malenka RC. An essential role for protein phosphatases in the induction of long-term depression in hippocampus. Science 1993;261: 1051–1055.

Otani S, Abraham WC. Inhibition of protein synthesis in the dentate gyrus, but not the entorhinal cortex, blocks maintenance of long-term potentiation in rats. Neuroscience Letters 1989;106:175–180.

Pavlov IP. Conditioned reflexes: an investigation of the physiological activity of the cerebral cortex. London: Oxford University Press, 1927.

Schwartz JH. Cognitive kinases. Proceedings of the National Academy of Sciences USA 1993;90:8310–8313.

Squire LR. Memory and brain. New York: Oxford University Press, 1987.

Thompson RF, Krupa DJ. Organization of memory traces in the mammalian brain. Annual Review of Neuroscience 1994; 17:519–49.

Thorndike EL. Animal intelligence: experimental studies. New York: Macmillan, 1911.

Chapter 21

Corbetta M, Miezin FM, Dobmeyer S, Shulman GL, Petersen SE. Attentional modulation of neural processing of shape, color, and velocity in humans. Science 1990; 248:1556–1559.

Crick R. The astonishing hypothesis: the scientific search for the soul. New York: Charles Scribner's Sons, 1994.

Damasio AR, Damasio H. Brain and language. Scientific American 1992;267: 88–95.

■ REFERENCES AND SUGGESTED READINGS

Fromkin V, Rodman R. An introduction to language. New York: Harcourt Brace Jovanovich, 1992.

Gardner H. The shattered mind. New York: Vintage Books, 1974.

Gardner RA, Gardner B. Teaching sign language to a chimpanzee. Science 1969; 165:664–672.

Gazzaniga MS. The bisected brain. New York: Appleton-Century-Crofts, 1970.

Geschwind N. Specializations of the human brain. Scientific American 1979;241: 180–199.

Geschwind N, Levitsky W. Human-brain: left-right asymmetries in temporal speech region. Science 1968;161:186–187.

Moran J, Desimone R. Selective attention gates visual processing in the extrastriate cortex. Science 1985;229:782–784.

Ojemann G, Mateer C. Human language cortex: localization of memory, syntax, and sequential motor-phoneme identification systems. Science 1979;205: 1401–1403.

Patterson FG. The gestures of a gorilla: language acquisition in another pongid. Brain and Language 1978;5:56–71.

Penfield W, Rasmussen T. The cerebral cortex of man. New York: Macmillan, 1950.

Petersen SE, Fox PT, Posner MI, Mintum M, Raichle ME. Positron emission tomographic studies of the cortical anatomy of single-word processing. Nature 1988;331:585–589.

Pinker S. The language instinct. New York: William Morrow & Co., 1994.

Posner MI, Petersen SE. The attention system of the human brain. Annual Review of Neuroscience 1990;13:25–42.

Posner MI, Raichle M. Images of mind. New York: Scientific American Library, 1994.

Posner MI, Snyder CRR, Davidson BJ. Attention and the detection of signals. Journal of Experimental Psychology: General 1980;109:160–174.

Rasmussen T, Milner B. The role of early left-brain injury in determining lateralization of cerebral speech functions. Annals of the New York Academy of Sciences 1977; 299:355–369.

Sokoloff L. Modeling metabolic processes in the brain in vivo. Annals of Neurology. Suppl. 1984;15:S1–S11.

Sperry RW. The great cerebral commissure. Scientific American 1964;210:42–52.

Spitzer H, Desimone R, Moran J. Increased attention enhances both behavioral and neuronal performance. Science 1988; 240:338–340.

Wurtz RH, Goldberg ME, Robinson DL. Brain mechanisms of visual attention. Scientific American 1982;246:124–135.

I N D E X